Nature's Gift to Neuroscience

In the 1960s, Sydney Brenner proposed to use the nematode worm *Caenorhabditis elegans* to discover the control mechanisms of animal development and to reveal how a small number of neurons generate different behaviours, giving birth to a vibrant community that uses this animal model for their studies. Brenner was aided in his aim by John Sulston, who mapped the *C. elegans* cell lineages – from a single cell to the multicellular adult – which transformed the field of developmental biology.

As a tribute to these two men, this book captures the perspectives of some of the early pioneers of the worm community, from Martin Chalfie, Robert Waterston and Donald Moerman to Catherine Rankin, Antony Stretton and John White. It also includes contributions from subsequent generations of the community, who explore the development and function of the *C. elegans* nervous system. This book features how this animal has become one of the best models for elucidating the biology of different sensory modalities and their complex behavioural outputs, or how this animal's survival strategies have contributed to our understanding of ageing and neurodegeneration. Thus, this volume documents the development of the *C. elegans* neuroscience field, from infancy to maturity.

The chapters in this book were originally published as a special issue of the *Journal of Neurogenetics*.

Chun-Fang Wu is Editor-in-Chief of the *Journal of Neurogenetics*. He has conducted neurobiological research in *Drosophila*, applying genetic, cell biological, physiological and behavioural techniques in the studies.

Joy Alcedo is Guest Editor for the *C. elegans* special issue of the *Journal of Neurogenetics*. Her research focuses on the sensory and neuromodulatory influences on *C. elegans* development and survival programs.

Nature's Gift to Neuroscience

A Tribute to Sydney Brenner and John Sulston

Edited by
Chun-Fang Wu and Joy Alcedo

CRC Press
Taylor & Francis Group
Boca Raton London New York

CRC Press is an imprint of the
Taylor & Francis Group, an **informa** business

First published 2022
by CRC Press
4 Park Square, Milton Park, Abingdon, Oxon, OX14 4RN

and by CRC Press
6000 Broken Sound Parkway NW, Suite 300, Boca Raton, FL 33487-2742

CRC Press is an imprint of Informa UK Limited

British Library Cataloguing-in-Publication Data
A catalogue record for this book is available from the British Library

ISBN13: 978-1-032-14521-1 (hbk)
ISBN13: 978-1-032-14522-8 (pbk)
ISBN13: 978-1-003-23975-8 (ebk)

DOI: 10.1201/9781003239758

Typeset in Minion Pro
by codeMantra

Publisher's Note
The publisher accepts responsibility for any inconsistencies that may have arisen during the conversion of this book from journal articles to book chapters, namely the inclusion of journal terminology.

Disclaimer
Every effort has been made to contact copyright holders for their permission to reprint material in this book. The publishers would be grateful to hear from any copyright holder who is not here acknowledged and will undertake to rectify any errors or omissions in future editions of this book.

Contents

Citation Information

The chapters in this book were originally published in the *Journal of Neurogenetics*, volume 34, issue 3–4 (2020). When citing this material, please use the original page numbering for each article, as follows:

Introduction
Nature's gift to neuroscience
Joy Alcedo, Yishi Jin, Douglas S. Portman, Veena Prahlad, David Raizen, Georgia Rapti, X.Z. Shawn Xu, Yun Zhang and Chun-Fang Wu
Journal of Neurogenetics, volume 34, issue 3–4 (2020) pp. 223–224

Chapter 1
My life with Sydney, 1961–1971
Antony O. W. Stretton
Journal of Neurogenetics, volume 34, issue 3–4 (2020) pp. 225–237

Chapter 2
John Sulston (1942–2018): a personal perspective
Robert H. Waterston and Donald G. Moerman
Journal of Neurogenetics, volume 34, issue 3–4 (2020) pp. 238–246

Chapter 3
A touching story
Martin Chalfie
Journal of Neurogenetics, volume 34, issue 3–4 (2020) pp. 247–250

Chapter 4
But can they learn? My accidental discovery of learning and memory in C. elegans
Catharine H. Rankin
Journal of Neurogenetics, volume 34, issue 3–4 (2020) pp. 251–254

Chapter 5
Of worms and men
John White
Journal of Neurogenetics, volume 34, issue 3–4 (2020) pp. 255–258

Chapter 6
A perspective on C. elegans *neurodevelopment: from early visionaries to a booming neuroscience research*
Georgia Rapti
Journal of Neurogenetics, volume 34, issue 3–4 (2020) pp. 259–272

Chapter 7
Neuronal specification in C. elegans*: combining lineage inheritance with intercellular signaling*
Antoine Barrière and Vincent Bertrand
Journal of Neurogenetics, volume 34, issue 3–4 (2020) pp. 273–281

Chapter 8
Molecular mechanisms governing axonal transport: a C. elegans *perspective*
Amruta Vasudevan and Sandhya P. Koushika
Journal of Neurogenetics, volume 34, issue 3–4 (2020) pp. 282–297

Chapter 9
C. elegans *MAGU-2/Mpp5 homolog regulates epidermal phagocytosis and synapse density*
Salvatore J. Cherra III, Alexandr Goncharov, Daniela Boassa, Mark Ellisman and Yishi Jin
Journal of Neurogenetics, volume 34, issue 3–4 (2020) pp. 298–306

Chapter 10
Synaptic remodeling, lessons from C. elegans
Andrea Cuentas-Condori and David M. Miller, 3rd
Journal of Neurogenetics, volume 34, issue 3–4 (2020) pp. 307–322

Chapter 11
What about the males? the C. elegans *sexually dimorphic nervous system and a CRISPR-based tool to study males in a hermaphroditic species*
Jonathon D. Walsh, Olivier Boivin and Maureen M. Barr
Journal of Neurogenetics, volume 34, issue 3–4 (2020) pp. 323–334

Chapter 12
Cell-type-specific promoters for C. elegans *glia*
Wendy Fung, Leigh Wexler and Maxwell G. Heiman
Journal of Neurogenetics, volume 34, issue 3–4 (2020) pp. 335–346

Chapter 13
C. elegans*: a sensible model for sensory biology*
Adam J. Iliff and X.Z. Shawn Xu
Journal of Neurogenetics, volume 34, issue 3–4 (2020) pp. 347–350

Chapter 14
Temperature signaling underlying thermotaxis and cold tolerance in Caenorhabditis elegans
Asuka Takeishi, Natsune Takagaki and Atsushi Kuhara
Journal of Neurogenetics, volume 34, issue 3–4 (2020) pp. 351–362

Chapter 15
Mechano-gated channels in C. elegans
Umar Al-Sheikh and Lijun Kang
Journal of Neurogenetics, volume 34, issue 3–4 (2020) pp. 363–368

Chapter 16
What can a worm learn in a bacteria-rich habitat?
He Liu and Yun Zhang
Journal of Neurogenetics, volume 34, issue 3–4 (2020) pp. 369–377

Chapter 17
C. elegans *aversive olfactory learning generates diverse intergenerational effects*
Ana Goncalves Pereira, Xicotencatl Gracida, Konstantinos Kagias and Yun Zhang
Journal of Neurogenetics, volume 34, issue 3–4 (2020) pp. 378–388

Chapter 18
Social and sexual behaviors in C. elegans*: the first fifty years*
Douglas S. Portman
Journal of Neurogenetics, volume 34, issue 3–4 (2020) pp. 389–394

Chapter 19
Small molecule signals mediate social behaviors in C. elegans
Caroline S. Muirhead and Jagan Srinivasan
Journal of Neurogenetics, volume 34, issue 3–4 (2020) pp. 395–403

Chapter 20
Intraguild predation between Pristionchus pacificus *and* Caenorhabditis elegans: *a complex interaction with the potential for aggressive behaviour*
Kathleen T. Quach and Sreekanth H. Chalasani
Journal of Neurogenetics, volume 34, issue 3–4 (2020) pp. 404–419

Chapter 21
Plasticity of pheromone-mediated avoidance behavior in C. elegans
YongJin Cheon, Hyeonjeong Hwang and Kyuhyung Kim
Journal of Neurogenetics, volume 34, issue 3–4 (2020) pp. 420–426

Chapter 22
Worms sleep: a perspective
David Raizen
Journal of Neurogenetics, volume 34, issue 3–4 (2020) pp. 427–429

Chapter 23
Cellular damage, including wounding, drives C. elegans *stress-induced sleep*
Desiree L. Goetting, Richard Mansfield, Rony Soto and Cheryl Van Buskirk
Journal of Neurogenetics, volume 34, issue 3–4 (2020) pp. 430–439

Chapter 24
Orcokinin neuropeptides regulate sleep in Caenorhabditis elegans
Madison Honer, Kristen Buscemi, Natalie Barrett, Niknaz Riazati, Gerald Orlando and Matthew D. Nelson
Journal of Neurogenetics, volume 34, issue 3–4 (2020) pp. 440–452

Chapter 25
Discriminating between sleep and exercise-induced fatigue using computer vision and behavioral genetics
Kelsey N. Schuch, Lakshmi Narasimhan Govindarajan, Yuliang Guo, Saba N. Baskoylu, Sarah Kim, Benjamin Kimia, Thomas Serre and Anne C. Hart
Journal of Neurogenetics, volume 34, issue 3–4 (2020) pp. 453–465

Chapter 26
The OptoGenBox – a device for long-term optogenetics in C. elegans
Inka Busack, Florian Jordan, Peleg Sapir and Henrik Bringmann
Journal of Neurogenetics, volume 34, issue 3–4 (2020) pp. 466–474

Chapter 27
Neuromodulators: an essential part of survival
Joy Alcedo and Veena Prahlad
Journal of Neurogenetics, volume 34, issue 3–4 (2020) pp. 475–481

Chapter 28
Neuroendocrine control of lipid metabolism: lessons from C. elegans
Supriya Srinivasan
Journal of Neurogenetics, volume 34, issue 3–4 (2020) pp. 482–488

Chapter 29
The discovery and consequences of the central role of the nervous system in the control of protein homeostasis
Veena Prahlad
Journal of Neurogenetics, volume 34, issue 3–4 (2020) pp. 489–499

Chapter 30
Host-microbe interactions and the behavior of Caenorhabditis elegans
Dennis H. Kim and Steven W. Flavell
Journal of Neurogenetics, volume 34, issue 3–4 (2020) pp. 500–509

Chapter 31
Neurogenetics of nictation, a dispersal strategy in nematodes
Heeseung Yang, Bo Yun Lee, Hyunsoo Yim and Junho Lee
Journal of Neurogenetics, volume 34, issue 3–4 (2020) pp. 510–517

Chapter 32
Regulatory systems that mediate the effects of temperature on the lifespan of Caenorhabditis elegans
Byounghun Kim, Jongsun Lee, Younghun Kim and Seung-Jae V. Lee
Journal of Neurogenetics, volume 34, issue 3–4 (2020) pp. 518–526

Chapter 33
The contribution of C. elegans *neurogenetics to understanding neurodegenerative diseases*
Joseph J. H. Liang, Issa A. McKinnon and Catharine H. Rankin
Journal of Neurogenetics, volume 34, issue 3–4 (2020) pp. 527–548

Chapter 34
A journey to 'tame a small metazoan organism', seen through the artistic eyes of C. elegans *researchers*
Eleni Gourgou, Alexandra R. Willis, Sebastian Giunti, Maria J. De Rosa, Amanda G. Charlesworth, Mirella Hernandez Lima, Elizabeth Glater, Sonja Soo, Bianca Pereira, Kübra Akbaş, Anushka Deb, Madhushree Kamak, Mark W. Moyle, Annika Traa, Aakanksha Singhvi, Surojit Sural and Eugene Jennifer Jin
Journal of Neurogenetics, volume 34, issue 3–4 (2020) pp. 549–560

Notes on Contributors

Kübra Akbaş, New Jersey Institute of Technology, Newark, NJ, USA.

Joy Alcedo, Department of Biological Sciences, Wayne State University, Detroit, MI, USA.

Umar Al-Sheikh, Department of Neurobiology and Department of Neurosurgery of the First Affiliated Hospital, Zhejiang University School of Medicine, China. NHC and CAMS Key Laboratory of Medical Neurobiology, MOE Frontier Science Center for Brain Research and Brain-Machine Integration, School of Brain Science and Brain Medicine, Zhejiang University, China.

Maureen M. Barr, Department of Genetics and Human Genetics Institute of New Jersey, Rutgers University, Piscataway, NJ, USA.

Natalie Barrett, Department of Biology, Saint Joseph's University, Philadelphia, PA, USA.

Antoine Barrière, Aix Marseille University, CNRS, IBDM, Turing Center for Living Systems, Marseille, France.

Saba N. Baskoylu, Carney Institute for Brain Science, Brown University, Providence, RI, USA. Department of Neuroscience, Brown University, Providence, RI, USA.

Vincent Bertrand, Aix Marseille University, CNRS, IBDM, Turing Center for Living Systems, Marseille, France.

Daniela Boassa, National Center for Microscopy and Imaging Research, University of California San Diego, La Jolla, CA, USA.

Olivier Boivin, Department of Genetics and Human Genetics Institute of New Jersey, Rutgers University, Piscataway, NJ, USA.

Henrik Bringmann, Max Planck Institute for Biophysical Chemistry, Göttingen, Germany. Faculty of Biology, Department of Animal Physiology, University of Marburg, Germany.

Inka Busack, Max Planck Institute for Biophysical Chemistry, Göttingen, Germany. Faculty of Biology, Department of Animal Physiology, University of Marburg, Germany.

Kristen Buscemi, Department of Biology, Saint Joseph's University, Philadelphia, PA, USA.

Sreekanth H. Chalasani, Neurosciences Graduate Program, University of California, San Diego, CA, USA. Molecular Neurobiology Laboratory, Salk Institute for Biological Studies, La Jolla, CA, USA.

Martin Chalfie, Department of Biological Sciences, Columbia University, New York, NY, USA.

Amanda G. Charlesworth, Department of Molecular Genetics, University of Toronto, Toronto, ON, Canada.

YongJin Cheon, Department of Brain and Cognitive Sciences, DGIST, Daegu, Republic of Korea.

Salvatore J. Cherra III, Section of Neurobiology, Division of Biological Sciences, University of California San Diego, La Jolla, CA, USA. Department of Neuroscience, University of Kentucky College of Medicine, Lexington, KY, USA.

Andrea Cuentas-Condori, Cell and Developmental Biology, Vanderbilt University, Nashville, TN, USA.

Anushka Deb, Tata Institute of Fundamental Research-TIFR, Mumbai, India.

Maria J. De Rosa, Instituto de Investigaciones Bioquımicas Bahıa Blanca-INIBIBB, Bahıa Blanca, Argentina.

Mark Ellisman, National Center for Microscopy and Imaging Research, University of California San Diego, La Jolla, CA, USA. Department of Neurosciences, School of Medicine, University of California San Diego, La Jolla, CA, USA.

Steven W. Flavell, Department of Brain and Cognitive Sciences, Picower Institute for Learning and Memory, Massachusetts Institute of Technology, Cambridge, MA, USA.

Wendy Fung, Department of Genetics, Blavatnik Institute, Harvard Medical School and Boston Children's Hospital, Boston, MA, USA.

Sebastian Giunti, Instituto de Investigaciones Bioquımicas Bahıa Blanca-INIBIBB, Bahıa Blanca, Argentina.

Elizabeth Glater, Pomona College, Claremont, CA, USA.

Desiree L. Goetting, Department of Biology, California State University Northridge, Los Angeles, CA, USA.

Alexandr Goncharov, Section of Neurobiology, Division of Biological Sciences, University of California San Diego, La Jolla, CA, USA.

Eleni Gourgou, Department of Mechanical Engineering, University of Michigan, Ann Arbor, MI, USA.

Lakshmi Narasimhan Govindarajan, Carney Institute for Brain Science, Brown University, Providence, RI, USA.Department of Cognitive, Linguistic and Psychological Sciences, Brown University, Providence, RI, USA.

Xicotencatl Gracida, Department of Organismic and Evolutionary Biology, Center for Brain Science, Harvard University, Cambridge, MA, USA.

Yuliang Guo, Carney Institute for Brain Science, Brown University, Providence, RI, USA. Department of Cognitive, Linguistic and Psychological Sciences, Brown University, Providence, RI, USA.

Anne C. Hart, Carney Institute for Brain Science, Brown University, Providence, RI, USA. Department of Neuroscience, Brown University, Providence, RI, USA.

Maxwell G. Heiman, Department of Genetics, Blavatnik Institute, Harvard Medical School and Boston Children's Hospital, Boston, MA, USA.

Mirella Hernandez Lima, Department of Mechanical Engineering, University of Michigan, Ann Arbor, MI, USA.

Madison Honer, Department of Biology, Saint Joseph's University, Philadelphia, PA, USA.

Hyeonjeong Hwang, Department of Brain and Cognitive Sciences, DGIST, Daegu, Republic of Korea.

Adam J. Iliff, Department of Molecular and Integrative Physiology, Life Sciences Institute, University of Michigan, Ann Arbor, MI, USA.

Eugene Jennifer Jin, Neurobiology Section, Division of Biological Sciences, University of California, San Diego, La Jolla, CA, USA.

Yishi Jin, Section of Neurobiology, Division of Biological Sciences, University of California San Diego, La Jolla, CA, USA. Department of Neurosciences, School of Medicine, University of California San Diego, La Jolla, CA, USA. Department of Cellular and Molecular Medicine, School of Medicine, University of California San Diego, La Jolla, CA, USA.

Florian Jordan, Max Planck Institute for Biophysical Chemistry, Göttingen, Germany.

Konstantinos Kagias, Department of Organismic and Evolutionary Biology, Center for Brain Science, Harvard University, Cambridge, MA, USA.

Madhushree Kamak, Tata Institute of Fundamental Research-TIFR, Mumbai, India.

Lijun Kang, Department of Neurobiology and Department of Neurosurgery of the First Affiliated Hospital, Zhejiang University School of Medicine, Zhejiang, China. NHC and CAMS Key Laboratory of Medical Neurobiology, MOE Frontier Science Center for Brain Research and Brain-Machine Integration, School of Brain Science and Brain Medicine, Zhejiang University, Zhejiang, China.

Byounghun Kim, Department of Biological Sciences, Korea Advanced Institute of Science and Technology, Daejeon, South Korea.

Dennis H. Kim, Division of Infectious Diseases, Boston Children's Hospital, and Department of Pediatrics, Harvard Medical School, Boston, MA, USA.

Kyuhyung Kim, Department of Brain and Cognitive Sciences, DGIST, Daegu, Republic of Korea. Korea Brain Research Institute (KBRI), Daegu, Republic of Korea.

Sarah Kim, Carney Institute for Brain Science, Brown University, Providence, RI, USA. Department of Neuroscience, Brown University, Providence, RI, USA.

Younghun Kim, Department of Biological Sciences, Korea Advanced Institute of Science and Technology, Daejeon, South Korea.

Benjamin Kimia, Carney Institute for Brain Science, Brown University, Providence, RI, USA. School of Engineering, Brown University, Providence, RI, USA.

Sandhya P. Koushika, Department of Biological Sciences, Tata Institute of Fundamental Research, Mumbai, India.

Atsushi Kuhara, Graduate School of Natural Science, Konan University, Kobe, Japan. Institute for Integrative Neurobiology, Konan University, Kobe, Japan. Faculty of Science and Engineering, Konan University, Kobe, Japan. AMED-PRIME, Japan Agency for Medical Research and Development, Tokyo, Japan.

Bo Yun Lee, Department of Biophysics and Chemical Biology, Seoul National University, Seoul, South Korea.

Jongsun Lee, Department of Biological Sciences, Korea Advanced Institute of Science and Technology, Daejeon, South Korea.

Junho Lee, Department of Biological Sciences, Seoul National University, Seoul, South Korea.

Seung-Jae V. Lee, Department of Biological Sciences, Korea Advanced Institute of Science and Technology, Daejeon, South Korea.

Joseph J. H. Liang, Djavad Mowafaghian Centre for Brain Health, University of British Columbia, Vancouver, Canada.

He Liu, Department of Organismic and Evolutionary Biology, Center for Brain Science, Harvard University, Cambridge, MA, USA.

Richard Mansfield, Department of Biology, California State University Northridge, Los Angeles, CA, USA.

Issa A. McKinnon, Djavad Mowafaghian Centre for Brain Health, University of British Columbia, Vancouver, Canada.

David M. Miller, 3rd, Neuroscience Program, Vanderbilt University, Nashville, TN, USA.

Donald G. Moerman, Department of Zoology, University of British Columbia, Vancouver, BC, USA.

Mark W. Moyle, School of Medicine, Yale University, New Haven, CT, USA.

Caroline S. Muirhead, Department of Biology and Biotechnology, Worcester Polytechnic Institute, Worcester, MA, USA.

Matthew D. Nelson, Department of Biology, Saint Joseph's University, Philadelphia, PA, USA.

Gerald Orlando, Department of Biology, Saint Joseph's University, Philadelphia, PA, USA.

Ana Goncalves Pereira, Department of Organismic and Evolutionary Biology, Center for Brain Science, Harvard University, Cambridge, MA, USA.

Bianca Pereira, Department of Biological Sciences, Wayne State University, Detroit, MI, USA.

Douglas S. Portman, Department of Biomedical Genetics, Del Monte Institute of Neuroscience, School of Medicine and Dentistry, University of Rochester, Rochester, NY, USA.

Veena Prahlad, Department of Biology, Aging Mind and Brain Initiative, and Iowa Neuroscience Institute, University of Iowa, Iowa City, IA, USA.

Kathleen T. Quach, Neurosciences Graduate Program, University of California, San Diego, CA, USA. Molecular Neurobiology Laboratory, Salk Institute for Biological Studies, La Jolla, CA, USA.

David Raizen, Department of Neurology, University of Pennsylvania School of Medicine, Philadelphia, PA, USA.

Catharine H. Rankin, Djavad Mowafaghian Centre for Brain Health, University of British Columbia, Vancouver, Canada. Department of Psychology, University of British Columbia, Vancouver, Canada.

Georgia Rapti, European Molecular Biology Laboratory, Unit of Developmental Biology, Heidelberg, Germany.

Niknaz Riazati, Department of Biology, Saint Joseph's University, Philadelphia, PA, USA.

Peleg Sapir, Max Planck Institute for Biophysical Chemistry, Göttingen, Germany.

Kelsey N. Schuch, Carney Institute for Brain Science, Brown University, Providence, RI, USA. Department of Molecular Biology, Cell Biology and Biochemistry, Brown University, Providence, RI, USA.

Thomas Serre, Carney Institute for Brain Science, Brown University, Providence, RI, USA. Department of Cognitive, Linguistic and Psychological Sciences, Brown University, Providence, RI, USA.

Aakanksha Singhvi, Fred Hutchinson Cancer Research Center, Seattle, WA, USA.

Sonja Soo, Department of Neurology and Neurosurgery, McGill University, Montreal, QC, Canada.

Rony Soto, Department of Biology, California State University Northridge, Los Angeles, CA, USA.

Jagan Srinivasan, Department of Biology and Biotechnology, Worcester Polytechnic Institute, Worcester, MA, USA.

Supriya Srinivasan, Department of Neuroscience and Dorris Neuroscience Center, The Scripps Research Institute, La Jolla, CA, USA.

Antony O. W. Stretton, University of Wisconsin-Madison, Madison, WI, USA.

Surojit Sural, Department of Mechanical Engineering, University of Michigan, Ann Arbor, MI, USA.

Natsune Takagaki, Graduate School of Natural Science, Konan University, Kobe, Japan. Institute for Integrative Neurobiology, Konan University, Kobe, Japan.

Asuka Takeishi, Neural Circuit of Multisensory Integration RIKEN Hakubi Research Team, RIKEN Cluster for Pioneering Research (CPR), RIKEN Center for Brain Science (CBS), Wako, Japan.

Annika Traa, Department of Neurology and Neurosurgery, McGill University, Montreal, QC, Canada.

Cheryl Van Buskirk, Department of Biology, California State University Northridge, Los Angeles, CA, USA.

Amruta Vasudevan, Department of Biological Sciences, Tata Institute of Fundamental Research, Mumbai, India.

Jonathon D. Walsh, Department of Genetics and Human Genetics Institute of New Jersey, Rutgers University, Piscataway, NJ, USA.

Robert H. Waterston, Department of Genome Sciences, University of Washington School of Medicine, Seattle, WA, USA.

Leigh Wexler, Department of Genetics, Blavatnik Institute, Harvard Medical School and Boston Children's Hospital, Boston, MA, USA.

John White, Laboratory of Cell and Molecular Biology, University of Wisconsin, Madison, WI, USA.

Alexandra R. Willis, Department of Molecular Genetics, University of Toronto, Toronto, ON, Canada.

Chun-Fang Wu, Department of Biology, University of Iowa, Iowa City, IA, USA.

X.Z. Shawn Xu, Life Sciences Institute and Department of Molecular & Integrative Physiology, University of Michigan, Ann Arbor, MI, USA.

Heeseung Yang, Department of Biological Sciences, Seoul National University, Seoul, South Korea.

Hyunsoo Yim, Department of Biological Sciences, Seoul National University, Seoul, South Korea.

Yun Zhang, Department of Organismic and Evolutionary Biology, Center for Brain Science, Harvard University, Cambridge, MA, USA.

INTRODUCTION

Nature's gift to neuroscience

In 1963, Sydney Brenner wrote a proposal to Max Perutz on tackling the next 'big questions' in molecular biology. Beginning with the postulate that simple organisms have many of the features that control development and physiology in more complex organisms, Brenner proposed 'to tame a small metazoan' to discover the 'control mechanisms' of development. Boldly, he proposed to identify and trace the lineage of every cell in a nematode worm. He was joined in this goal by John Sulston, who painstakingly produced the worm *C. elegans* cell lineages from single cell to adult. The product of this work, which was completed in the storied halls of the Medical Research Council's Laboratory of Molecular Biology (MRC LMB), was transformative at the time and remains a valuable resource to this day. Among the many contributions of the lineage project, it led to the discovery of mechanisms of apoptosis with H. Robert Horvitz. For their work, Brenner, Sulston, and Horvitz shared the 2002 Nobel Prize in Physiology or Medicine.

Their work and the work of many others who joined their quest identified and refined the pertinent questions in molecular genetics and development, questions that Brenner wrote to Perutz needed to be defined before progress can be made. In the process, Brenner, Sulston, their F1s and succeeding generations generated 'key unitary steps' in the development and function of different animal organ systems, including those of the nervous system. John White, a Brenner graduate student, mapped the first neural connectome with the help of Eileen Southgate and Nichol Thomson. In this issue, White discusses how the *C. elegans* and human nervous systems are fundamentally similar. He proposes a roadmap into reconstructions of more complex nervous systems, using concepts defined in the worm.

This issue also contains the personal perspectives of several pioneers from the early days of the *C. elegans* community. Martin Chalfie details how a drive with Horvitz led him to a poster by Sulston on touch-insensitive mutants at the 'First International *C. elegans* Meeting', which hooked Chalfie on touch receptor neurons—his 'Hershey Heaven' that led him to pursue diverse fields. Robert Waterston and Donald Moerman beautifully captures Sulston's spirit and passion for science, which Chalfie also highlights. Antony Stretton vividly describes his magical time in the early days of the LMB, with Brenner as a postdoctoral mentor. Catherine Rankin charmingly recounts her initial foray into *C. elegans* learning and memory after working on *Aplysia*.

This issue honors Brenner and Sulston for their contributions to the field of neuroscience. As Brenner had hoped in the 1960s, *C. elegans* has provided the foundation for many areas in neuroscience. This issue explores the *C. elegans* contributions to neuronal and glial development and circuit formation and their plasticity and diverse functions in health and disease. As new questions are defined, studies into *C. elegans* behavioral and physiological outputs in response to sensory inputs have also flourished. Here we also have a collection of articles that chronicles *C. elegans* sensory biology, aversive olfactory learning, aggression, social and sexual behaviors, sleep, survival programs, aging and neurodegenerative diseases. Indeed, *C. elegans* paved the way for the genetic study of aging, including how neurons might modulate aging or how age-dependent diseases affect the nervous system. We predict that *C. elegans* will continue to pave the way for solving 'big questions' in neuroscience, perhaps ushering the birth of more fields or re-birth of existing fields.

Brenner and Sulston's contributions resonate beyond scientific concepts. They championed a tradition of open access and data sharing, which permeates the *C. elegans* community and contributes to its power in advancing scientific progress. Thus, we hope that the work and personal perspectives highlighted here serve as resources and inspiration to present and future generations of the *C. elegans* community and their friends in other neurogenetics community and beyond.

Disclosure statement

No potential conflict of interest was reported by the author(s).

ORCID

Joy Alcedo http://orcid.org/0000-0002-5279-6640
Yishi Jin http://orcid.org/0000-0002-9371-9860
Douglas S. Portman http://orcid.org/0000-0003-2686-8839
David Raizen http://orcid.org/0000-0001-5935-0476
Georgia Rapti http://orcid.org/0000-0003-4836-8640
Shawn Xu http://orcid.org/0000-0001-7102-5035
Yun Zhang http://orcid.org/0000-0002-7631-858X
Chun-Fang Wu http://orcid.org/0000-0002-4973-2021

Joy Alcedo

Yishi Jin

Douglas S. Portman

Yun Zhang

Veena Prahlad

Chun-Fang Wu

David Raizen

Georgia Rapti

X.Z. Shawn Xu

Part I

The early years of *C. elegans* neurogenetics

My life with Sydney, 1961–1971

Antony O. W. Stretton

ABSTRACT

During the 1961–1971 decade, Sydney Brenner made several significant contributions to molecular biology—showing that the genetic code is a triplet code; discovery of messenger RNA; colinearity of gene and protein; decoding of chain terminating codons; and then an important transition: the development of the nematode *Caenorhabditis elegans* into the model eucaryote genetic system that has permeated the whole of recent biology.

What a joy, what a privilege, to have worked with Sydney Brenner at the Laboratory of Molecular Biology (LMB) for 10 years, between 1961 and 1971 (Figure 1). I worked with him on colinearity and decoding the amber chain termination codon, and then switched to nematodes and neurobiology. He recruited me because he was looking for a protein chemist to help with the colinearity project, which had been one of his major obsessions for years. He had been the internal examiner for my Cambridge PhD oral defense in 1960. I had worked on amino acid sequences in hemoglobin; I guess he must have liked my work, and he also figured (rightly) that I was unlikely to give him any lip. In what follows, I will try to bring out what struck me most about the character of this brilliant, complicated, irreverent, and hilarious man, and the original science he pioneered during those 10 years. I include some autobiography for background.

In fact, I first met Sydney in 1957, when I was a PhD student with Vernon Ingram, an organic chemist in the MRC Unit for the Study of Molecular Structure of Biological Systems, housed in the Cavendish Laboratory, the Physics Department of Cambridge University. Typically, MRC Units are small, highly specialized groups, and this one, with Max Perutz as its head, was focused on solving the structure of complicated molecules like proteins by X-ray crystallography; it was housed in the Cavendish Laboratory because William Bragg, the famous X-ray crystallographer, believed in Max. The Unit included John Kendrew and Hugh Huxley, who was Kendrew's PhD student; Francis Crick was another early member. Soon Max realized he needed chemists to make heavy atom derivatives of hemoglobin to help with the 'phase problem' in X-ray crystallography, and Vernon was one of these. Vernon then went on to do his famous experiments showing that sickle cell hemoglobin differed from normal hemoglobin by a single amino acid substitution (Ingram, 1956), and he became known as the 'father of molecular medicine'.

My initial combined lab and office space was a 3′ X 3′ section of a lab bench; a little later in the year, the entire unit moved into The Hut (Figure 2), an asbestos structure in the courtyard next to the bike racks, and I got half a lab bench to myself. The X-ray machines used by Max Perutz, John Kendrew and their colleagues stayed in the basement of the Cavendish, but they all had their offices in the Hut. Francis Crick was also a member of the lab, and the scientific interests of the group continued to broaden, largely due to Crick's interest in DNA and the genetic code. Sydney had joined him in 1956. It was an amazingly talented group!

Every morning at 11 am, the whole unit met for coffee (excellent coffee made by Leslie Barnett) in the entrance lobby. I was spellbound by Francis' and Sydney's brilliance. My training had been in organic chemistry, and I had never learned genetics. My closest contact with this new world had been Professor Sir Alexander Todd's superb lectures on the chemical structure of nucleic acids, but that was pure chemistry, with no reference to the importance of the nucleotide sequence. I had a very steep learning curve, and I kept my mouth shut for over a year.

Francis and Jim Watson had recently (Watson & Crick, 1953) published their seminal paper on DNA structure, and Francis was crystallizing his thoughts on the genetic code (see his glorious paper entitled 'On Protein Synthesis' (Crick, 1958)). At the MRC Unit, most of the coffee-time talk was about the Central Dogma and the genetic code. That was the influence that had taken Vernon to sickle cell hemoglobin. In the autumn of 1958, Vernon moved to the

Biology Department at MIT, and took me with him to finish my PhD research.

1957 was the year when Sydney and Francis had invited a bunch of illustrious American visitors, and they were going

Figure 1. Sydney as I knew him in the 1960s (MRC Laboratory of Molecular Biology).

to solve the colinearity problem. There was Seymour Benzer, who had just analyzed a huge number of mutations of the rII gene of bacteriophage T4, showing that they mapped as a linear structure, just like DNA and proteins (Benzer, 1961). But did the sequences really correspond? Were the mutations in DNA arranged in the same order as the amino acid changes they produced in the protein? That was the colinearity problem. No-one had yet identified the rII protein, and it seemed to be synthesized at low levels, making it hard to isolate in large enough amounts for protein chemistry. So there was a determined search for the right gene-protein pair. The issue of colinearity seems so clear and obvious these days, but its demonstration was a crucial test of the growing construct we now call molecular biology. Back then it was not the only possibility—for example, the work of Perutz and Kendrew reinforced the prevalent idea that proteins were intricate three-dimensional structures; the transition from a one-dimensional polypeptide chain to a complicated three-dimensional structure was not obvious. Maybe that is what the genes controlled. Anfinsen and Haber showed that RNase A could be reduced to break the disulfide bonds, then completely unfolded in urea and when the urea was removed it refolded, the disulfide bonds reformed, and it regained its full enzymatic activity (Anfinsen & Haber, 1961). This was taken as a 'don't worry' demonstration that the amino acid sequence itself can generate the right folding, but is this a fluke—what about other, larger proteins?

Seymour brought his graduate student Sewell Champe. Seymour also brought his wife, Dotty. What a lovely, sweet woman. She and Seymour were crazily in love, and everyone loved them for it. George Streisinger, another phage geneticist, came with his wife Lotte, a superb potter; they too were a wonderful example of a good marriage. Mahlon Hoagland, who had worked with Paul Zamecnik on the role of soluble RNA (tRNA) in protein synthesis was there. Paul Doty, the DNA physical chemist from Harvard, was there too. Much of the year was spent in talk, searching for the right protein with which to map the results of mutation. I remember Seymour eventually becoming very impatient with the (to him) distorted ratio of talk to action, and he started to work with Vernon on another abnormal hemoglobin, hemoglobin D (Benzer, Ingram, & Lehmann, 1958). He wanted, at least, to learn Vernon's techniques in protein chemistry. The year ended without colinearity being solved. But I was able to observe a pretty wide variety of intellectual styles of these very successful scientists. There are many 'right' styles, and that was a useful lesson for a first-year graduate student. I was very lucky!

During that year, I did not get to know Sydney as well as I did later, but it was obvious that he was absolutely brilliant. I also began to see that he had a fantastic sense of humor. At one big lab party at his house, he was talking with Ann Cullis, a young, elegant postgraduate physicist who was working with Max as his assistant, and I overheard her saying to Sydney 'I'm sorry, I'm no good at small talk' and Sydney replied 'Alright then, let's talk big!' I think he swiped that line from the Marx Brothers, like his reply to the question 'Do you have a rubber band, please?', 'Sorry, no. But I do have a string orchestra.' At that time, Sydney was a heavy smoker, and at each pull he puffed out his cheeks like Zephyr in Botticelli's 'Birth of Venus' and then took the smoke deep into his lungs. Fascinating! Was he deliberately increasing the absorptive surface area to get a better nicotine high? I should have asked him.

With Vernon I worked on sequencing the delta chain of human hemoglobin A_2, a minor normal hemoglobin controlled by a different gene from the major hemoglobin, hemoglobin A (Ingram & Stretton, 1961). After my PhD, I stayed at MIT for a year, and was appointed as an Instructor, so I had to teach. I found that I liked it, which was a surprise, since at the MRC lab, teaching and those who did it were regarded as rather despicable lower forms of life and their commitment to science was held in doubt. Talk about an attitude!

In 1961 I went back to Cambridge to join Sydney, and I occupied my old bench in the Hut! Already the new Hills Road lab in the Addenbrooke's Hospital complex was under construction, and we moved in during the Spring of 1962 (Figure 3). It was now named the Laboratory of Molecular

Figure 2. The Hut (Photo credit Hans Boye; MRC Laboratory of Molecular Biology).

Biology and besides the X-ray crystallographers and the molecular geneticists (Crick and Brenner), the new group included Fred Sanger and his associates who moved from the Biochemistry Department, and Hugh Huxley. These were the members of the Governing Board (Figure 4); Max was Chairman of the Board, and there were the three Divisions, Structural Studies (headed by Kendrew), Molecular Genetics (Crick and Brenner) and Protein Chemistry (Sanger). Each division had many junior appointees, and I was one of them.

I had a bay (two benches) in a lab next to Sydney's, and I was incredibly lucky to have two technicians assigned to me, Rita Fishpool and Eileen Southgate, both of whom had been trained by Vernon, so they knew all his peptide chemistry techniques. They were both very intelligent and they liked to work hard, and they picked up new techniques very quickly. That was an era when technicians were a separate class—their names did not appear as authors on papers, and they were not expected to participate in scientific discussions. At the Cavendish, they had a separate canteen from the faculty and students. This was the quintessence of the English class system. In the new lab, there was only one canteen, definitely a step in the right direction, but still the technicians usually chose to sit at separate tables. The whole system was in transition, and Rita and Eileen both agreed with me when I said that if they had been born a generation later they would have attended university instead of technical college and would almost certainly have gone on to earn PhDs. They were really good!

Having been in the States for three years, I saw the class system from the outside for the first time and saw it for what it was—EVIL! I was born in Rugby, and at age 13 went to Rugby School, a top English public school, as a day boy, and what a weird experience that was. We 'town boys' were a 10% minority, and most of the rest of the boys were upper middle class or better—they had to be for their parents to afford the fees. We were not. Plus, we came to the school speaking with the local accent—Roogbee — but that soon got knocked out of us. Almost all the boarders were embryonic Tories, and most of them truly despised the town boys for what we were. This was the first time I had experienced prejudice, although from my reading I conclude that it is nothing, nothing at all, like what too many people do to people with different colored skin. But it was not all bad: we were lucky in that we lived at home with our parents and had a relatively normal life. We were also extremely lucky for the education we received. The teachers were fantastic—talented men who loved their subjects and taught them extremely well. I must mention two who made the most difference to me: R.W. Stott, who taught chemistry, mostly inorganic, and took us to see blast furnaces and steel works, and Tim Tosswill who taught English and took it upon himself to make sure the scientists were civilized by getting us to read novels and poetry, and making us write an essay every week—the first one I got back I felt pretty good about because he gave me a 9, but then I found out that this was out of 20, not 10; he rarely gave anyone more than 9 and I wonder who could have earned a 20—maybe Shakespeare or Jane Austen? These two men were inspired and inspiring and are the reason I still love chemistry and books and reading.

1961 was a magic time at the LMB. That was the year that Sydney, together with his research student Alice Orgel, had worked on acridine mutagenesis, and had come to the conclusion that the mechanism was totally different from that of previous mutagens, like 5-bromouracil, that induced base substitutions and therefore discrete amino acid exchanges in the encoded protein (Orgel & Brenner, 1961).

Figure 3. The new Laboratory of Molecular Biology ca. 1962 (MRC Laboratory of Molecular Biology).

Acridines intercalate between base pairs in DNA and produce shifts in the reading frame. Francis and Sydney realized that this could be used to prove that the code was a triplet code. The acridine mutants of the rII gene in phage T4 could be divided into two sets, which they arbitrarily labeled + and −, on the basis that + mutants could suppress− mutants, and vice versa. Interpretation: one set of mutants cause a base insertion, the other set a base deletion, so when they are paired the reading frame is restored, and the protein is normal except for the region between the two mutations, where the amino acid sequence is altered (Streisinger & his colleagues, working on the lysozyme gene/protein in phage T4, later proved that this interpretation was correct (Terzaghi *et al.*, 1966)). Francis and Sydney realized that if the code is a triplet code, then 3 + or 3− mutants in the gene could restore the reading frame, and this is what they and their colleagues showed to be the case (Crick, Barnett, Brenner, & Watts-Tobin, 1961).

1961 was also the year when Sydney went to CalTech for a few weeks with Jacob and Meselson and came back having discovered messenger RNA, almost contemporaneously with Jim Watson's group (Brenner *et al.*, 1961; Gros *et al.*, 1961). Then Nirenberg's revolutionary discovery that poly-U encoded polyphenylalanine (Nirenberg *et al.*, 1961) set the whole coding field ablaze, and the identity of the codons quickly fell into place as the use of synthetic polynucleotides was extended by the Nirenberg, Ochoa and Khorana labs.

Another vivid memory: at the end of the summer of 1961, shortly after I had re-joined the lab, Francis attended the International Congress of Biochemistry in Moscow, and at the first morning coffee session after he got back he was full of excitement, bubbling over with his eagerness to tell us all about the results he had heard at the conference from Marshall Nirenberg's discovery, with Matthaei, that in a cell free system poly U led to the synthesis of polyphenylalanine (Nirenberg *et al.*, 1961). The first codon! Most impressive to me was that Francis immediately acknowledged that this result completely destroyed his comma-less code (Crick, Griffith, & Orgel, 1957): this elegant theory had predicted that the 64 possible nucleotide triplets would generate only 20 different sense codons, one for each of the 20 amino acids found in protein, if the nucleotide sequence is read in non-overlapping groups of 3, and that only one of the three possible reading frames was used—out-of-phase triplets were nonsense. A strong prediction was that homopolymers should also be nonsense, since all three reading frames are identical. Francis immediately saw that his theory was completely wrong, and the amazing thing was that he rejoiced! This is the way science is meant to be done! If an experiment disproves your theory, that is a major advance in your understanding. It is still enormous fun to read the argument that led to the comma-less code. While I was still teaching, I liked to use this example to tell undergraduates about the relative roles of theory and experiment in the advancement of science, and the way a brilliant man like Francis reacted.

Most of my first year back in Cambridge was spent trying to find a good protein for the colinearity problem. We knew that there were other groups on the trail and they had chosen smallish proteins which would make the protein chemistry easier—Cy Levinthal at MIT was working on alkaline phosphatase with Frank Rothman and Alan Garen, Charles Yanofsky at Stanford had a large group working on tryptophan synthetase. I started to work on beta-galactosidase from *E. coli*, but that was too big, I tried to find the rII protein from phage, but no joy, and I also worked on a promising small *E. coli* protein, galactokinase. Sydney had suggested using a strain of *E. coli* with an episome carrying another copy of the galactokinase gene to increase the level of the enzyme, which would help with the protein purification, and I asked him if he was sure that the two genes were identical. 'They are the same', he declared, so I went ahead, made extracts of big batches of the bacteria, and did ammonium sulfate cuts, and then a big DEAE-cellulose column. The enzyme assays showed two distinct peaks of enzyme, and when I showed Sydney, he was utterly silent—a most unusual condition for him! He knew that I knew that SB could BS!

But then Sydney had a brilliant insight about the T4 phage head protein. He knew that the head protein was the most abundant protein synthesized after phage infection. Dick Epstein and Charley Steinberg and their colleagues had isolated 'amber' mutants of phage, including some in the head protein gene (Epstein *et al.*, 1963). These were suppressible mutants that were normal on one strain of bacteria (su^+), but mutant on another (su^-). Amber mutants can occur in any gene, and on su^- host bacteria they are almost always complete nulls when they occur in an essential gene. They were believed to be 'nonsense' mutations that generated a codon that did not encode an amino acid; we now know that the su^+ host has a mutated t-RNA that can read the nonsense codon as an amino acid (Goodman, Abelson, Landy, Brenner, & Smith, 1968). Sydney's insight was that they are probably not just nonsense, but also chain termination signals: if they simply did not insert an amino acid into the growing chain, then the ribosomes would most likely get stuck, leading to a depletion of available ribosomes for continuing protein synthesis, whereas if they terminated the growing chain then the ribosomes could recycle. Experiments showed that protein synthesis was not slowed down in head protein amber mutants, so the prediction was that the amber mutants should release fragments of the head protein. Different mutants should produce fragments of different size. Howard Dintzis had shown that proteins are synthesized from the N-terminus (Dintzis, 1961), so the prediction was that a series of amber mutants of the head protein gene should make N-terminal fragments of different lengths. If the protein was digested with trypsin, then the different length fragments should produce different numbers of tryptic peptides; simply by counting the peptides in the different mutants, the fragments could be put in sequence of increasing size, without having to determine the amino acid sequence of the protein. Fred Sanger had done some preliminary experiments showing that the head protein tryptic peptides could be identified if single radioactive amino acids were used to label phage-infected bacteria, and that different radioactive amino acids generated different patterns of peptides. The first experiment to test these thoughts were done by Anand Sarabhai, Sydney's graduate student, and Sydney immediately asked me to join the project. Anand had to write his PhD thesis, and could not stay longer in England after his thesis defense because his family in Ahmedabad, India, owned a big textile factory, and needed Anand back home to help run the business. We had a series of 10 mutants, some isolated by Dick Epstein's student Antoinette Bolle, and some isolated by Sydney; the peptides labeled by different radioactive amino acids showed that the truncated proteins made by each mutant could be ordered into a series of increasing size, and the genetic map showed that the mutant sites were arranged in the same order. Colinearity! We wrote a paper, and Sydney submitted it to Nature, hoping it would be published in the last issue of 1963 to gain a year's priority (Sydney was very competitive), but it just missed and was published in the first issue of 1964 (Sarabhai, Stretton, Brenner, & Bolle, 1964). Within weeks, Yanofsky's group published their finding of colinearity from the analysis of amino acid substitutions in tryptophan synthetase (Yanofsky, Carlton, Guest, Helinski, & Henning, 1964).

Sydney wrote: 'it is ironic that a good deal of the genetic code had already been determined by biochemical methods before the classical problem of colinearity was solved' (Brenner, 1971). Sydney and Francis and their friends from the RNA Tie Club had wanted to solve the code by clever experiments involving the use of mutagens with known nucleotide changes and linking the amino acid exchanges they produced; in addition, comparing the amino acid sequences in proteins where the reading frame had been shifted between a pair of + and − acridine mutations should provide valuable information. Together, these would present an intriguing and rewarding cryptography problem completely in the spirit of the early people who thought about the code. They were much chagrined that most of the codons were in fact decoded by the 'Bloody Biochemists' (Sydney's label) using protein translation of synthetic nucleotides in cell-free systems. This was a time when the new 'molecular biologists' were deeply resented by much of the rest of the world of biology, especially by the biochemists. 'Frank Young, and Related Problems in Biochemistry' was the title of a fictitious seminar in biochemistry that Sydney made up as a mild retaliatory joke (Frank Young was the contemporary Prof. of Biochemistry in Cambridge, and there was no love lost between him and the MRC gang).

But what was the amber codon? Sydney and I pursued parallel paths, he doing genetics and I protein chemistry. I sequenced a novel head protein peptide that uniquely appeared in only one of the amber mutants (H36) and found that it was an N-terminal fragment of a peptide present in the wild type (Stretton & Brenner, 1965). This confirmed the prediction for the mode of action of the amber mutation. The chain termination occurred at a site that encoded glutamine in the wild type. When grown on a su^+ host, serine was inserted at the termination site, and the rest of the protein—the C-terminal part—was completed. Sam Kaplan joined us, and showed that other suppressors inserted different amino acids, and with different efficiencies (Kaplan, Stretton, & Brenner, 1965). I went on to show that when H36 was reverted by 4-amino purine, a mutagen that induces transitions, the mutant site now encoded either glutamine or tryptophan (the tryptophan revertant is temperature-sensitive) (Stretton, Kaplan, & Brenner, 1966). Spontaneous revertants generated tyrosine at the amber site (Stretton & Brenner, 1967). By now, much of the code had been solved, and the codons for most amino acids were known (or at least strongly predicted—the 'pure code' folks were hard to convince of the accuracy of translation of synthetic polynucleotides in cell-free systems). The amino acids connected by mutation led to the assignment of UAG as the amber codon. Very similar results were obtained by Weigert and Garen in the alkaline phosphatase gene and protein of *E. coli*. (Weigert & Garen, 1965)

One amazing thing we underlings observed was the interaction between Sydney and Francis. They shared an office, and would spend several hours each morning talking, just the two of them. Both, of course, were utterly brilliant, but it was hard to avoid the us/them division. In the Harvard Medical School Neurobiology Department I saw another great example of the synergy coming from the collaboration between equals in David Hubel and Torsten Wiesel, and yet another in Ed Furshpan and Dave Potter and I was lucky to experience it myself when Ed Kravitz and I worked together: deep pleasure and satisfaction. Sydney loved to talk, and every Saturday morning he would hold court in the media kitchen, over coffee. Much of the talk was about current experiments and exploring ideas, but he also told jokes and dipped into his deep store of anecdotes—they would often be repeats, but that didn't matter because they were so entertaining. Occasionally he would tell stories about his student days in South Africa, for example a group of like-minded students would gather to watch old Marx Brothers movies, and the rule was that no-one was allowed to laugh, otherwise they would miss some of the jokes. He also told of expeditions into the countryside by jeep where there were no finished roads and of getting stuck in mud and having to drag the jeeps out with a winch bolted to the front fender. He was wonderfully entertaining, and he liked to turn everything into a story and deposit it into his memory bank. His memory was amazing—he read the literature very widely and retained a huge fraction of what he read. I never saw him prepared with notes when he gave a lecture—he seemed to have gone over what he wanted to say in his mind, and that was enough. Most people can't do this—certainly I can't—and Sydney did not seem to have any inkling of how other minds might need to work. Not that he was ungenerous—he readily shared his thinking—often, talking through what he was thinking was his way of testing its validity, and I felt very lucky when he sometimes did that with me, one-on-one, or with a larger group over coffee.

After the code was solved, a lot of people were looking for something else to do. Some felt it was all over—solving the code was like the Holy Grail, but no more questing when you have found it! Sydney was not one of these. He moved right on. Choosing what to do next was always exhilarating for him—in his interviews with Lewis Wolpert (Brenner, 2001) he says that he is aware that the end game (filling in the details) is not very interesting to him, and he is much better at the beginning game. He was looking for the right system to solve more big problems in biology, and to him the big problems were development and the nervous system. He read widely and toyed with several different organisms before making his choice. He considered *Caulobacter* and various protozoa because they showed differentiation within a single cell; *Drosophila* was attractive because the genetics was already sophisticated for development and the nervous system, but the nervous system was still complicated; plus, his close friend Seymour Benzer had chosen that; and finally nematodes, leading to the final selection of *Caenorhabditis elegans*. It can be grown on Petri dishes, eating bacteria, it is small and transparent, it has a short generation time; it is a self-fertilizing hermaphrodite but occasionally throws off males that can mate with hermaphrodites, which have fully-formed female sex organs that they use occasionally when they happen to run into a male. Could a genetic system be developed? This was crucial to Sydney's scientific approach, so showing that mutants could be isolated and mapped was his top priority. He was most interested in the nervous system, so he isolated many mutants that had defective behavior (mostly all called *unc*, for uncoordinated), but initially he was interested in any marker that could be used to make a genetic map, so he also isolated morphological mutants (most of which were called *dpy*, for dumpy). The first genetic map, based on the work of Sydney and his long-term technician, Muriel Wigby, was published in 1974, He used mutations to define 6 linkage groups corresponding to the 6 chromosomes. This was Sydney's first research publication in the field (Brenner, 1974).

Another very important feature that attracted Sydney to nematodes was the relative simplicity of their nervous system. Sydney was aware of the classical work of Richard Goldschmidt (1908) on the large parasitic nematode *Ascaris lumbricoides* (now called *Ascaris suum*). Goldschmidt showed that the neurons were small in number and were individually identifiable—this was the first clear example of the concept of 'identified neurons' in any organism; in this case they were identified purely on the basis of their morphology and their relative positions within the ganglia of the central nervous system. Many of the nerve cells in *A. suum* are large, so there was a real possibility of doing electrophysiology. *C. elegans*, on the other hand, was a good system to do genetics, and small enough for electron microscopy, so working out the wiring diagram of all the neurons and their synapses seemed feasible, but at that time it seemed too small to do electrophysiology. Sydney's thought was that these two nematodes were close enough that they could be used as models of each other, taking advantage of the large cells of *Ascaris* for electrical recordings and the small size of *C. elegans* for electron microscopy and genetics. This was a wild guess, and the amazing thing is that although DNA sequencing has now shown that *A. suum* and *C. elegans* diverged about 500 million years ago (about the same time that each diverged from humans!) (Vanfleteren *et al.*, 1994), the structure of the nervous system is highly conserved apart from the huge difference in size. It took many years to show this to be largely true (Nanda & Stretton, 2010).

Sydney soon recruited Nichol Thomson, a highly skilled electron microscopist who had been Lord Rothschild's technician, and Nichol found that *C. elegans* was easy to fix and section; he could produce unbroken series of sections over long lengths of the worm, providing the raw material for

Figure 4. The Governing Board of the LMB in 1968 (MRC Laboratory of Molecular Biology). Left to right: Hugh Huxley, John Kendrew, Max Perutz, Francis Crick, Fred Sanger, Sydney Brenner.

reconstruction of the nervous system. John White, an engineer with expertise in electronics and computing, joined Sydney in 1969. Sydney's wish was to automate the analysis of serial electron micrographs by using computer graphics analysis, and John White knew computers. They acquired a new British computer, called a Modular 1, and Sydney became addicted, teaching himself to program in machine language. Apart from John, the rest of us were mystified; Sydney still loved to talk about what was on his mind, but this time we could not follow. John was a little impatient: 'why doesn't he stick to what he is good at?' Eventually John implemented the automated system, but some of the profiles in the sections were so complicated or distorted that they were hard for the computer to follow, and the human eye-brain system, mostly that of my former technician Eileen Southgate (whom I had trained to follow neurons through serial light microscope sections of *A. suum*) was the winner. The outcome was a huge paper, *The Mind of the Worm* (White, Southgate, Thomson, & Brenner, 1986), giving a description of the anatomy of each of the 302 neurons and their synaptic connections.

Another early recruit was John Sulston, trained in chemistry; he became a hero for Sydney when he discovered how to freeze *C. elegans* so that they could later be thawed and brought back to life, and the growing numbers of mutant stocks no longer had to be kept alive by serial propagation—an extremely important time-saver (see Brenner, 1974). John Sulston went on to make many important discoveries, including tracing the lineage of all the cells in the worm from egg to adult (Sulston, Schierenberg, White, & Thomson, 1983; Sulston & Horvitz, 1977), leading the consortium that sequenced the genome of *C. elegans* (*C. elegans* Sequencing Consortium, 1998), the first eucaryote to be sequenced, and then the human genome (his book *A Common Thread* with Georgina Ferry (Sulston & Ferry, 2002) is a delicious account of the battle between science in the public interest and the privatization of science for profit by corporate America, represented by Craig Venter).

My own path after the code was also an evolution. Originally, I wanted to use my protein chemistry experience to isolate and characterize the acetylcholine receptor, and I read a lot of pharmacology. Sydney and Francis thought I needed to be retrained if I was going into a new field, and they arranged for me to spend a year in the best neurobiology department in the US at the time, built by Steve Kuffler at Harvard Medical School. Steve and his colleagues thought I should start by taking the summer Woods Hole Neurobiology course, taught mostly by Ed Furshpan and Dave Potter, who, among their scientific achievements, had also mastered the art of teaching cellular electrophysiology to medical students. This was mostly a lab course—total immersion in the intricacies of intracellular recording for between 12 and 15 h a day, following their super clear lectures every morning. Seymour Benzer was one of the other students! I loved it! It was refreshing to dig deep in a field that I did not know at all, so far away from chemistry. Besides the Furshpan/Potter lectures, there were Saturday morning lectures by other members of the Neuro department, usually lasting at least 3 h. One of these was by Ed Kravitz, a biochemist who had just made a functional map of individual identified neurons in a lobster ganglion, and had assayed single cells, showing that inhibitors contain GABA and excitors do not. I was captivated, and when the time came to move to Harvard Med School at the end of the summer, I joined his lab for 15 months. Professionally, it was the best year I have ever had! Intellectually and emotionally Ed and I really connected. For the first time in my scientific life I was treated as an equal. We are still best friends.

We wanted to see if the cellular morphology of an individual identified neuron taken from different animals was identical. The question was analogous to the one Fred Sanger had asked about protein structure: is the structure unique, or is there microheterogeneity, with a 'center of gravity' of structure, which is similar but not identical in different individuals? To answer this, we extended the work of Ed Furshpan and Jaime Alvarez who had injected dyes from microelectrodes into individual neurons. Unfortunately, the dyes they used did not extend to the ends of the neurites, so we set out to find a better dye, and after trying many, found that Procion Yellow, which is fluorescent, could be traced to the outermost extent of the neural processes (Stretton & Kravitz, 1968). The answer to the question about reproducibility of structure was that the branching pattern was characteristic but not identical, like the way an oak tree in winter can be recognized even though the branches do not take identical paths. This was true of many different examples of identified neurons. Does this mean that the behavior controlled by the circuit that includes each neuron is variable, or does it mean that there are compensatory changes in other neurons so that the activity of the circuit is conserved? The lobster nervous system, simple as it is relative to vertebrate nervous systems, is still too complicated to try to answer this question, but I already knew about nematodes and Goldschmidt from Sydney, so when I returned to the LMB, I decided to do circuit analysis in *Ascaris suum*. Nematodes are much simpler than lobsters! I had already learned the necessary techniques, and Sydney was supportive. I chose to work on the motor nervous system and the control of locomotion, and that meant I had to describe the neuroanatomy of the nerve cords. This was part of the nervous system that Goldschmidt had not described, but Rosenbluth (1965) had shown that neuromuscular synapses are made in the dorsal and ventral nerve cords. But there was a lot of the worm to study—the head is a tiny fraction of the worm. Rita Fishpool and Eileen Southgate were assigned to me again, and together we traced neuronal processes through many thousands of serial sections (adult *A. suum* is typically 25 cm long!) of several individual worms, both female and male. We found that *A. suum* neurons are large enough that their processes can be traced through serial 10 μm sections by light microscopy, and we produced a map of the motor nervous system. But 3 years after I went back to Cambridge, my appointment ran out, and I was on the job market. I have been in the Zoology Department (now Department of Integrative Biology) of the University of Wisconsin-Madison ever since, and was lucky to have fantastic students and postdocs, and we did the electrophysiology, biochemistry, electron microscopy, and mass spectrometry (to characterize the multiplicity of neuropeptides that make the functional circuitry much more complicated than we ever imagined) (Davis & Stretton, 1996; Konop *et al.*, 2015a,b; Knickelbine *et al.*, 2018).

About the *C. elegans*/*A. suum* comparison: once again Sydney's instincts turned out to be right. The neurons of *C. elegans* and *A. suum* are essentially scale models of each other, with only apparently minor differences: *C. elegans* is a miniature version of *A. suum*. Considering the huge evolutionary divergence between them, the enormous difference in their size, and the different specializations likely to be responsible for their different lifestyles (free-living versus parasitic), this is remarkable, and even more remarkable when electrophysiological properties and neurotransmitter content of analogous neurons in the two species became known.

For many years, due to its experimental advantages at the time (large cells as microelectrode targets), the *Ascaris* electrophysiological data were the only ones available. The key was the commissures, which are dorso-ventral neural processes of single neurons (in other more complex nervous systems, commissures refer to bundles of neuronal processes linking symmetrical left/right structures); they were first described by Hesse (1892) but with no knowledge of function. We traced them through serial sections and found that they are branches of motorneurons with distinctive patterns of branching into the dorsal and ventral nerve cords: morphologically they fall into 5 distinct classes of identified motorneurons, recurring in multiple copies along the length of the worm. The motorneurons thus became additional 'identified neurons', joining the 162 neurons in the head ganglia reported by Goldschmidt (1908). The commissures were crucial in enabling us to do electrophysiological and chemical experiments. John Walrond and Ira Kass studied neuromuscular transmission from single identified motorneurons and found which were excitatory and which inhibitory (Stretton *et al.*, 1978). John also showed that there was wired-in reciprocal inhibition between dorsal and ventral musculature (Walrond & Stretton, 1985). In 1999, Richmond & Jorgensen reported a neuromuscular preparation from *C. elegans*, and characterized the postsynaptic responses produced in muscle by ACh and GABA.

Ralph Davis made the first intracellular recordings from nematode neurons, recording directly from commissures (Davis & Stretton, 1989a). He showed that motorneurons do not propagate action potentials, but have unusually high membrane resistance, which allows signals to propagate passively over long distances (the space constant, lambda, which is the distances over which the membrane potential falls exponentially to 1/e of its original value, is of the order of one centimeter). He also showed that intercellular synaptic transmission is analog, not digital—transmitter release is graded with presynaptic membrane potential (Davis & Stretton, 1989b), confirming a deduction made by del Castillo *et al.* (1963) based on muscle membrane potential changes in the presence of curare and neostigmine. It was nearly a decade until Shawn Lockery and his colleagues used patch electrodes to record from *C. elegans* neurons and showed that they had similar membrane properties (Goodman, Hall, Avery, & Lockery, 1998). The high membrane resistance in such small cells means that the cells are essentially iso-potential. Then, tonic synaptic transmission was also found in *C. elegans* (Liu, Hollopeter, & Jorgensen, 2009). Sydney was right!

The commissures also allowed the identification of the neurotransmitter used by excitatory motorneurons. del Castillo and colleagues had used pharmacological experiments to show that acetylcholine (ACh) was excitatory, and GABA was inhibitory (del Castillo, De Mello, & Morales, 1963, 1964). Carl Johnson extended this to the level of identified motorneurons: he dissected commissures of single motorneurons and showed that excitors, but not inhibitors, contained choline acetyltransferase, the biosynthetic enzyme for ACh (Johnson & Stretton, 1985). Carl then used a GABA-specific antibody and found that it labels inhibitory, but not excitatory, motorneurons (Johnson & Stretton, 1987). In *C. elegans*, Duerr, Han, Fields, and Rand (2008) found that choline acetyltransferase is localized in excitatory motorneurons, and McIntire, Jorgensen, Kaplan, and Horvitz (1993) identified GABA in inhibitory motorneurons. In addition to their electophysiological similarities, *A. suum* and *C. elegans* use the same classical chemical transmitters. Sydney was right again!

Meanwhile, Jim Angstadt showed that inhibitory motorneurons acted like voltage-controlled oscillators—above a certain threshold, the cells develop membrane potential oscillations with frequency controlled by the size of the input from excitatory motorneurons (Angstadt & Stretton, 1989). Both dorsal and ventral inhibitory motorneurons do this, and Ralph Davis showed that the dorsal and ventral inhibitors oscillate in antiphase. We were extremely excited by this because it brought together all of our analysis of circuitry, both at the electron microscope level (due to Pat Desnoyers and Judy Donmoyer—(see Johnson & Stretton, 1980) and the electrophysiological level. We finally had an explanation for the alternating and periodic contractions of dorsal and ventral muscle that occur during the locomotory behavior of the worm. Champagne flowed! But not for long. Jeff Meade recorded from inhibitory neurons in semi-intact behaving worms and found that the oscillations did not correspond to the waves of excitation and contraction in the musculature (Davis & Stretton, 1996). Our model was wrong: there was something missing from the circuit description. Some people (names withheld) thought this was a failure on our part, but to the contrary we had learned something important, and we knew we had to add to our model. Neuromodulators were an obvious candidate: in other systems, most known neuromodulators are neuropeptides, so we started a new line of research to explore the existence, diversity and activity of peptides in *A. suum*. For me this was particularly satisfying, since it meant going back to my old love, peptide chemistry, but with a quiver of new techniques that were much more sensitive and much faster.

First, we used antibodies against known peptides known from other organisms, both vertebrates and invertebrates. This was Paisarn Sithigorngul's project, and he quickly discovered a wide range of peptide-like immunoreactivities (Sithigorngul, Stretton, & Cowden, 1990); Paisarn was also a wizard at making monoclonal antibodies, and he isolated some highly specific anti-peptide antibodies (Sithigorngul, Cowden, & Stretton, 1996). Cindy Cowden and I did the first chemical isolations of *A, suum* FMRFamide-like peptides using an antibody against FMRFamide generously provided by Eve Marder and Ron Calabrese (Cowden, Stretton, & Davis, 1989; Cowden & Stretton, 1993, 1985; Marder, Calabrese, Nusbaum, & Trimmer, 1987). Sequencing was done by Edman degradation, and then the peptides could be synthesized in large enough amounts for activity determination. We were lucky that the UW-Madison Biotechnology Center had expertise in both sequencing and synthesis.

Ralph Davis did electrophysiology on identified motorneurons and found that endogenous peptides had potent effects (Davis & Stretton, 2001). For example, AF1, the first peptide we isolated, selectively reduces the resistance of inhibitory, but not excitatory, motorneurons, and disrupts signal propagation between the nerve cords; it also paralyses the worm (Cowden *et al.*, 1989). Different peptides had different physiological activity: sometimes individual peptides had complex activity, being excitatory on some targets and inhibitory on others, and sometimes with different temporal effects (Davis & Stretton, 2001).

Art Edison and Lynn Messinger brought molecular cloning into the lab (Edison, Messinger, & Stretton, 1997), and Jen Nanda extended this to *in situ* hybridization so that the expression of peptide-encoding transcripts could be assigned to identified neurons (Nanda & Stretton, 2010). Joanne Yew identified and sequenced new peptides by mass spectrometry (MS) (Yew *et al.*, 2005). Kari Andersen discovered how to dissect identified neuronal cell bodies, and Jessie Jarecki performed single cell MS—the cells are big enough that neuropeptides can be characterized and sequenced by MS, all from a single cell (Jarecki *et al.*, 2010), and with no constraint imposed by relatedness to known peptides. This work was continued by Molly Sygulla and Chris Konop (Konop, Knickelbine, Sygulla, Vestling, & Stretton, 2015a,b), and by Jenny Knickelbine, my last graduate student (Knickelbine *et al.*, 2018). We were building up a robust map of the neurons that synthesize different neuropeptides, robust because it relies on 3 independent techniques, antibody staining, *in situ* hybridization, and single cell MS, all carried out directly on normal unmodified cells.

We also continued to explore bioactivity, either by observing peptide-induced effects on locomotory behavior (Reinitz *et al.*, 2000, 2011) or effects on muscle contraction (Knickelbine *et al.*, 2018). We were surprised to find that almost all the peptides we isolated from *A. suum* affect locomotion and/or muscle contraction. What does this mean—is the way locomotory behavior is controlled as complicated as this seems to imply? We badly need to characterize receptors and match their K_D to the local concentration of each peptide at the receptor site, so that we can assess whether or not cross-talk between peptides encoded by different genes is a serious issue. It would be great to have an optical readout of extracellular peptide concentration at each receptor site, and the way these levels change in time and space, as well as the way each receptor responds. Most peptide receptors are G-protein-coupled receptors (GPCRs) and may themselves be subject to modulation by secondary ligands. The possibilities are as mind-boggling as allosteric control of enzyme activity by compounds unrelated to the catalytic chemistry. It calls for the mind-set of a biochemist—do more and more and more experiments to discover the intricacies of the fiddling done by natural selection. I am appalled at my naiveté when I thought, in 1967, that I was going to 'solve' the *A. suum* nervous system in 5 years; being in a Zoology Department for 47 years, and exposure to modern evolutionary biology, helped me gain a little humility!

Chris Konop extended the single cell work to RNA-Seq, and was quantifying differences in the levels of individual peptides in identified neurons, which had been previously suggested by our experiments using MS, immunocytochemistry and *in situ* hybridization, which are all only semi-quantitative techniques. But then my funding dried up and I had to close the lab, so in 2018 I retired; I was 82, so I am not complaining (much).

Returning to the issue of the *A. suum*/*C. elegans* comparison, the neuropeptide sequences found in *C. elegans* by Chris Li's and by Liliane Schoofs' laboratories, are in many cases identical, or closely sequence-related to the *A. suum* sequences (Husson, Mertens, Janssen, Lindemans, & Schoofs, 2007; Li, 2006; Rosoff, Doble, Price, & Li, 1993; Van Bael *et al.*, 2018). When the analysis was extended by Paul McVeigh and colleagues (McVeigh *et al.*, 2005, 2008) to other species of nematodes by genome and transcript data-mining (genome sequencing spearheaded by Mark Blaxter and colleagues—Kumar, Koutsovoulos, Kaur, & Blaxter, 2012), a similar result emerged. Some peptides are completely conserved among a wide range of nematodes; others clearly fall into sequence-related families. This implies a strong selective pressure on these sequences, but the conservation applies only to the neuropeptide sequences themselves - neuropeptides are cleaved out of precursor proteins by special proteases, and the parts of the precursor protein not destined to be neuropeptides generally show little sequence similarity. The nature of this selective pressure is not yet clear. But once again, this sequence conservation fits right in with Sydney's insight about the *A. suum*/*C. elegans* comparison.

So far in this story, I have brought out those features of the nervous systems of *A. suum* and *C. elegans* where Sydney's leap of faith was rewarded. But eventually, at some level, the identity has to break down—after all, they are different species. Their locomotory behavior is certainly different—in both worms, locomotion is driven by propagating body waves, either forwards or backwards. In adult *A. suum*, for forward-moving waves, the body generates 3 waveforms (of differing wavelength in different parts of the body) whereas in *C. elegans* there is slightly more than one waveform in the whole body. In addition, in *A. suum*, forward propagating waves drive the worm forwards, whereas in *C. elegans* they drive it backwards. We still do not understand these differences, and we are constantly watching for differences that might help. Faced with this issue, parasitologists shout 'Of course they are different!'. We agree, but we want an explanation. One idea we really like is that it is different cellular localization of neuropeptide expression (and/or the cellular expression of neuropeptide receptors) that can lead to behavioral differences. As explained above, in *A. suum* we feel confident about the peptide localizations because we have used three independent techniques, all of which identify the cells that contain either the peptide-encoding transcript or the peptide itself. In *C. elegans*, the localization has mostly been indirect, using DNA constructs that include GFP marker sequences and peptide-gene specific control sequences (Kim & Li, 2004). The results show that in most cases, based on the available evidence, the expression of homologous peptides is different in the two worms (Sithigorngul et al., 2011; Nanda & Stretton, 2010; Jarecki *et al.*, 2010; Konop *et al.*, 2015a,b). It is an open question whether the expression pattern seen with the DNA constructs is always faithful, i.e. whether the control sequences on the construct are complete. It is well known that control sequences can be located far from the gene being controlled, so it is possible that not all were included.

I like to think that Sydney would have been pleased by what we found in *A. suum*, but he had long since moved on, and had mostly lost any deep interest in explaining nematode behavior; the world of neuromodulation seems to be so intricate and detailed, a level that in general, over and over again, Sydney openly said he found boring, so he may well have been largely unaware of what we had been doing. I would have loved to have had a conversation with him about the evolutionary implications of what we found. Sydney certainly had already embraced the comparative approach when he proposed the *A. suum/C. elegans* comparison, and evolution was his passion at the end.

Since I left Cambridge, I only saw Sydney three times, twice when he visited Madison, and once when I re-visited the LMB. While he was still writing them, I used to relish his articles in Current Biology (Loose End… .False Starts – 2009), pure Sydney, still playful, still the master word spinner, and still the *enfant terrible*. I owe him a lot—especially for his generosity in sending me to Woods Hole and Harvard Medical School, which changed my life, as well as for his support in becoming a Fellow of King's College, Cambridge for 6 years. I loved working with him, loved his brilliance and his wit, and his devotion to excellence in science. What an extraordinary mind. Now that I have retired my activated state (Professor*) is rapidly decaying and I no longer meet undergraduates, many of whom used to ask me for advice on where to go to graduate school. I always told them that they should seek out someone with a mind that they admire, and with which their own minds resonate. This is what I felt about Sydney—he enriched my life, and I am eternally grateful. I miss his scientific grasp, as well as his wicked wit.

Acknowledgement

I am very grateful to Philippa Claude for her insightful critique.

References

Anfinsen, C.B., & Haber, E. (1961). Studies on the reduction and reformation of protein disulfide bonds. *The Journal of Biological Chemistry, 236*, 1361–1363.

Angstadt, J.D., & Stretton, A.O. (1989). Slow active potentials in ventral inhibitory motor neurons of the nematode Ascaris. *Journal of Comparative Physiology. A, Sensory, Neural, and Behavioral Physiology, 166*(2), 165–177. doi:10.1007/BF00193461

Benzer, S., Ingram, V.M., & Lehmann, H. (1958). Three varieties of human haemoglobin D. *Nature, 182* (4639), 852–854. doi:10.1038/182852a0

Benzer, S. (1961). On the topography of the genetic fine structure. *Proceedings of the National Academy of Sciences of the United States of America, 47*(3), 403–415. doi:10.1073/pnas.47.3.403

Brenner, S., Jacob, F., & Meselson, M. (1961). An unstable intermediate carrying information from genes to ribosomes for protein synthesis. *Nature, 190*, 576–581. doi:10.1038/190576a0

Brenner, S., Stretton, A.O., & Kaplan, S. (1965). Genetic code: The 'nonsense' triplets for chain termination and their suppression. *Nature, 206*(988), 994–998. doi:10.1038/206994a0

Brenner, S. (1971). Nonsense mutants and the genetic code. *JAMA, 218*(7), 1023–1026. doi:10.1001/jama.1971.03190200055011

Brenner, S. (1974). The genetics of *Caenorhabditis elegans*. *Genetics, 77*(1), 71–94.

Brenner, S. (2001). *My Life in Science, as told to Lewis Wolpert*. (Errol C. Friedberg and Eleanor Lawrence, Eds.). London, UK: Faculty of 1000 Ltd.

Cowden, C., Stretton, A.O., & Davis, R.E. (1989). AF1, a sequenced bioactive neuropeptide isolated from the nematode *Ascaris suum*. *Neuron, 2*(5), 1465–1473. doi:10.1016/0896-6273(89)90192-x

Cowden, C., & Stretton, A.O. (1993). AF2, an *Ascaris* neuropeptide: Isolation, sequence, and bioactivity. *Peptides, 14*(3), 423–430. doi:10.1016/0196-9781(93)90127-3

Cowden, C., & Stretton, A.O. (1995). Eight novel FMRFamide-like neuropeptides isolated from the nematode *Ascaris suum*. *Peptides, 16* (3), 491–500. doi:10.1016/0196-9781(94)00211-n

Crick, F.H.C., Barnett, L., Brenner, S., & Watts-Tobin, R.J. (1961). General nature of the genetic code for proteins. *Nature, 192*, 1227–1232. doi:10.1038/1921227a0

Crick, F.H.C. (1958). On protein synthesis. *Symposia of the Society for Experimental Biology, 12*, 138–163.

Crick, F.H., Griffith, J.S., Orgel, L.E. (1957). Codes without commas. *Proceedings of the National Academy of Sciences of the United States of America, 43* (5), 416–421. PMID:16590032. doi:10.1073/pnas.43.5.416., &

Davis, R.E., & Stretton, A.O. (1989a). Passive membrane properties of motorneurons and their role in long-distance signaling in the nematode Ascaris. *The Journal of Neuroscience : The Official Journal of the Society for Neuroscience, 9*(2), 403–14. doi:10.1523/JNEUROSCI.09-02-00403.1989

Davis, R.E., & Stretton, A.O. (1989b). Signaling properties of *Ascaris motorneurons*: Graded active responses, graded synaptic transmission, and tonic transmitter release. *The Journal of Neuroscience : The Official Journal of the Society for Neuroscience, 9*(2), 415–425. doi:10.1523/JNEUROSCI.09-02-00415.1989

Davis, R.E., & Stretton, A.O. (1996). The motornervous system of *Ascaris*: Electrophysiology and anatomy of the neurons and their control by neuromodulators. *Parasitology, 113* (S1), S97–S117. doi:10.1017/S0031182000077921

Davis, R.E., & Stretton, A.O. (2001). Structure-activity relationships of 18 endogenous neuropeptides on the motor nervous system of the nematode Ascaris suum. *Peptides, 22* (1), 7–23. doi:10.1016/S0196-9781(00)00351-X

del Castillo, J., De Mello, W.C., & Morales, T. (1963). The physiological role of acetylcholine in the neuromuscular system of *Ascaris lumbricoides*. *Archives Internationales de Physiologie et de Biochimie, 71*, 744–757.

del Castillo, J., De Mello, W.C., & Morales, T. (1964). Inhibitory action of gamma-aminobutyric acid (GABA) on Ascaris muscle. *Experientia, 20*(3), 141–143. doi:10.1007/BF02150701

Dintzis, H.M. (1961). Assembly of the peptide chains of hemoglobin. *Proceedings of the National Academy of Sciences of the United States of America, 47*, 247–261. doi:10.1073/pnas.47.3.247

Duerr, J.S., Han, H.P., Fields, S.D., & Rand, J.B. (2008). Identification of major classes of cholinergic neurons in the nematode *Caenorhabditis elegans*. *The Journal of Comparative Neurology, 506*(3), 398–408. doi:10.1002/cne.21551

Edison, A.S., Messinger, L.A., & Stretton, A.O. (1997). afp-1: A gene encoding multiple transcripts of a new class of FMRFamide-like neuropeptides in the nematode *Ascaris suum*. *Peptides, 18* (7), 929–935. doi:10.1016/s0196-9781(97)00047-8

Epstein, R.H., Bolle, A., Steinberg, C.M., Kellenberger, E., Boy de la Tour, E., Chevalley, R., … Lielausis, A. (1963). Physiological studies of conditional lethal mutants of bacteriophage T4D. *Cold Spring Harbor Symposia on Quantitative Biology., 28*(0), 375–394. doi:10.1101/SQB.1963.028.01.053

The *C. elegans* Sequencing Consortium. (1998). Genome sequence of the nematode *C. elegans*: A platform for investigating biology. *Science, 282*, 2012–2018. doi:10.1126/science.282.5396.2012

Goldschmidt, R. (1908). Das Nervensystem von *Ascaris lumbricoides* und *Ascaris megalocephala*. 1. *Zeitschrift fur Wissenschaftliche Zoologie, 90*, 73–136.

Goodman, H.M., Abelson, J., Landy, A., Brenner, S., & Smith, J.D. (1968). Amber suppression: A nucleotide change in the anticodon of a tyrosine transfer RNA. *Nature, 217*(5133), 1019–1024. doi:10.1038/2171019a0

Goodman, M.B., Hall, D.H., Avery, L., & Lockery, S.R. (1998). Active currents regulate sensitivity and dynamic range in *C. elegans* neurons. *Neuron, 20*(4), 763–772. doi:10.1016/S0896-6273(00)81014-4

Gros, F., Hiatt, H., Gilbert, W., Kurland, C.G., Risebrough, R.W., & Watson, J.D. (1961). Unstable ribonucleic acid revealed by pulse labelling of *Escherichia coli*. *Nature, 190*, 581–585. doi:10.1038/190581a0

Hesse, R. (1892). Uber das Nervensystem von *Ascaris lumbricoides* und *Ascaris megalocephala*. *Zeitschrift für wissenschaftliche Zoologie. Abteilung A, 90*, 73–136.

Husson, S.J., Mertens, I., Janssen, T., Lindemans, M., & Schoofs, L. (2007). Neuropeptidergic signaling in the nematode *Caenorhabditis elegans*. *Progress in Neurobiology*, *82*(1), 33–55. doi:10.1016/j.pneurobio.2007.01.006

Ingram, V.M., & Stretton, A.O. (1961). Human haemoglobin A_2: Chemistry, genetics and evolution. *Nature*, *190*, 1079–1084. doi:10.1038/1901079a0

Ingram, V.M. (1956). A specific chemical difference between the globins of normal human and sickle-cell anaemia haemoglobin. *Nature*, *178*(4537), 792–794. doi:10.1038/178792a0

Jarecki, J.L., Andersen, K., Konop, C.J., Knickelbine, J.J., Vestling, M.M., & Stretton, A.O. (2010). Mapping neuropeptide expression by mass spectrometry in single dissected identified neurons from the dorsal ganglion of the nematode *Ascaris suum*. *ACS Chemical Neuroscience*, *1*(7), 505–519. doi:10.1021/cn1000217

Johnson, C.D., & Stretton, A.O. (1985). Localization of choline acetyltransferase within identified motoneurons of the nematode Ascaris. *The Journal of Neuroscience : The Official Journal of the Society for Neuroscience*, *5*(8), 1984–1992. doi:10.1523/JNEUROSCI.05-08-01984.1985

Johnson, C.D., & Stretton, A.O. (1987). GABA-immunoreactivity in inhibitory motor neurons of the nematode Ascaris. *The Journal of Neuroscience : The Official Journal of the Society for Neuroscience*, *7*(1), 223–235. doi:10.1523/JNEUROSCI.07-01-00223.1987

Johnson, C.D., & Stretton, A.O.W. (1980). Neural control of locomotion in *Ascaris*: Anatomy, electrophysiology, and biochemistry. *In Nematodes as biological models* (Vol. 1, 159–195). New York: Academic Press.

Kaplan, S., Stretton, A.O., & Brenner, S. (1965). Amber suppressors: Efficiency of chain propagation and suppressor specific amino acids. *Journal of Molecular Biology*, *14*(2), 528–533. doi:10.1016/S0022-2836(65)80202-9

Kim, K., & Li, C. (2004). Expression and regulation of an FMRFamide-related neuropeptide gene family in *Caenorhabditis elegans*. *The Journal of Comparative Neurology*, *475* (4), 540–550. doi:10.1002/cne.20189

Knickelbine, J.J., Konop, C.J., Viola, I.R., Rogers, C.B., Messinger, L.A., Vestling, M.M., & Stretton, A.O.W. (2018). Different bioactive neuropeptides are expressed in two sub-classes of GABAergic RME nerve ring motorneurons in *Ascaris suum*. *ACS Chemical Neuroscience*, *9*(8), 2025–2040. doi:10.1021/acschemneuro.7b00450

Konop, C.J., Knickelbine, J.J., Sygulla, M.S., Vestling, M.M., & Stretton, A.O. (2015a). Different neuropeptides are expressed in different functional subsets of cholinergic excitatory motorneurons in the nematode *Ascaris suum*. *ACS Chemical Neuroscience*, *6* (6), 855–870. doi:10.1021/cn5003623

Konop, C.J., Knickelbine, J.J., Sygulla, M.S., Wruck, C.D., Vestling, M.M., & Stretton, A.O. (2015b). Mass spectrometry of single GABAergic somatic motorneurons identifies a novel inhibitory peptide, As-NLP-22, in the nematode *Ascaris suum*. *Journal of the American Society for Mass Spectrometry*, *26*(12), 2009–2023. doi:10.1007/s13361-015-1177-z

Kumar, S., Koutsovoulos, G., Kaur, G., & Blaxter, M., (2012). Toward 959 nematode genomes. *Worm*, *1*(1), 42–50. doi:10.4161/worm.19046

Li, C. (2006). The ever-expanding neuropeptide gene families in the nematode *Caenorhabditis elegans*. *Parasitology*, *131*(S1), S109–S127. doi:10.1017/S0031182005009376

Li, C., & Kim, K. (2008). Neuropeptides. *WormBook*, *25*, 1–36.

Liu, Q., Hollopeter, G., & Jorgensen, E.M. (2009). Graded synaptic transmission at the *Caenorhabditis elegans* neuromuscular junction. *Proceedings of the National Academy of Sciences of the United States of America*, *106*(26), 10823–10828. doi:10.1073/pnas.0903570106

Marder, E., Calabrese, R.L., Nusbaum, M.P., & Trimmer, B. (1987). Distribution and partial characterization of FMRFamide-like peptides in the stomatogastric nervous systems of the rock crab, Cancer borealis, and the spiny lobster, *Panulirus interruptus*. *The Journal of Comparative Neurology*, *259*(1), 150–163. doi:10.1002/cne.902590111

McIntire, S.L., Jorgensen, E., Kaplan, J., & Horvitz, H.R. (1993). The GABAergic nervous system of *Caenorhabditis elegans*. *Nature*, *364*(6435), 337–341. doi:10.1038/364337a0

McVeigh, P., Alexander-Bowman, S., Veal, E., Mousley, A., Marks, N.J., & Maule, A.G. (2008). Neuropeptide-like protein diversity in phylum Nematoda. *International Journal for Parasitology*, *38*(13), 1493–1503. doi:10.1016/j.ijpara.2008.05.006

McVeigh, P., Leech, S., Mair, G.R., Marks, N.J., Geary, T.G., & Maule, A.G. (2005). Analysis of FMRFamide-like peptide (FLP) diversity in phylum Nematoda. *International Journal for Parasitology*, *35*(10), 1043–1060. doi:10.1016/j.ijpara.2005.05.010

Nanda, J.C., & Stretton, A.O.W. (2010). In situ hybridization of neuropeptide-encoding transcripts afp-1, afp-3, and afp-4 in neurons of the nematode *Ascaris suum*. *The Journal of Comparative Neurology*, *518*(6), 896–910. doi:10.1002/cne.22251

Nirenberg, M.W., & Matthaei, J.H. (1961). The dependence of cell-free protein synthesis in *E. coli* upon naturally occurring or synthetic polyribonucleotides. *Proceedings of the National Academy of Sciences of the United States of America*, *47*, 1588–1602. doi:10.1073/pnas.47.10.1588

Orgel, A., & Brenner, S. (1961). Mutagenesis of Bacteriophage T4 by acridines. *Journal of Molecular Biology*, *3*, 762–768. doi:10.1016/s0022-2836(61)80081-8

Reinitz, C.A., Herfel, H.G., Messinger, L.A., & Stretton, A.O. (2000). Changes in locomotory behavior and cAMP produced in Ascaris suum by neuropeptides from *Ascaris suum* or *Caenorhabditis elegans*. *Molecular and Biochemical Parasitology*, *111* (1), 185–197. doi:10.1016/s0166-6851(00)00317-0

Reinitz, C.A., Pleva, A.E., & Stretton, A.O. (2011). Changes in cyclic nucleotides, locomotory behavior, and body length produced by novel endogenous neuropeptides in the parasitic nematode Ascaris suum. *Molecular and Biochemical Parasitology*, *180* (1), 27–34. doi:10.1016/j.molbiopara.2011.08.001

Richmond, J.E., & Jorgensen, E.M. (1999). One GABA and two acetylcholine receptors function at the C. elegans neuromuscular junction. *Nature Neuroscience*, *2*(9), 791–797. doi:10.1038/12160

Rosenbluth, J. (1965). Ultrastructure of somatic muscle cells in *Ascaris lumbricoides*. II. Intermuscular junctions, neuromuscular junctions, and glycogen stores. *The Journal of Cell Biology*, *26*(2), 579–591. doi:10.1083/jcb.26.2.579

Rosoff, M.L., Doble, K.E., Price, D.A., & Li, C. (1993). The *flp-1* propeptide is processed into multiple, highly similar FMRFamide-like peptides in *Caenorhabditis elegans*. *Peptides*, *14*(2), 331–338. doi:10.1016/0196-9781(93)90049-m

Sarabhai, A.S., Stretton, A.O., Brenner, S., & Bolle, A., (1964). Co-linearity of the gene with the polypeptide chain. *Nature*, *201*, 13–17. doi:10.1038/201013a0

Sithigorngul, P., Cowden, C., & Stretton, A.O. (1996). Heterogeneity of cholecystokinin/gastrin-like immunoreactivity in the nervous system of the nematode *Ascaris suum*. *The Journal of Comparative Neurology*, *370*(4), 427–442. doi:10.1002/(SICI)1096-9861(19960708)370:4<427::AID-CNE2>3.0.CO;2-6

Sithigorngul, P., Jarecki, J.L., & Stretton, A.O. (2011). A specific antibody to neuropeptide AF1 (KNEFIRFamide) recognizes a small subset of neurons in *Ascaris suum*: Differences from *Caenorhabditis elegans*. *The Journal of Comparative Neurology*, *519*(8), 1546–1561. doi:10.1002/cne.22584

Sithigorngul, P., Stretton, A.O., & Cowden, C. (1990). Neuropeptide diversity in *Ascaris*: An immunocytochemical study. *The Journal of Comparative Neurology*, *294*(3), 362–376. doi:10.1002/cne.902940306

Stretton, A.O., & Brenner, S. (1965). Molecular consequences of the amber mutation and its suppression. *Journal of Molecular Biology*, *12*, 456–465. doi:10.1016/s0022-2836(65)80268-6

Stretton, A.O., & Brenner, S. (1967). Spontaneous revertants of amber mutants. *Journal of Molecular Biology.*, *28*, 137–139.

Stretton, A.O., Fishpool, R.M., Southgate, E., Donmoyer, J.E., Walrond, J.P., Moses, J.E., & Kass, I.S. (1978). Structure and physiological activity of the motoneurons of the nematode Ascaris. *Proceedings of the National Academy of Sciences of the United States of America*, *75* (7), 3493–3497. doi:10.1073/pnas.75.7.3493

Stretton, A.O., Kaplan, S., & Brenner, S. (1966). Nonsense codons. *Cold Spring Harbor Symposia on Quantitative Biology, 31*, 173–179. doi: 10.1101/sqb.1966.031.01.025

Stretton, A.O.W., & Kravitz, E.A. (1968). Neuronal Geometry: Determination with a technique of intracellular dye injection. *Science (New York, N.Y.), 162* (3849), 132–134. doi:10.1126/science.162.3849.132

Sulston, J., & Ferry, G. (2002). *The common thread*. Washington, DC: Joseph Henry Press.

Sulston, J.E., & Horvitz, H.R. (1977). Post-embryonic cell lineages of the nematode, *Caenorhabditis elegans. Developmental Biology, 56*(1), 110–156. doi:10.1016/0012-1606(77)90158-0

Sulston, J.E., E., Schierenberg, J.G., White, & J.N., Thomson, (1983). The embryonic cell lineage of the nematode *Caenorhabditis elegans. Developmental Biology., 100*(1), 64–19. doi:10.1016/0012-1606(83)90201-4

Terzaghi, E., Okada, Y., Streisinger, G., Emrich, J., Inouye, M., & Tsugita, A. (1966). Change of a sequence of amino acids in phage T4 lysozyme by acridine-induced mutations. *Proceedings of the National Academy of Sciences of the United States of America, 56*(2), 500–507. doi:10.1073/pnas.56.2.500

Van Bael, S., Zels, S., Boonen, K., Beets, I., Schoofs, L., & Temmerman, L. (2018). A *Caenorhabditis elegans* mass spectrometric resource for neuropeptidomics. *Journal of the American Society for Mass Spectrometry, 29*(5), 879–889. doi:10.1007/s13361-017-1856-z

Vanfleteren, J.R., Van de Peer, Y., Blaxter, M.L., Tweedie, S.A., Trotman, C., Lu, L., … Moens, L. (1994). Molecular genealogy of some nematode taxa as based on cytochrome c and globin amino acid sequences. *Molecular Phylogenetics and Evolution, 3*(2), 92–101. doi:10.1006/mpev.1994.1012

Walrond, J.P., & Stretton, A.O. (1985). Reciprocal inhibition in the motor nervous system of the nematode Ascaris: Direct control of ventral inhibitory motoneurons by dorsal excitatory motoneurons. *The Journal of Neuroscience : The Official Journal of the Society for Neuroscience, 5*(1), 9–15. doi:10.1523/JNEUROSCI.05-01-00009.1985

Watson, J.D., & Crick, F.H.C. (1953). Molecular structure of nucleic acids; a structure for deoxyribose nucleic acid. *Nature, 171*(4356), 737–738. doi:10.1038/171737a0

Weigert, M.G., & Garen, A. (1965). Base composition of nonsense codons in *E. coli*. Evidence from amino-acid substitutions at a tryptophan site in alkaline phosphatase. *Nature, 206*(4988), 992–994. doi: 10.1038/206992a0

White, J.G., Southgate, E., Thomson, J.N., & Brenner, S. (1986). The structure of the nervous system of the nematode *Caenorhabditis elegans. Philosophical Transactions of the Royal Society of London. Series B, Biological Sciences, 314*(1165), 1–340. doi:10.1098/rstb.1986.0056

Yanofsky, C., Carlton, B.C., Guest, J.R., Helinski, D.R., & Henning, U. (1964). On the colinearity of gene structure and protein structure. *Proceedings of the National Academy of Sciences of the United States of America, 51*, 266–272. doi:10.1073/pnas.51.2.266

Yew, J.Y., Kutz, K.K., Dikler, S., Messinger, L., Li, L., & Stretton, A.O. (2005). Mass spectrometric map of neuropeptide expression in Ascaris suum. *The Journal of Comparative Neurology, 488*(4), 396–413. doi:10.1002/cne.20587]

John Sulston (1942–2018): a personal perspective

Robert H. Waterston and Donald G. Moerman

ABSTRACT
John Sulston changed the way we do science, not once, but three times – initially with the complete cell lineage of the nematode *Caenorhabditis elegans*, next with completion of the genome sequences of the worm and human genomes and finally with his strong and active advocacy for open data sharing. His contributions were widely recognized and in 2002 he received the Nobel Prize in Physiology and Medicine.

The cell lineage

John got involved in the lineage almost by accident. Joining the renowned MRC Laboratory of Molecular Biology in Cambridge in 1968 as a staff scientist in Sydney Brenner's group, he set out to discover neurotransmitters in the small nematode *C. elegans*. Sydney had settled on *C. elegans* for exhaustive studies of its development and nervous system because of its small size, simple nervous system and potential for genetic analysis. Others had already established that acetylcholine and GABA were important transmitters, so John elected to look for catecholamines. In his final months as a post-doc in Leslie Orgel's lab at the Salk Institute he had taken a summer course in neurobiology and learned procedures for detecting catecholamines using formaldehyde-induced-fluorescence (FIF). Unfortunately, he found these methods didn't work in the small worm because diffusion spread the signal, preventing identification of specific cells that might be catecholamine positive. To overcome this problem, he devised a clever way of freeze-drying the worms that prevented the spread of signal and allowed him to pinpoint the source of the signal. He then set off to find mutants directly that lacked the FIF signal, since he had no way of knowing what behavioral changes might result from catecholamine deficits. He adapted his methods so that he could screen thousands of animals – he was even in these early days undeterred by scale – and indeed found multiple mutants in several genes that altered the FIF signal, with some eliminating the signal altogether (Sulston, Dew, & Brenner, 1975). To his disappointment, he could find no obvious behavioral deficit. Subtle alterations in behavior would eventually be identified using those mutants but not until many years later. Those mutants could be reanalyzed because worms survive freezing and long-term storage, a protocol that John ported from mammalian tissue culture labs and adapted for the worm.

To complete his study, he needed to place the FIF-positive neurons in the larger nervous system. He stained worms with DAPI to reveal their nuclei and to his surprise he found that the adult worms had more cells in their nervous systems than newly hatched L1's. This finding contradicted the prevailing view that worms hatched with their full complement of somatic cells, yet he was able to show conclusively that the number of neurons in the ventral cord rose from 15 just after hatching to 57 by the end of the first larval stage. How did these new cells arise? Others in the lab were studying the embryonic lineage using differential interference contrast (DIC) microscopy – the transparency of the worm allowed the eggs to be observed live under the microscope and DIC revealed the cells and their nuclei as shades of grey in a finely shadowed field, like craters on the moon – and John started using the same methods on the newly hatched larvae. Initial attempts, however, came to naught since, unlike the eggs, the L1's were mobile, making tracking of the cells near impossible. To overcome this, John came up with a simple solution: place a small dollop of *E. coli* in the middle of the slide and the L1's stayed in one place, happily grazing. After an exciting weekend following L1's, John had the basic lineage of the extra ventral cord neurons figured out. Strikingly, the new neurons arose from the 15 precursors in a stereotypic pattern, with each precursor giving rise to a replacement for the parent cell and 3, 4 or 5 neurons. The variation in neuron progeny arose because of what was recognized as programmed cell death in some lineages, whereas in others those cells survived (J. E. Sulston, 1976).

In addition to the neurons added to the ventral cord, John noted other dividing cells and working with Bob Horvitz, then a recently arrived post-doc at the LMB, he worked out the full post-embryonic lineage of both the hermaphrodite and male (Sulston & Horvitz, 1977), with the exception of the somatic gonad, which he graciously handed over to Judith Kimble, a graduate student with David Hirsh at the time. He followed up these fundamental advances over the next few years with a variety of studies with colleagues. He was also called upon to resolve a dispute centered on the embryonic lineage of the intestine, but generally stayed away from any work on the embryonic lineage. In part this reluctance came from a desire not to interfere with those who were working on it, albeit unsuccessfully, and in part because the embryonic lineage was going to be much more difficult.

By 1981, however, with a sense of wanting to complete unfinished business, John decided to take on the embryonic lineage. He devised methods to help him succeed where others had failed. He used gossamers in the eyepiece and a sliding stage to track a nucleus of interest; he made quick sketches of the cells he was following and devised a color scheme to reflect their depth in the embryo (Figure 1). He grew to recognize all the cells in the 28 cell embryo, just prior to gastrulation, so that he could begin each day a couple hours into embryogenesis. He realized that he could put an embryo at 4 degrees overnight and pick up the next day where he had left off. But most of all he brought incredible powers of concentration and acute observation to the problem. He basically locked himself in a dark room each day to avoid any distractions, famously wearing a hole in the floor where his chair moved on the floor. But in a year and half, he had completed what many at the outset had viewed as impossibly difficult, perhaps quixotic, tracing the origins of the 558 final cells and the 113 cell deaths (Sulston, Schierenberg, White, & Thomson, 1983).

The final challenge was establishing the identity of cell types produced by the lineage. For some cells, such as those spread out along the length of the body, this task was straightforward, since the anatomy of the worm was increasingly well known. But for the neuronal cells in the head, John painstakingly followed the cells in a single embryo, fixed the embryo and then using serial sections in electron microscopy identified the features in each cell that allowed him to link his lineage to the neurons that John White was characterizing in the adult animal (White, Southgate, Thomson, & Brenner, 1986). With that task complete, John knew the origin of every cell in the animal throughout the life cycle.

The lineage proved in many respects bewildering. While the post-embryonic motor neurons of the ventral cord derived from a clear pattern across the fifteen precursors, the derivation of cell types in the embryo defied logic. While 80 of the 81 muscle cells derived from what had classically been defined as mesodermal lineages, the remaining cell was the product of the neuroectodermal AB founder. Some branches of the AB lineage (ABala, ABalpp) produced only neurons and glia, other branches gave rise to a variety of cell types, with a commitment to a neuronal cell type only at the last division. And some neurons arose from the otherwise completely mesodermal/hypodermal C founder cell. The mapping of the three-fold symmetric pharyngeal cells onto the binary lineage produced even more puzzling patterns, with seemingly cells recruited willy-nilly to generate the third occurrence of a specific cell type. It was clear, as Francois Jacob has written, evolution is a 'tinkerer, not an engineer.' (Jacob, 1977). For his work deciphering the lineage John shared the Nobel prize in Physiology and Medicine in 2002.

Genomes: the map

With the lineage completed, John made the startling choice to build a genome map of *C. elegans*, rather than use his knowledge of the lineage to study development. He had watched others spend weeks, months and even years trying to track down the DNA that encoded a gene that had given rise to an interesting mutant phenotype. John realized before almost anyone, how powerful a map would be in letting researchers find a gene in the enormity of a genome. The idea is deceptively simple in concept and was being pursued independently in yeast by Maynard Olson. Genomic libraries are collections of random fragments from across the genome. By repeatedly sampling from such a library, two clones may eventually be found that overlap. With ever deeper sampling, more overlaps will be found, until with enough redundancy, the overlapping clones will reconstitute the entire genome. At least, this was the theory.

Beyond the sheer numbers required for a genome as simple as the 100 Mb worm genome, the challenge was to find a method to characterize the cloned fragments well enough to enable the detection of unambiguous overlaps. Not having any first-hand knowledge of the then new molecular biology techniques, John worked with Jon Karn and Sydney to develop a strategy that would efficiently produce a unique 'fingerprint' of genomic clones that exploited the ability of restriction endonucleases to cleave DNA at specific sequences (Figure 2). Progress was steady but slow as John grew clones, made DNA, cleaved with restrictions enzymes, recorded the size patterns and searched for possible overlaps. But things sped up considerably when Alan Coulson joined John. Alan had worked with Fred Sanger on the development of DNA sequencing methods and their application to small genomes. Fred was retiring and Alan found just the challenge he was looking for in John's project. They divided up the tasks, with Alan focused on data production and methods development and John focused on developing the methods and software to speed data entry and analysis. This meant learning how to program in Fortran with some early encouragement from Roger Staden. Soon they had fingerprinted hundreds and then thousands of clones. In back-to-back papers with Olson in 1985, they published their initial map after fingerprinting more than 7000 clones, a truly astounding number at the time (Coulson, Sulston, Brenner, & Karn, 1986; Olson *et al.*, 1986).

The analysis of the data, however, revealed a problem; the clones were not a random sampling of the genome. Instead, some regions, like the rDNA repeat locus, were overrepresented in the libraries and about 10–20% of the

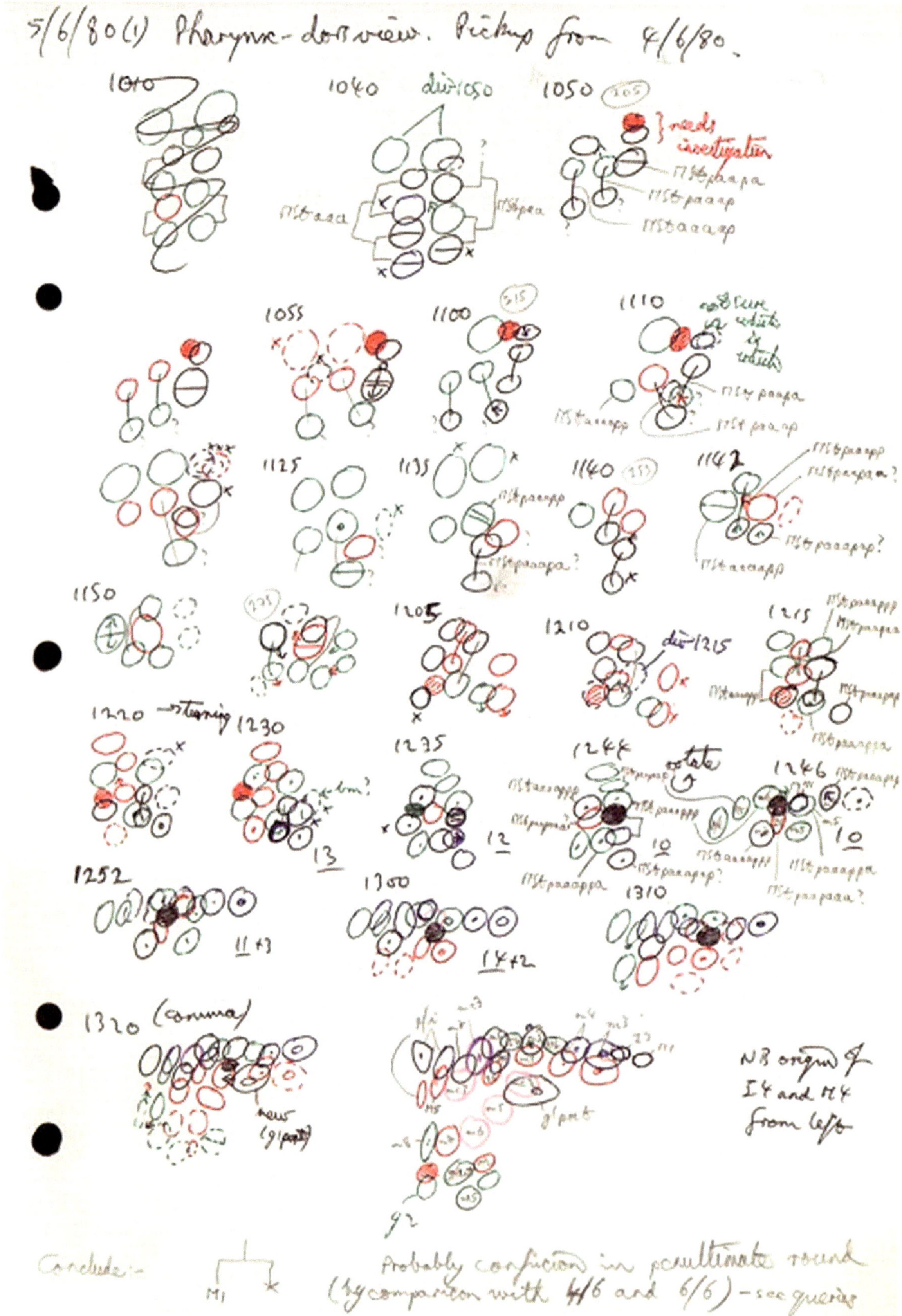

Figure 1. One of hundreds of drawings John used to trace the lineage (Gitschier, 2006). In this case he is tracing a cluster of cells at successive time points to investigate the progeny of the solid red cell at the top. The hatch marks are cell deaths. When asked how he arrived at such a scheme John responded 'Desperation'.

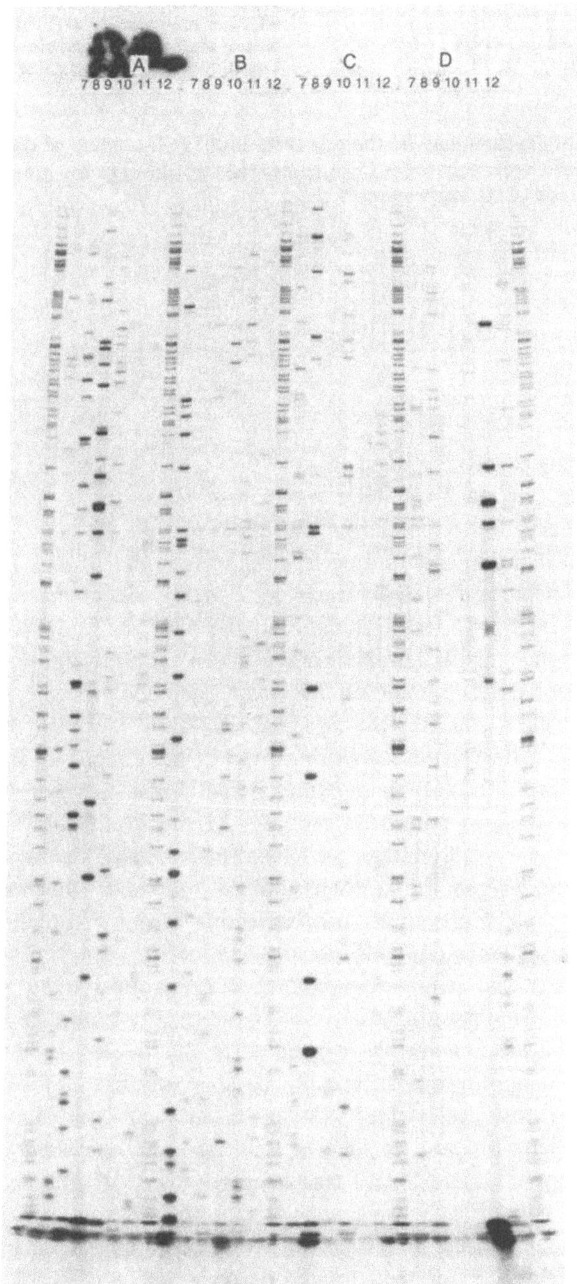

Figure 2. An example of a fingerprinting gel used in the mapping of the *C. elegans* genome. Lanes contain different samples with distinct restriction enzyme digested bands, while every seventh lane contains markers (Coulson *et al.*, 1986).

genome was not represented at all. The result was a badly fragmented map, with about 700 contigs (or sets of overlapping clones), rather than the hoped for 6 segments, with one for each chromosome. Nonetheless, the map was proving extremely useful to the worm community, simplifying their gene searches, especially as the contigs became associated with specific locations on the genetic map (see later).

John and Alan (the two were so closely associated in communications about the map that a technician in the Waterston lab thought they were one person – John N. Allen) tried new vectors, modified bacterial hosts and targeted searching in even more complex libraries, all with little improvement in genome coverage. Fortunately, students in Maynard Olson's laboratory were developing yeast based cloning methods, called Yeast Artificial Chromosomes (YACs), that not only offered the possibility of cloning larger genome fragments (the larger the pieces, the easier the puzzle) but also a eukaryotic host with at least different biases than those of *E.* coli (Burke, Carle, & Olson, 1987). At the time Bob Waterston had the good fortune to be on sabbatical at the LMB and located in John's lab, spending at least some of his time trying to find solutions to the gap problem. After some discussion, including sessions in the local pub, they agreed that Bob should pursue making worm genomic libraries in YACs upon returning to St. Louis. That cemented a close collaboration that extended over the next 15 years and the sequencing of two genomes, first the worm and then the human.

The YAC libraries turned out to have excellent representation of the missing regions. While the YACs failed to give good fingerprints, both because of their size and the difficulties in purifying them away from the yeast genome, their large size and the advanced state of the fingerprint map meant that individual YACs could be hybridized against a grid of *E. coli* clones representing the 700 contigs. YACs spanning the gaps would hybridize to clones from two or more contigs. Over the next couple years the map zipped up with almost no gaps beyond those between chromosomes (Coulson, Waterston, Kiff, Sulston, & Kohara, 1988, Coulson *et al.*, 1991). They selected 1000 YACs representing the whole genome, arrayed the clones on grids and shared them with the community. At last the worm genome was captured in an ordered array of clones.

Genomes: sequence

With the map in hand and the buzz about the initiation of the Human Genome Project, sequencing the worm genome was an obvious next step. Indeed, the idea of sequencing the worm genome had probably been there almost from the start. When John outlined his ideas for the mapping project in a one page note in the November 1984 issue of the Worm Breedeer's Gazette, there was a cryptic phrase – 'of providing information about the genome as a whole' – that might suggest he was thinking about it even then . Certainly, when he attended an early meeting that considered the Human Genome Project and when he appeared before the Albert's committee that was developing plans for sequencing the human genome (National Research Council *et al.*, 1988), John must have considered it. But he was focused on the map and, having watched Fred Sanger's efforts on the bacteriophage lambda genome and Bart Barrell's later efforts on various viral genomes, John knew better than most the challenges that would be faced in tackling a genome as large as the worm's 100 Mb genome. Even as the map was closing John avoided any commitment.

That all changed in early 1989, when Bob Horvitz got word that Jim Watson, who by then had assumed the leadership of the HGP, was considering leaving *C. elegans* off the list model organisms to include as testing grounds for sequencing strategies in the NIH's program. Horvitz recognized the disadvantages his lab (and other worm labs) would face if, for example, the fly genome were sequenced and the worm was left with only a map. He urged John to spearhead a worm sequencing effort and to approach Watson to convince him that the worm map provided the perfect starting point for a genome sequencing project. As luck would have it, the biennial international worm meeting was being held at CSHL in May of that year and a meeting was arranged with Watson, who, despite becoming director of the newly formed National Center for Human Genome Research, was still director of the CSHL. At that fateful meeting Watson agreed to entertain a grant proposal from Sulston and Waterston for a pilot project to sequence the worm genome. John famously described his feelings after the meeting of having a sense that the jail door had just slammed shut; he knew he was saying goodbye to his style of science of working alone or with one or two collaborators. Instead, he would have to direct others, a position he had studiously avoided, even avoiding taking on graduate students unless they were so independent that they wouldn't need his direction at all.

Once committed, John focused his energies on figuring out how to sequence the worm genome. He had never sequenced DNA himself, so he got Alan to get him started (although Alan had played such an important role with Sanger in getting DNA sequencing methods to work, he had no interest in more sequencing and continued to work on the worm map, closing the last few gaps and providing clones to investigators around the world). John tested various strategies, and rapidly concluded that shotgun sequencing would be most effective at the scale needed for the worm, where, analogous to the mapping effort, random clones from a genomic library are sequenced and then the overlapping sequences from different clones are used to stitch the genome back together. Shotgun sequencing, devised by Sanger for sequencing lambda and still in use by Barrell for the several hundred kilobase viral genomes, had fallen out of favor as a strategy, partly because the high redundancy required to get closure made it seem wasteful and partly, because even after high redundancy, there were regions that were incomplete, either as gaps or as places where because of technical limitations the sequence quality was too poor and thus needed more directed attention. The effort to fix, or edit, these regions, called 'finishing', and make the sequence almost error-free often equaled or even exceeded the effort required to get the initial shotgun data. But John recognized that the simplicity of shogun sequencing, at least in its initial phase, lent itself to scaling up in ways that other strategies didn't (and most of those other strategies still faced the problem of 'finishing').

The other major decision at the outset was how to actually get the sequence from clones, with the choice being between radioactive labeling, which was the standard, or using fluorescent labels on newly minted, but unproven sequencing instruments. A tour of the few labs using the new instruments convinced John that they were the future. The basic strategy was then set.

Of course, simply doing 500 times more of what Barrell was doing on the couple hundred thousand kilobase viral genomes wasn't going to work. A key part of the plan was to improve each of the multiple steps in sequencing to make them cheaper and easier to do at scale, all the while maintaining the high quality standards of the best small projects. John, along with the worm sequencing groups attacked one step after another, either by incorporating advances of others or by devising new protocols and computer programs. Throughput doubled yearly, with about half coming from process improvements and the other half from increasing the group size. By the end of the second year, it was clear that the three-year pilot phase with a target of 3 Mb of completed, 'finished' sequence was within reach and a scale up to complete the remaining 97 Mb was feasible (Wilson *et al.*, 1994). Finding the funding to do it was another thing.

Watson had made clear to John at the outset that while the NIH would fund his effort in the pilot phase, he would not fund further efforts. While costs were dropping, the costs of a full assault on the worm genome would be far more than the British Medical Research Council could afford. After some searching around, including an exploration of joining a private company, with Watson's help John was invited to submit a proposal to the Wellcome Trust, with the condition that John commit to begin sequencing the human genome as well. Though John knew this would mean still more time spent directing science, to get the worm sequence done he agreed to do it.

The worm sequencing proceeded apace and by 1998, the worm sequencing groups published a nearly complete version of the genome with more than 97 Mb completed, the first genome of any animal (*C. elegans Sequencing Consortium*, 1998). Over the next few years the last bits were sewn up and the worm sequence was complete telomere to telomere for all six chromosomes, with the exception of highly repetitive regions that were refractory to all available method (Hillier *et al.*, 2005). More recently, others, using long read sequencing, resolved even these highly repetitive regions (Yoshimura *et al.*, 2019).

With the backing of the Wellcome Trust, John went on to play a major role in the completion of the human genome sequence. The Sanger Centre/Institute contributed more completed sequence than any other group and John played a pivotal role in keeping the project on course, despite the challenges it faced. He also was a vigorous advocate of keeping the project truly international. Because that story has been well told in 'The Common Thread', written by John and Georgina Ferry in another example of John's success in collaborations, it needn't be retold here (Sulston & Ferry, 2002). With the successful completion of the human genome sequence assured, John stepped down as Director of what had become the Sanger Institute. He helped Alan with the map a bit, but became increasingly involved in bringing the benefits of the genome and science more broadly to the wider community, especially to populations often ignored by developed countries.

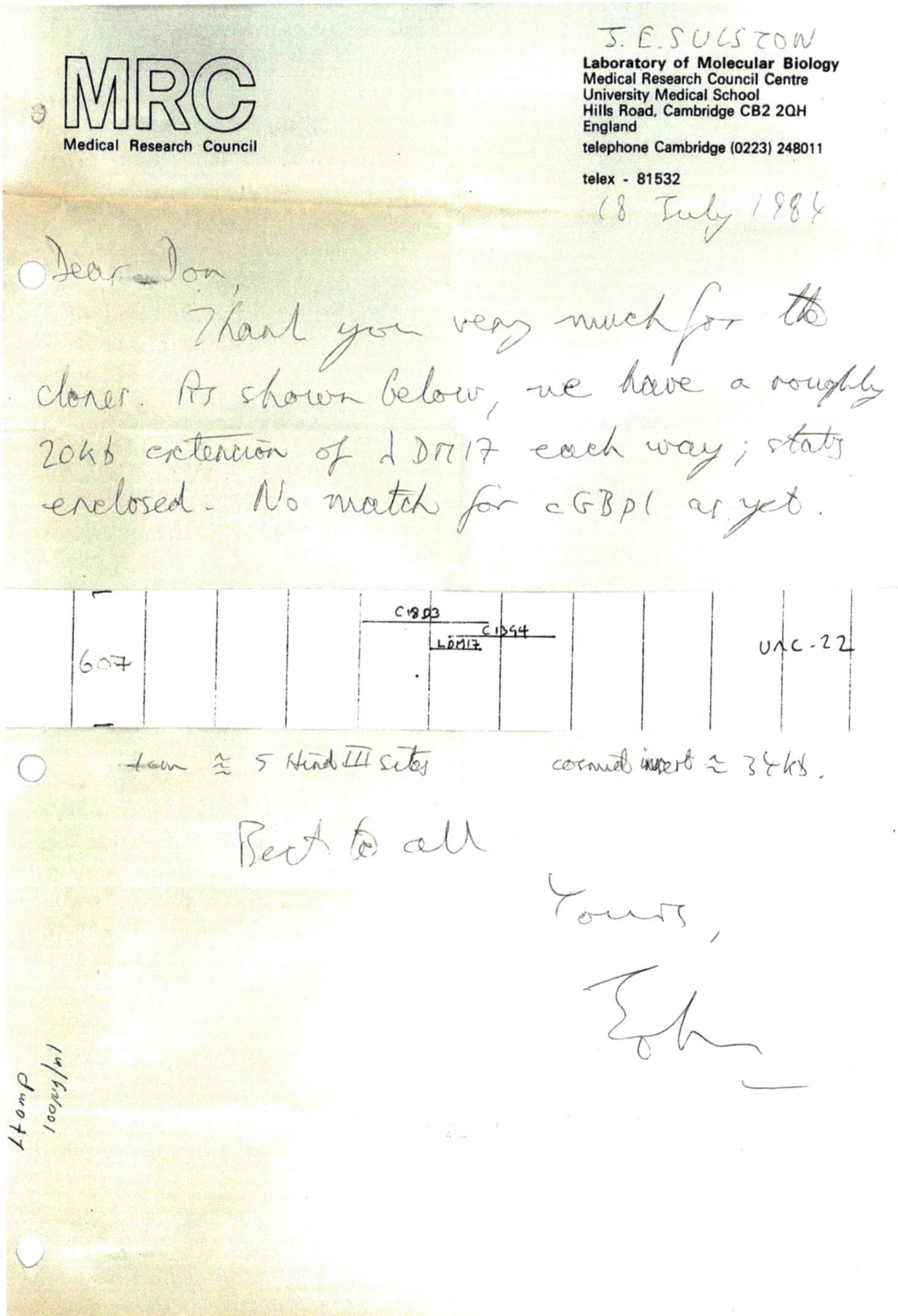

J.E. SULSTON

MRC
Medical Research Council

Laboratory of Molecular Biology
Medical Research Council Centre
University Medical School
Hills Road, Cambridge CB2 2QH
England
telephone Cambridge (0223) 248011

telex - 81532

18 July 1984

Dear Don,

Thank you very much for the clones. As shown below, we have a roughly 20kb extension of λDM17 each way; stats enclosed. No match for cGBp1 as yet.

Best to all

Yours,

John

Figure 3. A letter to one of the authors, who had sent John and Alan a lambda clone (LDM17) containing about 15 kb of the genetically defined *unc-22* gene. They returned two cosmids (C18D3 and C13G4) providing a 20 kb extension each way beyond the lambda clone. Even in these early stages, overlaps of the clone were found and thus served to anchor the physical map on the genetic map.

Open data

During the Human Genome Project John was a vocal and vigorous advocate of rapid, unrestricted data release. One result was the adoption of the so-called 'Bermuda Principles' by the international Human Genome Project, principles that served the project remarkably well despite the potential for abuse (Maxson Jones, Ankeny, & Cook-Deegan, 2018). In turn these principles have stood as a beacon to other fields, speeding the advance of science. Rapid release of data, especially for community resource projects, has become the norm in genomics and beyond. Open online community supported data archives and the eventual availability of open access journals are two extensions of this approach. In a real sense these principles of open access represent how John thought about and did science.

By nature John had a strong sense of community. While working alone at times, he also collaborated unselfishly throughout his career. With Horvitz, he discovered the post-embryonic cell lineage and found the first lineage mutants (Horvitz & Sulston, 1980). With Hedgecock, he investigated

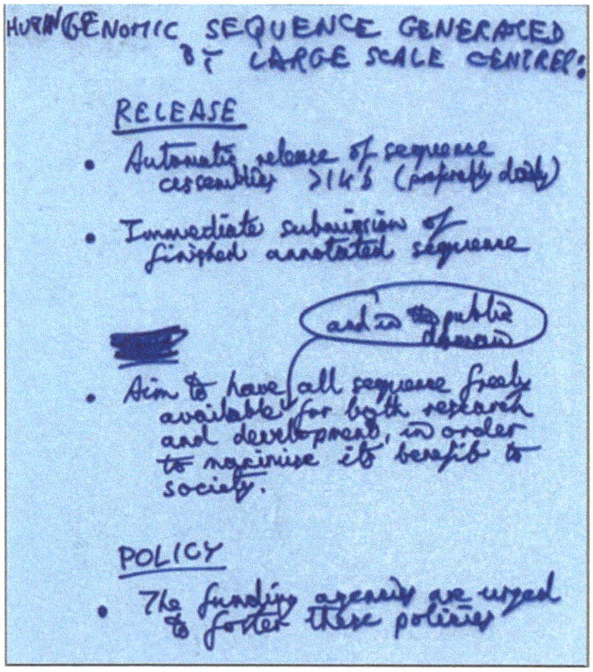

Figure 4. A photograph of the poster sheet at the Bermuda meeting in 1996 where John had written down the principles of data release as agreed to by the attendees (Maxson Jones *et al.*, 2018).

programmed cell death in the worm (Hedgecock, Sulston, & Thomson, 1983). With John White, he linked the lineage to cell types and then explored plasticity using laser ablation (Sulston & White, 1980). With Marty Chalfie he investigated touch sensitivity/mechanosensation in the worm (Chalfie & Sulston, 1981). With Alan Coulson and Bob Waterston he built the worm physical map (see above). He ran the Sanger Centre in that same spirit, recruiting talented people and then working with them, rather than directing them. This spirit extended to the entire Sanger staff: John continued to load samples on sequencing gels, both keeping him in the lab and including staff in his ring of collaborators. He also instilled in them a sense of purpose, akin to the parable about the whistling worker shouldering a heavy load of bricks for Chartres, who, when asked why he was so happy, replied, 'I'm building a cathedral.'

That sense of community and collaboration pervaded the worm mapping effort. John saw the physical map as a collaboration between his and Alan's work to establish clone overlaps and the entire *C. elegans* community as they used mutants to discover genes of consequence. When members of the community would locate their genes on the genetic map, John and Alan would ship the group clones that spanned the interval, all prepublication. Once the gene was located in that interval it became another anchor between the genetic and clone maps, creating a true genome map. Or if someone had cloned a gene by homology or transposon tagging, they would send their cloned DNA to John and Alan, who would find its location on the clone map and by inference the genetic map, creating still more landmarks (Figure 3). The tradition in the worm community, fostered by the Worm Breeder's Gazette, was one of willing sharing of data and ideas before publication, with the understanding that individuals would not unfairly exploit that prepublication access. They exploited the emerging ability to communicate across computers by posting the map on host computers in Cambridge, Boston, St. Louis and Los Angeles and allowing people to access the data via dialup connections. By the time the nearly completed map was presented at the biennial worm meeting at CSH in 1989, they could paste almost all the pieces in order along each of the six chromosomes.

That same spirit extended into the genome sequencing project. At the outset, with a clone-by-clone strategy, the scale was so small that John knew the effort would have little impact on most labs. Nonetheless, John informed the community at regular intervals precisely what clones were being worked on and he corresponded and shared sequence with labs working on genes in that area. As the effort scaled up, with multiple projects going on in various stages of completion, John explored ways to make the data available to any who wanted it. One problem early on was that the initial assemblies of the raw sequence were very unreliable and could have been misleading. Also, the central databases would only accept completed sequences, with a low error rate. Inevitably that meant a delay, sometimes quite long, between when there was assembled sequence with some errors and gaps but still very useful and when it could be submitted and released through the central databases.

Fortunately, two things happened that allowed the sequencing labs to present their data to the community in a timely manner. First, Phil Green developed software that made even initial assemblies quite reliable (Ewing, Hillier, Wendl, & Green, 1998). Second, the internet had matured to the point where the project could post sequence on their computers and anyone across the world could access it. The community was cautioned to use the unfinished sequences carefully since they contained errors and were subject to change. With those changes in place, John began posting the worm sequences online on a regular basis. Members of the community knew of the plans and regularly checked for updates on their region of interest or for homologs of their favorite genes.

When the Human Genome Project was moving more seriously into sequencing in the mid-90's, questions arose about just when sequencing centers would release their data and whether the genes they discovered could be patented. The example provided by the worm project proved crucial in policy debate that followed and in a meeting in Bermuda the members of the international HGP adopted a policy whereby centers would 1) eschew patenting; 2) release sequence daily; and 3) make the entire sequence freely available in the public domain without constraints (Figure 4) (Maxson Jones *et al.*, 2018). To facilitate coordination it also followed the worm precedent of assigning specific regions of the genome to centers, thus avoiding different centers targeting the same region. The policy stimulated immediate application of the sequence by making it available almost immediately to those searching for human disease genes.

The commitment to the policy was tested when the company Celera, in addition to patenting its own human genome sequences, incorporated all the available (and as yet unpublished) public sequence into its assembly, which the company sought to sell to commercial users and public labs. Despite this breach of scientific etiquette, with John's steadfast support, the HGP stood by its policy and today the human genome sequence is freely available to all users across the globe. The success of the policy in stimulating research has led to the adoption of similar policies in other realms of science (Maxson Jones *et al.*, 2018).

After John stepped down from his leadership role at the Sanger Centre, he continued his work furthering the openness of science and working to be sure its benefits were enjoyed by peoples of nations wealthy or poor. He published several papers on the benefits of open data. He served, first as a member and later as president, of the British Human Genetics Commission (HGC), a government advisory body that investigated issues such as direct-to-consumer genetic testing services and the protection of personal genetic information in insurance and employment. In 2001 he worked with Oxfam, an organization that he had long supported, visiting treatment programs for HIV/AIDS in South Africa; he later became the titular head for the charity's 'Cut the Cost of Medicines' campaign to reform the trade rules on patented medicines to benefit developing countries . With the bioethicist John Harris from the University of Manchester, they jointly founded The Institute for Science, Ethics and Innovation (iSEI). One important outcome of this collaboration was 'Who Owns Science? The Manchester Manifesto', an attempt to address the drawbacks of the existing system of intellectual property (University of Manchester. Institute for Science & Innovation 2009). A further major piece of work was the Royal Society report 'People and the Planet' (Sulston *et al.*, 2012). He chaired a 21-member working group of globally diverse contributors, addressing the impact of human population on the global environment.

Conclusion

John's work continues to impact much of biology daily. John demonstrated formally Virchow's dictum 'omnis cellula e cellula' for a single animal. In the worm, the lineage has facilitated the identification of major pathways that drive development and recently, using single cell RNA-sequencing methods, the gene expression changes that occur in most of the branches and into terminal differentiation have been uncovered. Yet the worm lineage has stood apart for many years as a powerful but singular example of the complex relationship of lineages to cell fate. Without more powerful analytical tools, it was impossible to follow complex developmental patterns in all but the simplest organisms. Other major model systems don't have fixed lineages and until recently the numbers of cells made tracing the cell divisions impossibly difficult. The situation has changed with the advent of molecular tags that vary during development (Frieda *et al.*, 2017; McKenna *et al.*, 2016; Salipante, Kas, McMonagle, & Horwitz, 2010) and, remarkably, lineage tracing can be applied to vertebrates. With single cell expression data sets for an increasing number of organisms during development, it is increasingly feasible to reconstruct lineages based on the expression similarities between mother and daughter cells and at the same time discover their differences. Similarly, single cell methods for measuring open chromatin when applied to embryos are providing insight into the regulatory elements that are used as cell divisions proceed.

John's pioneering efforts in genomics have had even broader impact. Although much of the biological community was skeptical of the value of genome sequences at the outset of the HGP, the utility of the sequences of the worm, of yeast, of human and of mouse genomes in those early days rapidly converted most skeptics to believers, creating a huge demand for more genome sequences. In turn that led to revolutionary advances in sequencing technology, with genome sequences of countless numbers of organisms available today. Also, DNA sequencing has become a cheap and quantitative assay for many biological phenomena, including community composition, RNA expression, chromatin accessibility and conformation, evolutionary and ecological studies and a continually growing list. Like molecular biology before it, genome science has pervaded most of biology.

John's commitment to keep genome information free, without patent restrictions, and to release data from resource focused projects before publication and as soon as practically possible has been followed widely, not just by the HGP, but in most genome scale projects. Biologists are so accustomed to having all the data online that it is difficult to even imagine an alternative world where sequence was held privately and only released to users with conditions. But today with the capture of individual human sequences by large companies and possibly governments to study the effects of human variation on health and other traits, biologists may be called to fight this battle yet again. How we regulate the ownership of data may be one of the most important political questions of our era (Harari, 2018). The rapid data release of so-called precompetitive data by the genomics field has been held up as a model in other scientific fields and adopted by them.

With all of these accomplishments John remained a warm and unassuming individual. He enjoyed growing his own fruit and vegetables in a delightfully wild garden and brought his cheese sandwich to work daily. He loved to share a pint in the pub with friends on a Friday evening, regaling the assemblage with anecdotes and stories or working through a difference of opinion. Punting down the Cam for a lab outing with John or fireworks in his garden on Guy Fawkes day are events that all who took part treasure. He cycled wherever possible and was proud that the solar panels he installed made his home energy neutral, even if it meant having to take a cold shower on occasion. Most of the money that he received through prizes, including the Nobel Prize he gave away to charitable causes.

John was an international treasure. In the worm community we joked that he should always travel with at least four bodyguards to handle all possible contingencies. As we have tried to illustrate in this perspective, so often he was the agent of change at a critical juncture in science. He will be sorely missed as a scientist, as a colleague and as a friend.

Disclosure statement

No potential conflict of interest was reported by the author(s).

ORCID

Robert H. Waterston http://orcid.org/0000-0001-5956-5215

References

Burke, D.T., Carle, G.F., & Olson, M.V. (1987). Cloning of large segments of exogenous DNA into yeast by means of artificial chromosome vectors. *Science, 236*(4803), 806–812.

C. elegans Sequencing Consortium. (1998). Genome sequence of the nematode *C. elegans*: A platform for investigating biology. *Science, 282*(5396), 2012–2018.DOI: 10.1126/science.282.5396.2012

Chalfie, M., & Sulston, J. (1981). Developmental genetics of the mechanosensory neurons of *Caenorhabditis elegans. Developmental Biology, 82*(2), 358–370.

Coulson, A., Kozono, Y., Lutterbach, B., Shownkeen, R., Sulston, J., & Waterston, R. (1991). YACs and the *C. elegans* Genome. *BioEssays: News and Reviews in Molecular, Cellular and Developmental Biology, 13*(8), 413–417.

Coulson, A., Sulston, J., Brenner, S., & Karn, J. (1986). Toward a physical map of the genome of the nematode *Caenorhabditis elegans. Proceedings of the National Academy of Sciences of the United States of America, 83*(20), 7821–7825.

Coulson, A., Waterston, R., Kiff, J., Sulston, J., & Kohara, Y. (1988). Genome linking with yeast artificial chromosomes. *Nature, 335*(6186), 184–186.

Ewing, B., Hillier, L., Wendl, M.C., & Green, P. (1998). Base-calling of automated sequencer traces using phred. I. accuracy assessment. *Genome Research, 8*(3), 175–185.

Frieda, K.L., Linton, J.M., Hormoz, S., Choi, J., Chow, K.-H.K., Singer, Z.S., Budde, M.W., Elowitz, M.B., & Cai, L. (2017). Synthetic recording and in situ readout of lineage information in single cells. *Nature, 541*(7635), 107–111.

Gitschier, J. (2006). Knight in common armor: An interview with sir John Sulston. *PLoS Genetics, 2*(12), e225. doi:10.1371/journal.pgen.0020225

Harari, Y.N. (2018). *21 lessons for the 21st Century*. Washington, DC: Random House.

Hedgecock, E., Sulston, J., & Thomson, J. (1983). Mutations affecting programmed cell deaths in the nematode *Caenorhabditis elegans. Science, 220*(4603), 1277–1279. doi:10.1126/science.6857247

Hillier, L.W., Coulson, A., Murray, J.I., Bao, Z., Sulston, J.E., & Waterston, R.H. (2005). Genomics in *C. Elegans*: So many genes, such a little worm. *Genome Research, 15*(12), 1651–1660.

Horvitz, H.R., & Sulston, J.E. (1980). Isolation and genetic characterization of cell-lineage mutants of the nematode *Caenorhabditis elegans. Genetics, 96*(2), 435–454.

Jacob, F. (1977). Evolution and tinkering. *Science, 196*(4295), 1161–1166.

Maxson Jones, K., Ankeny, R.A., & Cook-Deegan, R. (2018). The Bermuda Triangle: The pragmatics, policies, and principles for data sharing in the history of the human genome project. *Journal of the History of Biology, 51*(4), 693–805.

McKenna, A., Findlay, G.M., Gagnon, J.A., Horwitz, M.S., Schier, A.F., & Shendure, J. (2016). Whole-organism lineage tracing by combinatorial and cumulative genome editing. *Science, 353*(6298), aaf7907. doi:10.1126/science.aaf7907

National Research Council. Division on Earth and Life Studies, Commission on Life Sciences, and Committee on Mapping and Sequencing the Human Genome. (1988). *Mapping and Sequencing the Human Genome*. Washington, DC: National Academies Press.

Olson, M.V., Dutchik, J.E., Graham, M.Y., Brodeur, G.M., Helms, C., Frank, M., MacCollin, M., Scheinman, R., & Frank, T. (1986). Random-clone strategy for genomic restriction mapping in yeast. *Proceedings of the National Academy of Sciences of the United States of America, 83*(20), 7826–7830.

Salipante, S.J., Kas, A., McMonagle, E., & Horwitz, M.S. (2010). Phylogenetic analysis of developmental and postnatal mouse cell lineages. *Evolution & Development, 12*(1), 84–94.

Sulston, J.E. (1976). Post-embryonic development in the ventral cord of *Caenorhabditis elegans. Philosophical Transactions of the Royal Society of London B: Biological Sciences, 275*(938), 287–297.

Sulston, J., Dew, M., & Brenner, S. (1975). Dopaminergic neurons in the nematode *Caenorhabditis elegans. The Journal of Comparative Neurology, 163*(2), 215–226.

Sulston, J., & Ferry, G. (2002). *The common thread: A story of science, politics, ethics, and the human genome*. Washington, DC: Random House.

Sulston, J.E., & Horvitz, H.R. (1977). Post-embryonic cell lineages of the nematode, *Caenorhabditis elegans. Developmental Biology, 56*(1), 110–156.

Sulston, J.E., Schierenberg, E., White, J.G., & Thomson, J.N. (1983). The embryonic cell lineage of the nematode *Caenorhabditis elegans. Developmental Biology, 100*(1), 64–119.

Sulston, J.E., & White, J.G. (1980). Regulation and cell autonomy during postembryonic development of *Caenorhabditis elegans. Developmental Biology, 78*(2), 577–597.

Sulston, J., Bateson, P., Biggar, N., Fang, C., Cavenaghi, S., Cleland, J., & Cohen, J. (2012). *People and the planet. Royal Society Science Policy Centre Report 01/12*. London: The Royal Society.

University of Manchester. Institute for Science, Ethics and Innovation. (2009). *Who Owns Science? The Manchester Manifesto*. Manchester, UK: Institute for Science, Ethics and Innovation, The University of Manchester.

White, J.G., Southgate, E., Thomson, J.N., & Brenner, S. (1986). The structure of the nervous system of the nematode *Caenorhabditis elegans. Philosophical Transactions of the Royal Society of London. Series B, Biological Sciences, 314*(1165), 1–340.

Wilson, R., Ainscough, R., Anderson, K., Baynes, C., Berks, M., Bonfield, J., … Cooper, J. (1994). 2.2 Mb of contiguous nucleotide sequence from chromosome III of *C. elegans. Nature, 368*(6466), 32–38.

Yoshimura, J., Ichikawa, K., Shoura, M.J., Artiles, K.L., Gabdank, I., Wahba, L., … Schwarz, E.M. (2019). Recompleting the *Caenorhabditis elegans* Genome. *Genome Research, 29*(6), 1009–1022.

A touching story

Martin Chalfie

ABSTRACT
A slide taped to a window at the Woods Hole Marine Biology Laboratory was my first introduction to the touch receptor neurons of the nematode *Caenorhabditis elegans*. Studying these cells as a postdoc with Sydney Brenner gave me a chance to work with John Sulston on a fascinating set of neurons. I would never have guessed then that 43 years later I would still be excited about learning their secrets.

On 13 April 1977 Bob Horvitz and I drove to the Woods Hole Marine Biology Laboratory for what has since been known as the First International Worm Meeting (its actual title was 'Caenorhabditis elegans Workshop or The Worm Breeders' Confabulation'). I was just about to begin my postdoc with Sydney Brenner at the MRC Laboratory of Molecular Biology (LMB) and fully intended to study neurotransmitters in the worm. On the ride to Woods Hole, however, Bob urged me to take a look at some work that John Sulston was reporting on at the meeting, in part because John was not going to continue with the project and Bob thought I might be interested. I did not realize at the time that this suggestion was going to affect my entire scientific career.

What John described at the meeting was his discovery of a set of touch-sensing cells as well as several genes that were needed for their activity. He had been scheduled to give a poster of this work, but he had not bothered to make a poster. Instead he taped a slide showing an electron micrograph from one of the mutants onto a nearby window, so if you were interested in the work, you had to squint at the slide illuminated by the afternoon sun. Despite these less than optimal viewing conditions, I was hooked and immediately decided to change my postdoctoral project. The chance to look at an understudied sensory activity (and one I knew nothing about) and to manipulate that activity genetically was a powerful incentive. I was being offered a unique opportunity. John was astonishingly generous in letting me be part of the project.

The discovery of these cells (now called the touch receptor neurons, TRNs) and the mutants was the unexpected result of experiments John had done while studying dopaminergic neurons in *C. elegans* (Sulston, Dew, & Brenner, 1975). John had visualized dopaminergic neurons using a technique called formaldehyde-induced fluorescence. He even used this method to identify mutants that lacked dopamine, but this procedure was slow and labor-intensive, and not helped by John's initially throwing away the mutant plates and having to retrieve them from the garbage. John suspected that the neurons were mechanosensory rather than chemosensory because electron microscopic reconstructions being done in the lab at that time showed the cells having sensory dendrites that were embedded in the animal's cuticle, whereas other, presumably chemosensory, neurons had dendrites that poked through the cuticle. John reasoned that he could identify mutations affecting these cells with embedded dendrites by looking for animals that were touch insensitive.

At the time, electron microscopists, like Nichol Thomson, used eyebrow hairs glued to toothpicks to maneuver thin sections onto microscopy grids. John adapted this tool to do a simple test for touch sensitivity: if an animal moved away from the hair when it was drawn across the animal near the head or tail, the animal was sensitive to touch. Animals defective in touch did not move when touched with the hair, but moved normally otherwise. John used this test to screen for mutants that were insensitive to touch. In all, he found mutations in eight genes, now called *mec-1–mec-8* (for *mec*hanosensory abnormal genes 1–8).

Unfortunately, none of the mutants had defects in the dopaminergic cells. [John's hypothesis about the dopaminergic neurons being mechanosensory was, however, correct. Several years later, Beth Sawin in Bob Horvitz's lab showed that they sensed the texture of the bacterial lawn (Sawin, Ranganathan, & Horvitz, 2000).] Several of John's mutants, however, had defects in a different set of neurons. These neurons, six in all, each had a very long neurite that extended anteriorly from the cell body and was found adjacent to, but not embedded within, the cuticle. Intriguingly, the neurite was packed with large-diameter microtubules. No other cells had these microtubules. More importantly, mutations in one of the genes (*mec-3*) eliminated the cells and those in another gene (*mec-7*) eliminated the microtubules. The slide John taped to the window at Woods Hole was of a *mec-7* mutant.

Compared to the usual mutant phenotypes studied in the lab at the time (e.g. Long, Dumpy, Uncoordinated, Roller, Blister), touch-insensitivity was quite subtle. In fact, Sydney expressed considerable doubt that touch insensitivity was a real phenotype. His disbelief is ironic, because he had actually identified mutants defective in touch genes in his original genetic experiments (Brenner, 1974 and pers. comm.). He had not recognized that the mutants were insensitive to touch, but had isolated the strains because he felt that they were sluggish. Touch-insensitive mutants are, indeed, sluggish on lawns of bacteria when compared to wild-type animals, but this difference probably results from the inability of mutant animals to respond, as wild-type animals do, to the tap and subsequent vibration produced when their plate is slapped onto the stage of the dissecting microscope. John, going through Sydney's mutant collection, found that two of the early sluggish mutations *e75* (in *mec-2*) and *e398* (in *mec-8*) produced touch insensitivity. I still marvel at Sydney's ability to identify sluggishness as a definable phenotype. To me, this phenotype is considerably subtler than touch insensitivity.

I arrived in England soon after the Woods Hole meeting, but had planned to take a two-month vacation in Scotland and Norway before starting my postdoc. Before leaving for Scotland, I visited Cambridge briefly and had just enough time to find a place to stay, look around the lab, and have my first conversation with Sydney. We met one evening in the coffee room on the second floor of the LMB. A former postdoc just starting his assistant professorship at Harvard Medical School had been excited about my working with Sydney, but warned me that Sydney quickly decided if people were worth talking to or not, often when he first met them. As a result, I was somewhat intimidated seeing him that first time. Sydney asked about my plans, and I told him that instead of looking at worm neurotransmitters as I had described in my postdoc application, I was going to work on touch sensitivity. I am not entirely sure what he thought about my changing projects (we actually talked very little about science), but he approved of my leaving because no bench space was available at that moment. Everything was going smoothly, when Sydney asked if I had any questions. I had read his 1974 *Genetics* paper (Brenner, 1974) and thought that I understood most of it. I did not, however, know how he had derived the statistics that were found in one of his figures, and so I asked him for an explanation. He looked annoyed, shook his head, and refused to answer the question. In fact, that was the end of the meeting. At that moment I was sure that my scientific career was over.

I later learned that Sydney took a very hands-off approach to most of the people working in his group, and this attitude fostered our development as independent scientists (although a few people found the independence troubling). Everyone worked on their own project, although we often collaborated and we were always talking with each other. In many cases Sydney was not involved in the project, and if he was not involved, he would not want to be considered a coauthor. In fact, of the seven papers that came out of my postdoc, Sydney was a co-author only on one of them. As a result of this independence, I found that I rarely talked with Sydney about my work, about once a year. Many years later, I remarked to Phil Anderson, a friend and fellow postdoc in Sydney's lab, that I had been Sydney's postdoc for five years and talked to him about my work five times during that period. Phil said that he, too, had talked with Sydney five times in five years, and then we simultaneously said to each other that the longest scientific conversation we had had with Sydney about our work was when he was in bed in the hospital following his motorcycle accident and couldn't walk away. Several months after talking with Phil, I told another friend and fellow postdoc, Cynthia Kenyon, about our conversation. Cynthia was astonished, 'You talked to Sydney five times? I was also at the LMB for five years, and I only talked to him twice.'

Although I sometimes regret not having spoken more with Sydney, I feel that the independence he encouraged and insisted on was an important step in my development as a scientist. And I and others learned much about the way he thought just listening to him in the coffee room where he would often hold forth. I especially remember one day, probably in 1980, when Sydney, returning from giving a series of lectures in an Eastern Bloc country, told us that he had gotten a terrific exchange rate. We were a bit surprised at this remark, because we knew that these Soviet countries had artificial exchange rates that were not very favorable to Westerners. We asked Sydney to explain. He told us that he had finished his lectures on a Friday, was paid in cash in the local currency, but had to remain in the country until the next flight back to England, which was on Monday. His hosts had arranged for his meals and lodging over the weekend at a nearby hotel. Unfortunately, when Sydney went down to dinner that night, he discovered that the dining room came complete with a live orchestra, and the music was awful. After listening for a few moments, Sydney approached the orchestra leader, pulled all the local currency, which was worthless to him, out of his pocket, and asked, 'If I give you all this money, will you leave and not return until after I've gone on Monday?' The orchestra leader readily agreed, money exchanged hands, and soon the orchestra had packed and left. Sydney returned to his table to eat his meal in quiet. A few moments later, however, one of the other diners approached Sydney's table and said, 'I am a German businessman, and I have been staying at this hotel for a week and had to endure that horrible music. Please let me contribute.' He then gave Sydney (West) German marks. Others came up and offered Sydney Swiss francs, American dollars, etc. In all, Sydney said he made out quite well.

I returned to the lab from my vacation and was given a postage stamp-sized desk and a microscope for my experiments. This situation was quite normal; no one had any room. I was going to start looking for additional touch mutants, but before doing my first mutagenesis Bob Horvitz gave me his 'demonstration collection' of various *C. elegans* mutants, so I could see the range of phenotypes available at the time. I decided to test these animals for touch sensitivity and was surprised to find not only the known touch-insensitive strain in the collection, but also an additional strain. This latter strain had an *unc-22* mutation, which caused the animals to have random contractions of the body wall muscle (as a result the animals were called twitchers). This surprising, non-muscle phenotype (touch insensitivity) appeared to have been indirect, arising from the habituation of the touch response that these random contractions elicited, since I subsequently found that twitchers made by mutations in other genes were also touch insensitive. I was quite happy making this minor discovery so early.

A more substantial finding and one that gave me my first worm publication (Chalfie & Thomson, 1979) resulted from looking at electron micrographs of the touch sensing cells and a fortuitous conversation with Jonathan Hodgkin. One day early in my study of the TRNs and touch sensitivity, I went to the room where the serial section electron micrographs were stored to look at the TRNs. When I started looking at the cells, however, I realized that I had no idea what to look for. Since the cells had prominent microtubules and since I did not want to return to the lab and admit I did not know what I was doing, I decided to count the microtubules in the images. After counting microtubules in several hundred micrographs, I made a graph of the number versus the position of the cells in the series. The resulting graph looked like a mountain range with the values rising and falling all along the length of the neurite. I had no idea what this pattern meant, but decided I had spent sufficient time looking at the cells, so I could return to the lab without hanging my head. After all, I had produced a figure from all my efforts. On the way back, I ran into Jonathan Hodgkin, who ask me what I had been doing. After I showed him my graph, he asked me a great question that I should have asked myself: Is this what was supposed to happen? I said I didn't know but would find out. I learned that all microtubules were thought to start at the cell body, but if this was true, I should have seen a steadily decreasing number of microtubules in sections that were further and further from the cell body, not a mountain range. I then worked with Nichol Thomson, the master microscopist in the lab, and showed that the microtubules were relatively short compared to the length of the cells and distributed throughout the neurites (Chalfie & Thomson, 1979). Later, Dennis Bray, who I had met in Cambridge, showed a similar phenomenon in rat neurons (Bray & Bunge, 1981).

The publishing of my paper on microtubule length also showed me Jonathan's true nature. He thought my ordering a minimal number of reprints ridiculous and wasteful, since people at LMB expected others to make copies of their papers. I didn't really expect many people wanted copies of my paper, but I ordered them because I needed some reprints for my fellowship renewal (and I also wanted a printed copy). After a while I started getting requests and (although I did not realize it at the time) quite a few of them came from prominent professors at Stanford, including Linus Pauling. I sent out the reprints to all these people only to find later that these requests had originated with Jonathan, who, motivated by our betting how many requests would come in, arranged to have his friend Mariana Wolfner, a graduate student at Stanford at that time, acquire reprint request cards and send them to me. I have been wary of him ever since.

My main work was the genetic analysis of the touch mutants, and I eventually discovered three new genes (*mec-9*, *mec-10*, and *mec-12*), the last, like *mec-7*, being needed for the prominent microtubules seen only in the TRNs. This research meant that I had the great good fortune to work with John Sulston. I have written already about John being an astonishing experimentalist and humanist who had a great influence on my science and my life (Chalfie, 2018) and won't repeat myself here. A glimpse of John's generosity and kindness, however, can be seen in how our names are listed in our first paper on the touch mutants (Chalfie & Sulston, 1981). Before I had arrived in the lab, John had briefly mentioned his early work with the mutants in his paper on the dopamine neurons (Sulston *et al.*, 1975). When I started working on the project, he asked for my thoughts on whether he should write up his remaining data on the touch mutants or wait and combine his data with mine. I told him that I would prefer the latter option, and that is what happened. We did quite a lot more work on the project and I believe the final paper is more complete because of that, but John's selflessness, his lack of possessiveness, made a lasting impression. When it came time to write the paper, he not only wanted his name second, but seeing that I wanted to be listed without my middle initial, he told me to remove his middle initial to make our names more parallel.

Working with John was fun (as was enjoying a beer with him at the pub at the Frank Lee Centre near the lab). He approached research with a straight-forward understanding and clarity that I always envied. And he always came up with unique ways of solving problems. At one point, we did a series of experiments to show that the TRNs were, indeed, the touch-sensing cells. John, using the laser microscope developed by John White, would kill the TRNs in different combinations, confirm that the affected cell nuclei were missing, blind the samples, and give them to me to test for touch sensitivity. The experiments went very smoothly until one day when John and I disagreed about one of the animals. I said it was touch sensitive in the tail and he said it could not be because it lacked both tail TRNs. After a considerable amount of discussion, we both retested the animal for touch sensitivity and John reexamined it under the Nomarski microscope. We agreed that the animal was touch sensitive, but surprisingly, it seemed to lack both cells. We were puzzled, but John was determined to find an answer. He asked Nichol Thomson to do an electron microscopic 'autopsy' on the worm. Nichol took pictures of the worm at different positions along the animal's length and found that one TRN neurite was, indeed, still there. Apparently, the laser ablations had killed one cell but had only destroyed the nucleus in the other. The enucleated cell, however, appeared to function perfectly well.

During my time at the LMB, people would often talk about 'Hershey Heaven.' This phrase came from Nobel Laureate Alfred Hershey's idea of scientific heaven, which was 'to have one experiment that works, and keep doing it all the time' (Judson, 1980). On the face of it doing the same experiment repeatedly does not sound that appealing, but people interpreted Hershey Heaven to mean that every time you did the experiment, you would learn something new. To me, the TRNs are Hershey Heaven. During my time in Cambridge, John and I not only developed the first genetics system to study mechanosensation, but the study of the TRNs led me to investigate microtubule structure and function (Chalfie & Thomson, 1979, 1982), connectomics (Chalfie *et al.*, 1985), control of cell lineages (Chalfie, Horvitz, & Sulston, 1981), and neuronal outgrowth (Chalfie, Thomson, & Sulston, 1983). Subsequently, we have studied the molecular basis of mechanosensory transduction and its modulation, neuronal differentiation and specification, neuronal degeneration, and neuronal outgrowth. This work led to the discovery of novel transcription factors, cholesterol-binding proteins, chaperones, and channel proteins. And working on the cells and their gene expression motivated me to investigate GFP as a biological marker (Chalfie, Tu, Euskirchen, Ward, & Prasher, 1994). Together Sydney and John gave me an incredible gift.

Disclosure statement

No potential conflict of interest was reported by the author(s).

Acknowledgements

I want to thank Tulle Hazelrigg, Jonathan Hodgkin, and Bob Horvitz for their comments on the manuscript.

ORCID

Martin Chalfie http://orcid.org/0000-0002-9079-7046

References

Bray, D., & Bunge, M.B. (1981). Serial analysis of microtubules in cultured rat sensory axons. *Journal of Neurocytology, 10* (4), 589–605.

Brenner, S. (1974). The genetics of *Caenorhabditis elegans. Genetics, 77* (1), 71–94.

Chalfie, M. (2018). John Sulston (1942–2018). *Cell, 173* (4), 809–812.

Chalfie, M., & Sulston, J. (1981). Developmental genetics of the mechanosensory neurons of *Caenorhabditis elegans. Developmental Biology, 82* (2), 358–370.

Chalfie, M., & Thomson, J.N. (1979). Organization of neuronal microtubules in the nematode *Caenorhabditis elegans. The Journal of Cell Biology, 82* (1), 278–289.

Chalfie, M., & Thomson, J.N. (1982). Structural and functional diversity in the neuronal microtubules of *Caenorhabditis elegans. The Journal of Cell Biology, 93* (1), 15–23.

Chalfie, M., Horvitz, H.R., & Sulston, J.E. (1981). Mutations that lead to reiterations in the cell lineages of. *Cell, 24* (1), 59–69.

Chalfie, M., Sulston, J.E., White, J.G., Southgate, E., Thomson, J.N., & Brenner, S. (1985). The neural circuit for touch sensitivity in *Caenorhabditis elegans. The Journal of Neuroscience: The Official Journal of the Society for Neuroscience, 5* (4), 956–964.

Chalfie, M., Thomson, J.N., & Sulston, J.E. (1983). Induction of neuronal branching in *Caenorhabditis elegans. Science, 221* (4605), 61–63.

Chalfie, M., Tu, Y., Euskirchen, G., Ward, W.W., & Prasher, D.C. (1994). Green fluorescent protein as a marker for gene expression. *Science, 263* (5148), 802–805.

Judson, H.F. (1980). *The eighth day of creation: Makers of the revolution in biology.* New York: Simon and Schuster.

Sawin, E.R., Ranganathan, R., & Horvitz, H.R. (2000). *C. elegans* locomotory rate is modulated by the environment through a dopaminergic pathway and by experience through a serotonergic pathway. *Neuron, 26* (3), 619–631.

Sulston, J., Dew, M., & Brenner, S. (1975). Dopaminergic neurons in the nematode *Caenorhabditis elegans. The Journal of Comparative Neurology, 163*(2), 215–226.

But can they learn? My accidental discovery of learning and memory in *C. elegans*

Catharine H. Rankin

ABSTRACT
I did not set out to study *C. elegans*. My undergraduate and graduate training was in Psychology. My postdoctoral work involved studying learning and memory in 1 mm diameter juvenile *Aplysia californica*. As a starting Assistant Professor when I attempted to continue my studies on Aplysia I encountered barriers to carrying out that work; at about the same time I was introduced to *Caenorhabditis elegans* and decided to investigate whether they could learn and remember. My laboratory was the first to demonstrate conclusively that *C. elegans* could learn and in the years since then my lab and many others have demonstrated that *C. elegans* is capable of a variety of forms of learning and memory.

Looking back, my research trajectory into *C. elegans* happened entirely by chance as a result of the convergence of potential disasters and fortuitous moments. My undergraduate and graduate degrees were in Psychology. For my postdoctoral research, I studied the development of learning and memory in juvenile *Aplysia californica* with Dr. Thomas Carew in the Psychology Department at Yale University. We focused on the simplest forms of learning that primarily require only a single stimulus (single or repeated presentations): non-associative habituation, dishabituation and sensitization. Baby *Aplysia* are about 1 mm in diameter and had not previously been studied behaviorally and so I had to develop techniques to hold them still under the microscope, stimulate them and study their learning. When I began my first faculty position in the Psychology Department at the University of British Columbia in August of 1987, I started buying equipment to set up a lab to continue to study learning in juvenile *Aplysia*. Several things interrupted my forward progress – the mariculture facility that supplied the baby *Aplysia* was moving to a new location and closing for a year so no *Aplysia* would be available, and one of the vendors who was building some equipment for me demanded payment in advance and several months later ran away with a significant proportion of my already small start-up funds. Both events were potential disasters for a junior Assistant Professor, and it could have been tempting to give up! At the same time these disasters occurred I was trying to decide on the brand of dissecting microscope to purchase to equip my laboratory, and no one in the Psychology department then knew much about different brands of microscopes. I began asking everyone I met about microscopes and one day an undergraduate student I met told me he had worked in another lab with a number of different kinds of dissecting scopes and I should get in touch with the researcher in charge of that laboratory and get a recommendation. I did just that – I called Ann Rose in Medical Genetics at UBC and asked her about microscopes. She graciously extended an invitation for me come by her lab to look at the microscopes she used and I accepted her offer. When I visited her lab, she was very generous with her time and told me about the genetic research she was doing on the microscopic, 1 mm long nematode *C. elegans*. I was excited that her worms were the same size as my baby *Aplysia* and so I asked for some to use to test out all of the loaner microscopes that the salespeople had left for me. She agreed and sent me back to my office with a plate of worms.

At my office my first graduate student, Cathy Chiba, and I spent a lovely afternoon looking at the worms through different microscopes. Luckily the last colloquium I attended at Yale was by Robert Horvitz from MIT who gave a great talk summarizing the research on the cell lineage, the mapping of the nervous system and 'all things worm', so I could play the knowledgeable professor and tell her everything I knew about the worms. There was a moment while Cathy and I were admiring the worms and contemplating what was known about them when we looked at each other and asked 'Can they learn?' The next day I called Ann back and asked her that question. She said she didn't think so but had a binder filled with a print-out of the complete bibliography of all of the papers ever published on *C. elegans* (even then, in the pre-Wormbase days, the worm community came up with ways to share resources). We rushed over, borrowed the binder, worked our way through it and to our excitement found nothing pertaining to the subject. From here, a tentative direction for my future research was born. We naively took on the challenge of studying whether these microscopic nematodes could learn, without any experience in working with these organisms.

Again I went to see Ann and asked if I could learn basic worm techniques from her and try some experiments on my own, as the equipment I had purchased to study 1 mm *Aplysia* could easily be adapted to study 1 mm *C. elegans*, and it was going to be at least a year until I had *Aplysia* to work on. She taught us basic protocols for working with *C. elegans* such as preparing NGM plates, the *E. coli* food we rear them on and the techniques with which to move them to new plates. In the very old days, before the development of platinum picks, worm researchers used wooden sticks that they carved into a sharp tip with a scalpel – that was the technique she taught us. Today there is an alcohol lamp by every bench to sterilize platinum picks- in those days there was a supply of wooden sticks, a sharp scalpel and as a result there were wood shavings all over the floors (it was a relief a year or so later to switch to platinum picks).

Cathy and I got to work on *C. elegans* in February 1988 and quickly found that worms could learn! We wanted to start with habituation, a simple form of learning defined as a decrease in responding to a repeated stimulus. The neural circuits for the response to touching the head or the tail of the worm had been determined by Chalfie *et al.* (1985) and seemed like a good place to start. We decided we needed to find a mechanosensory stimulus that could be delivered automatically, as the odds of using a hair or an eyelash to touch worms at that same place on the head or tail at the same intensity 30–60 times at a 10 s inter-stimulus interval without unconscious experimenter bias was not realistic, thus ruling out head touch and tail touch. We chose to study the tap reversal response which uses the same mechanosensory neurons as the touch circuits. In response to a tap to the side of the agar filled petri plate on which they live worms stop what they are doing and crawl backwards for a short distance (I think of it as a type of startle response). The apparatus we used to deliver our taps consistently was an electromagnetic tapper that had been used by Ken Lukowiak's lab in Calgary to tap *Aplysia* siphons! Once we had a reliable stimulus we set to work and in a short period of time we found that *C. elegans* could show all of the simple forms of learning I studied in *Aplysia* (habituation, dishabituation and sensitization) as well as long-term memory for habituation. We had an abstract ready in the spring for a poster at the Annual Meeting of the Society for Neuroscience that fall. We presented our poster at SfN in Toronto in November 1988. At the meeting the major response to my poster from the invertebrate learning community was skepticism about why I would switch to this microscopic worm in which you could not do electrophysiology. This work was the basis of the first paper to conclusively demonstrate learning and memory in *C. elegans* (Rankin, Beck, & Chiba, 1990).

In late 1988 Ann told me about the worm meetings that happened every two years and so I submitted an abstract to the 1989 meeting at Cold Spring Harbor and was excited when my abstract was chosen for a talk. My first experience of the meeting (attended by ~360 people) was that I had landed on another planet – everyone was talking a language I did not understand for most of the sessions- I see from the notes I tried to take that there were talks on cell lineage, vulva formation, sex determination, genes that regulate development, etc. and I understood almost nothing. Mark Edgley gave an update on mapping and reported that 893 (!) genes had been physically mapped, and that there were ~1300 strains available. There was a neurobiology oriented session that I did understand some of: Cori Bargmann, then a post-doc in the Horvitz lab, gave a wonderful talk on laser ablation of different chemosensory neurons and Jim Thomas gave a great talk on osmotic avoidance. I was also part of this session and eventually it was my turn to speak. I had not spoken with many worm people before the meeting and was not aware of the consensus at the time that, based on the determinate development and what looked like a fixed wiring diagram, *C. elegans* would not be able to learn. Sydney Brenner had even commented on that publicly (which he later unhappily confirmed for me). Therefore, I did not realize that my findings would send a shock wave through the community. I got up, gave my talk and went from not knowing very many people to being the belle of the ball! After my talk, many people wanted to connect and were excited that the worm had now joined the ranks of animals that had demonstrated the ability to learn and remember. One of the most interesting conversations I had was with Dick Russell who had written a wonderful paper with Ed Hedgecock on thermosensation in the worm (Hedgecock & Russell, 1975). In that paper they reported that if placed on a thermal gradient *C. elegans* would return to the temperature at which they had been cultivated with food and avoid temperatures where they experienced no food. I had read the paper and shared my thoughts with him – my interpretation of his findings was that the worms might be learning and remembering the temperature and its relationship to the presence of bacteria and asked him if they had thought about it as learning. He laughed and told me that they had indeed thought about that, but since the consensus at the time was that *C. elegans* couldn't learn they had not mentioned that possibility. He was delighted they were wrong and that I, as a psychologist who understood learning, had conclusively shown that worms could learn and remember experiences.

Over the next several years I chose not to return to *Aplysia* research, but to stay with *C. elegans* and received much support from the worm community – Jim Thomas from the University of Washington invited me to come down to Seattle when Marty Chalfie gave a talk and I went to dinner with them. Marty was very helpful with suggesting mutant strains I could test to try to get insights into the neural circuit underlying the tap response and co-authored an SfN abstract on that topic with me. Several years later he came by the lab to give my graduate student Stephen Wicks lessons on how to laser ablate neurons. Don Moerman and Ann Rose at UBC always had time to answer my worm-related questions and make suggestions for genes to test for their possible involvement in tap-habituation. Two years later, at the first Wisconsin-Madison International worm meeting Ikue Mori was just finishing a post-doc and was returning to Japan to set up her own lab to study this thermotaxis phenomenon and we had a wonderful conversation about thermal learning and how to study it.

In the 30 years since then I have continued to study learning and memory in *C. elegans* and challenge many assumptions that were once held about behavioral plasticity in the worm. In the beginning of my explorations I thought by studying the simplest form of learning (habituation) in a small neural circuit (laser ablation studies suggested 5 sensory neurons, 4 pairs of interneurons with modulation by 3 additional neurons (one pair of sensory neurons and one interneuron; Wicks & Rankin, 1995) that habituation would be easy to figure out mechanistically. That turned out to be incorrect! My lab has gone on to show that *C. elegans* can express multiple kinds of memory including short-term, intermediate-term (Li *et al.*, 2013) and long-term memory (Beck & Rankin, 1995; Rose, Kaun, & Rankin, 2002; Rose, Kaun, Chen, & Rankin, 2003); that habituation could have an associative component in which worm memory is enhanced by a chemosensory cue that is present at both training and testing (Lau, Timbers, McEwan, Bozorgmehr, & Rankin, 2013; Rankin, 2000). Early experience can also determine the strength of synapses in adult animals as worms deprived of mechanosensory stimulation show decreased responses to tap and have fewer presynaptic vesicles in the tap sensory neurons and fewer *glr-1* glutamate receptors on command interneurons (Rose *et al.*, 2005). Our studies indicate that changes in behavior as a result of experience are the norm in *C. elegans*. An underlying assumption made by learning researchers was that habituation of all components of a response would be the same, and all would be mediated by a single cellular mechanism. Most studies of habituation measure a single response component (most commonly response frequency, next common a response magnitude measure) to assess learning. However, from our earliest work we found evidence that there are multiple mechanisms underlying habituation that can be genetically dissociated. In Rankin and Broster (1992) we hypothesized that habituation produced by stimuli delivered at different frequencies would be mediated by different mechanisms. A genetic analysis performed much later by Ardiel *et al.* (2018) found evidence of genes that affected habituation at high and low frequencies differently, confirming our hypothesis. The development of the Multi-Worm Tracker (Swierczek, Giles, Rankin, & Kerr, 2011) allowed us to carry out much larger screens of genes and measure many response characteristics simultaneously. This led us to the conclusion that habituation of different response components (frequency, latency, duration, speed) were at least somewhat independent measures and could be mediated by different genes (Ardiel, Yu, Giles, & Rankin, 2017; McDiarmid *et al.*, 2020). Ardiel *et al.* (2017) showed that each response component responded uniquely to repeated noxious stimulation, and that allowed the worm to adapt its overall behavior to sculpt its response pattern in a way that would enhance its ability to escape a potentially toxic environment while still responding to an acute noxious stimulus. Our hypotheses about different response components showing different patterns of plasticity have recently been supported by research studying habituation in zebrafish (Randlett *et al.*, 2019; reviewed in McDiarmid, Yu, & Rankin, 2019).

In retrospect, these accounts embody the ethos of *C. elegans* research: challenging old assumptions and proving the impossible possible. From these serendipitous beginnings, with the help and support of the *C. elegans* community, the field of learning and memory in *C. elegans* began. Since then the field has grown rapidly, and many researchers have expanded our knowledge about the types of learning and memory *C. elegans* can show and identified genetic mechanisms for many of them. It is quite astounding that the worm shows so much behavioral plasticity and can learn so many different things. The take-home message from this research on a microscopic worm with only ~300 neurons, that lives about 2 weeks is that it is highly adaptive for organisms to learn from their experience and to use that experience to guide their behavior. The breadth of the abilities of *C. elegans* to change its behavior as a result of experience is a testament to the importance of behavioral plasticity and learning in survival.

Acknowledgements

None of this could have been achieved without the help of many talented graduate and undergraduate students who have brought their questions, their insights and their curiosity to my lab and propelled the research into new directions. I have had continuous support from the Natural Sciences and Engineering Research Council of Canada for this work. Without the amazing worm community's advice, guidance and resources this work would not have been possible. Specific thanks for comments on this manuscript to Troy McDiarmid, Alex Yu, Lexis Kepler and Joseph Liang.

Disclosure statement

No potential conflict of interest was reported by the author(s).

ORCID

Catharine H. Rankin http://orcid.org/0000-0002-1781-0654

References

Ardiel, E.L., McDiarmid, T.A., Timbers, T.A., Lee, K.C.Y., Safaei, J., Pelech, S.L., & Rankin, C.H. (2018). Insights into the molecular mechanisms underlying interstimulus interval-dependent habituation in *Caenorhabditis elegans*. *Proceedings of the Royal Society B*, *285*(1891), 20182084. doi:10.1098/rspb.2018.2084

Ardiel, E.L., Yu, A.J., Giles, A.C., & Rankin, C.H. (2017). Habituation as an adaptive shift in response strategy mediated by neuropeptides. *Nature Press Journals: Science of Learning*, *2*(9), 1–10.

Beck, C.D.O., & Rankin, C.H. (1995). Heat shock disrupts long-term memory consolidation in *Caenorhabditis elegans*. *Learning & Memory*, *2*(3–4), 161–177. doi:10.1101/lm.2.3-4.161

Chalfie, M., Sulston, J.E., White, J.G., Southgate, E., Thomson, J.N., & Brenner, S. (1985). The neural circuit for touch sensitivity in *Caenorhabditis elegans*. *The Journal of Neuroscience: The Official*

Journal of the Society for Neuroscience, *5*(4), 956–964. doi:10.1523/JNEUROSCI.05-04-00956.1985

Hedgecock, E.M., & Russell, R.L. (1975). Normal and mutant thermotaxis in the nematode *Caenorhabditis elegans*. *Proceedings of the National Academy of Sciences of the United States of America*, *72*(10), 4061–4065. doi:10.1073/pnas.72.10.4061

Lau, H.L., Timbers, T.A., McEwan, A.H., Bozorgmehr, T., & Rankin, C.H. (2013). Genetic dissection of memory for associative and non-associative learning in *Caenorhabditis elegans*. *Genes, Brain, and Behavior*, *12*(2), 210–223. doi:10.1111/j.1601-183X.2012.00863.x

Li, C., Rose, J., Timbers, T.A., McEwan, A., Bozorgmehr, T., & Rankin, C.H. (2013). The FMRFamide-related neuropeptide FLP-20 is required in the mechanosensory neurons during memory for massed training in C. elegans. *Learning & Memory*, *20*(2), 103–108. doi:10.1101/lm.028993.112

McDiarmid, T.A., Belmadani, M., Liang, J., Meili, F., Mathews, E.A., Mullen, G.P., … Rankin, C.H. (2020). Systematic phenomics analysis of autism-associated genes reveals parallel networks underlying reversible impairments in habituation. *Proceedings of the National Academy of Sciences of the United States of America*, *117*(1), 656–667. doi:10.1073/pnas.1912049116

McDiarmid, T.A., Yu, A.J., & Rankin, C.H. (2019). Habituation is more than learning to ignore: Multiple mechanisms serve to facilitate shifts in behavioral strategy. *BioEssays*, *41*(9), 1900077. doi:10.1002/bies.201900077

Randlett, O., Haesemeyer, M., Forkin, G., Shoenhard, H., Schier, A.F., Engert, F., & Granato, M. (2019). Distributed plasticity drives visual habituation learning in larval zebrafish. *Current Biology*, *29*(8), 1337–1345.e4. doi:10.1016/j.cub.2019.02.039

Rankin, C.H. (2000). Context conditioning in habituation in the nematode *C. elegans*. *Behavioral Neuroscience*, *114*(3), 496–505. doi:10.1037/0735-7044.114.3.496

Rankin, C.H., Beck, C.D.O., & Chiba, C.M. (1990). *Caenorhabditis elegans*: A new model system for the study of learning and memory. *Behavioural Brain Research*, *37*(1), 89–92. doi:10.1016/0166-4328(90)90074-O

Rankin, C.H., & Broster, B.S. (1992). Factors affecting habituation and recovery from habituation in *C. elegans*. *Behavioral Neuroscience*, *106*(2), 239–249. doi:10.1037/0735-7044.106.2.239

Rose, J.K., Kaun, K.R., Chen, S.H., & Rankin, C.H. (2003). Glutamate receptor trafficking underlies long-term memory in *C. elegans*. *The Journal of Neuroscience*, *23*(29), 9595–9600. doi:10.1523/JNEUROSCI.23-29-09595.2003

Rose, J.K., Kaun, K.R., & Rankin, C.H. (2002). A New group-training procedure for habituation demonstrates that presynaptic glutamate release contributes to long-term memory in *Caenorhabditis elegans*. *Learning & Memory*, *9*(3), 130–137. doi:10.1101/lm.46802

Rose, J.K., Sangha, S., Rai, S., Norman, K., & Rankin, C.H. (2005). Decreased sensory stimulation reduces behavioral responding, retards development, and alters neuronal connectivity in *Caenorhabditis elegans*. *The Journal of Neuroscience: The Official Journal of the Society for Neuroscience*, *25*(31), 7159–7168. doi:10.1523/JNEUROSCI.1833-05.2005

Swierczek, N., Giles, A., Rankin, C.H., & Kerr, R. (2011). High-throughput behavioral analysis in *C. elegans*. *Nature Methods*, *8*(7), 592–598. doi:10.1038/nmeth.1625

Wicks, S.R., & Rankin, C.H. (1995). Integration of mechanosensory stimuli in *Caenorhabditis elegans*. *The Journal of Neuroscience*, *15*(3), 2434–2444. doi:10.1523/JNEUROSCI.15-03-02434.1995

Of worms and men

John White

ABSTRACT
Following the spectacular success of molecular genetics in deciphering the genetic code in the 1960s, several of its leading practitioners felt sufficiently emboldened to use their newly acquired skills to move on and study that most enigmatic of biological organs – the brain. Sydney Brenner's approach was to focus on *Caenorhabditis elegans*, a nematode that is genetically tractable, has a nervous system that generates a rich repertoire of behaviours yet is small enough to allow anatomical reconstructions with ultrastructural precision. Through force of personality and some inspired pioneering studies, Brenner managed to ignite a bonfire of enthusiasm for this organism, which has resulted in its nervous system becoming the best understood of that in any organism. Initially, many were skeptical that this rather strange structure with just a few hundred neurons would yield insights that were relevant to vertebrate nervous systems. However, fifty years on we know that the basic repertoire of molecular components of worm and human nervous systems are remarkably similar. Furthermore, worms have a similar diversity of these components rather than a primitive sub-set. It appears that the fundamental difference in a vertebrate nervous system is a huge expansion of the neural units that comprise a basic brain such as that exemplified in *C. elegans*.

When Sydney Brenner chose *Caenorhabditis elegans* as a model organism with which to study the nervous system using the tools of molecular genetics, he hoped that there would be sufficient structural and functional conservation among different nervous systems. Thus any knowledge acquired from *C. elegans* would provide insights into the function and development of vertebrate nervous systems. This was an act of faith on his part that was not shared by many mainstream neurobiologists at the time. Now, after more than 50 years of study, a wealth of knowledge has been gathered on the structure, function and development of the *C. elegans* nervous system, so I thought that this might be an opportune time to compare this miniscule brain with that of humans.

Remarkably, genomic studies have revealed that both *C. elegans* and humans have a total complement of around 20,000 genes as defined by DNA sequences that could encode proteins (Hillier *et al.*, 2005; Ezkurdia *et al.*, 2014). However, a human nervous system, with a total complement of neurons estimated at 86 billion (Herculano-Houzel, 2009), has to be vastly more complex than the nervous system of *C. elegans* with its complement of just 302 neurons (White, Southgate, Thomson, & Brenner, 1986). Yet, apparently, there does not appear to be a concomitant increase in the genetic information required to specify the human brain.

Although we like to think of the human brain as being the pinnacle in the evolution of nervous systems, it is by no means the biggest. Large mammals, such as elephants, have brains with a similar overall organization to those of humans but are considerably larger. In general, there is a rough correlation between brain and body size (Herculano-Houzel, 2009). The probable reason for this correlation is that vertebrate brains are organized as a collection of somatotopic maps (Saladin, 2012). The afferent pathways of sensory fields from the surface of an animal map to a planar homuncular representation within the cortex. The mapping is coherent but not necessary linear - regions on the skin of a vertebrate that have a high density of receptors, such as the digits of a human, map into relatively larger areas than do regions with fewer receptors, such as the back. There are similar coherent maps from the motor cortex via efferent pathways to body muscles. The large surface area of large mammals implies larger maps in the brain, assuming that the density of sensory receptors and motor neurons is similar across species. The structure of the *C. elegans* nervous system also exhibits striking somatotopic maps of the sensory receptors in the head to the nerve ring and from motor neurons in the nerve ring to the muscles in the head (White, 2018). Somatotopic maps therefore appear to be a conserved organizational feature of nervous systems across large scales of size.

It is interesting to see how the number of neurons in a nervous system might scale with the surface area. An adult *C. elegans* has a surface area of about 0.04 sq mm and has 302 neurons in its nervous system. An adult human has a surface area of around 2 sq m, which is therefore about 50,000,000 larger than that of *C. elegans*. Given the 302- neuron complement of *C. elegans*, scaling by surface area would predict

Figure 1. Sydney and Jim Watson in Singapore 2015.

around 15 billion neurons in humans, which intriguingly is in the same order of magnitude as the estimate of 86 billion neurons in the human brain. However, scaling by volume would predict around 38,250 billion human neurons - around three orders of magnitude too many. It therefore seems reasonable to suggest that the organization of nervous systems into collections of 2-dimensional somatotopic maps, representing the receptive fields of sensory receptors together with maps of motor neurons onto muscles, is the reason why the number of neurons (and hence size of brain) scales with the surface body area of an animal.

When the structure of the *C. elegans* nervous system was determined, it became apparent that the component neurons could be grouped into 118 clearly defined anatomical classes in the hermaphrodite (White *et al.*, 1986). More recently, an independent classification of neurons based solely on the combination of expressed transcription factors revealed a practically identical grouping of neuron classes (Hobert, Glenwinkel, & White, 2016). The size and complexity of vertebrate brains make the identification of neuron classes difficult in these animals. The cerebellum is the most regular structure in the vertebrate brain and has been classically described of being made up of just 5 neuron classes (Masland, 2004). This is undoubtedly an underestimate, but nevertheless a surprising observation. Recent studies using single cell RNA sequencing have demonstrated that there are around 34 classes of neuron in the hypothalamus, based on identified, unique combinations of transcription factor expression (Chen, Wu, Jiang, & Zhang, 2017). These are striking observations that reveal that, although vertebrate nervous systems have around 9 orders of magnitude more neurons than *C. elegans*, the complexity of these nervous systems in terms of the diversity of cell types is similar. These observations also show that the 302 neuron complement of *C. elegans* must be close to the smallest number of neurons possible while maintaining representatives of all the 118 classes.

Since Brenner first mooted *C. elegans* as a favorable organism in which to study a nervous system, a wealth of information has accrued on the anatomy, cell biology, function and genetic specification of the 302 neurons that control this little worm. In general, it has been found that most features are strikingly similar to humans. For example, the same neurotransmitters are generally used in the same contexts. In both worms and humans, acetylcholine is the main excitatory and GABA the main inhibitory neurotransmitter at neuromuscular junctions (Pereira *et al.*, 2015). Surprisingly, conservation also occurs at higher levels of functionality: serotonin is associated with feelings of well-being in humans (Hariri & Holmes, 2006) and can de-stress worms causing them to browse, feed and lay eggs if applied exogenously (Horvitz, Chalfie, Trent, Sulston, & Evans, 1982); dopamine is associated with the initiation of sequences of motor movement in both humans (Sveinbjornsdottir, 2016) and worms (Hills, Brockie, & Maricq, 2004).

One possible explanation for the apparent increase in capability of human over worm nervous systems is that humans could have a much larger repertoire of neurotransmitter receptors that can be used in different contexts. In addition, it may be that greater use is made of humoral signaling via neuropeptides or neurotransmitters acting on G-protein coupled receptors (GPCRs) to modulate the functions of neural circuits. Now that both the human and *C. elegans* genomes have been sequenced, it is possible to compare the total complement of GPCRs in the two organisms. Perhaps contrary to expectations, it was found that there are around 1300 putative GPCRs in *C. elegans* as compared to 400 in humans (Hobert, 2013). Many of these GPCRs are chemoreceptors (Robertson & Thomas, 2006) while others are peptide receptors, so it is likely that *C. elegans* can detect a larger repertoire of internal and external chemical signals than humans. Indeed, it has now become apparent that there is a parallel pattern of connections within the *C. elegans* nervous system that is not realized from the anatomical reconstructions, but in which long range diffusion of neuropeptides and some neurotransmitters link pre- and postsynaptic partners (Bentley *et al.*, 2016). In summary, these observations suggest that, at least in terms of the repertoire of molecular components, the *C. elegans* nervous system is similar or even more complex than that of humans (Hobert, 2013).

In terms of sensory perception, *C. elegans* lacks the high-bandwidth visual and auditory receptor channels of vertebrates but is nevertheless richly endowed with sensory receptors of many different modalities (Bargmann, 2006; Goodman, 2006). Furthermore, many of these sensory inputs have been shown to be used for memory and learning (Ardiel & Rankin, 2010). The movement of worms on planar substrates has been studied and has started to yield insights into how motor neurons coordinate body muscles to produce locomotion and turns (Zhen & Samuel, 2015). Recently, automated computer analyses of movements of wild type and mutant animals have shown that it is possible to describe locomotory behavior by a time sequence of combinations of four postures and thereby predict a novel neuron function (Brown, Yemini, Grundy, Jucikas, & Schafer, 2013). The natural habitat of *C. elegans* is in a 3-dimensional matrix of soil particles, so it is likely that much more will be revealed about worm behavior when studied in something closer to its natural habitat than planar surfaces, although this will be technically challenging (White, 2018).

Now that the synaptic interconnections of all neurons (i.e., the connectome) of both sexes of *C. elegans* have been determined (Cook *et al.*, 2019), and given the wealth of behavioral and cell biological information that has been accrued, it is worth considering what needs to be done in order to claim that we understand how the nervous system of this organism mediates its dynamic interactions with the environment, i.e., how it works. Probably the most convincing way of doing this is to simulate a complete worm in a computer along the physical properties of the milieu it inhabits (Gleeson, Lung, Grosu, Hasani, & Larson, 2018). If such a simulation is able to reproduce all the normal behaviors exhibited by wild-type and mutant worms in a variety of environmental contexts, then perhaps we will understand how the brain of this small animal generates patterns of behavior that are optimized for its survival and reproduction in its natural habitat. It could well be even more enlightening to study the *development* of the nervous system by computer simulation. This may not be an unrealistic fantasy. All the somatic cells in *C. elegans* are known, as is their lineal history (Sulston, Schierenberg, White, & Thomson, 1983). Molecular mechanisms for the guidance of outgrowing nerve processes are being uncovered (Gujar, Sundararajan, Stricker, & Lundquist, 2018). New optical techniques are starting to provide insights into the dynamics of brain assembly (Rapti, Li, Shan, Lu, & Shaham, 2017). These developments suggest that in the not-too-distant future it may be possible to consolidate this information with a knowledge of the receptors, channels, and adhesion molecules expressed on each cell to simulate the patterns of cell divisions, migrations, differentiations, process outgrowth, and synaptogenesis that make up the process of neurogenesis in *C. elegans* (Alicea, 2020). An exciting prospect would be to apply this strategy to vertebrate nervous systems, given that their cellular makeup and diversity is similar to that of *C. elegans*. This would enable a virtual vertebrate connectome to be obtained by computer simulation of the process of neurogenesis rather than by brute-force reconstruction of electron micrographs of serial sections, which is a formidable undertaking for a nervous system containing 86 billion neurons.

It may be instructive to consider whether some of the central topics in the studies of the human brain have relevance in the context of *C. elegans*. One prominent issue is the nature of consciousness. There are numerous definitions of what consciousness means but they are generally of the form: *the state of being aware of one's surroundings and self.* The responses of *C. elegans* to external stimuli have been extensively studied and have been shown to elicit responses that are appropriate to the stimulus (Bargmann, 2006). So, in this sense a *C. elegans* is certainly conscious. Indeed, I would argue that an autonomously driven car is also conscious. Rather like the concept of vitalism, which has fallen into disuse because of current knowledge of cell physiology,

Figure 2. Sydney at MRC LMB 40th anniversary celebrations in 1987.

consciousness may not be a useful concept for understanding the function of the nervous system of *C. elegans*. A concept derived from studies of higher nervous systems that is better defined is neural plasticity. At a gross level this is commonly seen as the ability of victims to eventually recover lost faculties following a stroke that kills off areas of the brain; however, it is likely that neural plasticity is a manifestation of the basic process of learning (Sweatt, 2016). There are well defined instances of neuronal plasticity during the development of the *C. elegans* nervous system and the detailed mechanisms are beginning to be understood at the molecular level (Jin & Qi, 2018), suggesting that this organism may have much to offer in the search for an understanding of this basic mechanism of cognition.

Creativity is an attribute that many would uniquely ascribe to human brains. However, as Brenner often commented, natural selection and replication of randomly induced variation is the only theory underpinning biology. This mechanism has produced all the rich panoply of life in the world apparently without the involvement of a nervous system. Nevertheless, a version of this process may well work at a neural level. A random change to the strength of a synapse from a sensory receptor could confer advantages to the animal by sensitizing or desensitizing the perception of a particular signal and become consolidated as a "memory" (Peymen *et al.*, 2019).

Modification of the strength of a synapse is generally accepted as the basic cellular mechanism by which the brain stores data, i.e., memories. *C. elegans* has around 7000 synapses (Cook *et al.*, 2019) whereas estimates for the human brain are around 10^{15} (Drachman, 2005), i.e., a factor of around 10^{11} greater. I have argued that the *C. elegans* nervous system has a similar level of basic molecular complexity to a human brain. Given that the genomes in the two organisms can code for a similar number of proteins, it is highly unlikely that genetic mechanisms are specifying the 10^{15} synapses in the human brain at a detailed level. More likely is that human brains are massively expanded versions of a basic brain as exemplified by that in *C. elegans*. Such an expansion would require very little extra genetic information. This level of expansion brings about a huge data storage capacity, which allows many types of memory (e.g., images, sounds, sequences of muscle activations) to be stored and recalled by will or external context. This facility enables mimicry and thereby the transfer of learned experiences, "memes" (Dawkins, 1976), by interhuman communication. In this way memories, repackaged as knowledge, can be transmitted horizontally rather than by vertical inheritance, thereby generating all the manifestations of human culture.

The nervous system of *C. elegans* is arguably the best characterized and understood of that of any animal. Brenner's vision that its study would reveal how genes specify a brain is coming to fruition. Even more than he dared hoped, it is becoming apparent that the *C. elegans* nervous system has remarkable parallels with the human brain and is providing revealing insights into how this enigmatic organ functions.

Disclosure statement

No potential conflict of interest was reported by the author(s).

References

Alicea, B. (2020). Raising the connectome: The emergence of neuronal activity and behavior in *Caenorhabditis elegans* (2020). *Front. Cell. Neurosci*, 15 September 2020. doi:10.3389/fncel.2020.524791

Ardiel, E.L., & Rankin, C.H. (2010). An elegant mind: Learning and memory in *Caenorhabditis elegans*. *Learning & Memory*, *17*(4), 191–201. doi:10.1101/lm.960510

Bargmann, C. (2006). Chemosensation in C. elegans. *WormBook*, 1–29. doi:10.1895/wormbook.1.123.1

Bentley, B., Branicky, R., Barnes, C.L., Chew, Y.L., Yemini, E., Bullmore, E.T., … Schafer, W.R. (2016). The multilayer connectome of *Caenorhabditis elegans*. *PLoS Computational Biology*, *12* (12), e1005283. doi:10.1371/journal.pcbi.1005283

Brown, A.E.X., Yemini, E.I., Grundy, L.J., Jucikas, T., & Schafer, W.R. (2013). A dictionary of behavioral motifs reveals clusters of genes affecting *Caenorhabditis elegans* locomotion. *Proceedings of the National Academy of Sciences of the United States of America*, *110*(2), 791–796. doi:10.1073/pnas.1211447110

Chen, R., Wu, X., Jiang, L., & Zhang, Y. (2017). Single-cell RNA-seq reveals hypothalamic cell diversity. *Cell Reports*, *18*(13), 3227–3241. doi:10.1016/j.celrep.2017.03.004

Cook, S.J., Jarrell, T.A., Brittin, C.A., Wang, Y., Bloniarz, A.E., Yakovlev, M.A., … Emmons, S.W. (2019). Whole-animal connectomes of both *Caenorhabditis elegans* sexes. *Nature*, *571*(7763), 63–71. doi:10.1038/s41586-019-1352-7

Dawkins, R. (1976). *The selfish gene*. Oxford, UK: Oxford University Press.

Drachman, D. (2005). Do we have brain to spare? *Neurology*, *64*(12), 2004–2005. doi:10.1212/01.WNL.0000166914.38327.BB

Ezkurdia, I., Juan, D., Rodriguez, J.M., Frankish, A., Diekhans, M., Harrow, J., … Tress, M.L. (2014). Multiple evidence strands suggest that there may be as few as 19,000 human protein-coding genes. *Human Molecular Genetics*, *23* (22), 5866–5878. doi:10.1093/hmg/ddu309

Gleeson, P., Lung, D., Grosu, R., Hasani, R., & Larson, S. (2018). c302: A multiscale framework for modelling the nervous system of *Caenorhabditis elegans*. *Philosophical Transactions of the Royal Society B: Biological Sciences*, *373* (1758), 20170379. doi:10.1098/rstb.2017.0379

Goodman, M. (2006). Mechanosensation. WormBook, 1–14. doi:10.1895/wormbook.1.62.1

Gujar, M.R., Sundararajan, L., Stricker, A., & Lundquist, E.A. (2018). Control of growth cone polarity, microtubule accumulation, and protrusion by UNC-6/Netrin and its receptors in *Caenorhabditis elegans*. *Genetics*, *210* (1), 235–255. doi:10.1534/genetics.118.301234

Hariri, A.R., & Holmes, A. (2006). Genetics of emotional regulation: The role of the serotonin transporter in neural function. *Trends in Cognitive Sciences*, *10* (4), 182–191. doi:10.1016/j.tics.2006.02.011

Herculano-Houzel, S. (2009). The human brain in numbers: A linearly scaled-up primate brain. *Frontiers in Human Neuroscience*, *3*, 31. doi:10.3389/neuro.09.031.2009

Hillier, L.W., Coulson, A., Murray, J.I., Bao, Z., Sulston, J.E., & Waterston, R.H. (2005). Genomics in *C. elegans*: So many genes, such a little worm. *Genome Research*, *15*(12), 1651–1660. doi:10.1101/gr.3729105

Hills, T., Brockie, P.J., & Maricq, A.V. (2004). Dopamine and glutamate control area-restricted search behavior in *Caenorhabditis elegans*. *The Journal of Neuroscience: The Official Journal of the Society for Neuroscience*, *24* (5), 1217–1225. doi:10.1523/JNEUROSCI.1569-03.2004

Hobert, O. (2013). The neuronal genome of *Caenorhabditis elegans*. *WormBook*, 1–106. doi:10.1895/wormbook.1.161.1

Hobert, O., Glenwinkel, L., & White, J. (2016). Revisiting neuronal cell type classification in *Caenorhabditis elegans*. *Current Biology*, *26* (22), R1197–R1203. doi:10.1016/j.cub.2016.10.027

Horvitz, H.R., Chalfie, M., Trent, C., Sulston, J., & Evans, P. (1982). Serotonin and octopamine in the nematode *Caenorhabditis elegans*. *Science*, *216* (4549), 1012–1014. doi:10.1126/science.6805073

Jin, Y., & Qi, Y.B. (2018). Building stereotypic connectivity: Mechanistic insights into structural plasticity from *C. elegans*. *Current Opinion in Neurobiology*, *48*, 97–105. doi:10.1016/j.conb.2017.11.005

Masland, R.H. (2004). Neuronal cell types. *Current Biology*, *14* (13), R497–R500. doi:10.1016/j.cub.2004.06.035

Pereira, L., Kratsios, P., Serrano-Saiz, E., Sheftel, H., Mayo, A.E., Hall, D.H., … Hobert, O. (2015). A cellular and regulatory map of the cholinergic nervous system of *C. elegans*. *eLife*, *4*. doi:10.7554/eLife.12432

Peymen, K., Watteyne, J., Borghgraef, C., Van Sinay, E., Beets, I., & Schoofs, L. (2019). Myoinhibitory peptide signaling modulates aversive gustatory learning in *Caenorhabditis elegans*. *PLoS Genetics*, *15*(2), e1007945. doi:10.1371/journal.pgen.1007945

Rapti, G., Li, C., Shan, A., Lu, Y., & Shaham, S. (2017). Glia initiate brain assembly through noncanonical Chimaerin-Furin axon guidance in *C. elegans*. *Nature Neuroscience*, *20*(10), 1350–1360. doi:10.1038/nn.4630

Robertson, H., & Thomas, J. (2006). The putative chemoreceptor families of C. elegans. *WormBook*, 1–12. doi:10.1895/wormbook.1.66.1[10.1895/wormbook.1.66.1]

Saladin, K.S. (2012). *Anatomy & physiology: The unity of form and function* (6th ed.). New York, NY: McGraw-Hill.

Sulston, J.E., Schierenberg, E., White, J.G., & Thomson, J.N. (1983). The embryonic cell lineage of the nematode *Caenorhabditis elegans*. *Developmental Biology*, *100*(1), 64–119. https://doi.org/10.1016/0012-1606.(83)90201-4 [Database] doi:10.1016/0012-1606(83)90201-4

Sveinbjornsdottir, S. (2016). The clinical symptoms of Parkinson's disease. *Journal of Neurochemistry*, *139*, 318–324. doi:10.1111/jnc.13691

Sweatt, J.D. (2016). Neural plasticity and behavior – Sixty years of conceptual advances. *Journal of Neurochemistry*, *139*, 179–199. doi:10.1111/jnc.13580

White, J. (2018). Clues to basis of exploratory behaviour of the *C. elegans* snout from head somatotropy. *Philosophical Transactions of the Royal Society B: Biological Sciences*, *373*(1758), 20170367. doi:10.1098/rstb.2017.0367

White, J.G., Southgate, E., Thomson, J.N., & Brenner, S. (1986). The structure of the nervous system of the nematode *Caenorhabditis elegans*. *Philosophical Transactions of the Royal Society of London. Series B, Biological Sciences*, *314*(1165), 1–340. doi:10.1098/rstb.1986.0056

Zhen, M., & Samuel, A.D.T. (2015). *C. elegans* locomotion: Small circuits, complex functions. *Current Opinion in Neurobiology*, *33*, 117–126. doi:10.1016/j.conb.2015.03.009

Part II

Nervous system development

OPEN ACCESS

A perspective on *C. elegans* neurodevelopment: from early visionaries to a booming neuroscience research

Georgia Rapti

ABSTRACT

The formation of the nervous system and its striking complexity is a remarkable feat of development. *C. elegans* served as a unique model to dissect the molecular events in neurodevelopment, from its early visionaries to the current booming neuroscience community. Soon after being introduced as a model, *C. elegans* was mapped at the level of genes, cells, and synapses, providing the first metazoan with a complete cell lineage, sequenced genome, and connectome. Here, I summarize mechanisms underlying *C. elegans* neurodevelopment, from the generation and diversification of neural components to their navigation and connectivity. I point out recent noteworthy findings in the fields of glia biology, sex dimorphism and plasticity in neurodevelopment, highlighting how current research connects back to the pioneering studies by Brenner, Sulston and colleagues. Multifaceted investigations in model organisms, connecting genes to cell function and behavior, expand our mechanistic understanding of neurodevelopment while allowing us to formulate emerging questions for future discoveries.

Introduction: from pioneer feats to a booming neuroscience community

The nervous system, the set of cells involved in perceiving external or internal stimuli and responding with animal behavior, displays an alluring degree of sophistication. Exploring what shapes its complexity is an endeavor deep-rooted in the descriptions of Golgi, Cajal and their predecessors from the 1800s to today's flourishing neuroscience community. Neurodevelopment follows similar patterns across organisms; cells commit to neural fates to generate neurons and glia, which migrate and extend processes to connect with target cells. It is shaped by cell death, fine-tuned by synaptic pruning, retains plasticity during development, and presents sex-dimorphism to support sex-specific behavior. Research in various models enables a sheer number of discoveries that connect genes to cells and behavior, revealing that neurodevelopment is driven by genetic pathways that are strikingly conserved. This fosters today's misconception that the major principles of neurodevelopment have been addressed. Yet, many questions remain unanswered. How does patterning, cell compartmentalization, and communication coordinate to establish connectivity? What mechanisms remodel the nervous system or drive differences across sexes or related species? How is connectivity shaped by glia, the underappreciated non-neuronal cells? Multifaceted investigations in genetically tractable models allow for the continuous formulation of emerging questions and discoveries.

The non-parasitic nematode *Caenorhabditis elegans*, a well-established genetic model, has been instrumental to key breakthroughs in neurodevelopment for decades. Its defined nervous system consists of 5000 synapses, 302 neurons in a hermaphrodite, 387 neurons in a male, 50 sex-shared ectodermal glia, and 6 associated mesodermal glia-like cells and other hypodermal cells. *C. elegans* was honed for the analysis of the nervous system by visionaries Sydney Brenner and John Sulston, among others. Brenner's and Sulston's early practices gave rise to everyday rituals in all *C. elegans* labs. Importantly, their discoveries and the community they pioneered enable numerous mechanistic findings in neurodevelopment. Studies in *C. elegans* that dissect neural patterning, guidance and connectivity reveal underlying conserved genes, opening doorways to explore complex nervous systems.

This edition is a tribute to Brenner and Sulston as pioneers of a model and a booming community, using *C. elegans* as a showcase of nervous system biology. We compiled manuscripts that discuss the emerging themes in *C. elegans* neurodevelopment and reflect on established and open questions. Here, I delineate key neurodevelopment aspects, featuring the powers of *C. elegans* neuroscience research in light of Brenner's and Sulston's contributions (Figure 1).

The C. elegans *nervous system: from cellular to anatomical and genome maps*

In the pre-Brenner years, *C. elegans* was initially used in research from the 1900s by Maupas, Nigon, and Dougherty

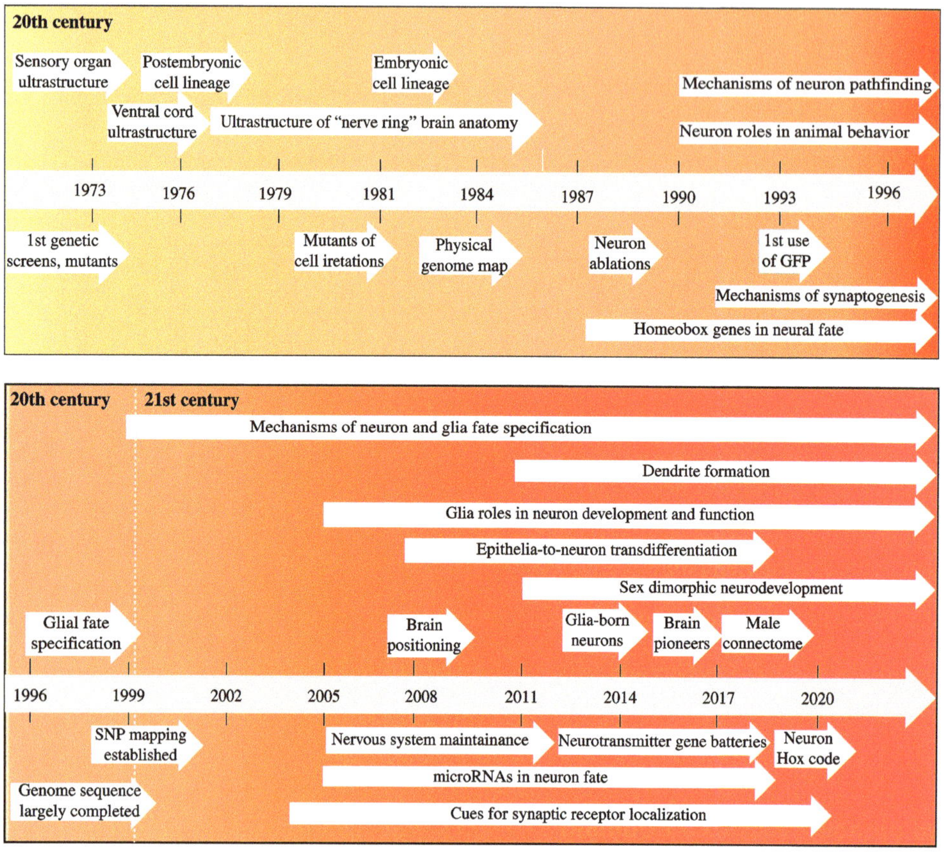

Figure 1. Timeline of some milestones in research of C. elegans neurodevelopment. This timeline presents a variety of milestones in the research of neurodevelopment, in cellular, genetic, genomic and mechanistic aspects. As all milestones of a research field cannot be presented, here I highlight early works that solidified different research directions. For details and citations of the events mentioned, see in text.

(Félix & Nigon, 2017). Later, Sydney Brenner chose *C. elegans* to study development and the nervous system. Brenner, Sulston, and their colleagues, contributed to the major steps that turn species into model organisms (Matthews & Vosshall, 2020). They proceeded from a vision of answering key questions to decrypting the animal's genetic and cellular maps and inspiring a research community to dissect them.

Setting his mind on studying animal development, Brenner chose '… a multicellular organism which has a short life cycle, can be easily cultivated, and is small enough to be handled in large numbers, like a micro-organism. It should have relatively few cells, so that exhaustive studies of lineage and patterns can be made, and should be amenable to genetic analysis.' (Brenner, 1963 MRC funding proposal). He further proposed '…to dissect the genetic specification of a nervous system in much the same way as was done for biosynthetic pathways in bacteria or for bacteriophage assembly … what was needed was an experimental organism which was suitable for genetical study and in which one could determine the complete structure of the nervous system' (Brenner, 1974). Bravely following his lead, his trainees Sulston and White, addressed the extraordinary goal of determining the entire nervous system structure. Sulston devised methods to transfer and visualize animals with Nomarksi microscopy and then embarked on a remarkable venture, to map the *C. elegans* path from a single cell to an adult worm. This was seemingly impossible, since embryonic divisions are fast and poorly discernible away from the egg's surface. Day after day for a year and a half, Sulston sat in a dark room for 12 h to track each division and daughter cell in developing embryos. Sulston confesses it was '… a challenge in the jigsaw-puzzling sense to get it all,' (Check, 2002). 'It was hugely exciting, looking at those cells dividing for the first time and knowing that I could see, I could find out. … then there was the group thing … the lineage was something that people really wanted.' (Gitschier, 2006). Sulston's determination and advances, narrated by Martin Chalfie in Sulston's Obituary (Chalfie, 2018) provided a complete map of how a fertilized egg gives rise to a hermaphrodite (Sulston, Schierenberg, White, & Thomson, 1983; Sulston & Horvitz, 1977), the first road map to study metazoan development at the cellular level.

Meanwhile, John White invested his programming expertise into reconstructing the ultrastructure of the *C. elegans* nervous system. White recalled that '… the project was ridiculously ambitious, given the computer hardware available at the time (1970). Yet, with the courage of innocence we forged ahead' (White, 2013). Together with the 'remarkable electron microscopist' Nichol Thomson and his 'meticulous technician' Eileen Southgate, they first reconstructed the animal's ventral nerve cord and proceeded in tracing neurites of the central neuropil, also known as the nerve ring. Their decade-long labor culminated in the ultrastructural analysis of an entire nervous system, which serves as a reference for all studies of neural circuitry and connectivity (White, Southgate, Thomson, & Brenner, 1976, 1986).

The tremendous amount of neurite trajectories and connections is incorporated into the *WormAtlas*, a *C. elegans* anatomy database generated by David Hall and colleagues, and complemented by recent nervous system reconstructions.

While generating *C. elegans* maps, cell identity was functionally linked to development using cell ablations. White developed laser ablation protocols, employed by Chalfie, Sulston, Bargmann, and others to correlate cell function to nervous system structure and animal behavior (Bargmann & Horvitz, 1991; Chalfie *et al.*, 1985; J. E. Sulston & White, 1980). Meanwhile, Brenner and Sulston connected cellular maps to genetics by establishing landmark physical genetic maps of the *C. elegans* genome, a prerequisite for its sequencing (Brenner, 1974; Sulston & Brenner, 1974). Sulston also spearheaded the sequencing of the animal's genome with Bob Waterston heading the sequencing effort at the Genome Sequencing Center (Washington University of St Louis). *C. elegans* was the first multicellular organism to have its genome sequenced (C. elegans sequencing consortium, 1998; Waterston & Sulston, 1995), giving the first complete genetic content required to build a nervous system (Bargmann, 1999).

Early studies of the genetics of the C. elegans *nervous system*

'The relationship between genes and development is unknown', Sulston and Horvitz wrote (Horvitz & Sulston, 1980). However, their work opened doors to numerous studies of *C. elegans* development that combined overturned this statement. Prior to sequencing the *C. elegans* genome, Brenner, Sulston and Horvitz systematically isolated and analyzed mutations affecting animal physiology and behavior to reveal developmental mechanisms. Brenner's expertise in phage genetics proved insightful in exploiting the value of mutational analysis. This was facilitated by the clonal propagation of hermaphrodites and the freezing protocol developed by Sulston, for strain storage without the need for continuous propagation. This advance allows maintenance and sharing of thousands of isolates from the early mutants to all engineered strains and the creation of consortia such as the Caenorhabditis Genetics Center.

The systematic generation of mutants was a productive conceptual leap for connecting animal and cell physiology to genetic information. Sequencing the *C. elegans* genome and the subsequently established methodologies for genetic mapping (Wicks, Yeh, Gish, Waterston, & Plasterk, 2001) enabled the identification of all mutations that impair nervous system development. Luckily, albeit unanticipated by Brenner, the external application of double-stranded RNA in *C. elegans* suppresses gene expression. Needless to say, genome sequencing allows for the recent CRISPR/Cas9 genome editing and reverse genetics by RNA interference (Dickinson & Goldstein, 2016; Kamath *et al.*, 2003). This array of unbiased and targeted gene manipulations allows for comprehensive research of neurodevelopment.

Core neurodevelopment events: birth and diversification, navigation and connectivity

Neurodevelopment proceeds through core processes of cell diversification, pathfinding, target selection, and connectivity (Figure 2). Investigations in *C. elegans* advance our molecular understanding of these events, pointing to principles that shape the prodigious complexity of nervous systems.

Generation and diversification of nervous system components

An early step in neurodevelopment is the allotment of ectodermal neural precursors. In *C. elegans* master regulators that specify the ectoderm founders remained unknown (Maduro, 2010). Neural specification is driven by transcription factors and cell interactions (Lee *et al.*, 2019; Stefanakis, Carrera, & Hobert, 2015). Early clues came from work by Sulston and colleagues on mutants of lineage iterations, creating supernumerary cells (Chalfie, Horvitz, & Sulston, 1981). It was demonstrated that upon commitment, the daughter cells of neuroblasts adopt distinct size and fate while in the absence of proneural transcription factors or specific kinases the daughter cells assume characteristics of their hypodermal sisters (Forrester, Dell, Perens, & Garriga, 1999; Frank, Baum, & Garriga, 2003; Singhvi & Garriga, 2009). Neural development is also shaped by cell death. Early work in *C. elegans* showed that of the 1090 somatic cells, 131 die before differentiating, most of which are in the ectodermal lineage (Sulston *et al.*, 1983). These also led to the identification of dedicated apoptotic factors driving these deaths (Horvitz, Shaham, & Hengartner, 1994; Shaham, 1998). It is now known that asymmetric divisions of neuroblasts and the death of their daughter cells, which shape the nervous system, depend on a complex array of transcription factors, kinases, GTPases, and contractile forces driving cell asymmetry (Cordes, Frank, & Garriga, 2006; Hirose & Horvitz, 2013; Metzstein, Hengartner, Tsung, Ellis, & Horvitz, 1996; Mishra, Wei, & Conradt, 2018; Ou, Stuurman, D'Ambrosio, & Vale, 2010; Teuliere, Cordes, Singhvi, Talavera, & Garriga, 2014).

Nervous system sophistication relies on diversification of its components, starting from the fates of daughter cells after neuroblast division. Diversification depends on patterning along the body axis and determination of subtype fate. The earliest mutants of neuronal specification affected homeobox genes (Finney, Ruvkun, & Horvitz, 1988; Way & Chalfie, 1988; White, Southgate, & Thomson, 1992). Neurodevelopmental roles of transcription factors are being refined over the years. Distinct transcription factors specify left-right neuron asymmetry (Lesch & Bargmann, 2010; Sarin, Antonio, Tursun, & Hobert, 2009) or regulate terminal differentiation of motoneuron, interneuron, or sensory modalities (Hobert, 2016; Kim, Kim, & Sengupta, 2010; Masoudi *et al.*, 2018; Poole, Bashllari, Cochella, Flowers, & Hobert, 2011; Satterlee *et al.*, 2001; Sengupta, Chou, & Bargmann, 1996). Interestingly, neurons of the same neurotransmitter identity are subject to regulation by shared factors, the *terminal selectors* (Gendrel, Atlas, & Hobert, 2016;

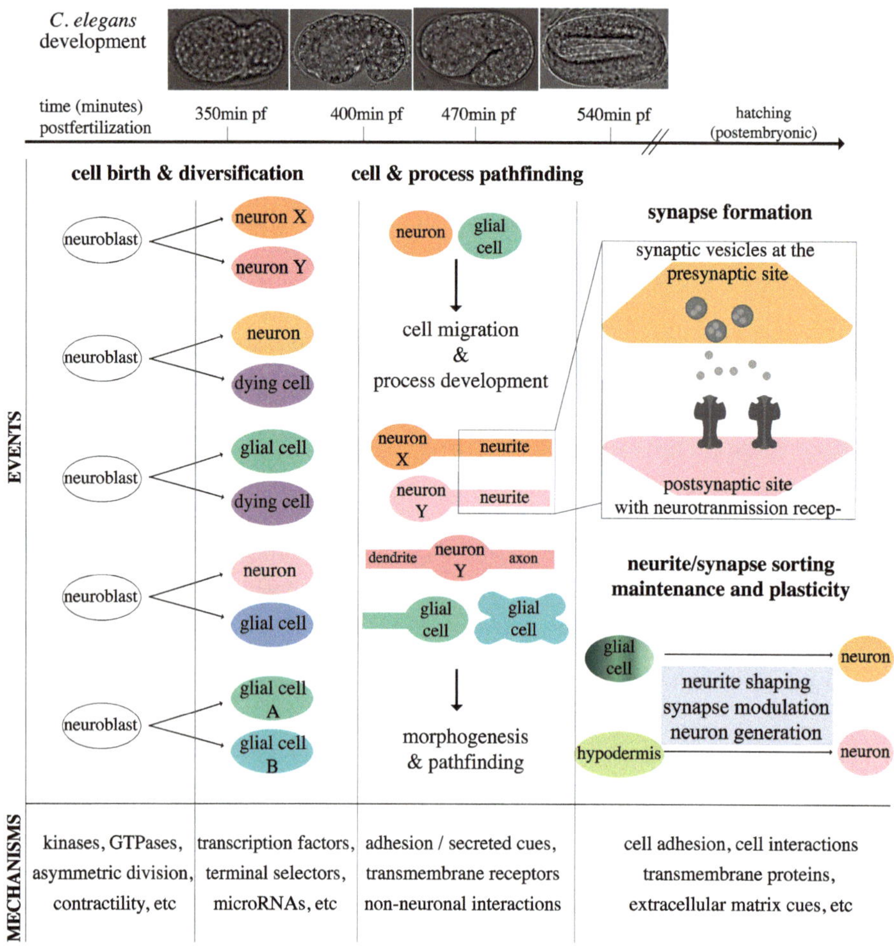

Figure 2. Core neurodevelopment processes that support the formation of the nervous system. The schematics summarizes selected events in nervous system formation, implicating cell autonomous factors, cell interactions and non cell-autonomous cues. Through development (from left to right), neuroblasts give rise to nervous system components, neurons and glia migrate, grow and diversify processes to then reach targets and generate synaptic connections. These connections are subject to maintenance and plasticity by mechanisms acting in neuronal or associated, non-neuronal cells, such as glial cells and hypodermal cells. Specifically, non-neuronal glia and hypodermal cells can generate postembryonic neurons, by division or transdifferentiation and in other instances they function for synapse maintenance or plasticity. For details and citations of the events and underlying mechanisms mentioned here, see in text.

Kratsios, Stolfi, Levine, & Hobert, 2012; Pereira *et al.*, 2015). Conversely, pan-neuronal identity is defined by redundant, parallel-acting cis-regulatory modules that direct expression to broad domains of the nervous system (Hobert, Carrera, & Stefanakis, 2010; Hobert & Kratsios, 2019). Each neuron subtype is now shown to present a unique combination of homeobox transcription factors (Reilly, Cros, Varol, Yemini, & Hobert, 2020). MicroRNAs are also shown to modulate fate; some can be subject to early priming in neuroblasts while others influence calcium signaling (Cochella & Hobert, 2012; Hsieh, Chang, & Chuang, 2012). Hence, new studies continue to uncover an increasing sophistication in mechanisms of neural differentiation.

Glia, non-neuronal ectodermal cells, also undergo diversification. This is regulated by transcription factors that segregate in non-neuronal daughter cells after progenitor division and act in glial cells (Labouesse, Hartwieg, & Horvitz, 1996) or by transcription factors that affect both glial and neuronal fate (Yoshimura, Murray, Lu, Waterston, & Shaham, 2008; Zhang, Noma, & Yan, 2020). Interestingly, the homeobox protein Prospero safeguards the postembryonic gene expression and function of glia (Wallace, Singhvi, Liang, Lu, & Shaham, 2016), suggesting that glia show maintenance and plasticity during the animal's life.

Altogether, specification of the nervous system relies on inherited factors and intercellular signaling (Bertrand, Bisso, Poole, & Hobert, 2011). In this edition, Barrière and Bertrand (2020) review how the nervous system diversifies in nematode species, through a combination of transcription factors and cell interactions through morphogen signaling. Sulston's cell lineaging and Brenner's mutants were key to analyzing cell interactions (Sulston *et al.*, 1983; Sulston & Horvitz, 1977) and their work in distant nematode species attempted to assign analogous neurons and features (Sulston, Dew, & Brenner, 1975). Barrière and Bertrand (2020) reflect on the evolvability of neural cell anatomy, cell fate, and wiring.

Nervous system morphogenesis: soma migrations and process pathfinding

Nervous system components migrate to adopt final positions in mature neuropils. Transcription factors and kinesin motors were early recognized to drive such migrations (Garriga, Guenther, & Horvitz, 1993; Wolf, Hung, Wightman, Way, & Garriga, 1998) while recent studies add the roles of actin, cytoskeletal scaffolds, and kinases (Levy-Strumpf & Culotti, 2007; Stringham & Schmidt, 2009; Tian *et al.*, 2015; Withee, Galligan, Hawkins, & Garriga, 2004). Cell interactions were implicated early in migration (Garriga, Desai, & Horvitz, 1993); they involve cell adhesion (Solecki, 2012), guidance cues (Sundararajan & Lundquist, 2012), morphogens, and the planar cell polarity pathway for rosette formation (Shah *et al.*, 2017). Recent research focuses on non-cell-autonomous factors of neuron migration, such as epidermal microRNAs (Pedersen *et al.*, 2013), heparan sulfate proteoglycans, and extracellular matrix (Saied-Santiago *et al.*, 2017; Tornberg *et al.*, 2011). Glial migrations are shown to depend on nutrient supply (Zhang, Ackley, & Yan, 2020) but remain far less studied.

Concurrently to or after migration, neurons grow neurites to reach their partners. The specification of neurites to become axons or dendrites and dendrite development, in particular, were not subject of early research. However, later work uncovered that axon-dendrite sorting is polarized by ankyrin and kinesin, while guidance cues and neuronal asymmetry define the site of axon formation (Adler, Fetter, & Bargmann, 2006; Maniar *et al.*, 2011). Glial-like mesodermal cells also specify certain axons through calcium signaling (Meng, Zhang, Jin, & Yan, 2016). Recent work reveals that the morphogenesis of dendrites and axons differ. Sensory dendrites form by retrograde extension upon extracellular attachment during neuronal migration (Heiman & Shaham, 2009). Some mechanosensory dendrites grow extensive arborization, driven by hypodermal cues, extracellular matrix, adhesion, and actin effectors (Dong, Liu, Howell, & Shen, 2013; Liu & Shen, 2012; Oren-Suissa, Hall, Treinin, Shemer, & Podbilewicz, 2010; Salzberg *et al.*, 2013; W. Zou *et al.*, 2018). Dendrites are also shaped by self-avoidance and axon-dendrite fasciculation (Chen, Hsu, Chang, & Pan, 2019; Smith, Watson, Vanhoven, Colón-Ramos, & Miller, 2012).

Axon formation driven by anterograde navigation was the subject of early scientific investigations. The earliest cues implicated in the process are the conserved Netrin, Robo, Semaphorin, Ephrin, and their receptors, now recognized as 'canonical guidance proteins' across organisms (Chédotal & Richards, 2010). Some of these were first identified in *C. elegans*. Brenner isolated mutants of the Netrin pathway and their roles in migration were characterized by Culotti, Wadsworth, Hall, and their colleagues (Culotti & Merz, 1998; Hedgecock, Culotti, & Hall, 1990; Keino-Masu *et al.*, 1996; Wadsworth, Bhatt, & Hedgecock, 1996). Soon after, Robo, Semaphorin, and TGF-β were implicated in pathfinding by the Bargmann and Culotti labs (Colavita & Culotti, 1998; Ikegami, Zheng, Ong, & Culotti, 2004; Roy, Zheng, Warren, & Culotti, 2000; Zallen, Yi, & Bargmann, 1998) while the roles of the fibroblast growth factor and ephrins were discovered later (Boulin, Pocock, & Hobert, 2006; Bulow & Hobert, 2004; Grossman, Giurumescu, & Chisholm, 2013). A rich array of studies highlights the roles of guidance cues in attraction or repulsion and identified their effectors, many conserved across organisms (Huang, Cheng, Tessier-Lavigne, & Jin, 2002; Xu & Quinn, 2012; Zheng, Diaz-Cuadros, & Chalfie, 2016). Axon pathfinding is also affected by factors driving cell migration, such as the Wnt morphogen, heparan sulfate proteoglycans, and the planar cell polarity pathway (Bülow *et al.*, 2008; Pan *et al.*, 2006; Sanchez-Alvarez *et al.*, 2011). It is now clear that guidance pathways are subject to significant crosstalk, with receptors and cues binding in more than one-to-one configurations (Fujisawa, Wrana, & Culotti, 2007; Rapti, Li, Shan, Lu, & Shaham, 2017; Yu, Hao, Lim, Tessier-Lavigne, & Bargmann, 2002). Downstream of guidance receptors, axons grow by dedicated growth cones with filopodia dependent on actin polymerization and GTPases (Demarco, Struckhoff, & Lundquist, 2012; Gujar, Sundararajan, Stricker, & Lundquist, 2018; Lebrand *et al.*, 2004; Lundquist, Herman, Shaw, & Bargmann, 1998; Sundararajan & Lundquist, 2012). The precision of axon paths is controlled by additional adhesion proteins, transcription factors, microRNAs, and the environment (Baum, Guenther, Frank, Pham, & Garriga, 1999; Pocock & Hobert, 2008; Poinat *et al.*, 2002; Schmitz, Kinge, & Hutter, 2007; Steimel *et al.*, 2010; Troemel, Sagasti, & Bargmann, 1999; Y. Zou, Chiu, Domenger, Chuang, & Chang, 2012). Future investigations will need to better trace how all these factors cooperate *in vivo*.

Notwithstanding the wide-ranging study of nervous system pathfinding, certain aspects remain less clear. How do the above pathways coordinate *in vivo*, in diverse contexts, in neurons and glia? What are their primary effects? How do they instruct the early formation of the nervous system? How are major axon tracts formed in the embryo and how do neurons and glia communicate in this process? Pioneers were proposed to drive the formation of the *C. elegans* ventral nerve cord, in early electron microscopy studies (Durbin, 1987), and their functional importance was shown with ablations (Hutter, 2003). Pioneers of the brain-like *nerve ring* remained elusive. It is now shown that brain pioneers cooperate with glia to drive hierarchical brain assembly, using diverse signaling cues (Kennerdell, Fetter, & Bargmann, 2009; Rapti *et al.*, 2017). This highlights the need to identify in the future the full array of neuron-glia interactions.

Nervous system connectivity: synapse formation and functional specification

Functional maturation and neural connectivity require protein localization in defined cell compartments. For example, sensory cilia consist of organelles with a defined proteome, including factors for extracellular vesicle release (Inglis, Blacque, & Leroux, 2009; Nechipurenko, Berciu, Sengupta, & Nicastro, 2017; Silva *et al.*, 2017; Wang *et al.*, 2014). Neurite polarity and compartmentalization are defined by finely-tuned anterograde and retrograde transport. Uncovering factors of axonal transport benefited early from Brenner's genetic screens; dozens of his isolated Unc mutants affect kinesins and interactors for anterograde or retrograde transport (Brenner, 1974). In this edition, Vasudevan and Koushika (2020) review the molecular mechanisms of protein trafficking in *C. elegans*, highlighting how polarized transport shapes neurons' forms, paths, and connectivity.

Upon navigation, neurites terminate their growth and form synapses to establish proper connectivity. Some Unc mutants from Brenner's screens were shown to affect transcription factors that drive synaptic differentiation, sometimes in coordination with neurotransmitter signaling (Jin, 2005; Kratsios *et al.*, 2015; Miller *et al.*, 1992). Kinases, GTPases, and calcium mechanisms were also recognized early to regulate synaptogenesis (Crump, Zhen, Jin, & Bargmann, 2001; Rongo & Kaplan, 1999). Later studies dissected a rich array of synaptogenesis mechanisms including key signaling mechanisms. These include roles for gap junctions, insulins, and heparan sulfates (Grill *et al.*, 2007; Hung *et al.*, 2013; Lázaro-Peña, Díaz-Balzac, Bülow, & Emmons, 2018; Yeh *et al.*, 2009) as well as adhesion and scaffolding complexes, some of which act hierarchically (Dai *et al.*, 2006; Patel *et al.*, 2006; Philbrook *et al.*, 2018; Shen & Bargmann, 2003). On the postsynaptic end, dendrites are shown to differentiate functional spines, apposing presynaptic sites (Cuentas-Condori *et al.*, 2019; Philbrook *et al.*, 2018). Non-neuronal cells also influence synaptic connectivity; glia can affect postembryonic synapse localization (Colón-Ramos, Margeta, & Shen, 2007), and the epidermis acts to maintain peripheral synapses (Cherra, Goncharov, Boassa, Ellisman, & Jin, 2020; this edition). The full array of interactions driving synaptogenesis remains to be identified.

The establishment of functional connectivity culminates with neurotransmitter release and specialized localization and activity of neurotransmitter receptors. Some of Brenner's Unc mutants affected conserved proteins of synaptic vesicle formation and release (Richmond, Davis, & Jorgensen, 1999; Weimer *et al.*, 2003). Later studies identified roles of clathrin-mediated and clathrin-independent mechanisms in synaptic vesicle endocytosis (Gan & Watanabe, 2018). In addition to mutants affecting synaptic vesicles, some of Brenner's mutants affected neurotransmitter receptors and their chaperones (Eimer *et al.*, 2007). Precise connectivity depends on the localization of neurotransmitter receptors. In central synapses, receptor abundance depends on conserved cytoplasmic calcium- or clathrin-binding proteins, but the underlying cell interactions remain elusive (Burbea, Dreier, Dittman, Grunwald, & Kaplan, 2002; Hoerndli *et al.*, 2015). At the neuromuscular junctions, synaptic localization of neurotransmitter receptors depends on extracellular domain interactions (Gally, Eimer, Richmond, & Bessereau, 2004; Gendrel, Rapti, Richmond, & Bessereau, 2009; Pinan-Lucarré *et al.*, 2014). Precise nervous system activity also entails modulation of synaptic strength and gating of neurotransmitter receptors by auxiliary proteins (Boulin *et al.*, 2012; Lei, Mellem, Brockie, Madsen, & Maricq, 2017; R. Wang *et al.*, 2012). In addition to chemical synapses, the complex integration of neural connectivity is achieved by an array of electrical connections (Bhattacharya, Aghayeva, Berghoff, & Hobert, 2019; Schafer, 2018).

This morphogenetic and functional diversification supports a dynamic nervous system activity throughout development. Mechanisms of circuit integration that drive *C. elegans* behavior are reviewed elsewhere (Goodman & Sengupta, 2019; Whittaker & Sternberg, 2004).

Nervous system maturation, maintenance, and plasticity throughout animal life

After establishing its embryonic structure, the *C. elegans* nervous system shows plasticity and diversifies its components postembryonically. Although the pre-Brenner field considered that nematodes did not add somatic cells postembryonically, Sulston's lineages showed otherwise. Sulston recognized that some dopaminergic neurons are generated by postembryonic divisions, adding to embryonic cells. He also noticed instances of postembryonic transdifferentiation; certain epithelial cells differentiate into neurons or male glia change their morphology and interactions (Sulston *et al.*, 1983; Sulston & Horvitz, 1977; Walthall & Chalfie, 1988). These observations are corroborated by recent studies that dissect these transdifferentiation events. Epithelia-to-neuron (Y-to-PDA) transdifferentiation occurs through epigenetic mechanisms (Jarriault, Schwab, & Greenwald, 2008; Zuryn *et al.*, 2014) while tail glia (PHso1) transdifferentiate in male-specific neurons by cell-intrinsic mechanisms (Molina-García *et al.*, 2019). Recent research work also reveals an instance of glia-born neurons that was missed by the early cell lineaging. Specific sex-shared glia (AMso) divide postembryonically to generate male-specific neurons (MCM), which is essential for sexually dimorphic behavior (Sammut *et al.*, 2015).

Recent investigations reveal that nervous system plasticity also occurs at the connectome level. After its establishment, connectivity is plastic in response to internal or external states. Sulston described early that embryonic and postembryonic lineages showed different motoneurons, suggesting that circuit wiring changes developmentally (Sulston, 1976; Sulston *et al.*, 1983; Sulston & Horvitz, 1977). Indeed, specific motoneurons undergo a switch in neuronal polarity, presynaptic and postsynaptic regions. In this edition, Cuentas-Condori and Miller (2020), review the mechanisms of this synaptic remodeling: the underlying mechanisms of transcription regulation, downstream cascades of protein recycling, microtubule dynamics, cell death, and extracellular interactions. Intriguingly, some implicated factors were initially isolated in Brenner's screens (Brenner, 1974). Cuentas-Condori and Miller (2020) highlight early and recent research work that shapes our understanding of synapse refinement.

The developmental plasticity of *C. elegans* connectivity is not limited to this specific switch. Recent ultrastructural studies by Zhen and colleagues describe global developmental changes in connectivity; they show that sensory and motor pathways gain new connections, while decision-making circuitry is maintained, and the brain becomes progressively more modular and feedforward (Witvliet *et al.*, 2020). The nervous system also changes in response to environmental stimuli. During the harsh environment-induced dauer stage, specific sensory neurons remodel their axon arborization, which retracts when animals return to a favorable environment (Schroeder *et al.*, 2013). The *C. elegans* electrical connectome is also dynamic, as gap junctions show striking changes in environment-induced diapause (Bhattacharya *et al.*, 2019). Last but not least, neural responses also change with aging as reviewed elsewhere (Melentijevic *et al.*, 2017; Stein & Murphy, 2012).

Besides its plasticity, the nervous system maintains its overall, embryonically-established structure throughout post-embryonic life. This requires mechanisms that safeguard neuronal fate, through transcription factor autoregulation, as well as the position and axonal integrity through fibroblast growth factor (FGF) and immunoglobulin-domain signaling (Bénard, Blanchette, Recio, & Hobert, 2012; Bénard & Hobert, 2009; Bulow & Hobert, 2004). Maintenance of glia is less studied but depends on FGF receptor and solute carrier factors (Shao, Watanabe, Christensen, Jorgensen, & Colón-Ramos, 2013).

Along with the mechanisms for maintaining neural cell bodies and processes, additional mechanisms are in place to maintain connectivity. In this edition, Cherra *et al.* (2020), present how the epidermis regulates synapse density at neuromuscular junctions. An MPP5 factor controls the localization of an immunoglobulin-domain, adhesion protein that regulates CED-1-dependent phagocytosis (Cherra *et al.*, 2020; Cherra & Jin, 2016). The CED-1 role in synapse engulfment comes a long way from its function in cell corpse engulfment, recognized by the Horvitz group (Zhou, Hartwieg, & Horvitz, 2001). Cherra and Jin (2016) suggest that the epidermis acts like glia for synapse elimination, at the *C. elegans* neuromuscular junctions that lack glia associations.

Glia in nervous system development; studying key roles of non-neuronal components

Glia are non-neuronal cells, abundant in complex nervous systems. They were long considered as connective tissue ('glue') providing trophic support to neurons and were neglected in quest of neuronal signaling. Glia are now implicated in nervous system development and function, coming closer to the center of attention. Glia communicate with neurons' chemical and electrical connections, regulate nutrients, neurite morphogenesis, and connectivity. They are linked to neurological disorders including epilepsy, autism spectrum disorders, and Alzheimer's (Allen & Lyons, 2018; Zuchero & Barres, 2015).

C. elegans glia associate with the sensory organs and axon-rich neuropils and were first described as 'nervous system support cells', in early studies by Sulston, Brenner, White, and Ward. (Shaham, 2015; Sulston *et al.*, 1983; Sulston & Horvitz, 1977; Ward, Thomson, White, & Brenner, 1975; White *et al.*, 1986). Twenty-five bilateral pairs of *C. elegans* ectoderm-derived glia come in different flavors. Twelve pairs of sheath glia wrap around axons or sensory endings (ADEsh, AMsh, CEPsh, ILsh, OLLsh, OLQsh PDEsh, PHsh). Some associate with synapses (CEPsh), while 13 pairs (socket glia) associate with sheath glia, generating pores for neurons to access the environment (ADEso, AMso, CEPso, ILo, OLLso, OLQso, PDEso, PHso). Early on, *C. elegans* glia were proposed to drive migration and to phagocytose dying cells (Sulston *et al.*, 1983), yet they were understudied for decades.

Contrary to vertebrate glia, *C. elegans* glia appear dispensable for trophic support of mature neurons. Shai Shaham, who pioneered *C. elegans* glia research, recognized that the cell autonomy of many cell death events suggested glia may not be crucial for neuron survival in this animal (Shaham & Horvitz, 1996; Shaham, pers comm). This conjecture was verified by the Shaham group for a number of glia-neuron associations (Bacaj, Tevlin, Lu, & Shaham, 2008; Katz, Corson, Iwanir, Biron, & Shaham, 2018; Shaham, 2015), enabling a unique experimental setting to uncouple trophic support from glia-neuron functional interactions. Taking advantage of this knowledge, recent studies dissect glia functions and interactions, which are key in shaping the *C. elegans* nervous system.

C. elegans glia are now implicated in numerous aspects of neural development: in morphogenesis of sensory dendrites and microvilli, axon pathfinding and initiation of brain assembly, synapse positioning and regulation of neurotransmission, male-specific generation of neurons, mechanosensation, animal longevity, and sleep (Bacaj *et al.*, 2008; Frakes *et al.*, 2020; Johnson, Fernandez-Abascal, Wang, Wang, & Bianchi, 2020; Low *et al.*, 2019; Perens & Shaham, 2005; Rapti *et al.*, 2017; Sammut *et al.*, 2015; Singhvi *et al.*, 2016; Wallace *et al.*, 2016; Yin *et al.*, 2017; Yoshimura *et al.*, 2008). *C. elegans* glia present heterogeneity in fate and function. Specific glia (CEPsh) communicate with several axons through distinct pathways (Rapti *et al.*, 2017) while dorsal and ventral glia of the same subtype (CEPsh) employ partly different transcription factors to regulate their fate (Yoshimura *et al.*, 2008). Other *C. elegans* glia (Amsh) are plastic and respond to external or internal conditions. They remodel their morphology together with their associated dendrites and show dynamic changes of gene expression, in different developmental stages or in response to temperature changes, starvation, or osmotic stress (Fung, Wexler, & Heiman, 2020; Lee, Procko, Lu, & Shaham, 2020; Procko, Lu, & Shaham, 2011). These glia (AMsh) also dynamically maintain their functional fate (Wallace *et al.*, 2016). Whether heterogeneity and plasticity apply to all *C. elegans* glia remains to be defined. Neuron-glia interactions remain under intense investigation and were reviewed often (Shaham, 2015; Singhvi & Shaham, 2019). Many questions persist in the quest of glial fate, morphogenesis, circuit formation and function.

For most of the 20th century, *C. elegans* glia remained obscure and vertebrate glia were studied as a by-product of recording neuronal connections. Today, different *C. elegans* glia are more or less studied. One of the reasons is that glia operate beyond the reach of tools designed to probe electrical signals. In a chicken or the egg problem, the lack of knowledge on glia hinders the identification of tools required to study them and the lack of tools hampers glial functional dissection. Luckily, recent research focuses on glia as equal protagonists of the nervous system. In this edition, the Heiman group contributes a resource manuscript presenting reagents to drive expression in different *C. elegans* glia (Fung *et al.*, 2020). Using previous literature, transcriptomics and new analysis, they present drivers for glia expression, comparing their specificity and robustness in relation to the animal's states. Such efforts, highlighting glia tools, facilitate the study of diverse glia to dissect aspects of their development and function. Given the numerous glial subtypes and functions, dissecting their biology will enable a more comprehensive view of nervous system complexity.

Sexual dimorphism in nervous system development

The remarkable intricacy of the nervous system defines the diversity of its components within and across sexes, to drive sex-specific behaviors allowing selective pressure against speciation. Pioneering work by Sulston, White, and colleagues provided maps of nervous system anatomy and connectome in the *C. elegans* hermaphrodite and spotlighted the lesser-studied male, describing a series of sexual variations in cell birth, elimination, or transformation (Sulston *et al.*, 1983; Sulston & Horvitz, 1977). Research of the male biology was facilitated by mutations that cause sex transformation, as studied by Brenner and Hodgkin (Hodgkin, Horvitz, & Brenner, 1979). Sexual diversification of the nervous system relies on the birth or death of sex-specific neurons, and the sex-dimorphic plasticity of sex-shared components. Early investigation highlighted that sex-dimorphic cell birth gives rise to sex-specific neurons, like the hermaphrodite HSN that innervate vulval muscle or the male-specific, sensory and pheromone-secreting CEM, while their opposite-sex counterparts undergo programmed cell death (Silva *et al.*, 2017; Sulston, Albertson, & Thomson, 1980, Sulston *et al.*, 1983; Sulston & Horvitz, 1977; Wang *et al.*, 2014; Ward *et al.*, 1975). Recent studies highlight glia that contribute to nervous system dimorphism: head and tail glia (AMso, PHso1) divide or differentiate, respectively, to produce the male interneuron MCM and the ciliated neuron PHD, which are required for male-specific behaviors (Molina-García *et al.*, 2019; Sammut *et al.*, 2015). Sex-shared neurons can also undergo sex-specific synaptic pruning or axon branching (Bayer & Hobert, 2018; Hart & Hobert, 2018; Oren-Suissa, Bayer, & Hobert, 2016).

In this edition, Walsh, Boivin, and Barr (2020) review the nervous system of *C. elegans* males and the mechanisms driving sexually dimorphic development and plasticity. They present challenges and strategies to explore nervous system sex-dimorphism (Walsh *et al.*, 2020). In addition, recent mapping of the male connectome by Emmons and colleagues will facilitate the dissection of sex-shared nervous system components that give rise to sex-specific behaviors and the plasticity of sex-related variations in neurodevelopment (Cook *et al.*, 2019).

Looking back and ahead; reflections beyond *C. elegans* neural development

The early completion of the *C. elegans* cell lineage, connectome, and genome sequence by Brenner, Sulston, and subsequent *C. elegans* researchers give the impression that neuroscientists discovered the instructions to build a nervous system (C.I. Bargmann, 1999). Until today, our comprehension of how the nervous system develops grows by leaps and bounds, through the continuous work of a booming neuroscience community (Figure 1). The spirit of open science fosters these tremendous advances. Data communication and reagent sharing started with platforms such as the Worm Breeder's Gazette, and International Worm Meetings. They culminate in consortia and databases such as the WormBook, WormAtlas, WormBase, CGC (Caenorhabditis Genetic Center), MMP (Million Mutation Project), modERN, modENCODE, and others. Sulston and his colleagues were early advocates of open science, an attribute highlighted when Sulston and Waterston led the public effort of the International Human Genome Consortium. Besides succeeding in this monumental achievement that altered the world, Sulston and Waterston insisted on making the data and reagents publicly available, years before publication. Sulston, Brenner, and their close colleagues and successors deserve our admiration, for pioneering scientific work in *C. elegans* and the first metazoan genome as well as for their scientific commitment that inspires *C. elegans* researchers to this day to make great strides toward understanding the intricacies of nervous system development and function.

Brenner and Sulston shared with Horvitz the 2002 Nobel Prize in Physiology or Medicine 'for their discoveries concerning genetic regulation of organ development and programmed cell death.' In their Nobel Prize interview they reflected on future research: 'We are drowning in an ocean of data but we are starving for knowledge … you must have a theoretical framework to embed this … what we are going to need is human intelligence,' highlights Brenner, and 'we should use our creativity' adds Sulston. 'If we understand the worm, we understand life. Which of course we're nowhere near.', Sulston further contemplates (The Guardian, 2002, John Sulston interview: One man and his worm).

Disclosure statement

No potential conflict of interest was reported by the author(s).

Funding

G.R. is supported by the European Molecular Biology Laboratory.

References

Adler, C.E., Fetter, R.D., & Bargmann, C.I. (2006). UNC-6/Netrin induces neuronal asymmetry and defines the site of axon formation. *Nature Neuroscience*, *9*(4), 511–518. doi:10.1038/nn1666

Allen, N.J., & Lyons, D.A. (2018). Glia as architects of central nervous system formation and function. *Science (New York, N.Y.)*, *362*(6411), 181–185. doi:10.1126/science.aat0473

Amruta Vasudevan & Sandhya P. Koushika (2020). Molecular mechanisms governing axonal transport: a *C. elegans* perspective, *Journal of Neurogenetics*, DOI: 10.1080/01677063.2020.1823385

Andrea Cuentas-Condori & David M. Miller, 3rd (2020). Synaptic remodeling, lessons from C. elegans, *Journal of Neurogenetics*, DOI: 10.1080/01677063.2020.1802725.

Bacaj, T., Tevlin, M., Lu, Y., & Shaham, S. (2008). Glia are essential for sensory organ function in *C. elegans*. *Science (New York, N.Y.).)*, *322*(5902), 744–747. doi:10.1126/science.1163074

Bargmann, C.I. (1999). Looking back, looking ahead. *Nature Neuroscience*, *2*(5), 389. doi:10.1038/8049

Bargmann, C.I., & Horvitz, H.R. (1991). Chemosensory neurons with overlapping functions direct chemotaxis to multiple chemicals in *C. elegans*. *Neuron*, *7*(5), 729–742. doi:10.1016/0896-6273(91)90276-6

Barrière, A., & Bertrand, V. (2020). Neuronal specification in *C. elegans*: Combining lineage inheritance with intercellular signaling. *Journal of Neurogenetics*. Advance online publication. doi:10.1080/01677063.2020.1781850

Baum, P.D., Guenther, C., Frank, C.A., Pham, B.V., & Garriga, G. (1999). The *Caenorhabditis elegans* gene ham-2 links Hox patterning to migration of the HSN motor neuron. *Genes & Development*, *13*(4), 472–483. doi:10.1101/gad.13.4.472

Bayer, E.A., & Hobert, O. (2018). Past experience shapes sexually dimorphic neuronal wiring through monoaminergic signalling. *Nature*, *561*(7721), 117–138. doi:10.1038/s41586-018-0452-0

Bénard, C.Y., Blanchette, C., Recio, J., & Hobert, O. (2012). The secreted immunoglobulin domain proteins ZIG-5 and ZIG-8 cooperate with L1CAM/SAX-7 to maintain nervous system integrity. *PLoS Genetics*, *8*(7), e1002819. doi:10.1371/journal.pgen.1002819

Bénard, C.Y., & Hobert, O. (2009). Looking beyond development: Maintaining nervous system architecture. *Current Topics in Developmental Biology*, *87*, 175–194. doi:10.1016/S0070-2153(09)01206-X

Bertrand, V., Bisso, P., Poole, R.J., & Hobert, O. (2011). Notch-dependent induction of left/right asymmetry in *C. elegans* interneurons and motoneurons. *Current Biology: CB*, *21*(14), 1225–1231. doi:10.1016/j.cub.2011.06.016

Bhattacharya, A., Aghayeva, U., Berghoff, E.G., & Hobert, O. (2019). Plasticity of the electrical connectome of *C. elegans*. *Cell*, *176*(5), 1174–1189. doi:10.1016/j.cell.2018.12.024

Boulin, T., Pocock, R., & Hobert, O. (2006). A novel Eph receptor-interacting IgSF protein provides *C. elegans* motoneurons with midline guidepost function . *Current Biology: Cb*, *16*(19), 1871–1883. doi:10.1016/j.cub.2006.08.056

Boulin, T., Rapti, G., Briseño-Roa, L., Stigloher, C., Richmond, J.E., Paoletti, P., & Bessereau, J.-L. (2012). Positive modulation of a Cys-loop acetylcholine receptor by an auxiliary transmembrane subunit. *Nature Neuroscience*, *15*(10), 1374–1381. doi:10.1038/nn.3197

Brenner, S. (1974). The genetics of *Caenorhabditis elegans*. *Genetics*, *77*(1), 71–94. doi:10.1002/cbic.200300625

Bulow, H.E., & Hobert, O. (2004). Differential sulfations and epimerization define heparan sulfate specificity in nervous system development. *Neuron*, *41*(5), 723–736. doi:10.1016/S0896-6273(04)00084-4

Bülow, H.E., Tjoe, N., Townley, R.A., Didiano, D., van Kuppevelt, T.H., & Hobert, O. (2008). Extracellular sugar modifications provide instructive and cell-specific information for axon-guidance choices. *Current Biology: CB*, *18*(24), 1978–1985. doi:10.1016/j.cub.2008.11.023

Burbea, M., Dreier, L., Dittman, J.S., Grunwald, M.E., & Kaplan, J.M. (2002). Ubiquitin and AP180 regulate the abundance of GLR-1 glutamate receptors at postsynaptic elements in *C. elegans*. *Neuron*, *35*(1), 107–120. doi:10.1016/S0896-6273(02)00749-3

C. elegans sequencing consortium. (1998). Genome sequence of the nematode *C. elegans*: A platform for investigating biology. The *C. elegans* Sequencing Consortium. *Science*, *282*(5396), 2012–2018. doi:10.1126/science.282.5396.2012

Chalfie, M. (2018). John Sulston (1942–2018). *Cell*, *173*(4), 809–812. doi:10.1016/j.cell.2018.04.024

Chalfie, M., Horvitz, H.R., & Sulston, J.E. (1981). Mutations that lead to reiterations in the cell lineages of *C. elegans*. *Cell*, *24*(1), 59–69. doi:10.1016/0092-8674(81)90501-8

Chalfie, M., Sulston, J.E., White, J.G., Southgate, E., Thomson, J.N., & Brenner, S. (1985). The neural circuit for touch sensitivity in *Caenorhabditis elegans*. *The Journal of Neuroscience: The Official Journal of the Society for Neuroscience*, *5*(4), 956–964. doi:10.1523/JNEUROSCI.05-04-00956.1985

Check, E. (2002). Worm cast in starring role for Nobel prize. *Nature*, *419*(6907), 548–549. doi:10.1038/419548a

Chédotal, A., & Richards, L.J. (2010). Wiring the brain: The biology of neuronal guidance. *Cold Spring Harbor Perspectives in Biology*, *2*(6), a001917 doi:10.1101/cshperspect.a001917

Chen, C.H., Hsu, H.W., Chang, Y.H., & Pan, C.L. (2019). Adhesive L1CAM-robo signaling aligns growth cone F-actin dynamics to promote axon-dendrite fasciculation in *C. elegans*. *Developmental Cell*, *49*(3), 490–491. doi:10.1016/j.devcel.2018.10.028

Cherra, S.J., Goncharov, A., Boassa, D., Ellisman, M., & Jin, Y. (2020). *C. elegans* MAGU-2/Mpp5 homolog regulates epidermal phagocytosis and synapse density. *Journal of Neurogenetics*. Advance online publication. doi:10.1080/01677063.2020.1726915

Cherra, S.J., & Jin, Y. (2016). A two-immunoglobulin-domain transmembrane protein mediates an epidermal-neuronal interaction to maintain synapse density. *Neuron*, *89*(2), 325–336. doi:10.1016/j.neuron.2015.12.024

Cochella, L., & Hobert, O. (2012). Embryonic priming of a miRNA locus predetermines postmitotic neuronal left/right asymmetry in *C. elegans*. *Cell*, *151*(6), 1229–1242. doi:10.1016/j.cell.2012.10.049

Colavita, A., & Culotti, J.G. (1998). Suppressors of ectopic UNC-5 growth cone steering identify eight genes involved in axon guidance in *Caenorhabditis elegans*. *Developmental Biology*, *194*(1), 72–85. doi:10.1006/dbio.1997.8790

Colón-Ramos, D.A., Margeta, M.A., & Shen, K. (2007). Glia promote local synaptogenesis through UNC-6 (netrin) signaling in *C. elegans*. *Science (New York, N.Y.)*, *318*(5847), 103–106. doi:10.1126/science.1143762

Cook, S.J., Jarrell, T.A., Brittin, C.A., Wang, Y., Bloniarz, A.E., Yakovlev, M.A., ... Emmons, S.W. (2019). Whole-animal connectomes of both *Caenorhabditis elegans* sexes. *Nature*, *571*(7763), 63–71. doi:10.1038/s41586-019-1352-7

Cordes, S., Frank, C.A., & Garriga, G. (2006). The *C. elegans* MELK ortholog PIG-1 regulates cell size asymmetry and daughter cell fate in asymmetric neuroblast divisions. *Development (Cambridge, England)*, *133*(14), 2747–2756. doi:10.1242/dev.02447

Crump, J.G., Zhen, M., Jin, Y., & Bargmann, C.I. (2001). The SAD-1 kinase regulates presynaptic vesicle clustering and axon termination. *Neuron*, *29*(1), 115–129. doi:10.1016/S0896-6273(01)00184-2

Cuentas-Condori, A., Mulcahy, B., He, S., Palumbos, S., Zhen, M., & Miller, D.M. (2019). *C. elegans* neurons have functional dendritic spines. *eLife*, *8*, e47918. doi:10.7554/eLife.47918

Culotti, J.G., & Merz, D.C. (1998). DCC and netrins. *Current Opinion in Cell Biology*, *10*(5), 609–613. doi:10.1016/S0955-0674(98)80036-7

Dai, Y., Taru, H., Deken, S.L., Grill, B., Ackley, B., Nonet, M.L., & Jin, Y. (2006). SYD-2 Liprin-alpha organizes presynaptic active zone formation through ELKS. *Nature Neuroscience*, *9*(12), 1479–1487. doi:10.1038/nn1808

Demarco, R.S., Struckhoff, E.C., & Lundquist, E.A. (2012). The Rac GTP exchange factor TIAM-1 acts with CDC-42 and the guidance receptor UNC-40/DCC in neuronal protrusion and axon guidance. *PLoS Genetics*, *8*(4), e1002665. doi:10.1371/journal.pgen.1002665

Dickinson, D.J., & Goldstein, B. (2016). CRISPR-based methods for *Caenorhabditis elegans* genome engineering. *Genetics*, *202*(3), 885–901. doi:10.1534/genetics.115.182162

Dong, X., Liu, O.W., Howell, A.S., & Shen, K. (2013). An extracellular adhesion molecule complex patterns dendritic branching and morphogenesis. *Cell*, *155*(2), 296–307. doi:10.1016/j.cell.2013.08.059

Durbin, R.M. (1987). *Studies on the development and organisation of the nervous system of Caenorhabditis elegans*. Cambridge: King's College, University of Cambridge.

Eimer, S., Gottschalk, A., Hengartner, M., Horvitz, H.R., Richmond, J., Schafer, W.R., & Bessereau, J.-L. (2007). Regulation of nicotinic receptor trafficking by the transmembrane Golgi protein UNC-50. *The EMBO Journal*, *26*(20), 4313–4323. http://www.ncbi.nlm.nih.gov/entrez/query.fcgi?db=pubmed&cmd=Retrieve&dopt=AbstractPlus&list_uids=17853888 doi:10.1038/sj.emboj.7601858

Félix, M.-A., & Nigon, M.V. (2017). History of research on *C. elegans* and other free-living nematodes as model organisms. *WormBook: The Online Review of C. elegans Biology*. doi:10.1895/wormbook.1.1

Finney, M., Ruvkun, G., & Horvitz, H.R. (1988). The *C. elegans* cell lineage and differentiation gene unc-86 encodes a protein with a homeodomain and extended similarity to transcription factors. *Cell, 55*(5), 757–769. doi:10.1016/0092-8674(88)90132-8

Forrester, W.C., Dell, M., Perens, E., & Garriga, G. (1999). A *C. elegans* Ror receptor tyrosine kinase regulates cell motility and asymmetric cell division. *Nature, 400*(6747), 881–885. doi:10.1038/23722

Frakes, A.E., Metcalf, M.G., Tronnes, S.U., Bar-Ziv, R., Durieux, J., Gildea, H.K., … Dillin, A. (2020). Four glial cells regulate ER stress resistance and longevity via neuropeptide signaling in *C. elegans*. *Science, 367*(6476), 436–440. doi:10.1126/science.aaz6896

Frank, C.A., Baum, P.D., & Garriga, G. (2003). HLH-14 is a *C. elegans* Achaete-Scute protein that promotes neurogenesis through asymmetric cell division. *Development (Cambridge, England)), 130*(26), 6507–6518. doi:10.1242/dev.00894

Fujisawa, K., Wrana, J.L., & Culotti, J.G. (2007). The slit receptor EVA-1 coactivates a SAX-3/Robo mediated guidance signal in *C. elegans*. *Science (New York, N.Y.), 317*(5846), 1934–1938. doi:10.1126/science.1144874

Fung, W., Wexler, L., & Heiman, M.G. (2020). Cell-type-specific promoters for *C. elegans* glia. *Journal of Neurogenetics*. Advance online publication. doi:10.1080/01677063.2020.1781851

Gally, C., Eimer, S., Richmond, J.E., & Bessereau, J.-L. (2004). A transmembrane protein required for acetylcholine receptor clustering in *Caenorhabditis elegans*. *Nature, 431*(7008), 578–582. doi:10.1038/nature02893

Gan, Q., & Watanabe, S. (2018). Synaptic vesicle endocytosis in different model systems. *Frontiers in Cellular Neuroscience, 12*(, 171. doi:10.3389/fncel.2018.00171

Garriga, G., Desai, C., & Horvitz, H.R. (1993). Cell interactions control the direction of outgrowth, branching and fasciculation of the HSN axons of *Caenorhabditis elegans*. *Development (Cambridge, England), 117*(3), 1071–1087.

Garriga, G., Guenther, C., & Horvitz, H.R. (1993). Migrations of the *Caenorhabditis elegans* HSNs are regulated by egl-43, a gene encoding two zinc finger proteins. *Genes & Development, 7*(11), 2097–2109. doi:10.1101/gad.7.11.2097

Gendrel, M., Atlas, E.G., & Hobert, O. (2016). A cellular and regulatory map of the GABAergic nervous system of *C. elegans*. *eLife, 5*, e17686. doi:10.7554/eLife.17686

Gendrel, M., Rapti, G., Richmond, J.E., & Bessereau, J.-L. (2009). A secreted complement-control-related protein ensures acetylcholine receptor clustering. *Nature, 461*(7266), 992–996. doi:10.1038/nature08430

Gitschier, J. (2006). Knight in common armor: An interview with Sir John Sulston. *PLoS Genetics, 2*(12), e225. doi:10.1371/journal.pgen.0020225

Goodman, M.B., & Sengupta, P. (2019). How *caenorhabditis elegans* senses mechanical stress, temperature, and other physical stimuli. *Genetics, 212*(1), 25–51. doi:10.1534/genetics.118.300241

Grill, B., Bienvenut, W.V., Brown, H.M., Ackley, B.D., Quadroni, M., & Jin, Y. (2007). *C. elegans* RPM-1 regulates axon termination and synaptogenesis through the Rab GEF GLO-4 and the Rab GTPase GLO-1. *Neuron, 55*(4), 587–601. doi:10.1016/j.neuron.2007.07.009

Grossman, E.N., Giurumescu, C.A., & Chisholm, A.D. (2013). Mechanisms of ephrin receptor protein kinase-independent signaling in amphid axon guidance in *Caenorhabditis elegans*. *Genetics, 195*(3), 899–913. doi:10.1534/genetics.113.154393

Gujar, M.R., Sundararajan, L., Stricker, A., & Lundquist, E.A. (2018). Control of growth cone polarity, microtubule accumulation, and protrusion by UNC-6/netrin and its receptors in *Caenorhabditis elegans*. *Genetics, 210*(1), 235–255. doi:10.1534/genetics.118.301234

Hart, M.P., & Hobert, O. (2018). Neurexin controls plasticity of a mature, sexually dimorphic neuron. *Nature, 553*(7687), 165–170. doi:10.1038/nature25192

Hedgecock, E.M., Culotti, J.G., & Hall, D.H. (1990). The unc-5, unc-6, and unc-40 genes guide circumferential migrations of pioneer axons and mesodermal cells on the epidermis in *C. elegans*. *Neuron, 4*(1), 61–85. doi:10.1016/0896-6273(90)90444-K

Heiman, M.G., & Shaham, S. (2009). DEX-1 and DYF-7 establish sensory dendrite length by anchoring dendritic tips during cell migration. *Cell, 137*(2), 344–355. doi:10.1016/j.cell.2009.01.057

Hirose, T., & Horvitz, H.R. (2013). An Sp1 transcription factor coordinates caspase-dependent and -independent apoptotic pathways. *Nature, 500*(7462), 354–358. doi:10.1038/nature12329

Hobert, O. (2016). Terminal selectors of neuronal identity. *Current Topics in Developmental Biology, 116*, 455–475. doi:10.1016/bs.ctdb.2015.12.007

Hobert, O., Carrera, I., & Stefanakis, N. (2010). The molecular and gene regulatory signature of a neuron. *Trends in Neurosciences, 33*(10), 435–445. doi:10.1016/j.tins.2010.05.006

Hobert, O., & Kratsios, P. (2019). Neuronal identity control by terminal selectors in worms, flies, and chordates. *Current Opinion in Neurobiology, 56*, 97–105. doi:10.1016/j.conb.2018.12.006

Hodgkin, J., Horvitz, H.R., & Brenner, S. (1979). Nondisjunction mutants of the nematode *Caenorhabditis elegans*. *Genetics, 91*(1), 67–94.

Hoerndli, F.J., Wang, R., Mellem, J.E., Kallarackal, A., Brockie, P.J., Thacker, C., … Maricq, A.V. (2015). Neuronal activity and CaMKII regulate kinesin-mediated transport of synaptic AMPARs. *Neuron, 86*(2), 457–474. doi:10.1016/j.neuron.2015.03.011

Horvitz, H.R., Shaham, S., & Hengartner, M.O. (1994). The genetics of programmed cell death in the nematode *Caenorhabditis elegans*. *Cold Spring Harbor Symposia on Quantitative Biology, 59*, 377–385. doi:10.1101/sqb.1994.059.01.042

Horvitz, H.R., & Sulston, J.E. (1980). Isolation and genetic characterization of cell-lineage mutants of the nematode *Caenorhabditis elegans*. *Genetics, 96*(2), 435–454.

Hsieh, Y.W., Chang, C., & Chuang, C.F. (2012). The MicroRNA mir-71 inhibits calcium signaling by targeting the TIR-1/Sarm1 adaptor protein to control stochastic L/R neuronal asymmetry in *C. elegans*. *PLoS Genetics, 8*(8), e1002864. doi:10.1371/journal.pgen.1002864

Huang, X., Cheng, H.J., Tessier-Lavigne, M., & Jin, Y. (2002). MAX-1, a novel PH/MyTH4/FERM domain cytoplasmic protein implicated in netrin-mediated axon repulsion. *Neuron, 34*(4), 563–576. doi:10.1016/S0896-6273(02)00672-4

Hung, W.L., Hwang, C., Gao, S., Liao, E.H., Chitturi, J., Wang, Y., … Zhen, M. (2013). Attenuation of insulin signalling contributes to FSN-1-mediated regulation of synapse development. *The EMBO Journal, 32*(12), 1745–1760. doi:10.1038/emboj.2013.91

Hutter, H. (2003). Extracellular cues and pioneers act together to guide axons in the ventral cord of *C. elegans*. *Development (Cambridge, England), 130*(22), 5307–5318. doi:10.1242/dev.00727

Ikegami, R., Zheng, H., Ong, S.-H., & Culotti, J. (2004). Integration of semaphorin-2A/MAB-20, ephrin-4, and UNC-129 TGF-beta signaling pathways regulates sorting of distinct sensory rays in *C. elegans*. *Developmental Cell, 6*(3), 383–395. doi:10.1016/S1534-5807(04)00057-7

Inglis, P.N., Blacque, O.E., & Leroux, M.R. (2009). Functional genomics of intraflagellar transport-associated proteins in *C. elegans*. *Methods in Cell Biology, 93*, 267–304. doi:10.1016/S0091-679X(08)93014-4

Jarriault, S., Schwab, Y., & Greenwald, I. (2008). A *Caenorhabditis elegans* model for epithelial-neuronal transdifferentiation. *Proceedings of the National Academy of Sciences of the United States of America, 105*(10), 3790–3795. doi:10.1073/pnas.0712159105

Jin, Y. (2005). Synaptogenesis. *WormBook : The online review of C. elegans biology*. doi:10.1895/wormbook.1.44.1

Johnson, C.K., Fernandez-Abascal, J., Wang, Y., Wang, L., & Bianchi, L. (2020). The Na+-K+-ATPase is needed in glia of touch receptors for responses to touch in *C. elegans*. *Journal of Neurophysiology, 123*(5), 2064–2074. doi:10.1152/jn.00636.2019

Kamath, R.S., Fraser, A.G., Dong, Y., Poulin, G., Durbin, R., Gotta, M., … Ahringer, J. (2003). Systematic functional analysis of the *Caenorhabditis elegans* genome using RNAi. *Nature, 421*(6920), 231–237. doi:10.1038/nature01278

Katz, M., Corson, F., Iwanir, S., Biron, D., & Shaham, S. (2018). Glia modulate a neuronal circuit for locomotion suppression during sleep in *C. elegans*. *Cell Reports*, *22*(10), 2575–2583. doi:10.1016/j.celrep.2018.02.036

Keino-Masu, K., Masu, M., Hinck, L., Leonardo, E.D., Chan, S.S.Y., Culotti, J.G., & Tessier-Lavigne, M. (1996). Deleted in Colorectal Cancer (DCC) encodes a netrin receptor. *Cell*, *87*(2), 175–185. doi:10.1016/S0092-8674(00)81336-7

Kennerdell, J.R., Fetter, R.D., & Bargmann, C.I. (2009). Wnt-Ror signaling to SIA and SIB neurons directs anterior axon guidance and nerve ring placement in *C. elegans*. *Development (Cambridge, England)*, *136*(22), 3801–3810. doi:10.1242/dev.038109

Kim, K., Kim, R., & Sengupta, P. (2010). The HMX/NKX homeodomain protein MLS-2 specifies the identity of the AWC sensory neuron type via regulation of the ceh-36 Otx gene in *C. elegans*. *Development (Cambridge, England)*, *137*(6), 963–974. doi:10.1242/dev.044719

Kratsios, P., Pinan-Lucarré, B., Kerk, S.Y., Weinreb, A., Bessereau, J.L., & Hobert, O. (2015). Transcriptional coordination of synaptogenesis and neurotransmitter signaling. *Current biology: CB*, *25*(10), 1282–1295. doi:10.1016/j.cub.2015.03.028

Kratsios, P., Stolfi, A., Levine, M., & Hobert, O. (2012). Coordinated regulation of cholinergic motor neuron traits through a conserved terminal selector gene. *Nature Neuroscience*, *15*(2), 205–214. doi:10.1038/nn.2989

Labouesse, M., Hartwieg, E., & Horvitz, H.R. (1996). The *Caenorhabditis elegans* LIN-26 protein is required to specify and/or maintain all non-neuronal ectodermal cell fates. *Development (Cambridge, England)*, *122*(9), 2579–2588.

Lázaro-Peña, M.I., Díaz-Balzac, C.A., Bülow, H.E., & Emmons, S.W. (2018). Synaptogenesis is modulated by heparan sulfate in *Caenorhabditis elegans*. *Genetics*, *209*(1), 195–208. doi:10.1534/genetics.118.300837

Lebrand, C., Dent, E.W., Strasser, G.A., Lanier, L.M., Krause, M., Svitkina, T.M., … Gertler, F.B. (2004). Critical role of Ena/VASP proteins for filopodia formation in neurons and in function downstream of netrin-1. *Neuron*, *42*(1), 37–49. doi:10.1016/S0896-6273(04)00108-4

Lee, I.H., Procko, C., Lu, Y., & Shaham, S. (2020). Induced nervous system remodeling and behavior. *BioRxiv*. doi:10.1101/2020.06.03.127894

Lee, J., Taylor, C.A., Barnes, K.M., Shen, A., Stewart, E.V., Chen, A., … Shen, K. (2019). A Myt1 family transcription factor defines neuronal fate by repressing non-neuronal genes. *eLife*, *8*, e46703. doi:10.7554/eLife.46703

Lei, N., Mellem, J.E., Brockie, P.J., Madsen, D.M., & Maricq, A.V. (2017). NRAP-1 is a presynaptically released NMDA receptor auxiliary protein that modifies synaptic strength. *Neuron*, *96*(6), 1303–1316.e6. doi:10.1016/j.neuron.2017.11.019

Lesch, B.J., & Bargmann, C.I. (2010). The homeodomain protein hmbx-1 maintains asymmetric gene expression in adult *C. elegans* olfactory neurons. *Genes & Development*, *24*(16), 1802–1815. http://genesdev.cshlp.org/cgi/doi/10.1101/gad.1932610 doi:10.1101/gad.1932610

Levy-Strumpf, N., & Culotti, J.G. (2007). VAB-8, UNC-73 and MIG-2 regulate axon polarity and cell migration functions of UNC-40 in *C. elegans*. *Nature Neuroscience*, *10*(2), 161–168. doi:10.1038/nn1835

Liu, O.W., & Shen, K. (2012). The transmembrane LRR protein DMA-1 promotes dendrite branching and growth in *C. elegans*. *Nature Neuroscience*, *15*(1), 57–63. doi:10.1038/nn.2978

Low, I.I.C., Williams, C.R., Chong, M.K., McLachlan, I.G., Wierbowski, B.M., Kolotuev, I., & Heiman, M.G. (2019). Morphogenesis of neurons and glia within an epithelium. *Development*, *146*(4), dev171124. doi:10.1242/dev.171124

Lundquist, E.A., Herman, R.K., Shaw, J.E., & Bargmann, C.I. (1998). UNC-115, a conserved protein with predicted LIM and actin-binding domains, mediates axon guidance in *C. elegans*. *Neuron*, *21*(2), 385–392. doi:10.1016/S0896-6273(00)80547-4

Maduro, M.F. (2010). Cell fate specification in the *C. elegans* embryo. *Developmental Dynamics: An Official Publication of the American Association of Anatomists*, *239*(5), 1315–1329. doi:10.1002/dvdy.22233

Maniar, T.A., Kaplan, M., Wang, G.J., Shen, K., Wei, L., Shaw, J.E., … Bargmann, C.I. (2011). UNC-33 (CRMP) and ankyrin organize microtubules and localize kinesin to polarize axon-dendrite sorting. *Nature Neuroscience*, *15*(1), 48–56. doi:10.1038/nn.2970

Masoudi, N., Tavazoie, S., Glenwinkel, L., Ryu, L., Kim, K., & Hobert, O. (2018). Unconventional function of an Achaete-Scute homolog as a terminal selector of nociceptive neuron identity. *PLoS Biology*, *16*(4), e2004979. doi:10.1371/journal.pbio.2004979

Matthews, B.J., & Vosshall, L.B. (2020). How to turn an organism into a model organism in 10 "easy" steps. *The Journal of Experimental Biology*, *223*(Suppl 1), jeb218198. doi:10.1242/jeb.218198

Melentijevic, I., Toth, M.L., Arnold, M.L., Guasp, R.J., Harinath, G., Nguyen, K.C., … Driscoll, M. (2017). *C. elegans* neurons jettison protein aggregates and mitochondria under neurotoxic stress. *Nature*, *542*(7641), 367–371. doi:10.1038/nature21362

Meng, L., Zhang, A., Jin, Y., & Yan, D. (2016). Regulation of neuronal axon specification by glia-neuron gap junctions in *C. elegans*. *eLife*, *5*, e19510. doi:10.7554/eLife.19510

Metzstein, M.M., Hengartner, M.O., Tsung, N., Ellis, R.E., & Horvitz, H.R. (1996). Transcriptional regulator of programmed cell death encoded by *Caenorhabditis elegans* gene ces-2. *Nature*, *382*(6591), 545–547. doi:10.1038/382545a0

Miller, D.D., Shen, M.M., Shamu, C.E., Burglin, T.R., Ruvkun, G., Dubois, M.L., … Wilson, L. (1992). *C. elegans* unc-4 gene encodes a homeodomain protein that determines the pattern of synaptic input to specific motor neurons David. *Nature*, *355* (6363), 841–845. doi:10.1038/355841a0

Mishra, N., Wei, H., & Conradt, B. (2018). *Caenorhabditis elegans* ced-3 caspase is required for asymmetric divisions that generate cells programmed to die. *Genetics*, *210*(3), 983–998. doi:10.1534/genetics.118.301500

Molina-García, L., Cook, S., Kim, B., Bonnington, R., Sammut, M., O'Shea, J., … Poole, R. (2019). A direct glia-to-neuron natural transdifferentiation ensures nimble sensory-motor coordination of male mating behaviour. BioRxiv, 285320. doi:10.1101/285320

Nechipurenko, I.V., Berciu, C., Sengupta, P., & Nicastro, D. (2017). Centriolar remodeling underlies basal body maturation during ciliogenesis in *Caenorhabditis elegans*. *eLife*, *6*, e25686. doi:10.7554/eLife.25686

Oren-Suissa, M., Bayer, E.A., & Hobert, O. (2016). Sex-specific pruning of neuronal synapses in *Caenorhabditis elegans*. *Nature*, *533*(7602), 206–211. doi:10.1038/nature17977

Oren-Suissa, M., Hall, D.H., Treinin, M., Shemer, G., & Podbilewicz, B. (2010). The fusogen EFF-1 controls sculpting of mechanosensory dendrites. *Science (New York, N.Y.).)*, *328*(5983), 1285–1288. doi:10.1126/science.1189095

Ou, G., Stuurman, N., D'Ambrosio, M., & Vale, R.D. (2010). Polarized myosin produces unequal-size daughters during asymmetric cell division. *Science (New York, N.Y.)*, *330*(6004), 677–680. doi:10.1126/science.1196112

Pan, C.L., Howell, J.E., Clark, S.G., Hilliard, M., Cordes, S., Bargmann, C.I., & Garriga, G. (2006). Multiple Wnts and Frizzled receptors regulate anteriorly directed cell and growth cone migrations in *Caenorhabditis elegans*. *Developmental Cell*, *10*(3), 367–377. doi:10.1016/j.devcel.2006.02.010

Patel, M.R., Lehrman, E.K., Poon, V.Y., Crump, J.G., Zhen, M., Bargmann, C.I., & Shen, K. (2006). Hierarchical assembly of presynaptic components in defined *C. elegans* synapses. *Nature Neuroscience*, *9*(12), 1488–1498. doi:10.1038/nn1806

Pedersen, M.E., Snieckute, G., Kagias, K., Nehammer, C., Multhaupt, H.A.B., Couchman, J.R., & Pocock, R. (2013). An epidermal microRNA regulates neuronal migration through control of the cellular glycosylation state. *Science (New York, N.Y.)*, *341*(6152), 1404–1408. doi:10.1126/science.1242528

Pereira, L., Kratsios, P., Serrano-Saiz, E., Sheftel, H., Mayo, A.E., Hall, D.H., … Hobert, O. (2015). A cellular and regulatory map of the cholinergic nervous system of *C. elegans*. *eLife*, *4*, e12432. doi:10.7554/eLife.12432

Perens, E.A., & Shaham, S. (2005). *C. elegans* daf-6 encodes a patched-related protein required for lumen formation. *Developmental Cell*, *8*(6), 893–906. doi:10.1016/j.devcel.2005.03.009

Philbrook, A., Ramachandran, S., Lambert, C.M., Oliver, D., Florman, J., Alkema, M.J., … Francis, M.M. (2018). Neurexin directs partner-specific synaptic connectivity in *C. elegans*. *eLife*, *7*, e35692. doi:10.7554/eLife.35692

Pinan-Lucarré, B., Tu, H., Pierron, M., Cruceyra, P.I., Zhan, H., Stigloher, C., … Bessereau, J.L. (2014). *C. elegans* Punctin specifies cholinergic versus GABAergic identity of postsynaptic domains. *Nature*, *511*(7510), 466–470. doi:10.1038/nature13313

Pocock, R., & Hobert, O. (2008). Oxygen levels affect axon guidance and neuronal migration in *Caenorhabditis elegans*. *Nature Neuroscience*, *11*(8), 894–900. doi:10.1038/nn.2152

Poinat, P., De Arcangelis, A., Sookhareea, S., Zhu, X., Hedgecock, E.M., Labouesse, M., & Georges-Labouesse, E. (2002). A conserved interaction between beta1 integrin/PAT-3 and Nck-interacting kinase/MIG-15 that mediates commissural axon navigation in *C. elegans*. *Current Biology*, *12*(8), 622–631. doi:10.1016/S0960-9822(02)00764-9

Poole, R.J., Bashllari, E., Cochella, L., Flowers, E.B., & Hobert, O. (2011). A Genome-Wide RNAi screen for factors involved in neuronal specification in *Caenorhabditis elegans*. *PLoS Genetics*, *7*(6), e1002109. doi:10.1371/journal.pgen.1002109

Procko, C., Lu, Y., & Shaham, S. (2011). Glia delimit shape changes of sensory neuron receptive endings in *C. elegans*. *Development*, *138*(7), 1371–1381. doi:10.1242/dev.058305

Rapti, G., Li, C., Shan, A., Lu, Y., & Shaham, S. (2017). Glia initiate brain assembly through noncanonical Chimaerin-Furin axon guidance in *C. elegans*. *Nature Neuroscience*, *20*(10), 1350–1360. doi:10.1038/nn.4630

Reilly, M.B., Cros, C., Varol, E., Yemini, E., & Hobert, O. (2020). Unique homeobox codes delineate all the neuron classes of *C. elegans*. *Nature*, *584*(7822), 595–601. doi:10.1038/s41586-020-2618-9

Richmond, J.E., Davis, W.S., & Jorgensen, E.M. (1999). UNC-13 is required for synaptic vesicle fusion in *C. elegans*. *Nature Neuroscience*, *2*(11), 959–964. http://www.ncbi.nlm.nih.gov/entrez/query.fcgi?db=pubmed&cmd=Retrieve&dopt=AbstractPlus&list_uids=10526333 doi:10.1038/14755

Rongo, C., & Kaplan, J.M. (1999). CaMKII regulates the density of central glutamatergic synapses in vivo. *Nature*, *402*(6758), 195–199. doi:10.1038/46065

Roy, P.J., Zheng, H., Warren, C.E., & Culotti, J.G. (2000). mab-20 encodes Semaphorin-2a and is required to prevent ectopic cell contacts during epidermal morphogenesis in *Caenorhabditis elegans*. *Development (Cambridge, England)*, *127*(4), 755–767.

Saied-Santiago, K., Townley, R.A., Attonito, J.D., Da Cunha, D.S., Díaz-Balzac, C.A., Tecle, E., & Bülow, H.E. (2017). Coordination of heparan sulfate proteoglycans with wnt signaling to control cellular migrations and positioning in *Caenorhabditis elegans*. *Genetics*, *206*(4), 1951–1967. doi:10.1534/genetics.116.198739

Salzberg, Y., Díaz-Balzac, C.A., Ramirez-Suarez, N.J., Attreed, M., Tecle, E., Desbois, M., … Bülow, H.E. (2013). Skin-derived cues control arborization of sensory dendrites in *Caenorhabditis elegans*. *Cell*, *155*(2), 308–320. doi:10.1016/j.cell.2013.08.058

Sammut, M., Cook, S.J., Nguyen, K.C.Q., Felton, T., Hall, D.H., Emmons, S.W., … Barrios, A. (2015). Glia-derived neurons are required for sex-specific learning in *C. elegans*. *Nature*, *526*(7573), 385–390. doi:10.1038/nature15700

Sanchez-Alvarez, L., Visanuvimol, J., McEwan, A., Su, A., Imai, J.H., & Colavita, A. (2011). VANG-1 and PRKL-1 cooperate to negatively regulate neurite formation in *Caenorhabditis elegans*. *PLoS Genetics*, *7*(9), e1002257 doi:10.1371/journal.pgen.1002257

Sarin, S., Antonio, C., Tursun, B., & Hobert, O. (2009). The *C. elegans* Tailless/TLX transcription factor nhr-67 controls neuronal identity and left/right asymmetric fate diversification. *Development (Cambridge, England))*, *136*(17), 2933–2944. doi:10.1242/dev.040204

Satterlee, J.S., Sasakura, H., Kuhara, A., Berkeley, M., Mori, I., & Sengupta, P. (2001). Specification of thermosensory neuron fate in *C. elegans* requires ttx-1, a homolog of otd/Otx. *Neuron*, *31*(6), 943–956. doi:10.1016/S0896-6273(01)00431-7

Schafer, W.R. (2018). The worm connectome: Back to the future. *Trends in Neurosciences*, *41*(11), 763–765. doi:10.1016/j.tins.2018.09.002

Schmitz, C., Kinge, P., & Hutter, H. (2007). Axon guidance genes identified in a large-scale RNAi screen using the RNAi-hypersensitive *Caenorhabditis elegans* strain nre-1(hd20) lin-15b(hd126). *Proceedings of the National Academy of Sciences of the United States of America*, *104*(3), 834–839. doi:10.1073/pnas.0510527104

Schroeder, N.E., Androwski, R.J., Rashid, A., Lee, H., Lee, J., & Barr, M.M. (2013). Dauer-specific dendrite arborization in *C. elegans* is regulated by KPC-1/Furin. *Current biology: CB*, *23*(16), 1527–1535. doi:10.1016/j.cub.2013.06.058

Sengupta, P., Chou, J.H., & Bargmann, C.I. (1996). odr-10 Encodes a seven transmembrane domain olfactory receptor required for responses to the odorant diacetyl. *Cell*, *84*(6), 899–909. doi:10.1016/S0092-8674(00)81068-5

Shah, P.K., Tanner, M.R., Kovacevic, I., Rankin, A., Marshall, T.E., Noblett, N., … Colavita, A. (2017). PCP and SAX-3/Robo pathways cooperate to regulate convergent extension-based nerve cord assembly in *C. elegans*. *Developmental Cell*, *41*(2), 195–203.e3. doi:10.1016/j.devcel.2017.03.024

Shaham, S. (1998). Identification of multiple *Caenorhabditis elegans* caspases and their potential roles in proteolytic cascades. *The Journal of Biological Chemistry*, *273*(52), 35109–35117. doi:10.1074/jbc.273.52.35109

Shaham, S. (2015). Glial development and function in the nervous system of *Caenorhabditis elegans*. *Cold Spring Harbor Perspectives in Biology*, *7*(4), a020578 doi:10.1101/cshperspect.a020578

Shai Shaham & H. Robert Horvitz (1996). An Alternatively Spliced *C. elegans* ced-4 RNA Encodes a Novel Cell Death Inhibitor, *Cell*, 201-208. DOI: 10.1016/S0092-8674(00)80092-6

Shao, Z., Watanabe, S., Christensen, R., Jorgensen, E.M., & Colón-Ramos, D.A. (2013). Synapse location during growth depends on glia location. *Cell*, *154*(2), 337–350. doi:10.1016/j.cell.2013.06.028

Shen, K., & Bargmann, C.I. (2003). The immunoglobulin superfamily protein SYG-1 determines the location of specific synapses in *C. elegans*. *Cell*, *112*(5), 619–630. doi:10.1016/S0092-8674(03)00113-2

Silva, M., Morsci, N., Nguyen, K.C.Q., Rizvi, A., Rongo, C., Hall, D.H., & Barr, M.M. (2017). Cell-specific α-tubulin isotype regulates ciliary microtubule ultrastructure, intraflagellar transport, and extracellular vesicle biology. *Current Biology: CB*, *27*(7), 968–980. doi:10.1016/j.cub.2017.02.039

Singhvi, A., & Garriga, G. (2009). Asymmetric divisions, aggresomes and apoptosis. *Trends in Cell Biology*, *19*(1), 1–7. doi:10.1016/j.tcb.2008.10.004

Singhvi, A., Liu, B., Friedman, C.J., Fong, J., Lu, Y., Huang, X.-Y., & Shaham, S. (2016). A glial K/Cl transporter controls neuronal receptive ending shape by chloride inhibition of an rGC. *Cell*, *165*(4), 936–948. doi:10.1016/j.cell.2016.03.026

Singhvi, A., & Shaham, S. (2019). Glia-neuron interactions in *Caenorhabditis elegans*. *Annual Review of Neuroscience*, *42*(1), 149–168. doi:10.1146/annurev-neuro-070918-050314

Smith, C.J., Watson, J.D., Vanhoven, M.K., Colón-Ramos, D.A., & Miller, D.M. (2012). Netrin (UNC-6) mediates dendritic self-avoidance. *Nature Neuroscience*, *15*(5), 731–737. doi:10.1038/nn.3065

Solecki, D.J. (2012). Sticky situations: Recent advances in control of cell adhesion during neuronal migration. *Current Opinion in Neurobiology*, *22*(5), 791–798. doi:10.1016/j.conb.2012.04.010

Stefanakis, N., Carrera, I., & Hobert, O. (2015). Regulatory logic of pan-neuronal gene expression in *C. elegans*. *Neuron*, *87*(4), 733–750. doi:10.1016/j.neuron.2015.07.031

Steimel, A., Wong, L., Najarro, E.H., Ackley, B.D., Garriga, G., & Hutter, H. (2010). The Flamingo ortholog FMI-1 controls pioneer-dependent navigation of follower axons in *C. elegans*. *Development*, *137*(21), 3663–3673. doi:10.1242/dev.054320

Stein, G.M., & Murphy, C.T. (2012). The intersection of aging, longevity pathways, and learning and memory in *C. elegans*. *Frontiers in Genetics*, *3*, 259. doi:10.3389/fgene.2012.00259

Stringham, E.G., & Schmidt, K.L. (2009). Navigating the cell: UNC-53 and the navigators, a family of cytoskeletal regulators with multiple roles in cell migration, outgrowth and trafficking. *Cell Adhesion & Migration*, *3*(4), 342–346. doi:10.4161/cam.3.4.9451

Sulston, J., Dew, M., & Brenner, S. (1975). Dopaminergic neurons in the nematode *Caenorhabditis elegans*. *The Journal of Comparative Neurology*, *163*(2), 215–226. doi:10.1002/cne.901630207

Sulston, J.E. (1976). Post-embryonic development in the ventral cord of *Caenorhabditis elegans*. *Philos Trans R Soc Lond B Biol Sci*, *275*(938), 287–297. doi:10.1098/rstb.1976.0084

Sulston, J.E., Albertson, D.G., & Thomson, J.N. (1980). The *Caenorhabditis elegans* male: Postembryonic development of nongonadal structures. *Developmental Biology*, *78*(2), 542–576. doi:10.1016/0012-1606(80)90352-8

Sulston, J.E., & Brenner, S. (1974). The DNA of *Caenorhabditis elegans*. *Genetics*, *77*(1), 95–104.

Sulston, J.E., & Horvitz, H.R. (1977). Post-embryonic cell lineages of the nematode, *Caenorhabditis elegans*. *Developmental Biology*, *56* (1), 110–156. doi:10.1016/0012-1606(77)90158-0

Sulston, J.E., Schierenberg, E., White, J.G., & Thomson, J.N. (1983). The embryonic cell lineage of the nematode *Caenorhabditis elegans*. *Developmental Biology*, *100*(1), 64–119. doi:10.1016/0012-1606(83)90201-4

Sulston, J.E., & White, J.G. (1980). Regulation and cell autonomy during postembryonic development of *Caenorhabditis elegans*. *Developmental Biology*, *78*(2), 577–597. doi:10.1016/0012-1606(80)90353-X

Sundararajan, L., & Lundquist, E.A. (2012). Transmembrane proteins UNC-40/DCC, PTP-3/LAR, and MIG-21 control anterior-posterior neuroblast migration with left-right functional asymmetry in *Caenorhabditis elegans*. *Genetics*, *192*(4), 1373–1388. doi:10.1534/genetics.112.145706

Teuliere, J., Cordes, S., Singhvi, A., Talavera, K., & Garriga, G. (2014). Asymmetric neuroblast divisions producing apoptotic cells require the cytohesin GRP-1 in *Caenorhabditis elegans*. *Genetics*, *198*(1), 229–247. doi:10.1534/genetics.114.167189

Tian, D., Diao, M., Jiang, Y., Sun, L., Zhang, Y., Chen, Z., … Ou, G. (2015). Anillin regulates neuronal migration and neurite growth by linking RhoG to the actin cytoskeleton. *Current Biology: Cb*, *25*(9), 1135–1145. doi:10.1016/j.cub.2015.02.072

Tornberg, J., Sykiotis, G.P., Keefe, K., Plummer, L., Hoang, X., Hall, J.E., … Bülow, H.E. (2011). Heparan sulfate 6-O-sulfotransferase 1, a gene involved in extracellular sugar modifications, is mutated in patients with idiopathic hypogonadotrophic hypogonadism. *Proceedings of the National Academy of Sciences of the United States of America*, *108*(28), 11524–11529. doi:10.1073/pnas.1102284108

Troemel, E.R., Sagasti, A., & Bargmann, C.I. (1999). Lateral signaling mediated by axon contact and calcium entry regulates asymmetric odorant receptor expression in *C. elegans*. *Cell*, *99*(4), 387–398. http://www.ncbi.nlm.nih.gov/entrez/query.fcgi?db=pubmed&cmd=Retrieve&dopt=AbstractPlus&list_uids=10571181 doi:10.1016/S0092-8674(00)81525-1

Wadsworth, W.G., Bhatt, H., & Hedgecock, E.M. (1996). Neuroglia and pioneer neurons express UNC-6 to provide global and local netrin cues for guiding migrations in *C. elegans*. *Neuron*, *16*(1), 35–46. doi:10.1016/S0896-6273(00)80021-5

Wallace, S.W., Singhvi, A., Liang, Y., Lu, Y., & Shaham, S. (2016). PROS-1/Prospero is a major regulator of the glia-specific secretome controlling sensory-neuron shape and function in *C. elegans*. *Cell Reports*, *15*(3), 550–562. doi:10.1016/j.celrep.2016.03.051

Walsh, J.D., Boivin, O., & Barr, M.M. (2020). What about the males? The *C. elegans* sexually dimorphic nervous system and a CRISPR-based tool to study males in a hermaphroditic species. Advance online publication. *Journal of Neurogenetics*. doi:10.1080/01677063.2020.1789978

Walthall, W.W., & Chalfie, M. (1988). Cell-cell interactions in the guidance of late-developing neurons in *Caenorhabditis elegans*. *Science (New York, N.Y.)*, *239*(4840), 643–645. doi:10.1126/science.3340848

Wang, J., Silva, M., Haas, L.A., Morsci, N.S., Nguyen, K.C.Q., Hall, D.H., & Barr, M.M. (2014). *C. elegans* ciliated sensory neurons release extracellular vesicles that function in animal communication. *Current Biology: CB*, *24*(5), 519–525. doi:10.1016/j.cub.2014.01.002

Wang, R., Mellem, J.E., Jensen, M., Brockie, P.J., Walker, C.S., Hoerndli, F.J., … Maricq, A.V. (2012). The SOL-2/Neto auxiliary protein modulates the function of AMPA-subtype ionotropic glutamate receptors. *Neuron*, *75*(5), 838–850. doi:10.1016/j.neuron.2012.06.038

Ward, S., Thomson, N., White, J.G., & Brenner, S. (1975). Electron microscopical reconstruction of the anterior sensory anatomy of the nematode *Caenorhabditis elegans*.?2UU. *The Journal of Comparative Neurology*, *160*(3), 313–337. doi:10.1002/cne.901600305

Waterston, R., & Sulston, J. (1995). The genome of *Caenorhabditis elegans*. *Proceedings of the National Academy of Sciences of the United States of America*, *92*(24), 10836–10840. doi:10.1073/pnas.92.24.10836

Way, J.C., & Chalfie, M. (1988). mec-3, a homeobox-containing gene that specifies differentiation of the touch receptor neurons in *C. elegans*. *Cell*, *54*(1), 5–16. doi:10.1016/0092-8674(88)90174-2

Weimer, R.M., Richmond, J.E., Davis, W.S., Hadwiger, G., Nonet, M.L., & Jorgensen, E.M. (2003). Defects in synaptic vesicle docking in unc-18 mutants. *Nature Neuroscience*, *6*(10), 1023–1030. doi:10.1038/nn1118

White, J.G. (2013). Getting into the mind of a worm–a personal view. *WormBook*. doi:10.1895/wormbook.1.158.1

White, J.G., Southgate, E., & Thomson, J.N. (1992). Mutations in the *Caenorhabditis elegans* unc-4 gene alter the synaptic input to ventral cord motor neurons. *Nature*, *355*(6363), 838–841. doi:10.1038/355838a0

White, J.G., Southgate, E., Thomson, J.N., & Brenner, S. (1976). The structure of the ventral nerve cord of *Caenorhabditis elegans*. *Philosophical Transactions of the Royal Society of London. Series B, Biological Sciences*, *275*(938), 327–348. doi:10.1098/rstb.1976.0086

White, J.G., Southgate, E., Thomson, J.N., & Brenner, S. (1986). The structure of the nervous system of the nematode *Caenorhabditis elegans*. *Philosophical Transactions of the Royal Society of London. Series B, Biological Sciences*, *314*(1165), 1–340. doi:10.1098/rstb.1986.0056

Whittaker, A.J., & Sternberg, P.W. (2004). Sensory processing by neural circuits in *Caenorhabditis elegans*. *Current Opinion in Neurobiology*, *14*(4), 450–456. doi:10.1016/j.conb.2004.07.006

Wicks, S.R., Yeh, R.T., Gish, W.R., Waterston, R.H., & Plasterk, R.H. (2001). Rapid gene mapping in *Caenorhabditis elegans* using a high density polymorphism map. *Nature Genetics*, *28*(2), 160–164. doi:10.1038/88878

Withee, J., Galligan, B., Hawkins, N., & Garriga, G. (2004). *Caenorhabditis elegans* WASP and Ena/VASP proteins play compensatory roles in morphogenesis and neuronal cell migration. *Genetics*, *167*(3), 1165–1176. doi:10.1534/genetics.103.025676

Witvliet, D., Mulcahy, B., Mitchell, J.K., Meirovitch, Y., Berger, D.K., Wu, Y., … Zhen, M. (2020). Connectomes across development reveal principles of brain maturation in *C. elegans*. *BioRxiv*. doi:10.1101/2020.04.30.066209

Wolf, F.W., Hung, M.S., Wightman, B., Way, J., & Garriga, G. (1998). vab-8 is a key regulator of posteriorly directed migrations in *C. elegans* and encodes a novel protein with kinesin motor similarity. *Neuron*, *20*(4), 655–666. doi:10.1016/S0896-6273(00)81006-5

Xu, Y., & Quinn, C.C. (2012). MIG-10 functions with ABI-1 to mediate the UNC-6 and SLT-1 axon guidance signaling pathways. *PLoS Genetics*, *8*(11), e1003054. doi:10.1371/journal.pgen.1003054

Yeh, E., Kawano, T., Ng, S., Fetter, R., Hung, W., Wang, Y., & Zhen, M. (2009). *Caenorhabditis elegans* innexins regulate active zone differentiation. *The Journal of Neuroscience: The Official Journal of the Society for Neuroscience*, *29*(16), 5207–5217. doi:10.1523/JNEUROSCI.0637-09.2009

Yin, J.A., Gao, G., Liu, X.J., Hao, Z.Q., Li, K., Kang, X.L., … Cai, S.Q. (2017). Genetic variation in glia-neuron signalling modulates ageing rate. *Nature*, *551*(7679), 198–203. doi:10.1038/nature24463

Yoshimura, S., Murray, J.I., Lu, Y., Waterston, R.H., & Shaham, S. (2008). mls-2 and vab-3 Control glia development, hlh-17/Olig expression and glia-dependent neurite extension in *C. elegans*. *Development*, *135*(13), 2263–2275. doi:10.1242/dev.019547

Yu, T.W., Hao, J.C., Lim, W., Tessier-Lavigne, M., & Bargmann, C.I. (2002). Shared receptors in axon guidance: SAX-3/Robo signals via UNC-34/enabled and a netrin-independent UNC-40/DCC function. *Nature Neuroscience*, *5*(11), 1147–1154. doi:10.1038/nn956

Zallen, J.A., Yi, B.A., & Bargmann, C.I. (1998). The conserved immunoglobulin superfamily member SAX-3/Robo directs multiple aspects of axon guidance in *C. elegans*. *Cell*, *92*(2), 217–227. doi:10.1016/S0092-8674(00)80916-2

Zhang, A., Ackley, B.D., & Yan, D. (2020). Vitamin B12 regulates glial migration and synapse formation through isoform-specific control of PTP-3/LAR PRTP expression. *Cell Reports*, *30*(12), 3981–3988.e3. doi:10.1016/j.celrep.2020.02.113

Zhang, A., Noma, K., & Yan, D. (2020). Regulation of gliogenesis by lin-32/Atoh1 in *Caenorhabditis elegans*. *G3 (Bethesda, Md.)*, *10*(9), 3271–3278. doi:10.1534/g3.120.401547

Zheng, C., Diaz-Cuadros, M., & Chalfie, M. (2016). GEFs and Rac GTPases control directional specificity of neurite extension along the anterior-posterior axis. *Proceedings of the National Academy of Sciences of the United States of America*, *113*(25), 6973–6978. doi:10.1073/pnas.1607179113

Zhou, Z., Hartwieg, E., & Horvitz, H.R. (2001). CED-1 is a transmembrane receptor that mediates cell corpse engulfment in *C. elegans*. *Cell*, *104*(1), 43–56. doi:10.1016/S0092-8674(01)00190-8

Zou, W., Dong, X., Broederdorf, T.R., Shen, A., Kramer, D.A., Shi, R., … Shen, K. (2018). A dendritic guidance receptor complex brings together distinct actin regulators to drive efficient F-actin assembly and branching. *Developmental Cell*, *45*(3), 362–375.e3. doi:10.1016/j.devcel.2018.04.008

Zou, Y., Chiu, H., Domenger, D., Chuang, C.F., & Chang, C. (2012). The lin-4 microRNA targets the LIN-14 transcription factor to inhibit netrin-mediated axon attraction. *Science Signaling*, *5*(228), ra43. doi:10.1126/scisignal.2002437

Zuchero, J.B., & Barres, B.A. (2015). Glia in mammalian development and disease. *Development (Cambridge, England)*, *142*(22), 3805–3809. doi:10.1242/dev.129304

Zuryn, S., Ahier, A., Portoso, M., White, E.R., Morin, M.C., Margueron, R., & Jarriault, S. (2014). Sequential histone-modifying activities determine the robustness of transdifferentiation. *Science (New York, N.Y.)*, *345*(6198), 826–829. doi:10.1126/science.1255885

Neuronal specification in *C. elegans*: combining lineage inheritance with intercellular signaling

Antoine Barrière and Vincent Bertrand

ABSTRACT

The nervous system is composed of a high diversity of neuronal types. How this diversity is generated during development is a key question in neurobiology. Addressing this question is one of the reasons that led Sydney Brenner to develop the nematode *C. elegans* as a model organism. While there was initially a debate on whether the neuronal specification follows a 'European' model (determined by ancestry) or an 'American' model (determined by intercellular communication), several decades of research have established that the truth lies somewhere in between. Neurons are specified by the combination of transcription factors inherited from the ancestor cells and signaling between neighboring cells (especially Wnt and Notch signaling). This converges to the activation in newly generated postmitotic neurons of a specific set of terminal selector transcription factors that initiate and maintain the differentiation of the neuron. In this review, we also discuss the evolution of these specification mechanisms in other nematodes and beyond.

Introduction

Bilaterian animals can display very sophisticated behaviors, mirrored by the cellular complexity of their nervous system that is composed by a high diversity of neuronal types. A major subject in neurobiology is to understand the origin of this diversity during development. This question was of great interest to Sydney Brenner and was one of the reasons why he chose to develop the nematode *C. elegans* as a model organism (Brenner, 1974). *C. elegans* indeed presents several advantages for the study of nervous system development. Its nervous system is relatively simple with only 302 neurons in the adult hermaphrodite (White, Southgate, Thomson, & Brenner, 1986). The animal is transparent, allowing researchers to follow nervous system development directly *in vivo* with single cell resolution. Finally, the life cycle is short (three days at 25 °C) and the brood size is large, making *C. elegans* a great system to identify genes involved in nervous system development via genetic screens.

The establishment of the lineage of *C. elegans* by John Sulston and colleagues showed that neurons are produced by an invariant series of cell divisions (Sulston & Horvitz, 1977; Sulston, Schierenberg, White, & Thomson, 1983). There was a debate among early *C. elegans* researchers as to whether the fate of a neuron is established in a 'European' way or 'American' way (Brown, 2003). In a 'European' model, the identity of a neuron is determined by the ancestry of the cell in a vertical manner, while in an 'American' model it is determined by the cellular environment in a horizontal manner.

In this review, we discuss the molecular and cellular mechanisms at the basis of neuronal specification in *C. elegans* showing that neurons are generated via a combination of lineage inheritance (vertical) and intercellular signaling (horizontal). We then present how terminal selector transcription factors trigger neuronal differentiation and maintain the differentiated state of neurons following specification. Finally, we discuss to what extent these neuronal specification programs are conserved in other nematodes and beyond.

Neuronal specification: a combination of lineage inheritance and intercellular signaling

In *C. elegans*, the majority of neurons is generated during embryogenesis by series of asymmetric divisions oriented along the anteroposterior axis (Sulston *et al.*, 1983). Many transcription factors that affect neuronal cell fate specification have been identified, via forward genetic screens or reverse RNAi screens (reviewed in Hobert, 2010). Analysis of their expression patterns suggests that, during development, a neuronal progenitor goes through a series of transient regulatory states defined by the expression of different combinations of transcription factors (see for example Bertrand & Hobert, 2009; Poole, Bashllari, Cochella, Flowers, & Hobert, 2011; Sarafi-Reinach, Melkman, Hobert, & Sengupta, 2001). These observations have been recently generalized using single-cell RNA sequencing of most embryonic cells (Packer *et al.*, 2019), showing that following each division of a neuronal progenitor, the two daughter cells acquire transcription profiles different from each other and from their mother cell.

This transition from one regulatory state to another is triggered by the asymmetric division process. In *C. elegans*, the Wnt/β-catenin asymmetry pathway (a specialized Wnt/β-catenin pathway) controls asymmetric divisions oriented along the anteroposterior axis (reviewed in Bertrand, 2016; Phillips & Kimble, 2009; Sawa & Korswagen, 2013). This pathway regulates many asymmetric divisions in the early embryo (Kaletta, Schnabel, & Schnabel, 1997; Lin, Hill, & Priess, 1998) and also controls the terminal asymmetric divisions of neuronal progenitors during neurulation (epidermal enclosure) (Figure 1(A); Bertrand & Hobert, 2009; Gordon & Hobert, 2015; Murgan *et al.*, 2015). The Wnt/β-catenin asymmetry pathway controls the expression of target genes in the daughter cells via the TCF transcription factor POP-1 and its transcriptional coactivator SYS-1, a β-catenin (Kidd, Miskowski, Siegfried, Sawa, & Kimble, 2005; Lin *et al.*, 1998; Lin, Thompson, & Priess, 1995; Phillips, Kidd, King, Hardin, & Kimble, 2007). The pathway modulates the activity of POP-1 and SYS-1 by regulating the nuclear export of POP-1 (Lo, Gay, Odom, Shi, & Lin, 2004; Rocheleau *et al.*, 1999) and the degradation of SYS-1 (Huang, Shetty, Robertson, & Lin, 2007; Phillips *et al.*, 2007). Following asymmetric division, high levels of SYS-1 relative to POP-1 in the posterior nucleus lead to the formation of a POP-1:SYS-1 complex that activates transcription (Huang *et al.*, 2007; Kidd *et al.*, 2005; Phillips *et al.*, 2007). In the anterior nucleus, SYS-1 levels are low and POP-1 mostly free of SYS-1 represses transcription.

Studies of *cis*-regulatory regions have established how the lineage history (transcription factors inherited from the mother cell) and the asymmetric division cue (Wnt/β-catenin asymmetry pathway) are integrated to generate novel regulatory states in the daughter cells. For example, in the larval neuroectodermal T lineage, the Hox transcription factor NOB-1 and the Pbx transcription factor CEH-20 are inherited from the mother cell and cooperate with the POP-1:SYS-1 complex in the posterior daughter to activate the expression of the Meis transcription factor gene *psa-3* by directly binding to its *cis*-regulatory regions (Arata *et al.*, 2006). Similarly, in the AIY neuron lineage of the embryo, the LIM-homeodomain transcription factor TTX-3, inherited from the mother cell (SMDD/AIY mother), cooperates with POP-1:SYS-1 to induce the transcription of the homeodomain gene *ceh-10* only in the posterior daughter (AIY neuron), by binding to its *cis*-regulatory regions (Bertrand & Hobert, 2009; Figure 1(B), division generating the SMDD and AIY neurons). While direct target genes activated in the posterior daughter cell contain POP-1 binding sites and are activated by the POP-1:SYS-1 complex, genes activated in the anterior daughter are not regulated via POP-1 binding sites. For example, in the AIY neuron lineage of the embryo, it has been observed that POP-1, in the absence of SYS-1, activates the expression of an anterior target gene *ttx-3* by forming a complex with another transcription factor protein REF-2 (a Zic factor) and binding to a REF-2 binding site in the *ttx-3 cis*-regulatory regions (Murgan *et al.*, 2015; Figure 1(B), division generating the SMDD/AIY mother and the SIAD/SIBV mother). Whether POP-1 activates transcription in a similar manner in anterior daughters of other neuronal lineages remains to be established. These data illustrate how combining lineage ancestry (via inherited transcription factors) with intercellular signaling (via the asymmetric Wnt pathway) generates neuronal type diversity in the nervous system of *C. elegans* (Figure 1(A)). Secreted Wnt ligands can play an instructive role in the asymmetric divisions regulated by the Wnt/β-catenin asymmetry pathway. For example, it

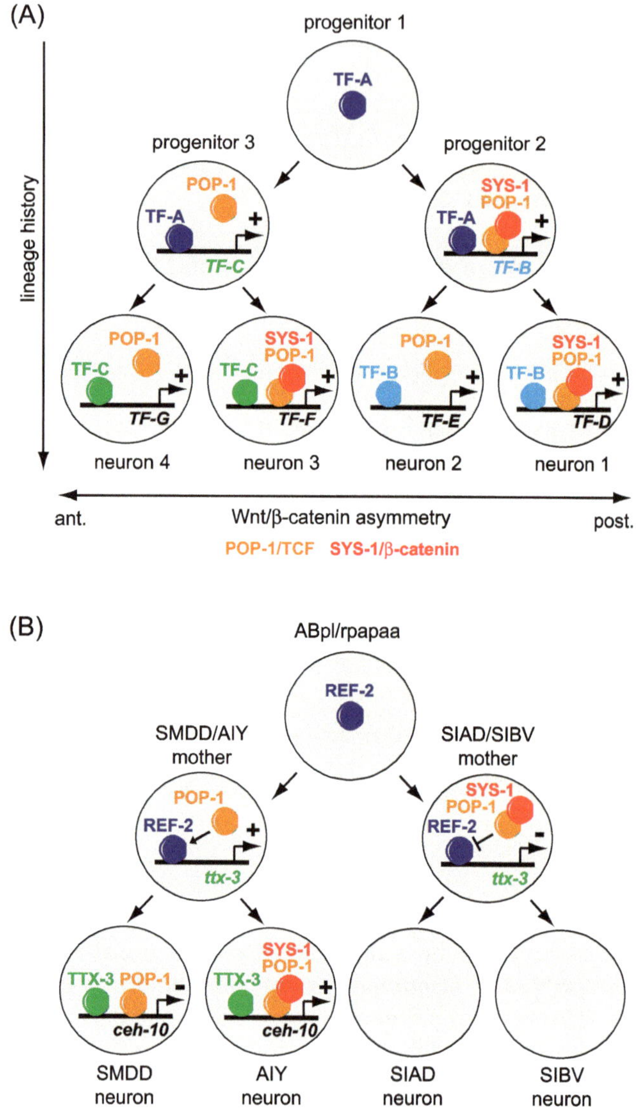

Figure 1. Generation of neuronal diversity by the combination of lineage history and Wnt/β-catenin asymmetry. (A) Following each asymmetric division, the Wnt/β-catenin asymmetry pathway cooperates with transcription factors (TFs) inherited from the mother cell to activate the expression of different transcription factors in the anterior and posterior daughter cells. (B) Example of the AIY neuron lineage (ABpl/rpapaa lineage). In the first cell division, the POP-1 protein interacts with the REF-2 protein to activate the transcription of the *ttx-3* gene in the anterior daughter (SMDD/AIY mother) via a REF-2 binding site present in the *ttx-3* promoter. This activation is blocked by SYS-1 in the posterior daughter (SIAD/SIBV mother). In the second cell division, the TTX-3 protein and the POP-1:SYS-1 complex activate the transcription of the *ceh-10* gene in the posterior daughter (AIY neuron) via TTX-3 and POP-1 binding sites present in the *ceh-10* promoter. In the anterior daughter (SMDD neuron), POP-1 without SYS-1 represses *ceh-10* expression.

has been established that for the endomesoderm precursor cell EMS or the larval blast cell T, Wnt ligands, secreted from a posterior source, control the division orientation and the asymmetry of daughter cell fates (Goldstein, Takeshita, Mizumoto, & Sawa, 2006). Interestingly, it has been recently observed that three Wnt ligands (CWN-1, CWN-2 and MOM-2), coming from the posterior of the embryo, regulate in an instructive manner the terminal asymmetric divisions of embryonic neuronal progenitors (Kaur *et al.*, 2020). This illustrates the importance of intercellular signaling during neuronal development in *C. elegans*.

Another intercellular signaling pathway playing an important role in neuronal cell fate specification in *C. elegans* is the Notch pathway. This pathway is involved in the generation of left-right asymmetries in the nervous system. While the nervous system of *C. elegans* is essentially bilateral symmetric, some neuronal left-right pairs, which are symmetric at the morphological level, display some functional and molecular asymmetries (Hobert, Johnston, & Chang, 2002). In the case of the pair of ASE taste neurons, the left and right cells express different sets of chemoreceptors that allow sensing different chemicals (Pierce-Shimomura, Faumont, Gaston, Pearson, & Lockery, 2001; Yu, Avery, Baude, & Garbers, 1997). The difference between left and right ASE is induced in the very early embryo (four-cell stage) by a Notch signal, coming from the posterior P lineage, which represses the expression of the T-box transcription factor genes *tbx-37* and *tbx-38* in the blastomere ABp, giving rise to the right ASE, but not in the blastomere ABa, giving rise to the left ASE (Poole & Hobert, 2006). Another example is the pairs of AIY interneurons and SMDD, SIAD and SIBV motor neurons, which act in a circuit controlling navigation behavior downstream of ASE. In these neurons, the bHLH gene *hlh-16* is expressed in a left-right asymmetric manner and is important for their correct axonal projections (Bertrand, Bisso, Poole, & Hobert, 2011). This asymmetry is established independently of the ASE asymmetry by a later Notch signal (during gastrulation) coming from the left mesoderm, which increases *hlh-16* expression on the left side. Another example of left-right asymmetry is the pair of AWC olfactory neurons: the odorant receptor gene *str-2* is activated in only one of the two AWC neurons in a stochastic manner with no left-right bias. Communication between the two AWC neurons, involving calcium signaling but not Notch, ensures that only one AWC expresses *str-2* (Troemel, Sagasti, & Bargmann, 1999).

To conclude, neuronal cell fate specification in *C. elegans* involves both lineage transmission (inherited transcription factors) and intercellular signaling (such as Wnt or Notch). It therefore follows neither a pure 'European' nor a pure 'American' model, but is instead a mix of both.

Neuronal differentiation: initiation and maintenance of terminal neuronal fate by terminal selectors

The function of a neuron relies on the specific set of terminal differentiation genes (or effector genes) that it expresses such as neurotransmitter receptors, ion channels or neurotransmitter synthesis pathway genes. In the *C. elegans* nervous system, the expression of many neuron type-specific terminal differentiation genes has been mapped. For example the cholinergic, GABAergic or glutamatergic neurons have been systematically identified (Gendrel, Atlas, & Hobert, 2016; Pereira *et al.*, 2015; Serrano-Saiz *et al.*, 2013). For several neuron types, the analysis of *cis*-regulatory regions of their specific terminal differentiation genes have revealed that they are often regulated by a common set of transcription factors, called terminal selectors (Hobert, 2008). These transcription factors also frequently autoregulate their expression, therefore maintaining the type identity of the neuron throughout the life of the animal. One example is the cholinergic interneuron AIY where the homeodomain transcription factors TTX-3 and CEH-10 directly activate and maintain the expression of a large battery of terminal differentiation genes: the choline acetyltransferase *cha-1*, the acetylcholine vesicular transporter *unc-17*, the neurotransmitter receptors *ser-2*, *gar-2* and *mod-1*, etc. (Figure 2; Wenick & Hobert, 2004). Other examples of terminal selectors include the two homeodomain transcription factors MEC-3 and UNC-86 in touch receptor neurons (Zhang *et al.*, 2002), the zinc finger transcription factor CHE-1 in the taste neuron ASE (Etchberger *et al.*, 2007) or the COE transcription factor UNC-3 in cholinergic motorneurons (Kratsios, Stolfi, Levine, & Hobert, 2011). In addition, many other transcription factors controlling the expression of terminal differentiation genes in various neuronal types have been identified (for a comprehensive review see Hobert, 2016).

The connection between these terminal differentiation programs and earlier cell fate specification events, such as asymmetric divisions, has been less characterized. The link has been elucidated in the case of the cholinergic interneuron AIY (Figure 2; Bertrand & Hobert, 2009; Murgan

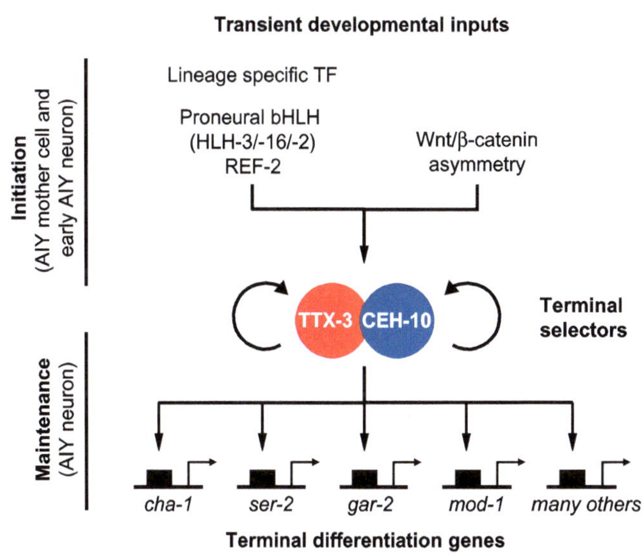

Figure 2. Connection of early specification events to the terminal differentiation program in the AIY interneuron. During embryonic development, transient developmental inputs (lineage specific transcription factors and the Wnt/β-catenin asymmetry pathway) initiate the expression of the terminal transcription factors TTX-3 and CEH-10 in the AIY lineage. TTX-3 and CEH-10 subsequently maintain their expression and activate a large battery of type specific terminal differentiation genes.

et al., 2015): the expression of the AIY terminal selector genes *ttx-3* and *ceh-10* is set up in the early postmitotic AIY neuron by two types of transient developmental inputs. One is the specific transcription factors inherited from the lineage history: proneural bHLH factors (HLH-3, HLH-16), their binding partner HLH-2 and the Zic transcription factor REF-2. The other input is the Wnt/β-catenin asymmetry pathway, which regulates the terminal asymmetric divisions in the AIY lineage. These lineage specific transcription factors and the Wnt/β-catenin asymmetry pathway are directly integrated at the level of the *cis*-regulatory regions of *ttx-3* and *ceh-10* to establish their coexpression specifically in the early AIY neurons. The lineage specific transcription factors (HLH-3, HLH-16, HLH-2 and REF-2) and the Wnt pathway effectors (POP-1 and SYS-1) subsequently disappear from the AIY neuron, and TTX-3 and CEH-10 expression is then maintained via a positive autoregulatory loop, where they directly bind their own *cis*-regulatory regions and positively regulate their expression. In the postmitotic AIY neuron, TTX-3 and CEH-10 directly activate and maintain the expression of a large battery of terminal differentiation genes, therefore, determining the type-specific function of the AIY neuron. This illustrates how terminal selector transcription factors can connect early specification events to the terminal differentiation of the neuron. Whether a similar regulatory logic applies for the activation of other terminal selectors remains to be established.

Evolution of neuronal specification programs in nematodes

Neuronal determination mechanisms based on lineage and cell neighborhood could constrain neuronal development and affect its flexibility and evolvability. Indeed, there was, early on, an impression of very high conservation of the nervous system across the vast phylogenetic distances of nematodes. In one of the seminal papers by John Sulston and Sydney Brenner, it was noted that: 'The nervous system of the nematode *Ascaris lumbricoides* appears to be very similar to that of *C. elegans*. The homology of the catecholaminergic neurons lends support to the possibility that, notwithstanding their great disparity in size, there is a strong conservation of the properties of the nervous system between these two nematodes' (Sulston, Dew, & Brenner, 1975). It is indeed tempting to try to assign almost one-to-one correspondence of the 302 *C. elegans* neurons (Sulston & Horvitz, 1977; Sulston *et al.*, 1983) to the 298 neurons found in *Ascaris* (Stretton *et al.*, 1992). But this apparent morphological conservation could underlie pervasive functional divergence. Due to the difficulty of duplicating genetic screens performed in *C. elegans* in other nematode species, there has been limited number of experiments analyzing the evolution of neuronal specification mechanisms. Overall, studies suggest that the neuroanatomy of nematodes is not as highly conserved as previously described (Han, Boas, & Schroeder, 2015; Figure 3(A)), although the ability to assign single-cell homology across great evolutionary distances is a critical factor to assess conservation and variation in neuronal fates.

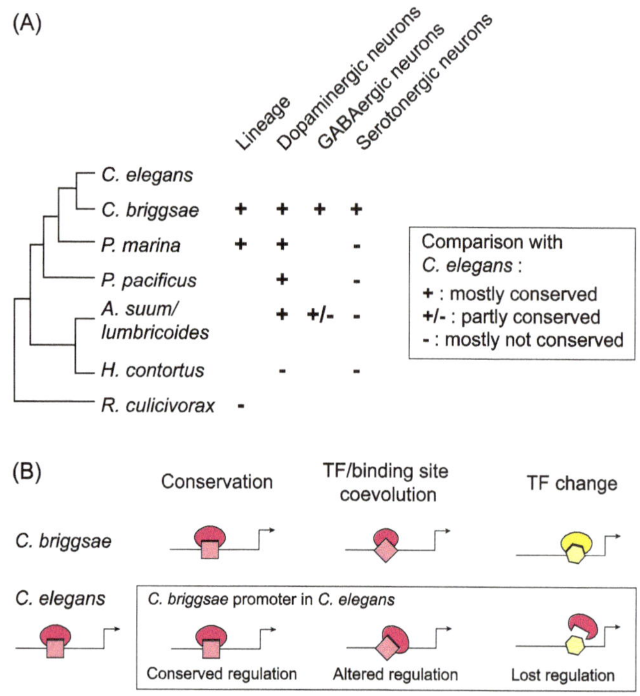

Figure 3. Conservation and divergence of neurodevelopment in nematodes. (A) Phylogeny of species discussed in this review, showing conservation or divergence of cell lineage or neurotransmitter cellular patterns, compared with *C. elegans*, at different phylogenetic distances. (B) Possible scenarios of transcriptional evolution. If a transcription factor and its target *cis*-regulatory element are conserved between *C. elegans* and *C. briggsae*, there is no change in regulation and no difference in expression pattern is observed when the *C. briggsae cis*-regulatory element is placed in the *C. elegans trans*-regulatory environment. If there was coevolution between *trans*-regulatory factors and *cis*-regulatory elements, the *C. briggsae cis*-regulatory element placed in the *C. elegans trans*-regulatory environment will give an altered expression pattern. Finally, if there was a change in transcription factors binding to a *cis*-regulatory element, the regulation might be completely lost when the *C. briggsae cis*-regulatory element is placed in the *C. elegans trans*-regulatory environment.

Variations in neuroanatomy and lineage

Gross neuroanatomical comparisons across many nematode species, most not amenable to manipulation, revealed extensive variability in number and time of birth of ventral nerve cord neurons and dye-filling sensory neurons across multiple clades (Han *et al.*, 2015). Finer analyses, comparing individual neurons between *C. elegans* and *Pristionchus pacificus* by serial electron microscopy, in the paryngeal (Bumbarger, Riebesell, Rodelsperger, & Sommer, 2013) and olfactory (Hong *et al.*, 2019) nervous systems, found remarkable conservation of neuron homology, as defined by cell body position and neurite anatomy, but also extensive rewiring and changes in fate, e.g. from interneuron to motor neuron.

Embryonic lineages were established for other species in the *Caenorhabditis* genus, namely *C. briggsae* (Zhao *et al.*, 2008), *C. remanei* and *C. brenneri* (Memar *et al.*, 2019). For these three species, embryonic development until the start of muscle contraction is identical, but for some small variations in the timing of division, and one cell escaping programmed cell death to become part of the pharynx in *C. briggsae* and *C. remanei* (Memar *et al.*, 2019). Accordingly, the number and identity of neurons appear to be perfectly conserved in the species investigated. However, once comparisons reach outside the *Caenorhabditis* genus, like with the marine nematode *Pellioditis marina* (Houthoofd *et al.*, 2003), developmental differences become apparent: while cell divisions maintain a strong homology (95.5%), final fate is less well conserved; only 85% of cells acquiring a neuronal fate in *P. marina* are also neurons in *C. elegans*. The difference is made up by cells acquiring pharyngeal, epidermal or programmed cell death fates, for a final count of 194 neurons at muscle contraction compared with 244 in *C. elegans*. In the more distantly related nematode *Romanomermis culicivorax* (Schulze & Schierenberg, 2009), the lineage is more generally divergent, and early development proceeds in a more monoclonal fashion. Cell positions are also markedly different, with neuronal precursors arranged as rings around the embryo, suggesting possible changes in cell–cell interactions.

Variations in neurotransmitter identity patterns

One of the first tools used to compare the nervous system across species is the Falck-Hillarp method, where formaldehyde reacts with catecholamines to form fluorophores. Using this method, Sulston *et al.* (1975) compared dopamine neurons between the yet to be described *C. elegans* nervous system and *Ascaris lumbricoides*, and identified strong similarities between the two species in the head and body, but noted the presence of a pair of dopaminergic neurons in the *A. lumbricoides* tail without equivalent in *C. elegans*. A later study, looking at dopaminergic neurons in rhabditid nematodes, including the *Caenorhabditis* genus, found perfect conservation of pattern and easy assignation of one-to-one homology (Rivard *et al.*, 2010). This constant pattern of dopaminergic neurons, although striking, might give an impression of almost perfect conservation. However, in the more distant specie *Haemonchus contortus*, dopamine was not detected in head neurons, but only in commissures (Rao, Forrester, Keller, & Prichard, 2011) suggesting a diverging pattern of dopaminergic neurons.

Similar studies targeting serotonergic neurons reveal that they are also highly conserved within the *Caenorhabditis* genus; however, in more distantly related species, only the serotonergic NSM neurons were found to be consistently conserved (Rivard *et al.*, 2010), including with *Ascaris suum* (Johnson, Reinitz, Sithigorngul, & Stretton, 1996) and *H. contortus* (Rao *et al.*, 2011). Finally, some GABAergic neurons were found to be so well conserved between *C. elegans* and *A. suum*, that clear homologs were identified for the four RMEs and DVB (Guastella, Johnson, & Stretton, 1991), and for DDs and VDs in the ventral nerve cord (Johnson & Stretton, 1987); however, three pairs of head neurons found in *A. suum* have no clear homologs in *C. elegans* (Guastella *et al.*, 1991).

Taken together these data reveal that, while neuronal types seem relatively well conserved in the *Caenorhabditis* genus, there are important variations at greater evolutionary distances.

Variations in transcriptional regulation

The variations in neuronal cell fates observed between nematodes suggest that the transcriptional mechanisms that specify neuronal identity have changed. In addition, even if neuronal types are conserved, the detailed transcriptional regulation of their identity may have evolved. These variations can happen at the level of the *trans*-acting factors, such as transcription factors or miRNA, and at the level of *cis*-regulatory regions (Figure 3(B)). Strikingly, several *trans* regulators of neuronal fate were found to vary across nematode species. The miRNA *lsy-6*, triggering the left-right asymmetry of ASE neurons in *C. elegans*, is absent from the genome of *P. pacificus* (Ahmed *et al.*, 2013). The transcription factor *che-1* involved in specifying ASE fate and restricted to that cell pair in *C. elegans*, is expressed in an additional pair of neurons, homolog of ASG, in *P. pacificus* (Hong *et al.*, 2019). For the nuclear receptor *odr-7*, involved in specifying AWA neuron cilia and expressed exclusively in that neuron in *C. elegans*, the *P. pacificus* ortholog is expressed in the homologs of AWC and ADF neurons, but not in the homolog of AWA (Hong *et al.*, 2019). The downstream terminal differentiation gene *odr-3* (a G-protein subunit) also diverged extensively in its expression compared with *C. elegans* (Hong *et al.*, 2019).

In addition to variations at the level of transcription factors, changes are also observed at the level of *cis*-regulatory elements. One tool of choice to study *cis*-regulatory changes between *C. elegans* and *C. briggsae*, which have one-to-one neuron homology, is to swap transcriptional reporters between species (Gordon & Ruvinsky, 2012; Figure 3(B)). If the regulatory mechanisms are conserved, the expression pattern would stay the same; if changes happened in *cis*-regulatory elements or in *trans* factors, the expression patterns driven by the reporters would vary. For example, in *C. elegans*, the transcription factor CEH-6 regulates *lin-11* expression in the interneuron RIC by binding to an intronic enhancer; in *C. briggsae*, the endogenous *lin-11* enhancer fails to drive expression in RIC, while reciprocal promoter swaps do drive expression in RIC (Amon & Gupta, 2017). Interestingly, an independent comparative RNAi screen found evidence for functional divergence of the CEH-6 transcription factor between *C. elegans* and *C. briggsae* (Verster, Ramani, McKay, & Fraser, 2014). This illustrates one example of functional divergence both in *cis* and in *trans* in the transcriptional regulation of neuronal fate. Over the years, different teams have performed promoter swaps for multiple genes, including some genes involved in regulation of neuronal fates. While for some genes like the transcription factors *ceh-24* (Harfe & Fire, 1998), *pag-3* (Aamodt *et al.*, 2000), *mec-3* (Xue, Finney, Ruvkun, & Chalfie, 1992), and *odr-7* (Colosimo, Tran, & Sengupta, 2003) the regulatory mechanisms appear to be conserved between *C. elegans* and *C. briggsae*, they diverged for others: regulation of the chemoreceptor *srsx-3* and of its putative transcription factor *lim-4* diverged between *C. elegans* and *C. briggsae* (Nokes *et al.*, 2009); the *C. briggsae* promoter of *gcy-5*, a chemoreceptor, does not drive expression in ASE neurons while its *C. elegans* ortholog does (Etchberger, Flowers, Poole, Bashllari, & Hobert, 2009); and the *C. briggsae* promoter of another chemoreceptor *odr-10* lacks a repressive UNC-3 binding site present in its *C. elegans* ortholog, and drives expression in one extra neuron pair (Kim, Colosimo, Yeung, & Sengupta, 2005). A detailed study of the promoter of the vesicular GABA transporter *unc-47* also revealed compensatory coevolution of *cis* and *trans* regulatory factors between *C. elegans* and *C. briggsae* (Barriere, Gordon, & Ruvinsky, 2012). More generally, this trend of cryptic changes in transcriptional regulation was found to be pervasive, when testing multiple neuronal genes from several *Caenorhabditis* species in the *C. elegans trans*-regulatory context (Barriere & Ruvinsky, 2014). Even though the final expression pattern of many genes may be conserved within *Caenorhabditis* nematodes, in accordance with conservation of fine anatomy and neurotransmitter neuronal fates, these promoter swaps reveal extensive developmental systems drift (True & Haag, 2001): while purifying selection maintains the final phenotype, the molecular mechanisms underlying its specification are less constrained and are able to diverge, provided that the output stays constant. *Cis*-regulatory elements diverged over time, but coevolved with their upstream transcription factors to maintain their outputs.

In conclusions, several scenarios of transcriptional evolution between *C. elegans* and *C. briggsae* are observed (Figure 3(B)): conservation of transcription factors and *cis*-regulatory elements (e.g. *ceh-24*); coevolution of *trans*-acting factors and *cis*-regulatory elements to maintain the final expression pattern (e.g. *unc-47*); and changes in regulatory mechanisms inducing variations in the final expression pattern (e.g. *lin-11*).

Beyond nematodes

Cis-regulatory elements involved in neuronal specification are usually not conserved over greater evolutionary distances, between phyla (Ruvinsky & Ruvkun, 2003). However, some degree of functional conservation can be observed at the level of the transcription factors that regulate neuronal cell fate acquisition. For example, proneural bHLH transcription factors (such as Achaete-scute or NeuroD family members) that play an important role in the early steps of neuronal specification in *C. elegans* (Frank, Baum, & Garriga, 2003; Hallam, Singer, Waring, & Jin, 2000; Krause *et al.*, 1997; Murgan *et al.*, 2015; Poole *et al.*, 2011), are also at the top of the hierarchy in the neuronal determination process in other animals (Baker & Brown, 2018). Conservation is also found at downstream steps. For example, in *C. elegans*, the dopamine pathway genes are activated by the Ets transcription factor AST-1 and the homeodomain transcription factor CEH-43 (Doitsidou *et al.*, 2013; Flames & Hobert, 2009), a role conserved in mice where their orthologs, Etv1 and Dlx2, are activators of the dopaminergic fate, specifically in the olfactory bulb. In addition, it has been recently observed that the Zic transcription factor REF-2 represses dopaminergic fate in *C. elegans*, a function conserved in the mouse where Zic1 and Zic2 repress the dopaminergic phenotype in the olfactory bulb (Tiveron *et al.*, 2017). Another interesting example is the role of COE transcription factors in cholinergic motor neuron differentiation. In *C. elegans*, the COE transcription factor UNC-3 regulates the expression of many terminal differentiation genes of cholinergic motor neurons including the cholinergic gene battery (Kratsios *et al.*, 2011). This role is conserved in the ascidian *Ciona intestinalis*, where the sole COE transcription factor controls the cholinergic fate of motor neurons (Kratsios *et al.*, 2011). However, while the COE transcription factor Ebf2 also regulates some aspects of axial motor neuron development in mice, it does not control their cholinergic fate (Catela *et al.*, 2019), suggesting both conservation and divergence in COE transcription factor functions. Some degree of conservation can also be observed over long evolutionary distances at cellular and signaling levels. For example, asymmetric divisions are widely used in diverse metazoans to generate neuronal diversity (Hartenstein & Stollewerk, 2015). In addition, the Wnt pathway, which controls neuronal progenitors asymmetric divisions in *C. elegans*, also plays a role in the regulation of neural stem cells asymmetric divisions in the mouse cortex (Chenn & Walsh, 2002; Delaunay, Cortay, Patti, Knoblauch, & Dehay, 2014; Kalani *et al.*, 2008; Woodhead, Mutch, Olson, & Chenn, 2006).

Conclusions

Since the seminal works of Sydney Brenner and John Sulston, decades of studies of nervous system development in *C. elegans* have established that neuronal diversity is generated via the combination of two types of information: lineage history (carried by inherited transcription factors) and intercellular signaling (especially Wnt and Notch signaling). Neuronal specification in *C. elegans* is therefore a mix between the lineage-based 'European' way and the interaction-based 'American' way. This information is integrated at the level of terminal selector transcription factors that activate and maintain the expression of large batteries of terminal differentiation genes in postmitotic neurons, and are therefore responsible for the acquisition and maintenance of neuronal type identities. While neuronal types are relatively well conserved within *Caenorhabditis* nematodes, there is more variability in distant nematodes. The evolution of neuronal specification involves changes at both the *cis*-regulatory element and *trans*-acting factor levels. However, some aspects of neuronal specification mechanisms are also conserved over large evolutionary distances. There is no doubt that, in the future, studies of nematodes, which offer an amazing level of cellular and molecular resolutions, will continue to deepen our knowledge of the mechanisms at the basis of nervous system development and evolution.

Acknowledgements

This review is part of a series of articles commemorating the contributions of Sydney Brenner and John Sulston to neuroscience.

Disclosure statement

The authors report no conflict of interest.

Funding

Work in our laboratory is funded by the Agence Nationale de la Recherche [ANR-14-CE11-0001, ANR-11-LABX-0054 and ANR-17-ERC2-0018] and the Fondation pour la Recherche Médicale [DEQ20180339160].

ORCID

Antoine Barrière http://orcid.org/0000-0001-8566-417X
Vincent Bertrand http://orcid.org/0000-0002-9036-2544

References

Aamodt, E., Shen, L., Marra, M., Schein, J., Rose, B., & McDermott, J.B. (2000). Conservation of sequence and function of the pag-3 genes from *C. elegans* and *C. briggsae*. *Gene*, *243*(1–2), 67–74. doi:10.1016/S0378-1119(99)00560-0

Ahmed, R., Chang, Z., Younis, A.E., Langnick, C., Li, N., Chen, W., … Dieterich, C. (2013). Conserved miRNAs are candidate post-transcriptional regulators of developmental arrest in free-living and parasitic nematodes. *Genome Biology and Evolution*, *5*(7), 1246–1260. doi:10.1093/gbe/evt086

Amon, S., & Gupta, B.P. (2017). Intron-specific patterns of divergence of lin-11 regulatory function in the *C. elegans* nervous system. *Developmental Biology*, *424*(1), 90–103. doi:10.1016/j.ydbio.2017.02.005

Arata, Y., Kouike, H., Zhang, Y., Herman, M.A., Okano, H., & Sawa, H. (2006). Wnt signaling and a Hox protein cooperatively regulate psa-3/Meis to determine daughter cell fate after asymmetric cell division in *C. elegans*. *Developmental Cell*, *11*(1), 105–115. doi:10.1016/j.devcel.2006.04.020

Baker, N.E., & Brown, N.L. (2018). All in the family: Proneural bHLH genes and neuronal diversity. *Development*, *145*(9), dev159426. doi:10.1242/dev.159426

Barriere, A., Gordon, K.L., & Ruvinsky, I. (2012). Coevolution within and between regulatory loci can preserve promoter function despite evolutionary rate acceleration. *PLoS Genetics*, *8*(9), e1002961. doi:10.1371/journal.pgen.1002961

Barriere, A., & Ruvinsky, I. (2014). Pervasive divergence of transcriptional gene regulation in Caenorhabditis nematodes. *PLoS Genetics*, *10*(6), e1004435. doi:10.1371/journal.pgen.1004435

Bertrand, V. (2016). β-Catenin-driven binary cell fate decisions in animal development. *Wiley Interdisciplinary Reviews-Developmental Biology*, *5*(3), 377–388. doi:10.1002/wdev.228

Bertrand, V., Bisso, P., Poole, R.J., & Hobert, O. (2011). Notch-dependent induction of left/right asymmetry in *C. elegans* interneurons and motoneurons. *Current Biology*, *21*(14), 1225–1231. doi:10.1016/j.cub.2011.06.016

Bertrand, V., & Hobert, O. (2009). Linking asymmetric cell division to the terminal differentiation program of postmitotic neurons in *C. elegans*. *Developmental Cell*, *16*(4), 563–575. doi:10.1016/j.devcel.2009.02.011

Brenner, S. (1974). The genetics of *Caenorhabditis elegans*. *Genetics*, *77*(1), 71–94.

Brown, A. (2003). *In the beginning was the worm*. New York: Columbia University Press.

Bumbarger, D.J., Riebesell, M., Rodelsperger, C., & Sommer, R.J. (2013). System-wide rewiring underlies behavioral differences in predatory and bacterial-feeding nematodes. *Cell*, *152*(1–2), 109–119. doi:10.1016/j.cell.2012.12.013

Catela, C., Correa, E., Wen, K., Aburas, J., Croci, L., Consalez, G.G., & Kratsios, P. (2019). An ancient role for collier/Olf/Ebf (COE)-type transcription factors in axial motor neuron development. *Neural Development*, *14*(1), 2. doi:10.1186/s13064-018-0125-6

Chenn, A., & Walsh, C.A. (2002). Regulation of cerebral cortical size by control of cell cycle exit in neural precursors. *Science (New York, N.Y.)*, *297*(5580), 365–369. doi:10.1126/science.1074192

Colosimo, M.E., Tran, S., & Sengupta, P. (2003). The divergent orphan nuclear receptor ODR-7 regulates olfactory neuron gene expression via multiple mechanisms in *Caenorhabditis elegans*. *Genetics*, *165*(4), 1779–1791.

Delaunay, D., Cortay, V., Patti, D., Knoblauch, K., & Dehay, C. (2014). Mitotic spindle asymmetry: A Wnt/PCP-regulated mechanism generating asymmetrical division in cortical precursors. *Cell Reports*, *6*(2), 400–414. doi:10.1016/j.celrep.2013.12.026

Doitsidou, M., Flames, N., Topalidou, I., Abe, N., Felton, T., Remesal, L., … Hobert, O. (2013). A combinatorial regulatory signature controls terminal differentiation of the dopaminergic nervous system in *C. elegans*. *Genes & Development*, *27*(12), 1391–1405. doi:10.1101/gad.217224.113

Etchberger, J.F., Flowers, E.B., Poole, R.J., Bashllari, E., & Hobert, O. (2009). Cis-regulatory mechanisms of left/right asymmetric neuron-subtype specification in *C. elegans*. *Development (Cambridge, England)*, *136*(1), 147–160. doi:10.1242/dev.030064

Etchberger, J.F., Lorch, A., Sleumer, M.C., Zapf, R., Jones, S.J., Marra, M.A., … Hobert, O. (2007). The molecular signature and cis-regulatory architecture of a *C. elegans* gustatory neuron. *Genes & Development*, *21*(13), 1653–1674. doi:10.1101/gad.1560107

Flames, N., & Hobert, O. (2009). Gene regulatory logic of dopamine neuron differentiation. *Nature*, *458*(7240), 885–889. doi:10.1038/nature07929

Frank, C.A., Baum, P.D., & Garriga, G. (2003). HLH-14 is a *C. elegans* achaete-scute protein that promotes neurogenesis through asymmetric cell division. *Development (Cambridge, England)*, *130*(26), 6507–6518. doi:10.1242/dev.00894

Gendrel, M., Atlas, E.G., & Hobert, O. (2016). A cellular and regulatory map of the GABAergic nervous system of *C. elegans*. *eLife*, *5*, e17686. doi:10.7554/eLife.17686

Goldstein, B., Takeshita, H., Mizumoto, K., & Sawa, H. (2006). Wnt signals can function as positional cues in establishing cell polarity. *Developmental Cell*, *10*(3), 391–396. doi:10.1016/j.devcel.2005.12.016

Gordon, K.L., & Ruvinsky, I. (2012). Tempo and mode in evolution of transcriptional regulation. *PLoS Genetics*, *8*(1), e1002432. doi:10.1371/journal.pgen.1002432

Gordon, P.M., & Hobert, O. (2015). A competition mechanism for a homeotic neuron identity transformation in *C. elegans*. *Developmental Cell*, *34*(2), 206–219. doi:10.1016/j.devcel.2015.04.023

Guastella, J., Johnson, C.D., & Stretton, A.O. (1991). GABA-immunoreactive neurons in the nematode Ascaris. *Journal of Comparative Neurology*, *307*(4), 584–597. doi:10.1002/cne.903070406

Hallam, S., Singer, E., Waring, D., & Jin, Y. (2000). The *C. elegans* NeuroD homolog cnd-1 functions in multiple aspects of motor neuron fate specification. *Development*, *127*(19), 4239–4252.

Han, Z., Boas, S., & Schroeder, N.E. (2015). Unexpected variation in neuroanatomy among diverse nematode species. *Frontiers in Neuroanatomy*, *9*, 162. doi:10.3389/fnana.2015.00162

Harfe, B.D., & Fire, A. (1998). Muscle and nerve-specific regulation of a novel NK-2 class homeodomain factor in *Caenorhabditis elegans*. *Development (Cambridge, England)*, *125*(3), 421–429.

Hartenstein, V., & Stollewerk, A. (2015). The evolution of early neurogenesis. *Developmental Cell*, *32*(4), 390–407. doi:10.1016/j.devcel.2015.02.004

Hobert, O. (2008). Regulatory logic of neuronal diversity: Terminal selector genes and selector motifs. *Proceedings of the National Academy of Sciences of the United States of America*, *105*(51), 20067–20071. doi:10.1073/pnas.0806070105

Hobert, O. (2010). Neurogenesis in the nematode Caenorhabditis elegans. WormBook, 1–24. doi:10.1895/wormbook.1.12.2

Hobert, O. (2016). A map of terminal regulators of neuronal identity in *Caenorhabditis elegans. Wiley Interdisciplinary Reviews-Developmental Biology, 5*(4), 474–498. doi:10.1002/wdev.233

Hobert, O., Johnston, R.J., Jr., & Chang, S. (2002). Left-right asymmetry in the nervous system: The *Caenorhabditis elegans* model. *Nature Reviews. Neuroscience, 3*(8), 629–640. doi:10.1038/nrn897

Hong, R.L., Riebesell, M., Bumbarger, D.J., Cook, S.J., Carstensen, H.R., Sarpolaki, T., … Sommer, R.J. (2019). Evolution of neuronal anatomy and circuitry in two highly divergent nematode species. *eLife, 8*, e47155. doi:10.7554/eLife.47155

Houthoofd, W., Jacobsen, K., Mertens, C., Vangestel, S., Coomans, A., & Borgonie, G. (2003). Embryonic cell lineage of the marine nematode Pellioditis marina. *Developmental Biology, 258*(1), 57–69. doi:10.1016/S0012-1606(03)00101-5

Huang, S., Shetty, P., Robertson, S.M., & Lin, R. (2007). Binary cell fate specification during *C. elegans* embryogenesis driven by reiterated reciprocal asymmetry of TCF POP-1 and its coactivator beta-catenin SYS-1. *Development (Cambridge, England), 134*(14), 2685–2695. doi:10.1242/dev.008268

Johnson, C.D., Reinitz, C.A., Sithigorngul, P., & Stretton, A.O. (1996). Neuronal localization of serotonin in the nematode *Ascaris suum. The Journal of Comparative Neurology, 367*(3), 352–360. doi:10.1002/(SICI)1096-9861(19960408)367:3 < 352::AID-CNE3 > 3.0.CO;2-4

Johnson, C.D., & Stretton, A.O. (1987). GABA-immunoreactivity in inhibitory motor neurons of the nematode Ascaris. *The Journal of Neuroscience, 7*(1), 223–235. doi:10.1523/JNEUROSCI.07-01-00223.1987

Kalani, M.Y.S., Cheshier, S.H., Cord, B.J., Bababeygy, S.R., Vogel, H., Weissman, I.L., … Nusse, R. (2008). Wnt-mediated self-renewal of neural stem/progenitor cells. *Proceedings of the National Academy of Sciences of the United States of America, 105*(44), 16970–16975. doi:10.1073/pnas.0808616105

Kaletta, T., Schnabel, H., & Schnabel, R. (1997). Binary specification of the embryonic lineage in *Caenorhabditis elegans. Nature, 390*(6657), 294–298. doi:10.1038/36869

Kaur, S., Melenec, P., Murgan, S., Bordet, G., Recouvreux, P., Lenne, P.F., & Bertrand, V. (2020). Wnt ligands regulate the asymmetric divisions of neuronal progenitors in *C. elegans* embryos. *Development, 147*(7), dev183186. doi:10.1242/dev.183186

Kidd, A.R., 3rd, Miskowski, J.A., Siegfried, K.R., Sawa, H., & Kimble, J. (2005). A beta-catenin identified by functional rather than sequence criteria and its role in Wnt/MAPK signaling. *Cell, 121*(5), 761–772. doi:10.1016/j.cell.2005.03.029

Kim, K., Colosimo, M.E., Yeung, H., & Sengupta, P. (2005). The UNC-3 Olf/EBF protein represses alternate neuronal programs to specify chemosensory neuron identity. *Developmental Biology, 286*(1), 136–148. doi:10.1016/j.ydbio.2005.07.024

Kratsios, P., Stolfi, A., Levine, M., & Hobert, O. (2011). Coordinated regulation of cholinergic motor neuron traits through a conserved terminal selector gene. *Nature Neuroscience, 15*(2), 205–214. doi:10.1038/nn.2989

Krause, M., Park, M., Zhang, J.M., Yuan, J., Harfe, B., Xu, S.Q., … Fire, A. (1997). A *C. elegans* E/daughterless bHLH protein marks neuronal but not striated muscle development. *Development, 124*(11), 2179–2189.

Lin, R., Hill, R.J., & Priess, J.R. (1998). POP-1 and anterior-posterior fate decisions in *C. elegans* embryos. *Cell, 92*(2), 229–239. doi:10.1016/S0092-8674(00)80917-4

Lin, R., Thompson, S., & Priess, J.R. (1995). pop-1 encodes an HMG box protein required for the specification of a mesoderm precursor in early *C. elegans* embryos. *Cell, 83*(4), 599–609. doi:10.1016/0092-8674(95)90100-0

Lo, M.C., Gay, F., Odom, R., Shi, Y., & Lin, R. (2004). Phosphorylation by the beta-catenin/MAPK complex promotes 14-3-3-mediated nuclear export of TCF/POP-1 in signal-responsive cells in *C. elegans. Cell, 117*(1), 95–106. doi:10.1016/S0092-8674(04)00203-X

Memar, N., Schiemann, S., Hennig, C., Findeis, D., Conradt, B., & Schnabel, R. (2019). Twenty million years of evolution: The embryogenesis of four Caenorhabditis species are indistinguishable despite extensive genome divergence. *Developmental Biology, 447*(2), 182–199. doi:10.1016/j.ydbio.2018.12.022

Murgan, S., Kari, W., Rothbächer, U., Iché-Torres, M., Mélénec, P., Hobert, O., & Bertrand, V. (2015). Atypical transcriptional activation by TCF via a Zic transcription factor in *C. elegans* neuronal precursors. *Developmental Cell, 33*(6), 737–745. doi:10.1016/j.devcel.2015.04.018

Nokes, E.B., Van Der Linden, A.M., Winslow, C., Mukhopadhyay, S., Ma, K., & Sengupta, P. (2009). Cis-regulatory mechanisms of gene expression in an olfactory neuron type in *Caenorhabditis elegans. Developmental Dynamics, 238*(12), 3080–3092. doi:10.1002/dvdy.22147

Packer, J.S., Zhu, Q., Huynh, C., Sivaramakrishnan, P., Preston, E., Dueck, H., … Murray, J.I. (2019). A lineage-resolved molecular atlas of *C. elegans* embryogenesis at single-cell resolution. *Science, 365*(6459), eaax1971. doi:10.1126/science.aax1971

Pereira, L., Kratsios, P., Serrano-Saiz, E., Sheftel, H., Mayo, A.E., Hall, D.H., … Hobert, O. (2015). A cellular and regulatory map of the cholinergic nervous system of *C. elegans. eLife, 4*, e12432. doi:10.7554/eLife.12432

Phillips, B.T., Kidd, A.R., 3rd, King, R., Hardin, J., & Kimble, J. (2007). Reciprocal asymmetry of SYS-1/beta-catenin and POP-1/TCF controls asymmetric divisions in *Caenorhabditis elegans. Proceedings of the National Academy of Sciences of the United States of America, 104*(9), 3231–3236. doi:10.1073/pnas.0611507104

Phillips, B.T., & Kimble, J. (2009). A new look at TCF and beta-catenin through the lens of a divergent *C. elegans* Wnt pathway. *Developmental Cell, 17*(1), 27–34. doi:10.1016/j.devcel.2009.07.002

Pierce-Shimomura, J.T., Faumont, S., Gaston, M.R., Pearson, B.J., & Lockery, S.R. (2001). The homeobox gene lim-6 is required for distinct chemosensory representations in *C. elegans. Nature, 410*(6829), 694–698. doi:10.1038/35070575

Poole, R.J., Bashllari, E., Cochella, L., Flowers, E.B., & Hobert, O. (2011). A genome-wide RNAi screen for factors involved in neuronal specification in *Caenorhabditis elegans. PLoS Genetics, 7*(6), e1002109. doi:10.1371/journal.pgen.1002109

Poole, R.J., & Hobert, O. (2006). Early embryonic programming of neuronal left/right asymmetry in *C. elegans. Current Biology, 16*(23), 2279–2292. doi:10.1016/j.cub.2006.09.041

Rao, V.T., Forrester, S.G., Keller, K., & Prichard, R.K. (2011). Localisation of serotonin and dopamine in Haemonchus contortus. International Journal for Parasitology, *41*(2), 249–254. doi:10.1016/j.ijpara.2010.09.002

Rivard, L., Srinivasan, J., Stone, A., Ochoa, S., Sternberg, P.W., & Loer, C.M. (2010). A comparison of experience-dependent locomotory behaviors and biogenic amine neurons in nematode relatives of *Caenorhabditis elegans. BMC Neuroscience, 11*, 22. doi:10.1186/1471-2202-11-22

Rocheleau, C.E., Yasuda, J., Shin, T.H., Lin, R., Sawa, H., Okano, H., … Mello, C.C. (1999). WRM-1 activates the LIT-1 protein kinase to transduce anterior/posterior polarity signals in *C. elegans. Cell, 97*(6), 717–726. doi:10.1016/S0092-8674(00)80784-9

Ruvinsky, I., & Ruvkun, G. (2003). Functional tests of enhancer conservation between distantly related species. *Development (Cambridge, England), 130*(21), 5133–5142. doi:10.1242/dev.00711

Sarafi-Reinach, T.R., Melkman, T., Hobert, O., & Sengupta, P. (2001). The Lin-11 LIM homeobox gene specifies olfactory and chemosensory neuron fates in C. elegans. *Development, 128*(17), 3269–3281.

Sawa, H., & Korswagen, H.C. (2013). *Wnt signaling in C. elegans.* WormBook, 1–30. doi:10.1895/wormbook.1.7.2

Schulze, J., & Schierenberg, E. (2009). Embryogenesis of *Romanomermis culicivorax*: An alternative way to construct a nematode. *Developmental Biology, 334*(1), 10–21. doi:10.1016/j.ydbio.2009.06.009

Serrano-Saiz, E., Poole, R.J., Felton, T., Zhang, F., De La Cruz, E.D., & Hobert, O. (2013). Modular control of glutamatergic neuronal identity in *C. elegans* by distinct homeodomain proteins. *Cell, 155*(3), 659–673. doi:10.1016/j.cell.2013.09.052

Stretton, A., Donmoyer, J., Davis, R., Meade, J., Cowden, C., & Sithigorngul, P. (1992). Motor behavior and motor nervous system function in the nematode *Ascaris suum*. *The Journal of Parasitology*, *78*(2), 206–214. doi:10.2307/3283468

Sulston, J.E., Dew, M., & Brenner, S. (1975). Dopaminergic neurons in the nematode *Caenorhabditis elegans*. Journal of Comparative Neurology, *163*(2), 215–226. doi:10.1002/cne.901630207

Sulston, J.E., & Horvitz, H.R. (1977). Post-embryonic cell lineages of the nematode, *Caenorhabditis elegans*. *Developmental Biology*, *56*(1), 110–156. doi:10.1016/0012-1606(77)90158-0

Sulston, J.E., Schierenberg, E., White, J.G., & Thomson, J.N. (1983). The embryonic cell lineage of the nematode *Caenorhabditis elegans*. *Developmental Biology*, *100*(1), 64–119. doi:10.1016/0012-1606(83)90201-4

Tiveron, M.-C., Beclin, C., Murgan, S., Wild, S., Angelova, A., Marc, J., … Cremer, H. (2017). Zic-proteins are repressors of dopaminergic forebrain fate in mice and *C. elegans*. *The Journal of Neuroscience*, *37*(44), 10611–10623. doi:10.1523/JNEUROSCI.3888-16.2017

Troemel, E.R., Sagasti, A., & Bargmann, C.I. (1999). Lateral signaling mediated by axon contact and calcium entry regulates asymmetric odorant receptor expression in *C. elegans*. *Cell*, *99*(4), 387–398. doi:10.1016/S0092-8674(00)81525-1

True, J.R., & Haag, E.S. (2001). Developmental system drift and flexibility in evolutionary trajectories. *Evolution & Development*, *3*(2), 109–119. doi:10.1046/j.1525-142x.2001.003002109.x

Verster, A.J., Ramani, A.K., McKay, S.J., & Fraser, A.G. (2014). Comparative RNAi screens in *C. elegans* and *C. briggsae* reveal the impact of developmental system drift on gene function. *PLoS Genetics*, *10*(2), e1004077. doi:10.1371/journal.pgen.1004077

Wenick, A.S., & Hobert, O. (2004). Genomic cis-regulatory architecture and trans-acting regulators of a single interneuron-specific gene battery in *C. elegans*. *Developmental Cell*, *6*(6), 757–770. doi:10.1016/j.devcel.2004.05.004

White, J.G., Southgate, E., Thomson, J.N., & Brenner, S. (1986). The structure of the nervous system of the nematode *Caenorhabditis elegans*. Philosophical Transactions of the Royal Society B: Biological Sciences, *314*(1165), 1–340. doi:10.1098/rstb.1986.0056

Woodhead, G.J., Mutch, C.A., Olson, E.C., & Chenn, A. (2006). Cell-autonomous beta-catenin signaling regulates cortical precursor proliferation. *The Journal of Neuroscience*, *26*(48), 12620–12630. doi:10.1523/JNEUROSCI.3180-06.2006

Xue, D., Finney, M., Ruvkun, G., & Chalfie, M. (1992). Regulation of the mec-3 gene by the *C. elegans* homeoproteins UNC-86 and MEC-3. *The EMBO Journal*, *11*(13), 4969–4979. doi:10.1002/j.1460-2075.1992.tb05604.x

Yu, S., Avery, L., Baude, E., & Garbers, D.L. (1997). Guanylyl cyclase expression in specific sensory neurons: A new family of chemosensory receptors. *Proceedings of the National Academy of Sciences of the United States of America*, *94*(7), 3384–3387. doi:10.1073/pnas.94.7.3384

Zhang, Y., Ma, C., Delohery, T., Nasipak, B., Foat, B.C., Bounoutas, A., … Chalfie, M. (2002). Identification of genes expressed in *C. elegans* touch receptor neurons. *Nature*, *418*(6895), 331–335. doi:10.1038/nature00891

Zhao, Z., Boyle, T.J., Bao, Z., Murray, J.I., Mericle, B., & Waterston, R.H. (2008). Comparative analysis of embryonic cell lineage between *Caenorhabditis briggsae* and *Caenorhabditis elegans*. *Developmental Biology*, *314*(1), 93–99. doi:10.1016/j.ydbio.2007.11.015

Molecular mechanisms governing axonal transport: a *C. elegans* perspective

Amruta Vasudevan and Sandhya P. Koushika

ABSTRACT
Axonal transport is integral for maintaining neuronal form and function, and defects in axonal transport have been correlated with several neurological diseases, making it a subject of extensive research over the past several years. The anterograde and retrograde transport machineries are crucial for the delivery and distribution of several cytoskeletal elements, growth factors, organelles and other synaptic cargo. Molecular motors and the neuronal cytoskeleton function as effectors for multiple neuronal processes such as axon outgrowth and synapse formation. This review examines the molecular mechanisms governing axonal transport, specifically highlighting the contribution of studies conducted in *C. elegans*, which has proved to be a tractable model system in which to identify both novel and conserved regulatory mechanisms of axonal transport.

Introduction

Axons are often the longest processes emerging from neuronal cell bodies, known to extend up to several metres in length for vertebrate neurons (Campenot & Eng, 2000). With most protein synthesis occurring in neuronal cell bodies, mechanisms for transport of material from the soma to the tips of axons and dendrites are essential (Morfini, Burns, Stenoien, & Brady, 2012). The earliest known experimental evidence for the existence of axonal transport comes from seminal work in the sciatic nerve by Weiss and Hiscoe (1948), wherein axoplasmic material accumulated at sites of surgical constriction that moved along the axon at a rate of ~1–2mm/day upon removal of the constriction (Morfini *et al.*, 2012; Weiss & Hiscoe, 1948). Subsequently, several studies examined the nature of axoplasmic transport, employing radioactive tracers to label proteins, lipids, and sterol content to monitor their flow through the axon (Morfini *et al.*, 2012). They reported that material was transported in both anterograde and retrograde directions, at different rates. These experiments identified two predominant "modes" of transport along neuronal processes; i) "Fast axonal transport" occurring at ~250–400mm/day, which is discussed in greater detail in the subsequent sections, and ii) "Slow axonal transport" shown to be ~1mm/day (Morfini *et al.*, 2012). Discovery of the molecular motors responsible for anterograde and retrograde transport (Maday, Twelvetrees, Moughamian, & Holzbaur, 2014), followed by their detailed structural characterization, has greatly improved the understanding of transport mechanisms employed by cells. It is now widely known that the Kinesin and Dynein family of motors transport intracellular organelles along microtubules in the anterograde and retrograde directions respectively (Gennerich & Vale, 2009). While several of these characterization studies have been conducted in cultured neurons *ex vivo*, *C. elegans* has proved to be an invaluable model system in both the elucidation of novel molecular mechanisms governing axonal transport, as well as molecular characterization of individual proteins involved. *C. elegans* is a ~1mm long, free-living, non-parasitic nematode, with a short life cycle (3 days from egg to adult at 25 °C), which makes it an attractive model system in which to conduct forward genetic screens (Brenner, 1974). One of its strengths is its transparency, that allows for the application of live imaging techniques in intact animals. In this review, we discuss the principles of axonal transport and highlight contributions made using *C. elegans*, which address key questions in this area.

General models of fast axonal transport

Microtubule-dependent motor proteins responsible for most fast axonal transport in neurons largely belong to the Kinesin or Dynein superfamily (Morfini *et al.*, 2012). Kinesins are ATPases that walk towards the plus ends of microtubules in a hand-over-hand motion, with each motor head taking 16 nm steps for every molecule of ATP hydrolysed (Gennerich & Vale, 2009). Cytoplasmic dynein, a member of the AAA family of ATPases, drives transport towards the minus ends of microtubules, using an inchworm-like movement with occasional hand-over-hand mode of stepping (Bhabha, Johnson, Schroeder, & Vale, 2016; Gennerich & Vale, 2009). Studies on intracellular transport across diverse cell types have revealed common underlying principles governing microtubule-based transport, such as i) cargo-specific mechanisms of motor recruitment and transport, ii) interactions between multiple motors on the cargo surface, and iii) navigation of the cargo-motor complex through obstacles. Several of these principles have been found to apply to axonal transport. Neurons, being polarized cells with distinct axonal and dendritic compartments, additionally exhibit region-specific regulation of cargo transport. These principles are discussed below.

Cargo-specific mechanisms of motor recruitment and transport

Neurons have diverse cargo such as synaptic vesicles, mitochondria, endosomes, peroxisomes, lysosomes, and autophagosomes. These cargoes exhibit distinct mechanisms of transport regulation, often linked to their unique functions within neurons. The regulation of cargo transport occurs through a) initiation of cargo motility, b) processivity of movement, or c) halting of motion at specific locations. Signalling endosomes have been shown to trigger Dynein-mediated retrograde transport upon activation and endocytosis of Trk receptors (Maday *et al.*, 2014). Synaptic vesicles, which often have to travel long distances to be delivered to distal presynaptic sites, regulate their processive motility by facilitating dimerization of the anterograde motor UNC-104/Kinesin-3 on the cargo surface (Klopfenstein, Tomishige, Stuurman, & Vale, 2002; Klopfenstein & Vale, 2004). Mitochondria, which have known roles in calcium homeostasis, are immobilized in axons at regions of high local calcium concentration through Miro, a calcium binding mitochondrial Rho GTPase that functions to recruit Kinesin-1 to the mitochondrial outer membrane (Guo *et al.*, 2005). These studies collectively suggest that different neuronal cargo have evolved distinct cargo-specific mechanisms of transport regulation to promote their respective functions. Molecular mechanisms underlying cargo-specific regulation of transport have been covered extensively in the following review (Maday *et al.*, 2014).

Interactions between multiple motors on the cargo surface

Intracellular cargoes are typically transported by multiple motors (Mallik & Gross, 2009). An individual cargo can be transported by multiple identical motors, different anterograde motors, or multiple opposing motors present on its surface. The ensemble of motors on a given cargo surface determine its motility characteristics. Presence of multiple identical motors has been shown to increase motor stall force, run length, cargo velocity and processivity of cargo motion *in vitro* (Gross, Vershinin, & Shubeita, 2007; Holzbaur & Goldman, 2010). Recruitment of multiple types of anterograde motors to the cargo surface can affect the velocity of cargo transport *in vivo* (Prevo, Scholey, & Peterman, 2017). *In vitro* studies also show that cargo transported by a combination of fast and slow motors display speeds intermediate between the speeds of the individual motors, and suggest that cargo transport velocity is dominated by the slower motors (Holzbaur & Goldman, 2010).

Several neuronal cargoes such as mitochondria, endosomes and lysosomes exhibit bidirectional transport along axons (Maday *et al.*, 2014). *In vitro* studies have proposed mechanisms by which opposing motors present on the cargo surface give rise to bidirectional transport behaviour (Hancock, 2018). The "tug-of-war" model proposes that opposing motors compete with each other to determine the net direction of cargo transport, while the "motor coordination model" proposes that regulators on the cargo surface ensure that motors for one direction of transport do not interact with microtubules when motors for the opposite direction are engaged (Hancock, 2018). Opposing motors have also been proposed to function cooperatively and promote transport in both directions (Brady, Pfister, & Bloom, 1990). Studies conducted in *C. elegans* neurons have helped uncover mechanisms of crosstalk between opposing motors, and are discussed in detail in subsequent sections.

Navigation of the cargo-motor complex through obstacles

The cargo-motor complex encounters numerous obstacles during the course of transport, in the form of molecular crowding or complex cytoskeletal geometries, leading to stalling of cargo (Holzbaur & Goldman, 2010; Sabharwal & Koushika, 2019). Crowding agents such microtubule-associated proteins can reduce the velocity and processivity of the cargo-motor complex (Vershinin, Carter, Razafsky, King, & Gross, 2007). Stalled cargo, actin-rich regions, and cytoskeletal intersections also pose as physical obstacles to cargo transport *in* vitro and *in vivo* (Ross, Shuman, Holzbaur, & Goldman, 2008; Sabharwal & Koushika, 2019; Schroeder, Mitchell, Shuman, Holzbaur, & Goldman, 2010). *In vitro* studies further suggest that levels of motors on the cargo surface determine how cargo traverse such obstacles (Ross *et al.*, 2008; Schroeder *et al.*, 2010), which have significant implications for axonal transport.

Region-specific regulation of cargo transport

Different regions of neurons employ distinct mechanisms of regulation of cargo transport. Axons and dendrites typically differ from each other in the polarity orientation of the underlying microtubules. Axons have microtubules laid out with their plus-ends towards growth cones while dendrites display mixed or minus-end-out orientation of microtubules (Rao & Baas, 2018). The polarity orientation of microtubules, and differential distribution of post-translational modifications among opposite polarity microtubules, inhibits transport of dendritic cargoes in axons and ensures removal of axonal cargoes from dendrites (Tas *et al.*, 2017). Subcompartments of axons, such as the axon initial segment (AIS), the distal tip and presynaptic sites, also exhibit distinct mechanisms of regulation of cargo transport. The AIS can specifically inhibit anterograde transport of dendritic cargo and trigger dynein-mediated retrograde transport back to dendrites (Leterrier, 2018). Distal tips of axons can efficiently activate dynein-mediated retrograde transport through microtubule plus-end binding proteins (Maday *et al.*, 2014), while presynaptic sites can either promote or inhibit cargo transport through neuronal activity (Qu, Kumar, Blockus, Waites, & Bartolini, 2019; Shakiryanova, Tully, & Levitan, 2006).

Regulators of anterograde axonal transport

In neurons, members of the Kinesin-1, Kinesin-2 and Kinesin-3 families significantly contribute to the anterograde axonal transport of neuronal cargo (Maday *et al.*, 2014). Several *C. elegans* studies have contributed to the discovery of Kinesins and Kinesin-like motor proteins *in vivo* (Otsuka *et al.*, 1991; Siddiqui, 2002). Techniques such as live fluorescence imaging and unbiased genetic screens have contributed to the elucidation of mechanisms governing anterograde axonal transport in *C. elegans* (Allen *et al.*, 2008; Brenner, 1974; Byrd *et al.*, 2001; Culotti & Russell, 1978; Kumar *et al.*, 2010; Mondal, Ahlawat, Rau, Venkataraman, & Koushika, 2011; Murthy, Bhat, & Koushika, 2011; Shakir, Fukushige, Yasuda, Miwa, & Siddiqui, 1993).

In the late 1980s, most of the Kinesin proteins identified and characterized across species, were found to regulate chromosome translocation and spindle formation during mitosis and meiosis (Endow, 1991), and the role of kinesins in *in vivo* anterograde axonal transport was poorly understood. Early contribution of studies in the nematode *C. elegans* has been the mapping and characterization of mutants isolated from forward genetic screens, several alleles of which were isolated by Sydney Brenner (1974). Studies conducted on recessive *unc-104* alleles, one isolated from a forward genetic screen for uncoordinated locomotion in *C. elegans* (Brenner, 1974), revealed that *unc-104* encoded a Kinesin family protein (Otsuka *et al.*, 1991). Ultrastructural analysis of neurons of an *unc-104* mutant isolated by Brenner (1974) revealed that the distribution of synaptic vesicles was severely perturbed, while the distribution of other membrane-bound organelles such as mitochondria and secretory granules was unaffected (Hall & Hedgecock, 1991). This study was the first to demonstrate that different kinesin paralogs can regulate the anterograde axonal transport of distinct intracellular organelles *in vivo*. UNC-104/KIF1A was later reported to mediate the anterograde transport of dense-core vesicles (DCVs) in *C. elegans* neurons (Goodwin, Sasaki, & Juo, 2012; Zahn *et al.*, 2004). UNC-104/IMAC in *D. melanogaster* (Pack-Chung, Kurshan, Dickman, & Schwarz, 2007), and KIF1A/Kinesin-3 in mammalian neurons (Aizawa *et al.*, 1992; Okada, Yamazaki, Sekine-Aizawa, & Hirokawa, 1995; Yonekawa *et al.*, 1998; Zhao *et al.*, 2001) were identified to regulate the anterograde transport of synaptic vesicles, suggesting that the role of the UNC-104/Kinesin-3 motor in anterograde transport of this cargo is conserved across multiple model systems.

The Kinesin-1 family members are known to mediate anterograde axonal transport of mitochondria (Mandal & Drerup, 2019). Analysis of other complementation groups isolated from Brenner's screen (Brenner, 1974) identified *unc-116* as the *C. elegans* homolog of mammalian Kinesin heavy chain (Patel, Thierry-Mieg, & Mancillas, 1993), and *vab-8* as a novel Kinesin-like motor regulating neuronal cell migration (Wolf, Hung, Wightman, Way, & Garriga, 1998). Subsequently, multiple Kinesin family motors were characterized and demonstrated to transport distinct cargo in other *C. elegans* cells, including ciliated sensory neurons in the head (Siddiqui, 2002).

In addition to the identification of neuronal Kinesin family motors, structural and molecular analysis of individual motors helped uncover additional mechanisms of regulation of transport in *C. elegans* neurons. The Pleckstrin Homology domain of UNC-104 was shown to be necessary for binding the cargo membrane (Klopfenstein & Vale, 2004), while its CC1-FHA domain was necessary for the dimerization and activation of UNC-104 *in vivo*, with implications for the anterograde transport of vesicles (Yue *et al.*, 2013). The latter study helped address a debate in the field as to whether UNC-104/Kinesin-3 functions as a monomer or dimer *in vivo* (Al-Bassam *et al.*, 2003; Kikkawa, Okada, & Hirokawa, 2000; Okada *et al.*, 1995; Tomishige, Klopfenstein, & Vale, 2002). While defects in the distribution and transport of Kinesin-3 cargo have been correlated to the progression of neurodegenerative diseases (Siddiqui & Straube, 2017), a study in *C. elegans* recently reported that mutations in *unc-104* associated with hereditary spastic paraplegia caused an increased association of UNC-104 with microtubules, an increase in its anterograde velocity and the anterograde transport of synaptic vesicles *in vivo* (Chiba *et al.*, 2019). These studies indicate that both a significant increase and decrease in UNC-104-mediated anterograde transport can perturb neuronal function and organismal behaviour. Anterograde motors are known to exist in autoinhibited states when not bound to microtubules or their respective cargo (Hammond, Blasius, Soppina, Cai, & Verhey, 2010; Hammond *et al.*, 2009). Similarly, in *C. elegans* motor neurons, UNC-104 is shown to exist in an autoinhibited state wherein its motor domain is bound to its stalk, thereby unable to interact with microtubules (Niwa *et al.*, 2016). This auto-inhibition is relieved by synaptic-vesicle-bound ARL-8, which suggests that the motility of UNC-104 is activated upon recruitment to cargo (Niwa *et al.*, 2016). It has earlier been proposed that Kinesin motors likely get degraded upon transport to nerve terminals (Dahlstrom, Pfister, & Brady, 1991; Li, Pfister, Brady, & Dahlström, 1999). Consistent with this hypothesis, UNC-104 has been shown to be degraded through a ubiquitin-dependent pathway, upon loss of specific binding to synaptic vesicles in *C. elegans* touch receptor neurons (Kumar *et al.*, 2010). This is the first *in vivo* study to demonstrate the degradation of Kinesin-3 family motors upon detachment from cargo, and has implications for cargo-specific regulation of the fate of motors (Figure 1(c)).

In vitro experiments report that an increase in the number of engaged motors significantly improves the run length of beads (Beeg *et al.*, 2008), indicating that factors governing

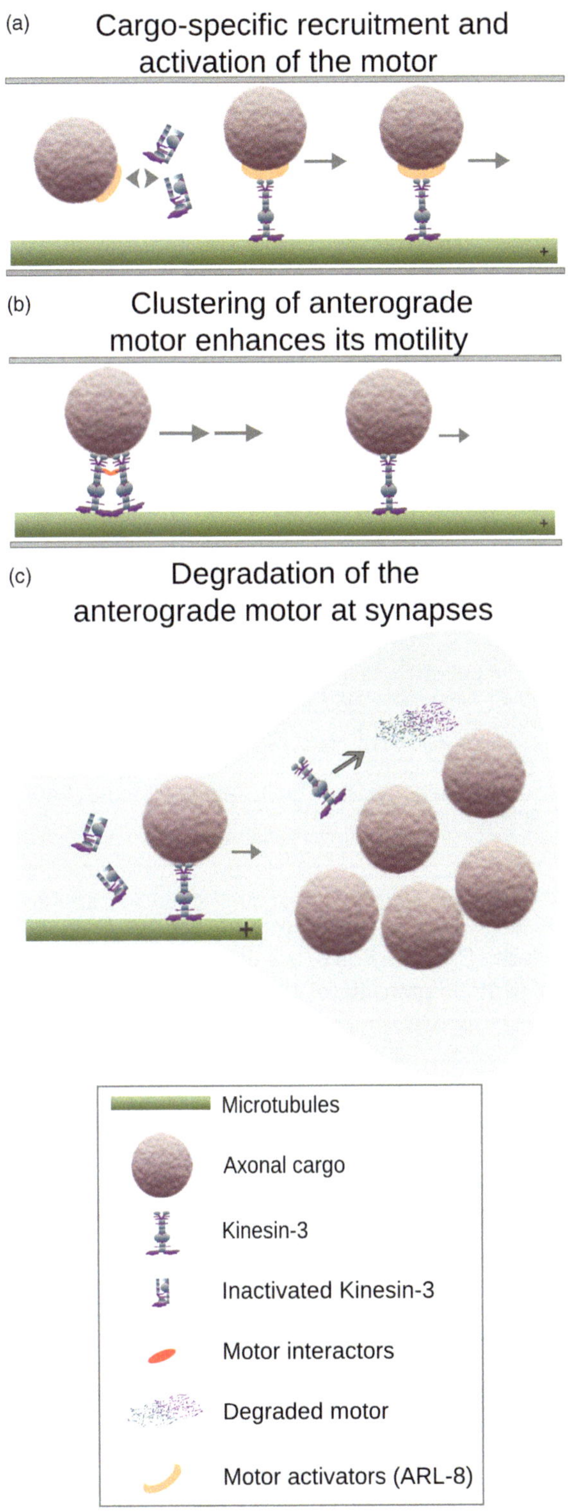

Figure 1. Mechanisms regulating anterograde axonal transport. a) Factors such as ARL-8 promote the activation and recruitment of the anterograde motor UNC-104/Kinesin-3 to synaptic vesicles. b) Factors known to interact with the anterograde motor UNC-104/Kinesin-3, such as SYD-2/Liprin-α, function to promote its motility *in vivo*. c) UNC-104/Kinesin-3 is degraded at synapses upon detachment from synaptic cargo.

motor recruitment and attachment to cargo surfaces can play a crucial role in regulating the distance traversed by cargo within neuronal processes. Several such factors have been identified from experiments conducted in *C. elegans* (Figure 1(a)). A list of several factors with known roles in anterograde axonal transport modulation have been included in Table 1.

Regulation of UNC-104/kinesin-3

RAB-3 (a member of the Rab family of small GTPases, involved in synaptic neurotransmission) is recruited to synaptic vesicles by AEX-3 (*C. elegans* ortholog of mammalian MADD family, a GEF for Rab GTPases), which was first isolated from a forward genetic screen in *C. elegans* for altered defecation behaviour (Iwasaki, Staunton, Saifee, Nonet, & Thomas, 1997; Thomas, 1990). AEX-3 thus regulates RAB-3 distribution, but not the distribution of other SV-associated proteins (Iwasaki *et al.*, 1997; Iwasaki & Toyonaga, 2000). Studies in cultured hippocampal neurons show that DENN/MADD functions to recruit the anterograde motors KIF1A and KIF1Bβ to vesicles carrying Rab3 (Niwa, Tanaka, & Hirokawa, 2008). Unlike the above mechanism, in *C. elegans*, ARL-8 (encodes a cargo-specific small GTPase) has been shown to recruit UNC-104 to synaptic vesicles (Klassen *et al.*, 2010), with both BLOS-9 (encodes the BORC subunit 8) (Niwa *et al.*, 2017) and SAM-4 (encodes a cargo-specific GEF) (Zheng *et al.*, 2014) regulating the interaction between UNC-104 and ARL-8. A novel role for active zone proteins SYD-2/Liprin-α (an active zone scaffolding protein) and UNC-10/RIM1 (an active zone protein, likely involved in priming of synaptic vesicles) in the recruitment of UNC-104 to RAB-3 containing synaptic vesicles, showed that they functioned redundantly with the motor's own PH domain in binding to axonal cargo (Bhan *et al.*, 2019). This observation provides a potential mechanism by which proteins involved in synapse formation or function may regulate motor recruitment and axonal transport of synaptic vesicles.

In vitro measurements of run lengths of Kinesin-3 motors are much shorter (~1-10μm) than cargo run lengths observed *in vivo* (~15–30 μm), which can be attributed to the presence of factors that function to enhance the motility characteristics of molecular motors *in vivo* (Beeg *et al.*, 2008; Sood *et al.*, 2018; Verbrugge, van den Wildenberg & Peterman, 2009). Some of these factors, identified from mammalian and *C. elegans* studies, have been listed in Table 1. Investigations conducted in *C. elegans* neurons revealed that SYD-2/Liprin-α (Wagner *et al.*, 2009) and LIN-2 (a MAGUK protein containing a CaMKII-homologous domain) function to cluster and increase the processive motility and velocity of UNC-104 (Figure 1(b)) (Wu, Muthaiyan Shanmugam, Bhan, Huang, & Wagner, 2016). CASY-1 (a calsyntenin family protein) was discovered to regulate UNC-104-mediated synaptic vesicle transport specifically in GABAergic motor neurons, likely by enhancing the motility of UNC-104 (Thapliyal *et al.*, 2018), thereby promoting anterograde transport. Bimolecular Fluorescence Complementation (BiFC), a technique that tests for the direct interaction between two proteins by fusing one to the N-terminal fragment and the other to the C-terminal fragment of Venus fluorescent protein and examining reconstituted fluorescence, identified that neuronal adaptors and interactors of UNC-104 regulate its subcellular distribution (Hsu,

Table 1. Factors regulating anterograde axonal transport.

Factors regulating motor activation/recruitment	
Kinesin-1 complex (UNC-116, KLC-1, KLC-2)	JIP-1/JIP1 (Matsuda *et al.*, 2001), Huntingtin (Gauthier *et al.*, 2004), MIRO-1/MIRO (Guo *et al.*, 2005), TRAK-1/TRAK1/2 (van Spronsen *et al.*, 2013), UNC-14 (Sakamoto *et al.*, 2005), UNC-76/FEZ1 (Fujita *et al.*, 2007), RanBP2 (Cho *et al.*, 2007), Syntabulin (Cai, Gerwin, & Sheng, 2005), BORC (Guardia, Farías, Jia, Pu, & Bonifacino, 2016), ARL-8/Arl8 (Farías *et al.*, 2017), SKIP (Farías *et al.*, 2017)
Kinesin-2	RAB4 (Dey, Banker, & Ray, 2017)
UNC-104/KIF1A	SAM-4/Myrlysin (Zheng *et al.*, 2014), BLOS-9/Blos9 (Niwa *et al.*, 2017), ARL-8/Arl8 (Klassen *et al.*, 2010), AEX-3/MADD (Niwa *et al.*, 2008), UNC-16/JIP3 (Choudhary *et al.*, 2017), SYD-2/Liprin-α (Bhan *et al.*, 2019), UNC-10/RIM1 (Bhan *et al.*, 2019)
Factors promoting motor motility	
Kinesin-1 complex (UNC-116, KLC-1, KLC-2)	MAP7D2 (Pan *et al.*, 2019), MARK/PAR1 Kinase (Mandelkow, Thies, Trinczek, Biernat, & Mandelkow, 2004)
UNC-104/KIF1A	MAP2 (Gumy *et al.*, 2017), SEPT9 (Karasmanis *et al.*, 2018), DCX (Monroy *et al.*, 2020), MAP9 (Monroy *et al.*, 2020), DYLT-1/DYNLT1 (Chen *et al.*, 2019), DYRB-1/DYNLRB1 (Chen *et al.*, 2019), DNC-1/DCTN1 (Chen *et al.*, 2019), DNC-5/DCTN5 (Chen *et al.*, 2019), DNC-6/DCTN6 (Chen *et al.*, 2019), SYD-2/Liprin-α (Wagner *et al.*, 2009), LIN-2/CASK (Wu *et al.*, 2016), PTL-1/Tau (Tien *et al.*, 2011), CASY-1/Calsyntenin-1 (Thapliyal *et al.*, 2018)
Factors inhibiting motor motility	
Kinesin-1 complex (UNC-116, KLC-1, KLC-2)	PTL-1/Tau (Dixit *et al.*, 2008), MAP2 (Gumy *et al.*, 2017), MAP9 (Monroy *et al.*, 2020), SEPT9 (Karasmanis *et al.*, 2018)
UNC-104/KIF1A	KBP (Kevenaar *et al.*, 2016), PTL-1/Tau (Dixit *et al.*, 2008)
Cargo	
Kinesin-1 complex (UNC-116, KLC-1, KLC-2)	APP-vesicles (Kamal, Stokin, Yang, Xia, & Goldstein, 2000), BDNF-vesicles (Colin *et al.*, 2008), Mitochondria (Patel *et al.*, 1993), Lysosomes (Guardia *et al.*, 2016)
Kinesin-2	Lysosomes (Brown *et al.*, 2005), Choline acetyltransferase vesicles (Ray *et al.*, 1999)
UNC-104/KIF1A	Synaptic vesicles (Hall & Hedgecock, 1991), Dense-Core vesicles (Zahn *et al.*, 2004)

Moncaleano, & Wagner, 2011). UNC-104 bound to UNC-16/JIP-3 predominantly localizes to the neuronal cell soma, while UNC-104 bound to DNC-1, a component of the dynactin complex, largely localizes to the distal tips of neurons, and UNC-104 bound to SYD-2/Liprin-α distributes along the neuronal process (Hsu *et al.*, 2011). These interactions have been suggested to influence the velocity and persistence times of UNC-104, with implications for axonal transport of synaptic vesicles (Hsu *et al.*, 2011).

Regulation of kinesin-1

The anterograde transport of mitochondria is mediated by UNC-116 and KLC-2 (Patel *et al.*, 1993; Sure *et al.*, 2018). UNC-16/JIP3, a JNK-interacting cargo-specific adapter, and UNC-76/FEZ1, a fasciculation and elongation protein zeta-1, regulate the anterograde transport of mitochondria in *C. elegans* touch receptor neurons (sensory neurons that respond to gentle touch) by regulating Kinesin-1 levels (Sure *et al.*, 2018). Additionally, phosphorylation of FEZ1 has been demonstrated to be important for its recruitment to the Kinesin-1 complex (Chua *et al.*, 2012) and the axonal transport and distribution of synaptic cargo such as SNB-1 (Synaptobrevin-1) and UNC-64/Syntaxin-1 (Butkevich *et al.*, 2016). *unc-69* (encodes the *C. elegans* homolog of mammalian SCOCO family of proteins) has been shown to function in the same genetic pathway as *unc-76/fez1* to regulate the distribution of SNB-1-containing synaptic vesicles, likely through direct interactions of UNC-69 with UNC-76, and recruitment to the Kinesin-1 complex (Chua *et al.*, 2012; Su *et al.*, 2006). This is consistent with reports from cultured cells where UNC-76/FEZ1 has been shown to physically interact with components of the Kinesin-1 complex, providing a direct mechanism for the regulation of motor motility (Blasius, Cai, Jih, Toret, & Verhey, 2007). *unc-14* (encodes a RUN domain protein) governs the distribution of SNB-1-containing synaptic vesicles in *C. elegans* neurons, likely through direct interactions with KLC-2 (kinesin light chain subunit) (Sakamoto *et al.*, 2005). These studies collectively show that anterograde transport mechanisms are largely conserved across vertebrate and invertebrate organisms.

Functional implications of anterograde axonal transport

Anterograde axonal transport is crucial for neuronal development and function (Morfini *et al.*, 2012). Studies conducted in *C. elegans* have helped identify several factors that link anterograde axonal transport to neuronal morphology. The Kinesin-1 subunit KLC-2, and VAB-8, a protein with Kinesin-like domains, play an important role in axon guidance and process outgrowth of *C. elegans* motor and touch sensory neurons (Lai & Garriga, 2004; Su *et al.*, 2006; Tsuboi, Hikita, Qadota, Amano, & Kaibuchi, 2005; Wolf *et al.*, 1998). AEX-3, a GEF for RAB-3, has been identified to be necessary for guidance of the *C. elegans* AVG neuron, and consequently, the organization of the entire Ventral Nerve Cord (Bhat & Hutter, 2016). It has been proposed that AEX-3 does so by regulating the anterograde trafficking of the Netrin receptor, UNC-5, mediated in part by UNC-104 (Bhat & Hutter, 2016).

C. elegans studies further demonstrate the importance of anterograde axonal transport in presynaptic assembly. Loss-of-function mutants of *unc-104* (Hall & Hedgecock, 1991; Miller *et al.*, 1996) and its genetic regulators *arl-8* (Klassen *et al.*, 2010; Niwa *et al.*, 2017) and *sam-4* (a putative GEF for ARL-8) (Niwa *et al.*, 2017; Zheng *et al.*, 2014) exhibit synapse assembly and transmission defects. Anterograde transport of synaptic vesicle proteins has been shown to be necessary for presynapse assembly in DD motoneurons after the first larval stage (Kurup, Yan, Goncharov, & Jin, 2015; Kurup, Yan, Kono, & Jin, 2017; Park *et al.*, 2011), and *en passant* presynaptic assembly in DA9 (Wu, Huo, Maeder, Feng, & Shen, 2013) and PDE (Lipton, Maeder, & Shen, 2018) neurons of *C. elegans*. UNC-104-mediated anterograde transport of integral autophagy proteins such as ATG-9 (Stavoe, Hill, Hall, & Colón-Ramos, 2016) is necessary for presynaptic assembly in *C. elegans* neurons, while KIF5B-mediated anterograde transport of Bassoon-containing vesicles is necessary for synaptic assembly in cultured mammalian neurons (Cai, Pan, & Sheng, 2007).

Reduction in UNC-104 activity has further been shown to accelerate age-dependent locomotion decline by perturbing synaptic vesicle transport and reducing synaptic transmission, while an increase in UNC-104 levels improves locomotion with age (Li *et al.*, 2016). Consistent with reports from *Drosophila* neurons (Fang, Soares, Teng, Geary, & Bonini, 2012), Kinesin-1-mediated anterograde axonal transport of mitochondria has been shown to protect against neurodegeneration in *C. elegans* neurons (Rawson *et al.*, 2014). The translocation of mitochondria has further been shown to be necessary for process regrowth in *C. elegans* neurons (Han, Baig, & Hammarlund, 2016). These studies collectively demonstrate the importance of anterograde axonal transport in neuronal development, synapse specification, and regulating age-dependent decline in neuronal function.

Regulators of retrograde axonal transport

Retrograde transport of neuronal cargo in axons is mediated largely by cytoplasmic Dynein (Gennerich & Vale, 2009). The Dynein motor complex comprises the heavy chain, light chain, light intermediate and intermediate chain subunits (Roberts, Kon, Knight, Sutoh, & Burgess, 2013). Following the discovery of the retrograde transport complex consisting of Dynein and its various accessory subunits, several *C. elegans* studies have helped uncover molecular mechanisms of regulation of retrograde transport *in vivo*. Characterization of *C. elegans* mutants isolated from a forward genetic screen for misaccumulation of Synaptobrevin (SNB-1) at the distal tip of *C. elegans* neurons, helped identify subunits of the neuronal Dynein complex and demonstrated that Dynein's recruitment to synaptic vesicles *in vivo* is only partially dependent on vesicular spectrin (Koushika *et al.*, 2004), an adaptor for Dynein (Muresan *et al.*, 2001). SNB-1 containing vesicles, returning to the cell body from the synapse in *C. elegans* touch receptor neurons, have been shown to lack RAB-3 but contain SNT-1 (Synaptotagmin) (Murthy *et al.*, 2011). Consistent with this observation, Dynein complex mutants show neuronal tip accumulation of SNB-1 and SNT-1 but not RAB-3 and SNG-1 (Synaptogyrin), additionally suggesting cargo specificity (Koushika *et al.*, 2004). Thus, retrograde transport in neurons is likely regulated by cargo-specific adaptors. A list of important factors regulating retrograde axonal transport can be found in Table 2.

Several *C. elegans* studies, employing a combination of classical genetics, molecular and biochemical approaches, helped identify various neuronal cell-type and cargo-specific regulators of the retrograde motor complex *in vivo*. The cyclin-dependent kinases PCT-1 and CDK-5 have been shown to act genetically upstream to components of the Dynein motor complex and inhibit Dynein by inactivating NUD-2 (*C. elegans* Nudel, a component of the Dynein complex) (Ou *et al.*, 2010). CDK-5 has further been proposed to inhibit Dynein-dependent transport of DCVs into the dendrites of DB motor neurons (Goodwin *et al.*, 2012). CDK-5, SYD-2 and SAD-1 have been proposed to inhibit Dynein-mediated transport of lysosomes into dendrites of *C. elegans* motor neurons (Edwards *et al.*, 2015). Another study identified that the cargo adaptor UNC-16/JIP-3 forms a complex with neuronal Kinesin-1 and DLI-1 (Dynein Light Intermediate chain subunit), to regulate the subcellular distribution of DLI-1 to neuronal distal tips (Figure 2(a)), and consequently regulate the retrograde transport of synaptic vesicle proteins SNB-1, SNT-1, and APL-1 (human APP-like protein) (Arimoto *et al.*, 2011). Tubulin mutations that increased the affinity of Dynein for microtubules caused Dynein-dependent mis-trafficking of synaptic vesicles to dendrites of *C. elegans* neurons (Hsu *et al.*, 2014). *C. elegans* tauopathy models expressing mutated human Tau

Table 2. Factors regulating retrograde axonal transport.

Activators	BICD-1/BICD2 (Splinter *et al.*, 2012), BICDR1 (Urnavicius *et al.*, 2018), HOOK1 (Olenick, Tokito, Boczkowska, Dominguez, & Holzbaur, 2016), HOOK3 (Olenick *et al.*, 2016), LIS-1/LIS1 (Splinter *et al.*, 2012), ERK1/2 (Mitchell *et al.*, 2012)
Adaptors	HAP1 (Li, Gutekunst, Hersch, & Li, 1998), TRAK-1/TRAK1/2 (van Spronsen *et al.*, 2013), RILP (Johansson *et al.*, 2007), UNC-16/JIP3 (Arimoto *et al.*, 2011), JIP4 (Montagnac *et al.*, 2009), Huntingtin (Caviston, Ross, Antony, Tokito, & Holzbaur, 2007)
Regulators of processive motility	Dynactin subunits (Arp1, Arp11, p24, p25, p27, p50, p62, $p150^{Glued}$) (Moughamian, Osborn, Lazarus, Maday, & Holzbaur, 2013), LIS-1/LIS1 (Shao *et al.*, 2013), PTL-1/Tau (Tien *et al.*, 2011)
Inhibitors	CDK-5/Cdk5 (Goodwin *et al.*, 2012; Ou *et al.*, 2010), PCT-1 (Ou *et al.*, 2010), SYD-2/ Liprin-α (Edwards *et al.*, 2015), PTL-1/Tau (Vershinin *et al.*, 2007)

(microtubule associated protein) show reduced anterograde and retrograde transport velocities of synaptic vesicles (Butler *et al.*, 2019). These mutations are proposed to perturb the direct interactions of Tau with Kinesins and components of the Dynein-Dynactin complex (Butler *et al.*, 2019). In summary, *C. elegans* studies have contributed significantly to the identification of inhibitors of retrograde axonal transport *in vivo* and provided insights into the interaction of the retrograde motor complex with the axonal microtubule cytoskeleton.

Cargo displaying bidirectional transport behaviour are proposed to be a consequence of "coordination" or "tug-of-war" between oppositely directed motors, in which cooperative behaviour of motors can also play key roles (Hancock, 2018). Consistent with reports from mammalian neurons, *C. elegans* studies have suggested the existence of regulatory crosstalk between anterograde and retrograde motors. Hypomorphic mutant alleles of components of the retrograde Dynein motor complex have been demonstrated to increase the anterograde bias in the axonal transport of synaptic cargo in *C. elegans* neurons, suggesting that the retrograde motor complex competes with the anterograde motor to regulate the direction of cargo transport (Kurup *et al.*, 2017). This is consistent with the "tug-of-war" model proposed in *in vitro* and cell culture studies (Hancock, 2018). Conversely, UNC-104 has also been reported to physically interact with several components of the Dynein-Dynactin complex, such as DYLT-1, DYRB-1, DNC-5 and DNC-6, which have been shown to affect the velocity and dwell times of UNC-104 along the neuronal process and the anterograde transport of SNB-1 in *C. elegans* neurons (Chen *et al.*, 2019). It has further been proposed that these interactions are necessary to activate UNC-104, promote anterograde axonal transport (Chen *et al.*, 2019), and likely function to localize components of the retrograde transport machinery to distal tips of neuronal processes (Figure 2(a)). This is consistent with mammalian studies, which have demonstrated direct interactions between Dynein and Kinesin light chains, proposing that this interaction is necessary for localising Dynein to the plus-ends of microtubules (Ligon, Tokito, Finklestein, Grossman, & Holzbaur, 2004). Collectively, findings in *C. elegans* support the model that opposing motors may also play a role in promoting each other's function.

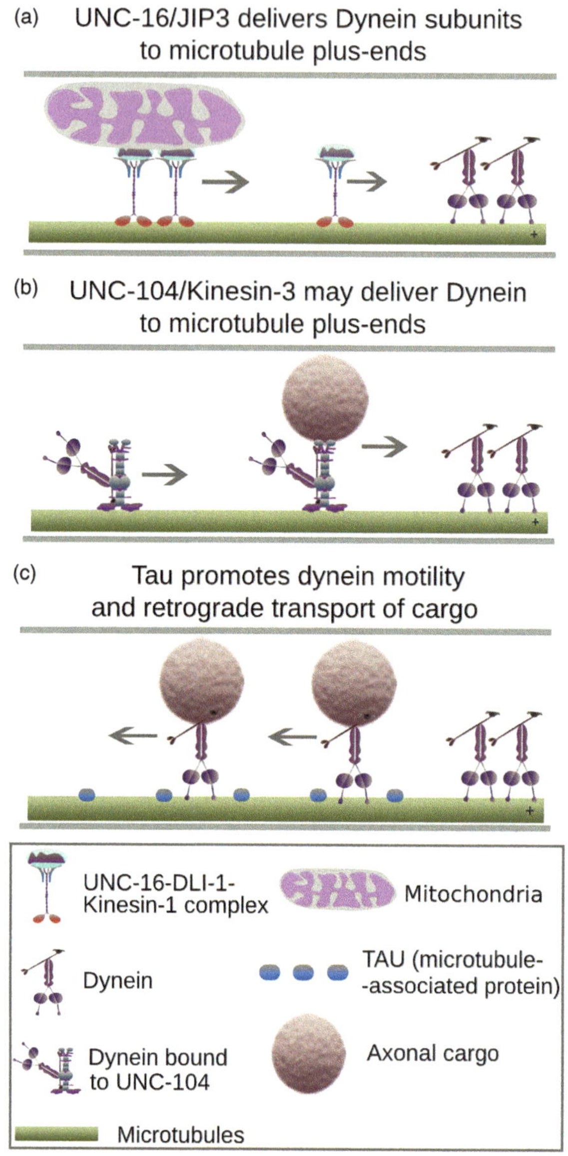

Figure 2. Mechanisms regulating retrograde axonal transport. a) Components of the dynein-dynactin complex are transported to microtubule plus-ends by anterograde motors such as Kinesin-1 through attachment to the adapter UNC-16/JIP-3. b) UNC-104/Kinesin-3 physically interacts with Dynein through its stalk domains, and may contribute to its delivery to microtubule plus-ends. c) Tau/PTL-1, an axonal microtubule associated protein, promotes the retrograde transport of cargo by enhancing the processivity of Dynein.

Functional implications of retrograde axonal transport

Several *C. elegans* studies have demonstrated the importance of retrograde transport in neuronal development. Dynein-mediated transport of synaptic vesicles, regulated by CDK-5, is necessary for the formation of new synapses after remodelling in the DD motoneurons of *C. elegans* (Park *et al.*, 2011). *C. elegans* BICD-1 (coiled-coil protein that is an accessory in the Dynein-Dynactin complex) is reported to regulate dendritic branching of the PVD neuron (Aguirre-Chen, Bulow, & Kaprielian, 2011). Studies also suggest that this dendritic branching is regulated by RAB-10, which functions to balance the activity of UNC-116 and Dynein to modulate the distribution of cargo such as the branching receptor DMA-1 in *C. elegans* neurons (Taylor, Yan, Howell, Dong, & Shen, 2015). Reduced levels of DNC-1 in *C. elegans* motoneurons increases the accumulation of autophagosomes at the distal tip, causes severe locomotion defects, and motor neuron degeneration, suggested to be important for disease progression in ALS (Ikenaka *et al.*, 2013). These studies collectively suggest that retrograde axonal transport is crucial for the distribution of neuronal cargo, and the maintenance of neuronal morphology, with implications for organismal behaviour.

Regulators of the neuronal cytoskeleton

The axonal cytoskeleton is composed of microtubule networks, actin assemblies, and intermediate filaments (Kapitein & Hoogenraad, 2011; Sainath & Gallo, 2015).

While molecular motors, such as Kinesins and Dyneins, predominantly use microtubules as tracks along which to transport neuronal cargo (Hirokawa & Noda, 2008), the other cytoskeletal elements also influence axonal transport (Morris & Hollenbeck, 1995; Perrot & Julien, 2009), which are discussed in the sections below.

Microtubules

Microtubules are cylindrically-shaped tubes made of polymers of αβ-tubulin GTP heterodimers. They can vary in their lattice organization and number of protofilaments, depending on the type of tubulin subunits involved, among other factors (Amos & Schlieper, 2005). They function as tracks for a majority of long-range axonal transport enabled by microtubule-dependent motors (Hirokawa & Noda, 2008). Several studies in *C. elegans* have demonstrated the importance of microtubule assembly in maintaining axonal transport. A detailed characterization of microtubule structures in *C. elegans* neurons revealed that most neurons contain 11-protofilament microtubules, with the exception of a) sensory cilia, which have both 13 and 11-protofilament microtubules and b) Touch Receptor Neurons, which contain 15-protofilament microtubules (Chalfie & Thomson, 1982). The distinct microtubule organization in touch receptor neurons is necessary for the organization and distribution of the DEG/ENaC channels, which are amiloride-sensitive, non-voltage gated Na^+ permeable channels, along the neuronal process (Bounoutas, O'Hagan, & Chalfie, 2009). Mutations in *tba-1*/α-tubulin (Baran *et al.*, 2010) and the *mec-7*/β-tubulin subunit (Kirszenblat, Neumann, Coakley, & Hilliard, 2013) have been shown to affect the distribution and transport of synaptic cargo in *C. elegans* neurons, while mutations in genes regulating microtubule acetylation (*mec-17*) and drugs perturbing microtubule polymerization (colchicine) have been demonstrated to affect the distribution and transport of neuronal ribosomes in touch receptor neurons of *C. elegans* (Noma, Goncharov, Ellisman, & Jin, 2017).

Microtubule stability, regulated by the nature of tubulin subunits, microtubule associated proteins and post-translational modifications, has been shown to influence axonal transport (Dubey, Ratnakaran, & Koushika, 2015). Mammalian studies show that microtubule nucleation and polymerisation is important for the transport of synaptic vesicles between successive boutons (Qu *et al.*, 2019). Likewise, microtubule dynamics has been shown to be necessary for maintaining synaptic vesicle transport during neuronal remodelling of DD motor neurons in *C. elegans* (Kurup *et al.*, 2015; 2017). It has further been shown to be regulated by TBA-1/α-tubulin, and DLK-1, a MAPKKK, known to regulate axonal transport by influencing the binding of the Kinesin-1 subunit KLC to its cargo (Horiuchi *et al.*, 2007; Kurup *et al.*, 2015, 2017). Conversely, studies in cultured neurons showed that pharmacological stabilization of dynamic microtubules promotes axonal transport (Dubey *et al.*, 2015). This suggests that the effect of microtubule dynamics on axonal transport of cargo is likely both a function of the developmental context and of molecular factors that are yet to be identified.

Microtubule-associated proteins

Microtubule associated proteins (MAPs) have been widely reported to influence microtubule-based transport of cargo in axons (Monroy *et al.*, 2018; 2020; Vershinin *et al.*, 2007). The effects fall into two classes (i) physically crowding the microtubule track (Dixit, Ross, Goldman, & Holzbaur, 2008; Vershinin *et al.*, 2007), and (ii) differentially influencing individual motors, for instance, promoting Kinesin-3 mediated transport and inhibiting Kinesin-1 mediated transport (Gumy *et al.*, 2017; Karasmanis *et al.*, 2018; Monroy *et al.*, 2020).

Studies conducted in *C. elegans* neurons identified that the microtubule associated protein ZYG-8 (member of the DCX (Doublecortin)-DCLK (Doublecortin-like kinase) subfamily) regulates the length, number, distribution and structural integrity of axonal microtubules (Bellanger *et al.*, 2012). These factors influence the transport properties of axonal cargo in *C. elegans* motor neurons (Yogev, Cooper, Fetter, Horowitz, & Shen, 2016). Microtubule lengths limit the run length of moving cargo, cargo pause at microtubule ends (Figure 3(c)), and a higher abundance of microtubules is correlated with shorter duration of pauses in the trajectories of moving cargo (Yogev *et al.*, 2016). This is the first *in vivo* demonstration of the effect of microtubule number and lengths on cargo distribution and transport in axons. PTL-1 (a microtubule-associated protein with Tau-like repeats) has been reported to promote Dynein-mediated axonal transport (Figure 2(b)), and the retrograde transport of UNC-104/Kinesin-3 and its associated cargo in *C. elegans* neurons (Tien, Wu, Hsu, Chang, & Wagner, 2011). The *C. elegans* PTRN-1, a member of the Calmodulin-regulated spectrin-associated protein (CAMSAP)/Patronin/Nezha family of microtubule minus-end binding proteins (Figure 3(e)), has been demonstrated to be necessary for microtubule polymerization, number, and organization, with implications for neuronal process outgrowth and cargo transport in *C. elegans* sensory neurons (Marcette, Chen, & Nonet, 2014; Richardson *et al.*, 2014). This is consistent with reports from mammalian studies which first identified the role of Patronin in stabilising microtubules by binding to their minus ends (Goodwin & Vale, 2010). Studies conducted in *C. elegans* neurons further demonstrated that mutations in *unc-116* (KHC-1/Kinesin-1), *unc-33* (CRMP) and *unc-44* (Ankyrin) affect the distribution and transport of synaptic vesicles, that are known to be transported by UNC-104/Kinesin-3, by affecting the polarity and orientation of the underlying microtubule network in dendrites (Maniar *et al.*, 2011; Yan *et al.*, 2013). Combining both *in vitro* and *in vivo* experiments, a recent study in *C. elegans* neurons demonstrated that exogenous expression of vertebrate KBP (a Kinesin-binding protein that may also associate with microtubules), which inhibits the interaction of KIF1A with microtubules *in vitro*, disrupts UNC-104-mediated synaptic vesicle transport *in vivo* (Kevenaar *et al.*, 2016). In summary, *C. elegans* studies, consistent with vertebrate literature, show

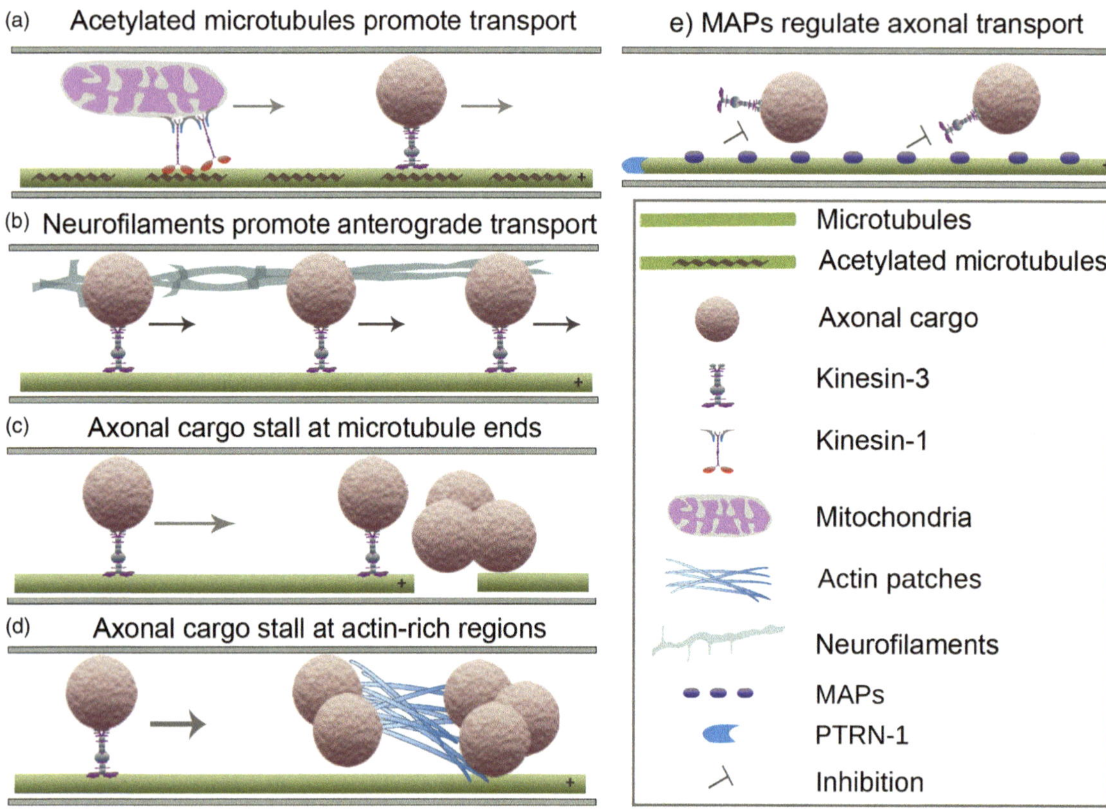

Figure 3. Role of the neuronal cytoskeleton in axonal transport. a) Acetylation of axonal microtubules by MEC-17 is important for the anterograde transport of neuronal cargo such as mitochondria. b) Neurofilaments stabilize axonal microtubules and promote the anterograde transport of synaptic vesicles. c) Axonal cargo stall at microtubule ends. d) Synaptic vesicles stall at actin patches, likely due to physical crowding. e) Microtubule minus-end binding protein PTRN-1 stabilises axonal microtubules. Certain microtubule-associated proteins can inhibit the association of anterograde motors with microtubules, thereby inhibiting anterograde axonal transport *in vivo*.

that MAPs can either promote or inhibit axonal transport by regulating microtubule assembly, stability, and interactions of molecular motors with microtubules.

Post-translational modifications

Post-translational modifications (PTMs) such as acetylation, polyglutamylation and detyrosination are markers for stable microtubules (Janke & Chloë Bulinski, 2011), and are often found enriched in axons as compared to dendrites (Hammond, Huang, *et al.*, 2010). PTMs have been reported to regulate the motility of microtubule-dependent motors such as Kinesin-1 (Balabanian, Berger, & Hendricks, 2017), and organelles such as mitochondria in cultured neurons (Magiera *et al.*, 2018).

The role of tubulin PTMs in regulating axonal transport is challenging to investigate in *C. elegans* neurons, as, unlike non-neuronal cells in culture, they have narrow geometries which make it difficult to observe cargo transport along individual microtubules. MEC-17, a *C. elegans* protein related to the Gcn5 histone acetyltransferases, is necessary for the formation of 15-protofilament microtubules in Touch Receptor Neurons, promotes neuronal process outgrowth, and maintains touch sensation (Akella *et al.*, 2010; Shida, Cueva, Xu, Goodman, & Nachury, 2010; Topalidou *et al.*, 2012). Absence of functional MEC-17 has been shown to cause the assembly of variable protofilament-number microtubules (11, 12, 13 & 15), perturb axonal transport, mitochondrial distribution, and cause microtubule instability, leading to axonal degeneration in *C. elegans* touch receptor neurons (Cueva, Hsin, Huang, & Goodman, 2012; Neumann & Hilliard, 2014). This is the first *in vivo* demonstration of the role of post-translational modifications in regulating protofilament numbers of microtubules, with direct implications for neuronal integrity and function.

Polyglutamylation of microtubules, regulated by a balance between the activities of the deglutamylase CCPP-1 and glutamylase TTLL-11 in *C. elegans*, has been shown to regulate microtubule doublet structure and stability, facilitate the recruitment of KLP-6 (a Kinesin-3 family motor) and limit the transport velocity of the anterograde motor OSM-3 (a homodimeric Kinesin-2 family motor) (O'Hagan *et al.*, 2011). Absence of CCPP-1 has been shown to cause progressive deterioration of cilia, suggesting that polyglutamylation-mediated microtubule stability and supported cargo transport are necessary for the maintenance of cilia (O'Hagan *et al.*, 2011; 2017; O'Hagan & Barr, 2012). The above studies suggest that microtubule PTMs in *C. elegans* neurons, like vertebrate neurons, play a direct role in microtubule assembly, stability, directing intracellular cargo transport, and maintaining neuronal structure and function, perhaps in part by facilitating the transport of neuronal cargo.

Functional implications of microtubule dynamics

Several *C. elegans* studies demonstrate the importance of microtubule assembly, dynamics, and potentially axonal transport for neuronal function. Loss-of-function mutations in tubulin subunits *mec-7* and *mec-12* and the tubulin acetyltransferase *mec-17* were shown to induce axon degeneration in *C. elegans* touch receptor neurons (Neumann & Hilliard, 2014). These mutants also displayed significant defects in the transport of mitochondria and synaptic vesicles, thus suggesting that microtubule stability and cargo transport are likely critical for preserving neuronal integrity (Neumann & Hilliard, 2014). PTL-1 (protein with Tau-like repeats), a microtubule associated protein known to regulate axonal transport, has been demonstrated to preserve neuronal structure (Chew, Fan, Gotz, & Nicholas, 2013). Loss of PTL-1 causes increased ectopic branching in mechanosensory and motor neurons in day 4 of adulthood, a phenotype only observed in wild type animals on day 8 (Chew *et al.*, 2013). These studies collectively suggest that microtubule stability/dynamics regulate neuronal structure and integrity, likely in part through the regulation of cargo transport.

Actin

Neuronal actin assemblies have been reported to assist long-distance transport of cargo in axons and dendrites (Venkatesh, Mathew, & Koushika, 2020). In neuronal processes, actin is present as waves, rings, polarized patches, trails, hotspots, bundled filopodial structures, and dendritic networks, several of which have been covered extensively in the following reviews (Roy, 2020; Venkatesh *et al.*, 2020). The first *in vivo* demonstration of actin trails and hot spots were made in the touch receptor neurons of *C. elegans* (Sood *et al.*, 2018). While stationary endosomes in vertebrate axons promote actin polymerization (Ganguly *et al.*, 2015; Hong, Qi, & Weaver, 2015), such observations have not been reported in *C. elegans* and *Drosophila* neurons, although both endosomes and synaptic vesicles are present at actin-rich regions (Sood *et al.*, 2018). Actin assemblies have been reported to serve as local "tracks" for the transport of cargo in vertebrate neurons (Venkatesh *et al.*, 2020). However, in both *C. elegans* and *Drosophila*, axonal actin-rich regions promote cargo stalling (Figure 3(d)), and regulate local cargo flux likely through physical crowding of the transport path (Sood et al., 2018). Thus *C. elegans* provides a readily accessible model system in which to investigate novel roles of the actin cytoskeleton in long-distance axonal transport.

Intermediate filaments

Intermediate filaments are cytoskeletal polymers know to interact with unassembled tubulin monomers to inhibit polymerization, and bind to microtubule networks to enhance their stability (Bocquet *et al.*, 2009; Chang & Goldman, 2004), both of which can influence axonal transport. Neurofilaments (NFs) are a special class of intermediate filaments that are expressed in neurons, and have been shown to provide structure to myelinated axons, control their diameter, and therefore axonal conductance (Yuan, Rao, Veeranna, & Nixon, 2017). In cultured neurons, neurofilament NF-L has been shown to modulate the anterograde and retrograde velocities of late endosomes and lysosomes (Perrot & Julien, 2009).

In *C. elegans*, intermediate filaments, such as IFP-1 and IFA-4, have been reported to stabilize microtubules in DD motor neurons, and inhibit synaptic vesicle transport during neuronal remodelling (Kurup, Li, Goncharov, & Jin, 2018). By contrast, another study in *C. elegans* identified and characterized a novel neurofilament protein, TAG-63, which has been demonstrated to promote the anterograde transport of synaptic cargo (Figure 3(b)) by regulating the run length, velocity and flux of UNC-104/Kinesin-3 and SNB-1-containing synaptic vesicles (Bhan *et al.*, 2019). This study is the first to report a direct role for neurofilament proteins in regulating the motility characteristics of axonal cargo *in vivo*. The exact nature of this regulation warrants further study, as it is not known whether it occurs via regulation of microtubule stability or direct interactions with the cargo-motor complex.

Perspective

Axonal transport is a critical process, central to neuronal function and maintenance. *In vitro* studies have provided a wealth of information about single and ensemble motor behaviours in different cytoskeletal geometries (Holzbaur & Goldman, 2010). Super resolution imaging techniques, such as single molecule localization microscopy, allow researchers to examine complex *in vivo* cytoskeletal geometries. Recent studies have succeeded in resolving individual microtubules in axons of cultured hippocampal neurons, using anti-tubulin nanobodies to stain microtubules, and a novel optical nanoscopy technique, named motor-PAINT, to assess the stability and orientation of individual microtubules (Mikhaylova *et al.*, 2015; Tas *et al.*, 2017). Such techniques, when applied to model organisms, can pave the way for investigating mechanisms by which motor-cargo complexes exhibit a preference for specific microtubules *in vivo*, for instance, to understand the role of post-translational modifications of microtubules in track selection by motors. Advanced microscopy techniques such as STORM (He *et al.*, 2016; Stewart & Shen, 2015) and Expansion microscopy (Yu *et al.*, 2020) have already begun to provide insights into the cytoskeletal architecture and synaptic organization of *C. elegans* neurons. These techniques allow investigators to translate the precision of *in vitro* measurements to *in vivo* systems.

Advances in imaging techniques help provide a physical understanding of axonal transport. However, the upstream regulators of axonal transport have been found to belong to diverse genetic pathways. Identification of such regulators would have been very difficult without the power of forward genetic screens. The traditional view is that the power of *C. elegans* as a model system is to identify novel regulators through unbiased forward genetic screens, and dissect genetic pathways through epistatic analyses with other characterized mutants. However, *C. elegans* provides two additional advantages, i) multiple *in vivo* developmental and functional contexts, and ii) diversity in the underlying cell biology between neurons. *C. elegans* provides neuronal systems exhibiting unique modes of development (e.g. stage-specific synaptic remodelling in DD motoneurons) and cell biological features (e.g. 15-protofilament microtubules in Touch Receptor Neurons). All studies discussed so far, beginning with the mutants first isolated by Brenner (1974) to current imaging advances, prove *C. elegans* to be a valuable model organism in which to address several open questions in the field of axonal transport.

Disclosure statement

No potential conflict of interest was reported by the author(s).

References

Aguirre-Chen, C., Bulow, H.E., & Kaprielian, Z. (2011). *C. elegans* bicd-1, homolog of the Drosophila dynein accessory factor Bicaudal D, regulates the branching of PVD sensory neuron dendrites. *Development (Cambridge, England)), 138*(3), 507–518. doi:10.1242/dev.060939

Aizawa, H., Sekine, Y., Takemura, R., Zhang, Z., Nangaku, M., & Hirokawa, N. (1992). Kinesin family in murine central nervous system. *The Journal of Cell Biology, 119*(5), 1287–1296. doi:10.1083/jcb.119.5.1287

Akella, J.S., Wloga, D., Kim, J., Starostina, N.G., Lyons-Abbott, S., Morrissette, N.S., ... Gaertig, J. (2010). MEC-17 is an alpha-tubulin acetyltransferase. *Nature, 467*(7312), 218–222. doi:10.1038/nature09324

Al-Bassam, J., Cui, Y., Klopfenstein, D., Carragher, B.O., Vale, R.D., & Milligan, R.A. (2003). Distinct conformations of the kinesin Unc104 neck regulate a monomer to dimer motor transition. *The Journal of Cell Biology, 163*(4), 743–753. doi:10.1083/jcb.200308020

Allen, P.B., Sgro, A.E., Chao, D.L., Doepker, B.E., Scott Edgar, J., Shen, K., & Chiu, D.T. (2008). Single-synapse ablation and long-term imaging in live *C. elegans*. *Journal of Neuroscience Methods, 173*(1), 20–26. doi:10.1016/j.jneumeth.2008.05.007

Amos, L. A., & Schlieper, D. (2005). Microtubules and Maps. In J. M. Squire & D. A. D. Parry (Eds.), Advances in Protein Chemistry (Vol. 71, pp. 257–298). Elsevier.

Arimoto, M., Koushika, S.P., Choudhary, B.C., Li, C., Matsumoto, K., & Hisamoto, N. (2011). The *Caenorhabditis elegans* JIP3 protein UNC-16 functions as an adaptor to link kinesin-1 with cytoplasmic dynein. *The Journal of Neuroscience: The Official Journal of the Society for Neuroscience, 31*(6), 2216–2224. doi:10.1523/JNEUROSCI.2653-10.2011

Balabanian, L., Berger, C.L., & Hendricks, A.G. (2017). Acetylated microtubules are preferentially bundled leading to enhanced kinesin-1 motility. *Biophysical Journal, 113*(7), 1551–1560. doi:10.1016/j.bpj.2017.08.009

Baran, R., Castelblanco, L., Tang, G., Shapiro, I., Goncharov, A., & Jin, Y. (2010). Motor neuron synapse and axon defects in a *C. elegans* alpha-tubulin mutant. *PLoS One, 5*(3), e9655. doi:10.1371/journal.pone.0009655

Beeg, J., Klumpp, S., Dimova, R., Gracià, R.S., Unger, E., & Lipowsky, R. (2008). Transport of beads by several kinesin motors. *Biophysical Journal, 94*(2), 532–541. doi:10.1529/biophysj.106.097881

Bellanger, J.-M., Cueva, J.G., Baran, R., Tang, G., Goodman, M.B., & Debant, A. (2012). The doublecortin-related gene *zyg-8* is a microtubule organizer in *Caenorhabditis elegans* neurons. *Journal of Cell Science, 125*(Pt 22), 5417–5427. doi:10.1242/jcs.108381

Bhabha, G., Johnson, G.T., Schroeder, C.M., & Vale, R.D. (2016). How dynein moves along microtubules. *Trends in Biochemical Sciences, 41*(1), 94–105. doi:10.1016/j.tibs.2015.11.004

Bhan, P., Muthaiyan Shanmugam, M., Wang, D., Bayansan, O., Chen, C., & Wagner, O.I. (2019). Characterization of TAG-63 and its role on axonal transport in *C. elegans*. *Traffic, 21*(2), 231–249. doi:10.1111/tra.12706

Bhat, J.M., & Hutter, H. (2016). Pioneer axon navigation is controlled by AEX-3, a guanine nucleotide exchange factor for RAB-3 in *Caenorhabditis elegans*. *Genetics, 203*(3), 1235–1247. doi:10.1534/genetics.115.186064

Blasius, T.L., Cai, D., Jih, G.T., Toret, C.P., & Verhey, K.J. (2007). Two binding partners cooperate to activate the molecular motor Kinesin-1. *The Journal of Cell Biology, 176*(1), 11–17. doi:10.1083/jcb.200605099

Bocquet, A., Berges, R., Frank, R., Robert, P., Peterson, A.C., & Eyer, J. (2009). Neurofilaments bind tubulin and modulate its polymerization. *The Journal of Neuroscience: The Official Journal of the Society for Neuroscience, 29*(35), 11043–11054. doi:10.1523/JNEUROSCI.1924-09.2009

Bounoutas, A., O'Hagan, R., & Chalfie, M. (2009). The multipurpose 15-protofilament microtubules in *C. elegans* have specific roles in mechanosensation. *Current Biology, 19*(16), 1362–1367. doi:10.1016/j.cub.2009.06.036

Brady, S.T., Pfister, K.K., & Bloom, G.S. (1990). A monoclonal antibody against kinesin inhibits both anterograde and retrograde fast axonal transport in squid axoplasm. *Proceedings of the National Academy of Sciences of the United States of America, 87*(3), 1061–1065. doi:10.1073/pnas.87.3.1061

Brenner, S. (1974). The genetics of *Caenorhabditis elegans*. *Genetics, 77*(1), 71–94.

Brown, C.L., Maier, K.C., Stauber, T., Ginkel, L.M., Wordeman, L., Vernos, I., & Schroer, T.A. (2005). Kinesin-2 is a motor for late endosomes and lysosomes: kinesin-2 is a late endosome motor. *Traffic, 6*(12), 1114–1124. doi:10.1111/j.1600-0854.2005.00347.x

Butkevich, E., Härtig, W., Nikolov, M., Erck, C., Grosche, J., Urlaub, H., ... Chua, J.J.E. (2016). Phosphorylation of FEZ1 by microtubule affinity regulating kinases regulates its function in presynaptic protein trafficking. *Scientific Reports, 6*(1), 26965. doi:10.1038/srep26965

Butler, V.J., Salazar, D.A., Soriano-Castell, D., Alves-Ferreira, M., Dennissen, F.J.A., Vohra, M., ... Kao, A.W. (2019). Tau/MAPT disease-associated variant A152T alters tau function and toxicity via impaired retrograde axonal transport. *Human Molecular Genetics, 28*(9), 1498–1514. doi:10.1093/hmg/ddy442

Byrd, D.T., Kawasaki, M., Walcoff, M., Hisamoto, N., Matsumoto, K., & Jin, Y. (2001). UNC-16, a JNK-signaling scaffold protein, regulates vesicle transport in *C. elegans*. *Neuron, 32*(5), 787–800. doi:10.1016/S0896-6273(01)00532-3

Cai, Q., Gerwin, C., & Sheng, Z.-H. (2005). Syntabulin-mediated anterograde transport of mitochondria along neuronal processes. *The Journal of Cell Biology, 170*(6), 959–969. doi:10.1083/jcb.200506042

Cai, Q., Pan, P.-Y., & Sheng, Z.-H. (2007). Syntabulin-kinesin-1 family member 5B-mediated axonal transport contributes to activity-dependent presynaptic assembly. *The Journal of Neuroscience: The Official Journal of the Society for Neuroscience, 27*(27), 7284–7296. doi:10.1523/JNEUROSCI.0731-07.2007

Campenot, R.B., & Eng, H. (2000). Protein synthesis in axons and its possible functions. *Journal of Neurocytology, 29* (11–12), 793–798. doi:10.1023/a:1010939307434

Caviston, J.P., Ross, J.L., Antony, S.M., Tokito, M., & Holzbaur, E.L.F. (2007). Huntingtin facilitates dynein/dynactin-mediated vesicle transport. *Proceedings of the National Academy of Sciences of the United States of America, 104*(24), 10045–10050. doi:10.1073/pnas.0610628104

Chalfie, M., & Thomson, J.N. (1982). Structural and functional diversity in the neuronal microtubules of *Caenorhabditis elegans*. *The Journal of Cell Biology, 93*(1), 15–23. doi:10.1083/jcb.93.1.15

Chang, L., & Goldman, R.D. (2004). Intermediate filaments mediate cytoskeletal crosstalk. *Nature Reviews. Molecular Cell Biology, 5*(8), 601–613. doi:10.1038/nrm1438

Chen, C., Peng, Y., Yen, Y., Bhan, P., Muthaiyan Shanmugam, M., Klopfenstein, D.R., & Wagner, O.I. (2019). Insights on UNC-104-

dynein/dynactin interactions and their implications on axonal transport in *Caenorhabditis elegans*. *Journal of Neuroscience Research*, *97*(2), 185–201. doi:10.1002/jnr.24339

Chew, Y.L., Fan, X., Gotz, J., & Nicholas, H.R. (2013). PTL-1 regulates neuronal integrity and lifespan in *C. elegans*. *Journal of Cell Science*, *126*(Pt 9), 2079–2091. doi:10.1242/jcs.jcs124404

Chiba, K., Takahashi, H., Chen, M., Obinata, H., Arai, S., Hashimoto, K., … Niwa, S. (2019). Disease-associated mutations hyperactivate KIF1A motility and anterograde axonal transport of synaptic vesicle precursors. *Proceedings of the National Academy of Sciences of the United States of America*, *116*(37), 18429–18434. doi:10.1073/pnas.1905690116

Choudhary, B., Kamak, M., Ratnakaran, N., Kumar, J., Awasthi, A., Li, C., … Koushika, S.P. (2017). UNC-16/JIP3 regulates early events in synaptic vesicle protein trafficking via LRK-1/LRRK2 and AP complexes. *PLoS Genetics*, *13*(11), e1007100. doi:10.1371/journal.pgen.1007100

Cho, K., Cai, Y., Yi, H., Yeh, A., Aslanukov, A., & Ferreira, P. A. (2007). Association of the Kinesin-Binding Domain of RanBP2 to KIF5B and KIF5C Determines Mitochondria Localization and Function. *Traffic*, *8*(12), 1722–1735. https://doi.org/10.1111/j.1600-0854.2007.00647.x

Chua, J.J.E., Butkevich, E., Worseck, J.M., Kittelmann, M., Gronborg, M., Behrmann, E., … Jahn, R. (2012). Phosphorylation-regulated axonal dependent transport of syntaxin 1 is mediated by a Kinesin-1 adapter. *Proceedings of the National Academy of Sciences of the United States of America*, *109*(15), 5862–5867. doi:10.1073/pnas.1113819109

Colin, E., Zala, D., Liot, G., Rangone, H., Borrell-Pagès, M., Li, X.-J., … Humbert, S. (2008). Huntingtin phosphorylation acts as a molecular switch for anterograde/retrograde transport in neurons. *The EMBO Journal*, *27*(15), 2124–2134. doi:10.1038/emboj.2008.133

Cueva, J.G., Hsin, J., Huang, K.C., & Goodman, M.B. (2012). Posttranslational acetylation of α-tubulin constrains protofilament number in native microtubules. *Current Biology*, *22*(12), 1066–1074. doi:10.1016/j.cub.2012.05.012

Culotti, J.G., & Russell, R.L. (1978). Osmotic avoidance defective mutants of the nematode *Caenorhabditis elegans*. *Genetics*, *90* (2), 243–256.

Dahlstrom, A.B., Pfister, K.K., & Brady, S.T. (1991). The axonal transport motor 'kinesin' is bound to anterogradely transported organelles: quantitative cytofluorimetric studies of fast axonal transport in the rat. *Acta Physiologica Scandinavica*, *141*(4), 469–476. doi:10.1111/j.1748-1716.1991.tb09107.x

Dey, S., Banker, G., & Ray, K. (2017). Anterograde transport of Rab4-associated vesicles regulates synapse organization in Drosophila. *Cell Reports*, *18*(10), 2452–2463. doi:10.1016/j.celrep.2017.02.034

Dixit, R., Ross, J.L., Goldman, Y.E., & Holzbaur, E.L.F. (2008). Differential regulation of dynein and kinesin motor proteins by tau. *Science (New York, N.Y.).)*, *319*(5866), 1086–1089. doi:10.1126/science.1152993

Dubey, J., Ratnakaran, N., & Koushika, S.P. (2015). Neurodegeneration and microtubule dynamics: death by a thousand cuts. *Frontiers in Cellular Neuroscience*, *9*, 343. doi:10.3389/fncel.2015.00343

Edwards, S.L., Morrison, L.M., Yorks, R.M., Hoover, C.M., Boominathan, S., & Miller, K.G. (2015). UNC-16 (JIP3) acts through synapse-assembly proteins to inhibit the active transport of cell soma organelles to *Caenorhabditis elegans* motor neuron axons. *Genetics*, *201*(1), 117–141. doi:10.1534/genetics.115.177345

Endow, S.A. (1991). The emerging kinesin family of microtubule motor proteins. *Trends in Biochemical Sciences*, *16* (6), 221–225. doi:10.1016/0968-0004(91)90089-E

Fang, Y., Soares, L., Teng, X., Geary, M., & Bonini, N.M. (2012). A novel Drosophila model of nerve injury reveals an essential role of nmnat in maintaining axonal integrity. *Current Biology*, *22*(7), 590–595. doi:10.1016/j.cub.2012.01.065

Farías, G.G., Guardia, C.M., De Pace, R., Britt, D.J., & Bonifacino, J.S. (2017). BORC/kinesin-1 ensemble drives polarized transport of lysosomes into the axon. *Proceedings of the National Academy of Sciences of the United States of America*, *114*(14), E2955–E2964. doi:10.1073/pnas.1616363114

Fujita, T., Maturana, A.D., Ikuta, J., Hamada, J., Walchli, S., Suzuki, T., … Kuroda, S. (2007). Axonal guidance protein FEZ1 associates with tubulin and kinesin motor protein to transport mitochondria in neurites of NGF-stimulated PC12 cells. *Biochemical and Biophysical Research Communications*, *361*(3), 605–610. doi:10.1016/j.bbrc.2007.07.050

Ganguly, A., Tang, Y., Wang, L., Ladt, K., Loi, J., Dargent, B., … Roy, S. (2015). A dynamic formin-dependent deep F-actin network in axons. *The Journal of Cell Biology*, *210*(3), 401–417. doi:10.1083/jcb.201506110

Gauthier, L.R., Charrin, B.C., Borrell-Pagès, M., Dompierre, J.P., Rangone, H., Cordelières, F.P., … Saudou, F. (2004). Huntingtin controls neurotrophic support and survival of neurons by enhancing BDNF vesicular transport along microtubules. *Cell*, *118*(1), 127–138. doi:10.1016/j.cell.2004.06.018

Gennerich, A., & Vale, R.D. (2009). Walking the walk: how kinesin and dynein coordinate their steps. *Current Opinion in Cell Biology*, *21*(1), 59–67. doi:10.1016/j.ceb.2008.12.002

Goodwin, P.R., Sasaki, J.M., & Juo, P. (2012). Cyclin-dependent kinase 5 regulates the polarized trafficking of neuropeptide-containing dense-core vesicles in *Caenorhabditis elegans* motor neurons. *The Journal of Neuroscience: The Official Journal of the Society for Neuroscience*, *32*(24), 8158–8172. doi:10.1523/JNEUROSCI.0251-12.2012

Goodwin, S.S., & Vale, R.D. (2010). Patronin regulates the microtubule network by protecting microtubule minus ends. *Cell*, *143*(2), 263–274. doi:10.1016/j.cell.2010.09.022

Gross, S.P., Vershinin, M., & Shubeita, G.T. (2007). Cargo transport: Two motors are sometimes better than one. *Current Biology*, *17*(12), R478–R486. doi:10.1016/j.cub.2007.04.025

Guardia, C.M., Farías, G.G., Jia, R., Pu, J., & Bonifacino, J.S. (2016). BORC functions upstream of kinesins 1 and 3 to coordinate regional movement of lysosomes along different microtubule tracks. *Cell Reports*, *17*(8), 1950–1961. doi:10.1016/j.celrep.2016.10.062

Gumy, L.F., Katrukha, E.A., Grigoriev, I., Jaarsma, D., Kapitein, L.C., Akhmanova, A., & Hoogenraad, C.C. (2017). MAP2 defines a pre-axonal filtering zone to regulate KIF1- versus KIF5-dependent cargo transport in sensory neurons. *Neuron*, *94*(2), 347.e7–362.e7. doi:10.1016/j.neuron.2017.03.046

Guo, X., Macleod, G.T., Wellington, A., Hu, F., Panchumarthi, S., Schoenfield, M., … Zinsmaier, K.E. (2005). The GTPase dMiro is required for axonal transport of mitochondria to Drosophila synapses. *Neuron*, *47*(3), 379–393. doi:10.1016/j.neuron.2005.06.027

Hall, D.H., & Hedgecock, M. (1991). Kinesin-related gene *unc-104* is required for axonal transport of synaptic vesicles in *C. elegans*. *Cell*, *65* (5), 837–847. doi:10.1016/0092-8674(91)90391-B

Hammond, J.W., Blasius, T.L., Soppina, V., Cai, D., & Verhey, K.J. (2010). Autoinhibition of the kinesin-2 motor KIF17 via dual intramolecular mechanisms. *The Journal of Cell Biology*, *189*(6), 1013–1025. doi:10.1083/jcb.201001057

Hammond, J.W., Cai, D., Blasius, T.L., Li, Z., Jiang, Y., Jih, G.T., … Verhey, K.J. (2009). Mammalian kinesin-3 motors are dimeric in vivo and move by processive motility upon release of autoinhibition. *PLoS Biology*, *7*(3), e1000072. doi:10.1371/journal.pbio.1000072

Hammond, J.W., Huang, C.-F., Kaech, S., Jacobson, C., Banker, G., & Verhey, K.J. (2010). Posttranslational modifications of tubulin and the polarized transport of kinesin-1 in neurons. *Molecular Biology of the Cell*, *21*(4), 572–583. doi:10.1091/mbc.e09-01-0044

Han, S.M., Baig, H.S., & Hammarlund, M. (2016). Mitochondria localize to injured axons to support regeneration. *Neuron*, *92*(6), 1308–1323. doi:10.1016/j.neuron.2016.11.025

Hancock, W. O. (2018). Mechanics of bidirectional cargo transport. In S. M. King (Ed.), *Dyneins: Structure, Biology and Disease* (pp. 152–171). Elsevier.

He, J., Zhou, R., Wu, Z., Carrasco, M.A., Kurshan, P.T., Farley, J.E., … Zhuang, X. (2016). Prevalent presence of periodic actin-spectrin-based membrane skeleton in a broad range of neuronal cell types and animal species. *Proceedings of the National Academy of Sciences of the United States of America*, *113*(21), 6029–6034. doi:10.1073/pnas.1605707113

Hirokawa, N., & Noda, Y. (2008). Intracellular transport and kinesin superfamily proteins, KIFs: Structure, function, and dynamics. *Physiological Reviews*, *88*(3), 1089–1118. doi:10.1152/physrev.00023.2007

Holzbaur, E.L., & Goldman, Y.E. (2010). Coordination of molecular motors: From in vitro assays to intracellular dynamics. *Current Opinion in Cell Biology*, *22*(1), 4–13. doi:10.1016/j.ceb.2009.12.014

Hong, N.H., Qi, A., & Weaver, A.M. (2015). PI(3,5)P2 controls endosomal branched actin dynamics by regulating cortactin-actin interactions. *The Journal of Cell Biology*, *210*(5), 753–769. doi:10.1083/jcb.201412127

Horiuchi, D., Collins, C.A., Bhat, P., Barkus, R.V., DiAntonio, A., & Saxton, W.M. (2007). Control of a kinesin-Cargo linkage mechanism by JNK pathway kinases. *Current Biology*, *17*(15), 1313–1317. doi:10.1016/j.cub.2007.06.062

Hsu, C.-C., Moncaleano, J.D., & Wagner, O.I. (2011). Sub-cellular distribution of UNC-104(KIF1A) upon binding to adaptors as UNC-16(JIP3), DNC-1(DCTN1/Glued) and SYD-2(Liprin-α) in *C. elegans* neurons. *Neuroscience*, *176*, 39–52. doi:10.1016/j.neuroscience.2010.12.044

Hsu, J.-M., Chen, C.-H., Chen, Y.-C., McDonald, K.L., Gurling, M., Lee, A., … Pan, C.-L. (2014). Genetic analysis of a novel tubulin mutation that redirects synaptic vesicle targeting and causes neurite degeneration in *C. elegans*. *PLoS Genetics*, *10*(11), e1004715. doi:10.1371/journal.pgen.1004715

Ikenaka, K., Kawai, K., Katsuno, M., Huang, Z., Jiang, Y.-M., Iguchi, Y., … Sobue, G. (2013). dnc-1/dynactin 1 knockdown disrupts transport of autophagosomes and induces motor neuron degeneration. *PLoS One*, *8*(2), e54511. doi:10.1371/journal.pone.0054511

Iwasaki, K., & Toyonaga, R. (2000). The Rab3 GDP/GTP exchange factor homolog AEX-3 has a dual function in synaptic transmission. *The EMBO Journal*, *19*(17), 4806–4816. doi:10.1093/emboj/19.17.4806

Iwasaki, K., Staunton, J., Saifee, O., Nonet, M., & Thomas, J.H. (1997). aex-3 encodes a novel regulator of presynaptic activity in *C. elegans*. *Neuron*, *18*(4), 613–622. doi:10.1016/S0896-6273(00)80302-5

Janke, C., & Chloë Bulinski, J. (2011). Post-translational regulation of the microtubule cytoskeleton: Mechanisms and functions. *Nature Reviews Molecular Cell Biology*, *12*(12), 773–786. doi:10.1038/nrm3227

Johansson, M., Rocha, N., Zwart, W., Jordens, I., Janssen, L., Kuijl, C., … Neefjes, J. (2007). Activation of endosomal dynein motors by stepwise assembly of Rab7-RILP-p150Glued, ORP1L, and the receptor betaIII spectrin. *The Journal of Cell Biology*, *176*(4), 459–471. doi:10.1083/jcb.200606077

Kamal, A., Stokin, G.B., Yang, Z., Xia, C.-H., & Goldstein, L.S.B. (2000). Axonal transport of amyloid precursor protein is mediated by direct binding to the kinesin light chain subunit of kinesin-I. *Neuron*, *28*(2), 449–459. doi:10.1016/S0896-6273(00)00124-0

Kapitein, L.C., & Hoogenraad, C.C. (2011). Which way to go? Cytoskeletal organization and polarized transport in neurons. *Molecular and Cellular Neurosciences*, *46*(1), 9–20. doi:10.1016/j.mcn.2010.08.015

Karasmanis, E.P., Phan, C.-T., Angelis, D., Kesisova, I.A., Hoogenraad, C.C., McKenney, R.J., & Spiliotis, E.T. (2018). Polarity of neuronal membrane traffic requires sorting of kinesin motor cargo during entry into dendrites by a microtubule-associated septin. *Developmental Cell*, *46*(2), 204–218.e7. doi:10.1016/j.devcel.2018.06.013

Kevenaar, J.T., Bianchi, S., van Spronsen, M., Olieric, N., Lipka, J., Frias, C.P., … Hoogenraad, C.C. (2016). Kinesin-binding protein controls microtubule dynamics and cargo trafficking by regulating kinesin motor activity. *Current Biology*, *26*(7), 849–861. doi:10.1016/j.cub.2016.01.048

Kikkawa, M., Okada, Y., & Hirokawa, N. (2000). 15 A resolution model of the monomeric kinesin motor, KIF1A. *Cell*, *100*(2), 241–252. doi:10.1016/S0092-8674(00)81562-7

Kirszenblat, L., Neumann, B., Coakley, S., & Hilliard, M.A. (2013). A dominant mutation in *mec-7/β-tubulin* affects axon development and regeneration in *Caenorhabditis elegans* neurons. *Molecular Biology of the Cell*, *24*(3), 285–296. doi:10.1091/mbc.E12-06-0441

Klassen, M.P., Wu, Y.E., Maeder, C.I., Nakae, I., Cueva, J.G., Lehrman, E.K., … Shen, K. (2010). An arf-like small G protein, ARL-8, promotes the axonal transport of presynaptic cargoes by suppressing vesicle aggregation. *Neuron*, *66*(5), 710–723. doi:10.1016/j.neuron.2010.04.033

Klopfenstein, D.R., Tomishige, M., Stuurman, N., & Vale, R.D. (2002). Role of phosphatidylinositol(4,5)bisphosphate organization in membrane transport by the Unc104 kinesin motor. *Cell*, *109*(3), 347–358. doi:10.1016/S0092-8674(02)00708-0

Klopfenstein, D.R., & Vale, R.D. (2004). The lipid binding pleckstrin homology domain in UNC-104 kinesin is necessary for synaptic vesicle transport in *Caenorhabditis elegans*□D □V. *Molecular Biology of the Cell*, *15* (8), 3729–3739. doi:10.1091/mbc.e04-04-0326

Koushika, S.P., Schaefer, A.M., Vincent, R., Willis, J.H., Bowerman, B., & Nonet, M.L. (2004). Mutations in *Caenorhabditis elegans* cytoplasmic dynein components reveal specificity of neuronal retrograde cargo. *The Journal of Neuroscience: The Official Journal of the Society for Neuroscience*, *24*(16), 3907–3916. doi:10.1523/JNEUROSCI.5039-03.2004

Kumar, J., Choudhary, B.C., Metpally, R., Zheng, Q., Nonet, M.L., Ramanathan, S., … Koushika, S.P. (2010). The *Caenorhabditis elegans* kinesin-3 motor UNC-104/KIF1A is degraded upon loss of specific binding to cargo. *PLoS Genetics*, *6*(11), e1001200. doi:10.1371/journal.pgen.1001200

Kurup, N., Li, Y., Goncharov, A., & Jin, Y. (2018). Intermediate filament accumulation can stabilize microtubules in *Caenorhabditis elegans* motor neurons. *Proceedings of the National Academy of Sciences of the United States of America*, *115*(12), 3114–3119. doi:10.1073/pnas.1721930115

Kurup, N., Yan, D., Goncharov, A., & Jin, Y. (2015). Dynamic microtubules drive circuit rewiring in the absence of neurite remodeling. *Current Biology*, *25*(12), 1594–1605. doi:10.1016/j.cub.2015.04.061

Kurup, N., Yan, D., Kono, K., & Jin, Y. (2017). Differential regulation of polarized synaptic vesicle trafficking and synapse stability in neural circuit rewiring in *Caenorhabditis elegans*. *PLoS Genetics*, *13*(6), e1006844. doi:10.1371/journal.pgen.1006844

Lai, T., & Garriga, G. (2004). The conserved kinase UNC-51 acts with VAB-8 and UNC-14 to regulate axon outgrowth in *C. elegans*. *Development*, *131*(23), 5991–6000. doi:10.1242/dev.01457

Leterrier, C. (2018). The axon initial segment: An updated viewpoint. *The Journal of Neuroscience*, *38*(9), 2135–2145. doi:10.1523/JNEUROSCI.1922-17.2018

Li, J.-Y., Pfister, K.K., Brady, S., & Dahlström, A. (1999). Axonal transport and distribution of immunologically distinct kinesin heavy chains in rat neurons. *Journal of Neuroscience Research*, *58* (2), 226–241. doi:10.1002/(SICI)1097-4547(19991015)58:2<226::AID-JNR3>3.0.CO;2-X

Li, L.-B., Lei, H., Arey, R.N., Li, P., Liu, J., Murphy, C.T., … Shen, K. (2016). The neuronal kinesin UNC-104/KIF1A is a key regulator of synaptic aging and insulin signaling-regulated memory. *Current Biology*, *26*(5), 605–615. doi:10.1016/j.cub.2015.12.068

Li, S.-H., Gutekunst, C.-A., Hersch, S.M., & Li, X.-J. (1998). Interaction of huntingtin-associated protein with dynactin $P150^{Glued}$. *The Journal of Neuroscience: The Official Journal of the Society for Neuroscience*, *18*(4), 1261–1269. doi:10.1523/JNEUROSCI.18-04-01261.1998

Ligon, L.A., Tokito, M., Finklestein, J.M., Grossman, F.E., & Holzbaur, E.L.F. (2004). A direct interaction between cytoplasmic dynein and kinesin I may coordinate motor activity. *The Journal of Biological Chemistry*, *279*(18), 19201–19208. doi:10.1074/jbc.M313472200

Lipton, D.M., Maeder, C.I., & Shen, K. (2018). Rapid assembly of presynaptic materials behind the growth cone in dopaminergic neurons is mediated by precise regulation of axonal transport. *Cell Reports*, *24*(10), 2709–2722. doi:10.1016/j.celrep.2018.07.096

Maday, S., Twelvetrees, A.E., Moughamian, A.J., & Holzbaur, E.L.F. (2014). Axonal transport: Cargo-specific mechanisms of motility and regulation. *Neuron, 84*(2), 292–309. doi:10.1016/j.neuron.2014.10.019

Magiera, M.M., Bodakuntla, S., Žiak, J., Lacomme, S., Marques Sousa, P., Leboucher, S., … Janke, C. (2018). Excessive tubulin polyglutamylation causes neurodegeneration and perturbs neuronal transport. *The EMBO Journal, 37*(23). doi:10.15252/embj.2018100440

Mallik, R., & Gross, S.P. (2009). Intracellular transport: How do motors work together? *Current Biology, 19*(10), R416–R418. doi:10.1016/j.cub.2009.04.007

Mandal, A., & Drerup, C.M. (2019). Axonal transport and mitochondrial function in neurons. *Frontiers in Cellular Neuroscience, 13*, 373. doi:10.3389/fncel.2019.00373

Mandelkow, E.-M., Thies, E., Trinczek, B., Biernat, J., & Mandelkow, E. (2004). MARK/PAR1 kinase is a regulator of microtubule-dependent transport in axons. *Journal of Cell Biology, 167*(1), 99–110. doi:10.1083/jcb.200401085

Maniar, T.A., Kaplan, M., Wang, G.J., Shen, K., Wei, L., Shaw, J.E., … Bargmann, C.I. (2011). UNC-33 (CRMP) and ankyrin organize microtubules and localize kinesin to polarize axon-dendrite sorting. *Nature Neuroscience, 15*(1), 48–56. doi:10.1038/nn.2970

Marcette, J.D., Chen, J.J., & Nonet, M.L. (2014). The *Caenorhabditis elegans* microtubule minus-end binding homolog PTRN-1 stabilizes synapses and neurites. *eLife, 3*, e01637. doi:10.7554/eLife.01637

Matsuda, S., Yasukawa, T., Homma, Y., Ito, Y., Niikura, T., Hiraki, T., … Nishimoto, I. (2001). c-Jun N-terminal kinase (JNK)-interacting protein-1b/islet-brain-1 scaffolds Alzheimer's amyloid precursor protein with JNK. *The Journal of Neuroscience: The Official Journal of the Society for Neuroscience, 21*(17), 6597–6607. doi:10.1523/JNEUROSCI.21-17-06597.2001

Mikhaylova, M., Cloin, B.M.C., Finan, K., van den Berg, R., Teeuw, J., Kijanka, M.M., … Kapitein, L.C. (2015). Resolving bundled microtubules using anti-tubulin nanobodies. *Nature Communications, 6*(1), 7933. doi:10.1038/ncomms8933

Miller, K.G., Alfonso, A., Nguyen, M., Crowell, J.A., Johnson, C.D., & Rand, J.B. (1996). A genetic selection for *Caenorhabditis elegans* synaptic transmission mutants. *Proceedings of the National Academy of Sciences of the United States of America, 93*(22), 12593–12598. doi:10.1073/pnas.93.22.12593

Mitchell, D.J., Blasier, K.R., Jeffery, E.D., Ross, M.W., Pullikuth, A.K., Suo, D., … Pfister, K.K. (2012). Trk activation of the ERK1/2 kinase pathway stimulates intermediate chain phosphorylation and recruits cytoplasmic dynein to signaling endosomes for retrograde axonal transport. *The Journal of Neuroscience, 32*(44), 15495–15510. doi:10.1523/JNEUROSCI.5599-11.2012

Mondal, S., Ahlawat, S., Rau, K., Venkataraman, V., & Koushika, S.P. (2011). Imaging in vivo neuronal transport in genetic model organisms using microfluidic devices. *Traffic, 12*(4), 372–385. doi:10.1111/j.1600-0854.2010.01157.x

Monroy, B.Y., Sawyer, D.L., Ackermann, B.E., Borden, M.M., Tan, T.C., & Ori-McKenney, K.M. (2018). Competition between microtubule-associated proteins directs motor transport. *Nature Communications, 9*(1), 1487. doi:10.1038/s41467-018-03909-2

Monroy, B.Y., Tan, T.C., Oclaman, J.M., Han, J.S., Simó, S., Niwa, S., … Ori-McKenney, K.M. (2020). A combinatorial MAP code dictates polarized microtubule transport. *Developmental Cell, 53*(1), 60–72.e4. doi:10.1016/j.devcel.2020.01.029

Montagnac, G., Sibarita, J.-B., Loubéry, S., Daviet, L., Romao, M., Raposo, G., & Chavrier, P. (2009). ARF6 interacts with JIP4 to control a motor switch mechanism regulating endosome traffic in cytokinesis. *Current Biology, 19*(3), 184–195. doi:10.1016/j.cub.2008.12.043

Morfini, G. A., Burns, M. R., Stenoien, D. L., & Brady, S. T. (2012). Axonal Transport. In S. T. Brady, G. J. Siegel, R. W. Albers & D. L. Price (Eds.), *Basic Neurochemistry* (pp. 146–164). Elsevier.

Morris, R.L., & Hollenbeck, P.J. (1995). Axonal transport of mitochondria along microtubules and F-actin in living vertebrate neurons. *The Journal of Cell Biology, 131*(5), 1315–1326. doi:10.1083/jcb.131.5.1315

Moughamian, A.J., Osborn, G.E., Lazarus, J.E., Maday, S., & Holzbaur, E.L.F. (2013). Ordered recruitment of dynactin to the microtubule plus-end is required for efficient initiation of retrograde axonal transport. *The Journal of Neuroscience: The Official Journal of the Society for Neuroscience, 33*(32), 13190–13203. doi:10.1523/JNEUROSCI.0935-13.2013

Muresan, V., Stankewich, M.C., Steffen, W., Morrow, J.S., Holzbaur, E.L.F., & Schnapp, B.J. (2001). Dynactin-dependent, dynein-driven vesicle transport in the absence of membrane proteins: A role for spectrin and acidic phospholipids. *Molecular Cell, 7* (1), 173–183. doi:10.1016/s1097-2765(01)00165-4

Murthy, K., Bhat, J.M., & Koushika, S.P. (2011). In vivo imaging of retrogradely transported synaptic vesicle proteins in *Caenorhabditis elegans* neurons. *Traffic, 12*(1), 89–101. doi:10.1111/j.1600-0854.2010.01127.x

Neumann, B., & Hilliard, M.A. (2014). Loss of MEC-17 leads to microtubule instability and axonal degeneration. *Cell Reports, 6*(1), 93–103. doi:10.1016/j.celrep.2013.12.004

Niwa, S., Lipton, D.M., Morikawa, M., Zhao, C., Hirokawa, N., Lu, H., & Shen, K. (2016). Autoinhibition of a neuronal kinesin UNC-104/KIF1A regulates the size and density of synapses. *Cell Reports, 16*(8), 2129–2141. doi:10.1016/j.celrep.2016.07.043

Niwa, S., Tanaka, Y., & Hirokawa, N. (2008). KIF1Bbeta- and KIF1A-mediated axonal transport of presynaptic regulator Rab3 occurs in a GTP-dependent manner through DENN/MADD. *Nature Cell Biology, 10*(11), 1269–1279. doi:10.1038/ncb1785

Niwa, S., Tao, L., Lu, S.Y., Liew, G.M., Feng, W., Nachury, M.V., & Shen, K. (2017). BORC regulates the axonal transport of synaptic vesicle precursors by activating ARL-8. *Current Biology, 27*(17), 2569–2578.e4. doi:10.1016/j.cub.2017.07.013

Noma, K., Goncharov, A., Ellisman, M.H., & Jin, Y. (2017). Microtubule-dependent ribosome localization in *C. elegans* neurons. *eLife, 6*, e26376. doi:10.7554/eLife.26376

O'Hagan, R., & Barr, M. (2012). Regulation of tubulin glutamylation plays cell-specific roles in the function and stability of sensory cilia. *Worm, 1*(3), 155–159. 10.4161/worm.19539

O'Hagan, R., Piasecki, B.P., Silva, M., Phirke, P., Nguyen, K.C.Q., Hall, D.H., … Barr, M.M. (2011). The tubulin deglutamylase CCPP-1 regulates the function and stability of sensory cilia in *C. elegans*. *Current Biology, 21*(20), 1685–1694. 10.1016/j.cub.2011.08.049

O'Hagan, R., Silva, M., Nguyen, K.C.Q., Zhang, W., Bellotti, S., Ramadan, Y.H., … Barr, M.M. (2017). Glutamylation regulates transport, specializes function, and sculpts the structure of cilia. *Current Biology, 27*(22), 3430.e6–3441.e6. doi:10.1016/j.cub.2017.09.066

Okada, Y., Yamazaki, H., Sekine-Aizawa, Y., & Hirokawa, N. (1995). The neuron-specific kinesin superfamily protein KIF1A is a unique monomeric motor for anterograde axonal transport of synaptic vesicle precursors. *Cell, 81* (5), 769–780. doi:10.1016/0092-8674(95)90538-3

Olenick, M.A., Tokito, M., Boczkowska, M., Dominguez, R., & Holzbaur, E.L.F. (2016). Hook adaptors induce unidirectional processive motility by enhancing the dynein-dynactin interaction. *The Journal of Biological Chemistry, 291*(35), 18239–18251. doi:10.1074/jbc.M116.738211

Otsuka, A.J., Jeyaprakash, A., García-Añoveros, J., Tang, L.Z., Fisk, G., Hartshorne, T., … Bornt, T. (1991). The *C. elegans unc-104* gene encodes a putative kinesin heavy chain-like protein. *Neuron, 6*(1), 113–122. doi:10.1016/0896-6273(91)90126-K

Ou, C.-Y., Poon, V.Y., Maeder, C.I., Watanabe, S., Lehrman, E.K., Fu, A.K.Y., … Shen, K. (2010). Two cyclin-dependent kinase pathways are essential for polarized trafficking of presynaptic components. *Cell, 141*(5), 846–858. doi:10.1016/j.cell.2010.04.011

Pack-Chung, E., Kurshan, P.T., Dickman, D.K., & Schwarz, T.L. (2007). A Drosophila kinesin required for synaptic bouton formation and synaptic vesicle transport. *Nature Neuroscience, 10*(8), 980–989. doi:10.1038/nn1936

Pan, X., Cao, Y., Stucchi, R., Hooikaas, P.J., Portegies, S., Will, L., … Hoogenraad, C.C. (2019). MAP7D2 localizes to the proximal axon and locally promotes kinesin-1-mediated cargo transport into the axon. *Cell Reports, 26*(8), 1988.e6–1999.e6. doi:10.1016/j.celrep.2019.01.084

Park, M., Watanabe, S., Poon, V.Y.N., Ou, C.-Y., Jorgensen, E.M., & Shen, K. (2011). CYY-1/cyclin Y and CDK-5 differentially regulate synapse elimination and formation for rewiring neural circuits. *Neuron, 70*(4), 742–757. doi:10.1016/j.neuron.2011.04.002

Patel, N., Thierry-Mieg, D., & Mancillas, J.R. (1993). Cloning by insertional mutagenesis of a cDNA encoding *Caenorhabditis elegans* kinesin heavy chain. *Proceedings of the National Academy of Sciences of the United States of America, 90*(19), 9181–9185. doi:10.1073/pnas.90.19.9181

Perrot, R., & Julien, J.-P. (2009). Real-time imaging reveals defects of fast axonal transport induced by disorganization of intermediate filaments. *FASEB Journal: Official Publication of the Federation of American Societies for Experimental Biology, 23*(9), 3213–3225. doi:10.1096/fj.09-129585

Prevo, B., Scholey, J.M., & Peterman, E.J.G. (2017). Intraflagellar transport: mechanisms of motor action, cooperation, and cargo delivery. *The FEBS Journal, 284*(18), 2905–2931. doi:10.1111/febs.14068

Qu, X., Kumar, A., Blockus, H., Waites, C., & Bartolini, F. (2019). Activity-dependent nucleation of dynamic microtubules at presynaptic boutons controls neurotransmission. *Current Biology, 29*(24), 4231–4240.e5. doi:10.1016/j.cub.2019.10.049

Rao, A.N., & Baas, P.W. (2018). Polarity sorting of microtubules in the axon. *Trends in Neurosciences, 41*(2), 77–88. doi:10.1016/j.tins.2017.11.002

Rawson, R.L., Yam, L., Weimer, R.M., Bend, E.G., Hartwieg, E., Horvitz, H.R., … Jorgensen, E.M. (2014). Axons degenerate in the absence of mitochondria in *C. elegans*. *Current Biology, 24*(7), 760–765. doi:10.1016/j.cub.2014.02.025

Ray, K., Perez, S.E., Yang, Z., Xu, J., Ritchings, B.W., Steller, H., & Goldstein, L.S.B. (1999). Kinesin-II Is required for axonal transport of choline acetyltransferase in Drosophila. *The Journal of Cell Biology, 147*(3), 507–518. doi:10.1083/jcb.147.3.507

Richardson, C.E., Spilker, K.A., Cueva, J.G., Perrino, J., Goodman, M.B., & Shen, K. (2014). PTRN-1, a microtubule minus end-binding CAMSAP homolog, promotes microtubule function in *Caenorhabditis elegans* neurons. *eLife, 3*, e01498. doi:10.7554/eLife.01498

Roberts, A.J., Kon, T., Knight, P.J., Sutoh, K., & Burgess, S.A. (2013). Functions and mechanics of dynein motor proteins. *Nature Reviews Molecular Cell Biology, 14*(11), 713–726. doi:10.1038/nrm3667

Ross, J.L., Shuman, H., Holzbaur, E.L.F., & Goldman, Y.E. (2008). Kinesin and dynein-dynactin at intersecting microtubules: Motor density affects dynein function. *Biophysical Journal, 94*(8), 3115–3125. doi:10.1529/biophysj.107.120014

Roy, S. (2020). Finding order in slow axonal transport. *Current Opinion in Neurobiology, 63*, 87–94. doi:10.1016/j.conb.2020.03.015

Sabharwal, V., & Koushika, S.P. (2019). Crowd control: Effects of physical crowding on cargo movement in healthy and diseased neurons. *Frontiers in Cellular Neuroscience, 13*, 470. doi:10.3389/fncel.2019.00470

Sainath, R., & Gallo, G. (2015). Cytoskeletal and signaling mechanisms of neurite formation. *Cell and Tissue Research, 359*(1), 267–278. doi:10.1007/s00441-014-1955-0

Sakamoto, R., Byrd, D.T., Brown, H.M., Hisamoto, N., Matsumoto, K., & Jin, Y. (2005). The *Caenorhabditis elegans* UNC-14 RUN domain protein binds to the kinesin-1 and UNC-16 complex and regulates synaptic vesicle localization. *Molecular Biology of the Cell, 16*(2), 483–496. doi:10.1091/mbc.e04-07-0553

Schroeder, H.W., Mitchell, C., Shuman, H., Holzbaur, E.L.F., & Goldman, Y.E. (2010). Motor number controls cargo switching at actin-microtubule intersections *in vitro*. *Current Biology, 20*(8), 687–696. doi:10.1016/j.cub.2010.03.024

Shakir, M.A., Fukushige, T., Yasuda, H., Miwa, J., & Siddiqui, S.S. (1993). *C. elegans* osm-3 gene mediating osmotic avoidance behaviour encodes a kinesin-like protein. *Neuroreport, 4*(7), 891–894. doi:10.1097/00001756-199307000-00013

Shakiryanova, D., Tully, A., & Levitan, E.S. (2006). Activity-dependent synaptic capture of transiting peptidergic vesicles. *Nature Neuroscience, 9*(7), 896–900. doi:10.1038/nn1719

Shao, C.-Y., Zhu, J., Xie, Y.-J., Wang, Z., Wang, Y.-N., Wang, Y., … Shen, Y. (2013). Distinct functions of nuclear distribution proteins LIS1, Ndel1 and NudCL in regulating axonal mitochondrial transport: Nud proteins regulate mitochondrial transport. *Traffic, 14*(7), 785–797. doi:10.1111/tra.12070

Shida, T., Cueva, J.G., Xu, Z., Goodman, M.B., & Nachury, M.V. (2010). The major alpha-tubulin K40 acetyltransferase alphaTAT1 promotes rapid ciliogenesis and efficient mechanosensation. *Proceedings of the National Academy of Sciences of the United States of America, 107*(50), 21517–21522. doi:10.1073/pnas.1013728107

Siddiqui, N., & Straube, A. (2017). Intracellular cargo transport by kinesin-3 motors. *Biochemistry. Biokhimiia, 82*(7), 803–815. doi:10.1134/S0006297917070057

Siddiqui, S.S. (2002). Metazoan motor models: Kinesin superfamily in *C. elegans*. *Traffic, 3*(1), 20–28. doi:10.1034/j.1600-0854.2002.30104.x

Sood, P., Murthy, K., Kumar, V., Nonet, M.L., Menon, G.I., & Koushika, S.P. (2018). Cargo crowding at actin-rich regions along axons causes local traffic jams. *Traffic, 19*(3), 166–181. doi:10.1111/tra.12544

Splinter, D., Razafsky, D.S., Schlager, M.A., Serra-Marques, A., Grigoriev, I., Demmers, J., … Akhmanova, A. (2012). BICD2, dynactin, and LIS1 cooperate in regulating dynein recruitment to cellular structures. *Molecular Biology of the Cell, 23*(21), 4226–4241. doi:10.1091/mbc.E12-03-0210

Stavoe, A.K.H., Hill, S.E., Hall, D.H., & Colón-Ramos, D.A. (2016). KIF1A/UNC-104 transports ATG-9 to regulate neurodevelopment and autophagy at synapses. *Developmental Cell, 38*(2), 171–185. doi:10.1016/j.devcel.2016.06.012

Stewart, E., & Shen, K. (2015). STORMing towards a clear picture of the cytoskeleton in neurons. *eLife, 4*. doi:10.7554/eLife.06235

Su, C. W., Tharin, S., Jin, Y., Wightman, B., Spector, M., Meili, D., Tsung, N., Rhiner, C., Bourikas, D., Stoeckli, E., Garriga, G., Horvitz, H. R., & Hengartner, M. O. (2006). The short coiled-coil domain-containing protein UNC-69 cooperates with UNC-76 to regulate axonal outgrowth and normal presynaptic organization in *Caenorhabditis elegans*. Journal of biology, 5(4), 9. https://doi.org/10.1186/jbiol39

Sure, G.R., Chatterjee, A., Mishra, N., Sabharwal, V., Devireddy, S., Awasthi, A., … Koushika, S.P. (2018). UNC-16/JIP3 and UNC-76/FEZ1 limit the density of mitochondria in *C. elegans* neurons by maintaining the balance of anterograde and retrograde mitochondrial transport. *Scientific Reports, 8*(1), 8938. doi:10.1038/s41598-018-27211-9

Tas, R.P., Chazeau, A., Cloin, B.M.C., Lambers, M.L.A., Hoogenraad, C.C., & Kapitein, L.C. (2017). Differentiation between oppositely oriented microtubules controls polarized neuronal transport. *Neuron, 96*(6), 1264–1271.e5. doi:10.1016/j.neuron.2017.11.018

Taylor, C.A., Yan, J., Howell, A.S., Dong, X., & Shen, K. (2015). RAB-10 regulates dendritic branching by balancing dendritic transport. *PLoS Genetics, 11*(12), e1005695. doi:10.1371/journal.pgen.1005695

Thapliyal, S., Vasudevan, A., Dong, Y., Bai, J., Koushika, S.P., & Babu, K. (2018). The C-terminal of CASY-1/Calsyntenin regulates GABAergic synaptic transmission at the *Caenorhabditis elegans* neuromuscular junction. *PLoS Genetics, 14*(3), e1007263. doi:10.1371/journal.pgen.1007263

Thomas, J.H. (1990). Genetic analysis of defecation in *Caenorhabditis elegans*. *Trends in Genetics, 6*, 175. doi:10.1016/0168-9525(90)90166-4

Tien, N.-W., Wu, G.-H., Hsu, C.-C., Chang, C.-Y., & Wagner, O.I. (2011). Tau/PTL-1 associates with kinesin-3 KIF1A/UNC-104 and affects the motor's motility characteristics in *C. elegans* neurons. *Neurobiology of Disease, 43*(2), 495–506. doi:10.1016/j.nbd.2011.04.023

Tomishige, M., Klopfenstein, D.R., & Vale, R.D. (2002). Conversion of UNC104/KIF1A kinesin into a processive motor after dimerization. *Science, 297*(5590), 2263–2267. doi:10.1126/science.1073386

Topalidou, I., Keller, C., Kalebic, N., Nguyen, K.C.Q., Somhegyi, H., Politi, K.A., ... Chalfie, M. (2012). Genetically separable functions of the MEC-17 tubulin acetyltransferase affect microtubule organization. *Current Biology*, *22*(12), 1057–1065. doi:10.1016/j.cub.2012.03.066

Tsuboi, D., Hikita, T., Qadota, H., Amano, M., & Kaibuchi, K. (2005). Regulatory machinery of UNC-33 Ce-CRMP localization in neurites during neuronal development in *Caenorhabditis elegans*. *Journal of Neurochemistry*, *95*(6), 1629–1641. doi:10.1111/j.1471-4159.2005.03490.x

Urnavicius, L., Lau, C.K., Elshenawy, M.M., Morales-Rios, E., Motz, C., Yildiz, A., & Carter, A.P. (2018). Cryo-EM shows how dynactin recruits two dyneins for faster movement. *Nature*, *554*(7691), 202–206. doi:10.1038/nature25462

van Spronsen, M., Mikhaylova, M., Lipka, J., Schlager, M.A., van den Heuvel, D.J., Kuijpers, M., ... Hoogenraad, C.C. (2013). TRAK/Milton motor-adaptor proteins steer mitochondrial trafficking to axons and dendrites. *Neuron*, *77*(3), 485–502. doi:10.1016/j.neuron.2012.11.027

Venkatesh, K., Mathew, A., & Koushika, S.P. (2020). Role of actin in organelle trafficking in neurons. *Cytoskeleton*, *77*(3–4), 97–109. doi:10.1002/cm.21580

Verbrugge, S., van den Wildenberg, S.M.J.L., & Peterman, E.J.G. (2009). Novel ways to determine kinesin-1's run length and randomness using fluorescence microscopy. *Biophysical Journal*, *97*(8), 2287–2294. doi:10.1016/j.bpj.2009.08.001

Vershinin, M., Carter, B.C., Razafsky, D.S., King, S.J., & Gross, S.P. (2007). Multiple-motor based transport and its regulation by Tau. *Proceedings of the National Academy of Sciences of the United States of America*, *104*(1), 87–92. doi:10.1073/pnas.0607919104

Wagner, O.I., Esposito, A., Kohler, B., Chen, C.-W., Shen, C.-P., Wu, G.-H., ... Klopfenstein, D.R. (2009). Synaptic scaffolding protein SYD-2 clusters and activates kinesin-3 UNC-104 in *C. elegans*. *Proceedings of the National Academy of Sciences of the United States of America*, *106*(46), 19605–19610. doi:10.1073/pnas.0902949106

Weiss, P., & Hiscoe, H.B. (1948). Experiments on the mechanism of nerve growth. *The Journal of Experimental Zoology*, *107*(3), 315–395. doi:10.1002/jez.1401070302

Wolf, F.W., Hung, M.-S., Wightman, B., Way, J., & Garriga, G. (1998). vab-8 Is a key regulator of posteriorly directed migrations in *C. elegans* and encodes a novel protein with kinesin motor similarity. *Neuron*, *20*(4), 655–666. doi:10.1016/S0896-6273(00)81006-5

Wu, G.-H., Muthaiyan Shanmugam, M., Bhan, P., Huang, Y.-H., & Wagner, O.I. (2016). Identification and characterization of LIN-2(CASK) as a regulator of kinesin-3 UNC-104(KIF1A) motility and clustering in neurons: Regulation of UNC-104 via LIN-2 and SYD-2. *Traffic*, *17*(8), 891–907. doi:10.1111/tra.12413

Wu, Y.E., Huo, L., Maeder, C.I., Feng, W., & Shen, K. (2013). The Balance between capture and dissociation of presynaptic proteins controls the spatial distribution of synapses. *Neuron*, *78*(6), 994–1011. doi:10.1016/j.neuron.2013.04.035

Yan, J., Chao, D.L., Toba, S., Koyasako, K., Yasunaga, T., Hirotsune, S., & Shen, K. (2013). Kinesin-1 regulates dendrite microtubule polarity in *Caenorhabditis elegans*. *eLife*, *2*, e00133. doi:10.7554/eLife.00133

Yogev, S., Cooper, R., Fetter, R., Horowitz, M., & Shen, K. (2016). Microtubule organization determines axonal transport dynamics. *Neuron*, *92*(2), 449–460. doi:10.1016/j.neuron.2016.09.036

Yonekawa, Y., Harada, A., Okada, Y., Funakoshi, T., Kanai, Y., Takei, Y., ... Hirokawa, N. (1998). Defect in synaptic vesicle precursor transport and neuronal cell death in KIF1A motor protein-deficient mice. *Journal of Cell Biology*, *141*(2), 431–441. doi:10.1083/jcb.141.2.431

Yu, C.-C., Barry, N.C., Wassie, A.T., Sinha, A., Bhattacharya, A., Asano, S., ... Boyden, E.S. (2020). Expansion microscopy of *C. elegans*. *eLife*, *9*, e46249. doi:10.7554/eLife.46249

Yuan, A., Rao, M.V., Veeranna., & Nixon, R.A. (2017). Neurofilaments and neurofilament proteins in health and disease. *Cold Spring Harbor Perspectives in Biology*, *9*(4), a018309. doi:10.1101/cshperspect.a018309

Yue, Y., Sheng, Y., Zhang, H.-N., Yu, Y., Huo, L., Feng, W., & Xu, T. (2013). The CC1-FHA dimer is essential for KIF1A-mediated axonal transport of synaptic vesicles in *C. elegans*. *Biochemical and Biophysical Research Communications*, *435*(3), 441–446. doi:10.1016/j.bbrc.2013.05.005

Zahn, T.R., Angleson, J.K., MacMorris, M.A., Domke, E., Hutton, J.F., Schwartz, C., & Hutton, J.C. (2004). Dense core vesicle dynamics in *Caenorhabditis elegans* neurons and the role of kinesin UNC-104: IDA-1:: GFP transport in *C. elegans* neurons. *Traffic*, *5*(7), 544–559. doi:10.1111/j.1600-0854.2004.00195.x

Zhao, C., Takita, J., Tanaka, Y., Setou, M., Nakagawa, T., Takeda, S., ... Hirokawa, N. (2001). Charcot-marie-tooth disease type 2A caused by mutation in a microtubule motor KIF1Bβ. *Cell*, *105*(5), 587–597. doi:10.1016/S0092-8674(01)00363-4

Zheng, Q., Ahlawat, S., Schaefer, A., Mahoney, T., Koushika, S.P., & Nonet, M.L. (2014). The vesicle protein SAM-4 regulates the processivity of synaptic vesicle transport. *PLoS Genetics*, *10*(10), e1004644. doi:10.1371/journal.pgen.1004644

C. elegans MAGU-2/Mpp5 homolog regulates epidermal phagocytosis and synapse density

Salvatore J. Cherra III, Alexandr Goncharov, Daniela Boassa, Mark Ellisman and Yishi Jin

ABSTRACT
Synapses are dynamic connections that underlie essential functions of the nervous system. The addition, removal, and maintenance of synapses govern the flow of information in neural circuits throughout the lifetime of an animal. While extensive studies have elucidated many intrinsic mechanisms that neurons employ to modulate their connections, increasing evidence supports the roles of non-neuronal cells, such as glia, in synapse maintenance and circuit function. We previously showed that *C. elegans* epidermis regulates synapses through ZIG-10, a cell-adhesion protein of the immunoglobulin domain superfamily. Here we identified a member of the Pals1/MPP5 family, MAGU-2, that functions in the epidermis to modulate phagocytosis and the number of synapses by regulating ZIG-10 localization. Furthermore, we used light and electron microscopy to show that this epidermal mechanism removes neuronal membranes from the neuromuscular junction, dependent on the conserved phagocytic receptor CED-1. Together, our study shows that *C. elegans* epidermis constrains synaptic connectivity, in a manner similar to astrocytes and microglia in mammals, allowing optimized output of neural circuits.

Introduction

Synapses enable the transmission and integration of information within the nervous system. Proper synaptic connectivity is essential to govern nervous system functions such as sensory perception, learning, and coordinated movement. Aberrant synaptic connections have been associated with a variety of neurological disorders. Synapse formation, elimination, and maintenance work together to ensure precision and plasticity of neuronal circuits throughout the lifetime of an animal. Many neuronal-intrinsic mechanisms involve various classes of cell surface proteins and intracellular signaling pathways that establish synaptic connections (Cherra & Jin, 2015; de Wit & Ghosh, 2016; Sudhof, 2018). Additional work has shown how extrinsic mechanisms involving non-neuronal cells, such as astrocytes and microglia, cooperate with neurons to modulate neuronal circuits (Allen & Eroglu, 2017; Chung, Allen, & Eroglu, 2015). It is now well established that astrocytes and microglia play an active role in the pruning of synaptic connections during development of the visual system in mice (Chung *et al.*, 2013; Stevens *et al.*, 2007). In *Drosophila*, glial cells also remove synapses and axonal components as a means to eliminate neuronal connections (Awasaki *et al.*, 2006; Fuentes-Medel *et al.*, 2009).

One common mechanism for non-neuronal cells to assist in the wiring and rewiring of circuits is through the phagocytosis pathway. The phagocytosis pathway is highly conserved throughout evolution and plays essential roles in the removal of cell corpses and cellular debris (Bangs, Franc, & White, 2000; Mangahas & Zhou, 2005). While the core machinery that orchestrates engulfment has been widely studied, additional mechanisms that modulate the initiation or target specificity of this pathway still remain largely unknown. The *C. elegans* motor circuit provides a simple model to understand how non-neuronal cells modulate synapse elimination via the phagocytosis pathway. Our previous work uncovered a cell surface protein, ZIG-10, which mediates an interaction between the epidermis and neurons. The cell-cell interaction mediated by ZIG-10 enables the elimination of synapses through the activation of the phagocytosis pathway (Cherra & Jin, 2016).

In this study we have further investigated how the phagocytosis pathway maintains optimal synaptic connectivity in the motor circuit. We have discovered that a member of the MPP5/Pals1 membrane-associated guanylate kinase (MAGUK) family, MAGU-2, regulates synapse density. MAGU-2 functions in the epidermis and regulates ZIG-10 localization. Furthermore, we show that motor neuron membrane transfer to the epidermis is dependent on the phagocytosis receptor,

CED-1. Our study highlights a new role for MAGUKs in regulating synaptic connectivity.

Results

Identification of MAGU-2 in ZIG-10 pathway

We have previously reported that maintenance of synaptic connectivity in the *C. elegans* motor circuit requires components of the phagocytosis pathway acting in the epidermis (Cherra & Jin, 2016). To further understand the regulation of phagocytosis-mediated synapse elimination, we screened for additional genes using a function-based assay. Ectopic expression of ZIG-10 in the ventral cord GABAergic motor neurons causes a reduction in GABAergic synapse number (Cherra & Jin, 2016) and enhances locomotor deficit caused by an acetylcholine receptor mutation, *acr-2(n2420)* (Jospin *et al.*, 2009) (Supplemental Figure 1(a–c)). We employed this ectopic ZIG-10 expression induced phenotype to identify intracellular signaling molecules using RNAi against genes that contained SH3 domains (Cherra & Jin, 2016). One candidate corresponded to the *magu-2* gene, knockdown of which decreased the locomotor defects in *zig-10(tm6127); acr-2(n2420)* animals that also ectopically expressed ZIG-10 in GABAergic neurons. The observed *magu-2(RNAi)* effect on locomotor defects was further verified using genetic null mutations in *magu-2* (Supplementary Figure 1(a)). Additionally, the *magu-2(gk218)* null mutation restored the number of GABAergic synapses to wild type levels in animals ectopically expressing ZIG-10 (Supplementary Figure 1(b–c)). Since loss of function in *magu-2* alone does not cause any defects in GABA synapse morphology (Supplementary Figure 1(b–c)), this suggests that MAGU-2 acts in a cellular context dependent on the ZIG-10 pathway.

MAGU-2 belongs to the membrane-associated guanylate kinase (MAGUK) family that contains a PDZ domain, an SH3 domain, and a guanylate kinase domain in its carboxy-terminus (Supplementary Figure 2(a)) (Hobert, 2013; Zhu, Shang, & Zhang, 2016). The family of MAGUKs modulate various forms of intercellular junctions by regulating protein localization (Zhu *et al.*, 2016). In mammals, CASK and PSD95 modulate intercellular junctions in the nervous system through their regulation of synaptic transmission and neurotransmitter receptor clustering. Based on homology, MAGU-2 belongs to the MAGUK subfamily of membrane palmitoylated proteins (MPP) and displays highest percent of total identity (~32%) with Mpp5/Pals1 (Supplementary Figure 2(b)). In the developing mouse nervous system, Mpp5/Pals1 is expressed in Schwann cells and Muller glia and is essential for establishing cell polarity (Ozcelik *et al.*, 2010; van Rossum *et al.*, 2006). Additionally, Mpp5/Pals1 regulates the maintenance of cerebellar progenitors (Park *et al.*, 2016) and axon sorting in the peripheral nervous system (Zollinger, Chang, Baalman, Kim, & Rasband, 2015). Retinal ganglion cell-specific Mpp5/Pals1 knockout mice display neuronal degeneration early during development, mimicking the clinical degeneration observed in Leber congenital amaurosis patients (Cho *et al.*, 2012; Park *et al.*, 2011). The *Drosophila* homolog, Stardust, acts as a scaffolding protein at epithelial intercellular junctions to establish cell polarity (Tepass, 2012). However, it is unclear whether the MPP subfamily also regulates synaptic connectivity.

MAGU-2 acts in the epidermis to regulate the synapse number of cholinergic motor neurons

To determine where MAGU-2 functions to modulate synaptic connectivity, we first investigated the *magu-2* gene structure to identify its promoter based on the information from AceView and Wormbase (Lee *et al.*, 2018; Thierry-Mieg & Thierry-Mieg, 2006). Both sources predicted *magu-2* to span more than 10 kb, producing two potential isoforms (Figure 1(a–b)), with the long isoform a being supported by a recent dataset based on single-cell RNA-seq (Cao *et al.*, 2017). We isolated RNA from a mixed population of wild type N2 animals. We then used isoform-specific forward primers and a common reverse primer to amplify each putative isoform. Using genomic DNA as a control, we found that only the shorter isoform b was detected in our assay as a 270 bp-band whereas 748 bp-band indicative of the longer isoform was not detected (Figure 1(d)). This analysis does not exclude the possibility that isoform a may be expressed at lower levels or in a very limited subset of tissues.

Based on our transcript analysis, we created a MAGU-2::GFP fusion protein driven by the 4 kb DNA sequence upstream of *magu-2* isoform b to determine where MAGU-2 was expressed (Figure 1(c)). We detected MAGU-2::GFP expression from embryo to adulthood (Figure 1(e–f)). MAGU-2::GFP was visible as a diffuse signal throughout the epidermis and in various cells in the head, pharyngeal muscle, and posterior intestine, but was not observed in neurons or muscle (Figure 1(g–h)). Since ZIG-10 acts in both epidermis and cholinergic neurons to regulate cholinergic synapse density (Cherra & Jin, 2016), we investigated whether *magu-2* affects cholinergic synapses. We analyzed two *magu-2* deletion mutants, *ok1059* and *gk218*, which both remove most of the gene, and therefore likely are null alleles (Figure 2(a–d); Supplementary Figure 2(a)). Both alleles displayed an increase in cholinergic synapses, similar to *zig-10(tm6127)* animals (Figure 2(d)). We observed that *zig-10(tm6127); magu-2(gk218)* double mutants showed no further increase in cholinergic synapses (Figure 2(d)). To assess whether the additional cholinergic synapses in *magu-2* mutants formed functional postsynaptic compartments, we assayed the animals' sensitivity to the cholinergic agonist, levamisole. We have previously shown that the excessive synapses observed in the *zig-10(tm6127)* animals caused hypersensitivity to levamisole, leading to more rapid paralysis than wild type animals (Cherra & Jin, 2016). The *magu-2(gk218)* mutants showed an increased sensitivity to levamisole similar to *zig-10(tm6127)* mutants (Figure 2(e)). Overall, these observations are consistent with MAGU-2 and ZIG-10 functioning in the same pathway.

Expression of MAGU-2::GFP in *magu-2(gk218)* animals was sufficient to restore cholinergic synapse density to wild type levels (Figure 2(h,m)), demonstrating that isoform b

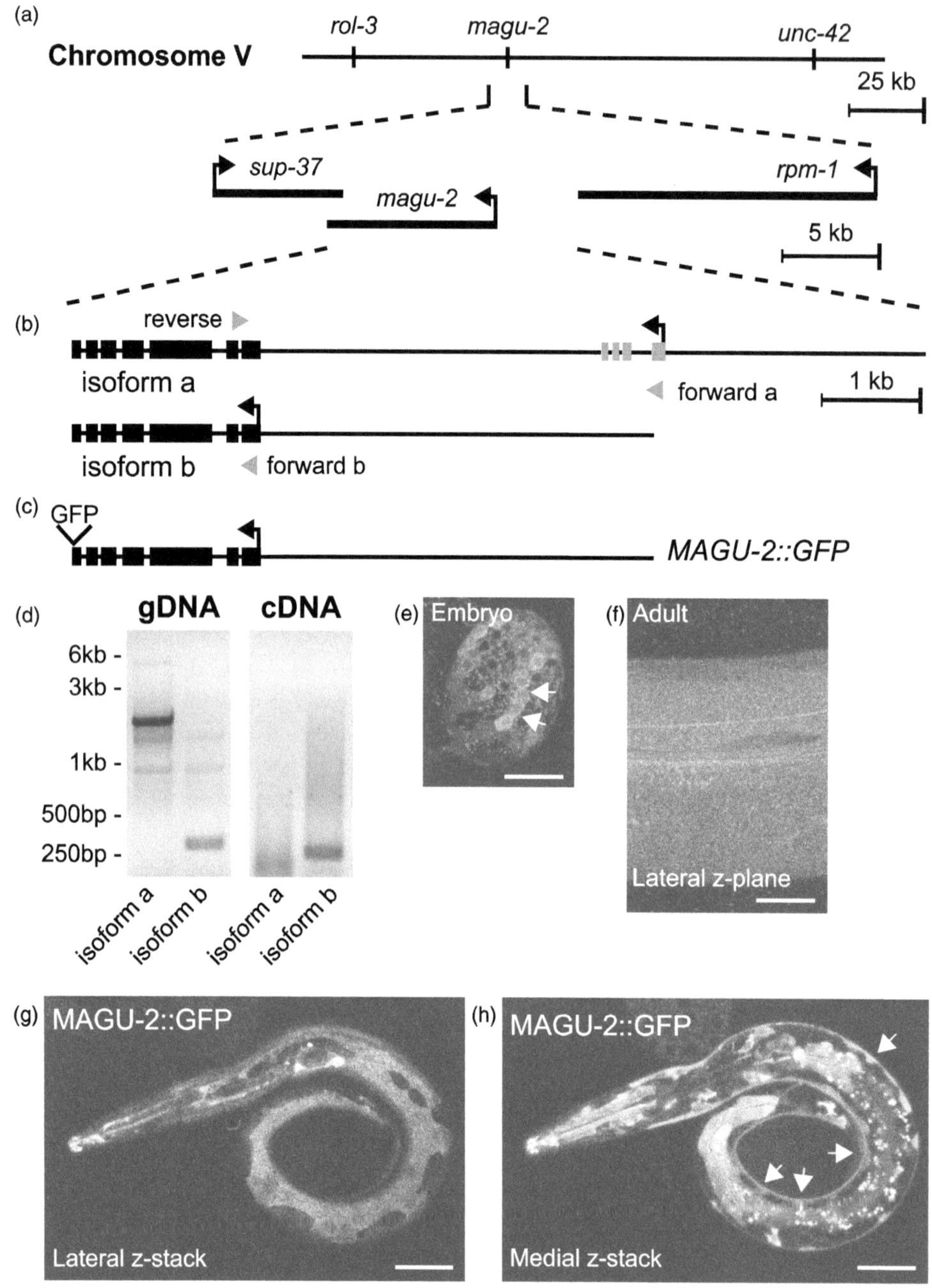

Figure 1. MAGU-2 is expressed in the epidermis. (a) Diagram of the *magu-2* genetic locus on chromosome V. The *magu-2* gene is flanked by the neighboring genes, *sup-37* and *rpm-1*. (b) Diagram of *magu-2* gene structure representing the predicted a and b isoforms. Gray arrowheads indicate primers used for isoform analysis. (c) Diagram of MAGU-2::GFP transgene. (d) Images of PCR products using *magu-2a* and *magu-2b* specific primers after gel electrophoresis. Genomic DNA (gDNA) from N2 was used as a PCR reaction control. Messenger RNAs from N2 were reverse transcribed into cDNA. PCR of *magu-2a* is expected to produce a 5.7 kb band from gDNA and 748 bp band from cDNA; *magu-2b* is expected to produce a 323 bp band from gDNA and a 270 bp band from cDNA. (e) MAGU-2::GFP expression in an embryo, between 150 and 300 min after fertilization. Arrows indicate cluster of cells expressing MAGU-2::GFP. (f) MAGU-2::GFP expression in a *z*-plane through the lateral epidermis of an adult animal. (g–h) Representative fluorescence images of L1 larva expressing MAGU-2::GFP. Scale bars are 20 μm. (g) Maximum projection through superficial layers of lateral aspect of animal showing MAGU-2 expression in epidermis, but not in neuronal commissures. (h) Maximum projection through medial layers around nerve cord showing MAGU-2 expression in epidermal ridge, but not in neuronal cell bodies. Arrows indicate epidermal ridge.

was sufficient to replace MAGU-2 function in the *gk218* mutant. We further addressed in which cells MAGU-2 was functioning to regulate synapse density using tissue-specific promoters to express MAGU-2 isoform b cDNA. Expression of MAGU-2 cDNA in the epidermis of *magu-2(gk218)* animals restored the density of cholinergic synapses to wild type levels (Figure 2(j,m)). However, expression of MAGU-2 in the nervous system or muscles was not sufficient to restore cholinergic synapse density (Figure 2(i,k,m)). Moreover, expressing mouse Mpp5 cDNA in the epidermis sufficiently restored synapse number to wild type levels (Figure 2(l–m)), indicating that mpp5 can function similarly to MAGU-2. Together, these data indicate that MAGU-2 is expressed and functions solely in the epidermis to modulate synaptic connectivity.

MAGU-2 affects phagocytosis in epidermis

To determine if MAGU-2 regulates epidermal phagocytosis, we investigated whether *magu-2(gk218)* animals displayed changes in the epidermal structures marked by GFP::FYVE.

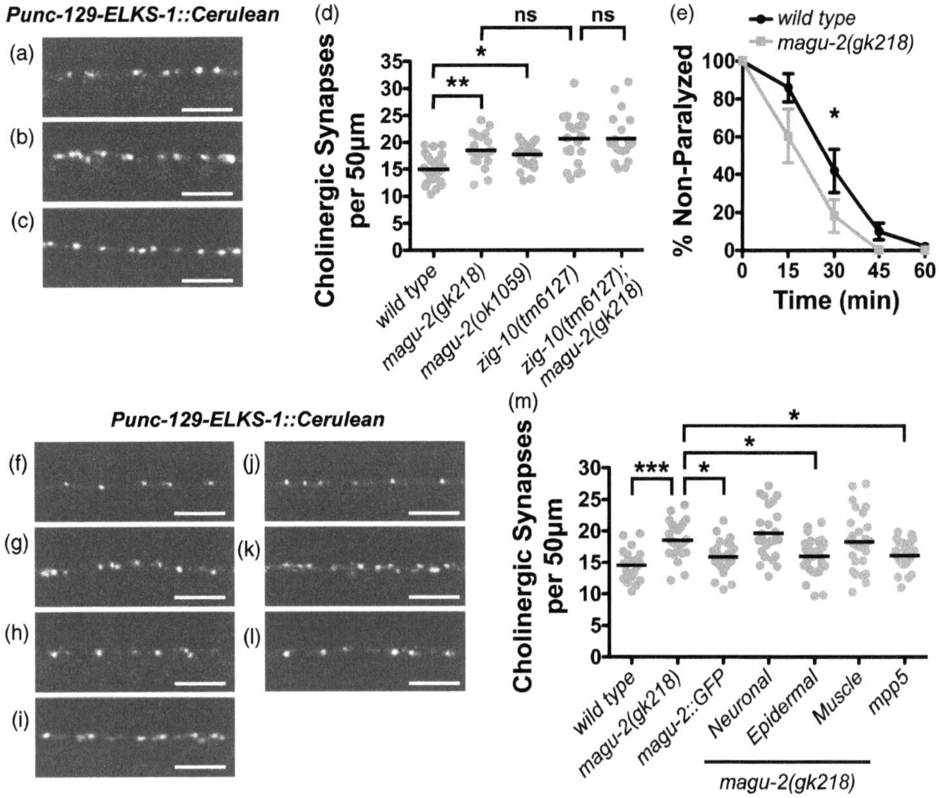

Figure 2. MAGU-2 functions in the epidermis to regulate cholinergic synapse density. Representative images of cholinergic synapses labeled by ELKS-1::Cerulean in wild type (a), *magu-2(gk218)* (b), and *magu-2(ok1059)* (c) animals. Scale bars are 5 μm. (d) Quantification of synapses from the indicated genotypes of animals expressing ELKS-1::Cerulean in cholinergic neurons. Gray dots represent individual animals; the black lines indicate the means. $*p < 0.05$; $**p < 0.01$; ns: not significant. (e) Quantification of paralysis over time in the presence of 1 mM levamisole; $*p < 0.05$. Representative images of cholinergic synapses labeled by ELKS-1::Cerulean in wild type (f), *magu-2(gk218)* (g), and *magu-2(gk218); MAGU-2::GFP* (h), *magu-2(gk218); Neuronal::MAGU-2* (i), *magu-2(gk218); Epidermal::MAGU-2* (j), *magu-2(gk218); Muscle::MAGU-2* (k), and *magu-2(gk218); Epidermal::mpp5* (l) animals. Scale bars are 5 μm. (m) Quantification of cholinergic synapses labeled by ELKS-1::Cerulean in wild type or *magu-2(0)* animals. Transgenic expression of MAGU-2 was achieved using its endogenous promoter, *rgef-1* promoter for neurons, *col-10* promoter for epidermis, or *myo-3* promoter for muscle. Mouse mpp5 cDNA was expressed using the *col-10* promoter. Gray dots represent individual animals; the black lines indicate the means; $*p < 0.05$; $***p < 0.001$.

FYVE domains associate with the membranes of endosomes and phagosomes in developing embryos of *C. elegans* (Yu, Lu, & Zhou, 2008). We have reported that both fluorescently-labeled FYVE domain and CED-1 expressed in the epidermis show colocalization with puncta labeled for presynaptic proteins (Cherra & Jin, 2016). Here, we analyzed the colocalization between GFP::FYVE structures in the epidermis and cholinergic synapses labeled by the presynaptic marker, mCherry::RAB-3. We observed that wild type animals displayed ~4 colocalization events per 100 micrometers where GFP::FYVE-labeled structures were associated with mCherry::RAB-3 expressed by cholinergic motor neurons (Figure 3(a–b)). In *magu-2(gk218)* animals, we found a significant decrease in the number of phagosomes near cholinergic synapses, similar to that of *ced-1(e1735)* animals (Figure 3(a–b)). These results indicate that MAGU-2 regulates epidermal phagocytosis.

To further evaluate the phagocytosis of neuronal materials by the epidermis, we performed electron microscopy analysis using the mini singlet oxygen generator (miniSOG) to label specific tissues. When stimulated by blue light, miniSOG produces singlet oxygen, and in the presence of diaminobenzadine (DAB), miniSOG oxidizes DAB to generate electron-dense osmiophilic polymers within nanometers of miniSOG localization (Shu *et al.*, 2011). We expressed miniSOG fused to ZIG-10 in cholinergic neurons to determine if cholinergic membranes were engulfed by the epidermis. On electron micrographs of the nerve cord, miniSOG::ZIG-10 signals were detected in a subset of neuronal processes, whose positions were consistent with being cholinergic axons and dendrites (Figure 3(c–d)). Importantly, in addition to the miniSOG signals found in the nerve cord, we also observed miniSOG-containing vesicles within the epidermis (Figure 3(c)). This suggested that the epidermis engulfs portions of cholinergic neuronal membranes. As further support for this idea, we analyzed cholinergic-driven miniSOG::ZIG-10 in the phagocytotic receptor *ced-1(e1735)* animals, which are deficient in executing phagocytosis (Hedgecock, Sulston, & Thomson, 1983; Zhou, Hartwieg, & Horvitz, 2001). In over 300 EM sections from two wild type animals, we found more than twenty miniSOG-labeled vesicles derived from cholinergic neurons in the epidermis. In the same number of sections from three *ced-1(e1735)* animals, we only observed miniSOG-labeled cholinergic axons or dendrites (Figure 3(d)). Thus, these data from miniSOG-mediated correlated-light-electron-microscopy analyses show that neuronal membranes are transferred to epidermis through the phagocytosis pathway.

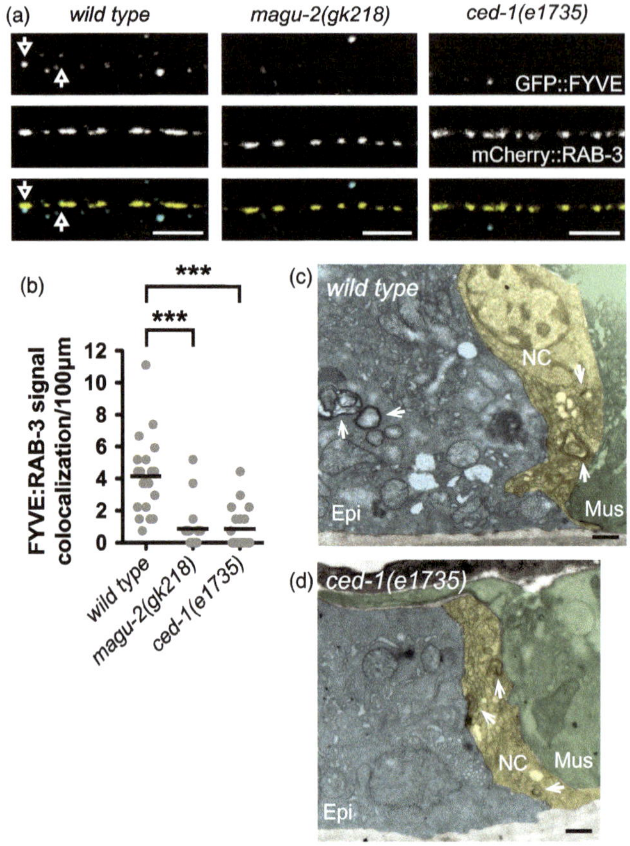

Figure 3. MAGU-2 modulates epidermal phagocytosis of neuronal membranes. (a) Representative single, 0.5μm-plane fluorescence images of wild type, *magu-2(gk218), and ced-1(e1735)* animals expressing GFP::FYVE to label epidermal phagosomes and mCherry::RAB-3 to label cholinergic synapses. Arrows indicate GFP::FYVE spots that colocalize with RAB-3::mCherry. Scale bars are 5 μm. (b) Quantification of colocalization between epidermal phagosomes and cholinergic synapses. Gray dots represent individual animals; the black lines indicate the means; ***<0.001. (c–d) Representative EM images of the ventral nerve cord in animals expressing miniSOG::ZIG-10 in cholinergic neurons. Scale bars are 500 nm. Cholinergic miniSOG-labeled structures are indicated by white arrows. Green shading indicates the muscle, blue indicates the epidermis, and yellow indicates the nerve cord. (c) Representative image of the ventral nerve cord flanked by muscle and epidermis. Cholinergic membranous material labeled by miniSOG::ZIG-10 is visible in the nerve cord and inside the epidermis in a wild type animal. (d) Representative image of the ventral cord in *ced-1(e1735)* animals, which show no miniSOG labeled structures inside the epidermis.

ZIG-10 localization depends on magu-2

MAGUKs regulate localization and surface expression of ion channels (El-Husseini *et al.*, 2000), glutamate receptors (Mi *et al.*, 2004), and N-cadherins (Wang *et al.*, 2014). Mutations in the extracellular Ig domain of ZIG-10 or RNAi-mediated knockdown displayed a similar phenotype to the *zig-10(tm6127)* mutation, suggesting that extracellular interactions and cell surface expression levels of ZIG-10 were essential to its function (Cherra & Jin, 2016). We hypothesized that MAGU-2 may regulate ZIG-10 expression or localization, similar to the interactions between cell adhesion proteins and other MAGUKs. To test this possibility, we generated a single-copy insertion of a functional GFP::ZIG-10 (Supplemental Figure 3). We then co-stained for GFP::ZIG-10 and the presynaptic active zone protein, UNC-10 in wild type and *magu-2(gk218)* animals harboring the GFP::ZIG-10 transgene. In *magu-2(gk218)* animals, there was a significant decrease in GFP::ZIG-10 near UNC-10-labeled synapses (Figure 4(a,b)). However, there was no change in GFP::ZIG-10 intensity in neuronal cell bodies (Figure 4(c,d)). These data suggest that MAGU-2 regulates ZIG-10 localization near synapses but does not alter ZIG-10 expression levels.

Discussion

The regulation of synaptic connections plays an essential role in promoting circuit function and robust signaling. Cell-cell signaling is a common mechanism to modulate synapse formation and maintenance. In mammals, *Drosophila*, and *C. elegans* non-neuronal cells play essential roles in controlling synapse number through multiple mechanisms (Allen, 2013; Cherra & Jin, 2015; Corty & Freeman, 2013). Phagocytosis-mediated synapse pruning requires cell-cell interactions that enable glia or other non-neuronal cells to remove synapses (Cherra & Jin, 2016; Chung *et al.*, 2013; MacDonald *et al.*, 2006). Here, we have presented a new approach for investigating cellular interactions by electron microscopy, using genetically encoded miniSOG enzyme that enables the labeling of specific proteins or cellular compartments for electron microscopic analysis (Shu *et al.*, 2011). miniSOG can be expressed in a tissue-specific or temporal manner to enable the analysis of discrete interactions, such as between the epidermis and neurons. This approach provides a complementary method for immuno-EM analysis of protein localization or cell-cell interactions, including the analysis of phagocytosis.

Within the nervous system, phagocytosis by glia or epidermal cells plays multiple roles in removing dead cells or debris, degrading axons after injury, and pruning axons or synapses to modulate neural circuits (Awasaki *et al.*, 2006; Cherra & Jin, 2016; MacDonald *et al.*, 2006; Rasmussen, Sack, Martin, & Sagasti, 2015; Stevens *et al.*, 2007). While the removal of apoptotic cells by phagocytes has been widely studied, the recognition and engulfment of synapses is not well understood. In addition to phagocyte-corpse interactions mediated by a phagocytosis receptor, CED-1/Draper/Megf, other cell surface proteins further modulate the phagocytosis process, such as integrin and the immunoglobulin domain superfamily member, ZIG-10 (Albert, Kim, & Birge, 2000; Cherra & Jin, 2016). Here, we uncovered a role for the Pals1/Mpp5 homolog, MAGU-2, in regulating non-neuronal phagocytosis and synapse density at the neuromuscular junction. MPP members establish cell polarity, participate in axon sorting, and reduce or prevent neuronal degeneration through the maintenance of cell-cell interactions (Cho *et al.*, 2012; van Rossum *et al.*, 2006; Zollinger *et al.*, 2015). Our data suggest that cell polarity and MPPs may also modulate synaptic connectivity.

We propose that MAGU-2 functions in the epidermis to maintain cellular interactions between the epidermis and neurons by regulating the localization of ZIG-10. In the presence of MAGU-2, ZIG-10 localizes near synapses and enables the execution of phagocytosis to reduce the number of cholinergic synapses at the neuromuscular junction. Since MPPs are expressed in brain regions where glial-mediated

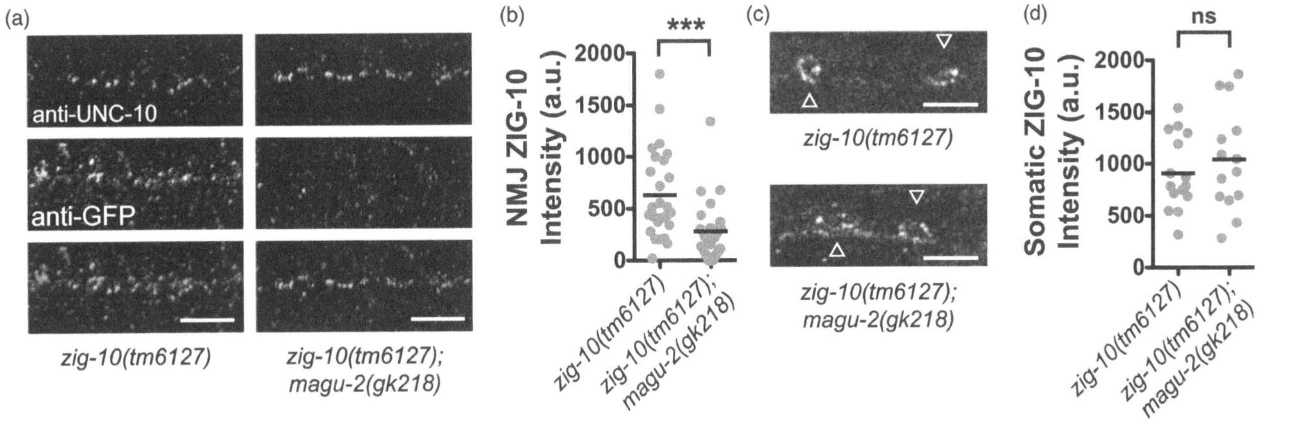

Figure 4. MAGU-2 regulates ZIG-10 localization. (a) Representative single, 0.5 μm-plane fluorescence images of the dorsal nerve cord in *zig-10(tm6127) and zig-10(tm6127); magu-2(gk218)* animals expressing a single-copy transgene of GFP::ZIG-10. Animals were stained for UNC-10 to label synapses and GFP to identify ZIG-10 localization. Scale bars are 5 μm. (b) Quantification of fluorescence intensity of GFP::ZIG-10 near synapses. Gray dots represent individual animals; the black lines indicate the means; ***$p < 0.001$. (c) Representative single, 0.5 μm-plane fluorescence images of neuronal cell bodies from *zig-10(tm6127) and zig-10(tm6127); magu-2(gk218)* animals stained for GFP::ZIG-10. Scale bars are 5 μm. White arrowheads indicate cell bodies. (d) Quantification of fluorescence intensity of GFP::ZIG-10 in neuronal cell bodies. Gray dots represent individual animals; the black lines indicate the means; ns: not significant.

synaptic pruning occurs (Clarke *et al.*, 2018; Zhang *et al.*, 2014), this raises the possibility that a conserved mechanism involving MPPs and phagocytosis may underlie synaptic pruning in mammals. Together these results provide new insights into how the phagocytosis pathway modulates synaptic connectivity.

Materials and methods

Strains and transgenes

All strains were maintained at 20 °C as previously described (Brenner, 1974). Transgenic animals were created by microinjection as previously described (Mello, Kramer, Stinchcomb, & Ambros, 1991). Single-copy insertion of GFP::ZIG-10 at *cxTi10882* on chromosome IV was generated using CRISPR/Cas9 and modified plasmids (Frokjaer-Jensen, Davis, Ailion, & Jorgensen, 2012; Takayanagi-Kiya, Zhou, & Jin, 2016). For further strain and transgene information, see Table S1 and Supplementary Figure 3.

mRNA Analysis for magu-2

To determine if *magu-2* produced both predicted isoforms, we isolated RNA from mixed stage worms using Trizol according to manufacturer's specifications. SuperScript First Strand Synthesis kit (Invitrogen) was used to generate cDNA. We performed RT-PCR using primers: 5′-TCGATACCACAGGCACTGT CC-3′ as *magu-2a* forward, 5′-CATTGCTGCAACATCTGG ACC-3′ as *magu-2b* forward, and 5′-TTTCTTCAACTT CAGCGAGTGG-3′ as the common *magu-2* reverse primer.

DNA plasmid construction

MAGU-2::GFP was constructed by PCR amplifying the 4 kb sequence upstream of *magu-2 b* as its promoter and cloning it into pCR8 (Invitrogen) to generate pCZGY3355. The genomic *magu-2* coding sequence was amplified by PCR from wild type worms and Gibson assembly was used to fuse GFP to the C-terminus of *magu-2* to generate MAGU-2::GFP (pCZGY3358). For tissue-specific rescue constructs, *magu-2* cDNA was cloned from wild type animals using SuperScript III First-Strand Synthesis System (Invitrogen). The splice leader 1 sequence was used with gene-specific primers that recognized the longest predicted mRNA to generate *magu-2* cDNA, which then was cloned into pCR8 (Invitrogen) to produce pCZGY3350. Mouse mpp5 cDNA was amplified from a mouse brain cDNA library and cloned into pCR8 to produce pCZGY3359. Generation of tissue-specific rescue constructs was performed using LR reactions (Invitrogen) with destination vectors containing tissue-specific promoters : rgef-1, col-10, unc-17b, myo-3 (Altun-Gultekin *et al.*, 2001; Cherra, & Jin, 2016; Okkema *et al.*, 1993; Rand, 2007). For single-copy insertion, Gibson assembly was used to clone *Pzig-10-gfp::zig-10, Prps-0-hygromycin resistance gene*, and homology arms around the *cxTi10882* locus on chromosome IV (pCZGY3354). See Table S2 for more information.

Whole-mount immunocytochemistry

To visualize GFP::ZIG-10 at near endogenous levels, CRISPR technology was used to insert into chromosome IV a single-copy of GFP::ZIG-10, which encodes the genomic locus of ZIG-10 with GFP fused directly after the signal peptide. The localization of GFP::ZIG-10 (*juSi333*) in the *zig-10(tm6127)* background in the presence or absence of the *magu-2(gk218)* allele was analyzed. Adult animals were washed 3 times in M9, followed by one 30-min wash in water. Animals were then placed on poly-L-lysine coated slides for freeze-crack and fixation with methanol and acetone as previously described (Duerr *et al.*, 1999). Following fixation, slides were washed in PBST and blocked with 5% goat serum. Samples were then incubated overnight with antibodies against GFP (1:500, A1112, RRID:AB_10073917, Invitrogen) and against UNC-10 (1:50, RRID:AB_10570332, Developmental Studies Hybridoma Bank). Samples were washed and incubated with secondary antibodies: goat anti-mouse Alexa 594 (RRID:AB_141372, Invitrogen) or goat anti-rabbit Alexa 488 (RRID:AB_143165, Invitrogen). After washing, samples were mounted with Vectashield mounting media (RRID:AB_2336789, Vector Labs).

MiniSOG photo-oxidation and electron microscopy

To investigate the cellular interactions between the epidermis and the nerve cord, we used miniSOG to label cholinergic neurons using tissue-specific promoters. miniSOG-labeled cellular membranes were visualized by EM after fixation and photo-oxidation. Adult animals were placed in 2% glutaraldehyde, 2% paraformaldehyde solution in 100 mM sodium cacodylate buffer. After the animals stopped moving, they were cut in half and incubated for 1 h on ice. The samples were washed in 100 mM sodium cacodylate buffer and then blocked in 50 mM glycine, 10 mM potassium cyanide, 20 mM aminotriazole in 100 mM sodium cacodylate buffer for 2 h on ice. The samples were placed in a MatTek culture dish containing ice-cold 2.5 mM oxygentated diaminobenzadine and 10 mM HCl in 100 mM sodium cacodylate buffer and illuminated with blue light using a Leica SPEII confocal microscope for 20 min. The samples were then washed in 100 mM sodium cacodylate buffer and post fixed overnight at 4 °C in 2% osmium tetroxide in 100 mM sodium cacodylate buffer. The following day samples were rinsed with ice-cold ddH_2O, dehydrated with ethanol and acetone, embedded in Durcupan, and baked for 3 days at 60 °C.

After embedding, 60-nm-thick serial sections were prepared for analysis. MiniSOG-labeled structures were identified by visual inspection of electron micrographs as vesicular structures appearing darker than background and that persist through at least two consecutive serial sections. For miniSOG::ZIG-10 expressed in cholinergic neurons, the number of miniSOG-labeled structures was counted in the nerve cord and in the epidermis by an individual blinded to the sample genotypes.

Fluorescent microscopy

All images were captured at 63× magnification using an LSM710 confocal microscope (Zeiss) using identical settings for each fluorescent protein marker. L4 animals were immobilized in M9 buffer by rolling animals such that their dorsal or ventral surface contacted the coverslip. All image analysis was performed using the Fiji distribution of NIH ImageJ.

For synapse number, dorsal synapses posterior to the vulva were imaged, and a single z-plane of 0.5-µm thickness was analyzed. Synapse density was determined using the ImageJ Analyze Particles function to count the number of synaptic puncta larger than 0.05 μm^2 over a defined length of nerve cord.

For colocalization of phagosomes with dorsal synapses posterior to the vulva, confocal images were manually inspected for colocalization between GFP::FYVE and mCherry::RAB-3, as defined by particles containing pixels from both the red and green channels. The number of colocalization events per length of nerve cord analyzed was then determined.

For quantifying ZIG-10 near the neuromuscular junctions, synapses were identified by eye as bright spots labeled by the UNC-10 antibody. GFP::ZIG-10 intensity was quantified by drawing a 2-µm-wide rectangle adjacent to the UNC-10-labeled nerve cord using ImageJ. The integrated density from the green channel was then measured inside of the rectangle. The rectangle was then moved 10 µm perpendicular to the nerve cord to measure the background integrated density. The background signal was subtracted from the GFP::ZIG-10 signal adjacent to the nerve cord to produce the measured intensity of ZIG-10 at the neuromuscular junction.

Levamisole sensitivity assay

L4 animals were moved to a fresh plate. The following day young adult animals were placed on plates containing 1 mM levamisole (Sigma). Animals were gently touched every 15 min for one hour to assess paralysis; animals that did not move after three touches were considered paralyzed.

Statistical analysis

All quantitative data are displayed as mean with individual data points represented as circles. For comparisons between two groups with normal distributions, Student's t-test was used. A Mann-Whitney test was used for comparisons between two groups that did not display Gaussian distribution. For comparisons between multiple groups, an ANOVA was used followed by post-hoc t-tests using a Bonferroni correction for multiple comparisons. A p values < 0.05 was considered statistically significant. Power analysis was performed to ensure that data was collected from a large enough sample size to provide a beta error ≤ 0.2.

Acknowledgements

We thank the members of the Jin laboratory for constructive comments. We thank J.S. Dittman for reporter lines. We appreciate Wormbase for genetic and genomic information.

Disclosure statement

No potential conflict of interest was reported by the author(s).

Funding

Some strains were provided by National BioResource Project (Dr. S. Mitani) and the *Caenorhabditis* Genetics Center, which is funded by the National Institutes of Health Office of Research Infrastructure Programs [P40-OD010440]. The electron micrographs were taken in the UCSD Cellular and Molecular Medicine Electron microscopy core facility, which is supported in part by National Institutes of Health Award [S10-OD023527]. This work was supported by grants from the National Institutes of Health to S.J.C. [K99-NS097638], D.B. [R01-GM086197], M.E. [P41-GM103412], and Y.J. [R01 and R37-NS035546].

ORCID

Salvatore J. Cherra http://orcid.org/0000-0003-1581-9150
Yishi Jin http://orcid.org/0000-0002-9371-9860

References

Albert, M.L., Kim, J.I., & Birge, R.B. (2000). alphavbeta5 integrin recruits the CrkII-Dock180-rac1 complex for phagocytosis of apoptotic cells. *Nature Cell Biology*, *2*, 899–905. doi:10.1038/35046549

Allen, N.J. (2013). Role of glia in developmental synapse formation. *Current Opinion in Neurobiology*, *23*, 1027–1033. doi:10.1016/j.conb.2013.06.004

Allen, N.J., & Eroglu, C. (2017). Cell biology of astrocyte-synapse interactions. *Neuron*, *96*, 697–708. doi:10.1016/j.neuron.2017.09.056

Altun-Gultekin, Z., Andachi, Y., Tsalik, E.L., Pilgrim, D., Kohara, Y., & Hobert, O. (2001). A regulatory cascade of three homeobox genes, ceh-10, ttx-3 and ceh-23, controls cell fate specification of a defined interneuron class in *C. elegans*. *Development (Cambridge, England)*, *128*, 1951–1969.

Awasaki, T., Tatsumi, R., Takahashi, K., Arai, K., Nakanishi, Y., Ueda, R., & Ito, K. (2006). Essential role of the apoptotic cell engulfment genes draper and ced-6 in programmed axon pruning during Drosophila metamorphosis. *Neuron*, *50*, 855–867. doi:10.1016/j.neuron.2006.04.027

Bangs, P., Franc, N., & White, K. (2000). Molecular mechanisms of cell death and phagocytosis in Drosophila. *Cell Death & Differentiation*, *7*, 1027–1034. doi:10.1038/sj.cdd.4400754

Brenner, S. (1974). The genetics of *Caenorhabditis elegans*. *Genetics*, *77*, 71–94.

Cao, J., Packer, J.S., Ramani, V., Cusanovich, D.A., Huynh, C., Daza, R., … Shendure, J. (2017). Comprehensive single-cell transcriptional profiling of a multicellular organism. *Science*, *357*, 661–667. doi:10.1126/science.aam8940

Cherra, S.J., 3rd, & Jin, Y. (2015). Advances in synapse formation: Forging connections in the worm. *Wiley Interdisciplinary Reviews: Developmental Biology*, *4*, 85–97. doi:10.1002/wdev.165

Cherra, S.J., 3rd, & Jin, Y. (2016). A two-immunoglobulin-domain transmembrane protein mediates an epidermal-neuronal interaction to maintain synapse density. *Neuron*, *89*, 325–336. doi:10.1016/j.neuron.2015.12.024

Cho, S.H., Kim, J.Y., Simons, D.L., Song, J.Y., Le, J.H., Swindell, E.C., … Kim, S. (2012). Genetic ablation of Pals1 in retinal progenitor cells models the retinal pathology of Leber congenital amaurosis. *Human Molecular Genetics*, *21*, 2663–2676. doi:10.1093/hmg/dds091

Chung, W.S., Allen, N.J., & Eroglu, C. (2015). Astrocytes control synapse formation, function, and elimination. *Cold Spring Harbor Perspectives in Biology*, *7*, a020370. doi:10.1101/cshperspect.a020370

Chung, W.S., Clarke, L.E., Wang, G.X., Stafford, B.K., Sher, A., Chakraborty, C., … Barres, B.A. (2013). Astrocytes mediate synapse elimination through MEGF10 and MERTK pathways. *Nature*, *504*, 394–400. doi:10.1038/nature12776

Clarke, L.E., Liddelow, S.A., Chakraborty, C., Munch, A.E., Heiman, M., & Barres, B.A. (2018). Normal aging induces A1-like astrocyte reactivity. *Proceedings of the National Academy of Sciences of the United States of America*, *115*, E1896–E1905. doi:10.1073/pnas.1800165115

Corty, M.M., & Freeman, M.R. (2013). Cell biology in neuroscience: Architects in neural circuit design: glia control neuron numbers and connectivity. *Journal of Cell Biology*, *203*, 395–405. doi:10.1083/jcb.201306099

de Wit, J., & Ghosh, A. (2016). Specification of synaptic connectivity by cell surface interactions. *Nature Reviews Neuroscience*, *17*, 4–35. doi:10.1038/nrn.2015.3

Duerr, J.S., Frisby, D.L., Gaskin, J., Duke, A., Asermely, K., Huddleston, D., … Rand, J.B. (1999). The cat-1 gene of *Caenorhabditis elegans* encodes a vesicular monoamine transporter required for specific monoamine-dependent behaviors. *The Journal of Neuroscience*, *19*, 72–84. doi:10.1523/JNEUROSCI.19-01-00072.1999

El-Husseini, A.E., Topinka, J.R., Lehrer-Graiwer, J.E., Firestein, B.L., Craven, S.E., Aoki, C., & Bredt, D.S. (2000). Ion channel clustering by membrane-associated guanylate kinases. Differential regulation by N-terminal lipid and metal binding motifs. *Journal of Biological Chemistry*, *275*, 23904–23910. doi:10.1074/jbc.M909919199

Frokjaer-Jensen, C., Davis, M.W., Ailion, M., & Jorgensen, E.M. (2012). Improved Mos1-mediated transgenesis in *C. elegans*. *Nature Methods*, *9*, 117–118. doi:10.1038/nmeth.1865

Fuentes-Medel, Y., Logan, M.A., Ashley, J., Ataman, B., Budnik, V., & Freeman, M.R. (2009). Glia and muscle sculpt neuromuscular arbors by engulfing destabilized synaptic boutons and shed presynaptic debris. *PLoS Biology*, *7*, e1000184. doi:10.1371/journal.pbio.1000184

Hedgecock, E.M., Sulston, J.E., & Thomson, J.N. (1983). Mutations affecting programmed cell deaths in the nematode *Caenorhabditis elegans*. *Science*, *220*, 1277–1279. doi:10.1126/science.6857247

Hobert, O. (2013). The neuronal genome of *Caenorhabditis elegans*. *WormBook*, Aug 13:1–106. doi:10.1895/wormbook.1.161.1

Jospin, M., Qi, Y.B., Stawicki, T.M., Boulin, T., Schuske, K.R., Horvitz, H.R., … Jin, Y. (2009). A neuronal acetylcholine receptor regulates the balance of muscle excitation and inhibition in *Caenorhabditis elegans*. *PLoS Biology*, *7*, e1000265. doi:10.1371/journal.pbio.1000265

Lee, R.Y.N., Howe, K.L., Harris, T.W., Arnaboldi, V., Cain, S., Chan, J., … Sternberg, P.W. (2018). WormBase 2017: molting into a new stage. *Nucleic Acids Research*, *46*, D869–D874. doi:10.1093/nar/gkx998

MacDonald, J.M., Beach, M.G., Porpiglia, E., Sheehan, A.E., Watts, R.J., & Freeman, M.R. (2006). The Drosophila cell corpse engulfment receptor Draper mediates glial clearance of severed axons. *Neuron*, *50*, 869–881. doi:10.1016/j.neuron.2006.04.028

Mangahas, P.M., & Zhou, Z. (2005). Clearance of apoptotic cells in *Caenorhabditis elegans*. *Seminars in Cell and Developmental Biology*, *16*, 295–306. doi:10.1016/j.semcdb.2004.12.005

Mello, C.C., Kramer, J.M., Stinchcomb, D., & Ambros, V. (1991). Efficient gene transfer in *C. elegans*: Extrachromosomal maintenance and integration of transforming sequences. *The EMBO Journal*, *10*, 3959–3970. doi:10.1002/j.1460-2075.1991.tb04966.x

Mi, R., Sia, G.M., Rosen, K., Tang, X., Moghekar, A., Black, J.L., McEnery, M., Huganir, R.L., & O'Brien, R.J. (2004). AMPA receptor-dependent clustering of synaptic NMDA receptors is mediated by Stargazin and NR2A/B in spinal neurons and hippocampal interneurons. *Neuron*, *44*, 335–349. doi:10.1016/j.neuron.2004.09.029

Okkema, P.G., Harrison, S.W., Plunger, V., Aryana, A., & Fire, A. (1993). Sequence requirements for myosin gene expression and regulation in *Caenorhabditis elegans*. *Genetics*, *135*, 385–404.

Ozcelik, M., Cotter, L., Jacob, C., Pereira, J.A., Relvas, J.B., Suter, U., & Tricaud, N. (2010). Pals1 is a major regulator of the epithelial-like polarization and the extension of the myelin sheath in peripheral nerves. *Journal of Neuroscience*, *30*, 4120–4131. doi:10.1523/JNEUROSCI.5185-09.2010

Park, B., Alves, C.H., Lundvig, D.M., Tanimoto, N., Beck, S.C., Huber, G., … Wijnholds, J. (2011). PALS1 is essential for retinal pigment epithelium structure and neural retina stratification. *Journal of Neuroscience*, *31*, 17230–17241. doi:10.1523/JNEUROSCI.4430-11.2011

Park, J.Y., Hughes, L.J., Moon, U.Y., Park, R., Kim, S.B., Tran, K., … Kim, S. (2016). The apical complex protein Pals1 is required to maintain cerebellar progenitor cells in a proliferative state. *Development*, *143*, 133–146. doi:10.1242/dev.124180

Rand, J.B. (2007). Acetylcholine. *WormBook*, Jan 30; 1–21.

Rasmussen, J.P., Sack, G.S., Martin, S.M., & Sagasti, A. (2015). Vertebrate epidermal cells are broad-specificity phagocytes that clear sensory axon debris. *The Journal of Neuroscience, 35*, 559–570. doi: 10.1523/JNEUROSCI.3613-14.2015

Shu, X., Lev-Ram, V., Deerinck, T.J., Qi, Y., Ramko, E.B., Davidson, M.W., ... Tsien, R.Y. (2011). A genetically encoded tag for correlated light and electron microscopy of intact cells, tissues, and organisms. *PLoS Biology, 9*, e1001041. doi:10.1371/journal.pbio.1001041

Stevens, B., Allen, N.J., Vazquez, L.E., Howell, G.R., Christopherson, K.S., Nouri, N., ... Barres, B.A. (2007). The classical complement cascade mediates CNS synapse elimination. *Cell, 131*, 1164–1178. doi:10.1016/j.cell.2007.10.036

Sudhof, T.C. (2018). Towards an understanding of synapse formation. *Neuron, 100*, 276–293. doi:10.1016/j.neuron.2018.09.040

Takayanagi-Kiya, S., Zhou, K., & Jin, Y. (2016). Release-dependent feedback inhibition by a presynaptically localized ligand-gated anion channel. *Elife, 5*, pii: e21734. doi:10.7554/eLife.21734

Tepass, U. (2012). The apical polarity protein network in Drosophila epithelial cells: Regulation of polarity, junctions, morphogenesis, cell growth, and survival. *Annual Review of Cell and Developmental Biology, 28*, 655–685. doi:10.1146/annurev-cellbio-092910-154033

Thierry-Mieg, D., & Thierry-Mieg, J. (2006). AceView: A comprehensive cDNA-supported gene and transcripts annotation. *Genome Biology, 7*, S12–S14. doi:10.1186/gb-2006-7-s1-s12

van Rossum, A.G., Aartsen, W.M., Meuleman, J., Klooster, J., Malysheva, A., Versteeg, I., ... Wijnholds, J. (2006). Pals1/Mpp5 is required for correct localization of Crb1 at the subapical region in polarized Muller glia cells. *Human Molecular Genetics, 15*, 2659–2672. doi:10.1093/hmg/ddl194

Wang, S.H., Celic, I., Choi, S.Y., Riccomagno, M., Wang, Q., Sun, L.O., ... Kolodkin, A.L. (2014). Dlg5 regulates dendritic spine formation and synaptogenesis by controlling subcellular N-cadherin localization. *The Journal of Neuroscience, 34*, 12745–12761. doi:10.1523/JNEUROSCI.1280-14.2014

Yu, X., Lu, N., & Zhou, Z. (2008). Phagocytic receptor CED-1 initiates a signaling pathway for degrading engulfed apoptotic cells. *PLoS Biology, 6*, e61. doi:10.1371/journal.pbio.0060061

Zhang, Y., Chen, K., Sloan, S.A., Bennett, M.L., Scholze, A.R., O'Keeffe, S., ... Wu, J.Q. (2014). An RNA-sequencing transcriptome and splicing database of glia, neurons, and vascular cells of the cerebral cortex. *The Journal of Neuroscience, 34*, 11929–11947. doi:10.1523/JNEUROSCI.1860-14.2014

Zhou, Z., Hartwieg, E., & Horvitz, H.R. (2001). CED-1 is a transmembrane receptor that mediates cell corpse engulfment in *C. elegans*. *Cell, 104*, 43–56. doi:10.1016/S0092-8674(01)00190-8

Zhu, J., Shang, Y., & Zhang, M. (2016). Mechanistic basis of MAGUK-organized complexes in synaptic development and signalling. *Nature Reviews Neuroscience, 17*, 209–223. doi:10.1038/nrn.2016.18

Zollinger, D.R., Chang, K.J., Baalman, K., Kim, S., & Rasband, M.N. (2015). The polarity protein Pals1 regulates radial sorting of axons. *The Journal of Neuroscience, 35*, 10474–10484. doi:10.1523/JNEUROSCI.1593-15.2015

OPEN ACCESS

Synaptic remodeling, lessons from *C. elegans*

Andrea Cuentas-Condori and David M. Miller, 3rd

ABSTRACT

Sydney Brenner's choice of *Caenorhabditis elegans* as a model organism for understanding the nervous system has accelerated discoveries of gene function in neural circuit development and behavior. In this review, we discuss a striking example of synaptic remodeling in the *C. elegans* motor circuit in which DD class motor neurons effectively reverse polarity as presynaptic and postsynaptic domains at opposite ends of the DD neurite switch locations. Originally revealed by EM reconstruction conducted over 40 years ago, DD remodeling has since been investigated by live cell imaging methods that exploit the power of *C. elegans* genetics to reveal key effectors of synaptic plasticity. Although synapses are also extensively rewired in developing mammalian circuits, the underlying remodeling mechanisms are largely unknown. Here, we highlight the possibility that studies in *C. elegans* can reveal pathways that orchestrate synaptic remodeling in more complex organisms. Specifically, we describe (1) transcription factors that regulate DD remodeling, (2) the cellular and molecular cascades that drive synaptic remodeling and (3) examples of circuit modifications in vertebrate neurons that share some similarities with synaptic remodeling in *C. elegans* DD neurons.

Introduction

As a pioneer in the use of phage genetics to unravel the fundamental mechanisms of gene expression, Sydney Brenner possessed an insightful understanding of how mutant analysis could be exploited to tackle more complex questions in biology. With the overarching goal of understanding how genes build the brain, he chose *Caenorhabditis elegans* because its simple nervous system could be fully described and because its rapid, 3-day life cycle facilitates genetic analysis (Brenner, 1974). Its small size also mattered, not only for the practical advantage of culturing large numbers of animals for mutant screens but also because Brenner understood that it would be necessary to use electron microscopy (EM) to define the 'wiring diagram' (Brenner, 1973). In an early step toward this goal, Brenner *et al.* published a description of serial section EM reconstruction of the adult ventral nerve cord (White, Southgate, Thomson, & Brenner, 1976). An accompanying analysis of the ventral cord cell lineage by John Sulston suggested that eight motor neuron classes were generated in two developmental periods (Sulston, 1976; Sulston & Horvitz, 1977), initially DA, DB and DD motor neurons in the embryo and then VA, VB, VC, VD and AS classes from a second wave of cell divisions late in the first larval stage. This finding was intriguing because it suggested that the motor circuit of newly hatched larvae with only three motor neuron types (DA, DB and DD) should differ from that of the adult with its full complement of eight motor neuron classes (DA, DB, DD, VA, VB, VC, VD, AS; Figure 1(A)).

Caenorhabditis elegans development involves four successive larval stages (L1 to L4) before adulthood. EM reconstruction of an early L1 larva yielded the unanticipated finding that the functional polarity of DD motor neurons is reversed in comparison to the adult (White, Albertson, & Anness, 1978). In the newly hatched L1, each of the six DD motor neurons innervates ventral muscles and also extends a circumferential commissure to the dorsal nerve cord to receive synaptic inputs from DA and DB motor neurons (Figure 1(B)). In the adult, this stereotypical DD morphology is retained but synaptic output is switched to dorsal muscles and input is provided by VA and VB motor neurons in the ventral nerve cord (Figure 1(C)). Reconstruction of an L2 larva revealed DD neurons with adult-like connectivity. Thus, DD synaptic remodeling was likely to occur during the L1 to L2 transition, when post-embryonic motor neurons are generated (Figure 1(D)). Importantly, the second member of the D class, ventral D (VD) motor neuron, develops post-embryonically and synapses onto ventral muscles while receiving input on the dorsal side (White *et al.*, 1976). In other words, VD neurons adopt the synaptic arrangement of early L1 DD neurons (Figures 1(B) and 2(A)).

To determine if the arrival of larval neurons was necessary for DD remodeling, White *et al.* (1978) reconstructed the dorsal and ventral nerve cords of a *lin-6* mutant that blocks post-embryonic cell division. This experiment showed

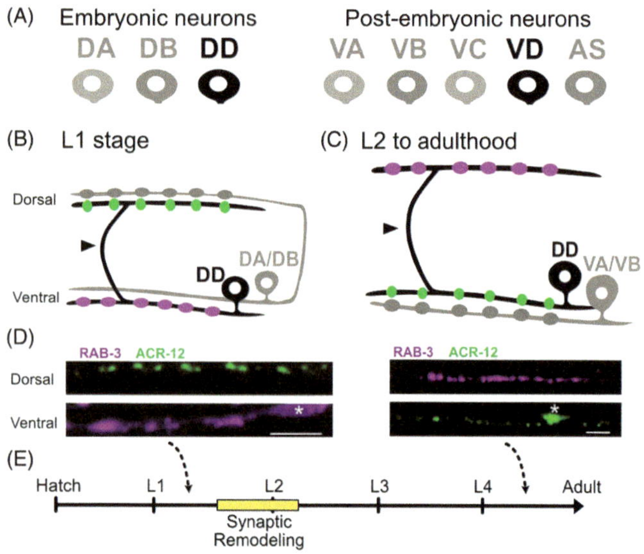

Figure 1. Dorsal D motor neurons undergo synaptic remodeling during early development. (A) (Left) The newly hatched L1 larva contains 3 classes of ventral cord motor neurons: DA, DB, DD. (Right) Five additional postembryonic motor neuron classes (VA, VB, VC, VC, AS) are added to the ventral cord during the L1 to L2 larval transition. (B) During the first larval stage (L1), DD motor neurons (black) provide output to body muscles at ventral presynaptic boutons (purple) and receive input from cholinergic DA/DB neurons (gray) through ACR-12 nACh receptors at postsynaptic terminals (green) on the dorsal side. Arrowhead points to commissure. (C) (Top) DD motor neurons (black) remodel to place presynaptic boutons (purple) on the dorsal side, and relocate postsynaptic terminals (green) to the ventral side for cholinergic input from VA/VB motor neurons (gray). Arrowhead points to commissure. D. DD presynaptic boutons labeled with mCherry::RAB-3 and DD postsynaptic terminals marked with ACR-12::GFP before in early L1 (Left) and after (Right) remodeling at L4 stage. Asterisk labels cell bodies. Scale bar = 10 μm. Images adapted from (He *et al.*, 2015). (E) DD neurons remodel over a 4-6-hour period that spans the transition from the L1 to L2 larval stages (yellow).

that the relocation of DD presynaptic boutons to the dorsal cord is not impaired by the absence of postembryonic ventral cord motor neurons. However, input from DA and DB neurons, which is typically eliminated in the wild type, persists on the dorsal process of DD neurons in the *lin-6* mutant background. These results suggest that post-embryonic neurons are not required for presynaptic DD remodeling, but may be necessary for further refinement of this circuit, (i.e. elimination of DA and DB inputs).

Remarkably, 20 years elapsed before the publication of a subsequent study of DD remodeling, a breakthrough paper from Yishi Jin's lab reporting the first use of a novel GFP synaptic marker to monitor DD remodeling (Hallam & Jin, 1998). This live-cell imaging approach, which has since been widely adopted for studies of synaptic morphogenesis and plasticity (Hendi, Kurashina, & Mizumoto, 2019), depended on several key technological advances. These include (1) methods for generating transgenic strains (Mello, Kramer, Stinchcomb, & Ambros, 1991), (2) cloning of genes specifically expressed in GABAergic neurons (e.g. *unc-25*; Eastman, Horvitz, & Jin, 1999; Mclntire, Jorgensen, & Horvitz, 1993a; Mclntire, Jorgensen, Kaplan, & Horvitz, 1993b) and (3) the discovery that GFP could be used for visualizing neuronal morphology *in vivo* (Chalfie, Tu, Euskirchen, Ward, & Prasher, 1994) and for tagging the presynaptic protein, Synaptobrevin/SNB-1 (Jorgensen *et al.*, 1995; M L Nonet, 1999). Hallam and Jin exploited these methods to produce a transgenic line that used the *unc-25* promoter (Eastman *et al.*, 1999; Michael L. Nonet, Saifee, Zhao, Rand, & Wei, 1998) for selective labeling of GABAergic presynaptic terminals with SNB-1::GFP. Direct observation confirmed the earlier prediction from EM reconstruction (White *et al.*, 1978) that DD presynaptic domains relocate from the ventral nerve cord to dorsal DD process during the L1 to L2 transition (Hallam & Jin, 1998; Figure 1).

The additional key prediction, that the postsynaptic apparatus is also relocated during DD remodeling, was not confirmed by live-cell imaging until almost 40 years after the original EM reconstruction. As noted above, in the early L1, the dorsal arm of each DD neuron is postsynaptic to cholinergic DA and DB motor neurons (Figure 1(B)). After remodeling in the L2, DD inputs are switched to the ventral nerve cord where they are postsynaptic to cholinergic VA and VB motor neurons (Figure 1(C); White *et al.*, 1978). With the determination that the nicotinic acetylcholine receptor (nAChR) subunit, ACR-12, is required for cholinergic activation of ventral cord GABAergic neurons (Cinar, Keles, & Jin, 2005; Petrash, Philbrook, Haburcak, Barbagallo, & Francis, 2013), it was possible to monitor the relocation of the DD postsynaptic domain in live animals. ACR-12::GFP localizes as discrete puncta on the dendrites of both GABAergic motor neurons (DDs and VDs) in apposition to cholinergic presynaptic terminals (Cuentas-Condori *et al.*, 2019; He *et al.*, 2015; Petrash *et al.*, 2013; Philbrook *et al.*, 2018). In DD neurons, ACR-12::GFP is initially positioned on the dorsal neurite at the L1 stage and then relocates to the ventral cord as remodeling ensues (He *et al.*, 2015; Howell, White, & Hobert, 2015; Figure 1(B,C)). Thus, as originally deduced from EM reconstruction in 1978, presynaptic and postsynaptic complexes are repositioned to opposite ends of the DD neuron and effectively switch locations during early larval development (White *et al.*, 1978; Figure 1).

By exploiting these and other synaptic markers, several laboratories have sought to elucidate the underlying molecular mechanisms that drive remodeling of the presynaptic and postsynaptic compartments in D-class motor neurons. Some of these findings were recently reviewed (Jin & Qi, 2018; Kurup & Jin, 2016). Here, we seek to integrate these studies in a historical narrative that highlights Sydney Brenner's role in launching this field of research and also incorporate new findings on the mechanism of synaptic disassembly. This review will discuss (1) Transcriptional pathways that regulate synaptic remodeling, (2) Downstream targets that mediate cell biological mechanisms for synaptic remodeling and the role of activity in these pathways, (3) The relevance of these findings to understanding synaptic remodeling in the mammalian nervous system.

Synaptic remodeling of D-type motor neurons is transcriptionally regulated

The stereotypical occurrence of DD remodeling during a specific developmental period (i.e. L1 stage larvae) points to regulation by a genetic program (White *et al.*, 1978). This

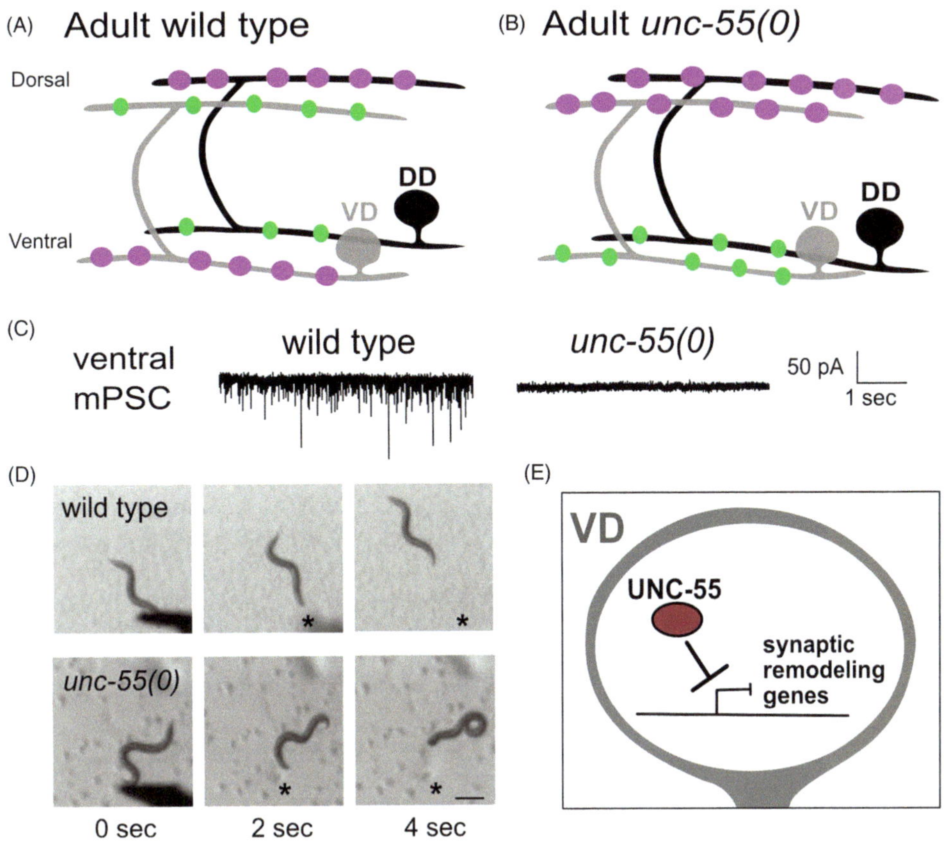

Figure 2. Ventral D (VD) motor neurons ectopically remodel in *unc-55* mutants. (A) In adult wild-type worms, DD (black) presynaptic boutons (magenta) are located on the dorsal side and postsynaptic terminals (green) are positioned on the ventral side. VD (gray) presynaptic terminals are positioned on the ventral side (magenta) whereas postsynaptic terminals are located on the dorsal side. (B) In adult *unc-55(0)* mutants, both DD (black) and VD (gray) presynaptic boutons (magenta) are located on the dorsal side. Postsynaptic terminals (green) of both DD and VD neurons are positioned on the ventral side. (C) (Left) Wild-type worms show robust miniature Post Synaptic Currents (mPSCs) in ventral muscles whereas mPSCs are not detected in *unc-55* mutants (Right). Adapted from (Petersen *et al.*, 2011). (D) Head touch (asterisk) evokes backward movement in the wild type (top) but unc-55 mutants coil ventrally with head touch (asterisk) due to absence of inhibitory GABAergic input on the ventral side. Scale bar = 250 μm. Adapted from (Petersen *et al.*, 2011). (E) The UNC-55/COUP-TF transcription factor functions in VD motor neurons to antagonize expression of synaptic remodeling genes.

idea was substantiated by the finding that the heterochronic protein, LIN-14, controls the timing of DD remodeling. The presynaptic marker, SNB-1::GFP, is precociously relocated to the dorsal nerve cord in *lin-14* mutants, suggesting that LIN-14 normally prevents the premature activation of the presynaptic remodeling program (Hallam & Jin, 1998; Figure 3(A)).

UNC-55/COUP-TF

Although LIN-14 was known to be nuclear-localized in 1998 (Hallam & Jin, 1998; Ruvkun & Giusto, 1989) and later confirmed to function as a transcription factor (Hristova, Birse, Hong, & Ambros, 2005), direct evidence of transcriptional control of the DD synaptic remodeling program first emerged from studies of the *unc-55* locus. As is the case for the majority of 'uncoordinated' or 'unc' loci, the original *unc-55* mutant alleles were isolated by Sydney Brenner (Brenner, 1974). *unc-55* mutants display a characteristic movement defect of strong ventral coiling during backward locomotion (Figure 2(D)). EM reconstruction by Leon Nawrocki, a postdoc with John White, revealed that VD class neurons adopt the DD pattern of dorsal synaptic output (Hardy, 1990; Figure 2(A,B)), a conclusion confirmed by

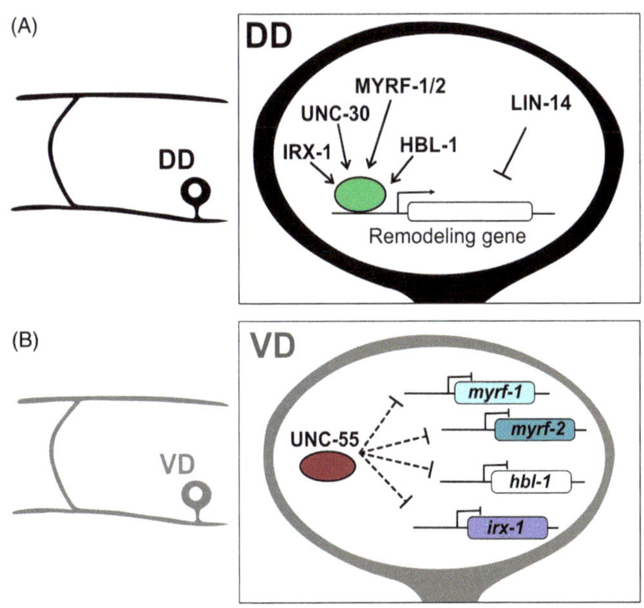

Figure 3. Transcriptional regulation of synaptic remodeling in D-type GABAergic motor neurons. (A) (Left) Morphology of DD motor neuron. (Right) Transcription factors IRX-1/Iroquois MYRF-1/MYRF-2 and HBL-1/Hunchback promote expression of DD remodeling genes, whereas LIN-14 antagonizes remodeling genes. (B) (Left) Morphology of VD motor neuron. (Right) The transcription factor UNC-55/COUP-TFII inhibits expression of IRX-1/Iroquois, HBL-1/Hunchback and MYRF-1/MYRF-2 in VD neurons to prevent ectopic synaptic remodeling.

immunostaining results (Walthall & Plunkett, 1995). The absence of inhibitory GABA synapses on the ventral side (Figure 2(B,C)) results in excess ventral cholinergic excitation and the consequent coiling phenotype (Figure 2(D); Miller-Fleming *et al.*, 2016; Thompson-Peer, Bai, Hu, & Kaplan, 2012; Walthall & Plunkett, 1995).

Molecular cloning revealed that *unc-55* encodes a member of the conserved COUP-TF family of transcription factors that is selectively expressed in VD but not DD motor neurons (Zhou & Walthall, 1998). As a likely transcriptional repressor, UNC-55 was proposed to function in VD neurons to block ectopic activation of the DD synaptic remodeling program (Shan, Kim, Li, & Walthall, 2005; Figure 2(E)). This idea is consistent with several lines of evidence: (1) In *unc-55* mutants, VD synapses are initially established with ventral muscles and then relocated to the dorsal nerve cord in a developmental sequence that mimics native DD remodeling (Petersen *et al.*, 2011; Thompson-Peer *et al.*, 2012); (2) Forced expression of UNC-55 in DD neurons antagonizes synaptic remodeling (Shan *et al.*, 2005); (3) UNC-55 controls expression of downstream effectors of the DD remodeling program including transcription factors IRX-1/Iroquois (Petersen *et al.*, 2011), MYRF1/2/Myelin Regulatory Factor (Meng et al., 2017; Yu et al., 2017) and HBL-1/Hunchback (Thompson-Peer *et al.*, 2012), the DEG/ENaC protein, UNC-8 (Miller-Fleming *et al.*, 2016), the Ig-domain protein OIG-1 (He *et al.*, 2015; Howell *et al.*, 2015) and regulators of cAMP homeostasis (Yu *et al.*, 2017).

IRX-1/Iroquois

The IRX-1/Iroquois transcription factor was initially identified by a gene expression profiling strategy that detected *unc-55*-regulated transcripts in VD neurons. A subsequent RNAi screen determined that IRX-1 is required for synaptic remodeling (Miller-Fleming *et al.*, 2020; Petersen *et al.*, 2011). IRX-1 is a member of the conserved Iroquois family of homeodomain transcription factors that includes mammalian homologs that specify neuronal identity in the developing nervous system (Cavodeassi, Modolell, & Gómez-Skarmeta, 2001; Houweling *et al.*, 2001). In *C. elegans*, IRX-1/Iroquois is expressed in DD motor neurons but is normally turned off by UNC-55 in the VD class (Figure 3). Ectopic expression of IRX-1/Iroquois in *unc-55* mutant VD neurons triggers the relocation of ventral VD presynaptic boutons to the dorsal nerve cord (Petersen *et al.*, 2011) and also the reciprocal removal of dorsal postsynaptic nACh receptors for reassembly on the ventral side (He *et al.*, 2015). In addition, forced expression of IRX-1/Iroquois in otherwise wild-type VD neurons is sufficient to induce the overall synaptic remodeling program (Petersen *et al.*, 2011). Thus, IRX-1/Iroquois appears to orchestrate native DD synaptic remodeling by regulating expression of key downstream effectors that control both presynaptic and postsynaptic plasticity (see below; He *et al.*, 2015; Petersen *et al.*, 2011). Although cell-specific RNAi of *irx-1* in DD neurons delays but does not block DD remodeling, the incomplete penetrance of this effect could be due to partial IRX-1 knockdown by RNAi (Petersen *et al.*, 2011).

HBL-1/Hunchback

An independent approach revealed that UNC-55 also negatively regulates expression of the HBL-1/Hunchback transcription factor in VD neurons (Figure 3). HBL-1 promotes the translocation of ventral presynaptic components to the dorsal side in *unc-55* mutant VDs and also drives synaptic remodeling in wild-type DD neurons (Thompson-Peer *et al.*, 2012). HBL-1 has not been tested, however, for a potential role in remodeling the DD postsynaptic apparatus (e.g. ACR-12, Figure 1). In addition, forced expression of HBL-1 in VD neurons is not sufficient to induce remodeling which suggests that HBL-1 could function downstream of IRX-1. Notably, the micro RNA, *mir-84*, antagonizes HBL-1 expression and this function is required to block the HBL-1-dependent precocious removal of DD presynaptic terminals in the early L1 larva in *mir-84* mutants (Thompson-Peer *et al.*, 2012).

MYRF1/2/Myelin Regulatory Factor

A forward genetic screen revealed that members of Myelin Regulatory Factor (MYRF) family of transcription factors mediate the translocation of ventral DD presynaptic domains to the dorsal nerve cord (Figure 3; J. Meng *et al.*, 2017). MYRF transcription factors are highly conserved and notably regulate myelination in mammals (Bujalka *et al.*, 2013). The *C. elegans* paralogs, MYRF-1 and MYRF-2, are expressed in DD neurons and function together in a heteromeric complex to regulate DD remodeling (J. Meng *et al.*, 2017). Interestingly, MYRF transcription factors localize to the ER where an N-terminal fragment is released by proteolytic cleavage for translocation to the nucleus to function as a transcription factor (Bujalka *et al.*, 2013; J. Meng *et al.*, 2017). MYRF-1 and MYRF-2 display a similar cell biological mechanism but the signal that activates the pathway in DD neurons is not known. Remodeling is blocked in a genetic background that selectively eliminates MYRF-1 function in DD neurons which suggests that MYRF-1/2 function is essential for DD remodeling (J. Meng *et al.*, 2017). It is still an open question whether MYRF-1/2 function is required for the removal of dorsal postsynaptic ACR-12 complexes in DD neurons and their reassembly on the ventral side. In the future it will be important to delineate the specific roles of MYRF-1/2 versus that of IRX-1/Iroquois in DD remodeling.

UNC-30/PITX

Finally, DD remodeling is also disrupted by mutations that disable the UNC-30/PITX transcription factor (Figure 3; Howell *et al.*, 2015). In wild-type L1 larvae prior to remodeling, DD presynaptic markers localize to the ventral side whereas postsynaptic components are limited to the dorsal nerve cord (Hallam & Jin, 1998; He *et al.*, 2015; Figure 1). In *unc-30* mutants, however, fluorescently labeled DD presynaptic and postsynaptic markers are observed in both ventral and dorsal nerve cords of early L1 larvae (Howell *et al.*, 2015). This effect of apparently precocious remodeling in *unc-30* mutants could arise in part from the dysregulation of LIN-14 and OIG-1, both of which antagonize DD remodeling (Hallam & Jin, 1998; He *et al.*, 2015; Howell *et al.*, 2015). Alternatively, the abnormal synaptic organization of *unc-30* mutant DD neurons could result from the fundamental role of UNC-30 in the differentiation of DD and VD neurons (Jin, Hoskins, & Horvitz, 1994; McIntire *et al.*, 1993b). In any case, the role of UNC-30 in DD remodeling is likely complex because UNC-30 is also necessary for IRX-1 expression which promotes remodeling (Petersen *et al.*, 2011).

To summarize, at least six transcription factors function together to specify both the timing and progression of synaptic remodeling in *C. elegans* GABAergic neurons. In this developmentally regulated program, some transcription factors promote synaptic remodeling (IRX-1/Iroquois HBL-1/Hunchback, MYRF-1/2/Myelin Regulatory Factor; Meng *et al.*, 2017; Petersen *et al.*, 2011; Thompson-Peer *et al.*, 2012), others function to prevent it (LIN-14, UNC-55/COUP-TF; Figure 3; Hallam & Jin, 1998; Walthall & Plunkett, 1995) and at least one (UNC-30/PITX) exerts the dual role of both promoting and antagonizing remodeling (Howell *et al.*, 2015; Petersen *et al.*, 2011). The surprising intricacy of this gene regulatory network in the *C. elegans* nervous system with its 302 neurons (White, Southgate, Thomson, & Brenner, 1986) points to the potentially daunting challenge of dissecting transcriptional control of synaptic remodeling in more complex nervous systems that could involve a larger number of additional transcription factors with specialized functions deriving from the evolutionary amplification of transcription factor families (Babu, Luscombe, Aravind, Gerstein, & Teichmann, 2004). Ultimately, a detailed understanding of the relationship between transcription factors and their targets (Petersen *et al.*, 2011; Yu *et al.*, 2017) will be key to determining how genetic programs orchestrate synaptic remodeling.

Cellular and molecular mechanisms that regulate DD remodeling

Neuronal activity promotes DD synaptic remodeling

Although several genetic programs regulate synaptic remodeling in D-type motor neurons (Figure 3), additional evidence suggests that neuronal activity also promotes remodeling (Figure 4(A)). For example, remodeling of the DD presynaptic domains is delayed by mutations that disable neurotransmitter release (e.g. *unc-13/Munc13* and *unc-18/Mun18*) and accelerated by mutations that increase neurotransmission (e.g. *tom-1/Tomosyn* and *slo-1/BK potassium channel*; Thompson-Peer *et al.*, 2012). Similarly, synaptic remodeling in the mammalian visual system is retarded but not blocked by sensory deprivation (i.e. absence of light) thus suggesting that both genetically encoded programs and neural activity promote remodeling in this circuit. (Kang *et al.*, 2013). Indeed, in *C. elegans*, DD expression of HBL-1/Hunchback, a transcription factor that promotes remodeling (Figure 3), is diminished in *unc-13* and *unc-18* mutants, which points to an activity-dependent mechanism for elevating HBL-1/Hunchback gene expression (Figure 4(B)). A downstream role for HBL-1/Hunchback is also suggested by the finding that a *hbl-1* mutant blocks the acceleration of DD presynaptic remodeling by *tom-1* and *slo-1* mutants (Thompson-Peer *et al.*, 2012). Additionally, optogenetic activation of L1 larval DD neurons drives precocious presynaptic remodeling (Miller-Fleming *et al.*, 2016). Although the activity-dependent mechanisms that promote HBL-1 expression are unknown, other downstream effectors of synaptic remodeling that depend on neuronal activity are beginning to emerge (Figure 4(D)).

The DEG/ENaC cation channel protein, UNC-8, promotes synaptic disassembly in an activity-dependent mechanism

A role for neural activity in GABA neuron synaptic remodeling is also supported by experiments showing that DEG/ENaC cation channel subunit, UNC-8, triggers presynaptic disassembly in a mechanism that depends on intracellular Ca^{++} (Miller-Fleming *et al.*, 2016). UNC-8 is expressed in DD neurons where it promotes remodeling but is turned off by UNC-55/COUP-TF in VD neurons to prevent ectopic removal of ventral VD presynaptic domains. UNC-55 regulation is likely indirect in this case as UNC-55 negatively regulates IRX-1 which in turn promotes UNC-8 expression (Miller-Fleming *et al.*, 2020; Figure 4(B,C,F)).

A reconstituted UNC-8 channel preferentially gates Na^{++} in Xenopus oocytes (Wang *et al.*, 2013) which suggests that the native UNC-8 DEG/ENaC channel could depolarize the GABA neurons in which it is expressed. In turn, this effect is predicted to enhance Ca^{++} import by presynaptic Voltage Gated Calcium Channels (VGCC), an idea supported by the additional findings that UNC-2/VGCC is required for UNC-8-dependent presynaptic disassembly (Figure 4(D); Miller-Fleming *et al.*, 2016). This model parallels an earlier finding in which a Drosophila Pickpocket/DEG/ENaC channel is synaptically localized in motor neurons where it elevates intracellular Ca^{++} to promote neurotransmitter release (Orr *et al.*, 2017; Younger, Mu, Tong, Pym, & Davis, 2013). The radically different outcome of synaptic destruction that arises from UNC-8 function in *C. elegans* depends on the serine/threonine phosphatase CalcineurinA/TAX-6 and its regulatory subunit CalcineurinB/CNB-1. Calcineurin/CaN is activated by intracellular Ca^{++} and genetic results suggest that it functions upstream of UNC-8. Thus, UNC-8/DEG/ENaC, UNC-2/VGCC and CaN/TAX-6 may constitute a positive feedback loop to amplify Ca^{++} influx (Figure 4(D)). In turn, elevated Ca^{++} might activate synaptically localized apoptotic components that have been previously shown to drive synapse removal and that likely function in the UNC-8 pathway (Figure 4(D); L. Meng *et al.*, 2015; Miller-Fleming *et al.*, 2016). Intriguingly, CaN is known to antagonize postsynaptic function by dephosphorylating AMPA receptors to promote their endocytic removal (Lee, Kameyama, Huganir, &

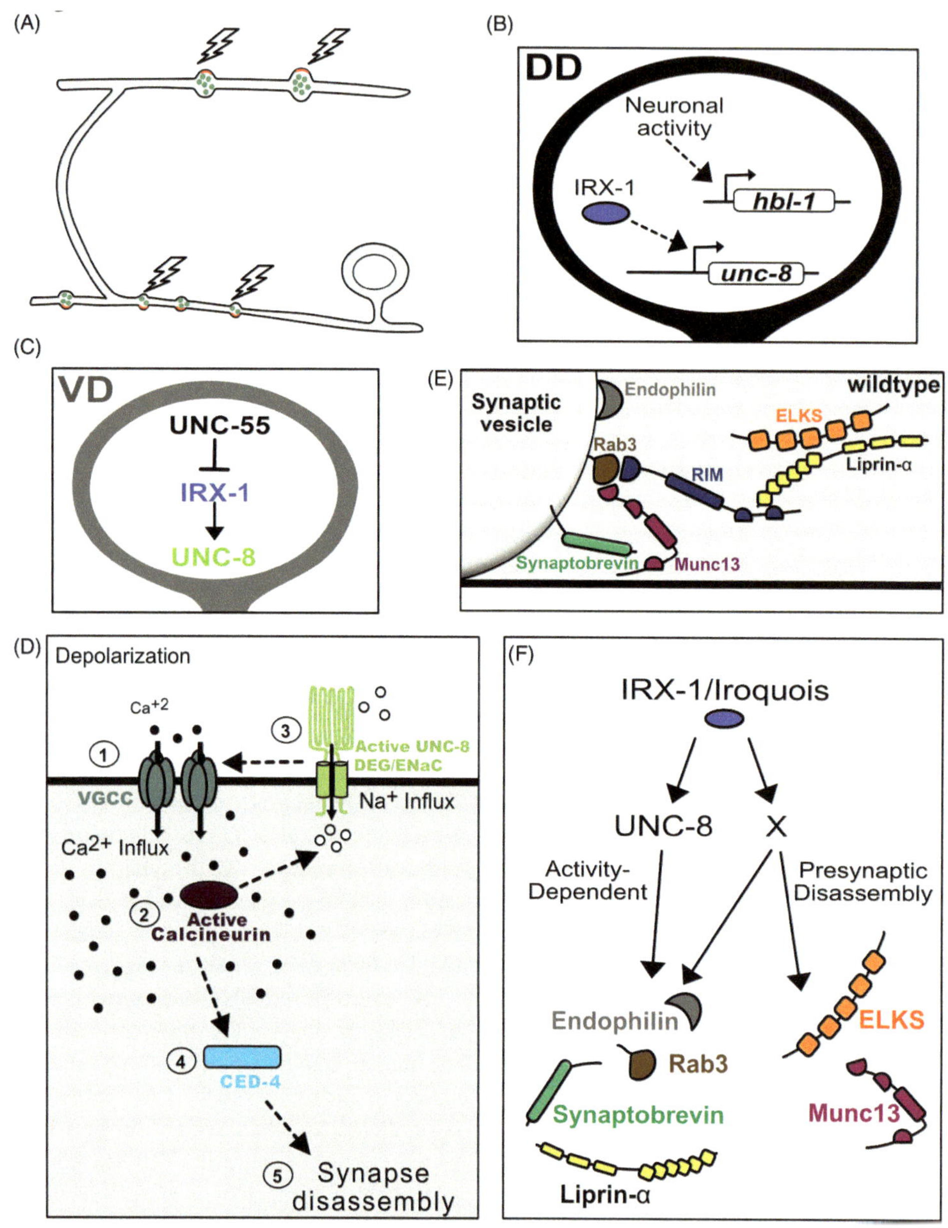

Figure 4. Neuronal activity promotes synaptic remodeling. (A) In DD neurons, neuronal activity promotes the disassembly of ventral presynaptic boutons as well as the formation of new terminals on the dorsal side. (B) Neuronal activity promotes expression of the pro-remodeling transcription factor HBL-1/Hunchback in DD neurons. The homeodomain transcription factor IRX-1/Iroquois/factor drives expression of the DEG/ENaC channel subunit UNC-8. (C) The transcriptional repressor UNC-55/COUP-TFII functions in VD neurons to block expression of the pro-remodeling transcription factor IRX-1/Iroquois and its downstream target UNC-8. (D) Proposed activity-dependent mechanism for synaptic remodeling (1) With depolarization, Voltage-gated Calcium-channels (VGCC) import Ca^{++} to (2) activate the Ca^{++}-dependent phosphatase Calcineurin/CaN (3) CaN functions upstream of ENaC/UNC-8, which mediates Na^{++} import, leading to further membrane depolarization and activation of VGCC. These signaling events trigger a positive feedback loop that elevates intracellular Ca^{++} for activation (4) of a cell-death pathway involving CED-4 and potentially additional downstream effectors (5) of presynaptic disassembly. (E) Presynaptic proteins Endophilin/UNC-57, Rab3, RIM/UNC-10, Munc13/UNC-13, Synaptobrevin/SNB-1, ELKS and Liprin-alpha/SYD-2 regulate synaptic vesicle fusion and neurotransmitter release. (F) The transcription factor IRX-1/Iroquois drives presynaptic disassembly by promoting expression of UNC-8/ENaC expression and an unknown target (X) to remove Endophilin, Rab3, Synaptobrevin and Liprin-α in an activity-dependent mechanism. IRX-1 also drives the elimination of active zone proteins ELKS and Munc13 that are not disassembled by the UNC-8 pathway.

Bear, 1998; Sanderson *et al.*, 2012; Winder, Mansuy, Osman, Moallem, & Kandel, 1998). It will be interesting to determine if presynaptic disassembly in remodeling GABA neurons could also depend on other CaN/TAX-6 targets that have yet to be identified at remodeling presynaptic terminals.

Notably, the *unc-8* locus was initially detected by a dominant allele, *unc-8(d)*, isolated in Sydney Brenner's original screen for uncoordinated mutants (Brenner, 1974). The *unc-8(d)* mutant DEG/ENaC channel is constitutively active and results in the degeneration of ventral cord DA/DB cholinergic neurons that normally drive movement (Shreffler & Wolinsky, 1997; Wang *et al.*, 2013). *unc-8* is abundantly expressed in DA/DB motor neurons but at a substantially lower levels in DDs which could account for the selective degeneration of DA/DB neurons in *unc-8(d)* mutants (Miller-Fleming *et al.*, 2020; Wang *et al.*, 2013). Interestingly, UNC-8 does not activate synaptic remodeling in DA/DB motor neurons but is sufficient to induce presynaptic disassembly in both DD and VD class GABAergic neurons (Miller-Fleming *et al.*, 2016). The differential effects of UNC-8 on motor neuron function and synaptic remodeling point to potential neuron-specific regulation of UNC-8 channel activity and intracellular localization.

Parallel-acting pathways dismantle the presynaptic apparatus in remodeling GABAergic neurons

Although mutations or pharmacologic treatments that inactivate UNC-8 channel function retard synaptic remodeling, disassembly of the presynaptic apparatus is not completely blocked thus suggesting that additional parallel acting pathways are likely involved (Miller-Fleming *et al.*, 2016). Recent work has shown that the transcription factor IRX-1/Iroquois promotes UNC-8 expression in DD neurons (Figure 4(B,F)). Genetic evidence suggests that IRX-1/Iroquois must also regulate other downstream effectors to orchestrate presynaptic disassembly. Of particular interest is the finding that the UNC-8 pathway selectively removes a subset of presynaptic markers (e.g. SNB-1/V-SNARE, RAB-3/RAB3) but does not target key effectors of synaptic vesicle fusion and release, UNC-13/Munc13 and ELKS-1/ELKS (Figure 4(E)), both of which are removed by the additional IRX-1/Iroquois regulated pathway (Figure 4(F); Miller-Fleming *et al.*, 2020). This finding is significant because it suggests that the mechanism of synaptic destruction in this case selectively targets specific components and thus likely involves molecularly distinct pathways that operate in tandem to disassemble the presynaptic apparatus. Experiments that identify downstream targets of IRX-1/Iroquois should be helpful for unraveling the mechanism of this effect. Because expression of IRX-1/Iroquois is controlled by UNC-55, previous studies that have identified UNC-55-regulated genes could also point to proteins that work downstream of IRX-1/Iroquois (Petersen *et al.*, 2011; Yu *et al.*, 2017).

GABA signaling promotes DD remodeling

Because GABA signaling can regulate neuronal plasticity *in vivo* (Afroz, Parato, Shen, & Smith, 2016; Deidda *et al.*, 2015; Hensch *et al.*, 1998; Wu *et al.*, 2012), several groups have explored the possibility that DD synaptic remodeling in *C. elegans* is also regulated by GABA. Across species, GABA synthesis depends on the Glutamic Acid Decarboxylase (GAD) enzyme which is encoded by the *unc-25* gene in *C. elegans* (Jin, Jorgensen, Hartwieg, & Horvitz, 1999). An early study determined that the density of dorsal GABAergic synapses, which are formed after DD remodeling, was not different between wild-type and *unc-25* mutants in young adults (Jin *et al.*, 1999) thus suggesting that GABA signaling is dispensable for DD remodeling. However, new findings from additional studies of this circuit, challenge this conclusion and suggest that GABA signaling regulates the progression of the DD remodeling program: (1) Optogenetic activation of DD neurons accelerates both presynaptic disassembly and dorsal reassembly of DD presynaptic domains synapses (Miller-Fleming, 2016; Miller-Fleming *et al.*, 2016). (2) Mutants that disrupt GABA synthesis (GAD/*unc-25*) or vesicular uptake (VGAT/*unc-47*) delay removal of ventral GABAergic synapses (Miller-Fleming, 2016; Thompson-Peer *et al.*, 2012). (3) GABA signaling accelerates DD remodeling in *tom-1* mutants in which neurotransmitter release is elevated (Miller-Fleming, 2016; Thompson-Peer *et al.*, 2012). (4) Disruption of GABA signaling during early development slows DD presynaptic remodeling (Han, Bellemer, & Koelle, 2015). Finally, (5) metabotropic GABA receptors (e.g. *gbb-1* and *gbb-2*) and the GABA re-uptake transporter, SNF-11, promote elimination of ventral synapses in remodeling GABAergic neurons (Miller-Fleming, 2016). The original finding that the number of dorsally placed DD synapses in *unc-25* mutants is indistinguishable from wild type at the adult stage (Jin *et al.*, 1999) suggests that additional parallel acting pathways are likely responsible for completing the assembly of DD synaptic boutons when GABA signaling is impaired during early larval development. An important question for the future is to determine if delayed DD remodeling, as occurs in mutants with disrupted GABA signaling, also perturbs overall function of the mature motor circuit. This possibility seems plausible given that multiple new postembryonic motor neurons are incorporated into the ventral cord during the period in which DD remodeling transpires (Sulston, 1976; White *et al.*, 1978; Figure 1).

cAMP levels regulate presynaptic remodeling in GABAergic neurons

Genomic experiments (ChIP-Seq) to reveal targets of UNC-30/PITX and UNC-55/COUP-TF, transcription factors that regulate GABA neuron synaptic remodeling, detected key effectors of cAMP metabolism, notably PDE-4 (phosphodiesterase) and ACY-1 (adenylate cyclase; Yu *et al.*, 2017). Genetic experiments that derive from these findings are consistent with the hypothesis that cAMP signaling promotes DD presynaptic remodeling (Figure 5). For example, VD motor neurons remodel ectopically in *unc-55* mutants (Figure 2) and this phenotype is correlated with reduced expression of PDE-4. This finding suggests that UNC-55 normally activates PDE-4 expression to limit cAMP levels and thus prevent VD remodeling (Figure 5(B)). Conversely, IRX-1/Iroquois, antagonizes PDE-4 expression (Yu *et al.*, 2017), an effect consistent with the role of IRX-1/Iroquois in promoting DD remodeling (Figure 5(A); Petersen *et al.*, 2011). Direct measurements with an *in vivo* FRET assay confirmed that cAMP levels are correlated with DD synaptic remodeling. The regulation of cAMP levels is apparently complex as other antagonists of the synaptic remodeling pathway, LIN-14 and OIG-1 (see below), also limit cAMP levels potentially by promoting PDE-4 expression (Figure 5). cAMP likely functions in combination with additional pathways because genetic mutants predicted to alter cAMP levels (*e.g. pde-4*) exert modest effects on synaptic remodeling (Yu *et al.*, 2017). For example, the proposed role of UNC-8 in elevating intracellular Ca^{++} to promote presynaptic disassembly (Miller-Fleming *et al.*, 2016) might also boost cAMP

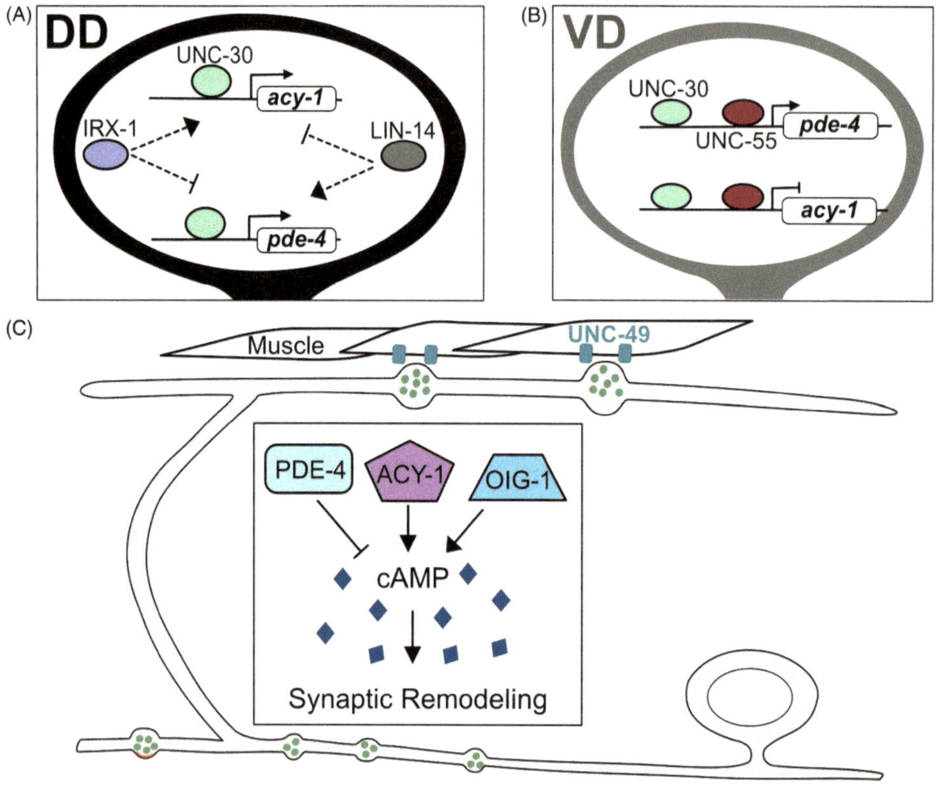

Figure 5. cAMP promotes synaptic remodeling. (A) Transcriptional control of biosynthetic (*acy-1*/adenylate cyclase) and metabolic (*pde-4*/phosphodiesterase) regulators of cAMP in DD neurons by IRX-1/Iroquois, UNC-30/PITX and LIN-14. (B) In VD neurons, UNC-30/PITX and UNC-55/COUP-TF promote expression of *pde-4*/phosphodiesterase and antagonize expression of *acy-1*/adenylate cyclase to prevent cAMP levels from exceeding a critical threshold that triggers presynaptic remodeling. (C) cAMP promotes the elimination of ventral presynaptic vesicles (green) and the localization of dorsal synaptic vesicles (green) adjacent to clusters of the postsynaptic UNC-49 GABAergic receptors (blue) in dorsal muscles in remodeling DD neurons. cAMP levels are reduced by PDE-4/phosphodiesterase and elevated by the ACY-1/adenylate cyclase and OIG-1/One-Ig-domain transmembrane protein.

levels since adenylate cyclase activity is Ca^{++} dependent (Halls & Cooper, 2011; Koch *et al.*, 2011). The downstream cell biological effects of cAMP are similarly uncharacterized but could potentially alter microtubule dynamics which is known to depend on cAMP signaling (Ghosh-Roy, Wu, Goncharov, Jin, & Chisholm, 2010) and to promote DD synaptic remodeling (see below; Kurup, Yan, Goncharov, & Jin, 2015).

Presynaptic remodeling depends on microtubule dynamics

DD neurons adopt a 'unipolar' morphology in which a single neurite maintains axonal and dendritic compartments in separate locations (Figure 1). Initially, in the early L1, the axonal neurotransmitter release machinery is restricted to a ventral region of the DD neurite proximal to the DD cell soma whereas the dendritic compartment is distally positioned in the dorsal segment of the DD neurite. With remodeling, these presynaptic (ventral) and postsynaptic (dorsal) domains exchange locations (Figure 1; White *et al.*, 1978). Interestingly, despite the switch in DD signaling polarity, microtubule (MT) orientation is not altered by remodeling. However, MT dynamics is elevated during this period and is required for DD remodeling (Figure 6(F); Kurup *et al.*, 2015).

Most MTs in DD neurons adopt the 'plus-end out' orientation both before and after remodeling (Figure 6; Kurup *et al.*, 2015). 'Plus-end' refers to the MT end to which more α and β tubulin dimers are added during MT growth and removed during MT shrinkage (Baas & Lin, 2011). The resultant 'dynamic instability' of MTs is characteristically elevated during cell biological events (e.g. cell division) in which the MT cytoskeleton is actively reorganized (Gardner, Zanic, & Howard, 2013). Several lines of evidence indicate that DD remodeling depends on MT dynamics. First, genetic mutations that stabilize MTs block DD remodeling and this effect can be partially relieved by treatment with the MT depolymerizing drug, nocodazole (Kurup *et al.*, 2015). Second, factors that regulate the transition from MT growth to shrinkage (i.e. 'catastrophe') such as the conserved kinase, DLK-1, and MT associated proteins, Kinesin-3/KLP-7 and Spastin/SPAS-1, promote DD remodeling (Kurup *et al.*, 2015). Third, the MT stabilizing role of intermediate filaments antagonizes remodeling (Figure 6(D,F); Kurup, Li, Goncharov, & Jin, 2018). A role for DLK-1 in synaptic remodeling is notable because DLK-1 also promotes axon regeneration in a cell biological mechanism that drives MT growth (Ghosh-Roy, Goncharov, Jin, & Chisholm, 2012). Additional unknown factors are likely required for activating MT dynamics in DD neurons, however, because a genetic ablation of DLK-1 activity results in only a slight delay in synaptic remodeling (Kurup *et al.*, 2015).

The plus-end motors, UNC-116/Kinesin1 and UNC-104/Kinesin3 function together to deliver presynaptic components to the dorsal neurite during remodeling (Figure

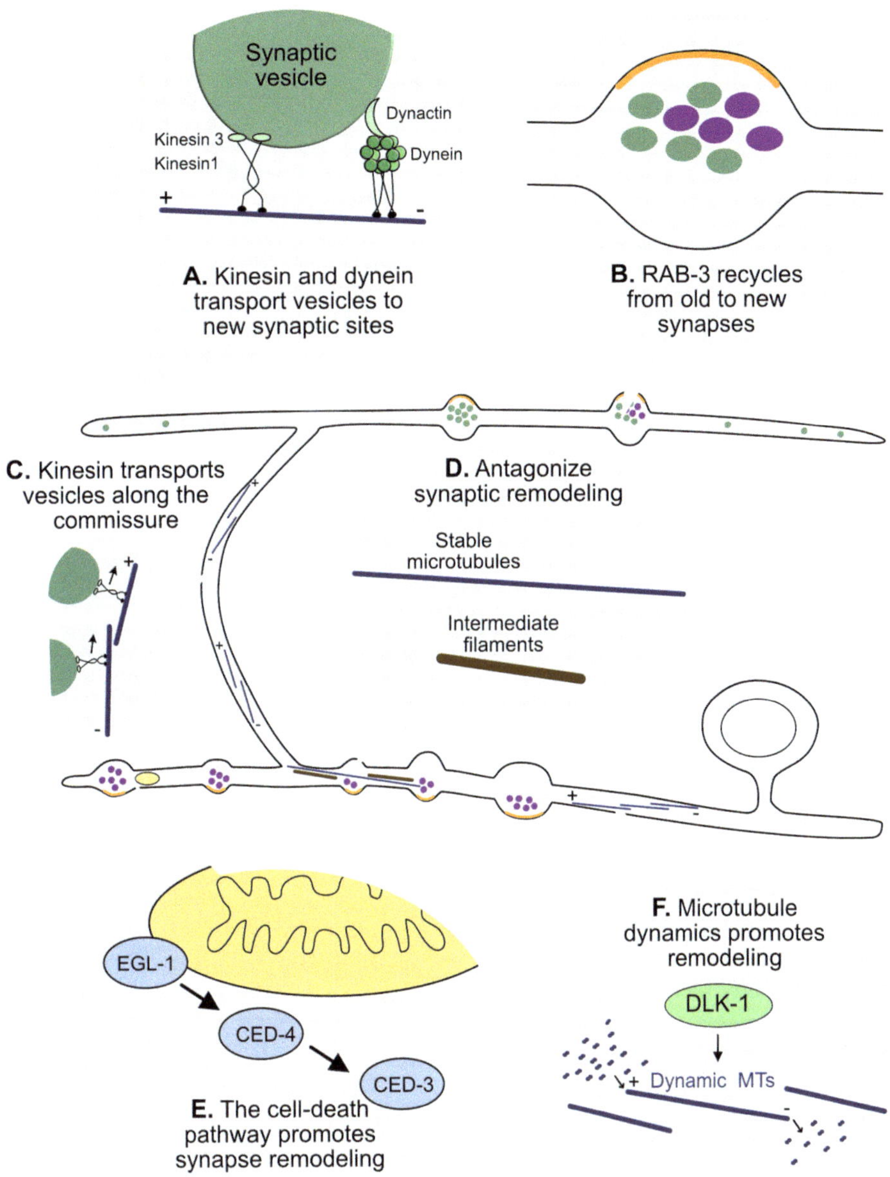

Figure 6. Cellular regulators of synaptic remodeling. (A) Anterograde transport of synaptic vesicles by motor proteins Kinesin1/UNC-116 and Kinesin3/UNC-104 on microtubules (blue) to the anterior distal tip of the dorsal DD neurite is opposed by the retrograde motor complex of Dynein/DHC-1 and Dynactin/DNC-4 that relocates synaptic vesicles to the posterior DD neurite (blue). (B) Experiments with photoconverted Dendra2::RAB-3 demonstrated that RAB-3 from old synaptic terminals (magenta) can be relocated to new dorsal synapses in remodeling DD neurons. (C) Kinesin1/UNC-116 transports synaptic vesicles along microtubules (blue) in the DD commissure. (D) Stable microtubules (blue), intermediate filaments (brown) and the kinase TTBK-3 antagonize synaptic remodeling in *tba-1(gf);dlk-1* double mutants (see text). (E) Cell-death pathway components, (EGL-1, CED-4, CED-3) associate with presynaptic mitochondria (yellow) to drive elimination of ventral synaptic terminals. (F) DLK-1 signaling promotes microtubule (blue) dynamics for synaptic remodeling.

6(A,C); Kurup, Yan, Kono, & Jin, 2017; Park *et al.*, 2011). Although MT dynamics is required for this kinesin-dependent function, the mechanistic basis for the effect is unknown (Kurup *et al.*, 2015). Intriguingly, an optogenetic experiment suggests that at least some of the cargo delivered by UNC-104/Kinesin1 to the dorsal side may include recycled components of the presynaptic apparatus. In this study, the synaptic vesicle protein, RAB-3, was tagged with Dendra, a photoconvertible GFP, to confirm its translocation from disassembled ventral DD synapses to nascent dorsal synapses during remodeling (Figure 6(B); Park *et al.*, 2011). In the future, it will be interesting to determine if additional presynaptic components are also recycled for reassembly at new DD synapses and to delineate the cell biological mechanism of potential endocytic events that are likely involved. The essential role for motor-dependent trafficking in DD remodeling is underscored by the finding that the cyclin-dependent kinase, CDK-5, functions upstream of UNC-104 and promotes remodeling (Park *et al.*, 2011). The molecular mechanism of CDK-5-dependent activation of UNC-104 is unknown.

The cell death pathway promotes remodeling of D-type motor neurons

Presynaptic Synaptobrevin/SNB-1 puncta are transiently localized to the axially projecting neurites of RME neurons in a remodeling event that mimics the sequential assembly and removal of DD presynaptic domains (L. Meng *et al.*, 2015). The RME remodeling phenotype was exploited in a genetic screen that revealed that members of the canonical cell-death pathway are required to remove transient RME presynaptic terminals. Interestingly, the apoptotic pathway is also involved in the removal of ventral presynaptic terminals during DD remodeling (Figure 6(E)). Additional genetic and imaging experiments suggested that apoptotic components are delivered to the presynaptic domain in association with mitochondria (L. Meng *et al.*, 2015) where CED-3/Caspase-mediated activation of the actin-severing protein, gelsolin, triggers presynaptic disassembly. Recent genetic results suggest that the apoptotic pathway may function in an activity-dependent mechanism of presynaptic disassembly that is triggered by transcriptionally regulated expression of the UNC-8/DEG/ENaC sodium channel subunit (Figure 4(D); Miller-Fleming *et al.*, 2016).

Regulation of postsynaptic remodeling in D-type motor neurons

Although both presynaptic and postsynaptic compartments are relocated in remodeling DD neurons (Figure 1), little is known of the postsynaptic mechanism in part because a reliable markers for the postsynaptic apparatus, ACR-12 and LEV-10, were only recently identified (Cuentas-Condori *et al.*, 2019; He, Cuentas-Condori, & Miller, 2019; Petrash *et al.*, 2013). ACR-12 encodes an α-subunit of a heteromeric nicotinic acetylcholine receptor (nAChR) in GABAergic motor neurons that also contains UNC-29, UNC-38, UNC-63 and LEV-1 AChR subunits (Philbrook *et al.*, 2018). GFP-marked ACR-12 and the auxiliary protein, LEV-10, localize to the postsynaptic compartments of DD and VD neurons in close apposition to presynaptic input from ventral cord cholinergic motor neurons (Figure 1(B,C); Cuentas-Condori *et al.*, 2019; He *et al.*, 2019; Petrash *et al.*, 2013; Philbrook *et al.*, 2018). Initially, in early L1 larvae, ACR-12::GFP localizes to the dorsal DD neurite but then disappears as remodeling ensues and nascent ACR-12::GFP puncta emerge on the ventral side (Figure 1(B,C); He *et al.*, 2015).

The Ig-domain protein, OIG-1, antagonizes DD synaptic remodeling

The translocation of dorsal ACR-12::GFP puncta to the ventral side is accelerated in *oig-1* mutants (He *et al.*, 2015; Howell *et al.*, 2015). The *oig-1* locus encodes a small protein of 137 amino acids with a single ImmunoGlobulin-(Ig) domain. OIG-1 is up-regulated by LIN-14 and the UNC-30/PITX transcription factor in early L1 larval DD neurons (Howell *et al.*, 2015) but is turned off in the late L1 by IRX-1/Iroquois as the DD remodeling program is activated (Figure 7(A); He *et al.*, 2015). Thus, OIG-1 normally functions to antagonize DD remodeling and is repressed by the transcription factor IRX-1/Iroquois to prevent this effect. OIG-1 also appears to antagonize ectopic remodeling in VD neurons. OIG-1 is highly expressed in the VD neurons throughout development due to repression of IRX-1 by

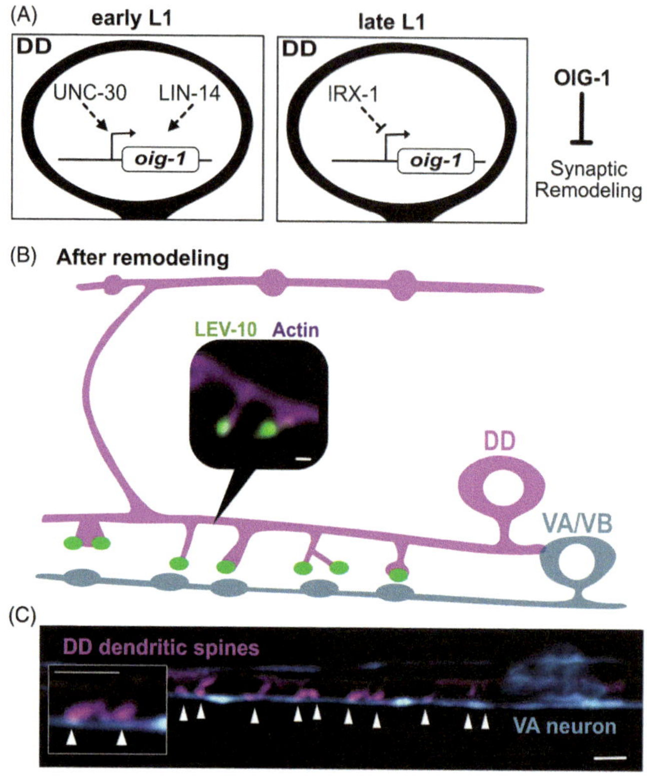

Figure 7. Postsynaptic remodeling. (A) The One-Ig-domain protein, OIG-1, is upregulated by the PITX/UNC-30 transcription factor in early L1 larval DD neurons but turned off by Iroquois/IRX-1 during the late L1 to prevent OIG-1 from antagonizing synaptic remodeling. (B) Graphical representation of dendritic spines protruding from the ventral postsynaptic neurite of a DD neuron and contacting presynaptic terminals of cholinergic VA/VB neurons. Inset shows a fluorescent image of the actin marker, LifeAct::mCherry (magenta), and the postsynaptic protein, LEV-10 (green) at the spine tip. Scale bar = 200 nm. Adapted from (Cuentas-Condori *et al.*, 2019) (C) Fluorescent image shows DD dendritic spines (magenta) projecting toward a presynaptic VA neuron (blue). Arrowheads denote sites of contact between postsynaptic spines and the VA process. Scale bar = 1 µm. Adapted from (Cuentas-Condori *et al.*, 2019).

UNC-55/COUP-TF (He *et al.*, 2015). OIG-1 expression in VD neurons may also depend on direct interaction of UNC-55 with the *oig-1* promoter (Howell *et al.*, 2015). In wild-type animals, ACR-12::GFP is exclusively localized to dorsal VD neurites but also appears on the ventral side in *oig-1* mutants (He *et al.*, 2015). Remarkably, genetic analysis indicates that OIG-1 also antagonizes remodeling of the presynaptic apparatus in both DD and VD neurons. For example, DD presynaptic markers (e.g. SNB-1::GFP) are precociously translocated to the ventral side in *oig-1* mutants (He *et al.*, 2015; Howell *et al.*, 2015). Notably, OIG-1 is the only known downstream effector that regulates both presynaptic and postsynaptic remodeling.

The OIG-1 mechanism of action is unclear. Although OIG-1 contains a canonical signal peptide and is secreted when over-expressed in transgenic animals (He *et al.*, 2015; Howell *et al.*, 2015), secretion is not required for its synaptic remodeling function (He *et al.*, 2015); when expressed at native levels, the endogenous OIG-1 protein is not secreted and shows an intracellular location (He *et al.*, 2019). An intracellular role is also consistent with the finding that OIG-1 expression in GABAergic neurons rescues the misplacement of ACR-12::GFP in *oig-1* mutants whereas forced expression of OIG-1 in nearby cholinergic ventral cord motor neurons does not complement the *oig-1* ectopic remodeling phenotype (He *et al.*, 2015; Howell *et al.*, 2015). The reported role of OIG-1 in limiting cAMP levels for presynaptic remodeling is similarly unknown (Yu *et al.*, 2017).

Caenorhabditis elegans GABAergic neurons have functional dendritic spines

In addition to the replacement of presynaptic components with the ACR-12 postsynaptic receptor, the remodeling mechanism transforms the initially oblong DD axonal compartments into dendritic spines that protrude from the ventral DD neurite (Figure 7(B,C); Cuentas-Condori *et al.*, 2019; Philbrook *et al.*, 2018). The possibility that DD neurons might display dendritic spines was first noted by John White *et al.* (1978, 1976, 1986) in EM reconstructions of the ventral cord. This idea is notable because dendritic spines are specialized postsynaptic structures that detect neurotransmitter release from presynaptic neurons. In mammalian neurons, spine morphogenesis is dynamic and responsive to stimuli correlated with learning and memory (Hering, Sheng, & Medical, 2001).

Recent studies have confirmed that dendritic spines in *C. elegans* DD and VD GABAergic motor neurons display key hallmarks of mammalian spines as they (1) are structurally defined by a dynamic actin cytoskeleton, (2) localize postsynaptic proteins in apposition to excitatory presynaptic terminals (Figure 7(B,C)), (3) localize ER and ribosomes, (4) display Ca^{++} transients evoked by presynaptic activity and, (5) respond to activity-dependent signals that modulate spine density (Cuentas-Condori *et al.*, 2019; Philbrook *et al.*, 2018). Interestingly, postsynaptic spine formation and maintenance requires the trans-synaptic adhesion protein Neurexin/NRX-1, which functions in presynaptic cholinergic motor neurons. Surprisingly, this Neurexin/NRX-1-dependent effect does not require the canonical trans-synaptic interacting partner, the membrane protein neuroligin/NLG-1 and thus likely interacts with an alternative component that is currently unknown (Philbrook *et al.*, 2018). The dramatic emergence of dendritic spines in remodeling DD neurons offers a unique opportunity to exploit the power of *C. elegans* genetics and live cell imaging to define evolutionarily ancient but shared mechanisms that drive spine morphogenesis.

Insights from synaptic remodeling in *C. elegans*

Sydney Brenner set out to understand how genes define the structure of the nervous system (Brenner, 1974). He selected *C. elegans* for this undertaking because he expected that its rapid life cycle would facilitate mutant analysis and its small size would allow the use of EM to define the wiring diagram. By design, his research strategy did not attempt to tackle directly the extraordinarily more difficult problem of unraveling how genes specify the brain. He chose instead to rely on the premise that all nervous systems are built with fundamental genetic programs and thus, that these secret plans could be divined much more easily in a nematode than in a more complex organism (Brenner, 1973).

As described above, DD motor neurons effectively reverse functional polarity with presynaptic (axonal) and postsynaptic (dendritic) compartments exchanging locations at opposite ends of each DD neuron (Figure 1). Similar examples of polarity reversal are currently unknown in other organisms. Remarkably, however, for neurons in multiple species, the asymmetric features that distinguish dendritic vs axonal compartments can be reallocated in response to injury. Axotomy, for example, can result in the transformation of an existing dendrite into an axon both in cultured neurons (Dotti & Banker, 1987) and in a living organism (Whitington & Sink, 2004). In addition, synapses are extensively relocated in the developing mammalian nervous system (De Paola *et al.*, 2006; Stettler, Yamahachi, Li, Denk, & Gilbert, 2006) and thus could depend on molecular pathways that also drive synaptic remodeling in DD neurons. Outlined below are additional examples of synaptic remodeling in more complex organisms for which studies of DD remodeling in *C. elegans* could be informative.

Remodeling of GABAergic neurons in the mammalian brain

In *C. elegans*, as in mammals, GABA-dependent inhibition determines circuit function (Lehmann, Steinecke, & Bolz, 2012; Pelkey *et al.*, 2017; Schuske, Beg, & Jorgensen, 2004). Strong conservation of key molecular determinants of GABAergic function, including the GABA biosynthetic enzyme, GAD (UNC-25), vesicular GABA transporter, VGAT (UNC-47) and GABA ionotropic (UNC-49) and metabotropic (GBB-1/2) receptors highlight striking molecular similarities and suggest that developmental mechanisms that control GABA-dependent circuit refinement might also be conserved (Jin *et al.*, 1999; Mclntire *et al.*, 1993a; 1993b). GABAergic neurons constitute about 20–30% of the mammalian cortex, typically provide inhibitory input to glutamatergic neurons and are structurally and functionally diverse (Hendry, Schwark, & Jones, 1987; Pelkey *et al.*, 2017; Sherwood *et al.*, 2010). Similar to DD neurons, some mammalian GABAergic neurons receive excitatory input through dendritic spines and others innervate target cells through *en-passant* boutons (Kawaguchi, Karube, & Kubota, 2006; Pelkey *et al.*, 2017). GABAergic interneurons can also be extensively refined during postnatal development. Of particular interest, the elimination of perisomatic inputs from GABAergic basket cells to glutamatergic pyramidal neurons depends on a mechanism that requires GABA signaling (Sullivan *et al.*, 2018; Wu *et al.*, 2012). The parallel role of GABA in promoting the removal of presynaptic termini in developing DD neurons in *C. elegans* could be indicative of shared cell biological pathways for synaptic remodeling (Miller-Fleming, 2016).

Activity-dependent removal of presynaptic domains in the mammalian visual circuit

In the developing mammalian visual circuit, retinal ganglion cells (RGCs) project to the thalamus to innervate geniculate neurons. Initially, presynaptic boutons are dispersed

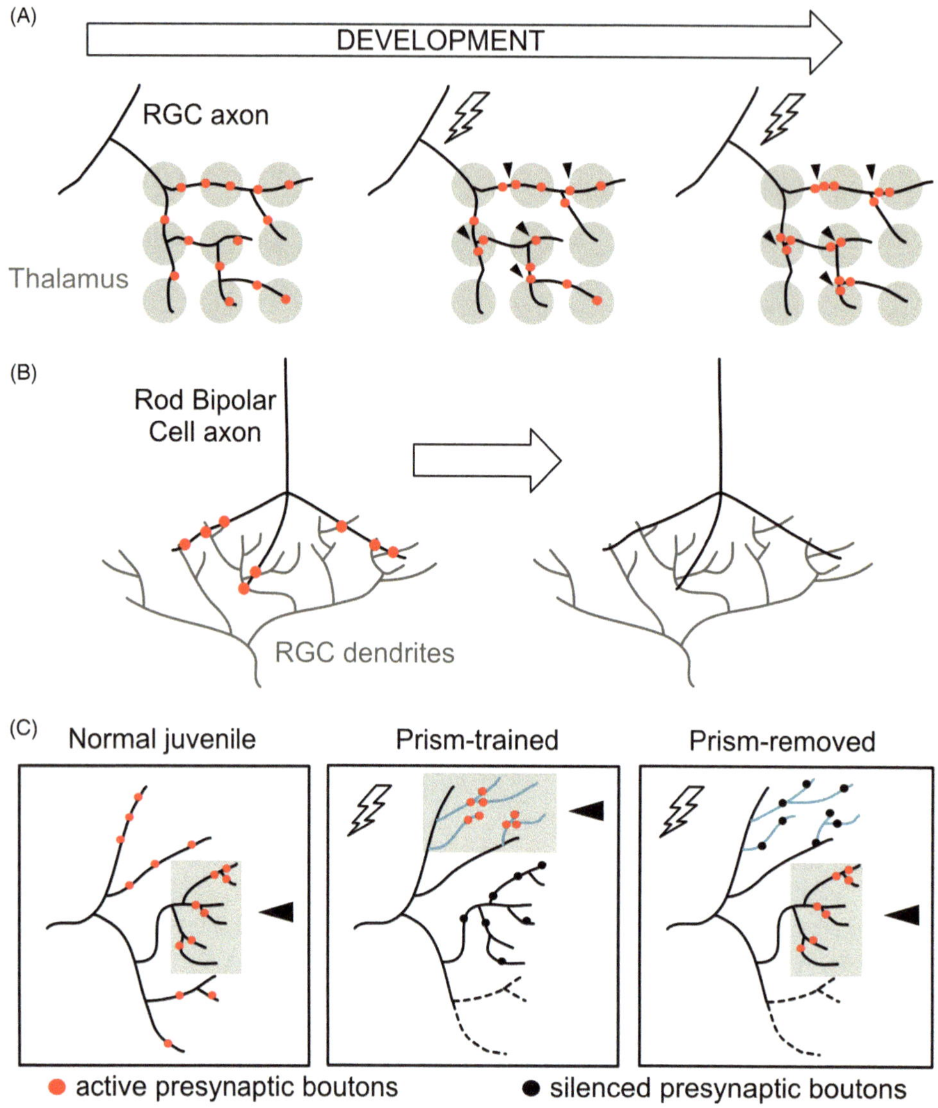

Figure 8. Presynaptic remodeling within intact axons in vertebrate circuits. (A) Initially, a Retinal Ganglion Cell (RGC) axon (black) innervates broadly multiple geniculate neurons in the thalamus (gray). Activity (lightning bolt) induces the relocation and clustering of RGC axons at proximal positions. (B) RGC dendrites (gray) receive input from Rod Bipolar cell axons (black). During early development, synaptic boutons (red) in Rod Bipolar Cells are eliminated while intact axonal trajectories are maintained. (C) Activity drives clustering of presynaptic boutons in the auditory circuit of barn owls. Normal (untrained) juveniles associate visual and auditory cues in the normal axonal region (gray box, arrowhead) where active synaptic boutons cluster (red). Prism-trained owls learn to associate auditory cues with an optically imposed object location. In this paradigm, active synaptic boutons (red) cluster in the adaptive region (gray box, arrowhead) whereas inactive boutons (black) remain in the normal region. After prisms are removed, active synaptic boutons (red) cluster at the normal region (gray box, arrowhead), whereas inactive boutons (black) remain in the adaptive zone.

throughout each RGC axon to synapse with multiple geniculate targets. This florid pattern of connectivity is then refined in a mechanism that eliminates distal RGC boutons while clustering others at proximal locations (Figure 8(A); Hong *et al.*, 2014). Importantly, RGC axonal projections remain intact as boutons are relocated and are not retracted until a later, separate pruning step (Hong & Chen, 2011). Inputs to RGCs from rod Bipolar Cells (BCs) in the retina are also eliminated from stable axonal-dendritic contacts during development (Figure 8(B); Morgan, Soto, Wong, & Kerschensteiner, 2011). Thus, synaptic remodeling within existing RGC and BC axons parallels refinement of the DD circuit in which presynaptic domains are repositioned to new locations within the DD neurite. As discussed above, at least one presynaptic protein, RAB-3, is recycled from old to new boutons in DD neurons (Park *et al.*, 2011). It will be interesting to determine whether RGCs similarly recycle existing synaptic material from distal regions to the newly formed bouton clusters. This possibility seems plausible because a related phenomenon occurs in the mature mammalian nervous system in which presynaptic components can be actively exchanged between *en-passant* synapses (Tsuriel *et al.*, 2006).

Synaptic remodeling in RGC axons is also activity-dependent as deprivation of sensory neuron input diminishes bouton clustering (Hong *et al.*, 2014). Similarly, reduced synaptic activity impairs DD remodeling by delaying both formation of new synapses in the dorsal DD neurite (Thompson-Peer *et al.*, 2012) as well as the elimination of old presynaptic domains on the ventral side (Miller-

Fleming *et al.*, 2016). The activity-dependent effect on DD presynaptic disassembly depends on the cell-autonomous roles of the UNC-8/DEG/ENaC cation channel and the serine-threonine phosphatase TAX-6/Calcineurin which function together to elevate intracellular Ca^{++} (Miller-Fleming *et al.*, 2016). It will be interesting to determine if similar components direct synaptic remodeling in RGCs.

Altered behavior in the barn owl involves the reallocation of the presynaptic apparatus

Anatomical and functional studies of the barn owl auditory localization pathway provide additional examples of activity-dependent remodeling that involves the coincident assembly of synapses in new locations as others are removed (Figure 7(C)). Juvenile owls, fitted with prisms that distort the visual field, learn to associate auditory cues with the imposed new optical location (Knudsen & Knudsen, 1989). This phenomenon is correlated with the expansion of axonal arbors into receptive fields associated with the learned behavior (Debello, Feldman, & Knudsen, 2001). Clustering of presynaptic boutons in these new adaptive zones is also enhanced in comparison to the normal receptive field but the overall number of synapses in each region is not significantly different. Thus, this mechanism appears to have effectively reduced the separation between adjacent presynaptic domains by balancing nascent assembly with synaptic elimination in nearby regions (Debello *et al.*, 2001; Mcbride, Rodriguez-Contreras, Trinh, Bailey, & Debello, 2008). Notably, synaptic remodeling in the *C. elegans* GABAergic circuit also involves the elimination of established synapses paired with assembly of presynaptic boutons in new locations (Figure 1). Interestingly, prism-trained owls retain the capacity to associate auditory cues with the normal visual field after the prisms are removed. This finding suggests that the adaptive synaptic clusters, which are maintained in trained animals, are functionally silenced with the restoration of normal visual cues (Mcbride & Debello, 2015). We speculate that this example of synaptic silencing in the barn owl auditory circuit could be potentially accomplished as in the *C. elegans* GABAergic neurons by the selective disassembly of key components such as ELKS or Munc13 that are required for neurotransmitter release but are not needed for the maintenance of synaptic structure (Figure 4; Liu *et al.*, 2014; Miller-Fleming *et al.*, 2020; Varoqueaux *et al.*, 2002). In the case of the barn owl, this surgically precise mechanism could facilitate an adaptive response to temporal cues while also maintaining the long-term capacity to restore normal visual input (Mcbride & Debello, 2015).

Conclusions and future prospects

Sydney Brenner was a supreme optimist. After concluding that all of the major questions about gene expression had been answered, he set out to deduce how genes determine behavior (Brenner, 1973). To simplify the problem of acquiring this information, he selected a small organism with a limited number of neurons and facile genetics. Although EM reconstruction ultimately revealed the complete wiring diagram of the *C. elegans* nervous system and, genetic analysis detected hundreds of 'unc' mutants that altered movement (Brenner, 1974; White *et al.*, 1986), Brenner's original goal of deducing the genomic logic of behavior has yet to be realized. Nevertheless, the tools that he developed and his vision of how they could be employed, inspired an army of enthusiastic peers (Emmons, *et al.*, 2015). Here we have featured their studies of synaptic plasticity, a dynamic facet of the nervous system that tunes circuit function. We have limited our focus to studies of a developmentally regulated remodeling event that alters the architecture of DD class GABAergic motor neurons in the *C. elegans* ventral nerve cord. Remarkably, DD presynaptic and postsynaptic domains swap locations at opposite ends of the DD neurite with no apparent alterations in external DD morphology during early development (Figure 1; Hallam & Jin, 1998; White *et al.*, 1978). The stereotypical timing of DD remodeling is indicative of a genetic program and multiple transcription factors that regulate DD rewiring have been identified (Figure 3). In fact, two of these transcription factors, UNC-30/PITX and UNC-55/COUP-TF, were initially detected as movement-defective alleles in Brenner's original genetic screen (Brenner, 1974; Howell *et al.*, 2015; Shan *et al.*, 2005). Although DD rewiring is subject to genetic control, the developmentally regulated translocation of the DD presynaptic domain is accelerated by neuron activity (Miller-Fleming *et al.*, 2016; Thompson-Peer *et al.*, 2012). Because the mechanism of this effect involves the voltage-gated calcium channel UNC-2/VGCC and the calcium-activated phosphatase, TAX-6/Calcineurin, elevated intracellular calcium is an attractive choice for a likely driver of presynaptic disassembly. Genetic experiments suggest that downstream calcium-dependent effects could target a canonical apoptotic pathway (Miller-Fleming *et al.*, 2016). The reported involvement of the CED-3/caspase-activated actin-severing protein, gelsolin, points to a critical role for the actin cytoskeleton in presynaptic remodeling but the mechanism of this effect is unknown (L. Meng *et al.*, 2015). Recent studies revealed the surprising finding that the activity-dependent remodeling pathway targets a subset of presynaptic components. At least two key regulators of neurotransmitter release, UNC-13/Munc13 and ELKS (Liu *et al.*, 2014; Varoqueaux *et al.*, 2002), are instead disassembled by a separate parallel-acting pathway regulated by the homeodomain transcription factor, IRX-1/Iroquois (Figure 4; Miller-Fleming *et al.*, 2020). The existence of distinct disassembly pathways is indicative of specific molecular interactions that selectively eliminate specific presynaptic components. RNA-Seq profiling experiments to identify IRX-1/Iroquois targets (Spencer *et al.*, 2014; Taylor *et al.*, 2019) could be useful for delineating the biochemical mechanism of these effects. This approach could also be useful for delineating the mechanism of postsynaptic remodeling, which is also regulated by IRX-1/Iroquois, but has not been extensively investigated. Of particular interest is the question of whether known regulators of presynaptic remodeling (Figures 4–6) are also involved in dismantling the postsynaptic apparatus (Figure 7). Only one downstream effector is known to regulate remodeling of both presynaptic and postsynaptic DD compartments, the small single-Ig domain protein OIG-1, but its role is mysterious and requires further investigation (He *et al.*, 2015, 2019; Howell *et al.*, 2015). Finally, future studies should continue to exploit an additional key strength of *C. elegans* as an experimental organism that was also recognized by Sydney Brenner; worms are transparent and small enough to fit on a microscope slide. One of the major challenges of delineating mechanisms that regulate synaptic dynamics in mammals (Figure 8) is the relative inaccessibility of developing circuits to live cell imaging (Südhof, 2018). By using new methods of bright, cell-specific fluorescent labeling of native proteins, it should be possible to monitor the dynamics of synaptic destruction and reassembly in time-lapse imaging experiments that avoid potential artifacts arising from multicopy transgenic arrays (He *et al.*, 2019; Hefel & Smolikove, 2019; Schwartz & Jorgensen, 2016).

Disclosure statement

No potential conflict of interest was reported by the author(s).

ORCID

Andrea Cuentas-Condori http://orcid.org/0000-0002-4847-0031
David M. Miller, 3rd http://orcid.org/0000-0001-9048-873X

References

Afroz, S., Parato, J., Shen, H., & Smith, S.S. (2016). Synaptic pruning in the female hippocampus is triggered at puberty by extrasynaptic GABA A receptors on dendritic spines. *eLife, 5*, e15106. doi:10.7554/eLife.15106

Baas, P.W., & Lin, S. (2011). Hooks and comets : The story of microtubule polarity orientation in the neuron. *Developmental Neurobiology, 71*(6), 403–418. doi:10.1002/dneu.20818

Babu, M.M., Luscombe, N.M., Aravind, L., Gerstein, M., & Teichmann, S.A. (2004). Structure and evolution of transcriptional regulatory networks. *Current Opinion in Structural Biology, 14*(3), 283–291. doi:10.1016/j.sbi.2004.05.004

Brenner, S. (1973). The genetics of behaviour. *British Medical Bulletin, 29*(3), 269–271. doi:10.1093/oxfordjournals.bmb.a071019

Brenner, S. (1974). The genetics of *Caenorhabditis elegans. Genetics, 77* (1), 71–94.

Bujalka, H., Koenning, M., Jackson, S., Perreau, V.M., Pope, B., Hay, C.M., … Emery, B. (2013). MYRF is a membrane-associated transcription factor that autoproteolytically cleaves to directly activate myelin genes. *PLoS Biology, 11*(8), e1001625. doi:10.1371/journal.pbio.1001625

Cavodeassi, F., Modolell, J., & Gómez-Skarmeta, J.L. (2001). The Iroquois family of genes: From body building to neural patterning. *Development (Cambridge, England), 128* (15), 2847–2855.

Chalfie, M., Tu, Y., Euskirchen, G., Ward, W., & Prasher, D. (1994). Green fluorescent protein as a marker for gene expression. *Science, 263*(5148), 802–805. doi:10.1126/science.8303295

Cinar, H., Keles, S., & Jin, Y. (2005). Expression profiling of GABAergic motor neurons in *Caenorhabditis elegans. Current Biology : CB, 15* (4), 340–346. https://doi.org/10.1016/j. doi:10.1016/j.cub.2005.02.025

Cuentas-Condori, A., Mulcahy, B., He, S., Palumbos, S., Zhen, M., & Miller, D.M. (2019). *C. elegans* neurons have functional dendritic spines. *eLife, 8*, e47918. doi:10.7554/eLife.47918

De Paola, V., Holtmaat, A., Knott, G., Song, S., Wilbrecht, L., Caroni, P., & Svoboda, K. (2006). Cell type-specific structural plasticity of axonal branches and boutons in the adult neocortex. *Neuron, 49* (6), 861–875. doi:10.1016/j.neuron.2006.02.017

Debello, W.M., Feldman, D.E., & Knudsen, E.I. (2001). Adaptive axonal remodeling in the midbrain auditory space map. *The Journal of Neuroscience : The Official Journal of the Society for Neuroscience, 21*(9), 3161–3174. doi:10.1523/JNEUROSCI.21-09-03161.2001

Deidda, G., Allegra, M., Cerri, C., Naskar, S., Bony, G., Zunino, G., … Cancedda, L. (2015). Early depolarizing GABA controls critical-period plasticity in the rat visual cortex. *Nature Neuroscience, 18*(1), 87–96. doi:10.1038/nn.3890

Dotti, C., & Banker, G. (1987). Experimentally induced alteration in the polarity of developin neurons. *Nature, 330*(6145), 254–256. doi:10.1038/330254a0

Eastman, C., Horvitz, H.R., & Jin, Y. (1999). Coordinated transcriptional regulation of the unc-25 glutamic acid decarboxylase and the unc-47 GABA vesicular transporter by the *Caenorhabditis elegans* UNC-30 homeodomain protein. *The Journal of Neuroscience, 19*(15), 6225–6234. doi:10.1523/JNEUROSCI.19-15-06225.1999

Emmons, S.W. (2015). The beginning of connectomics: A commentary on White *et al.* (1986) 'The structure of the nervous system of the nematode *Caenorhabditis elegans. Philosophical Transactions of the Royal Society B: Biological Sciences, 370*(1666), 20140309. doi:10.1098/rstb.2014.0309

Gardner, M.K., Zanic, M., & Howard, J. (2013). Microtubule catastrophe and rescue. *Current Opinion in Cell Biology, 25*(1), 14–19. doi:10.1016/j.ceb.2012.09.006

Ghosh-Roy, A., Goncharov, A., Jin, Y., & Chisholm, A.D. (2012). Kinesin-13 and tubulin posttranslational modifications regulate microtubule growth in axon regeneration. *Developmental Cell, 23*(4), 716–728. doi:10.1016/j.devcel.2012.08.010

Ghosh-Roy, A., Wu, Z., Goncharov, A., Jin, Y., & Chisholm, A.D. (2010). Calcium and cyclic AMP promote axonal regeneration in *Caenorhabditis elegans* and require DLK-1 kinase. *The Journal of Neuroscience : The Official Journal of the Society for Neuroscience, 30*(9), 3175–3183. doi:10.1523/JNEUROSCI.5464-09.2010

Hallam, S.J., & Jin, Y. (1998). lin-14 regulates the timing of synaptic remodelling in *Caenorhabditis elegans. Nature, 395*(6697), 78–82. doi:10.1038/25757

Halls, M.L., & Cooper, D.M.F. (2011). Regulation by Ca2+-signaling pathways of adenylyl cyclases. *Cold Spring Harbor Perspectives in Biology, 3*(1), a004143. doi:10.1101/cshperspect.a004143

Han, B., Bellemer, A., & Koelle, M.R. (2015). An evolutionarily conserved switch in response to GABA affects development and behavior of the locomotor circuit of *Caenorhabditis elegans. Genetics, 199* (4), 1159–1172. doi:10.1534/genetics.114.173963

Hardy, P.A. (1990). Genetic aspects of nervous system development. *Journal of Neurogenetics, 6*(3), 115–131. doi:10.3109/01677069009107105

He, S., Cuentas-Condori, A., & Miller, D.M. (2019). NATF (Native and Tissue-Specific Fluorescence): A strategy for brigth, tissue-specific GFP labeling of native proteins in *Caenorhabditis elegans. Genetics, 212*(2), 387–395. doi:10.1534/genetics.119.302063

Hefel, A., & Smolikove, S. (2019). Tissue-specific split sfGFP system for streamlined expression of GFP tagged proteins in the *Caenorhabditis elegans* germline. *G3 (Bethesda, Md.), 9*(6), 1933–1943. doi:10.1534/g3.119.400162

Hendi, A., Kurashina, M., & Mizumoto, K. (2019). Intrinsic and extrinsic mechanisms of synapse formation and specificity in *C. elegans. Cellular and Molecular Life Sciences, 76*(14), 2719–2738. doi:10.1007/s00018-019-03109-1

Hendry, S.H.C., Schwark, H.D., & Jones, E.G. (1987). Numbers and proportions of GABA-immunoreactive different areas of monkey cerebral cortex neurons. *Journal of Neuroscience, 7*, 1503–1519.

Hensch, T.K., Fagiolini, M., Mataga, N., Stryker, M.P., Baekkeskov, S., & Kash, S.F. (1998). Local GABA circuit control of experience-dependent plasticity in developing visual cortex. *Science (New York, N.Y.), 282*(5393), 1504–1509. doi:10.1126/science.282.5393.1504

He, S., Philbrook, A., McWhirter, R., Gabel, C.V., Taub, D.G., Carter, M.H., ... Miller, D.M. (2015). Transcriptional control of synaptic remodeling through regulated expression of an immunoglobulin superfamily protein. *Current Biology : CB*, *25*(19), 2541–2548. doi:10.1016/j.cub.2015.08.022

Hering, H., Sheng, M., & Medical, H.H. (2001). Dendritic spines : Structure, dynamics and regulaion. *Nature Reviews Neuroscience*, *2*(12), 880–888. doi:10.1038/35104061

Hong, Y.K., & Chen, C. (2011). Wiring and rewiring of the retinogeniculate synapse. *Current Opinion in Neurobiology*, *21*(2), 228–237. doi:10.1016/j.conb.2011.02.007

Hong, Y.K., Park, S.H., Litvina, E.Y., Morales, J., Sanes, J.R., & Chen, C. (2014). Refinement of the Retinogeniculate Synapse by Bouton Clustering. *Neuron*, *84*(2), 332–339. doi:10.1016/j.neuron.2014.08.059

Houweling, A.C., Dildrop, R., Peters, T., Mummenhoff, J., Moorman, A.F., Rüther, U., & Christoffels, V.M. (2001). Gene and cluster-specific expression of the Iroquois family members during mouse development. *Mechanisms of Development*, *107* (1-2), 169–174. doi:10.1016/S0925-4773(01)00451-8

Howell, K., White, J.G., & Hobert, O. (2015). Spatiotemporal control of a novel synaptic organizer molecule. *Nature*, *523*(7558), 83–87. doi:10.1038/nature14545

Hristova, M., Birse, D., Hong, Y., & Ambros, V. (2005). The *Caenorhabditis elegans* heterochronic regulator LIN-14 is a novel transcription factor that controls the developmental timing of transcription from the insulin/insulin-like growth factor gene ins-33 by direct DNA binding. *Molecular and Cellular Biology*, *25*(24), 11059–11072. doi:10.1128/MCB.25.24.11059-11072.2005

Jin, Y., Hoskins, R., & Horvitz, H.R. (1994). Control of type-D GABAergic neuron differentiation by *C. elegans* UNC-30 homeodomain protein. *Nature*, *372*(6508), 780–783. doi:10.1038/372780a0

Jin, Y., Jorgensen, E., Hartwieg, E., & Horvitz, H.R. (1999). The *Caenorhabditis elegans* gene unc-25 encodes glutamic acid decarboxylase and is required for synaptic transmission but not synaptic development. *The Journal of Neuroscience : The Official Journal of the Society for Neuroscience*, *19*(2), 539–548. doi:10.1523/JNEUROSCI.19-02-00539.1999

Jin, Y., & Qi, Y.B. (2018). Building stereotypic connectivity: Mechanistic insights into structural plasticity from *C. elegans*. *Current Opinion in Neurobiology*, *48*, 97–105. doi:10.1016/j.conb.2017.11.005

Jorgensen, E.M., Hartwieg, E., Schuske, K., Nonet, M.L., Jin, Y., & Horvitz, H.R. (1995). Defective recycling of synaptic vesicles in synaptotagmin mutants of *Caenorhabditis elegans*. *Nature*, *378*(6553), 196–199. doi:10.1038/378196a0

Kang, E., Durand, S., Leblanc, J.J., Hensch, T.K., Chen, C., & Fagiolini, M. (2013). Visual acuity development and plasticity in the absence of sensory experience. *The Journal of Neuroscience : The Official Journal of the Society for Neuroscience*, *33*(45), 17789–17796. doi:10.1523/JNEUROSCI.1500-13.2013

Kawaguchi, Y., Karube, F., & Kubota, Y. (2006). Dendritic branch typing and spine expression patterns in cortical nonpyramidal cells. *Cerebral Cortex (New York, N.Y. : 1991)*, *16*(5), 696–711. doi:10.1093/cercor/bhj015

Knudsen, E.I., & Knudsen, F. (1989). Vision calibrates sound localization in developing barn owls. *The Journal of Neuroscience : The Official Journal of the Society for Neuroscience*, *9*(9), 3306–3313. doi:10.1523/JNEUROSCI.09-09-03306.1989

Koch, S.M., Dela Cruz, C.G., Hnasko, T.S., Edwards, R.H., Huberman, A.D., & Ullian, E.M. (2011). Pathway-specific genetic attenuation of glutamate release alters select features of competition-based visual circuit refinement. *Neuron*, *71*(2), 235–242. doi:10.1016/j.neuron.2011.05.045

Kurup, N., & Jin, Y. (2016). Neural circuit rewiring: Insights from DD synapse remodeling. *Worm*, *5*(1), e1129486. doi:10.1080/21624054.2015.1129486

Kurup, N., Li, Y., Goncharov, A., & Jin, Y. (2018). Intermediate filament accumulation can stabilize microtubules in *Caenorhabditis elegans* motor neurons. *Proceedings of the National Academy of Sciences*, *115*(12), 3114–3119. doi:10.1073/pnas.1721930115

Kurup, N., Yan, D., Goncharov, A., & Jin, Y. (2015). Dynamic microtubules drive circuit rewiring in the absence of neurite remodeling. Current Biology : CB, *25*(12), 1594–1605. doi:10.1016/j.cub.2015.04.061

Kurup, N., Yan, D., Kono, K., & Jin, Y. (2017). Differential regulation of polarized synaptic vesicle trafficking and synapse stability in neural circuit rewiring in *Caenorhabditis elegans*. *PLOS Genetics*, *13*(6), e1006844. doi:10.1371/journal.pgen.1006844

Lee, H.K., Kameyama, K., Huganir, R.L., & Bear, M.F. (1998). NMDA induces long-term synaptic depression and dephosphorylation of the GluR1 subunit of AMPA receptors in hippocampus. *Neuron*, *21*(5), 1151–1162. doi:10.1016/S0896-6273(00)80632-7

Lehmann, K., Steinecke, A., & Bolz, J. (2012). GABA through the ages: Regulation of cortical function and plasticity by inhibitory interneurons. *Neural Plasticity*, *2012*, 892784. doi:10.1155/2012/892784

Liu, C., Bickford, L.S., Held, R.G., Nyitrai, H., Su, T.C., & Kaeser, P.S. (2014). The active zone protein family ELKS supports Ca2+ influx at nerve terminals of inhibitory hippocampal neurons. *The Journal of neuroscience : The official journal of the Society for Neuroscience*, *34*(37), 12289–12303. doi:10.1523/JNEUROSCI.0999-14.2014

Mcbride, T.J., & Debello, W.M. (2015). Input clustering in the normal and learned circuits of adult barn owls. *Neurobiology of Learning and Memory*, *121*, 39–51. doi:10.1016/j.nlm.2015.01.011

Mcbride, T.J., Rodriguez-Contreras, A., Trinh, A., Bailey, R., & Debello, W.M. (2008). Learning drives differential clustering of axodendritic contacts in the barn owl auditory system. *The Journal of Neuroscience : The Official Journal of the Society for Neuroscience*, *28*(27), 6960–6973. doi:10.1523/JNEUROSCI.1352-08.2008

Mclntire, S.L., Jorgensen, E., & Horvitz, H.R. (1993a). Genes required for GABA function in *Caenorhabditis elegans*. *Nature*, *364*(6435), 334–337. doi:10.1038/364334a0

Mclntire, S.L., Jorgensen, E., Kaplan, J., & Horvitz, H.R. (1993b). The GABAergic nervous system of *C. elegans*. *Nature*, *364*(6435), 337–414. doi:10.1038/364337a0

Mello, C., Kramer, J., Stinchcomb, D., & Ambros, V. (1991). Efficient gene transfer in *C. elegans*: Extrahcormosomal maintenance and integration of transforming sequences. *The Embo Journal*, *10*(12), 3959–3970. doi:10.1002/j.1460-2075.1991.tb04966.x

Meng, J., Ma, X., Tao, H., Jin, X., Witvliet, D., Mitchell, J., ... Qi, Y.B. (2017). Myrf ER-bound transcription factors drive *C. elegans* synaptic plasticity via cleavage-dependent nuclear translocation. *Developmental Cell*, *41*(2), 180–194.e7. doi:10.1016/j.devcel.2017.03.022

Meng, L., Mulcahy, B., Cook, S.J., Neubauer, M., Wan, A., Jin, Y., & Yan, D. (2015). The cell death pathway regulates synapse elimination through cleavage of gelsolin in *Caenorhabditis elegans* neurons. *Cell Reports*, *11*(11), 1737–1748. doi:10.1016/j.celrep.2015.05.031

Miller-Fleming, T.W. (2016). Molecular dissection of synaptic remodeling in GABAergic neurons.

Miller-Fleming, T.W., Cuentas-Condori, A., Palumbos, S., Manning, L., Richmond, J.R., & Miller, D.M. (2020). Transcriptional control of parallel-acting pathways that remove discrete presynaptic proteins in remodeling neurons. BioRxiv.

Miller-Fleming, T.W., Petersen, S.C., Manning, L., Matthewman, C., Gornet, M., Beers, A., ... Miller, D.M. (2016). The DEG/ENaC cation channel protein UNC-8 drives activity-dependent synapse removal in remodeling GABAergic neurons. *eLife*, *5*, e14599. doi:10.7554/eLife.14599

Morgan, J.L., Soto, F., Wong, R.O.L., & Kerschensteiner, D. (2011). Development of cell type-specific connectivity patterns of converging excitatory axons in the retina. *Neuron*, *71*(6), 1014–1021. doi:10.1016/j.neuron.2011.08.025

Nonet, M.L. (1999). Visualization of synaptic specializations in live *C. elegans* with synaptic vesicle protein-GFP fusions. *Journal of Neuroscience Methods*, *89*(1), 33–40. doi:10.1016/S0165-0270(99)00031-X

Nonet, M.L., Saifee, O., Zhao, H., Rand, J.B., & Wei, L. (1998). Synaptic transmission deficits in *Caenorhabditis elegans* synaptobrevin mutants. *The Journal of Neuroscience : The Official Journal of the*

Society for Neuroscience, *18*(1), 70–80. doi:10.1523/JNEUROSCI.18-01-00070.1998

Orr, B.O., Gorczyca, D., Younger, M.A., Jan, L.Y., Jan, Y.N., & Davis, G.W. (2017). Composition and control of a Deg/ENaC channel during presynaptic homeostatic plasticity. *Cell Reports*, *20*(8), 1855–1866. doi:10.1016/j.celrep.2017.07.074

Park, M., Watanabe, S., Poon, V.Y.N., Ou, C.Y., Jorgensen, E.M., & Shen, K. (2011). CYY-1/Cyclin Y and CDK-5 differentially regulate synapse elimination and formation for rewiring neural circuits. *Neuron*, *70*(4), 742–757. doi:10.1016/j.neuron.2011.04.002

Pelkey, K.A., Chittajallu, R., Craig, M.T., Tricoire, L., Wester, J.C., & Mcbain, X.C.J. (2017). Hippocampal GABAergic inhibitory interneurons. *Physiological Reviews*, *97* (4), 1619–1747. doi:10.1152/physrev.00007.2017

Petersen, S.C., Watson, J.D., Richmond, J.E., Sarov, M., Walthall, W.W., & Miller, D.M. (2011). A transcriptional program promotes remodeling of GABAergic synapses in *Caenorhabditis elegans*. *The Journal of Neuroscience : The Official Journal of the Society for Neuroscience*, *31*(43), 15362–15375. doi:10.1523/JNEUROSCI.3181-11.2011

Petrash, H.A., Philbrook, A., Haburcak, M., Barbagallo, B., & Francis, M.M. (2013). ACR-12 ionotropic acetylcholine receptor complexes regulate inhibitory motor neuron activity in *Caenorhabditis elegans*. *The Journal of Neuroscience : The Official Journal of the Society for Neuroscience*, *33*(13), 5524–5532. doi:10.1523/JNEUROSCI.4384-12.2013

Philbrook, A., Ramachandran, S., Lambert, C.M., Oliver, D., Florman, J., Alkema, M.J., … Francis, M.M. (2018). Neurexin directs partner-specific synaptic connectivity in *C. elegans*. *eLife*, *7*, e35692. doi:10.7554/eLife.35692

Ruvkun, G., & Giusto, J. (1989). The *Caenorhabditis elegans* heterochronic gene lin-14 encodes a nuclear protein that forms a temporal developmental switch. *Nature*, *338*(6213), 313–319. doi:10.1038/338313a0

Sanderson, J.L., Gorski, J.A., Gibson, E.S., Lam, P., Freund, R.K., Chick, W.S., & Dell'Acqua, M.L. (2012). Akap150-anchored calcineurin regulates synaptic plasticity by limiting synaptic incorporation of Ca2+-permeable AMPA receptors. *The Journal of Neuroscience : The Official Journal of the Society for Neuroscience*, *32*(43), 15036–15052. doi:10.1523/JNEUROSCI.3326-12.2012

Schuske, K., Beg, A.A., & Jorgensen, E.M. (2004). The GABA nervous system in *C. elegans*. *Trends in Neurosciences*, *27*(7), 407–414. doi:10.1016/j.tins.2004.05.005

Schwartz, M.L., & Jorgensen, E.M. (2016). SapTrap, a toolkit for high-throughput CRISPR/Cas9 gene modification in *Caenorhabditis elegans*. *Genetics*, *202*(4), 1277–1288. doi:10.1534/genetics.115.184275

Shan, G., Kim, K., Li, C., & Walthall, W.W. (2005). Convergent genetic programs regulate similarities and differences between related motor neuron classes in *Caenorhabditis elegans*. *Developmental Biology*, *280*(2), 494–503. doi:10.1016/j.ydbio.2005.01.032

Sherwood, C.C., Raghanti, M.A., Stimpson, C.D., Spocter, M.A., Uddin, M., Boddy, A.M., … Hof, P.R. (2010). Inhibitory interneurons of the human prefrontal cortex display conserved evolution of the phenotype and related genes. *Proceedings. Biological Sciences*, *277*(1684), 1011–1020. doi:10.1098/rspb.2009.1831

Shreffler, W., & Wolinsky, E. (1997). Genes controlling ion permeability in both motorneurons and muscle. *Behavior Genetics*, *27*(3), 211–221. doi:10.1023/A:1025605929373

Spencer, W.C., McWhirter, R., Miller, T., Strasbourger, P., Thompson, O., Hillier, L.W., … Miller, D.M. (2014). Isolation of specific neurons from *C. elegans* larvae for gene expression profiling. *PLoS One.*, *9*(11), e112102. doi:10.1371/journal.pone.0112102

Stettler, D.D., Yamahachi, H., Li, W., Denk, W., & Gilbert, C.D. (2006). Axons and synaptic boutons are highly dynamic in adult visual cortex. *Neuron*, *49*(6), 877–887. doi:10.1016/j.neuron.2006.02.018

Südhof, T. (2018). Towards an understanding of synapse formation. *Neuron*, *100*(2), 276–293. doi:10.1016/j.neuron.2018.09.040

Sullivan, C.S., Gotthard, I., Wyatt, E.V., Bongu, S., Mohan, V., Weinberg, R.J., & Maness, P.F. (2018). Perineuronal net protein neurocan inhibits NCAM/EphA3 repellent signaling in GABAergic interneurons. *Scientific Reports*, *8*(1), 1–15. doi:10.1038/s41598-018-24272-8

Sulston, J.E. (1976). Post-embryonic development in the ventral cord of *Caenorhabditis elegans*. *Philosophical Transactions of the Royal Society of London*, *275*(938), 287–297.

Sulston, J.E., & Horvitz, H.R. (1977). Post-embryonic cell lineages of the nematode, *Caenorhabditis elegans*. *Developmental Biology*, *56* (1), 110–156. doi:10.1016/0012-1606(77)90158-0

Taylor, S., Santpere, G., Reilly, M., Glenwinkel, L., Poff, A., McWhirther, R., … Miller, D.M. III, (2019). Expression profiling of the mature *C. elegans* nervous system by single-cell RNA-Sequencing. BioRxiv.

Thompson-Peer, K.L., Bai, J., Hu, Z., & Kaplan, J. (2012). HBL-1 patterns synaptic remodeling in *C. elegans*. *Neuron*, *73*(3), 453–465. doi:10.1016/j.neuron.2011.11.025

Tsuriel, S., Geva, R., Zamorano, P., Dresbach, T., Boeckers, T., Gundelfinger, E.D., … Ziv, N.E. (2006). Local sharing as a predominant determinant of synaptic matrix molecular dynamics. *PLoS Biology*, *4*(9), e271. doi:10.1371/journal.pbio.0040271

Varoqueaux, F., Sigler, A., Rhee, J., Brose, N., Enk, C., Reim, K., & Rosenmund, C. (2002). Total arrest of spontaneous and evoked synaptic transmission but normal synaptogenesis in the absence of Munc13-mediated vesicle priming. *Proceedings of the National Academy of Sciences*, *99*(13), 9037–9042. doi:10.1073/pnas.122623799

Walthall, W.W., & Plunkett, J.A. (1995). Genetic transformation of the synaptic pattern of a motoneuron class in *Caenorhabditis elegans*. *The Journal of Neuroscience : The Official Journal of the Society for Neuroscience*, *15*(2), 1035–1043. doi:10.1523/JNEUROSCI.15-02-01035.1995

Wang, Y., Matthewman, C., Han, L., Miller, T., Miller, D.M., & Bianchi, L. (2013). Neurotoxic unc-8 mutants encode constitutively active DEG/ENaC channels that are blocked by divalent cations. *The Journal of General Physiology*, *142*(2), 157–169. doi:10.1085/jgp.201310974

White, J.G., Albertson, D.G., & Anness, M. (1978). Connectivity changes in a class of motoneurone during the development of a nematode. *Nature*, *271* (5647), 764–766. doi:10.1038/271764a0

White, J.G., Southgate, E., Thomson, J.N., & Brenner, S. (1976). The structure of the ventral nerve cord of *Caenorhadbitis elegans*. *Philosophical Transactions of the Royal Society of London*, *275*: 327–348.

White, J.G., Southgate, E., Thomson, J.N., & Brenner, S. (1986). The Structure of the Nervous System of the Nematode *Caenorhabditis elegans*. *Philosophical Transactions of the Royal Society of London. Series B, Biological Sciences*, *314*(1165), 1–340. doi:10.1098/rstb.1986.0056

Whitington, P.M., & Sink, H. (2004). Development of a polar morphology by identified embryonic motoneurons. *International Journal of Developmental Neuroscience : The Official Journal of the International Society for Developmental Neuroscience*, *22* (1), 39–45. doi:10.1016/j.ijdevneu.2003.10.004

Winder, D.G., Mansuy, I.M., Osman, M., Moallem, T.M., & Kandel, E.R. (1998). Genetic and pharmacological evidence for a novel, intermediate phase of long-term potentiation suppressed by calcineurin. *Cell*, *92*(1), 25–37. doi:10.1016/S0092-8674(00)80896-X

Wu, X., Fu, Y., Knott, G., Lu, J., Di Cristo, G., & Huang, Z.J. (2012). GABA signaling promotes synapse elimination and axon pruning in developing cortical inhibitory interneurons. *The Journal of Neuroscience : The Official Journal of the Society for Neuroscience*, *32*(1), 331–343. doi:10.1523/JNEUROSCI.3189-11.2012

Younger, M.A., Mu, M., Tong, A., Pym, E.C., & Davis, G.W. (2013). A Presynaptic ENaC Channel Drives Homeostatic Plasticity. *Neuron*, *79* (6), 1183–1196. doi:10.1016/j.neuron.2013.06.048

Yu, B., Wang, X., Wei, S., Fu, T., Dzakah, E.E., Waqas, A., … Shan, G. (2017). Convergent Transcriptional Programs Regulate cAMP Levels in C. *elegans* GABAergic Motor Neurons. *Developmental Cell*, *43* (2), 212–215. doi:10.1016/j.devcel.2017.09.013

Zhou, H., & Walthall, W. (1998). UNC-55, an Orphan Nuclear Hormone Receptor, Orchestrates Synaptic Specificity among Two Classes of Motor Neurons in *Caenorhabditis elegans*. *The Journal of Neuroscience : The Official Journal of the Society for Neuroscience*, *18*(24), 10438–10444. doi:10.1523/JNEUROSCI.18-24-10438.1998

What about the males? the *C. elegans* sexually dimorphic nervous system and a CRISPR-based tool to study males in a hermaphroditic species

Jonathon D. Walsh, Olivier Boivin and Maureen M. Barr

ABSTRACT

Sexual dimorphism is a device that supports genetic diversity while providing selective pressure against speciation. This phenomenon is at the core of sexually reproducing organisms. *Caenorhabditis elegans* provides a unique experimental system where males exist in a primarily hermaphroditic species. Early works of John Sulston, Robert Horvitz, and John White provided a complete map of the hermaphrodite nervous system, and recently the male nervous system was added. This addition completely realized the vision of *C. elegans* pioneer Sydney Brenner: a model organism with an entirely mapped nervous system. With this 'connectome' of information available, great strides have been made toward understanding concepts such as how a sex-shared nervous system (in hermaphrodites and males) can give rise to sex-specific functions, how neural plasticity plays a role in developing a dimorphic nervous system, and how a shared nervous system receives and processes external cues in a sexually-dimorphic manner to generate sex-specific behaviors. In *C. elegans*, the intricacies of male-mating behavior have been crucial for studying the function and circuitry of the male-specific nervous system and used as a model for studying human autosomal dominant polycystic kidney disease (ADPKD). With the emergence of CRISPR, a seemingly limitless tool for generating genomic mutations with pinpoint precision, the *C. elegans* model system will continue to be a useful instrument for pioneering research in the fields of behavior, reproductive biology, and neurogenetics.

Introduction

Sydney Brenner championed the use of *C. elegans* as a model organism to study nervous system development and function (Brenner, 1974). The genetic amenability of the *C. elegans* hermaphrodite is perfect for classic forward genetic screens, with Brenner's first mutagenesis screen identifying nearly 100 genes, most of which were identified by movement defects. Brenner's (1974) vision for an animal model with a completely mapped nervous system was realized – in part – by the pioneering lineage studies of John Sulston and Robert Horvitz and the circuitry diagrams of John White (Sulston & Horvitz, 1977; Sulston, Schierenberg, White, & Thomson, 1983; White, Southgate, Thomson, & Brenner, 1986).

John Sulston also spotlighted the lesser studied sex of *C. elegans*: the male. Sulston and colleagues discovered that hermaphrodite and male *C. elegans* display striking sexual dimorphisms in development, programmed cell death, and the nervous system in hermaphrodite and male animals, with 959 and 1031 total cells in hermaphrodites and males, respectively, 302 neurons in hermaphrodites and 381 in males (Sulston & Horvitz, 1977; Sulston *et al.*, 1983; White *et al.*, 1986). A subset of neurons are ciliated at dendritic ends: 60 in hermaphrodites and 112 in males (52 male-specific) were identified (Perkins, Hedgecock, Thomson, & Culotti, 1986; Sulston, Albertson, & Thomson, 1980). Very recently, Barrios and Poole discovered four additional male-specific neurons that arise from glial cells (Molina-García *et al.*, 2019; Sammut *et al.*, 2015).

All these works laid the foundation for the wide body of literature that provides explanations for a large range of biological questions including those relevant to human development and disease. A key feature of this organism, one that makes it a hugely powerful model but also introduces some challenges, is that it is a hermaphroditic species with an extremely low incidence of males (0.2–0.5%). Brenner generated male stocks, a product of offspring resulting from male-mated hermaphrodites, which would result in 50% male progeny (Brenner, 1974). Jonathan Hodgkin identified high incidence of males (*him*) mutations that enable male stock maintenance without mating crosses (Hodgkin, Horvitz, & Brenner, 1979). These *him* strains are widely used and key to studying any aspect of *C. elegans* male biology.

In this article, we focus on new advances in understanding sexually dimorphic development and behavior in the *C. elegans* male nervous system made possible by the pioneering work of John Sulston. In addition, we present a CRISPR/Cas9 method to engineer a mutation in the *him-5* gene that will facilitate study of *C. elegans* male biology in this hermaphroditic species.

C. elegans comes in two sexes

C. elegans has two sexes: male and hermaphrodite. Studies of the two sexes have defined molecular mechanisms driving sexually dimorphic processes, including sex determination, differentiation, and development. However, investigations into genes influencing adult male behaviors have been limited. The *C. elegans* genome is comprised of approximately 20,122 protein-coding genes, of which approximately 6000 have no assigned function estimated by lack of gene ontology associations (https://wormbase.org//about/wormbase_release_WS274#0-10). Hence to fully understand the organism and its genome, it is essential to study both sexes.

C. elegans sex is determined by the X to autosome ratio. Hemizygous XO animals are males. XX animals are hermaphrodites, which are morphologically female whose germ line makes sperm prior to producing oocytes. This dosage dependent sex-determination is controlled through a transcription factor, TRA-1A that has multiple target loci (Berkseth, Ikegami, Arur, Lieb, & Zarkower, 2013; Hodgkin, 1987a; Zarkower & Hodgkin, 1992). The self-fertilizing hermaphrodite will produce approximately 300 offspring, of which 99.5–99.8% are XX hermaphrodites and 0.2–0.5% are XO males due to nondisjunction (Hodgkin *et al.*, 1979). The frequency of males can be increased by heat shock (Hodgkin, 1983), by crossing males with hermaphrodites, or by introducing mutations that increase chromosomal nondisjunction. Males crossed with hermaphrodites yield 50% XX hermaphrodites and 50% XO males with offspring counts as high as 1400 per mated hermaphrodite (Chasnov, 2013). Mutations that increase the frequency of nondisjunction (*him*; high incidence of males, (Hodgkin *et al.*, 1979)) allow the efficient production of male progeny even if males cannot sire cross progeny. *him-5* hermaphrodites generate 35% male progeny. *him-5* males exhibit wild-type mating efficiency and behavior. The ability of hermaphrodites to self-fertilize obviates the need to propagate by mating and enables study of male related physiology/behaviors. The hermaphroditic reproduction strategy is a powerful tool to isolate mutants defective in male development, fertility, or behavior.

The neurons and mechanisms of male mating behavior

The *C. elegans* male nervous system possesses 387 neurons to the hermaphrodite's 302. The cephalic and amphid sensory organs in the head are found in both males and hermaphrodites and display sexually dimorphic features. In the male cephalic organs, the cephalic socket creates a pore through which the cephalic male CEM cilia protrude and release ciliary extracellular vesicles to the environment (Akella *et al.*, 2019; Silva *et al.*, 2017; Wang *et al.*, 2014; Ward *et al.*, 1975; Ware *et al.*, 1975). In the hermaphrodite cephalic organs, the CEM neurons that are programmed to die and the cephalic socket cell is closed to the environment. In the L4 male, the bilateral amphid socket glia undergo a male-specific division to self-renew and to generate two mystery cells of the male (MCM). Hence, the male and hermaphrodite display sexual dimorphisms in sex-shared glia and in the male-specific nervous system (Molina-García *et al.*, 2019; Sammut *et al.*, 2015; Sulston *et al.*, 1980; White *et al.*, 1986). Among the sex-shared ventral cord neurons there are sex-specific motor neurons that arise from the same precursor cell lineages in males and hermaphrodites: The CA/CP motor neurons in the male and VCs in the hermaphrodite (Sulston & Horvitz, 1977; Sulston *et al.*, 1980). With the male ventral cord CA/CP neurons and the head CEM and MCM neurons as the exceptions, most of the male sex-specific neurons are located in the tail.

In a landmark series of papers, Sulston characterized the lineage and development of the non-gonadal structures of the *C. elegans* male (Sulston & Horvitz, 1977, 1981; Sulston & White, 1980; Sulston *et al.*, 1980). The male tail consists of an elongated bursa, fan, and proctodeum. The fan is composed of nine pairs of bilaterally arranged rays. The proctodeum houses two spicules and the gubernaculum. Additionally, there is a hook sensillum located anterior to the cloaca and a left-right pair of postcloacal sensillae. Each ray and hook sensillum contains two ciliated sensory neurons, an A type neuron whose ciliated sensory ending is embedded within the cuticle and is predicted to be mechanosensory, and a B type neuron whose cilium is exposed to the environment (with the exception of ray 6B) and is predicted to be chemosensory. Each ray has one glial structure cell (in contrast to other sensilla that have socket and sheath glial cells). Each spicule has two ciliated sensory neurons (SPV and SPD) and one proprioceptive SPC motor neuron.

Copulation behavior is one of the more ancient social behaviors, yet remarkably little is known about the molecular basis of sexual behaviors. *C. elegans* male mating behavior offers a powerful model to study the nervous system, genes, and behavior. In *C. elegans*, the male performs most of the overt sensory and motor behaviors that occur during mating. Male mating behavior is the most complex behavior in *C. elegans*. Although, intricate, male mating behavior can be broken down into simpler, stereotypical sub-steps: response, backing, turning, location of vulva, spicule insertion, and sperm transfer (Hodgkin, 1983; Liu & Sternberg, 1995). This raises the question of how these behaviors are genetically controlled. In his Worm Breeders' Gazette correspondence 'The Turn and the Screw' (http://www.wormbook.org/wli/wbg7.2p22/), Sulston *et al* describe the first laser ablations of *C. elegans* male-specific neuronal precursors to explore neuronal function in mating. This 1980s Worm Breeders' Gazette hardcopy version of an online BioRxiv preprint paved the way for Liu and Sternberg's seminal paper on the neuronal basis of *C. elegans* male mating behavior (Hodgkin, 1983; Liu & Sternberg, 1995).

Many of the male-specific neurons are associated with male-specific behaviors (Barr & Garcia, 2006; Barr, García, & Portman, 2018; Liu & Sternberg, 1995). Each behavioral sub-step is regulated by a specific set of male-specific neurons, which provides great power when experimentally identifying the functional relevance of behavioral phenotypes. The shared nervous system (found in hermaphrodites and males) also contributes to male behaviors (Emmons, 2018).

In this context, the shared nervous system may possess sexually dimorphic properties.

The sex-shared nervous system and male-specific remodeling

Early pioneering studies by Sulston and White provided a fundamental understanding of how a sexually dimorphic nervous system is constructed and wired. The *C. elegans* sexes share a large portion of the nervous system (294 sex-shared neurons), however, these neurons will adopt male or hermaphrodite functionality based on their genetic sex. These sex-specific differences can range from morphological changes, altering synapse partners, and even changing their response to environmental chemical cues (Portman, 2017). The hermaphrodite-specific lineages have been extensively studied (Hobert, 2010), but the male sex has been somewhat overlooked, largely due to the challenges that arise when pursuing male-specific questions in a predominantly hermaphroditic species. Recent serial electron microscopy reconstructions and completion of the whole animal connectomes of both sexes by Emmons and Hall have provided profound new insights regarding sexual dimorphisms in the nervous system (Cook *et al.*, 2019).

A prime example of morphological differences of sex-shared neurons and male-specific remodeling of the shared nervous system is the phasmid C PHC neuronal pair. In hermaphrodites these neurons are responsible for responses to harsh tail touch and innervate the pre-anal ganglion. In the male, these neurons extend into the tail during L4 stage, innervate male-specific motor neurons and get repurposed for male mating behavior (Serrano-Saiz *et al.*, 2017). Male-specific remodeling can also result from past experiences. DVB neurites in males get remodeled in response to copulatory behavior. Males that are more practiced in mating demonstrate increased mating success attributed to a higher chance of spicule insertion and sperm transfer (Hart & Hobert, 2018).

The shared phasmid B PHB phasmid neurons also undergo a sex-specific rewiring program. Sexual maturation leads to synaptic pruning of connections from this neuronal pair in both males and hermaphrodites (Oren-Suissa, Bayer, & Hobert, 2016). In adults, PHB neurons connect to the AVA interneuron pair in hermaphrodites but to the AVG interneuron in males (Cook *et al.*, 2019; Jarrell *et al.*, 2012; White *et al.*, 1986). As juveniles, both male and hermaphrodite PHB neurons synapse to AVA and AVG neurons, demonstrating sexually dimorphic pruning during sexual maturation (Oren-Suissa *et al.*, 2016). This sex specific pruning has been attributed to male-specific *unc-6* netrin expression in AVG interneurons (Weinberg, Berkseth, Zarkower, & Hobert, 2018).

Nervous system developmental plasticity in the male

Two cases of developmental plasticity in a sex-dependent manner comes from shared socket glia cells that – in only males – transdifferentiate into a male-specific neuron or that undergo a cell division to self-renew and to generate a neuron. Barrios and Poole discovered that the shared amphid socket glia cells divide in the L4 male to self-renew and to generate the MCM neurons that are required for sex specific learning (Sammut *et al.*, 2015). Shortly thereafter, Barrios and Poole discovered another case of male-specific neuronal generation, but this time via transdifferentiation of glial to neuronal identity. Between the 3rd and 4th larval stages in the male tail, phasmid socket 1 (PHso1) cells transdifferentiate into ciliated phasmid D PHD neurons that control male locomotion during mating (Molina-García *et al.*, 2019). Sulston noted that the male and hermaphrodite adult phasmid sensory organs were different (Sulston *et al.*, 1980), but did not observe this late transdifferentiation event in his lineage studies. This brings the anatomy of the *C. elegans* male to 387 neurons (93 of which are male-specific) and 90 glia (Molina-García *et al.*, 2019). The number of male-specific EF and DX interneurons varies between three and four depending on the worm, and therefore, some males have 385 neurons (Cook *et al.*, 2019).

Sex-specific responses to neurotransmitters and external cues as a result of dimorphic circuitry

Further dimorphisms are observed between the sexes when it comes to neurotransmitter identity (Hobert & Kratsios, 2019). AIM neurons switch to a male-specific neurotransmitter identity from glutamatergic to cholinergic at the L4 larval stage, whereas hermaphrodites maintain their glutamatergic identity (Pereira *et al.*, 2015). Environmental effects can also have an impact on sex-specific responses to chemicals. Sex-specific pruning can be disrupted by starvation of juvenile males, leading to an increase in sensitivity to chemical cues from the environment (Bayer & Hobert, 2018). The male-specific repression of the diacetyl receptor gene *odr-10* in AWA neurons prioritizes mating over feeding, demonstrating that sex-specific sensory receptor expression in a shared neuron (Ryan *et al.*, 2014). Defecation and foraging behaviors are examples of behavioral programs that are controlled/respond through sex-shared neural circuits that are dimorphic (Gruninger, Gualberto, LeBoeuf, & Garcia, 2006; LeBoeuf & Rene Garcia, 2017). Neural circuits that govern sex-specific behaviors are recently reviewed by Scott Emmons (Emmons, 2018).

A class of neurons with sex-shared and sex-specific implications in behavior and physiology: the ciliary extracellular vesicle releasing neurons

Extracellular vesicles or EVs are tiny membrane bound packets that carry proteins, lipids, metabolites, and nucleic acids and are secreted from cells in all kingdoms of life. Many human cell types secrete EVs into body fluids and these EVs can potentially signal between cells and tissues. For more information about EVs, we refer the readers to reviews on EV biogenesis and function (Beer & Wehman, 2017; Cicero, Lo Cicero, Stahl, & Raposo, 2015; EL

Andaloussi, Mäger, Breakefield, & Wood, 2013; Stahl & Raposo, 2019).

Extracellular vesicle releasing neurons (EVNs) are comprised of 21 male-specific and six sex-shared ciliated neurons that possess the remarkable ability to shed and release ciliary extracellular vesicles (EVs) into the environment or extracellular space that are involved in animal-to-animal communication (Figure 1(B)) (Wang *et al.*, 2014). The six sex-shared inner labial type 2 (IL2) neurons are involved in alcohol sensing, nictation behavior, and sensing blue light (Johnson *et al.*, 2017; Lee *et al.*, 2011; Zaslaver *et al.*, 2015). The IL2 neurons have no known function in mating behavior. The 21 male-specific neurons are 4 cephalic male CEM neurons, 16 RnB neurons (ray 1 to 9, excluding ray 6, left and right symmetry), and one HOB hook neuron (Barr & Sternberg, 1999; Barr *et al.*, 2001; Barrios, Nurrish, & Emmons, 2008). The CEM neurons respond to chemoattractants secreted by hermaphrodites for mating (Chasnov, So, Chan, & Chow, 2007). The HOB neuron is required for vulva location and the ray neurons are important for response and turning behavior (Liu & Sternberg, 1995). To date, these 27 ciliated neurons are the only known neurons to release GFP-tagged ciliary EVs (or any EVs) outside the worm. *C. elegans* does produce other EVs inside the body of the animal, including EVs important for gastrulation in the embryo, cuticle formation during larval development, extrinsic repair of injured neurons, and neuronal waste expulsion in the adult (Beer & Wehman, 2017; Kolotuev, Apaydin, & Labouesse, 2009; Liégeois, Benedetto, Garnier, Schwab, & Labouesse, 2006; Melentijevic *et al.*, 2017; Oren-Suissa, Gattegno, Kravtsov, & Podbilewicz, 2017; Wehman, Poggioli, Schweinsberg, Grant, & Nance, 2011).

The cilium is a site of EV shedding and environmental release (Figure 1(B)). *C. elegans* cilia possess a similar basic structure to that of other ciliated organisms. At the base is the periciliary membrane compartment (akin to the ciliary pocket in mammalian cells) and transition zone. The periciliary membrane compartment acts as a trafficking hub for sorting proteins between ciliary and cell body destinations (Kaplan *et al.*, 2012). The axoneme consists of the transition zone, middle and distal segments (Akella *et al.*, 2019; Doroquez, Berciu, Anderson, Sengupta, & Nicastro, 2014; Perkins *et al.*, 1986; Silva *et al.*, 2017). Environmental release of ciliary EVs is regulated by the conserved intraflagellar transport (IFT) machinery and EVN-specific ciliary resident proteins (O'Hagan *et al.*, 2017; Silva *et al.*, 2017; Wang *et al.*, 2014, 2015). The kinesin-3 protein KLP-6 and myristoylated coil-coil protein CIL-7 are exclusively expressed in the 27 EVNs and regulate ciliary EV release (Maguire *et al.*, 2015; Wang *et al.*, 2014). Ciliary kinesin-2 and kinesin-3 KLP-6, IFT components, tubulin, and tubulin glutamylases and deglutamylases are required for ciliary EV biogenesis (O'Hagan *et al.*, 2017; Silva *et al.*, 2017; Wang *et al.*, 2014). EVs may also function in neuron-glia communication through periciliary membrane compartment-released EVs at the base of CEM cilia, which get released into the pore of the cephalic sensory organ formed by sheath and socket cells (Akella *et al.*, 2020). The functional significance of polycystin regulation of mating behavior, interactions between the polycystins, the function of polycystins in EVs, and the function of EVs in mating behavior in general remain enigmatic and are currently being investigated.

Clinical relevance of mating behavior in *C. elegans*

Among the most important discoveries related to *C. elegans* and mating behavior, was that LOV-1 and PKD-2, orthologs of human polycystin-1 (PC1) and polycystin-2 (PC2), respectively, are involved in male mating behavior and are localized to cilia of male-specific neurons (Barr & Sternberg, 1999; Barr *et al.*, 2001). LOV-1 and PC1 share a similar architecture as an 11 transmembrane receptor with a large extracellular domain. PKD-2 is a transient receptor protein (TRP) polycystin channel (Hanaoka *et al.*, 2000; Nauli *et al.*, 2003; Sharif-Naeini *et al.*, 2009). *lov-1* and *pkd-2* are co-expressed and act in the 21 male-specific ciliated EVNs. LOV-1 and PKD-2 localize to cilia and are secreted in ciliary extracellular vesicles. In EVs isolated from mixed cultures, PKD-2 and other cargo are required to elicit male tail chasing and circling (Silva *et al.*, 2017; Wang *et al.*, 2014). In neurons, *lov-1* and *pkd-2* are required for sex drive, response to mate contact, and vulva location (Barr & Sternberg, 1999; Barr *et al.*, 2001; Barrios *et al.*, 2008). Remarkably, ciliary extracellular vesicles (EVs) carrying polycystin-2 are released upon mechanical stimulation of the sensory cilia of male and are targeted at the vulva of his mating partner. This is the first report of directional transfer of activity-evoked EVs from one animal to another (Wang, Nikonorova, Gu, Sternberg, & Barr, 2020).

Autosomal Dominant Polycystic Kidney Disease (ADPKD) is a common, life threatening disease that affects 1/400–1/1000 individuals. ADPKD is caused by mutations in PKD1 and PKD2, which encode polycystin-1 and polycystin-2 (PC1 and PC2) (Boucher & Sandford, 2004; Igarashi & Somlo, 2002). Remarkably, the function of the polycystins remains an enigma 25 years after their cloning and 17 years after their discovery on renal primary cilia. Besides cilia, PC1 and PC2 are also found in other subcellular locations including extracellular vesicles (EVs) of human urine. Urinary EVs can be used as biomarkers of renal disease (Karpman, Ståhl, & Arvidsson, 2017) including ADPKD (Hogan *et al.*, 2015). Whether these polycystin-carrying EVs are of ciliary origin and what role EVs play in healthy and diseased kidneys remains unknown. The *C. elegans* ADPKD model is a beautiful example of the utility of model organisms in studying the underlying mechanisms of human disease. In *C. elegans*, knockouts of either *lov-1* or *pkd-2* polycystin results in a mating behavior defect but is lethal in mice (Barr *et al.*, 2001; Boulter *et al.*, 2001; Wu *et al.*, 2000). In *C. elegans* and mammals the polycystins LOV-1/PC1 and PKD-2/PC2 act in the same genetic pathway and function in a sensory capacity (Barr *et al.*, 2001; Yoder, Hou, & Guay-Woodford, 2002).

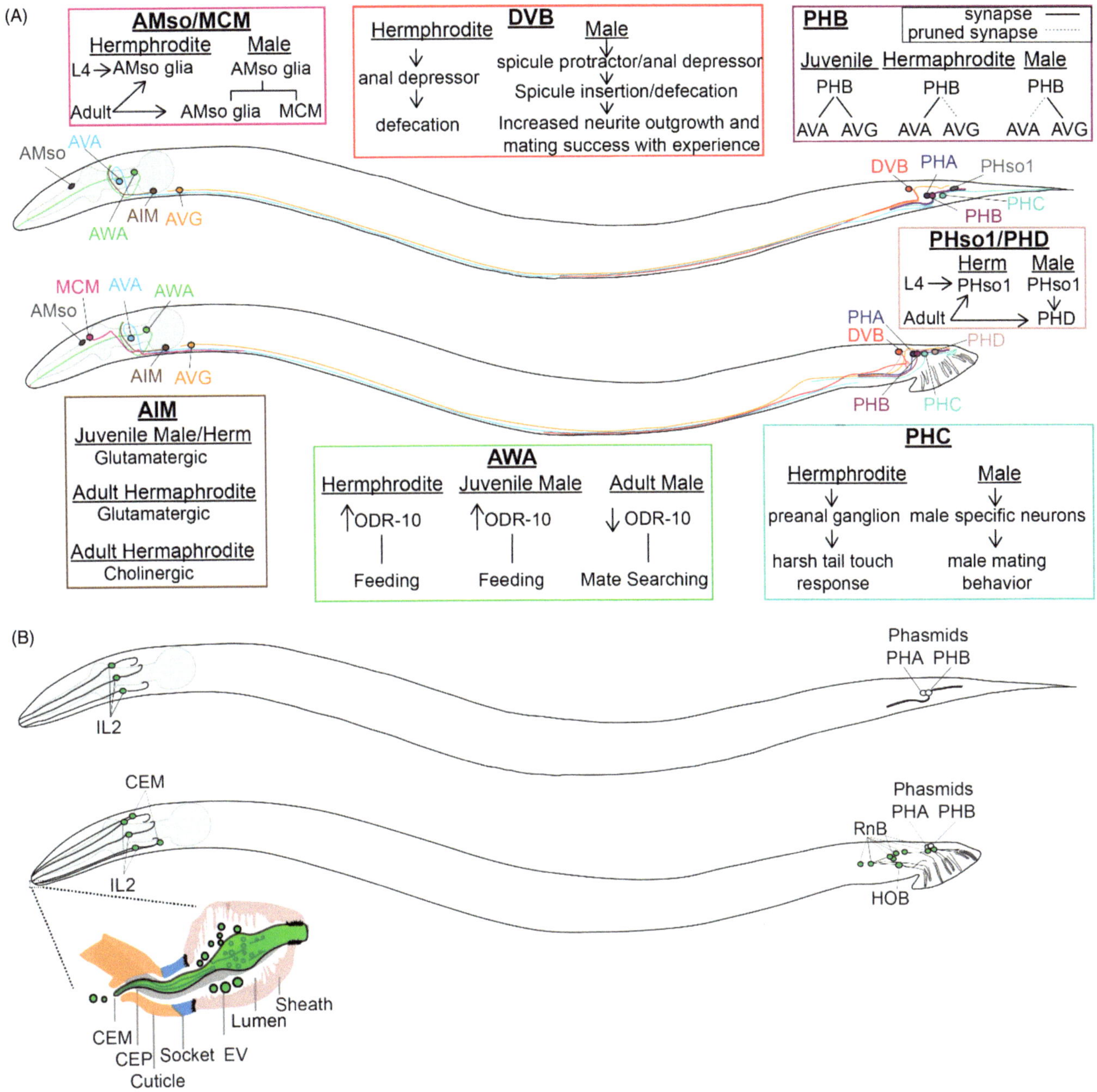

Figure 1. Abbreviated wiring diagram of sexually dimorphic neurons (A) Sex-shared neurons that are sexually dimorphic after sexual maturation of the male. Includes an example of each concept described in the article. Top worm is hermaphrodite and the bottom worm is male. Neuron cell bodies are depicted as circles, neurites as lines, and glia depicted as ovals (AMso and PMso1). Pharynx is depicted in the head for reference. Text-filled boxes represent the sexually dimorphic property of the neuron. (B) Extracellular vesicle releasing neurons (EVNs) in males (bottom) and hermaphrodites (top). Pharynx and phasmids are shown as landmarks for head and tail, respectively. The structure of a CEM cilium is shown in detail (bottom). EVs are seen inside and outside the worm represented by circles on the CEM cilium.

The complications that arise when studying male-specific biology in a hermaphroditic species

C. *elegans* is a powerful model organism that has been used by biologists for over half a century (Brenner, 1974; Fatt & Dougherty, 1963; Nigon & Dougherty, 1949). Although, a dioecious species, *C. elegans* strongly favors self-reproduction over mating and the incidence of males in a wild-type population is approximately 0.2–0.5% (Hodgkin, 1983). Male biology is a subject of interest for researchers, and studying genes specifically expressed in the male presents some unique challenges. The first generation of heat-shock generated 'male stocks' can have stress-induced epigenetic changes that persist for multiple generations (Klosin, Casas, Hidalgo-Carcedo, Vavouri, & Lehner, 2017; Zhou, He, Deng, Pang, & Tang, 2019). This dilemma is compounded when experimenting with genes that result in male mating deficiencies when mutated. When trying to make male-specific CRISPR reporters or knockouts, it seems intuitive to do this in a high incidence of males strain such as *him-5*. The *him-5* mutation increases the rate of male progeny to 35% but does not increase brood size so this comes at the cost of a decrease in hermaphrodite progeny (Hodgkin, 1983). This deficit of hermaphrodites is especially problematic when the CRISPR mutation efficiency is below a certain threshold. In our experience, we have not been able to successfully create a CRISPR-generated fluorescent protein fusion in *him-5* background, whereas we have no issues using N2 (widely used hermaphrodite lab strain). To further complicate the use of the *him-5* genetic background for generating CRISPR-induced mutations, *him-5* affects repair mechanism choice as a negative regulator of meiotic non-homologous end joining (NHEJ) (Macaisne, Kessler, & Yanowitz, 2018; Mateo *et al.*, 2016). This could severely reduce the efficiency of homology directed recombination (HDR) in *him-5* strains for CRISPR-mediated recombination. The obvious solution to this issue is to cross in the desired *Him* strain after generating CRISPR-induced mutations in the N2 background. However, this strategy has its own limitations that can be burdensome and time consuming. Examples include when genes are closely linked to the *Him* allele, when crossing mating deficient strains, or when dealing with strains containing multiple mutations that must all be homozygous after a cross. There are current methods for generating males in higher incidence that do not rely on crossing or heat shock to generate 'male stocks.' These strategies, such as male-producing RNAi and using transgenes to masculinize hermaphrodites, provide their own unique challenges and advantages.

RNAi has been used as a temporary means to generate males. *him-14* RNAi (also known as 'male food') was first described by Darrell Killian and Jane Hubbard in 2001 as 'effective and reversible, but not efficient' in a Worm Breeder's Gazette correspondence 'RNAi feeding to produce males'. A systematic comparison of other Him inducing RNAi targets was later performed, identifying RNAi induced knockdowns of several genes that were more efficient at generating males (Timmons *et al.*, 2014). These *Him* inducing gene knockdowns are useful for cases where a temporary need for males is sufficient or where reversibility is required. Due to the temporary nature of this method, other approaches would be more appropriate for experiments that require permanent production of males.

Masculinization of hermaphrodites is an alternative strategy for studying the development of sex-based differences in *C. elegans* and can serve a different purpose than studying wild-type males. Specific loss of function or gain-of-function alleles can effectively masculinize hermaphrodites (Hodgkin, 1987b; Hodgkin & Brenner, 1977). These alleles range in phenotype from mild masculinization with no male behaviors to nearly complete masculinization with pseudo-males demonstrating mating behavior but with low fertility (Hodgkin, 1987b; Schedl, Graham, Barton, & Kimble, 1989). Feminization of hermaphrodites is also possible (Barton, Schedl, & Kimble, 1987; Hodgkin, 1987b; Hodgkin & Brenner, 1977; Schedl *et al.*, 1989). This has been very useful at uncovering essential biological and behavioral aspects of sex differences in the *C. elegans* nervous system. Examples include using transgenics to study environmental sexual conditioning by masculinizing/feminizing the nervous system (Sakai *et al.*, 2013) and masculinizing/feminizing single neurons or subsets of neurons to study sexual differences in sex-shared neurons (Hart & Hobert, 2018; Oren-Suissa *et al.*, 2016). Phenotypic variation among these mutants, however, necessitates studying male biology in genetically male animals to fully understand the development of the male nervous system and male-specific behaviors.

Here, we propose a different approach for addressing some of the challenges associated with studying male-specific biology in a hermaphroditic species: a CRISPR/Cas-9 mutagenesis strategy, designed to effectively and reproducibly generate males in almost any genetic background.

The *him*CRISPR method

We used CRISPR/Cas9 technologies to induce the equivalent of the *him-5(e1490)* point mutation for the purposes of studying male specific physiology and/or behaviors. The *him-5(e1490)* mutant was identified in a forward genetic screen (Hodgkin *et al.*, 1979) and generates a high incidence of males at approximately 35% as a result of a single point mutation of the splice acceptor site of intron 3 (Figure 2, Supporting Information Figure 1) (Meneely, McGovern, Heinis, & Yanowitz, 2012). This method relies on CRISPR-induced homology-directed repair (HDR) of our target locus (*him-5*) with the use of single-stranded oligonucleotides (ssODN) as the repair donor containing approximately 70 bp of homology (35 bp on each side of edits) (Paix *et al.*, 2014). Our ssODN donor is 161 bp long and the homology arms of the construct are located within 100 bp of the cut site. We injected a mix of Cas9 protein, crRNA, tracrRNA, ssODN, and a pRF4 (*rol-6*, dominant roller) co-injection marker to screen for transformants (Dokshin, Ghanta, Piscopo, & Mello, 2018). In addition to the *him-5(e1490)* point mutation, the ssODN includes other silent mutations to prevent re-cutting post-recombination and to generate a restriction enzyme site for screening for successful editing. Young adult N2 worms are injected with the injection mix. F1 rollers were singled 3–4 days after injection and allowed to lay eggs. The F1 rollers were then genotyped via single-worm PCR and restriction enzyme digest to identify successful edits. F2s that were genotyped homozygous or heterozygous were singled onto multiple plates. F3s with a high incidence of males were identified as homozygous for the *him-5* CRISPR-induced mutation. Homozygotes of the gene of interest are confirmed through genotyping lysates from a pool of F3 from each *him* F2 plate, and the mutation were sequenced by Sanger sequencing of PCR products (Supporting Information Figure 2). In this article, two independent isolates *him-5(my80)* and *him-5(my81)* were phenotypically analyzed and found to recapitulate the *him-5(e1490)* phenotype.

Our *him*CRISPR approach is very efficient. With the small group of F_1 progeny selected from one round of injections, 8/11 had mutation events occur at only one of the *him-5* alleles while 1/11 had mutation events at both. In other words, approximately 82% of F1 progeny genotyped for the expected mutation event (Supporting Information Figure 1). This is within the range of expected efficiencies as reported previously (Dokshin *et al.*, 2018). In some cases, NHEJ recombinants may be identified, in addition or

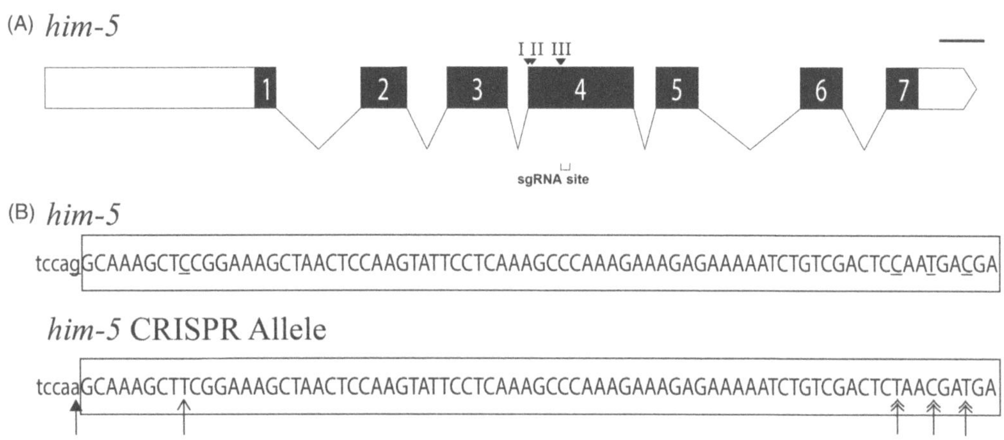

Figure 2. Genomic structure of *him-5* and mutagenesis strategy. (A) Genomic structure of *him-5* depicting the 7 exons in black. Black arrows indicate the sites of nucleotide changes introduced in the CRISPR mutagenesis specified by Roman numerals I – III (I = e1490 CAG → CAA mutation; III = silent mutation introducing HindIII restriction site; III = Silent mutations to prevent re-cutting of the sgRNA site). sgRNA target site is indicated by the bracket below exon 4. Scale bar is 100 bp (B) Nucleotide-level depiction of the site of mutagenesis for the *him-5* CRISPR allele. Box surrounds exon 4. Filled black arrow indicates the splice site mutation that is the causal mutation for the *him-5(e1490)* high incidence of males (AG -> AA). Single-lined arrow indicates the silent point mutation that introduces a BamHI restriction enzyme site. The three doublelined arrows indicate the silent mutations introduced to disrupt the sgRNA target site to prevent re-cutting after HDR has occurred.

instead of HDR recombinants, that produce indels resulting in higher incidence of males when using this method. In this case, it is up to experimenter discretion to use this variant over the well characterized *him-5(e1490)* mutation. Our motivation for the *him*CRISPR method was to generate an efficient way to acquire high incidence of males and replicate the phenotype of the widely used *him-5(e1490)* splice-site mutation strain.

Characterizing *him*-5 CRISPR strains

We analyzed brood size, male progeny count, male mating efficiency, and male mating behaviors using standard assays (Hodgkin, 1983; Liu & Sternberg, 1995; Meneely *et al.*, 2012). Quantification of brood size and number of male progeny were used to characterize our CRISPR strains (Meneely *et al.*, 2012). Brood size and incidence of males assay was performed by placing single hermaphrodites on a seeded plate, allowed to lay eggs for 24 h, then moved to a new plate. This was done over the course of 5 days or until the hermaphrodite was sperm depleted ($N = 15$). Our CRISPR-generated strains produce males at frequencies that are comparable to both our *him-5(e1490)* strain and the expected frequency described in the literature (Figure 3(A)) (Meneely *et al.*, 2012). Brood size was also comparable to *him-5(e1490)* and *N2* (Figure 3(A)).

Mating efficiency data was collected according to Hodgkins (Hodgkin, 1983). Mating efficiency assays were conducted by placing 6 males and 6 hermaphrodites onto a mating plate for 24 h. Males were then removed from the plate and hermaphrodites were allowed to lay eggs. After each day, cross progeny and total progeny were counted and the mated hermaphrodites were moved to a new plate. These counts were done over the course of 5 days. Three independent mating plates were analyzed for each genotype for three independent replicates ($N = 9$, 27 total males per genotype). Test strains were mated to *dpy*-17 and *unc-52* hermaphrodites and cross progeny were compared between groups. We use these two strains to distinguish between varying levels of mating defects as each strain introduces a different level of difficulty for the males. *unc-52* worms are extremely Unc (uncoordinated) and adults do not move at all except for slight head movements for feeding. This makes them much easier mating targets than *dpy-17* hermaphrodite partners. Using these two strains can give us a measure of just how mating deficient a strain is. With the mating efficiency assay, a failure to sire cross progeny may reflect a defect in mating behavior but could just as likely be caused by defects in male tail development, sperm, or general locomotion. Both *him-5(my80)* and *him-5(my81)* strains mated at similar efficiencies to the wild-type *him-5(e1490)* when mated with either *unc-52* or *dpy-17* (Figure 3(B)).

Finally, mating behavior assays were performed according to Sternberg (Liu & Sternberg, 1995) and the data was used to calculate response and vulva location efficiencies. For the mating assays, day one adult virgin males (isolated at the L4 stage away from hermaphrodites) were placed on mating plates with 20 day-one *unc-31* hermaphrodites (isolated at L4 away from males). Mating was observed for 5 min and each mating step was observed and recorded (response, backing, turning, vulva location, spicule insertion, sperm transfer). Response and vulva location efficiencies were calculated and compared to *him-5(e1490)* as the wild-type control, and *pkd-2(sy606); him-5(e1490)* which is defective in response and location of vulva (Lov) defective. Both *him-5(my80)* and *him-5(my81)* exhibited normal mating behavior with respect to response and location of vulva when compared to *him-5(e1490)* (Figure 3(C,D)). Our analysis of the *him*CRISPR demonstrates that *him-5(my80)* and *him-5(my81)* are indistinguishable from the original *him-5(e1490)* strain that is commonly accepted as a wild-type control for mating assays.

Discussion

CRISPR/Cas9-mediated mutagenesis is an effective alternative for increasing the incidence of male progeny in *C.*

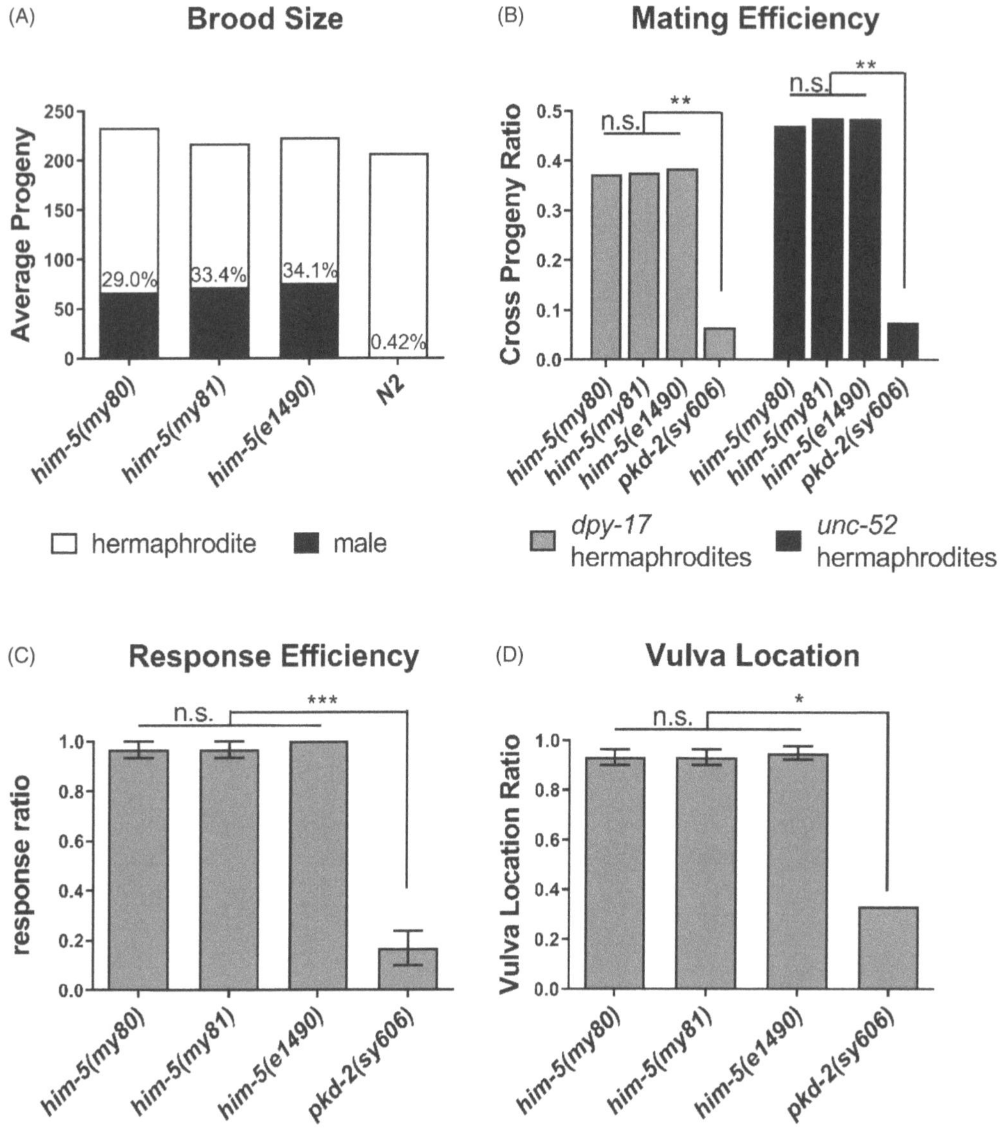

Figure 3. *him*CRISPR strains *him-5(my80)* and *him-5(my81)* display the same sex related characteristics as *him-5(e1490)*. (A) Total progeny of singled hermaphrodites was counted over 5 days ($N = 15$, 3 independent replicates). *him-5(my80)* and *him-5(my81)* male percentage were not significantly different from him-5(e1490). *him-5(my80)*, *him-5(my81)*, and *him-5(e1490)* were significantly different from N2 $p < .001$ (nonparametric one-way ANOVA, Dunn's *post hoc* analysis). (B) Mating efficiency was assayed by comparing the ratio of cross progeny to total progeny in the F1 generation from a cross with either dpy-17 or unc-52 hermaphrodites. *him-5(my80)*, *him-5(my81)*, and *him-5(e1490)* exhibited statistically similar mating efficiency, whereas they all performed significantly better than pkd-2(sys606);*him-5(e1490)* ($p < .01$), a mating deficient strain (non-parametric one-way ANOVA, Dunn's *post hoc* analysis). C&D) Response and vulva location efficiency were measured in mating behavior assays. The two graphs represent different measures of the same experiment. For *him-5(my80)*, *him-5(my81)*, and *him-5(e1490)* $N = 30$. For pkd-2(sys606);*him-5(e1490)*$N = 30$ for response, however, only one male responded during the course of the assays, therefore, for the vulva location measure, $N = 1$. Neither response nor vulva location differed among *him-5(my80)*, *him-5(my81)*, and *him-5(e1490)* and were all significantly different from pkd-2(sys606);*him-5(e1490)*(Response $p < .001$, location of vulva $p < .05$, non-parametric one-way ANOVA, Dunn's *post hoc* analysis).

elegans. Our CRISPR-generated *him*-5 strains are phenotypically indistinguishable from CB1490 [*him-5(e1490)*] with regard to brood size, mating efficiency, and male response and vulva location behaviors. The *him*CRISPR method can be adapted for the other commonly used *him* strain, *him-8*, by simply altering the sgRNA and homology repair construct to target that locus. We use *him-5(e1490)* for most of our male-related experiments, therefore, this was the subject of our investigation for validating this method. This *him*CRISPR method is a great way to circumvent experimental challenges when studying males in a hermaphroditic species by being able to generate any *Him* strain regardless of genetic background.

This tool is uniquely suited to address challenges of studying male-specific biology in a hermaphroditic species. One such challenge presents when genes are directly involved in male mating success, thus, loss-of-function mutation strains that are not already in a *him-5* background will make genetic crosses complicated and time consuming at best. With this *him*CRISPR strategy, mutations that result in loss or lack of male-mating success will be less problematic since common strategies for generating high incidence of males relies on mating. Mutations that are closely linked on the same chromosome as *him-5* (or *him-8*) will no longer require reliance on recombination frequencies. Finding recombinants can be cumbersome and time consuming when the genes are closely linked (1% linkage or less). Another challenge we have faced is generating fluorescently tagged reporters of male-specific genes using CRISPR. Given that we have had better success with generating fluorescent protein-tagged CRISPR mutations in the N2 background, we generate male stocks via heat shock to examine gene expression patterns before generating permanent male-producing lines through crossing. With this tool, we will be able to induce both the *him-5* CRISPR allele and the male-specific fluorescent-tagged protein of interest in multiplex. This can also be a tool for addressing concerns regarding phenotypic discrepancies in genetically different backgrounds, which is an issue with commonly used N2 male-stocks (Zhao, Wang, Poole, & Gems, 2019). Heat shock alone, as a means of generating males, can be a cause for phenotypic differences through stress induced genetic changes (Klosin *et al.*, 2017; Melnick *et al.*, 2019). In addition to circumventing these specific challenges, it would be interesting to consider this tool as a way to study male evolution by generating males in other hermaphroditic *Caenorhabditis* and/or other nematode species (Cutter, Morran, & Phillips, 2019).

Mating behavior is the most complex behavior exhibited by *C. elegans*. As with any *C. elegans* behavioral experiment, male mating assays are influenced by many variables including environment, experimental bias, and day-to-day differences in behavioral responses (http://www.wormbook.org/chapters/www_behavior/behavior.html). For reproducibility, controls (wild-type *him-5* and a mating defective mutant such as *pkd-2; him-5*) are essential. If wild-type *him-5* males do not behave normally, then assays for the day are terminated. To avoid experimental bias, the observer should be blinded to the genotype of the animals being assays. Environmental variations include culture conditions, transfer of males to mating plates, ambient temperature, humidity, and time of day when assays are performed.

With the complexity of mating behavior in general, slight differences in how assays are performed may have profound impact on male behavioral output and its interpretation. A striking example comes from studies looking at the effect of aging on male reproductive abilities. Hart and Hobert demonstrated an increase in mating success in day 3 males over day 1 males that correlated with remodeling of the DVB neurites and repeated sensory input over time (Hart & Hobert, 2018). This study focused on spicule insertion and sperm transfer and found that day 1 males had a higher incidence of spicule prodding but lower incidence of spicule protraction whereas day 3 males demonstrated the inverse, thus were better at mating (Hart & Hobert, 2018). Mating efficiency was not measured. In contrast, others report a clear decline in mating success in day 3 or older males (Chatterjee *et al.*, 2013; Guo, Navetta, Gualberto, & García, 2012). To assess age-associated decline in mating, Chatterjee *et al.* (2013) used mating efficiency, mating response, and vulva location assays in day 1 and day 5 males (only assessed mating efficiency at day 3). This study also used a ring of chemical repellent (garlic) to trap males and reduce escapees, which is not commonly used. Guo *et al.* (2012) measured mating potency, which is a variation on the efficiency assay. In the mating potency assay, single males are mated with temperature-sensitive *pha-1* hermaphrodites at the restrictive temperature to ensure only cross progeny can be produced. The proportion of males that can produce any cross progeny are scored. This contrasts with the mating efficiency assay that utilizes Dpy and Unc hermaphrodites, mates hermaphrodites with multiple males per mating plate, and scores all progeny as a proportion of cross to total progeny. The ease of the mating potency assay is that one does not have to count cross progeny, rather one simply identifies whether there are any cross progeny on the plate. Although, these approaches are reaching for the same outcome, differences in the assays such as hermaphrodite substrates (moving versus nonmoving) and the nature of the readout can influence behavioral assay interpretations. Guo *et al.* (2012) also used spicule insertion and sperm transfer as a measure of male-mating success and concluded that male mating was significantly reduced in day 3 males (Guo *et al.*, 2012). Depending on specific assays and parameters, one can draw different conclusions regarding the impact of aging on male virility.

In Hodgkin (1983), an attempt to create a standardized measure of mating efficiency was proposed by calculating a tiered system of efficiencies (i.e. a 0–4 scale, 4 being perfect mating efficiency 0 being no detectable mating efficiency). Using those criteria, the negative control we use for mating assays, *pkd-2(sy606)* would be a 3 on that scale (10–30% of WT), which is considered efficient mating, even though it is vulva location defective and has a significantly lower production of cross progeny. These examples demonstrate how the complexity of behavior in a biological system is difficult to standardize, and great care must be taken when comparing results from multiple studies.

The pioneering work of John Sulston *et al.* led to the completion of the entire *C. elegans* connectome in both sexes and paved the way to understand how the male-specific nervous system and sex-shared nervous system to integrate organismal survival with reproductive function. New evidence of nervous system developmental plasticity in the male reveals an unexplored area of male-specific *C. elegans* physiology. The two incidences of glia to neuron transformation during sex maturation raises a fundamental question: to what extent is the *C. elegans* nervous system plastic and capable of rewiring and/or regenerating? Sex-specific responses based on dimorphic properties of sex-shared neural circuits reveals yet another layer of development that provides fundamental insights into mechanisms of sex-specific behavior and neural physiology.

Acknowledgements

The authors thank current Barr lab members Jyothi Shilpa Akella, Kade Power, Inna Nikonorova, and especially Juan Wang for discussions, insights, and helpful feedback on this work; Arantza Barrios for feedback and clarifications on the MCM and PHD neurons; Brittany Klimek for her help with genotyping. We especially thank Helen Ushakov who expertly does our microinjections and Gloria Androwski who keeps our lab running; Barth Grant for early advice regarding CRISPR techniques; and Wormbase for quick responses and for being awesome. We also thank the National BioResource Project (Tokyo

Women's Medical College, Tokyo, Japan) and *Caenorhabditis* Genetics Center (CGC) for strains.

Disclosure statement

No potential conflict of interest was reported by the author(s).

Funding

The CGC is supported by the National Institutes of Health - Office of Research Infrastructure Programs [P40 OD010440]. This work was funded by the National Institutes of Health (NIH) awards DK059418 and DK116606 to MMB and NIH Institutional Research and Academic Career Development Award (IRACDA) K12GM093854 to JDW.

ORCID

Jonathon D. Walsh http://orcid.org/0000-0002-4277-7657
Maureen M. Barr http://orcid.org/0000-0003-4483-2952

References

Akella, J.S., Carter, S.P., Nguyen, K., Tsiropoulou, S., Moran, A.L., Silva, M., ... Blacque, O.E. (2020). Ciliary Rab28 and the BBSome negatively regulate extracellular vesicle shedding. *eLife*, *9*:e50580 doi: 10.7554/eLife.50580

Akella, J.S., Silva, M., Morsci, N.S., Nguyen, K.C., Rice, W.J., Hall, D.H., & Barr, M.M. (2019). Cell type-specific structural plasticity of the ciliary transition zone in *C. elegans*. *Biology of the Cell*, *111*, 95–107. doi:10.1111/boc.201800042

Barr, M.M., DeModena, J., Braun, D., Nguyen, C.Q., Hall, D.H., & Sternberg, P.W. (2001). The *Caenorhabditis elegans* autosomal dominant polycystic kidney disease gene homologs lov-1 and pkd-2 act in the same pathway. *Current Biology*, *11*, 1341–1346. doi:10.1016/S0960-9822(01)00423-7

Barr, M.M. and Garcia, L.R. Male mating behavior (June 19, 2006), WormBook, ed. The C. elegans Research Community, WormBook, doi/10.1895/wormbook.1.78.1, http://www.wormbook.org.

Barr, M.M., García, L.R., & Portman, D.S. (2018). Sexual dimorphism and sex differences in *Caenorhabditis elegans* neuronal development and behavior. *Genetics*, *208*, 909–935. doi:10.1534/genetics.117.300294

Barr, M.M., & Sternberg, P.W. (1999). A polycystic kidney-disease gene homologue required for male mating behaviour in *C. elegans*. *Nature*, *401*, 386–389. doi:10.1038/43916

Barrios, A., Nurrish, S., & Emmons, S.W. (2008). Sensory regulation of *C. elegans* male mate-searching behavior. *Current Biology*, *18*(23), 1865–1871. doi:10.1016/j.cub.2008.10.050

Barton, M.K., Schedl, T.B., & Kimble, J. (1987). Gain-of-function mutations of fem-3, a sex-determination gene in *Caenorhabditis elegans*. *Genetics*, *115*(1), 107–119.

Bayer, E.A., & Hobert, O. (2018). Past experience shapes sexually dimorphic neuronal wiring through monoaminergic signalling. *Nature*, *561*, 117–121. doi:10.1038/s41586-018-0452-0

Beer, K.B., & Wehman, A.M. (2017). Mechanisms and functions of extracellular vesicle release in vivo—What we can learn from flies and worms. *Cell Adhesion & Migration*, *11*, 135–150. doi:10.1080/19336918.2016.1236899

Berkseth, M., Ikegami, K., Arur, S., Lieb, J.D., & Zarkower, D. (2013). TRA-1 ChIP-seq reveals regulators of sexual differentiation and multilevel feedback in nematode sex determination. *Proceedings of the National Academy of Sciences of the United States of America*, *110*, 16033–16038. doi:10.1073/pnas.1312087110

Boucher, C., & Sandford, R. (2004). Autosomal dominant polycystic kidney disease (ADPKD, MIM 173900, PKD1 and PKD2 genes, protein products known as polycystin-1 and polycystin-2). *European Journal of Human Genetics*, *12*, 347–354. doi:10.1038/sj.ejhg.5201162

Boulter, C., Mulroy, S., Webb, S., Fleming, S., Brindle, K., & Sandford, R. (2001). Cardiovascular, skeletal, and renal defects in mice with a targeted disruption of the Pkd1 gene. *Proceedings of the National Academy of Sciences of the United States of America*, *98*, 12174–12179. doi:10.1073/pnas.211191098

Brenner, S. (1974). The genetics of *Caenorhabditis elegans*. *Genetics*, *77*(1), 71–94.

Chasnov, J.R. (2013). The evolutionary role of males in *C. elegans*. *Worm*, *2*(1), e21146. doi:10.4161/worm.21146

Chasnov, J.R., So, W.K., Chan, C.M., & Chow, K.L. (2007). The species, sex, and stage specificity of a Caenorhabditis sex pheromone. *Proceedings of the National Academy of Sciences of the United States of America*, *104*, 6730–6735. doi:10.1073/pnas.0608050104

Chatterjee, I., Ibanez-Ventoso, C., Vijay, P., Singaravelu, G., Baldi, C., Bair, J., ... Singson, A. (2013). Dramatic fertility decline in aging *C. elegans* males is associated with mating execution deficits rather than diminished sperm quality. *Experimental Gerontology*, *48*, 1156–1166. doi:10.1016/j.exger.2013.07.014

Cicero, A.L., Lo Cicero, A., Stahl, P.D., & Raposo, G. (2015). Extracellular vesicles shuffling intercellular messages: For good or for bad. *Current Opinion in Cell Biology*, *35*, 69–77. doi:10.1016/j.ceb.2015.04.013

Cook, S.J., Jarrell, T.A., Brittin, C.A., Wang, Y., Bloniarz, A.E., Yakovlev, M.A., ... Emmons, S.W. (2019). Whole-animal connectomes of both *Caenorhabditis elegans* sexes. *Nature*, *571*, 63–71. doi:10.1038/s41586-019-1352-7

Cutter, A.D., Morran, L.T., & Phillips, P.C. (2019). Males, outcrossing, and sexual selection in Caenorhabditis nematodes. *Genetics*, *213*(1), 27–57. doi:10.1534/genetics.119.300244

Dokshin, G.A., Ghanta, K.S., Piscopo, K.M., & Mello, C.C. (2018). Robust genome editing with short single-stranded and long, partially single-stranded DNA donors in *Caenorhabditis elegans*. *Genetics*, *210*, 781–787. doi:10.1534/genetics.118.301532

Doroquez, D.B., Berciu, C., Anderson, J.R., Sengupta, P., & Nicastro, D. (2014). A high-resolution morphological and ultrastructural map of anterior sensory cilia and glia in *Caenorhabditis elegans*. *eLife*, *3*, e01948. doi:10.7554/eLife.01948

EL Andaloussi, S., Mäger, I., Breakefield, X.O., & Wood, M.J.A. (2013). Extracellular vesicles: Biology and emerging therapeutic opportunities. *Nature Reviews. Drug Discovery*, *12*, 347–357. doi:10.1038/nrd3978

Emmons, S.W. (2018). Neural circuits of sexual behavior in *Caenorhabditis elegans*. Annual Review of Neuroscience, *41*, 349–369. doi:10.1146/annurev-neuro-070815-014056

Fatt, H.V., & Dougherty, E.C. (1963). Genetic control of differential heat tolerance in two strains of the nematode *Caenorhabditis elegans*. *Science (New York, N.Y.)*, *141*, 266–267. doi:10.1126/science.141.3577.266

Gruninger, T.R., Gualberto, D.G., LeBoeuf, B., & Garcia, L.R. (2006). Integration of male mating and feeding behaviors in *Caenorhabditis elegans*. *Journal of Neuroscience*, *26*(1), 169–179. doi:10.1523/JNEUROSCI.3364-05.2006

Guo, X., Navetta, A., Gualberto, D.G., & García, L.R. (2012). Behavioral decay in aging male *C. elegans* correlates with increased cell excitability. *Neurobiology of Aging*, *33*, 1483.e5–e23. doi:10.1016/j.neurobiolaging.2011.12.016

Hanaoka, K., Qian, F., Boletta, A., Bhunia, A.K., Piontek, K., Tsiokas, L., ... Germino, G.G. (2000). Co-assembly of polycystin-1 and -2 produces unique cation-permeable currents. *Nature*, *408*, 990–994. doi:10.1038/35050128

Hart, M.P., & Hobert, O. (2018). Neurexin controls plasticity of a mature, sexually dimorphic neuron. *Nature*, *553*, 165–170. doi:10.1038/nature25192

Hobert O. Neurogenesis in the nematode Caenorhabditis elegans (October 4, 2010), WormBook, ed. The C. elegans Research Community, WormBook, doi/10.1895/wormbook.1.12.2, http://www.wormbook.org.

Hobert, O., & Kratsios, P. (2019). Neuronal identity control by terminal selectors in worms, flies, and chordates. *Current Opinion in Neurobiology*, *56*, 97–105. doi:10.1016/j.conb.2018.12.006

Hodgkin, J. (1983). Male phenotypes and mating efficiency in *Caenorhabditis elegans*. *Genetics*, *103*(1), 43–64.

Hodgkin, J. (1987a). Primary sex determination in the nematode *C. elegans*. *Development*, *101*(Suppl) 5–16.

Hodgkin, J. (1987b). Sex determination and dosage compensation in *Caenorhabditis elegans*. *Annual Review of Genetics*, *21*(1), 133–154. doi:10.1146/annurev.ge.21.120187.001025

Hodgkin, J., Horvitz, H.R., & Brenner, S. (1979). Nondisjunction mutants of the nematode *Caenorhabditis elegans*. *Genetics*, *91*(1), 67–94.

Hodgkin, J.A., & Brenner, S. (1977). Mutations causing transformation of sexual phenotype in the nematode *Caenorhabditis elegans*. *Genetics*, *86*(2 Pt. 1), 275–287.

Hogan, M.C., Bakeberg, J.L., Gainullin, V.G., Irazabal, M.V., Harmon, A.J., Lieske, J.C., … Ward, C.J. (2015). Identification of biomarkers for PKD1 using urinary exosomes. *Journal of the American Society of Nephrology*, *26*, 1661–1670. doi:10.1681/ASN.2014040354

Igarashi, P., & Somlo, S. (2002). Genetics and pathogenesis of polycystic kidney disease. *Journal of the American Society of Nephrology*, *13*(9), 2384–2398. doi:10.1097/01.asn.0000028643.17901.42

Jarrell, T.A., Wang, Y., Bloniarz, A.E., Brittin, C.A., Xu, M., Thomson, J.N., … Emmons, S.W. (2012). The connectome of a decision-making neural network. *Science (New York, N.Y.)*, *337*, 437–444. doi:10.1126/science.1221762

Johnson, J.R., Edwards, M.R., Davies, H., Newman, D., Holden, W., Jenkins, R.E., … Barclay, J.W. (2017). Ethanol stimulates locomotion via a G-signaling pathway in IL2 neurons in. *Genetics*, *207*, 1023–1039. doi:10.1534/genetics.117.300119

Kaplan, O.I., Doroquez, D.B., Cevik, S., Bowie, R.V., Clarke, L., Anna, A.W., … Blacque, O.E. (2012). Endocytosis genes facilitate protein and membrane transport in *C. elegans* sensory cilia. *Current Biology*, *22*, 451–460. doi:10.1016/j.cub.2012.01.060

Karpman, D., Ståhl, A.-L., & Arvidsson, I. (2017). Extracellular vesicles in renal disease. *Nature Reviews. Nephrology*, *13*, 545–562. doi:10.1038/nrneph.2017.98

Klosin, A., Casas, E., Hidalgo-Carcedo, C., Vavouri, T., & Lehner, B. (2017). Transgenerational transmission of environmental information in *C. elegans*. *Science*, *356*, 320–323. doi:10.1126/science.aah6412

Kolotuev, I., Apaydin, A., & Labouesse, M. (2009). Secretion of Hedgehog-related peptides and WNT during *Caenorhabditis elegans* development. *Traffic (Copenhagen, Denmark)*, *10*, 803–810. doi:10.1111/j.1600-0854.2008.00871.x

LeBoeuf, B., & Rene Garcia, L. (2017). *Caenorhabditis elegans* male copulation circuitry incorporates sex-shared defecation components to promote intromission and sperm transfer. *G3 (Bethesda, MD.)*, *7*, 647–662. doi:10.1534/g3.116.036756

Lee, H., Choi, M.-K., Lee, D., Kim, H.-S., Hwang, H., Kim, H., … Lee, J. (2011). Nictation, a dispersal behavior of the nematode *Caenorhabditis elegans*, is regulated by IL2 neurons. *Nature Neuroscience*, *15*(1), 107–112. doi:10.1038/nn.2975

Liégeois, S., Benedetto, A., Garnier, J.-M., Schwab, Y., & Labouesse, M. (2006). The V0-ATPase mediates apical secretion of exosomes containing Hedgehog-related proteins in *Caenorhabditis elegans*. *The Journal of Cell Biology*, *173*, 949–961. doi:10.1083/jcb.200511072

Liu, K.S., & Sternberg, P.W. (1995). Sensory regulation of male mating behavior in *Caenorhabditis elegans*. *Neuron*, *14*(1), 79–89. doi:10.1016/0896-6273(95)90242-2

Macaisne, N., Kessler, Z., & Yanowitz, J.L. (2018). Meiotic double-strand break proteins influence repair pathway utilization. *Genetics*, *210*, 843–856. doi:10.1534/genetics.118.301402

Maguire, J.E., Silva, M., Nguyen, K.C.Q., Hellen, E., Kern, A.D., Hall, D.H., & Barr, M.M. (2015). Myristoylated CIL-7 regulates ciliary extracellular vesicle biogenesis. *Molecular Biology of the Cell*, *26*, 2823–2832. doi:10.1091/mbc.E15-01-0009

Mateo, A.-R.F., Kessler, Z., Jolliffe, A.K., McGovern, O., Yu, B., Nicolucci, A., … Derry, W.B. (2016). The p53-like protein CEP-1 is required for meiotic fidelity in *C. elegans*. *Current Biology*, *26*, 1148–1158. doi:10.1016/j.cub.2016.03.036

Melentijevic, I., Toth, M.L., Arnold, M.L., Guasp, R.J., Harinath, G., Nguyen, K.C., … Driscoll, M. (2017). *C. elegans* neurons jettison protein aggregates and mitochondria under neurotoxic stress. *Nature*, *542*, 367–371. doi:10.1038/nature21362

Melnick, M., Gonzales, P., Cabral, J., Allen, M.A., Dowell, R.D., & Link, C.D. (2019). Heat shock in *C. elegans* induces downstream of gene transcription and accumulation of double-stranded RNA. *PLoS One*, *14*(4), e0206715. doi:10.1371/journal.pone.0206715

Meneely, P.M., McGovern, O.L., Heinis, F.I., & Yanowitz, J.L. (2012). Crossover distribution and frequency are regulated by him-5 in *Caenorhabditis elegans*. *Genetics*, *190*, 1251–1266. doi:10.1534/genetics.111.137463

Molina-García, L., Kim, B., Cook, S.J., Bonnington, R., O'Shea, J., Sammut, M., … Poole, R.J. (2019). A direct glia-to-neuron natural transdifferentiation ensures nimble sensory-motor coordination of male mating behaviour. *bioRxiv*. 285320. doi:10.1101/285320

Nauli, S.M., Alenghat, F.J., Luo, Y., Williams, E., Vassilev, P., Li, X., … Zhou, J. (2003). Polycystins 1 and 2 mediate mechanosensation in the primary cilium of kidney cells. *Nature Genetics*, *33*(2), 129–137. doi:10.1038/ng1076

Nigon, V., & Dougherty, E.C. (1949). Reproductive patterns and attempts at reciprocal crossing of *Rhabditis elegans* maupas, 1900, and Rhabditis briggsae Dougherty and nigon, 1949 (Nematoda: Rhabditidae). *The Journal of Experimental Zoology*, *112*, 485–503. doi:10.1002/jez.1401120307

O'Hagan, R., Silva, M., Nguyen, K.C.Q., Zhang, W., Bellotti, S., Ramadan, Y.H., … Barr, M.M. (2017). Glutamylation regulates transport, specializes function, and sculpts the structure of cilia. *Current Biology*, *27*, 3430–3441.e6. doi:10.1016/j.cub.2017.09.066

Oren-Suissa, M., Bayer, E.A., & Hobert, O. (2016). Sex-specific pruning of neuronal synapses in *Caenorhabditis elegans*. *Nature*, *533*, 206–211. doi:10.1038/nature17977

Oren-Suissa, M., Gattegno, T., Kravtsov, V., & Podbilewicz, B. (2017). Extrinsic repair of injured dendrites as a paradigm for regeneration by fusion in *Caenorhabditis elegans*. *Genetics*, *206*(1), 215–230. doi:10.1534/genetics.116.196386

Paix, A., Wang, Y., Smith, H.E., Lee, C.-Y.S., Calidas, D., Lu, T., … Seydoux, G. (2014). Scalable and versatile genome editing using linear DNAs with microhomology to Cas9 Sites in *Caenorhabditis elegans*. *Genetics*, *198*, 1347–1356. doi:10.1534/genetics.114.170423

Pereira, L., Kratsios, P., Serrano-Saiz, E., Sheftel, H., Mayo, A.E., Hall, D.H., … Hobert, O. (2015). A cellular and regulatory map of the cholinergic nervous system of *C. elegans*. *eLife*, *4*, 4. doi:10.7554/eLife.12432

Perkins, L.A., Hedgecock, E.M., Thomson, J.N., & Culotti, J.G. (1986). Mutant sensory cilia in the nematode *Caenorhabditis elegans*. *Developmental Biology*, *117*, 456–487. doi:10.1016/0012-1606(86)90314-3

Portman, D.S. (2017). Sexual modulation of sex-shared neurons and circuits in *Caenorhabditis elegans*. *Journal of Neuroscience Research*, *95*(1–2), 527–538. doi:10.1002/jnr.23912

Ryan, D.A., Miller, R.M., Lee, K., Neal, S.J., Fagan, K.A., Sengupta, P., & Portman, D.S. (2014). Sex, age, and hunger regulate behavioral prioritization through dynamic modulation of chemoreceptor expression. *Current Biology*, *24*, 2509–2517. doi:10.1016/j.cub.2014.09.032

Sakai, N., Iwata, R., Yokoi, S., Butcher, R.A., Clardy, J., Tomioka, M., & Iino, Y. (2013). A Sexually conditioned switch of chemosensory behavior in *C. elegans*. *PLos One*, *8*, e68676. doi:10.1371/journal.pone.0068676

Sammut, M., Cook, S.J., Nguyen, K.C.Q., Felton, T., Hall, D.H., Emmons, S.W., … Barrios, A. (2015). Glia-derived neurons are required for sex-specific learning in *C. elegans*. *Nature*, *526*, 385–390. doi:10.1038/nature15700

Schedl, T., Graham, P.L., Barton, M.K., & Kimble, J. (1989). Analysis of the role of tra-1 in germline sex determination in the nematode *Caenorhabditis elegans*. *Genetics*, *123*, 755–769.

Serrano-Saiz, E., Pereira, L., Gendrel, M., Aghayeva, U., Bhattacharya, A., Howell, K., … Hobert, O. (2017). A neurotransmitter atlas of the *Caenorhabditis elegans* male nervous system reveals sexually dimorphic neurotransmitter usage. *Genetics*, *206*, 1251–1269. doi:10.1534/genetics.117.202127

Sharif-Naeini, R., Folgering, J.H.A., Bichet, D., Duprat, F., Lauritzen, I., Arhatte, M., … Honoré, E. (2009). Polycystin-1 and -2 dosage regulates pressure sensing. *Cell*, *139*, 587–596. doi:10.1016/j.cell.2009.08.045

Silva, M., Morsci, N., Nguyen, K.C.Q., Rizvi, A., Rongo, C., Hall, D.H., & Barr, M.M. (2017). Cell-specific α-tubulin isotype regulates ciliary microtubule ultrastructure, intraflagellar transport, and extracellular vesicle biology. *Current Biology*, *27*, 968–980. doi:10.1016/j.cub.2017.02.039

Stahl, P.D., & Raposo, G. (2019). Extracellular vesicles: Exosomes and microvesicles, integrators of homeostasis. *Physiology (Bethesda, MD.)*, *34*, 169–177. doi:10.1152/physiol.00045.2018

Sulston, J.E., Albertson, D.G., & Thomson, J.N. (1980). The *Caenorhabditis elegans* male: Postembryonic development of nongonadal structures. *Developmental Biology*, *78*, 542–576. doi:10.1016/0012-1606(80)90352-8

Sulston, J.E., & Horvitz, H.R. (1977). Post-embryonic cell lineages of the nematode, *Caenorhabditis elegans*. *Developmental Biology*, *56*(1), 110–156. doi:10.1016/0012-1606(77)90158-0

Sulston, J.E., & Horvitz, H.R. (1981). Abnormal cell lineages in mutants of the nematode *Caenorhabditis elegans*. *Developmental Biology*, *82*(1), 41–55. doi:10.1016/0012-1606(81)90427-9

Sulston, J.E., Schierenberg, E., White, J.G., & Thomson, J.N. (1983). The embryonic cell lineage of the nematode *Caenorhabditis elegans*. *Developmental Biology*, *100*(1), 64–119. doi:10.1016/0012-1606(83)90201-4

Sulston, J.E., & White, J.G. (1980). Regulation and cell autonomy during postembryonic development of *Caenorhabditis elegans*. *Developmental Biology*, *78*, 577–597. doi:10.1016/0012-1606(80)90353-X

Timmons, L., Luna, H., Martinez, J., Moore, Z., Nagarajan, V., Kemege, J.M., & Asad, N. (2014). Systematic comparison of bacterial feeding strains for increased yield of *Caenorhabditis elegans* males by RNA interference-induced non-disjunction. *FEBS Letters*, *588*, 3347–3351. doi:10.1016/j.febslet.2014.07.023

Wang, J., Kaletsky, R., Silva, M., Williams, A., Haas, L.A., Androwski, R.J., … Barr, M.M. (2015). Cell-specific transcriptional profiling of ciliated sensory neurons reveals regulators of behavior and extracellular vesicle biogenesis. *Current Biology*, *25*, 3232–3238. doi:10.1016/j.cub.2015.10.057

Wang, J., Nikonorova, I.A., Gu, A., Sternberg, P.W., & Barr, M.M. (2020). Polycystin-2 ciliary extracellular vesicle release and targeting. *Current Biology*, *30*, R1–R3.

Wang, J., Silva, M., Haas, L.A., Morsci, N.S., Nguyen, K.C.Q., Hall, D.H., & Barr, M.M. (2014). *C. elegans* ciliated sensory neurons release extracellular vesicles that function in animal communication. *Current Biology*, *24*, 519–525. doi:10.1016/j.cub.2014.01.002

Ward, S., Thomson, N., White, J. G., & Brenner, S. (1975). Electron microscopical reconstruction of the anterior sensory anatomy of the nematode Caenorhabditis elegans. The Journal of Comparative Neurology, 160(3), 313–337. DOI: 10.1002/cne.901600305

Ware, R. W., Clark, D., Crossland, K., & Russell, R. L. (1975). The nerve ring of the nematodeCaenorhabditis elegans: Sensory input and motor output. In The Journal of Comparative Neurology (Vol. 162, Issue 1, pp. 71–110). https://doi.org/10.1002/cne.901620106

Wehman, A.M., Poggioli, C., Schweinsberg, P., Grant, B.D., & Nance, J. (2011). The P4-ATPase TAT-5 inhibits the budding of extracellular vesicles in *C. elegans* embryos. Current Biology, *21*, 1951–1959. doi: 10.1016/j.cub.2011.10.040

Weinberg, P., Berkseth, M., Zarkower, D., & Hobert, O. (2018). Sexually dimorphic unc-6/netrin expression controls sex-specific maintenance of synaptic connectivity. *Current Biology*, *28*, 623–629.e3. doi:10.1016/j.cub.2018.01.002

White, J.G., Southgate, E., Thomson, J.N., & Brenner, S. (1986). The structure of the nervous system of the nematode *Caenorhabditis elegans*. *Philosophical Transactions of the Royal Society of London. Series B, Biological Sciences*, *314*, 1–340. doi:10.1098/rstb.1986.0056

Wu, G., Markowitz, G.S., Li, L., D'Agati, V.D., Factor, S.M., Geng, L., … Somlo, S. (2000). Cardiac defects and renal failure in mice with targeted mutations in Pkd2. *Nature Genetics*, *24*(1), 75–78. doi:10.1038/71724

Yoder, B.K., Hou, X., & Guay-Woodford, L.M. (2002). The polycystic kidney disease proteins, polycystin-1, polycystin-2, polaris, and cystin, are co-localized in renal cilia. *Journal of the American Society of Nephrology*, *13*, 2508–2516. doi:10.1097/01.ASN.0000029587.47950.25

Zarkower, D., & Hodgkin, J. (1992). Molecular analysis of the *C. elegans* sex-determining gene tra-1: A gene encoding two zinc finger proteins. *Cell*, *70*, 237–249. doi:10.1016/0092-8674(92)90099-X

Zaslaver, A., Liani, I., Shtangel, O., Ginzburg, S., Yee, L., & Sternberg, P.W. (2015). Hierarchical sparse coding in the sensory system of *Caenorhabditis elegans*. *Proceedings of the National Academy of Sciences of the United States of America*, *112*, 1185–1189. doi:10.1073/pnas.1423656112

Zhao, Y., Wang, H., Poole, R.J., & Gems, D. (2019). A fln-2 mutation affects lethal pathology and lifespan in *C. elegans*. *Nature Communications*, *10*(1), 5087. doi:10.1038/s41467-019-13062-z

Zhou, L., He, B., Deng, J., Pang, S., & Tang, H. (2019). Histone acetylation promotes long-lasting defense responses and longevity following early life heat stress. *PLoS Genetics*, *15*, e1008122. doi:10.1371/journal.pgen.1008122

OPEN ACCESS

Cell-type-specific promoters for *C. elegans* glia

Wendy Fung, Leigh Wexler and Maxwell G. Heiman

ABSTRACT

Glia shape the development and function of the *C. elegans* nervous system, especially its sense organs and central neuropil (nerve ring). Cell-type-specific promoters allow investigators to label or manipulate individual glial cell types, and therefore provide a key tool for deciphering glial function. In this technical resource, we compare the specificity, brightness, and consistency of cell-type-specific promoters for *C. elegans* glia. We identify a set of promoters for the study of seven glial cell types (*F16F9.3*, amphid and phasmid sheath glia; *F11C7.2*, amphid sheath glia only; *grl-2*, amphid and phasmid socket glia; *hlh-17*, cephalic (CEP) sheath glia; and *grl-18*, inner labial (IL) socket glia) as well as a pan-glial promoter (*mir-228*). We compare these promoters to promoters that are expressed more variably in combinations of glial cell types (*delm-1* and *itx-1*). We note that the expression of some promoters depends on external conditions or the internal state of the organism, such as developmental stage, suggesting glial plasticity. Finally, we demonstrate an approach for prospectively identifying cell-type-specific glial promoters using existing single-cell sequencing data, and we use this approach to identify two novel promoters specific to IL socket glia (*col-53* and *col-177*).

Introduction

Cell-type-specific promoters are the key to studying any individual cell in *C. elegans*. With such a promoter in hand, one can visualize the cell in live animals by expressing a soluble fluorescent protein; probe cell biology in real time by expressing fluorescent constructs that label subcellular structures or serve as biosensors of cellular activity; disrupt cell function by genetic ablation or cell-specific depletion of single proteins; and determine cell-autonomy of gene function by expressing a rescuing transgene in specific cells in an otherwise mutant background.

C. elegans glia illustrate the power of cell-specific promoters to open up a new area of study. The powerful approaches listed above have enabled the study of several glial cell types. In particular, the amphid sheath (AMsh) and cephalic sheath (CEPsh) glia have been studied in detail over the last two decades, while more recent work has begun to examine the amphid socket (AMso), phasmid socket (PHso1 and PHso2), and inner labial socket (ILso) glia (Singhvi & Shaham, 2019). In contrast, other glial cell types, for which no cell-type-specific promoters have been identified, remain largely unexplored.

In this technical resource, we summarize and compare the promoters that have been used to study distinct glial cell types in *C. elegans*. We evaluate a set of specific, bright, and consistent promoters for the study of seven glial cell types as well as a pan-glial driver. We note that some promoters exhibit interesting dynamics in their expression patterns that are suggestive of glial plasticity. Finally, we demonstrate an approach to identify new cell-type-specific glial promoters by taking advantage of recent single-cell sequencing experiments, and we use this approach to identify *col-53* and *col-177* as novel cell-specific promoters for ILso glia.

Materials and methods

Strains and plasmids

All strains were grown at 20 °C on nematode growth media (NGM) with *E. coli* OP50 bacteria (Brenner, 1974). Transgenic strains generated in this study were constructed in an N2 background. N2 animals were injected with 25–75 ng/μL per plasmid at a final concentration of 100 ng/μL DNA. Additional information on strains and plasmids used in this study are in Supplementary Table S1.

Fluorescence microscopy and image processing

L2 to L3 animals were selected based on size and the extent of vulva development on the day of imaging. Animals were washed and immobilized in M9 solution containing 50 mM sodium azide and mounted on 2% agarose pads. Image stacks were collected on a DeltaVision Core imaging system (Applied

Precision) with a UApo 40x/1.35 NA, PlanApo 60x/1.42 NA, or UPlanSApo 100x/1.40 NA oil immersion objective and a CoolSnap HQ2 camera. Images were subsequently deconvolved using Softworx (Applied Precision) and maximum intensity projections were generated in ImageJ. The brightness and contrast of each projection were linearly adjusted in Affinity Photo 1.7.3. Fluorescent signals were pseudo-colored, and merged images were generated using the Screen layer mode in Affinity Photo 1.7.3. When the whole animal is shown, a composite of multiple fields was manually stitched together. The maximum intensity projections of the stitched images were generated with the same optical stacks across the whole animal, except for *F16F9.3* and *grl-2*, for which different optical stacks were used to generate the maximum intensity projections of the head and tail, because the head and tail glia are in different planes. Additionally, for *grl-2*, the head was split into two regions and different optical stacks were used to generate the maximum intensity projections to resolve both the AMso glia and the excretory cells separately. All figures were assembled in Affinity Designer 1.7.3. For quantification of cell fluorescence expression, FIJI software (Schindelin *et al.*, 2012) was used to measure the integrated density of GFP or RFP in the cell body and the average of three background regions in the head was subtracted.

Assessment of glial marker consistency

To assess the consistency of each glial marker, L2 to L3 transgenic animals were washed and immobilized in M9 solution containing 50 mM sodium azide and mounted on 2% agarose pads. Fluorescently labeled glial cells were scored visually across optical stacks with a Deltavision Core imaging system (Applied Precision) and a PlanApo 60x/1.42 NA oil immersion objective.

Analysis of single-cell RNA sequencing data

To identify novel markers for ILso glia, we used the 'Differential Expression' feature on the VisCello data explorer (Packer *et al.*, 2019). We generated a list of genes that are highly enriched in the predicted ILso glia cluster compared to all other cell types in the single-cell RNA sequencing dataset. This was achieved by first selecting the following: Global dataset (Choose Sample), Cell type/subtype (Meta Class), and ILso (Group 1). Group 2 was left blank. Next, downsample cells was performed with the default settings (0.05 FDR cutoff, 200 Max Cell in subset, and 1000 Max Cell in background). Finally, differential expression (DE) analysis was run to generate a table of differentially expressed genes for the ILso glia cluster. A list of genes that are enriched in other glial types, including AMso glia (Supplementary Figure S2(C)), can be generated by replacing ILso with the glial type of interest (Group 1).

The graphical plots of *grl-18* and *col-53* or *col-177* expression (Figure 5(A), Supplementary Figure S1(A)) were generated using the online VisCello data explorer (https://cello.shinyapps.io/celegans_explorer/). The 'Cell Type' explorer was used to select the glia and excretory cell subset, and data was displayed using the 'UMAP-2D [Paper]' projection colored by cell type/subtype. Separate searches for *grl-18* and *col-53* or *col-177* were used to generate a plot of the glia and excretory cell subset showing the relative expression levels of each gene using the black_red palette for *grl-18* and viridis palette for *col-53* or *col-177.* Inset boxes in Figure 5(A) and Supplementary Figure S1(A) are magnifications of the predicted ILso cluster, which contains 163 cells.

Results and discussion

Survey of glial promoters used in the literature

We performed a survey of cell-specific promoters that have been used to label and manipulate *C. elegans* glia (Table 1). All *C. elegans* glia are associated with sense organs, mostly in the head and tail (Ward, Thomson, White, & Brenner, 1975). Hermaphrodites have 24 sense organs: 18 in the head (two amphid, AM; four cephalic, CEP; six inner labial, IL; six outer labial, OL), two in the anterior midbody (anterior deirid, ADE); two in the posterior midbody (posterior deirid, PDE), and two in the tail (phasmid, PH) (White, Southgate, Thomson, & Brenner, 1986). Each sense organ comprises one or more ciliated sensory neurons; one sheath glial cell; and, in most cases, one socket glial cell (the phasmid has two socket glia). The distal endings of the sheath and socket glia are arranged to form a tube-shaped epithelium that is continuous with the skin (Figure 1) (Low *et al.*, 2019; Ward *et al.*, 1975). The proximal portion of this tube is formed by the ending of the sheath glial cell, while the distal portion of the tube is formed by the ending of the socket glial cell. In each sense organ, sensory neurons protrude through this glial tube to directly access the environment (Figure 1). The sheath (sh) and socket (so) glia of most sense organs appear to represent distinct cell types. For example, the AMsh glia are a different cell type than the AMso glia or CEPsh glia based on morphology, molecular profile, and function (Mizeracka & Heiman, 2015). As discussed in detail below, each of these glial cell types expresses a distinct transcriptional reporter. The functional relevance of these striking transcriptional differences has not been fully established, but these transcriptional reporters serve as powerful tools to study and manipulate individual glial cell types. In addition to the glial cell types described above, there are also sex-specific glia in copulatory sense organs in the male tail, and glial-like functions can be performed by mesoderm-derived GLR cells and by the skin (Singhvi & Shaham, 2019; Yang & Chien, 2019). Here, we will focus exclusively on sheath and socket glia of sense organs that are present in both sexes.

Cell-type-specific promoters for *C. elegans* glia have been identified through three approaches: reverse genetics, forward genetics, and transcriptional profiling. In reverse genetics, the molecular identity of a gene motivates investigation of its expression pattern. The study of molecularly interesting gene families led to the first identification of promoters expressed specifically in glia, including the basic helix-loop-helix transcription factor gene *hlh-17* (McMiller & Johnson, 2005; Yoshimura, Murray, Lu, Waterston, & Shaham, 2008); the inositol triphosphate receptor gene *itr-1* (Gower *et al.*, 2001); the neurexin gene *itx-1* (Haklai-Topper *et al.*, 2011);

Table 1. Glial promoters used in the literature.

AMsh or PHsh		
F16F9.3	2 kb promoter	Bacaj *et al.* (2008); Braunreiter, Hamlin, & Lyman-Gingerich (2014); Heiman & Shaham (2009); Low *et al.* (2019); Mizeracka, Rogers, Shaham, Bulyk, & Heiman (2019); Oikonomou *et al.* (2011); Procko *et al.* (2012, 2011); Singhvi *et al.* (2016); Wallace *et al.* (2016); Yip & Heiman (2018)
vap-1 (developmentally regulated)	2797 bp or 5205 bp promoter	Bacaj *et al.* (2008); Ding *et al.* (2015); Oikonomou *et al.* (2011); Perens & Shaham (2005); Procko *et al.* (2011); Wallace *et al.* (2016); Wang *et al.* (2017) Wang, D'Urso, & Bianchi (2012, 2008)
T02B11.3	2.5 kb promoter	Bacaj *et al.* (2008); Grant, Matthewman, & Bianchi (2015); Mizeracka *et al.* (2019); Oikonomou, Perens, Lu, & Shaham (2012) Oikonomou *et al.* (2011); Procko *et al.* (2011); Wang *et al.* (2017, 2008)
fig-1	2.2 kb promoter	Bacaj *et al.* (2008); Fenk & de Bono (2015); Kage-Nakadai *et al.* (2016); Wallace *et al.* (2016); Yoshida *et al.* (2016)
ver-1 (thermo- and dauer-regulated)	2110 bp promoter	Popovici *et al.* (2002); Procko *et al.* (2012, 2011); Yoshida *et al.* (2016)
F53F4.13	650 bp promoter	Bacaj *et al.* (2008); Mizeracka *et al.* (2019); Singhvi *et al.* (2016)
F11C7.2	350 bp promoter	Bacaj *et al.* (2008); Wallace *et al.* (2016)
hmit-1.2 (osmo-regulated)	979 bp promoter	Kage-Nakadai *et al.* (2016, 2011)
lit-1	2.5 kb promoter	Oikonomou *et al.* (2011); Wallace *et al.* (2016)
grl-28	2553 bp promoter	Hao *et al.* (2006)
acd-1	2550 bp promoter, plus coding sequences	Wang *et al.* (2008)
F52E1.2	3.3 kb promoter	Ohkura & Bürglin (2011)
hmit-1.3	4883 bp promoter	Kage-Nakadai *et al.* (2011)
F58F9.6	4 kb promoter	Procko *et al.* (2012)
K02E11.4	800 bp promoter	Wallace *et al.* (2016)
R11D1.3	900 bp promoter	Wallace *et al.* (2016)
bgnt-1.1	recombineered fosmid	Timbers *et al.* (2016)
AMso or PHso		
itr-1	2.3 kb promoter (from between exon 1 and exon 2)	Gower *et al.* (2001); Han *et al.* (2013); Heiman & Shaham (2009); Mizeracka *et al.* (2019); Sammut *et al.* (2015); Tucker, Sieber, Morphew, & Han (2005); Wang *et al.* (2017)
grl-2	2981 bp promoter	Hao *et al.* (2006); Hunt-Newbury *et al.* (2007); Low *et al.* (2019); Mizeracka *et al.* (2019); Molina-García *et al.* (2018); Sammut *et al.* (2015)
lin-48	6.8 kb promoter	Johnson *et al.* (2001); Mizeracka *et al.* (2019); Molina-García *et al.* (2018)
grd-15	840 bp promoter	Hunt-Newbury *et al.* (2007); Timbers *et al.* (2016)
unc-53	3.4 kb promoter (from exon 8 to exon 13)	Stringham, Pujol, Vandekerckhove, & Bogaert (2002); Tucker *et al.* (2005)
alr-1	1 kb promoter	Tucker *et al.* (2005)
abt-4	2472 bp promoter	Hunt-Newbury *et al.* (2007)
grl-6	939 bp promoter	Hunt-Newbury *et al.* (2007)
srp-2	2833 bp promoter	Hunt-Newbury *et al.* (2007)
Combinations of AMsh, AMso, PHsh, and PHso		
daf-6	3 kb promoter	Kage-Nakadai *et al.* (2016); Moussaif & Sze (2009); Ohkura & Bürglin (2011); Perens & Shaham (2005); Wallace *et al.* (2016)
ztf-16	2.1 kb enhancer (from 4637 to 2536 bp upstream of ATG)	Procko *et al.* (2012); Sammut *et al.* (2015)
ttx-1	3.5 kb enhancer (from 11 to 7.5 kb upstream of ATG)	Procko *et al.* (2011)
CEPsh		
hlh-17	2.5 kb promoter or 1.9 kb promoter	Cianciulli *et al.* (2019); Colón-Ramos *et al.* (2007); Gibson *et al.* (2018); Ji *et al.* (2019); Katz, Corson, Iwanir, Biron, & Shaham (2018); McMiller & Johnson (2005); Mizeracka *et al.* (2019); Rapti *et al.* (2017); Sammut *et al.* (2015); Shao *et al.* (2013); Stout & Parpura (2011); Yoshida *et al.* (2016); Yoshimura *et al.* (2008)
swip-10	1.5 kb enhancer (from fifth intron, 1013 to 2504 bp downstream of ATG)	Hardaway *et al.* (2015)
twk-16	3412 bp promoter	Cianciulli *et al.* (2019)
ILso		
grl-18	2968 or 2962 bp promoter	Cebul *et al.* (2020); Hao *et al.* (2006); Mizeracka *et al.* (2019)
Combinations of ILsh, ILso, OLsh, and OLso		
itx-1	2977 bp promoter or 1.8 kb promoter	Haklai-Topper *et al.* (2011); Han *et al.* (2013); McQuary *et al.* (2016); Sammut *et al.* (2015)
delm-1	2 kb promoter	Han *et al.* (2013)
delm-2	1.4 kb promoter	Han *et al.* (2013)
Most (or many) glia		
ptr-10	300 bp promoter	Gibson *et al.* (2018); Hardaway *et al.* (2015); Katz *et al.* (2018); Oikonomou *et al.* (2011); Rapti *et al.* (2017); Sammut *et al.* (2015); Wallace *et al.* (2016); Yin *et al.* (2017); Yoshida *et al.* (2016); Yoshimura *et al.* (2008)
mir-228	2.2 kb promoter	Molina-García *et al.* (2018); Pierce *et al.* (2008); Rapti *et al.* (2017); Wallace *et al.* (2016); Wang *et al.* (2017); Yin *et al.* (2017)
kcc-3	3.9 kb promoter and 2.2 kb downstream sequences	McQuary *et al.* (2016); Singhvi *et al.* (2016); Tanis, Bellemer, Moresco, Forbush, & Koelle (2009); Yoshida *et al.* (2016)
pros-1	recombineered fosmid or 9 kb promoter	Kage-Nakadai *et al.* (2016); Wallace *et al.* (2016)
Unidentified combinations of glia		
grd-8	905 bp promoter	Hao *et al.* (2006)
grd-16	2866 bp promoter	Hao *et al.* (2006)
grl-1	287 bp promoter	Hao *et al.* (2006)
grl-6	940 bp promoter	Hao *et al.* (2006)
grl-8	2893 bp promoter	Hao *et al.* (2006)
grl-12	2989 bp promoter	Hao *et al.* (2006)
grl-13	2843 bp promoter	Hao *et al.* (2006)
grl-17	2869 bp promoter	Hao *et al.* (2006)
grl-29	1893 bp promoter	Hao *et al.* (2006)
rgba-1	2.1 kb promoter	Yin *et al.* (2017)
argk-1	1.2 kb promoter	McQuary *et al.* (2016)

Promoters with glial-specific expression were identified in the literature. Within each group, promoters are listed in order from most to least frequently used.

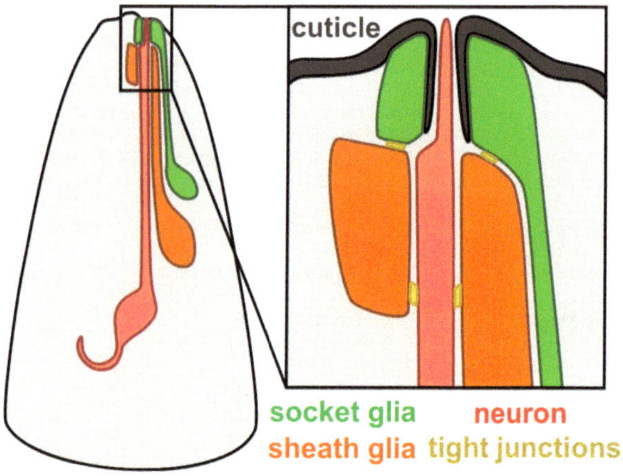

Figure 1. Schematic of a *C. elegans* sense organ in thehead. A typical *C. elegans* sense organ contains one or more sensory neurons (red) and two glial cells, called the sheath (orange) and socket (green). Most sensory neurons extend a ciliated dendritic ending through a tube-shaped pore formed by the sheath and socket glia. The socket glia secretes cuticle (gray) that forms an open pore through which chemosensory dendrite endings protrude, as shown, or a closed sheet into which mechanosensory dendrite endings are embedded (not shown). Tight junctions (yellow) are present between the neuron and sheath glia, the sheath and socket glia, and the socket glia and skin. See online version for color figure.

the VEGF receptor gene *ver-1* (Popovici, Isnardon, Birnbaum, & Roubin, 2002); the venom allergen protein gene *vap-1* (Bacaj, Tevlin, Lu, & Shaham, 2008); the ion channel genes *delm-1* and *delm-2* (Han *et al.*, 2013); the microRNA *mir-228* (Pierce *et al.*, 2008); and a group of nematode-specific genes with distant similarity to Hedgehog genes, including the 'groundhog' (*grd*) and 'ground-like' (*grl*) genes (Hao, Johnsen, Lauter, Baillie, & Bürglin, 2006). In a typical experiment, a putative promoter fragment from the gene of interest is inserted upstream of GFP in a transgene and the resulting expression pattern is examined. Expression in the AMsh (*vap-1*, *ver-1*) or CEPsh (*hlh-17*) glia is readily recognized due to the symmetry and distinctive morphology of these cells (Table 1). However, expression in the ILsh, ILso, OLsh, and OLso glia is harder to recognize due to the variable cell body positions and similar symmetry and morphology of these glial types (Bargmann & Avery, 1995). Thus, most promoters that label combinations of these cells, including *itx-1* and the *delm*, *grl*, and *grd* gene families, have not had their expression patterns precisely defined (Table 1).

As a second approach, forward genetics has led to the discovery of glial-expressed genes through the identification of mutants that disrupt neuronal or glial function. These include *daf-6* and *lit-1*, which affect glial morphogenesis and thus sensory neuron function (Oikonomou *et al.*, 2011; Perens & Shaham, 2005); *ttx-1* and *ztf-16*, which affect a glial-specific temperature response (Procko, Lu, & Shaham, 2012, 2011); and *kcc-3*, which affects the microenvironment that glia create around sensory neuron endings (Singhvi *et al.*, 2016; Yoshida *et al.*, 2016) (Table 1). These genes came under study due to their roles in AMsh glia, but several are also expressed in other classes of glia.

Finally, using transcriptional profiling approaches, existing glial markers have been used to 'bootstrap' the way to new markers, typically by using fluorescence-activated cell sorting to purify a glial population and then subjecting it to RNA profiling. This approach led to the identification of numerous AMsh markers, including *fig-1*, *F16F9.3*, *T02B11.3*, and others (Bacaj *et al.*, 2008; Wallace, Singhvi, Liang, Lu, & Shaham, 2016) (Table 1). The expression of these genes is highly specific to the AMsh, but they were missed by other approaches because they are not members of a conserved gene family and (with the exception of *fig-1*) have not been found to cause phenotypes when disrupted. The AMsh and PHsh are highly similar, and many AMsh markers are also expressed in the PHsh.

Recommended promoters for specific glial cell types

There is currently no consensus on which promoters are most appropriate for studying specific glia types. Therefore, we collected glial promoters from the literature and reexamined the specificity, brightness, and consistency of their expression patterns. Specificity requires that expression is restricted to individual glial cell types for easy visualization and for reducing off-target effects of genetic manipulations. Brightness allows fine structural details of glia to be resolved, which provides a tool for examining glia-neuron interactions and an efficient way to screen for mutants with glial morphology defects. Last, consistency assesses whether the promoter reliably labels the same number of cells across individuals. Based on these criteria, we selected five promoters to target seven glial cell types: the AMsh and PHsh; CEPsh; the AMso, PHso1 and PHso2; and ILso glia (Figure 2, Table 2). However, it is important to note that brightness and consistency are also affected by the structure of the transgene, including copy number and whether it is genomically integrated, as well as the fluorescent reporter used.

A pair of AMsh glia are part of the bilateral amphid sense organs in the head (Ward *et al.*, 1975). Each AMsh is associated with 12 sensory neurons that respond to diverse stimuli including chemical and mechanical cues, temperature, osmolarity, and pheromones (Bargmann, 2006). AMsh glia are the most well-studied *C. elegans* glia, with many highly specific markers identified (Bacaj *et al.*, 2008; Wallace *et al.*, 2016) (Table 1). We selected *F16F9.3* and *F11C7.2* as AMsh promoters that offer different advantages (Figure 2(A)). *F16F9.3* is reliably expressed in AMsh glia as well as the PHsh glia, their functional counterparts in the tail (24/25 animals with an integrated *F16F9.3*pro:mCherry transgene had 2/2 AMsh and 2/2 PHsh glia marked; 1/25 animals had 1/2 AMsh and 2/2 PHsh glia marked) (Figure 2(A), Supplementary Table S2). The strong expression of *F16F9.3* allows us to clearly visualize the tube-shaped pores in the sheath glia through which the sensory dendrite endings of the neurons protrude (Figure 2(B)). Other promoters, including *F53F4.13* and *T02B11.3*, also offer strong specific labeling of AMsh and PHsh. By comparison, because *F11C7.2* expression is restricted to the AMsh glia (Figure 2(A)), it offers the advantage of genetically manipulating AMsh glia without affecting PHsh glia (16/23 animals with an *F11C7.2*pro:GFP extrachromosomal array had 2/2 AMsh

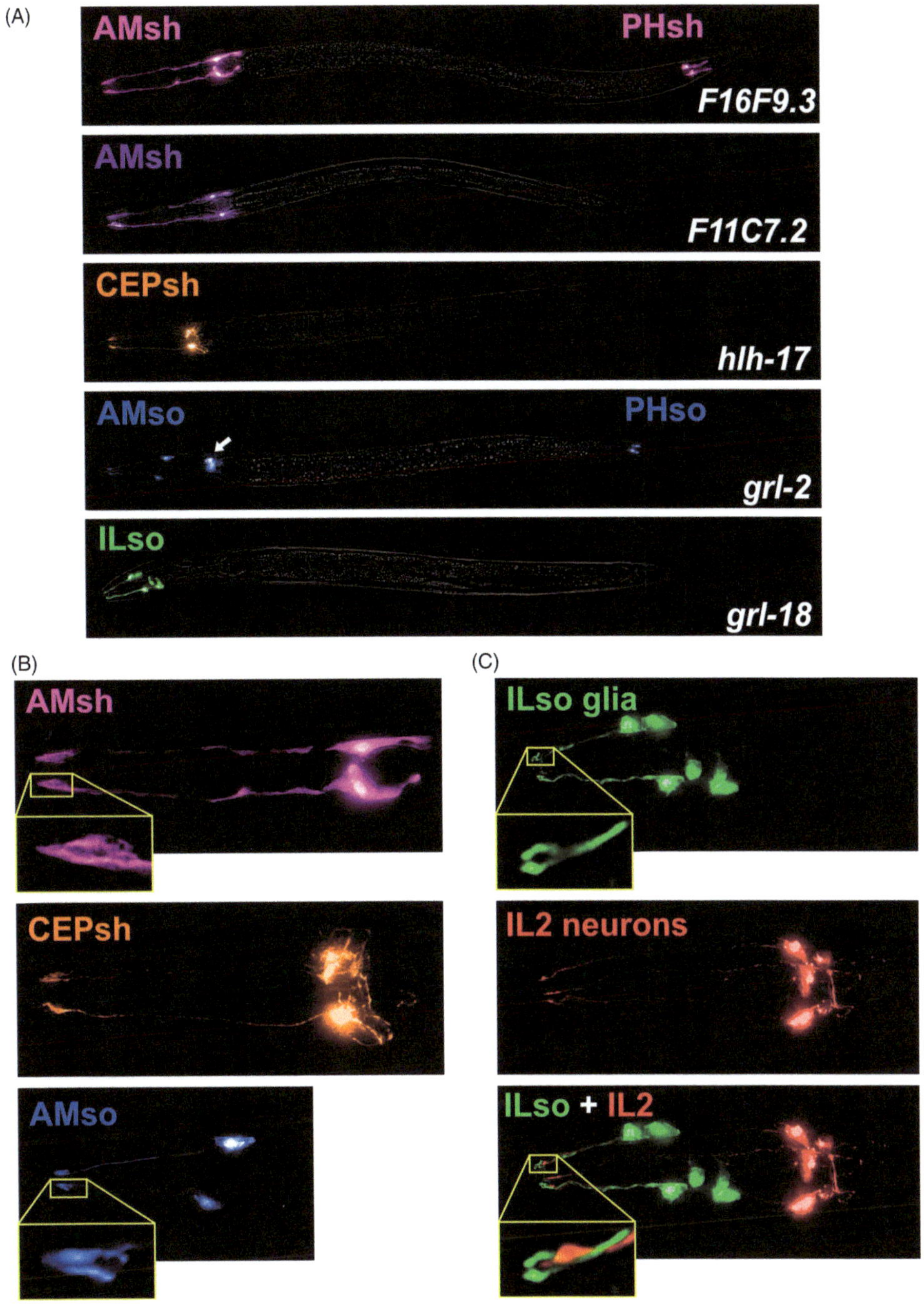

Figure 2. Cell-type-specific promoters for sheath and socket glia. (A) Merged brightfield images and pseudo-colored fluorescence projections of animals expressing fluorescent proteins under control of promoters selected for brightness, specificity, and consistency. *F16F9.3* (pink, AMsh and PHsh); *F11C7.2* (purple, AMsh only); *hlh-17* (orange, CEPsh); *grl-2* (blue, AMso, PHso1, and PHso2; also expressed in excretory duct and pore cells, white arrow); *grl-18* (green, ILso). (B) Magnified images of AMsh (pink, *F16F9.3*), CEPsh (orange, *hlh-17*), and AMso (blue, *grl-2*), showing that their brightness is sufficient to resolve fine structural details including the tube-like pores of the AMsh and AMso glia and the branch-like posterior processes of the CEPsh glia. (C) Head of an animal expressing *grl-18*pro:GFP (green, ILso glia) and *klp-6*pro:mCherry (red, IL2 neurons), demonstrating that the *grl-18* promoter labels the six ILso glia, identified by their processes which each form a pore for the ciliated dendritic ending of an IL2 sensory neuron. See online version for color figure.

glia and no PHsh glia marked, whereas 7/23 animals had 1/2 AMsh glia and no PHsh glia marked, Supplementary Table S2). In summary, *F16F9.3* has higher overall expression, while *F11C7.2* has higher specificity for AMsh glia.

Each of four CEPsh glia is associated with the fourfold symmetric cephalic sense organs in the head. Each CEPsh glial cell wraps around the dendrite of a mechanosensory neuron (CEP) and, in males, an additional pheromone-detecting neuron (CEM) (Ward *et al.*, 1975). We selected *hlh-17* as a cell-specific promoter for CEPsh glia (Figure 2(A)). We evaluated the expression pattern of a transcriptional reporter for *hlh-17* (McMiller & Johnson, 2005; Yoshimura *et al.*, 2008) and observed that it is restricted to CEPsh glia and is consistently expressed in all four cells (22/22 animals with an integrated *hlh-17*pro:GFP transgene had 4/4 CEPsh glia marked, Supplementary Table S2). Though the transcriptional reporter we examined uses a ~2.5 kb promoter fragment and appears to be highly specific to CEPsh glia, it is important to note that reporters containing a different promoter fragment of the *hlh-17* gene are expressed

Table 2. Recommended cell-type-specific glial promoters.

Gene	Promoter (bp)	Glia marked	Other cells marked	References
F16F9.3	2000	AMsh, PHsh	none	Bacaj *et al.* (2008)
F11C7.2	350	AMsh	none	Bacaj *et al.* (2008)
hlh-17	2500	CEPsh	none	McMiller & Johnson (2005); Yoshimura *et al.* (2008)
grl-2	2981	AMso, PHso1, PHso2	excretory duct, pore	Hao *et al.* (2006)
grl-18	2962	ILso	vulval epithelial cells	Cebul *et al.* (2020); Hao *et al.* (2006)
mir-228	2200	all (or nearly all) glia	seam cells, excretory cell	Pierce *et al.* (2008)

For each marker, the promoter column indicates the size (bp) of the genomic DNA fragment that is taken directly upstream of the gene translation start site, except *hlh-17* for which the 3′ end of the promoter fragment is 118 bp upstream of the translation start site.

in additional cells (Yoshimura *et al.*, 2008). The marker typically becomes brighter as animals reach adulthood but the extended anterior and posterior processes of the glia are visible at all larval stages. We can clearly observe the thin, branch-like posterior processes that wrap around the nerve ring – a feature that distinguishes CEPsh glia from other glia of *C. elegans* (Figure 2(B)). The ability to visualize these fine details makes it possible to investigate the role of CEPsh glia in shaping neuronal connections in the central neuropil, or nerve ring (Colón-Ramos, Margeta, & Shen, 2007; Ji *et al.*, 2019; Rapti, Li, Shan, Lu, & Shaham, 2017; Shao, Watanabe, Christensen, Jorgensen, & Colón-Ramos, 2013).

The two AMso glia wrap around the distal tips of a subset of amphid sensory neurons and guide them toward pores in the overlying cuticle so that they can access the external environment (Ward *et al.*, 1975). We selected *grl-2* to label and target AMso glia (Figure 2(A)). It is consistently expressed in AMso glia as well as the PHso glia in the tail (23/23 animals with an integrated *grl-2*pro:YFP transgene have 2/2 AMso glia, 2/2 PHso1, and 2/2 PHso2 glia marked, Supplementary Table S2). However, the marker is also expressed brightly in the excretory duct and pore cells (Hao *et al.*, 2006) (Figure 2(A), arrow), which makes it problematic to use *grl-2* for genetic perturbations, as disrupting the excretory cells can lead to lethality or defects in osmoregulation (Nelson & Riddle, 1984; Sundaram & Buechner, 2016). Although *grl-2* is less cell-type-specific than the AMsh or CEPsh promoters, it is bright (Figure 2(B)) and consistent and exhibits a more restricted expression pattern than other AMso glia markers that have been identified to date (Table 1). As previous work has shown, we found that *lin-48* offers more restricted expression in the phasmid glia (PHso1 but not PHso2) but is also expressed in the posterior intestine and additional cells in the head, including the excretory duct and unidentified neurons (Johnson, Fitzsimmons, Hagman, & Chamberlin, 2001). It will be important to identify new AMso glia markers whose expression is excluded from the excretory cells so that they can be used as tools to elucidate AMso glia biology and function in future studies.

Last, we selected the *grl-18* promoter for the six ILso glia that are part of the sixfold symmetric inner labial sense organs of the head (Figure 2(A)). In a previous study (Hao *et al.*, 2006), *grl-18* was tentatively identified as a potential marker for ILso or the OLso quadrant (dorsal and ventral, but not lateral) glia based on position and morphology. We find that *grl-18* is a highly specific marker for ILso glia. Each ILso glia wraps around the dendrite endings of two sensory neurons, called the IL1 and IL2 neurons (Ward *et al.*, 1975). Therefore, we examined animals co-expressing *grl-18*pro:GFP and the IL2 neuron-specific marker *klp-6*pro:mCherry (Figure 2(C)). We observed the ciliated endings of the IL2 dendrites protruding through the tube-like pores formed by the glia expressing *grl-18*pro:GFP, demonstrating that *grl-18* labels ILso glia. This is consistent with our previous observation that *grl-18*pro:GFP labels the lateral ILso glial cells that form specialized attachments to the BAG and URX neurons (Cebul, McLachlan, & Heiman, 2020). To assess the quality of this marker, we examined the expression pattern more carefully. We found that its expression is highly consistent (28/28 animals with an integrated *grl-18*pro:GFP transgene had 6/6 ILso glia marked, Supplementary Table S2). It is mostly specific to the ILso glia, however after animals enter the fourth larval (L4) stage, *grl-18*pro:GFP is also expressed in vulval epithelial cells. The marker is very bright, so it can be used to visualize the pore structure of the glia (Figure 2(C)) and screen for mutants with defects in morphology to investigate ILso glia development. It is also suitable for structured illumination microscopy, which requires greater brightness than conventional imaging (Cebul *et al.*, 2020). An inherent limitation is that the ILso cell body positions are variable (Bargmann & Avery, 1995) and often appear to overlap in a two-dimensional projection, making it difficult to show all six cells clearly in a single image. Overall, we have demonstrated that *grl-18* specifically labels ILso glia, and this opens up opportunities for studying ILso glia, which remain largely unexplored.

Based on their specificity, brightness, and consistency, the markers listed in Table 2 are especially useful for labeling and genetically manipulating the AMsh, PHsh, CEPsh, AMso, PHso1, PHso2, and ILso glia. We are not aware of promoters that specifically label the CEPso, OLsh, OLso, ADEsh, ADEso, PDEsh, or PDEso. In fact, little is known about these glia at all. Currently, the best approach for studying them may be to use promoters that are expressed in combinations of multiple glial cell types, as described further below.

Promoters that label multiple glial cell types

Pan-neuronal promoters, such as *rab-3*pro (Nonet *et al.*, 1997), have proven vital for the study of neuronal function and, similarly, the ability to drive expression in all glial cells is important for studying glial function. Several promoters that drive expression in many glial types have been identified (Table 1). We selected *mir-228* as a pan-glial promoter,

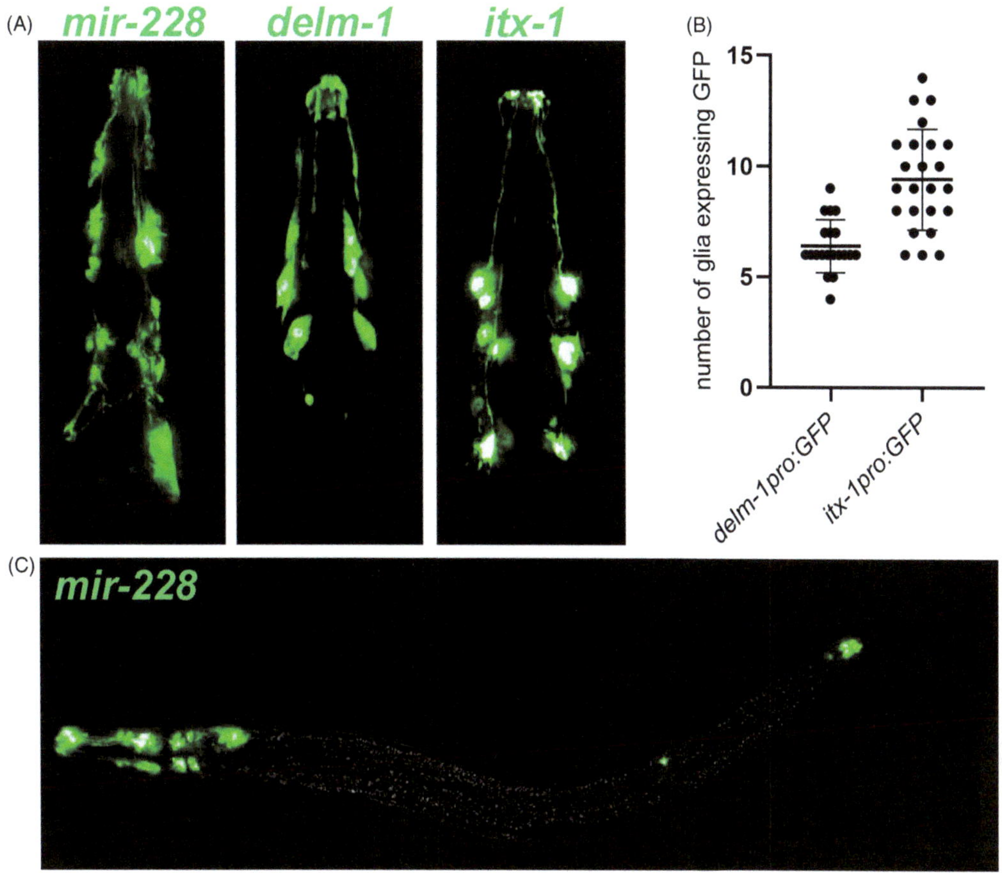

Figure 3. Promoters expressed in combinations of glia. (A) Images of head expression of GFP under control of the pan-glial promoter *mir-228* and the variable glial promoters *delm-1* and *itx-1*. (B) Quantification of the number of GFP-expressing glia observed with *delm-1* ($n = 20$) and *itx-1* ($n = 24$) extrachromosomal reporters. Error bars indicate standard deviation. $p < 0.0001$, unpaired t-test with Welch's correction. (C) Merged brightfield and fluorescence projections of a whole animal expressing pan-glial *mir-228*pro:GFP.

because it drives robust expression in most if not all glia (Figure 3(A,C)) (Pierce *et al.*, 2008). However, it is important to note that it is also expressed in some non-glial cells, including the hypodermal seam cells and the excretory cell.

Promoters expressed in other combinations of glia are also useful for studying glial biology. Curiously, some of these promoters exhibit more variable expression across individuals than the highly consistent cell-type-specific promoters described above. Two such promoters are *delm-1* and *itx-1*, both of which are expressed in combinations of glia in the head (Figure 3(A)) (Haklai-Topper *et al.*, 2011; Han *et al.*, 2013). While the identity of some of these glia is known (see Table 1), the expression patterns of these promoters have not been fully characterized. Although there is some overlap in their expression, as described previously (Han *et al.*, 2013), it is clear that *delm-1* and *itx-1* are expressed in different combinations of glia (Figure 3(A)). In contrast to the highly consistent cell-type-specific promoters described above, *delm-1*pro:GFP and *itx-1*pro:GFP extrachromosomal arrays exhibit greater variability from animal to animal, although each promoter shows a consistent range of expression (Figure 3(B)). *delm-1* is expressed in fewer glia and has less variability in expression than *itx-1* (Figure 3(B), Supplementary Table S2). A high degree of variability in GFP intensity between different glia within the same animal was often observed, particularly for *itx-1*. It is possible that differences in the threshold of detection introduced more variability into the number of GFP-expressing glia that could be discerned. Despite this variability, these promoters provide a useful tool for driving expression in combinations of glia that cannot be accessed with other promoters.

Plasticity in expression of glial promoters

The variability in expression of these promoters may reflect transgene artifacts or true biological differences across individuals. Indeed, the expression of some glial promoters has been shown to be dynamic across life stages, suggesting that glia exhibit surprising plasticity in response to environmental conditions and the internal state of the animal. We noted that the expression of *vap-1* in the AMsh is dependent upon developmental stage. Using a 2797 bp promoter, an integrated *vap-1*pro:RFP transgene exhibits absent or very faint expression in the AMsh prior to the L4 stage, but it then becomes robust in L4 animals and is upregulated further in adulthood (Figure 4(A,C)). These developmental changes in expression suggest that changes in glial function may occur during late larval development and maturation. In addition to developmental stage, *vap-1* expression is subject to further modulation through activity of amphid sensory neurons. DYF-7 is required to anchor the dendrite endings of sensory

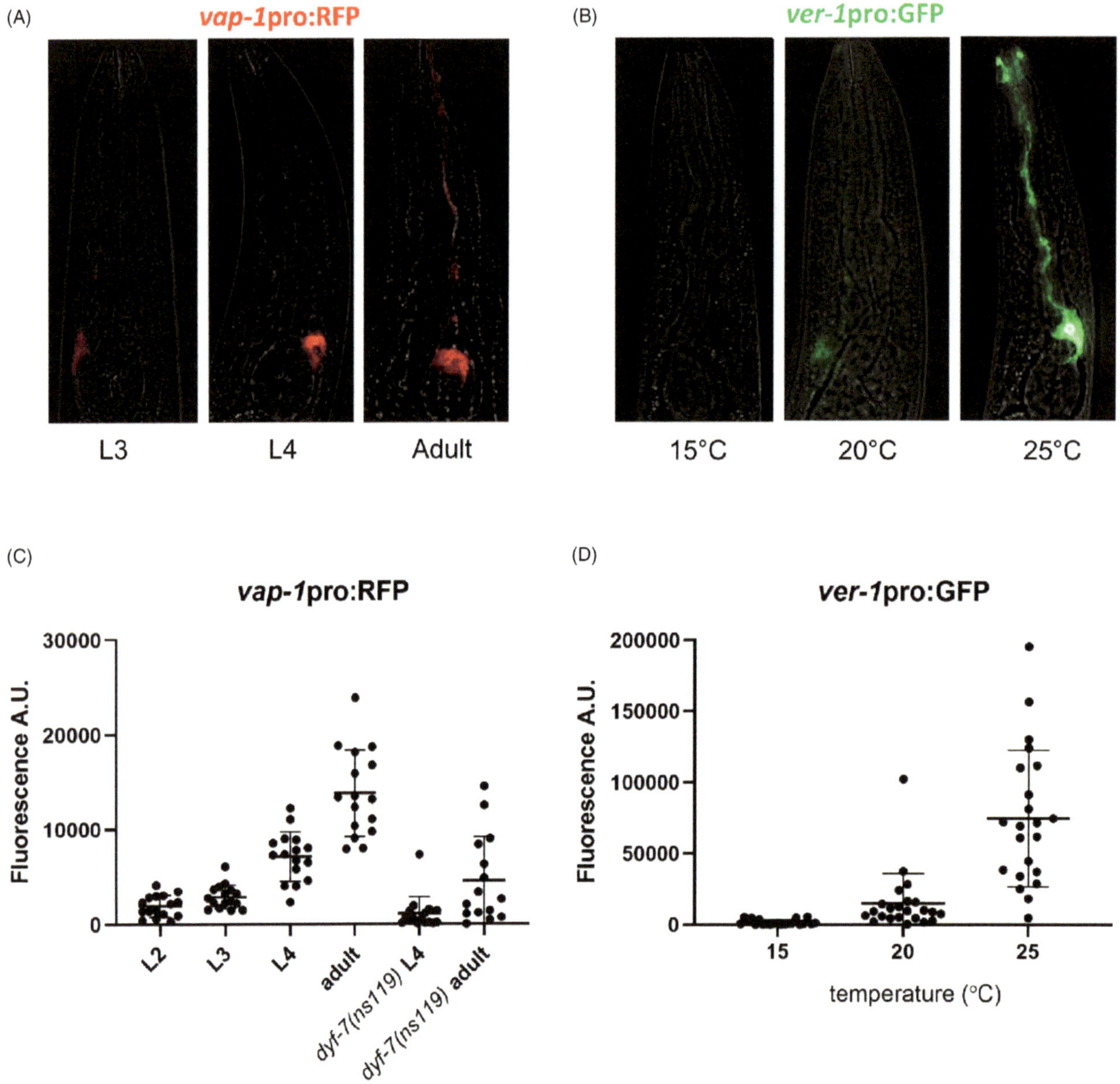

Figure 4. Plasticity in glial gene expression. (A) *vap-1*pro:RFP expression in AMsh glia in animals at the third larval (L3), fourth larval (L4), and 1-day adult life stages, showing developmental regulation of gene expression. (B) *ver-1*pro:GFP expression in AMsh glia in L4 animals grown at 15 °C, 20 °C, or 25 °C, showing thermoregulation of gene expression as previously described (Procko *et al.*, 2012, 2011). (C) Quantification of *vap-1*pro:RFP fluorescence in AMsh glia from second larval (L2) stage to 1-day adults and in *dyf-7(ns119)* L4 and adult animals, showing that sensory defects alter gene expression. $n = 15$–16 per group. $p < 0.05$, *dyf-7* L4 vs *dyf-7* adults; $p < 0.0001$, *dyf-7* adults vs wild-type adults; unpaired *t*-test with Welch's correction (D) Quantification of *ver-1*pro:GFP fluorescence in the AMsh glia in L4 animals from 15 °C to 25 °C. $n = 21$–23 per group. $p < 0.05$, 20 °C vs 15 °C; $p < 0.0001$, 20 °C vs 25 °C; Brown-Forsyth and Welch one-way ANOVA with Dunnett's multiple comparisons test.

neurons to the tip of the nose, and loss of function of this protein results in shortened dendrites that diminish the function of these sensory neurons (Heiman & Shaham, 2009; Low *et al.*, 2019). Loss of DYF-7 function due to the null allele *dyf-7(ns119)* causes severe defects in *vap-1* expression in AMsh glia, suggesting that input from the external environment may modulate AMsh gene expression (Figure 4(C)). Interestingly, while *vap-1*pro:RFP expression at the L4 stage is nearly lost in *dyf-7(ns119)* mutants, it is still upregulated to some degree in adults, albeit to a much lesser extent than in wild type (Figure 4(C)). These observations suggest that there may be multiple layers of plasticity in glial gene expression, both through developmental stage and sensory activity.

The expression of *ver-1* in AMsh glia has been previously reported to be modulated by temperature, increasing steadily from undetectable expression at 15 °C to very strong expression at 25 °C (Procko *et al.*, 2012, 2011). We were able to recapitulate these results (Figure 4(B,D)). Additionally, *ver-1* expression has also been shown to be induced by entry into dauer, a long-lived stress-resistant life stage, concomitant with remodeling of glia (Procko *et al.*, 2011).

Other examples of dynamic gene expression in glia include *hmit-1.2*, which is upregulated in AMsh glia in response to osmotic stress (Kage-Nakadai, Uehara, & Mitani, 2011), and *daf-6*, which is downregulated in AMsh glia during dauer (Moussaif & Sze, 2009). While these data provide evidence for glial plasticity, the functional significance of

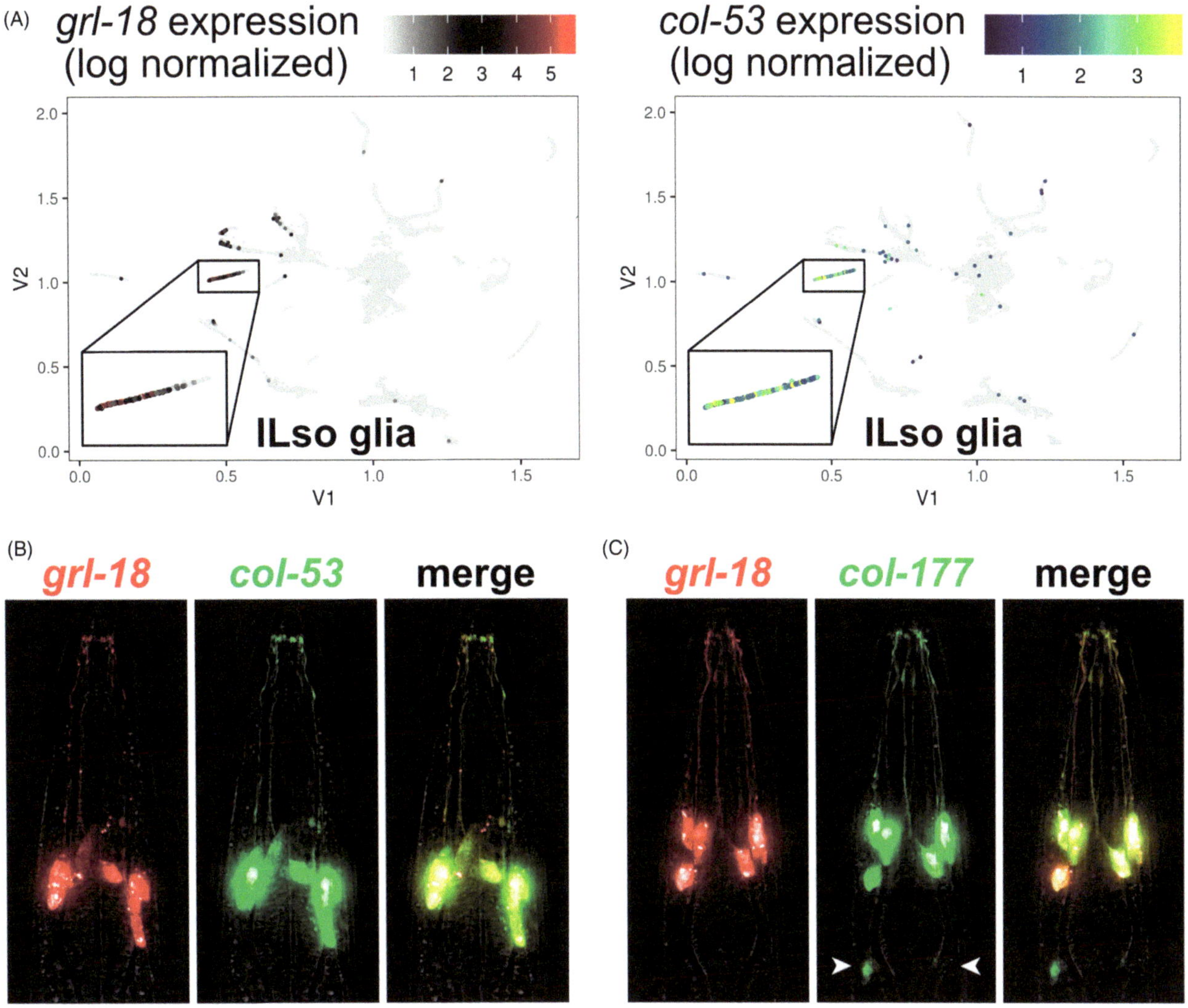

Figure 5. Prospective identification of novel cell-type-specific promoters for ILso glia. (A) Cell cluster plots illustrating the expression of *grl-18* and *col-53* in embryonic glia and excretory cells from Packer *et al.*, 2019. Each point represents an individual cell. The color indicates the relative expression level of each gene. For *grl-18*, low expression is black and high expression is red. For *col-53*, low is blue and high is green/yellow. Gray indicates no detected transcripts for the gene of interest. Inset box, the cell cluster predicted to include ILso glia. Cell cluster plot for *col-177* is in Supplementary Figure S1(A). (B, C) Merged brightfield and fluorescence images of an animal expressing *grl-18*pro:mApple together with (B) *col-53*pro:GFP or (C) *col-177*pro:GFP, showing that these markers are expressed in the same cells. *col-177* is also faintly expressed in cells with a glial morphology, tentatively identified as OLL socket glia in the head (arrowheads, see Supplementary Figure S1(B)) and ADE and PDE socket glia in the body. See online version for color figure.

these changes in gene expression for AMsh function and its interactions with neurons remain unclear. While all of these examples involve AMsh glia, it is likely that other glia may exhibit similar plasticity. Dynamic expression of promoters in other glia has yet to be identified and the full extent of glial plasticity in response to different environmental and state changes remains unknown.

Single-cell transcriptional profiling identifies new cell-type-specific promoters for glia

Discovery of glial cell-type-specific promoters has often relied on serendipity, and has been limited to study of one glial cell type at a time. The recent revolution in single-cell transcriptional profiling raises the possibility that promoters specific to each glial cell type can be identified systematically. Single-cell transcriptional profiling of embryos identified progenitor cells that give rise to eight major groups of tissues, including a group that contains glia and excretory cells. Packer *et al.* used existing markers from the literature to propose identities for clusters of cells corresponding to nine glial subtypes (Packer *et al.*, 2019).

In order to test whether this dataset could be used to prospectively identify novel cell-type-specific glial markers, we attempted to identify a novel marker for ILso glia. The ILso glia cluster is characterized by its enrichment in *grl-18* expression (Figure 5(A)). Using the VisCello data explorer (http://github.com/qinzhu/VisCello.celegans), we generated a list of additional genes that are highly enriched in the ILso glia cluster compared to the global dataset (see Methods). From the list, we selected the collagen genes, *col-53* and *col-177*, to characterize further. They are strong candidates to be markers for ILso glia, because they are highly expressed in the ILso glia cluster (Figure 5(A), Supplementary Figure S1(A)), like *grl-18*, and because socket glia secrete cuticle, which consists of collagen and other extracellular matrix proteins.

We created *col-53* and *col-177* reporters consisting of 733 bp or 2.75 kb promoter regions upstream of their respective translation start sites fused to GFP. We generated animals with a *col-53*pro:GFP or *col-177*pro:GFP extrachromosomal array and an integrated *grl-18pro*:mApple transgene. We found that *col-53* and *col-177* are each expressed in the same six glial cells as *grl-18* (Figure 5(B,C)), demonstrating that they are novel cell-type-specific promoters for ILso glia. For *col-177*, we found that it is also expressed in two additional cells in the head that sit posteriorly to the ILso glia and extend processes to the tip of the nose. Based on morphology, position, symmetry, and the association of OLL neuron sensory dendrites with their distal endings (Supplementary Figure S1(B)), we have tentatively identified these cells as the lateral OLL/Rso glia, for which there is currently no marker. We also observe faint expression in cells in the body whose morphology, position, and symmetry is consistent with the anterior and posterior deirid (ADE, PDE) socket glia.

Our findings provide proof-of-principle that single-cell transcript profiling enables prospective identification of novel cell-type-specific glial promoters. Indeed, we have also found that this approach can be used to identify promoters that are expressed in combinations of glia (*mam-5*, Supplementary Figure S2(A,B)) or that are likely to be expressed in other discrete glial types (AMso, Supplementary Figure S2(C)). Markers for the remaining unexplored glia will be essential for studies that systematically examine the roles of glia in the diverse sense organs of *C. elegans*. Furthermore, it will be extremely valuable to identify glial promoters that are expressed early in development or whose expression changes with environmental context, such as neuronal activity. Glial markers expressed in embryos will allow us to visualize and track the birth, migration, and morphogenesis of glia during nervous system assembly (Rapti *et al.*, 2017). Meanwhile, glial markers whose expression is regulated by the environment will provide insight on how glia adapt and potentially alter interactions with their associated neurons to drive appropriate behaviors. These rich datasets and tools will open up new avenues for discovery in the field of glial biology.

Acknowledgements

The authors thank Shai Shaham and members of the Shaham laboratory, Meera Sundaram, and John Murray for sharing information and advice on glial-specific promoters and for providing strains and plasmids, and John Murray for help using the VisCello data explorer. The authors thank members of the Heiman laboratory for unpublished work further characterizing candidate promoters, including Ian McLachlan (*itx-1*), Elizabeth Cebul (*grl-18, delm-1, delm-2*), and Karolina Mizeracka (*grl-2, lin-48*). Some strains were provided by the CGC, which is funded by National Institutes of Health Office of Research Infrastructure Programs (P40 OD010440).

Disclosure statement

No potential conflict of interest was reported by the author(s).

Funding

This work was supported by NIH [T32NS007473] to L.W. and NIH [R01NS112343] to M.G.H.

ORCID

Wendy Fung http://orcid.org/0000-0002-2446-8535
Leigh Wexler http://orcid.org/0000-0001-9616-5448
Maxwell G. Heiman http://orcid.org/0000-0002-2557-6490

References

Bacaj, T., Tevlin, M., Lu, Y., & Shaham, S. (2008). Glia are essential for sensory organ function in *C. elegans*. *Science (New York, N.Y.)*, *322* (5902), 744–747. doi:10.1126/science.1163074

Bargmann, C.I. (2006). Chemosensation in *C. elegans*. In The C. elegans Research Community (Ed.), *WormBook: The Online Review of C. elegans Biology* (1–29). doi:10.1895/wormbook.1.123.1

Bargmann, C.I., & Avery, L. (1995). Laser killing of cells in *Caenorhabditis elegans*. *Methods in Cell Biology*, *48*, 225–250. doi:10.1016/s0091-679x(08)61390-4

Brenner, S. (1974). The genetics of Caenorhabditis elegans. *Genetics*, *77* (1), 71–94

Braunreiter, K., Hamlin, S., & Lyman-Gingerich, J. (2014). Identification and characterization of a novel allele of *Caenorhabditis elegans* bbs-7. *PloS One*, *9* (12), e113737. doi:10.1371/journal.pone.0113737

Cebul, E.R., McLachlan, I.G., & Heiman, M.G. (2020). Dendrites with specialized glial attachments develop by retrograde extension using SAX-7 and GRDN-1. *Development (Cambridge, England)*, *147* (4), dev180448. doi:10.1242/dev.180448

Cianciulli, A., Yoslov, L., Buscemi, K., Sullivan, N., Vance, R.T., Janton, F., … Nelson, M.D. (2019). Interneurons regulate locomotion quiescence via cyclic adenosine monophosphate signaling during stress-induced sleep in *Caenorhabditis elegans*. *Genetics*, *213* (1), 267–279. doi:10.1534/genetics.119.302293

Colón-Ramos, D.A., Margeta, M.A., & Shen, K. (2007). Glia promote local synaptogenesis through UNC-6 (netrin) signaling in *C. elegans*. *Science (New York, N.Y.)*, *318* (5847), 103–106. doi:10.1126/science.1143762

Ding, G., Zou, W., Zhang, H., Xue, Y., Cai, Y., Huang, G., … Kang, L. (2015). In vivo tactile stimulation-evoked responses in *Caenorhabditis elegans* amphid sheath glia. *PloS One*, *10* (2), e0117114. doi:10.1371/journal.pone.0117114

Fenk, L.A., & de Bono, M. (2015). Environmental CO2 inhibits *Caenorhabditis elegans* egg-laying by modulating olfactory neurons and evokes widespread changes in neural activity. *Proceedings of the National Academy of Sciences of the United States of America*, *112* (27), E3525–3534. doi:10.1073/pnas.1423808112

Gibson, C.L., Balbona, J.T., Niedzwiecki, A., Rodriguez, P., Nguyen, K.C.Q., Hall, D.H., & Blakely, R.D. (2018). Glial loss of the metallo β-lactamase domain containing protein, SWIP-10, induces age- and glutamate-signaling dependent, dopamine neuron degeneration. *PLoS Genetics*, *14* (3), e1007269. doi:10.1371/journal.pgen.1007269

Gower, N.J., Temple, G.R., Schein, J.E., Marra, M., Walker, D.S., & Baylis, H.A. (2001). Dissection of the promoter region of the inositol 1,4,5-trisphosphate receptor gene, itr-1, in *C. elegans*: A molecular basis for cell-specific expression of IP3R isoforms. *Journal of Molecular Biology*, *306* (2), 145–157. doi:10.1006/jmbi.2000.4388

Grant, J., Matthewman, C., & Bianchi, L. (2015). A novel mechanism of pH buffering in *C. elegans* Glia: bicarbonate transport via the voltage-gated ClC Cl-channel CLH-1. *The Journal of Neuroscience : The Official Journal of the Society for Neuroscience*, *35* (50), 16377–16397. doi:10.1523/JNEUROSCI.3237-15.2015

Haklai-Topper, L., Soutschek, J., Sabanay, H., Scheel, J., Hobert, O., & Peles, E. (2011). The neurexin superfamily of *Caenorhabditis elegans*.

Gene Expression Patterns : GEP, 11 (1–2), 144–150. doi:10.1016/j.gep.2010.10.008

Han, L., Wang, Y., Sangaletti, R., D'Urso, G., Lu, Y., Shaham, S., & Bianchi, L. (2013). Two novel DEG/ENaC channel subunits expressed in glia are needed for nose-touch sensitivity in Caenorhabditis elegans. *The Journal of Neuroscience : The Official Journal of the Society for Neuroscience, 33* (3), 936–949. doi:10.1523/JNEUROSCI.2749-12.2013

Hao, L., Johnsen, R., Lauter, G., Baillie, D., & Bürglin, T.R. (2006). Comprehensive analysis of gene expression patterns of hedgehog-related genes. *BMC Genomics, 7*, 280. doi:10.1186/1471-2164-7-280

Hardaway, J.A., Sturgeon, S.M., Snarrenberg, C.L., Li, Z., Xu, X.Z.S., Bermingham, D.P., ... Blakely, R.D. (2015). Glial expression of the *Caenorhabditis elegans* gene swip-10 supports glutamate dependent control of extrasynaptic dopamine signaling. *The Journal of Neuroscience : The Official Journal of the Society for Neuroscience, 35* (25), 9409–9423. doi:10.1523/JNEUROSCI.0800-15.2015

Heiman, M.G., & Shaham, S. (2009). DEX-1 and DYF-7 establish sensory dendrite length by anchoring dendritic tips during cell migration. *Cell, 137* (2), 344–355. doi:10.1016/j.cell.2009.01.057

Hunt-Newbury, R., Viveiros, R., Johnsen, R., Mah, A., Anastas, D., Fang, L., ... Moerman, D.G. (2007). High-throughput in vivo analysis of gene expression in *Caenorhabditis elegans. PLoS Biology, 5* (9), e237. doi:10.1371/journal.pbio.0050237

Ji, T., Wang, K., Fan, J., Huang, J., Wang, M., Dong, X., ... Colón-Ramos, D.A. (2019). ADAMTS-family protease MIG-17 regulates synaptic allometry by modifying the extracellular matrix and modulating glia morphology during growth (preprint). *Neuroscience.* doi:10.1101/734830

Johnson, A.D., Fitzsimmons, D., Hagman, J., & Chamberlin, H.M. (2001). EGL-38 Pax regulates the ovo-related gene lin-48 during *Caenorhabditis elegans* organ development. *Development (Cambridge, England), 128*, 2857–2865.

Kage-Nakadai, E., Ohta, A., Ujisawa, T., Sun, S., Nishikawa, Y., Kuhara, A., & Mitani, S. (2016). *Caenorhabditis elegans* homologue of Prox1/Prospero is expressed in the glia and is required for sensory behavior and cold tolerance. *Genes Cells, 21* (9), 936–948. doi:10.1111/gtc.12394

Kage-Nakadai, E., Uehara, T., & Mitani, S. (2011). H+/myo-inositol transporter genes, hmit-1.1 and hmit-1.2, have roles in the osmoprotective response in *Caenorhabditis elegans. Biochemical and Biophysical Research Communications, 410* (3), 471–477. doi:10.1016/j.bbrc.2011.06.001

Katz, M., Corson, F., Iwanir, S., Biron, D., & Shaham, S. (2018). Glia modulate a neuronal circuit for locomotion suppression during sleep in *C. elegans. Cell Reports, 22* (10), 2575–2583. doi:10.1016/j.celrep.2018.02.036

Low, I.I.C., Williams, C.R., Chong, M.K., McLachlan, I.G., Wierbowski, B.M., Kolotuev, I., & Heiman, M.G. (2019). Morphogenesis of neurons and glia within an epithelium. *Development (Cambridge, England), 146* (4), dev171124. doi:10.1242/dev.171124

McMiller, T.L., & Johnson, C.M. (2005). Molecular characterization of HLH-17, a *C. elegans* bHLH protein required for normal larval development. *Gene, 356*, 1–10. doi:10.1016/j.gene.2005.05.003

McQuary, P.R., Liao, C.-Y., Chang, J.T., Kumsta, C., She, X., Davis, A., ... Hansen, M. (2016). *C. elegans* S6K mutants require a creatine-kinase-like effector for lifespan extension. *Cell Reports, 14* (9), 2059–2067. doi:10.1016/j.celrep.2016.02.012

Mizeracka, K., & Heiman, M.G. (2015). The many glia of a tiny nematode: studying glial diversity using *Caenorhabditis elegans. Wiley Interdisciplinary Reviews. Developmental Biology, 4* (2), 151–160. doi:10.1002/wdev.171

Mizeracka, K., Rogers, J.M., Shaham, S., Bulyk, M.L., & Heiman, M.G. (2019). Lineage-specific control of convergent cell identity by a Forkhead repressor (preprint). *Developmental Biology.* doi:10.1101/758508

Molina-García, L., Cook, S.J., Kim, B., Bonnington, R., Sammut, M., O'Shea, J., ... Poole, R.J. (2018). A direct glia-to-neuron natural transdifferentiation ensures nimble sensory-motor coordination of male mating behaviour (preprint). *Neuroscience.* doi:10.1101/285320

Moussaif, M., & Sze, J.Y. (2009). Intraflagellar transport/Hedgehog-related signaling components couple sensory cilium morphology and serotonin biosynthesis in *Caenorhabditis elegans. The Journal of Neuroscience : The Official Journal of the Society for Neuroscience, 29* (13), 4065–4075. doi:10.1523/JNEUROSCI.0044-09.2009

Nelson, F.K., & Riddle, D.L. (1984). Functional study of the *Caenorhabditis elegans* secretory-excretory system using laser microsurgery. *The Journal of Experimental Zoology, 231* (1), 45–56. doi:10.1002/jez.1402310107

Nonet, M.L., Staunton, J.E., Kilgard, M.P., Fergestad, T., Hartwieg, E., Horvitz, H.R., ... Meyer, B.J. (1997). *Caenorhabditis elegans rab-3* mutant synapses exhibit impaired function and are partially depleted of vesicles. *The Journal of Neuroscience : The Official Journal of the Society for Neuroscience, 17* (21), 8061–8073. doi:10.1523/JNEUROSCI.17-21-08061.1997

Ohkura, K., & Bürglin, T.R. (2011). Dye-filling of the amphid sheath glia: Implications for the functional relationship between sensory neurons and glia in *Caenorhabditis elegans. Biochemical and Biophysical Research Communications, 406* (2), 188–193. doi:10.1016/j.bbrc.2011.02.003

Oikonomou, G., Perens, E.A., Lu, Y., & Shaham, S. (2012). Some, but not all, retromer components promote morphogenesis of *C. elegans* sensory compartments. *Developmental Biology, 362* (1), 42–49. doi:10.1016/j.ydbio.2011.11.009

Oikonomou, G., Perens, E.A., Lu, Y., Watanabe, S., Jorgensen, E.M., & Shaham, S. (2011). Opposing activities of LIT-1/NLK and DAF-6/patched-related direct sensory compartment morphogenesis in *C. elegans. PLoS Biology, 9* (8), e1001121. doi:10.1371/journal.pbio.1001121

Packer, J.S., Zhu, Q., Huynh, C., Sivaramakrishnan, P., Preston, E., Dueck, H., ... Murray, J.I. (2019). A lineage-resolved molecular atlas of *C. elegans* embryogenesis at single-cell resolution. *Science, 365*(6459), eaax1971. doi:10.1126/science.aax1971

Perens, E.A., & Shaham, S. (2005). *C. elegans* daf-6 encodes a patched-related protein required for lumen formation. *Developmental Cell, 8* (6), 893–906. doi:10.1016/j.devcel.2005.03.009

Pierce, M.L., Weston, M.D., Fritzsch, B., Gabel, H.W., Ruvkun, G., & Soukup, G.A. (2008). MicroRNA-183 family conservation and ciliated neurosensory organ expression. *Evolution & Development, 10* (1), 106–113. doi:10.1111/j.1525-142X.2007.00217.x

Popovici, C., Isnardon, D., Birnbaum, D., & Roubin, R. (2002). *Caenorhabditis elegans* receptors related to mammalian vascular endothelial growth factor receptors are expressed in neural cells *Neuroscience Letters, 329* (1), 116–120. doi:10.1016/S0304-3940(02)00595-5

Procko, C., Lu, Y., & Shaham, S. (2011). Glia delimit shape changes of sensory neuron receptive endings in *C. elegans. Development (Cambridge, England), 138* (7), 1371–1381. doi:10.1242/dev.058305

Procko, C., Lu, Y., & Shaham, S. (2012). Sensory organ remodeling in *Caenorhabditis elegans* requires the zinc-finger protein ZTF-16. *Genetics, 190* (4), 1405–1415. doi:10.1534/genetics.111.137786

Rapti, G., Li, C., Shan, A., Lu, Y., & Shaham, S. (2017). Glia initiate brain assembly through noncanonical Chimaerin-Furin axon guidance in *C. elegans. Nature Neuroscience, 20* (10), 1350–1360. doi:10.1038/nn.4630

Sammut, M., Cook, S.J., Nguyen, K.C.Q., Felton, T., Hall, D.H., Emmons, S.W., ... Barrios, A. (2015). Glia-derived neurons are required for sex-specific learning in *C. elegans. Nature, 526* (7573), 385–390. doi:10.1038/nature15700

Schindelin, J., Arganda-Carreras, I., Frise, E., Kaynig, V., Longair, M., Pietzsch, T., ... Cardona, A. (2012). Fiji: an open-source platform for biological-image analysis. *Nature Methods, 9* (7), 676–682. doi:10.1038/nmeth.2019

Shao, Z., Watanabe, S., Christensen, R., Jorgensen, E.M., & Colón-Ramos, D.A. (2013). Synapse location during growth depends on glia location. *Cell, 154* (2), 337–350. doi:10.1016/j.cell.2013.06.028

Singhvi, A., Liu, B., Friedman, C.J., Fong, J., Lu, Y., Huang, X.-Y., & Shaham, S. (2016). A glial K/Cl transporter controls neuronal receptive ending shape by chloride inhibition of an rGC. *Cell, 165* (4), 936–948. doi:10.1016/j.cell.2016.03.026

Singhvi, A., & Shaham, S. (2019). Glia-neuron interactions in *Caenorhabditis elegans*. *Annual Review of Neuroscience*, *42*, 149–168. doi:10.1146/annurev-neuro-070918-050314

Stout, R.F., & Parpura, V. (2011). Voltage-gated calcium channel types in cultured *C. elegans* CEPsh glial cells. *Cell Calcium*, *50* (1), 98–108. doi:10.1016/j.ceca.2011.05.016

Stringham, E., Pujol, N., Vandekerckhove, J., & Bogaert, T. (2002). unc-53 controls longitudinal migration in *C. elegans*. *Development (Cambridge, England)*, *129*, 3367–3379.

Sundaram, M.V., & Buechner, M. (2016). The *Caenorhabditis elegans* excretory system: A model for tubulogenesis, cell fate specification, and plasticity. *Genetics*, *203* (1), 35–63. doi:10.1534/genetics.116.189357

Tanis, J.E., Bellemer, A., Moresco, J.J., Forbush, B., & Koelle, M.R. (2009). The potassium chloride cotransporter KCC-2 coordinates development of inhibitory neurotransmission and synapse structure in *Caenorhabditis elegans*. *The Journal of Neuroscience : The Official Journal of the Society for Neuroscience*, *29* (32), 9943–9954. doi:10.1523/JNEUROSCI.1989-09.2009

Timbers, T.A., Garland, S.J., Mohan, S., Flibotte, S., Edgley, M., Muncaster, Q., … Leroux, M.R. (2016). Accelerating gene discovery by phenotyping whole-genome sequenced multi-mutation strains and using the sequence kernel association test (SKAT). *PLoS Genetics*, *12* (8), e1006235. doi:10.1371/journal.pgen.1006235

Tucker, M., Sieber, M., Morphew, M., & Han, M. (2005). The *Caenorhabditis elegans* aristaless orthologue, alr-1, is required for maintaining the functional and structural integrity of the amphid sensory organs. *Molecular Biology of the Cell*, *16* (10), 4695–4704. doi:10.1091/mbc.e05-03-0205

Wallace, S.W., Singhvi, A., Liang, Y., Lu, Y., & Shaham, S. (2016). PROS-1/Prospero is a major regulator of the glia-specific secretome controlling sensory-neuron shape and function in *C. elegans*. *Cell Reports*, *15* (3), 550–562. doi:10.1016/j.celrep.2016.03.051

Wang, Y., Apicella, A., Lee, S.-K., Ezcurra, M., Slone, R.D., Goldmit, M., … Bianchi, L. (2008). A glial DEG/ENaC channel functions with neuronal channel DEG-1 to mediate specific sensory functions in *C. elegans*. *The EMBO Journal*, *27* (18), 2388–2399. doi:10.1038/emboj.2008.161

Wang, Y., D'Urso, G., & Bianchi, L. (2012). Knockout of glial channel ACD-1 exacerbates sensory deficits in a *C. elegans* mutant by regulating calcium levels of sensory neurons. *Journal of Neurophysiology*, *107* (1), 148–158. doi:10.1152/jn.00299.2011

Wang, W., Perens, E.A., Oikonomou, G., Wallace, S.W., Lu, Y., & Shaham, S. (2017). IGDB-2, an Ig/FNIII protein, binds the ion channel LGC-34 and controls sensory compartment morphogenesis in *C. elegans*. *Developmental Biology*, *430* (1), 105–112. doi:10.1016/j.ydbio.2017.08.009

Ward, S., Thomson, N., White, J.G., & Brenner, S. (1975). Electron microscopical reconstruction of the anterior sensory anatomy of the nematode *Caenorhabditis elegans*.?2UU. *The Journal of Comparative Neurology*, *160* (3), 313–337. doi:10.1002/cne.901600305

White, J.G., Southgate, E., Thomson, J.N., & Brenner, S. (1986). The structure of the nervous system of the nematode *Caenorhabditis elegans*. *Philosophical Transactions of the Royal Society of London. Series B, Biological Sciences*, *314* (1165), 1–340. doi:10.1098/rstb.1986.0056

Yang, W.-K., & Chien, C.-T. (2019). Beyond being innervated: the epidermis actively shapes sensory dendritic patterning. *Open Biology*, *9* (3), 180257. doi:10.1098/rsob.180257

Yin, J.-A., Gao, G., Liu, X.-J., Hao, Z.-Q., Li, K., Kang, X.-L., … Cai, S.-Q. (2017). Genetic variation in glia-neuron signalling modulates ageing rate. *Nature*, *551* (7679), 198–203. doi:10.1038/nature24463

Yip, Z.C., & Heiman, M.G. (2018). Ordered arrangement of dendrites within a *C. elegans* sensory nerve bundle. *eLife*, *7*, e35825. doi:10.7554/eLife.35825

Yoshida, A., Nakano, S., Suzuki, T., Ihara, K., Higashiyama, T., & Mori, I. (2016). A glial K(+) /Cl(-) cotransporter modifies temperature-evoked dynamics in *Caenorhabditis elegans* sensory neurons. *Genes, Brain, and Behavior*, *15* (4), 429–440. doi:10.1111/gbb.12260

Yoshimura, S., Murray, J.I., Lu, Y., Waterston, R.H., & Shaham, S. (2008). mls-2 and vab-3 Control glia development, hlh-17/Olig expression and glia-dependent neurite extension in *C. elegans*. *Development (Cambridge, England))*, *135*(13), 2263–2275. doi:10.1242/dev.019547

Part III

From inputs to outputs

C. elegans: a sensible model for sensory biology

Adam J. Iliff and X.Z. Shawn Xu

ABSTRACT

From Sydney Brenner's backyard to hundreds of labs across the globe, inspiring six Nobel Prize winners along the way, *Caenorhabditis elegans* research has come far in the past half century. The journey is not over. The virtues of *C. elegans* research are numerous and have been recounted extensively. Here, we focus on the remarkable progress made in sensory neurobiology research in *C. elegans*. This nematode continues to amaze researchers as we are still adding new discoveries to the already rich repertoire of sensory capabilities of this deceptively simple animal. Worms possess the sense of taste, smell, touch, light, temperature and proprioception, each of which is being studied in genetic, molecular, cellular and systems-level detail. This impressive organism can even detect less commonly recognized sensory cues such as magnetic fields and humidity.

Introduction

Sensory neurobiology exists at the interface between biology, chemistry and physics. Our sensory processes mediate our interaction with the physical world. It is incredible that life has evolved to detect a vast array of forces and molecules present in our universe. Organisms as seemingly disparate as nematodes and humans have much in common in regard to their sensory processes. The relationship of an organism with its external and internal environments arguably begins with the sensory inputs and ends with behavioral output. Perception occurs *via* sensory neurons that activate in response to specific stimuli. These cues act upon sensory transduction machinery expressed by the sensory neuron itself or in specialized structures that communicate with the sensory neuron. Some of these sensory structures have evolved into large complex organs, such as the mammalian eye. However, such complexity incorporating large numbers of cells is not required for a sophisticated sensory system. Even a tiny one millimeter long organism with a compact nervous system of only 302 neurons can detect a surprisingly vast and varied array of physical stimuli, such as mechanical forces, chemicals, light, temperature, humidity and electromagnetic fields (Figure 1). The evolution of these sensory modalities confers numerous benefits to survival, including the ability to find food and mates and to avoid hazard.

Mechanosensation

Caenorhabditis elegans research has proven indispensable in determining the transduction machinery underlying mechanical sensory modalities, such as touch and proprioception. For instance, the founding members of the mechanosensitive DEG/ENaC ion channel family were identified and characterized using genetic approaches with touch behavioral assays in *C. elegans*, which facilitated identification of vertebrate homologs (Driscoll & Chalfie, 1991; Goodman, 2006). The mechanosensitive TRPN/NOMPC channel TRP-4 transduces touch in CEP/ADE/PDE neurons and mediates stretch-triggered proprioception in the DVA neuron; this mechanosensitive channel is conserved between worms and other invertebrates and low vertebrates (Kang, Gao, Schafer, Xie, & Xu, 2010; Li, Feng, Sternberg, & Xu, 2006). In addition to DVA, some other neurons, such as ventral cord motor neurons, also mediate proprioception, but the underlying mechanosensitive channels remain to be identified (Wen *et al.*, 2012). As most molecules involved in sensory transduction are evolutionarily conserved (Hobert, 2013), identifying sensory receptors, transduction channels and related molecules involved in a mechanosensory process in *C. elegans* often enhances our molecular understanding of mechanosensation in more complex animals.

Chemosensation

Beyond mechanical stimuli, *C. elegans* also tastes and smells *via* multiple chemoreceptor families. Worm chemosensation has been studied extensively and the neural circuits and receptors responsible have been described in exquisite detail (Bargmann, 2006). Bargmann and colleagues led the identification of the large chemoreceptor family of G-protein couple receptors (GPCRs) with the discovery of the GPCR ODR-10 and its odorant ligand diacetyl (Sengupta, Chou, & Bargmann, 1996; Troemel, Chou, Dwyer, Colbert, &

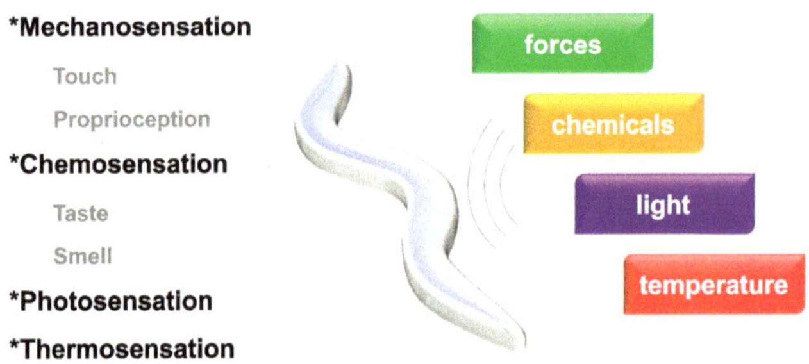

Figure 1. The primary sensory modalities in *Caenorhabditis elegans*. Worms possess a rich repertoire of sensory modalities, sensing mechanical forces, chemicals, light and temperature. In addition, worms can detect electric and magnetic fields and humility.

Bargmann, 1995). Of course, mammalian olfactory receptors are also known to be GPCRs (Buck & Axel, 1991). In addition, the transduction pathway is G protein signaling coupled to either cyclic nucleotide-gated (CNG) channels (TAX-4/TAX-2) or TRP channels (OSM-9/OCR-2 proteins). Both of these transduction mechanisms are also found in mammalian olfactory neurons. In addition to odorants and tastants, we now know that worms can detect a wide range of chemicals, including the physiologically relevant gases oxygen and carbon dioxide *via* a separate set of sensory neurons with guanylate cyclases as molecular sensors (Bargmann, 2006; Bretscher, Busch, & de Bono, 2008; Hallem & Sternberg, 2008). Worm sensory neurons also confer sensitivity to the pH and osmolarity of solutions (Bargmann, 2006; Wang, Li, Liu, Liu, & Xu, 2016), allowing worms to avoid harmful environments.

Photosensation

Surprisingly, these eyeless animals are not blind! That is, they can detect light, particularly the short wavelengths of light, in ways that are only now becoming understood. *C. elegans* exhibits escape behaviors in response to blue, violet and ultraviolet light stimulation (Edwards *et al.*, 2008; Liu *et al.*, 2010; Ward, Liu, Feng, & Xu, 2008). Detecting and responding to this potentially damaging light could promote survival. Transparent body permits light to reach internal photoreceptor cells. Light stimulation of photoreceptor neurons triggers escape responses. The protein responsible for photosensation in photoreceptor cells is unlike any found elsewhere in nature. That receptor (named LITE-1), which is a member of the invertebrate taste receptor family, was only recently classified as a *bona fide* photoreceptor, exhibiting a photon-capturing efficiency more than 10-fold greater than known photoreceptors such as rhodopsin (Gong *et al.*, 2016). LITE-1 thus defines a new class of photoreceptors. In addition to sensing light, worms can also sense light-generated chemicals, which can regulate feeding behavior (Bhatla & Horvitz, 2015).

Thermosensation

The ability to detect both heat and cold is well-established in *C. elegans*. Worms develop a preference for the temperature in which they are cultivated (Goodman, 2014; Mori & Ohshima, 1995). When exposed to a gradient of surface temperatures, worms crawl to the area with the same temperature in which they were cultivated (termed thermotaxis). This feat demonstrates that *C. elegans* can detect (and remember!) temperature. The TAX-2/TAX-4 CNG channels involved in chemosensation are also required for thermosensory transduction. AFD neurons are the major thermosensory neurons in *C. elegans* and express these CNG channels, as well as the three membrane receptor guanylate cyclases GCY-8, GCY-18 and GCY-23 that directly sense heat (Inada *et al.*, 2006). In addition, worms display noxious heat avoidance behavior that depends on several heat-sensitive neurons, such as AFD and FLP (Liu, Schulze, & Baumeister, 2012; Saro *et al.*, 2020). In addition to sensing heat, worms can respond to cooling. The TRP family channel TRPA-1 functions as a cold-sensitive channel in *C. elegans* and promotes cold-dependent longevity that is mediated by the intestine and IL1 neurons, both of which are cold-sensitive (Chatzigeorgiou *et al.*, 2010; Xiao *et al.*, 2013; Zhang *et al.*, 2018). Recently, a new type of cold-sensing receptor was discovered in *C. elegans* (Gong *et al.*, 2019). The glutamate receptor protein GLR-3 and its mammalian homolog GluK2 sense noxious cold and may be the long-sought noxious cold sensor (Gong *et al.*, 2019). The salt-sensing neuron ASER expresses the *glr-3* gene and requires GLR-3 for cold-sensing, indicating that ASER is also a cold-sensitive neuron.

Magnetosensation

C. elegans sensory neuroscience has moved beyond the five basic Aristotelian senses and generated some astounding discoveries. For instance, *C. elegans* can detect both components of the electromagnetic field. In the presence of an applied electric field, worms will orient their crawling toward the negative pole (Gabel *et al.*, 2007). More recently, Vidal-Gadea *et al.* (2015) showed that these worms are sensitive to the geomagnetic field and use this capability to orient their burrowing trajectory. Artificially reversing the direction of the Earth's magnetic field caused nematodes to burrow in the opposite direction. Worms might integrate geomagnetic information in order to promote navigation to favorable environments. While this behavior suggests nematodes possess magnetosensation, this ability in humans is not yet resolved.

Sensory integration

We have illustrated that *C. elegans* can sense many stimuli. How does its nervous system process all this information in order to behave properly? The neuronal connectivity of the entire nervous has been known for decades (White, Southgate, Thomson, & Brenner, 1986), but the dynamics of the embedded functional neural circuits have only begun to be elucidated. Downstream of sensory neurons, interneurons integrate the activity of multiple sensory neurons and coordinate a behavioral response. A number of studies have combined aversive and attractive stimuli to determine how behavioral responses are coordinated. That is, how the worm makes decisions (reviewed in Ghosh, Nitabach, Zhang, & Harris, 2017). Notably, such decisions are also subject to modulation by internal states, for example, nutritional status (Ghosh *et al.*, 2016). In one interesting case, information from mechanosensory neurons and thermosensory neurons combine to confer an emergent sensory modality, the ability to detect humidity (Russell, Vidal-Gadea, Makay, Lanam, & Pierce-Shimomura, 2014).

Concluding remarks

Our sensory systems bring the world out there into the world within. New discoveries of how worms sense these myriad stimuli are helping us understand our own sensory processes. The integration of multiple types of sensory information likely confers important advantages to survival and reproduction. Despite its simplicity, recent studies reveal that *C. elegans* can detect more of this universe than previously thought. It is even possible that worms can detect gravitational forces (Chen, Ko, Chuang, Bau, & Raizen, 2019). We invite the reader to ponder what else this amazing creature can do!

Acknowledgments

The authors thank Elizabeth Ronan for comments. Research in the Xu lab is supported by grants from the NIH (to X.Z.S.X).

Disclosure statement

No potential conflict of interest was by reported the authors.

References

Bargmann, C. (2006). Chemosensation in C. elegans. *WormBook*. Retrieved from doi:10.1895/wormbook.1.123.1

Bhatla, N., & Horvitz, H.R. (2015). Light and hydrogen peroxide inhibit C. elegans feeding through gustatory receptor orthologs and pharyngeal neurons. *Neuron*, *85*(4), 804–818. Retrieved from doi:10.1016/j.neuron.2014.12.061

Bretscher, A.J., Busch, K.E., & de Bono, M. (2008). A carbon dioxide avoidance behavior is integrated with responses to ambient oxygen and food in Caenorhabditis elegans. *Proceedings of the National Academy of Sciences*, *105*(23), 8044–8049. Retrieved from doi:10.1073/pnas.0707607105

Buck, L., & Axel, R. (1991). A novel multigene family may encode odorant receptors: A molecular basis for odor recognition. *Cell*, *65*(1), 175–187. Retrieved from doi:10.1016/0092-8674(91)90418-X

Chatzigeorgiou, M., Yoo, S., Watson, J.D., Lee, W.H., Spencer, W.C., Kindt, K.S., … Schafer, W.R. (2010). Specific roles for DEG/ENaC and TRP channels in touch and thermosensation in C. elegans nociceptors. *Nature Neuroscience*, *13*(7), 861–868. Retrieved from doi:10.1038/nn.2581

Chen, W.L., Ko, H., Chuang, H.S., Bau, H.H., & Raizen, D. (2019). Caenorhabditis elegans exhibits positive gravitaxis [Preprint]. *BioRxiv*, Retrieved from doi:10.1101/658229

Driscoll, M., & Chalfie, M. (1991). The mec-4 gene is a member of a family of Caenorhabditis elegans genes that can mutate to induce neuronal degeneration. *Nature*, *349*(6310), 588–593. doi:10.1038/349588a0

Edwards, S.L., Charlie, N.K., Milfort, M.C., Brown, B.S., Gravlin, C.N., Knecht, J.E., & Miller, K.G. (2008). A novel molecular solution for ultraviolet light detection in Caenorhabditis elegans. *PLoS Biology*, *6*(8), e198. Retrieved from doi:10.1371/journal.pbio.0060198

Gabel, C.V., Gabel, H., Pavlichin, D., Kao, A., Clark, D.A., & Samuel, A.D.T. (2007). Neural circuits mediate electrosensory behavior in Caenorhabditis elegans. *The Journal of Neuroscience*, *27*(28), 7586–7596. Retrieved from doi:10.1523/JNEUROSCI.0775-07.2007

Ghosh, D.D., Nitabach, M.N., Zhang, Y., & Harris, G. (2017). Multisensory integration in C. elegans. *Current Opinion in Neurobiology*, *43*, 110–118. Retrieved from doi:10.1016/j.conb.2017.01.005

Ghosh, D.D., Sanders, T., Hong, S., McCurdy, L.Y., Chase, D.L., Cohen, N., … Nitabach, M.N. (2016). Neural architecture of hunger-dependent multisensory decision making in C. elegans. *Neuron*, *92*(5), 1049–1062. Retrieved from doi:10.1016/j.neuron.2016.10.030

Gong, J., Liu, J., Ronan, E.A., He, F., Cai, W., Fatima, M., … Xu, X.Z.S. (2019). A cold-sensing receptor encoded by a glutamate receptor gene. *Cell*, *178*(6), 1375–1386.e11. Retrieved from doi:10.1016/j.cell.2019.07.034

Gong, J., Yuan, Y., Ward, A., Kang, L., Zhang, B., Wu, Z., … Xu, X.Z.S. (2016). The C. elegans taste receptor homolog LITE-1 is a photoreceptor. *Cell*, *167*(5), 1252–1263.e10. Retrieved from doi:10.1016/j.cell.2016.10.053

Goodman, M. (2006). Mechanosensation. *WormBook*. Retrieved from doi:10.1895/wormbook.1.62.1

Goodman, M.B. (2014). Thermotaxis navigation behavior. *WormBook*, 1–10. Retrieved from doi:10.1895/wormbook.1.168.1

Hallem, E.A., & Sternberg, P.W. (2008). Acute carbon dioxide avoidance in Caenorhabditis elegans. *Proceedings of the National Academy of Sciences of the United States of America*, *105*(23), 8038–8043. Retrieved from doi:10.1073/pnas.0707469105

Hobert, O. (2013). The neuronal genome of Caenorhabditis elegans. *WormBook*, 1–106. Retrieved from doi:10.1895/wormbook.1.161.1

Inada, H., Ito, H., Satterlee, J., Sengupta, P., Matsumoto, K., & Mori, I. (2006). Identification of guanylyl cyclases that function in thermosensory neurons of Caenorhabditis elegans. *Genetics*, *172*(4), 2239–2252. Retrieved from doi:10.1534/genetics.105.050013

Kang, L., Gao, J., Schafer, W.R., Xie, Z., & Xu, X.Z.S. (2010). C. elegans trp family protein trp-4 is a pore-forming subunit of a native mechanotransduction channel. *Neuron*, *67*(3), 381–391. Retrieved from doi:10.1016/j.neuron.2010.06.032

Li, W., Feng, Z., Sternberg, P.W., & Xu, X.Z.S. (2006). A C. elegans stretch receptor neuron revealed by a mechanosensitive TRP channel homologue. *Nature*, *440*(7084), 684–687. Retrieved from doi:10.1038/nature04538

Liu, J., Ward, A., Gao, J., Dong, Y., Nishio, N., Inada, H., … Xu, X.Z.S. (2010). C. elegans phototransduction requires a G protein-dependent cGMP pathway and a taste receptor homolog. *Nature Neuroscience*, *13*(6), 715–722. Retrieved from doi:10.1038/nn.2540

Liu, S., Schulze, E., & Baumeister, R. (2012). Temperature- and touch-sensitive neurons couple cng and trpv channel activities to control heat avoidance in Caenorhabditis elegans. *PLoS One*, *7*(3), e32360. Retrieved from doi:10.1371/journal.pone.0032360

Mori, I., & Ohshima, Y. (1995). Neural regulation of thermotaxis in Caenorhabditis elegans. *Nature*, *376*(6538), 344–348. Retrieved from doi:10.1038/376344a0

Russell, J., Vidal-Gadea, A.G., Makay, A., Lanam, C., & Pierce-Shimomura, J.T. (2014). Humidity sensation requires both mechanosensory and thermosensory pathways in Caenorhabditis elegans. *Proceedings of the National Academy of Sciences of the United States of America*, *111*(22), 8269–8274. Retrieved from doi:10.1073/pnas.1322512111

Saro, G., Lia, A.S., Thapliyal, S., Marques, F., Busch, K.E., & Glauser, D.A. (2020). Specific ion channels control sensory gain, sensitivity, and kinetics in a tonic thermonociceptor. *Cell Reports*, *30*(2), 397–408.e4. Retrieved from doi:10.1016/j.celrep.2019.12.029

Sengupta, P., Chou, J.H., & Bargmann, C.I. (1996). Odr-10 encodes a seven transmembrane domain olfactory receptor required for responses to the odorant diacetyl. *Cell*, *84*(6), 899–909. Retrieved from doi:10.1016/S0092-8674(00)81068-5

Troemel, E.R., Chou, J.H., Dwyer, N.D., Colbert, H.A., & Bargmann, C.I. (1995). Divergent seven transmembrane receptors are candidate chemosensory receptors in C. elegans. *Cell*, *83*(2), 207–218. Retrieved from doi:10.1016/0092-8674(95)90162-0

Vidal-Gadea, A., Ward, K., Beron, C., Ghorashian, N., Gokce, S., Russell, J., … Pierce-Shimomura, J. (2015). Magnetosensitive neurons mediate geomagnetic orientation in Caenorhabditis elegans. *eLife*, *4*, e07493. Retrieved from doi:10.7554/eLife.07493

Wang, X., Li, G., Liu, J., Liu, J., & Xu, X.Z.S. (2016). Tmc-1 mediates alkaline sensation in C. elegans through Nociceptive Neurons. *Neuron*, *91*(1), 146–154. Retrieved from doi:10.1016/j.neuron.2016.05.023

Ward, A., Liu, J., Feng, Z., & Xu, X.Z.S. (2008). Light-sensitive neurons and channels mediate phototaxis in C. elegans. *Nature Neuroscience*, *11*(8), 916–922. Retrieved from doi:10.1038/nn.2155

Wen, Q., Po, M.D., Hulme, E., Chen, S., Liu, X., Kwok, S.W., … Samuel, A.D.T. (2012). Proprioceptive coupling within motor neurons drives C. elegans forward locomotion. *Neuron*, *76*(4), 750–761. Retrieved from doi:10.1016/j.neuron.2012.08.039

White, J.G., Southgate, E., Thomson, J.N., & Brenner, S. (1986). The structure of the nervous system of the nematode Caenorhabditis elegans. *Philosophical Transactions of the Royal Society of London. Series B, Biological Sciences*, *314*(1165), 1–340. Retrieved from doi:10.1098/rstb.1986.0056

Xiao, R., Zhang, B., Dong, Y., Gong, J., Xu, T., Liu, J., & Xu, X.Z.S. (2013). A genetic program promotes C. elegans longevity at cold temperatures via a thermosensitive trp channel. *Cell*, *152*(4), 806–817. Retrieved from doi:10.1016/j.cell.2013.01.020

Zhang, B., Gong, J., Zhang, W., Xiao, R., Liu, J., & Xu, X.Z.S. (2018). Brain-gut communications via distinct neuroendocrine signals bidirectionally regulate longevity in C. elegans. *Genes & Development*, *32*(3–4), 258–270. Retrieved from doi:10.1101/gad.309625.117

Temperature signaling underlying thermotaxis and cold tolerance in *Caenorhabditis elegans*

Asuka Takeishi, Natsune Takagaki and Atsushi Kuhara

ABSTRACT

Caenorhabditis elegans has a simple nervous system of 302 neurons. It however senses environmental cues incredibly precisely and produces various behaviors by processing information in the neural circuit. In addition to classical genetic analysis, fluorescent proteins and calcium indicators enable *in vivo* monitoring of protein dynamics and neural activity on either fixed or free-moving worms. These analyses have provided the detailed molecular mechanisms of neuronal and systemic signaling that regulate worm responses. Here, we focus on responses of *C. elegans* against temperature and review key findings that regulate thermotaxis and cold tolerance. Thermotaxis of *C. elegans* has been studied extensively for almost 50 years, and cold tolerance is a relatively recent concept in *C. elegans*. Although both thermotaxis and cold tolerance require temperature sensation, the responsible neurons and molecular pathways are different, and *C. elegans* uses the proper mechanisms depending on its situation. We summarize the molecular mechanisms of the major thermosensory circuit as well as the modulatory strategy through neural and tissue communication that enables fine tuning of thermotaxis and cold tolerance.

Among various environmental cues, temperature is an unavoidable stimulus which animals are constantly exposed to. Each animal has an acceptable environmental temperature range that matches their living conditions. For instance, parasitic worms prefer hosts' body temperature, such as the human parasite *A. ceylanicum* which prefers around 38 °C (Bryant *et al.*, 2018), and some species have a tolerance for extreme temperature, such as *Diamesa kohshimai* that lives at the North Pole and survives even at 16°C (Kohshima, 1984). The environmental temperature is particularly important for ectotherms as it largely regulates their body temperature as well as affects all internal biochemical reactions. Animals thus have evolved behavioral strategies to seek an appropriate environmental temperature. Thermotaxis is the initial behavior choice for most animals to relocate themselves to the appropriate temperature circumstances when they encounter an unpreferable temperature environment. Species-specific thermotaxis mechanisms have been described (Garrity, Goodman, Samuel, & Sengupta, 2010; Glauser & Goodman, 2016; Hoffstaetter, Bagriantsev, & Gracheva, 2018); however, the mechanism of thermosensation and signal transduction in the neural network is not fully understood.

Caenorhabditis elegans has extremely sensitive and sophisticated thermosensory mechanisms and shows robust thermotaxis and temperature tolerant. It has thus been used as an attractive model system to reveal the neuronal and molecular basis of thermosensory signaling, and it was found that signaling transduction mechanisms are similar to phototransduction in vertebrates. In this review, we will discuss the mechanisms of temperature sensation, signal transduction and behavior determination that were revealed in *C. elegans*.

Thermotaxis behaviors in *C. elegans*

Animals migrate toward their preferable temperature on temperature gradient. Unlike mammals and other organisms that prefer a specific temperature, *C. elegans* prefer the past cultivation temperature (T_c) that they were grown in for several hours (Hedgecock & Russell, 1975; Mori & Ohshima, 1995). The behavior strategy of *C. elegans* is rather simple; they move forward when they are facing attractive cues, and worms change their run direction by making turns when they meet repellent cues (Albrecht & Bargmann, 2011; Pierce-Shimomura, Morse, & Lockery, 1999; Ryu & Samuel, 2002; Zariwala, Miller, Faumont, & Lockery, 2003). On the spatial temperature gradient around T_c (within 2°C), reversal and turning behaviors are suppressed and worms continue to run forward by tracking the isothermal gradient

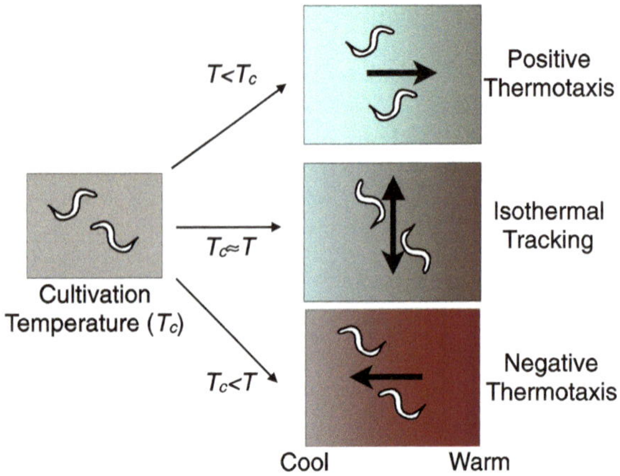

Figure 1. Thermotaxis behavior of *C. elegans* on the temperature gradient within the physiological temperature range. Worms memorize cultivation temperature (T_c) and migrate toward T_c on the temperature gradient. Negative thermotaxis (migration toward cooler side) and positive thermotaxis (migration toward warmer side) are observed when worms are placed warmer and cooler temperature gradient than T_c. They track isotherms around T_c.

(Isothermal tracking) (Figure 1) (Hedgecock & Russell, 1975; Luo, Clark, Biron, Mahadevan, & Samuel, 2006; Mori & Ohshima, 1995). When worms are at a warmer temperature gradient than T_c, they migrate toward cooler temperatures (negative thermotaxis) (Figure 1) (Clark, Gabel, Lee, & Samuel, 2007; Luo *et al.*, 2014; Ramot, MacInnis, Lee, & Goodman, 2008; Ryu & Samuel, 2002), and vice versa to migrate to the warmer side on a cooler temperature gradient than T_c (positive thermotaxis) (Figure 1) (Jurado, Kodama, Tanizawa, & Mori, 2010; Luo *et al.*, 2014; Mori & Ohshima, 1995; Ramot, MacInnis, Lee, *et al.*, 2008). During negative and positive thermotaxis, worms take a biased random walk: they reorient to a preferable direction by turning (Ramot, MacInnis, Lee, *et al.*, 2008). During negative thermotaxis, in addition to a biased turning direction, they show longer forward movement toward the cooler side and increase their turning when they are running toward the warmer side (Clark *et al.*, 2007; Luo *et al.*, 2014; Ramot, MacInnis, Lee, *et al.*, 2008; Ryu & Samuel, 2002).

Thermotaxis behavior depends on temperature stimuli, such as the shallowness of gradient and temperature range (Jurado *et al.*, 2010; Ramot, MacInnis, Lee, *et al.*, 2008), and the innate condition of animals: starved animals do not show negative thermotaxis (Chi *et al.*, 2007; Hedgecock & Russell, 1975; Kodama *et al.*, 2006; Mohri *et al.*, 2005; Ramot, MacInnis, Lee, *et al.*, 2008; Yamada & Ohshima, 2003). Thermotaxis behavior of worms is thus dynamic but also plastic.

Thermosensory neurons that sense physiological temperature

Worms are extremely sensitive to temperature change and can detect a temperature gradient as shallow as 0.01 °C/cm (Luo *et al.*, 2006). This precise thermo sensing is mainly mediated by a pair of sensory neurons, AFD neurons, together with other sensory neurons, AWC and ASI (Beverly, Anbil, & Sengupta, 2011; Kuhara *et al.*, 2008; Mori & Ohshima, 1995). Genetic or laser ablation of AFD suppresses all thermotaxis behaviors (Mori & Ohshima, 1995). *In vivo* calcium imaging was performed on AFD with genetically encoded calcium indicators and the results show AFD responds to temperature stimuli above temperature threshold (T^*) in a manner that depends on T_c (Clark, Biron, Sengupta, & Samuel, 2006; Kimura, Miyawaki, Matsumoto, & Mori, 2004). In the temperature range that AFD responds, calcium concentration in AFD raises in response to warming stimuli and drops in response to cooling (Clark *et al.*, 2006; Kimura *et al.*, 2004; Wasserman, Beverly, Bell, & Sengupta, 2011). These AFD responses are observed both in immobilized and free-moving worms, and phasic activity of AFD correlates with the temperature fluctuations caused by head movements on the thermal gradient plate (Clark *et al.*, 2006; Tsukada *et al.*, 2016; Venkatachalam *et al.*, 2016). Patch clamp experiments also show warming increases a non-selective cation current of AFD that results in depolarization, and cooling vice versa decreases cation current to induce hyperpolarization (Ramot, MacInnis, Goodman, 2008). Primary culture of isolated AFD neurons also shows temperature-dependent calcium responses (Kobayashi *et al.*, 2016), which indicates that the response of AFD is driven by cell autonomous mechanisms.

AWC and ASI, known as odor sensing neurons (Bargmann, Hartwieg, & Horvitz, 1993; Bargmann & Horvitz, 1991), also contribute to thermotaxis behaviors under specific conditions (Beverly *et al.*, 2011; Biron, Wasserman, Thomas, Samuel, & Sengupta, 2008; Kuhara *et al.*, 2008). T_c dependent activation of AWC and ASI is observed against warming stimuli; however, their responses are stochastic and unsimilar to phase-locked AFD responses (Beverly *et al.*, 2011; Biron *et al.*, 2008; Kuhara *et al.*, 2008). Consistent with the role of AWC or ASI activity in odor sensing, AWC and ASI induce or suppress turning, respectively (Beverly *et al.*, 2011; Biron *et al.*, 2008; Gray, Hill, & Bargmann, 2005). Thus, ablation of AWC results in prolonged forward movement during thermotaxis (Biron *et al.*, 2008). This effect of AWC or ASI ablation is minor on thermotaxis navigation (Beverly *et al.*, 2011; Biron *et al.*, 2008; Kuhara *et al.*, 2008) suggesting that their roles to mediate frequency of reorientation are modulatory.

Thermotransduction of AFD

As thermotaxis behavior is an attractive model to study mechanisms of thermosensation, neuronal plasticity and adaptation, the thermosensory mechanisms of AFD have been studied extensively. The molecular mechanism of thermotransduction is similar to mammalian photoreceptor cells that use the cyclic GMP mediated signaling pathway. AFD-specific receptor type guanylyl cyclases (AFD-rGCs), GCY-8, GCY-18 and GCY-23, that produce cGMP from GTP are essential for thermo transduction of AFD (Inada *et al.*, 2006; Yu, Avery, Baude, & Garbers, 1997) together with cGMP gated channels, TAX-2/TAX-4 (Figure 2) (Dusenbery, Sheridan, & Russell, 1975; Hedgecock & Russell, 1975; Mori & Ohshima, 1995). Mutants that lack all three AFD-rGCs or TAX-2/TAX-4 do not show the temperature-dependent

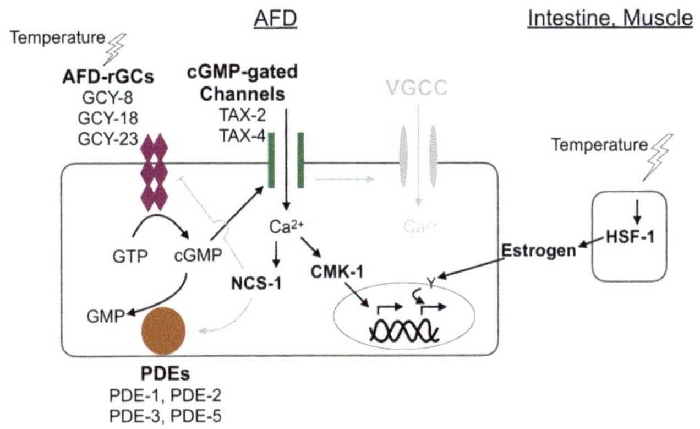

Figure 2. Thermosensory transduction of AFD. Temperature is sensed by AFD specific guanylyl cyclases (AFD-rGCs); GCY-8, GCY-18 and GCY-23. AFD-rGCs convert GTP to cGMP, which then opens cGMP-gated channels. Calcium influx induces CMK-1 localization change into nucleus which drives gene expression. cGMP is degraded by phosphodiesterases (PDEs). The involvement of voltage-gated calcium channels (VGCCs) and the feedback mechanism of NCS-1 are still unclear. Systemic estrogen signaling also mediates gene expression in AFD to regulate thermotaxis.

calcium response of AFD nor thermotaxis behaviors (Inada *et al.*, 2006; Kimura *et al.*, 2004). Single or double mutants of AFD-rGCs also show defects in thermotaxis behavior and calcium responses; however, the effect is subtle compared to the triple mutant, suggesting partial redundancy of these rGCs (Inada *et al.*, 2006; Takeishi *et al.*, 2016; Wang, O'Halloran, & Goodman, 2013; Wasserman *et al.*, 2011). Worm neurons express voltage-gated calcium channels but their contribution to thermosensation in AFD is unknown.

The family of transient receptor potential (TRP) channels are known as common thermo receptors in mammals and other animals (Clapham & Miller, 2011; Glauser & Goodman, 2016; Hoffstaetter *et al.*, 2018; Palkar, Lippoldt, & McKemy, 2015; Patapoutian, Peier, Story, & Viswanath, 2003; Voets, 2014). However, AFD does not express them and it has long been a mystery what functions as a thermosensory protein in AFD. As TAX-2/TAX-4 are expressed broadly even in non-thermosensory neurons, they are unlikely to be thermosensory proteins (Coburn, Mori, Ohshima, & Bargmann, 1998; Kimura *et al.*, 2004). Experiments to express *gcy-18* or *gcy-23* into non-thermosensory cells ectopically conferred thermosensitivity onto non-thermosensory neurons and muscles, which strongly supports that these rGCs function as thermosensory proteins (Takeishi *et al.*, 2016). Interestingly, the temperature threshold of responses of those AFD-rGC misexpressing cells were cell-type specific and independent of T_c (Takeishi *et al.*, 2016). These results suggest activation of rGCs and T^* setting depends on an intrinsic mechanism of each cell.

The level of cGMP is critical to set T^*, and T^* shifts to higher temperature when worms are exposed to cGMP analog, 8-Br-GMP (Wasserman *et al.*, 2011). Recent analysis with a genetically encoded cGMP indicator showed that cGMP concentration in the dendritic tip of AFD, where AFD-rGCs localize, increases upon warming temperature (Figure 3) (Aoki, Shiota, Nakano, & Mori, 2019; Woldemariam *et al.*, 2019). cGMP concentration is tightly regulated by synthesis via AFD-rGCs and degradation via PDEs (Aoki *et al.*, 2019; Woldemariam *et al.*, 2019). Among the six PDEs that the worm genome contains, PDE-1, PDE-2, PDE-3 and PDE-5 hydrolyse both cGMP and cAMP, while PDE-4 and PDE-6 are cAMP specific (Omori & Kotera, 2007; Wang *et al.*, 2013). Single mutants for these PDEs show defects in thermotaxis, but combined mutations of them show more severe defects, suggesting redundancy of these PDE functions to control cGMP levels and calcium responses of AFD (Aoki *et al.*, 2019; Woldemariam *et al.*, 2019). AWC and ASI also require TAX-2/TAX-4 for their thermosensory signaling (Beverly *et al.*, 2011; Kuhara *et al.*, 2008); however, thermosensory proteins and detailed thermotransduction mechanisms are still unknown.

Adaptation of AFD calcium response to T_c

How is the precise setting of T^* mediated in AFD? AFD completely shifts T^* to a new temperature after >5 h exposure (Yu *et al.*, 2014). Upon temperature shift from cold to hot, calcium imaging and patch clamp experiments show T^* of AFD shifts in two kinetics; a faster shift that happens in the time scale of minutes, and slow shift that takes hours to settle T^* to the final temperature (Hawk *et al.*, 2018; Ramot, MacInnis, & Goodman, 2008; Yu *et al.*, 2014). The fast shift is likely modulated by non-transcriptional mechanism, and the slow shift involves gene transcription.

Non-transcriptional modulation likely involves a concentration change of cGMP and/or calcium through AFD-rGC and PDE activity. In addition to AFD-rGCs and PDEs,

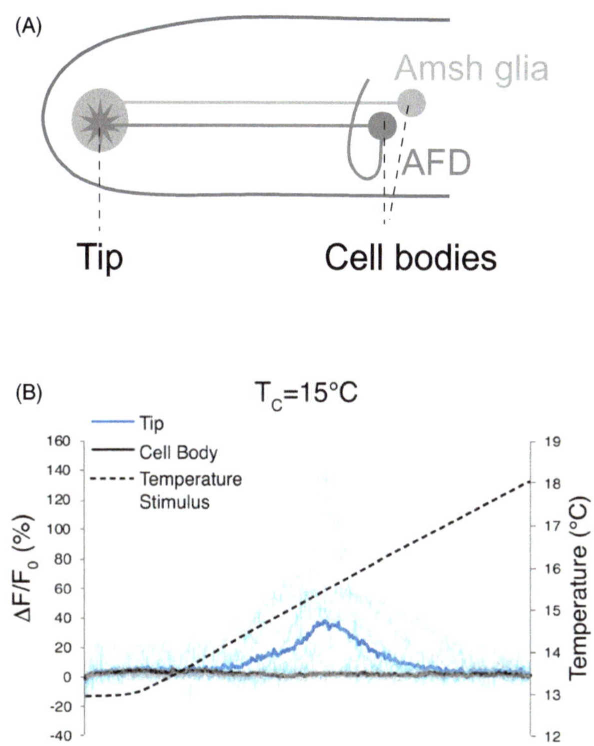

Figure 3. (A) Schematic of AFD neuron and amphid sheath glia (Amsh) in the head of AFD. AFD has finger-like villi on the tip of the dendrite that is surrounded by Amsh glia. (B) Dynamics of cGMP concentration in the tip and cell body of AFD. cGMP level was monitored by a genetically encoded cGMP indicator (FlincG) with warming stimuli. cGMP concentration increased upon warming stimuli in the tip where AFD-rGCs and cGMP-gated channels localize, but not in the cell body. Data source: Woldemariam *et al.* (2019).

neuronal calcium sensor 1 (NCS-1), a calcium-binding protein related to vertebrate guanylate cyclase-activating proteins (GCAPs), was identified as modulator of AFD T^* (Burgoyne, 2007; Wang *et al.*, 2013). Mutants for NCS-1 show increased T^* and defects in fast adaptation during warmer temperature shifts (Wang *et al.*, 2013). As higher cGMP levels result in higher T^* (Wasserman *et al.*, 2011), it is suggested that NCS-1 functions to regulate cGMP levels through calcium-dependent negative feedback that inhibits AFD-rGCs or activates PDEs. Patch clamp experiments also show NCS-1 inhibits voltage-activated outward currents of the voltage-gated K^+ channel (Wang *et al.*, 2013). A recent study shows SLO-1/SLO-2 K^+ channels contribute to modulate the kinetics of AFD adaptation (Aoki *et al.*, 2018), and these repolarization mechanisms also seem to have important roles in tuning AFD T^*.

Slower T^* shifts involve gene transcription that includes AFD-rGC expression level changes via calcium/calmodulin-dependent protein kinase 1 (CMK-1) (Yu *et al.*, 2014). The expression level of each AFD-rGC is higher at warmer temperatures (Yu *et al.*, 2014), which is consistent with the result that increased cGMP signaling results in higher T^*. In *cmk-1* mutants, temperature-dependent expression changes of AFD-rGCs are not observed, and T^* stays low in the warming shift experiment (Yu *et al.*, 2014).

In addition to CMK-1, the mutant of heat shock transcription factor-1 (HSF-1) also shows lower T^* in a shifting experiment to warmer temperatures and atactic thermotaxis after shifting to a warmer temperature (Sugi, Nishida, & Mori, 2011). HSF-1 is expressed and required in muscle or intestine to modulate the T^* of AFD through estrogen signaling and heat shock transcription factors (Sugi *et al.*, 2011). Thus, multiple AFD intrinsic and systemic pathways are involved in precise tuning of AFD T^*.

Thermotaxis regulation by downstream of AFD

AIY is the primary postsynaptic partner of AFD (White, Southgate, Thomson, & Brenner, 1986). Worms in which AIY is ablated or inhibited exhibit constant negative thermotaxis behavior independent of T_c (Luo *et al.*, 2014; Mori & Ohshima, 1995) and the constitutive cryophilic mechanism of AIY ablation is AFD dependent (Luo *et al.*, 2014). This result suggests AIY mediates this behavior toward warmer temperatures, and that there is a still unknown mechanism to conduct temperature information from AFD to the downstream negative thermotaxis pathway without AFD–AIY communication. The temperature threshold of the AIY response to warming stimuli is tightly synchronized with AFD responses (Hawk *et al.*, 2018). Also, the probability and amplitude of temperature-dependent calcium-responses in AIY is T_c dependent: higher probability at cooler temperature than T_c, and lower probability at warmer temperature than T_c (Table 1) (Hawk *et al.*, 2018). This result is consistent with thermotaxis behavior in AIY inhibited worms and confirms that abolishing AIY responses induces negative thermotaxis.

Table 1. Orientation of thermotaxis is determined by connectivity of AFD and AIY.

Genotype	AFD-AIY connectivity	Probability of AIY response	Behavior
ttx-3, AIY ablation, *dgk-1*, AFD::caPKC-1	AFD AIY	(increasing, low to high)	Negative thermotaxis
Wild type ($T>T_c$)	AFD → AIY		
Wild type ($T<T_c$)	AFD → AIY		Positive thermotaxis
pkc-1/ttx-4	AFD ➡ AIY		

More connectivity induces higher probability of responses of AIY. Lower probability of AIY responses directs worms toward cooler side, and vice versa. Genotype: overexpression of constitutively-active PKC-1 into AFD (AFD::caPKC-1).

A detailed analysis reveals that AFD T^* changes happen faster than thermotaxis temperature preference changes when worms are exposed to new temperature (Biron *et al.*, 2006; Hawk *et al.*, 2018). AIY also shows a slower T^* shift compared to AFD (Biron *et al.*, 2006; Hawk *et al.*, 2018). This divergence of adaptation timescale is mediated by the synaptic communication between AFD and AIY via PKC-1/TTX-4 and diacylglycerol kinase, DGK-3 (Biron *et al.*, 2006; Hawk *et al.*, 2018; Luo *et al.*, 2014; Okochi, Kimura, Ohta, & Mori, 2005). T_c-independent positive thermotaxis and negative thermotaxis was observed in the *pkc-1/ttx-4* mutant and in worms that overexpress constitutive-active PKC-1 into AFD (AFD::caPKC-1), respectively (Table 1) (Hawk *et al.*, 2018; Okochi *et al.*, 2005). The temperature response of AIY was altered in *pkc-1* mutants and AFD::caPKC-1, although the AFD response was the same as wild type indicating presynaptic plasticity of the AFD–AIY synapse modulates thermotaxis orientation (Hawk *et al.*, 2018). Worms also exhibit constitutive positive thermotaxis when ectopic gap junctions are overexpressed in AFD and AIY (Table 1) (Hawk *et al.*, 2018). These results suggest communication of AIY and AFD is important in determining the temperature preference (Hawk *et al.*, 2018; Narayan, Laurent, & Sternberg, 2011).

In addition to AIY, AIZ and RIA are shown to mediate thermotaxis behavior (Mori & Ohshima, 1995; Ouellette, Desrochers, Gheta, Ramos, & Hendricks, 2018). AIZ ablated worms show constitutive positive thermotaxis (Mori & Ohshima, 1995). Since ablation of AIZ has an opposite phenotype to AIY ablation, AIY is thought to be inhibitory to AIZ (Mori & Ohshima, 1995). The interneuron RIA receives signals both from AIY and AIZ. RIA-inhibited worms are atactic in isothermal tracking experiments on asymmetric temperature gradients, such as radial temperature gradients (Mori & Ohshima, 1995; Ouellette *et al.*, 2018; Tanizawa *et al.*, 2006). During isothermal tracking, head bends give oscillating temperature stimuli to thermosensory neurons, and downstream RIA neurons activate different axonal compartments of the nerve ring (Ouellette *et al.*, 2018). This function of RIA is likely to be important to detect precise temperature on the asymmetric temperature gradient and to decide the orientation of the run (Ouellette *et al.*, 2018).

AWC is also connected to AIY and AIZ, and ASI to AIZ. AFD and AWC are both glutaminergic onto AIY (Narayan *et al.*, 2011; Ohnishi, Kuhara, Nakamura, Okochi, & Mori, 2011). As AWC and ASI respond to other environmental stimuli that AFD does not respond to (Metaxakis, Petratou, & Tavernarakis, 2018), they might be tuning thermotaxis behavior by integrating those environmental and internal factors into thermotaxis signaling pathways within these neurons or downstream neurons.

Dendritic tip structure of AFD modulates thermotaxis

The dendritic tip of each sensory neuron of *C. elegans* has a unique shape (Figure 3(A)) (Doroquez, Berciu, Anderson, Sengupta, & Nicastro, 2014; Perkins, Hedgecock, Thomson, & Culotti, 1986). AFD has a finger-like villi ending that is composed of an actin-based structure around small microtubule-based cilium (Doroquez *et al.*, 2014; Perkins *et al.*, 1986), and AFD specific rGCs are located in the villi structure (Inada *et al.*, 2006; Nguyen, Liou, Hall, & Leroux, 2014). The villi structure of AFD is likely to be essential for AFD function, as defects in the structure are seen in AFD signaling mutants, including shortened villi in triple mutants of the AFD-rGCs (Satterlee *et al.*, 2001). In addition to signaling in AFD, amphid sheath glia surrounding AFD also respond to temperature and regulate villi structure of AFD by controlling extracellular ion concentration (Shaham, 2015; Singhvi *et al.*, 2016). Amphid sheath glia express K^+/Cl^- cotransporter (KCC)−3, and both thermotaxis behavior and temperature-dependent AFD calcium response is impaired in a mutant of *kcc-3* (Singhvi *et al.*, 2016; Yoshida *et al.*, 2016). The *kcc-3* mutant has shortened AFD villi which is caused by constitutive activation of GCY-8, one of the AFD-rGCs (Singhvi *et al.*, 2016). These results suggest KCC-3 in amphid sheath glia increases extracellular Cl^- ion concentration, which then inhibits cGMP production by GCY-8 in AFD to regulate thermotaxis behaviors. The effect of the Cl^- ion is likely to be only for GCY-8, and not for the other AFD-rGCs, GCY-23 and GCY-18.

Noxious temperature avoidance and escape

Worms have rather wide range of innocuous temperature. However, when worms encounter noxious heat or coldness, or rapid temperature change even within physiological temperature range, they show escaping or avoidance behavior (Glauser *et al.*, 2011; Mohammadi, Byrne Rodgers, Kotera, & Ryu, 2013; Wittenburg & Baumeister, 1999). When they detect noxious heat on their head, they use a withdrawal strategy that consists of reverse movement and a sharp turn (Wittenburg & Baumeister, 1999). They show accelerated forward movement when they feel heat to their tail and show probabilistic backward, forward or pausing behavior when feeling noxious stimuli in the mid body (Mohammadi *et al.*, 2013). These strategies are different from the biased random walk that they use for negative or positive thermotaxis (Schafer, 2012), suggesting that the response against noxious temperature is mediated via different mechanisms from thermotaxis. FLP, AFD, ASE, PHC and PVD neurons were identified to contribute to noxious heat or cold responses, and the responsible neurons depend on the stimuli (Chatzigeorgiou *et al.*, 2010; Gong *et al.*, 2019; Liu, Schulze, & Baumeister, 2012; Mohammadi *et al.*, 2013). PVD neurons mediate noxious cold responses through TRPA-1 (Chatzigeorgiou *et al.*, 2010). ASE mediates cold response through a kainate-type glutamate receptor, GLR-3 (Gong *et al.*, 2019). GLR-3 transmits temperature signal through trimeric G protein signaling, which is independent of its glutamate-gated channel function (Gong *et al.*, 2019). PVD, FLP, AFD and PHC neurons were shown to regulate responses against noxious heat (Chatzigeorgiou *et al.*, 2010; Liu *et al.*, 2012; Mohammadi *et al.*, 2013). TRPV channels play important roles in these responses: FLP and PHC require both *osm-9* and *ocr-2*, and PVD requires *ocr-2* (Liu *et al.*, 2012; Mohammadi *et al.*, 2013). The FLP-21/NPR-1-mediated neuropeptide signaling pathway was also shown to function synergically with TRPV channels for noxious heat escape (Glauser *et al.*, 2011). While TRPA-1 and GLR-3 have been tested for their direct cold sensitivity, no evidence showed that *C. elegans* TRPV and CNG channels can be directly activated by heat (Tobin *et al.*, 2002), although TRPV and a CNG channel in other species are known to function as heat sensor proteins (Caterina *et al.*, 1997; Finka, Cuendet, Maathuis, Saidi, & Goloubinoff, 2012).

Cold tolerance and acclimation

The mechanisms for temperature tolerance and acclimation are critical for animal survival. One mechanism that has been well studied involves changes in the fatty acid composition of cell membranes. An increase in the percentage of unsaturated fatty acids maintains cell-membrane fluidity at lower temperatures, which is required for cold tolerance in animals and plants (Murray, Hayward, Govan, Gracey, & Cossins, 2007; Savory, Sait, & Hope, 2011). Many cells accumulate sugars such as trehalose to prevent their cytosol from freezing (Goto, Takahashi, & Suzuki, 1993; Sakurai, 2012). These adaptive 'outputs' have been strongly focused on, however, the molecular systems that deal with temperature 'input' are only beginning to gain academic interest.

Caenorhabditis elegans has cold tolerance ability (Murray *et al.*, 2007). For example, wild-type animals grown at 15 °C are able to survive after transferred to 2 °C for 48 h, while most 20/25 °C-grown animals cannot survive at 2 °C (Figure 4) (Ohta, Ujisawa, Sonoda, & Kuhara, 2014; Ujisawa, Ohta, Okahata, Sonoda, & Kuhara, 2014). Cold tolerance is dynamically changeable within only 2–3 h of exposure to new cultivation temperature. For example, 25 C-cultivated animals do not survive at 2 °C, whereas 25 °C-cultivated animals can survive at 2 C after exposure to 15 °C for 3 h (Figure 4) (Ohta *et al.*, 2014). This phenomenon is defined as cold acclimation (Okahata, Wei, Ohta, & Kuhara, 2019).

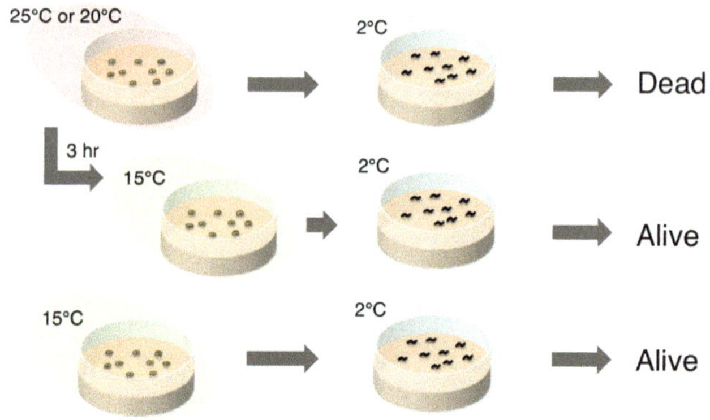

Figure 4. Cultivation temperature-dependent cold tolerance and acclimation. Wild-type animals cultivated at 25 °C cannot survive after cold stimulus at 2 °C for 48 h, while 15 °C-cultivated animals can survive after a similar stimulus. 25 °C-cultivated animals can survive at 2 °C after cultivation at 15 °C for 3 h.

G protein-coupled thermosensory signaling in ASJ for performing cold tolerance

Temperature information sensed by a specific sensory neuron is essential for cold tolerance. Mutants lacking the cGMP-gated channel TAX-4/TAX-2 in sensory neurons show abnormally increased cold tolerance. Abnormal *tax-4* cold tolerance is caused by ASJ neurons, which are known as light and pheromone sensing neurons (Figure 5) (Ohta *et al.*, 2014). Ablation of wild-type ASJ results in increased cold tolerance, suggesting ASJ negatively regulates cold tolerance (Figure 5) (Ohta *et al.*, 2014). Higher calcium concentration at warmer temperature suggests that ASJ acts as a thermosensory neuron, and TAX-4 is the primary channel for ASJ thermosensation. All of these suggest that the ASJ is a temperature sensing neuron in cold tolerance (Ohta *et al.*, 2014).

The molecular mechanism for phototransduction in ASJ has been identified (Figure 5). Light is received by a photoreceptor LITE-1 in ASJ, which sequentially activates Gα protein (GOA-1 and GPA-3), GC (DAF-11 and ODR-1) and a cGMP-gated channel TAX-4/TAX-2. PDE (PDE-1, 2 and 5) acts as an inhibitor of light signaling by hydrolyzing cGMP to 5′GMP (Figure 5) (Liu *et al.*, 2010). Differences and commonalities of the molecular components between light and temperature sensations have been identified (Ujisawa, Ohta, Uda-Yagi, & Kuhara, 2016). In cold tolerance, Gα (GOA-1, GPA-1 3), GC (DAF-11 and ODR-1) and PDE (PDE-1, 2, 3 and 5) are involved in thermosensation (Ujisawa *et al.*, 2016), in which GPA-1 and PDE-3 are required for thermosignaling specifically (Figure 5) (Ujisawa *et al.*, 2016). LITE-1 is the sole light receptor in ASJ, but LITE-1 is not required for ASJ temperature sensation and cold tolerance (Liu *et al.*, 2010; Ohta *et al.*, 2014; Ujisawa *et al.*, 2016), indicating that an unidentified temperature receptor functions in ASJ (Figure 5).

ADL thermosensory neuron in cold tolerance

ENDU-2 is homologous to human calcium-dependent endoribonuclease EndoU and its mutant *endu-2* displays increased tolerance to cold (Ohta *et al.*, 2014; Ujisawa *et al.*, 2018). The abnormal *endu-2* cold tolerance is caused by defective ENDU-2 in the ADL olfactory neuron and muscle cells. Higher calcium concentration at a warmer temperature reveals ADL acts as a thermosensory neuron, in which three TRP channel subunits OCR-1/OCR-2/OSM-9 and the KCNQ-type potassium channel subunits KQT-2/KQT-3 are required for temperature signaling (Okahata *et al.*, 2019; Ujisawa *et al.*, 2018). ADL thermal response requires ENDU-2 function in ADL or specific muscle cells, suggesting cell autonomous and non-cell autonomous functions of ENDU-2 (Figure 6) (Ujisawa *et al.*, 2018).

Downstream molecules of ENDU-2 have been identified by transcriptome analysis. One downstream gene is an apoptotic gene, *ced-3* encoding caspase (Ujisawa *et al.*, 2018). *ced-3* mutants exhibit decreased cold tolerance, which is caused by defective CED-3 functions in ADL and muscle cells (Ujisawa *et al.*, 2018). It is reported that CED-3 controls synaptic remodeling in neurons during development (Meng *et al.*, 2015). ENDU-2 plays a role as a negative

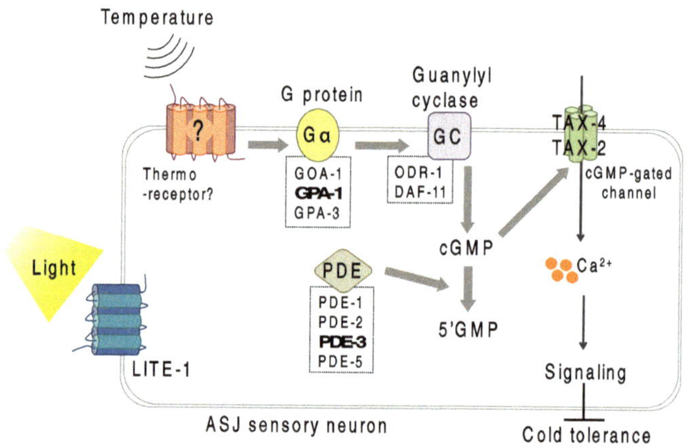

Figure 5. Temperature and light signaling model in ASJ sensory neuron regulating cold tolerance. Temperature- and light-signaling include both common and function-specific gene products. Molecular names shown in bold indicate those specific to temperature signaling. Additional Gα, GC, and PDE may also be required for temperature signaling because temperature response is not completely diminished in the mutant lacking the proteins shown in this model. Temperature receptor remains unidentified.

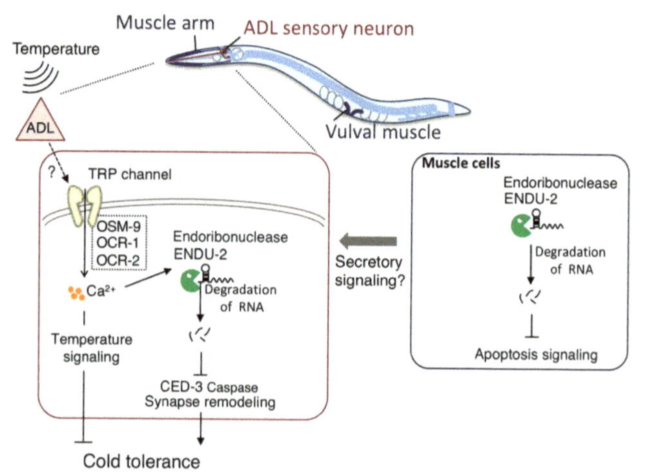

Figure 6. A model for ENDU-2-mediated cold tolerance. The ENDU-2 and TRP channels positively regulate neural activity of the ADL sensory neuron. ENDU-2 could negatively regulate CED-3-dependent synaptic remodeling and apoptotic signaling in both muscle cells and ADL may decrease ADL function.

regulator of CED-3-dependent synaptic remodeling (Figure 6). These imply that ENDU-2 is involved in cold tolerance and synaptic remodeling through the caspase pathway (Figure 6) (Ujisawa *et al.*, 2018).

Oxygen modulates thermosensitivity of ADL in cold acclimation

Cold acclimation is modulated by environmental oxygen concentration (Okahata *et al.*, 2019). A KCNQ potassium channel KQT-2 is critical for ADL thermosensation, and its mutant *kqt-2* decreases ADL thermosensitivity which causes supra normal cold acclimation (Figure 7). The supranormal cold acclimation of the *kqt-2* mutant is exaggerated when *kqt-2* mutants are cultivated on medium-size agar plates, compared with smaller plates (Okahata *et al.*, 2019). Oxygen signaling from URX oxygen-sensing neurons modulates ADL excitability for sensing the nematode pheromone, ascaroside, through chemical and electrical synaptic connections via RMG interneurons (Fenk & de Bono, 2017). Mutation of the URX oxygen receptor GCY-35 decreases cold acclimation and suppresses the supranormal cold acclimation and ADL thermal responsiveness in *kqt-2* mutants (Okahata *et al.*, 2019). Signaling of the URX-sensed oxygen and ADL-sensed temperature are processed in ADL to determine cold acclimation in a simple neural circuit (Figure 7). This simple neuronal circuit integrates two different sensory modalities, temperature and oxygen, and thus is an attractive model for studying integration and discrimination of multiple sensory signals.

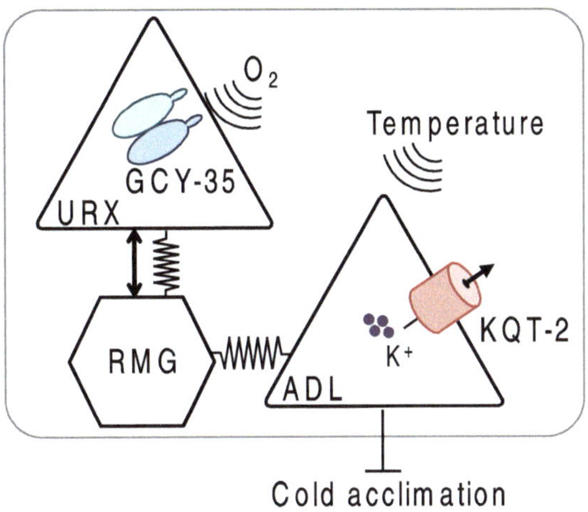

Figure 7. A model for the neural circuit integrating temperature and oxygen signaling. High O_2 level sensed by URX neurons through oxygen receptor GCY-35 transmits signals to ADL that travel via the RMG interneuron and then ADL activity may be inhibited, which negatively regulates cold acclimation.

Tissue networks for regulation of cold tolerance

Tissues downstream of thermosensory neurons for cold tolerance are identified (Figure 8) (Ohta *et al.*, 2014). Cold tolerance is regulated by molecules in the ageing pathway such as DAF-16/FOXO-type transcriptional factor and AGE-1/PI3 kinase, which play a role in insulin signaling (Savory *et al.*, 2011). Under temperature stimuli, ASJ synapses release insulin-like molecules INS-6 and DAF-28, which are received by the DAF-2/insulin receptor at the intestine and neurons (Ohta *et al.*, 2014). In the intestine, DAF-2 regulates downstream signal relevant molecules such as AGE-1 and AKT kinase/AKT-1, which inhibit nuclear translocation of the DAF-16. Therefore, a temperature signal negatively regulates DAF-16-dependent gene expression, resulting in increased tolerant ability to cold (Ohta *et al.*, 2014). Abnormal phospholipid saturation in a desaturase mutant causes decreased cold tolerance (Murray *et al.*, 2007). Glycerol accumulation in the intestine is also involved in cold tolerance (Liu, Xiao, Ji, Zhang, & Zou, 2017). Cold activates Gsα and protein kinase A (PKA) which regulates expression of a hormone-sensitive lipase HOSL-1, which then regulates glycerol accumulation resulting from lipid hydrolysis important

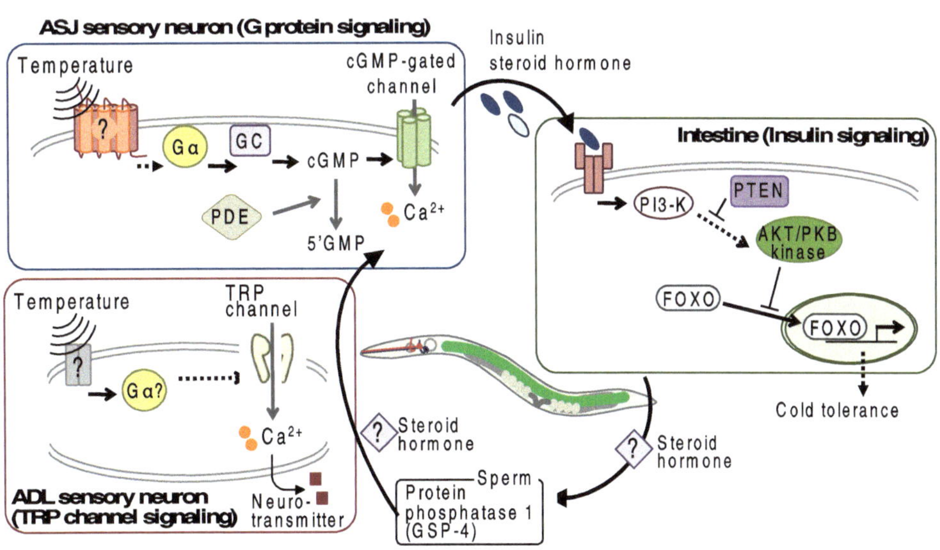

Figure 8. A tissue network model for performing cold tolerance. A tissue network including neurons, intestinal cells, and sperm controls cold tolerance. Temperature is sensed by the ASJ and ADL sensory neurons, then the ASJ releases insulin and steroid hormones that are received by the intestine. Intestinal cells, in turn, affect the sperm through nuclear hormone receptors, which affect ASJ temperature signalling by a yet-unidentified feedback system.

for cold tolerance (Liu *et al.*, 2017). These genes are likely the targets of DAF-16.

The DAF-2-mediated insulin pathway is also involved in a variety of stress-tolerance responses. As a cold stimulus is a stressful condition for *C. elegans*, stress responsive genes are likely to be involved in cold tolerance. However, cold tolerance is almost normal in various stress response mutants such as heat-shock factor (*hsf-1*), MAP kinase (*jnk-1*, *kgb-1* and *kgb-2*) and heat-shock proteins (*hsp-16.2, 16.41* and *16.48*) (Ohta *et al.*, 2014). Metabolic stress induced by dietary restriction is also involved in a cold-adaptive response that is abrogated by supplementation with choline (Klapper, Findeis, Koefeler, & Doring, 2016). The intestine itself is cold-sensitive and plays a central role in ageing and the stress response, in which a cold receptor TRP channel (TRPA-1) and its downstream protein kinase C (PKC-2) are essential (Xiao *et al.*, 2013), but *trpa-1* and *pkc-2* mutants show normal cold tolerance (Ohta *et al.*, 2014).

DNA microarray analysis of *daf-2* mutants in response to temperature stimuli was performed to identify the downstream genes for intestine insulin signaling, and it was found that expressions of sperm genes are altered (Sonoda, Ohta, Maruo, Ujisawa, & Kuhara, 2016). Mutation of the sperm-specific protein phosphatase PP1 (GSP-4) showed increased cold tolerance (Sonoda *et al.*, 2016). The temperature response of the ASJ neuron is decreased in a *gsp-4* mutant (Sonoda *et al.*, 2016). Genetic analyses demonstrate that the intestine affects gene expression in sperm, and sperm affects gene expression in the ASJ (Sonoda *et al.*, 2016). A model for the tissue network and a feedback system involved in cold tolerance is proposed (Figure 8). Temperature activates ASJ and ADL, and ASJ regulates the intestine through insulin and steroid-hormonal signaling. Intestine affects the sperm, which in turn affects ASJ thermosensory signaling, perhaps through unidentified secretory signaling (Sonoda *et al.*, 2016).

Multiple aspects for regulating cold tolerance

Long-chain acyl-CoA dehydrogenase (VLCAD) is required in mice for survival in the cold. VLCAD upregulates the *TMEM135* gene encoding a novel mouse protein. Over production of TMEM135 in *C. elegans* causes increased survival rate to cold stress (Exil *et al.*, 2010), and TMEM135 deficient *C. elegans* exhibits reduction of fat storage, mitochondria activity and longevity (Exil *et al.*, 2010). Low temperature experience introduces loss of pigmentation, a decrease in the size of the intestine and gonads and disruption of the vulva in worms (Robinson & Powell, 2016). Mutation in a G-protein coupled receptor FSHR-1 causes strong resistance to cold-shock with abnormal pigmentation (Robinson & Powell, 2016). PROS-1, a homeodomain transcription factor homologous to Drosophila prospero/mammalian Prox1, regulates development of glial cells surrounding head sensory neurons. Knockdown of *pros-1* gene causes abnormal morphology of sensory neurons and abnormal cold tolerance (Kage-Nakadai *et al.*, 2016).

PAQR-2 is homologous to the anti-diabetic mammalian protein known as adiponectin receptor AdipoR2, and belongs to the PAQR protein family (Devkota *et al.*, 2017). PAQR-2 and its partner IGLR-2 act as regulators of membrane homeostasis by increasing the proportion of unsaturated phospholipids in the plasma membrane, which enables worms to survive at low temperatures such as 15 °C (Pilon & Svensk, 2013; Svensk *et al.*, 2013, 2016). PAQR-2 may act cell-nonautonomously to maintain membrane fluidity through upregulating fatty acid desaturation (Bodhicharla, Devkota, Ruiz, & Pilon, 2018), wherein PAQR-2 is speculated to function as sensor of the low membrane fluidity caused by cold or increasing exogenous saturated fatty acids, SFAs (Devkota *et al.*, 2017).

Natural variation of cold tolerance and acclimation in nematode

Wild-type strains of *C. elegans* have been isolated from many areas. These naturally varying strains differ in cold tolerance and acclimation (Okahata *et al.*, 2016). For example, the AB1 strain from Australia rapidly acclimates to colder temperatures, whereas the CB4856 strain from Hawaii acclimates more slowly. Responsible polymorphisms for these variations have been mapped onto a specific region of chromosome I (Okahata *et al.*, 2016).

Nematodes are among the most abundant animals on earth, and some of them live in a particularly cold condition. A northern root-knot nematode, *Meloidogyne hapla*, lives in temperatures ranging from −15 °C to 27 °C. The J2 larval stage of *M. hapla* can strongly enhance its cold tolerance (Wu *et al.*, 2018). An Antarctic nematode, *Panagrolaimus sp.* DAW1, is a rare multicellular organism known to survive intracellular freezing on a routine basis (Wharton & Brown, 1991; Wharton & Ferns, 1995). DAW1 stores trehalose in cold-acclimated condition, and expression of trehalose synthesis genes *tps-2* and *lea-1* is upregulated after cold stimuli (Seybold, Wharton, Thorne, & Marshall, 2017). A parasitic nematode, *Marshallagia marshalli*, survived exposure to freezing temperatures below −30 °C (Carlsson, Irvine, Wilson, & Coulson, 2013). Entomopathogenic nematodes, *Steinernema feltiae* and *Heterorhabditis bacteriophora*, survive at −13 °C by cryoprotective dehydration (Ali & Wharton, 2013). It has been reported that parasitic and non-parasitic nematodes show differences in proliferation rates and cold tolerance (Anderson & Coleman, 1982), and these molecular and biochemical mechanisms remain largely unknown.

Concluding remarks

Caenorhabditis elegans has a simple neuronal circuit through which worms can produce variable temperature responses depending on their circumstances. One of the open questions in thermosensation of AFD is the mechanism of AFD-rGCs to control cGMP level. In mammalian and other systems, guanylyl cyclases generally function as dimers or trimers and are activated by conformation change (Potter, 2011). An expression level difference was observed among three AFD-rGCs (Yu *et al.*, 2014), however, it is still unknown whether their activation is regulated by dimerization or what function each AFD-rGC has. Further study of cGMP dynamics with recently published cGMP indicators will reveal insights into the mechanism to regulate precise cGMP levels that is likely to contribute to fast T^* shift of AFD (Aoki *et al.*, 2019; Woldemariam *et al.*, 2019). In addition, these studies will also give detailed cGMP-dependent sensory mechanisms that might be evolutionally conserved, as guanylyl cyclases are common receptors for environmental cues, such as odor and gas, in mammals (Maruyama, 2016). The temperature receptor(s) in cold tolerance remains unidentified. Trimeric G proteins are involved in thermosensation of ASJ, and upstream molecules for G proteins such as G protein-coupled receptors are candidates for the thermoreceptor in cold tolerance. Recently, it was reported that DEG-1, a degenerin/epithelial sodium channel (DEG/ENaC), probably acts as a temperature receptor in ASG sensory neuron for controlling cold tolerance (Takagaki *et al.*, 2020).

The molecular mechanisms of thermotaxis and cold tolerance are also not yet fully understood. For instance, pathways downstream of CMK-1 seem to have other factors in addition to the canonical CREB pathway and HSF-1 in thermotaxis (Yu *et al.*, 2014). It is also still unclear what molecular signal modulates thermosensitivity of AWC and ASI (Beverly *et al.*, 2011; Biron *et al.*, 2008; Kuhara *et al.*, 2008). Whether these molecules and neurons are also involved in the performance of cold tolerance is a simple question. Further identification of these signaling pathways will reveal the precise mechanism of neuronal plasticity and circuit communication that drives temperature responsive behaviors.

Acknowledgements

We thank P. Sengupta for comments on this manuscript.

Disclosure statement

No potential conflict of interest was reported by the author(s).

Funding

A.K. was supported by the Asahi Glass Foundation, the Takeda Science Foundation, the Suzuken Memorial Foundation, the Hirao Taro Foundation of KONAN GAKUEN for Academic Research, AMED Mechano Biology [19gm5810024h0003, 20gm5810024h0004], KAKENHI from JSPS and MEXT Japan [17K19410, 18H02484, 15H05928].

ORCID

Atsushi Kuhara http://orcid.org/0000-0003-2994-8658

References

Albrecht, D.R., & Bargmann, C.I. (2011). High-content behavioral analysis of *Caenorhabditis elegans* in precise spatiotemporal chemical environments. *Nature Methods*, *8*, 599–605. doi:10.1038/nmeth.1630

Ali, F., & Wharton, D.A. (2013). Cold tolerance abilities of two entomopathogenic nematodes, *Steinernema feltiae* and *Heterorhabditis bacteriophora*. *Cryobiology*, *66*(1), 24–29. doi:10.1016/j.cryobiol.2012.10.004

Anderson, R.V., & Coleman, D.C. (1982). Nematode temperature responses: A niche dimension in populations of bacterial-feeding nematodes. *Journal of Nematology*, *14*(1), 69–76.

Aoki, I., Shiota, M., Nakano, S., & Mori, I. (2019). cGMP dynamics underlie thermosensation in *C. elegans*. *bioRxiv*. 764571.

Aoki, I., Tateyama, M., Shimomura, T., Ihara, K., Kubo, Y., Nakano, S., & Mori, I. (2018). SLO potassium channels antagonize premature decision making in. *Communications Biology*, *1*(1), 123. doi:10.1038/s42003-018-0124-5

Bargmann, C.I., Hartwieg, E., & Horvitz, H.R. (1993). Odorant-selective genes and neurons mediate olfaction in *C. elegans*. *Cell*, *74*, 515–527. doi:10.1016/0092-8674(93)80053-H

Bargmann, C.I., & Horvitz, H.R. (1991). Chemosensory neurons with overlapping functions direct chemotaxis to multiple chemicals in *C. elegans*. *Neuron*, *7*, 729–742. doi:10.1016/0896-6273(91)90276-6

Beverly, M., Anbil, S., & Sengupta, P. (2011). Degeneracy and neuromodulation among thermosensory neurons contribute to robust thermosensory behaviors in *Caenorhabditis elegans*. *Journal of Neuroscience*, *31*, 11718–11727. doi:10.1523/JNEUROSCI.1098-11.2011

Biron, D., Shibuya, M., Gabel, C., Wasserman, S.M., Clark, D.A., Brown, A., ... Samuel, A.D.T. (2006). A diacylglycerol kinase modulates long-term thermotactic behavioral plasticity in *C. elegans*. *Nature Neuroscience*, *9*, 1499–1505. doi:10.1038/nn1796

Biron, D., Wasserman, S., Thomas, J.H., Samuel, A.D., & Sengupta, P. (2008). An olfactory neuron responds stochastically to temperature and modulates *Caenorhabditis elegans* thermotactic behavior. *Proceedings of the National Academy of Sciences of the United States of America*, *105*, 11002–11007. doi:10.1073/pnas.0805004105

Bodhicharla, R., Devkota, R., Ruiz, M., & Pilon, M. (2018). Membrane fluidity is regulated cell nonautonomously by *Caenorhabditis elegans* PAQR-2 and its mammalian homolog adipoR2. *Genetics*, *210*(1), 189–201. doi:10.1534/genetics.118.301272

Bryant, A.S., Ruiz, F., Gang, S.S., Castelletto, M.L., Lopez, J.B., & Hallem, E.A. (2018). A critical role for thermosensation in host seeking by skin-penetrating nematodes. *Current Biology*, *28*, 2338–2347.e2336.

Burgoyne, R.D. (2007). Neuronal calcium sensor proteins: Generating diversity in neuronal Ca^{2+} signalling. *Nature Reviews Neuroscience*, *8*, 182–193. doi:10.1038/nrn2093

Carlsson, A.M., Irvine, R.J., Wilson, K., & Coulson, S.J. (2013). Adaptations to the Arctic: Low-temperature development and cold tolerance in the free-living stages of a parasitic nematode from Svalbard. *Polar Biology*, *36*, 997–1005. doi:10.1007/s00300-013-1323-7

Caterina, M.J., Schumacher, M.A., Tominaga, M., Rosen, T.A., Levine, J.D., & Julius, D. (1997). The capsaicin receptor: A heat-activated ion channel in the pain pathway. *Nature*, *389*, 816–824. doi:10.1038/39807

Chatzigeorgiou, M., Yoo, S., Watson, J.D., Lee, W.-H., Spencer, W.C., Kindt, K.S., ... Schafer, W.R. (2010). Specific roles for DEG/ENaC and TRP channels in touch and thermosensation in *C. elegans* nociceptors. *Nature Neuroscience*, *13*, 861–868. doi:10.1038/nn.2581

Chi, C.A., Clark, D.A., Lee, S., Biron, D., Luo, L., Gabel, C.V., ... Samuel, A.D.T. (2007). Temperature and food mediate long-term thermotactic behavioral plasticity by association-independent mechanisms in *C. elegans*. *Journal of Experimental Biology*, *210*, 4043–4052. doi:10.1242/jeb.006551

Clapham, D.E., & Miller, C. (2011). A thermodynamic framework for understanding temperature sensing by transient receptor potential (TRP) channels. *Proceedings of the National Academy of Sciences of the United States of America*, *108*, 19492–19497. doi:10.1073/pnas.1117485108

Clark, D.A., Biron, D., Sengupta, P., & Samuel, A.D. (2006). The AFD sensory neurons encode multiple functions underlying thermotactic behavior in *Caenorhabditis elegans. Journal of Neuroscience, 26*, 7444–7451. doi:10.1523/JNEUROSCI.1137-06.2006

Clark, D.A., Gabel, C.V., Lee, T.M., & Samuel, A.D. (2007). Short-term adaptation and temporal processing in the cryophilic response of *Caenorhabditis elegans. Journal of Neurophysiology, 97*, 1903–1910. doi:10.1152/jn.00892.2006

Coburn, C.M., Mori, I., Ohshima, Y., & Bargmann, C.I. (1998). A cyclic nucleotide-gated channel inhibits sensory axon outgrowth in larval and adult *Caenorhabditis elegans*: A distinct pathway for maintenance of sensory axon structure. *Development (Cambridge, England), 125*, 249–258.

Devkota, R., Svensk, E., Ruiz, M., Ståhlman, M., Borén, J., & Pilon, M. (2017). The adiponectin receptor AdipoR2 and its *Caenorhabditis elegans* homolog PAQR-2 prevent membrane rigidification by exogenous saturated fatty acids. *PLoS Genetics, 13*, e1007004. doi:10.1371/journal.pgen.1007004

Doroquez, D.B., Berciu, C., Anderson, J.R., Sengupta, P., & Nicastro, D. (2014). A high-resolution morphological and ultrastructural map of anterior sensory cilia and glia in *Caenorhabditis elegans. eLife, 3*, e01948. doi:10.7554/eLife.01948

Dusenbery, D.B., Sheridan, R.E., & Russell, R.L. (1975). Chemotaxis-defective mutants of the nematode *Caenorhabditis elegans. Genetics, 80*, 297–309.

Exil, V.J., Silva Avila, D., Benedetto, A., Exil, E.A., Adams, M.R., Au, C., & Aschner, M. (2010). Stressed-induced TMEM135 protein is part of a conserved genetic network involved in fat storage and longevity regulation in *Caenorhabditis elegans. PLOS One, 5*(12), e14228. doi:10.1371/journal.pone.0014228

Fenk, L.A., & de Bono, M. (2017). Memory of recent oxygen experience switches pheromone valence in *Caenorhabditis elegans. Proceedings of the National Academy of Sciences of the United States of America, 114*, 4195–4200. doi:10.1073/pnas.1618934114

Finka, A., Cuendet, A.F., Maathuis, F.J., Saidi, Y., & Goloubinoff, P. (2012). Plasma membrane cyclic nucleotide gated calcium channels control land plant thermal sensing and acquired thermotolerance. *The Plant Cell, 24* , 3333–3348. doi:10.1105/tpc.112.095844

Garrity, P.A., Goodman, M.B., Samuel, A.D., & Sengupta, P. (2010). Running hot and cold: Behavioral strategies, neural circuits, and the molecular machinery for thermotaxis in *C. elegans* and *Drosophila. Genes and Development, 24*, 2365–2382. doi:10.1101/gad.1953710

Glauser, D.A., Chen, W.C., Agin, R., MacInnis, B.L., Hellman, A.B., Garrity, P.A., ... Goodman, M.B. (2011). Heat avoidance is regulated by transient receptor potential (TRP) channels and a neuropeptide signaling pathway in *Caenorhabditis elegans. Genetics, 188*(1), 91–103. doi:10.1534/genetics.111.127100

Glauser, D.A., & Goodman, M.B. (2016). Molecules empowering animals to sense and respond to temperature in changing environments. *Current Opinion in Neurobiology, 41*, 92–98. doi:10.1016/j.conb.2016.09.006

Gong, J., Liu, J., Ronan, E.A., He, F., Cai, W., Fatima, M., ... Xu, X.Z.S. (2019). A cold-sensing receptor encoded by a glutamate receptor gene. *Cell, 178*, 1375–1386. doi:10.1016/j.cell.2019.07.034

Goto, M., Takahashi, K., & Suzuki, C. (1993). Ecological study on the barnyard grass stem borer, *Enosima leucotaeniella* (RAGONOT) (Lepidoptera: Pyralidae): VIII. Seasonal changes of carbohydrate contents in overwintering larvae. *Applied Entomology and Zoology, 28*, 417–421. doi:10.1303/aez.28.417

Gray, J.M., Hill, J.J., & Bargmann, C.I. (2005). A circuit for navigation in *Caenorhabditis elegans. Proceedings of the National Academy of Sciences of the United States of America, 102*, 3184–3191.

Hawk, J.D., Calvo, A.C., Liu, P., Almoril-Porras, A., Aljobeh, A., Torruella-Suárez, M.L., ... Colón-Ramos, D.A. (2018). Integration of plasticity mechanisms within a single sensory neuron of *C. elegans* actuates a memory. *Neuron, 97*, 356–367.e354. doi:10.1016/j.neuron.2017.12.027

Hedgecock, E.M., & Russell, R.L. (1975). Normal and mutant thermotaxis in the nematode *Caenorhabditis elegans. Proceedings of the National Academy of Sciences of the United States of America, 72*, 4061–4065. doi:10.1073/pnas.72.10.4061

Hoffstaetter, L.J., Bagriantsev, S.N., & Gracheva, E.O. (2018). TRPs et al.: A molecular toolkit for thermosensory adaptations. *Pflügers Archiv – European Journal of Physiology, 470*, 745–759. doi:10.1007/s00424-018-2120-5

Inada, H., Ito, H., Satterlee, J., Sengupta, P., Matsumoto, K., & Mori, I. (2006). Identification of guanylyl cyclases that function in thermosensory neurons of *Caenorhabditis elegans. Genetics, 172*, 2239–2252. doi:10.1534/genetics.105.050013

Jurado, P., Kodama, E., Tanizawa, Y., & Mori, I. (2010). Distinct thermal migration behaviors in response to different thermal gradients in *Caenorhabditis elegans. Genes, Brain and Behavior, 9*(1), 120–127.

Kage-Nakadai, E., Ohta, A., Ujisawa, T., Sun, S., Nishikawa, Y., Kuhara, A., & Mitani, S. (2016). *Caenorhabditis elegans* homologue of Prox1/Prospero is expressed in the glia and is required for sensory behavior and cold tolerance. *Genes to Cells, 21*, 936–948. doi:10.1111/gtc.12394

Kimura, K.D., Miyawaki, A., Matsumoto, K., & Mori, I. (2004). The *C. elegans* thermosensory neuron AFD responds to warming. *Current Biology, 14*, 1291–1295. doi:10.1016/j.cub.2004.06.060

Klapper, M., Findeis, D., Koefeler, H., & Doring, F. (2016). Methyl group donors abrogate adaptive responses to dietary restriction in *C. elegans. Genes and Nutrition, 11*(1), 4. doi:10.1186/s12263-016-0522-4

Kobayashi, K., Nakano, S., Amano, M., Tsuboi, D., Nishioka, T., Ikeda, S., ... Mori, I. (2016). Single-cell memory regulates a neural circuit for sensory behavior. *Cell Reports, 14*(1), 11–21. doi:10.1016/j.celrep.2015.11.064

Kodama, E., Kuhara, A., Mohri-Shiomi, A., Kimura, K.D., Okumura, M., Tomioka, M., ... Mori, I. (2006). Insulin-like signaling and the neural circuit for integrative behavior in *C. elegans. Genes and Development, 20*, 2955–2960. doi:10.1101/gad.1479906

Kohshima, S. (1984). A novel cold-tolerant insect found in a Himalayan glacier. *Nature, 310*, 225–227. doi:10.1038/310225a0

Kuhara, A., Okumura, M., Kimata, T., Tanizawa, Y., Takano, R., Kimura, K.D., ... Mori, I. (2008). Temperature sensing by an olfactory neuron in a circuit controlling behavior of *C. elegans. Science, 320*, 803–807. doi:10.1126/science.1148922

Liu, F., Xiao, Y., Ji, X.L., Zhang, K.Q., & Zou, C.G. (2017). The cAMP-PKA pathway-mediated fat mobilization is required for cold tolerance in *C. elegans. Scientific Reports, 7*(1), 638. doi:10.1038/s41598-017-00630-w[PMC].

Liu, J., Ward, A., Gao, J., Dong, Y., Nishio, N., Inada, H., ... Xu, X.Z.S. (2010). *C. elegans* phototransduction requires a G protein-dependent cGMP pathway and a taste receptor homolog. *Nature Neuroscience, 13*, 715–722. doi:10.1038/nn.2540

Liu, S., Schulze, E., & Baumeister, R. (2012). Temperature- and touch-sensitive neurons couple CNG and TRPV channel activities to control heat avoidance in *Caenorhabditis elegans. PLOS One, 7*, e32360. doi:10.1371/journal.pone.0032360

Luo, L., Clark, D.A., Biron, D., Mahadevan, L., & Samuel, A.D. (2006). Sensorimotor control during isothermal tracking in *Caenorhabditis elegans. Journal of Experimental Biology, 209*, 4652–4662. doi:10.1242/jeb.02590

Luo, L., Cook, N., Venkatachalam, V., Martinez-Velazquez, L.A., Zhang, X., Calvo, A.C., ... Samuel, A.D.T. (2014). Bidirectional thermotaxis in *Caenorhabditis elegans* is mediated by distinct sensorimotor strategies driven by the AFD thermosensory neurons. *Proceedings of the National Academy of Sciences of the United States of America, 111*, 2776–2781. doi:10.1073/pnas.1315205111

Maruyama, I.N. (2016). Receptor guanylyl cyclases in sensory processing. *Frontiers in Endocrinology, 7*, 173. doi:10.3389/fendo.2016.00173[PMC].

Meng, L., Mulcahy, B., Cook, S.J., Neubauer, M., Wan, A., Jin, Y., & Yan, D. (2015). The cell death pathway regulates synapse elimination through cleavage of gelsolin in *Caenorhabditis elegans* neurons. *Cell Reports, 11*(11), 1737–1748. doi:10.1016/j.celrep.2015.05.031

Metaxakis, A., Petratou, D., & Tavernarakis, N. (2018). Multimodal sensory processing in *Caenorhabditis elegans. Open Biology, 8*, 180049.

Mohammadi, A., Byrne Rodgers, J., Kotera, I., & Ryu, W.S. (2013). Behavioral response of *Caenorhabditis elegans* to localized thermal stimuli. *BMC Neuroscience, 14*(1), 66. doi:10.1186/1471-2202-14-66

Mohri, A., Kodama, E., Kimura, K.D., Koike, M., Mizuno, T., & Mori, I. (2005). Genetic control of temperature preference in the nematode *Caenorhabditis elegans. Genetics, 169*, 1437–1450. doi:10.1534/genetics.104.036111

Mori, I., & Ohshima, Y. (1995). Neural regulation of thermotaxis in *Caenorhabditis elegans. Nature, 376*, 344–348. doi:10.1038/376344a0

Murray, P., Hayward, S.A., Govan, G.G., Gracey, A.Y., & Cossins, A.R. (2007). An explicit test of the phospholipid saturation hypothesis of acquired cold tolerance in *Caenorhabditis elegans. Proceedings of the National Academy of Sciences of the United States of America, 104*, 5489–5494. doi:10.1073/pnas.0609590104

Narayan, A., Laurent, G., & Sternberg, P.W. (2011). Transfer characteristics of a thermosensory synapse in *Caenorhabditis elegans. Proceedings of the National Academy of Sciences of the United States of America, 108*, 9667–9672. doi:10.1073/pnas.1106617108

Nguyen, P.A., Liou, W., Hall, D.H., & Leroux, M.R. (2014). Ciliopathy proteins establish a bipartite signaling compartment in a *C. elegans* thermosensory neuron. *Journal of Cell Science, 127*, 5317–5330.

Ohnishi, N., Kuhara, A., Nakamura, F., Okochi, Y., & Mori, I. (2011). Bidirectional regulation of thermotaxis by glutamate transmissions in *Caenorhabditis elegans. The EMBO Journal, 30*, 1376–1388. doi:10.1038/emboj.2011.13

Ohta, A., Ujisawa, T., Sonoda, S., & Kuhara, A. (2014). Light and pheromone-sensing neurons regulates cold habituation through insulin signalling in *Caenorhabditis elegans. Nature Communications*, 5(1), 4412. doi:10.1038/ncomms5412

Okahata, M., Ohta, A., Mizutani, H., Minakuchi, Y., Toyoda, A., & Kuhara, A. (2016). Natural variations of cold tolerance and temperature acclimation in *Caenorhabditis elegans. Journal of Comparative Physiology B, 186*, 985–998. doi:10.1007/s00360-016-1011-3

Okahata, M., Wei, A.D., Ohta, A., & Kuhara, A. (2019). Cold acclimation via the KQT-2 potassium channel is modulated by oxygen in *Caenorhabditis elegans. Science Advances, 5*, eaav3631. doi:10.1126/sciadv.aav3631

Okochi, Y., Kimura, K.D., Ohta, A., & Mori, I. (2005). Diverse regulation of sensory signaling by *C. elegans* nPKC-epsilon/eta TTX-4. *The EMBO Journal, 24*, 2127–2137.

Omori, K., & Kotera, J. (2007). Overview of PDEs and their regulation. *Circulation Research, 100*, 309–327.

Ouellette, M.H., Desrochers, M.J., Gheta, I., Ramos, R., & Hendricks, M. (2018). A gate-and-switch model for head orientation behaviors in *Caenorhabditis elegans. eNeuro, 5*, doi:10.1523/ENEURO.0121-18.2018

Palkar, R., Lippoldt, E.K., & McKemy, D.D. (2015). The molecular and cellular basis of thermosensation in mammals. *Current Opinion in Neurobiology, 34*, 14–19.

Patapoutian, A., Peier, A.M., Story, G.M., & Viswanath, V. (2003). ThermoTRP channels and beyond: Mechanisms of temperature sensation. *Nature Reviews Neuroscience, 4*, 529–539. doi:10.1038/nrn1141

Perkins, L.A., Hedgecock, E.M., Thomson, J.N., & Culotti, J.G. (1986). Mutant sensory cilia in the nematode *Caenorhabditis elegans. Developmental Biology, 117*, 456–487. doi:10.1016/0012-1606(86)90314-3

Pierce-Shimomura, J.T., Morse, T.M., & Lockery, S.R. (1999). The fundamental role of pirouettes in *Caenorhabditis elegans* chemotaxis. *The Journal of Neuroscience, 19*, 9557–9569. doi:10.1523/JNEUROSCI.19-21-09557.1999

Pilon, M., & Svensk, E. (2013). PAQR-2 may be a regulator of membrane fluidity during cold adaptation. *Worm, 2*, e27123. doi:10.4161/worm.27123

Potter, L.R. (2011). Guanylyl cyclase structure, function and regulation. *Cellular Signalling, 23*, 1921–1926. doi:10.1016/j.cellsig.2011.09.001

Ramot, D., MacInnis, B.L., & Goodman, M.B. (2008). Bidirectional temperature-sensing by a single thermosensory neuron in *C. elegans. Nature Neuroscience, 11*, 908–915. doi:10.1038/nn.2157

Ramot, D., MacInnis, B.L., Lee, H.C., & Goodman, M.B. (2008). Thermotaxis is a robust mechanism for thermoregulation in *Caenorhabditis elegans* nematodes. *Journal of Neuroscience, 28*, 12546–12557. doi:10.1523/JNEUROSCI.2857-08.2008

Robinson, J.D., & Powell, J.R. (2016). Long-term recovery from acute cold shock in *Caenorhabditis elegans. BMC Cell Biology, 17*, 2. doi:10.1186/s12860-015-0079-z

Ryu, W.S., & Samuel, A.D. (2002). Thermotaxis in *Caenorhabditis elegans* analyzed by measuring responses to defined thermal stimuli. *The Journal of Neuroscience, 22*, 5727–5733. doi:10.1523/JNEUROSCI.22-13-05727.2002

Sakurai, M. (2012). The functional mechanism of trehalose as a stress protectant from a viewpoint of its hydration property. *Cryobiology and Cryotechnology, 58*(1), 41–51.

Satterlee, J.S., Sasakura, H., Kuhara, A., Berkeley, M., Mori, I., & Sengupta, P. (2001). Specification of thermosensory neuron fate in *C. elegans* requires ttx-1, a homolog of otd/Otx. *Neuron, 31*, 943–956. doi:10.1016/s0896-6273(01)00431-7

Savory, F.R., Sait, S.M., & Hope, I.A. (2011). DAF-16 and Delta9 desaturase genes promote cold tolerance in long-lived *Caenorhabditis elegans* age-1 mutants. *PLOS One., 6*, e24550. doi:10.1371/journal.pone.0024550

Schafer, W.R. (2012). Tackling thermosensation with multidimensional phenotyping. *BMC Biology, 10*(1), 91. doi:10.1186/1741-7007-10-91

Seybold, A.C., Wharton, D.A., Thorne, M.A.S., & Marshall, C.J. (2017). Investigating trehalose synthesis genes after cold acclimation in the Antarctic nematode *Panagrolaimus* sp. *Biology Open, 6*, 1953–1959. doi:10.1242/bio.023341

Shaham, S. (2015). Glial development and function in the nervous system of *Caenorhabditis elegans. Cold Spring Harbor Perspectives in Biology, 7*, a020578. doi:10.1101/cshperspect.a020578

Singhvi, A., Liu, B., Friedman, C.J., Fong, J., Lu, Y., Huang, X.Y., & Shaham, S. (2016). A glial K/Cl transporter controls neuronal receptive ending shape by chloride inhibition of an rGC. *Cell, 165*, 936–948. doi:10.1016/j.cell.2016.03.026

Sonoda, S., Ohta, A., Maruo, A., Ujisawa, T., & Kuhara, A. (2016). Sperm affects head sensory neuron in temperature tolerance of *Caenorhabditis elegans. Cell Reports, 16*(1), 56–65. doi:10.1016/j.celrep.2016.05.078

Sugi, T., Nishida, Y., & Mori, I. (2011). Regulation of behavioral plasticity by systemic temperature signaling in *Caenorhabditis elegans. Nature Neuroscience, 14*, 984–992. doi:10.1038/nn.2854

Svensk, E., Devkota, R., Ståhlman, M., Ranji, P., Rauthan, M., Magnusson, F., ... Pilon, M. (2016). *Caenorhabditis elegans* PAQR-2 and IGLR-2 protect against glucose toxicity by modulating membrane lipid composition. *PLOS Genetics, 12*, e1005982. doi:10.1371/journal.pgen.1005982

Svensk, E., Ståhlman, M., Andersson, C.H., Johansson, M., Borén, J., & Pilon, M. (2013). PAQR-2 regulates fatty acid desaturation during cold adaptation in *C. elegans. PLOS Genetics, 9*, e1003801. doi:10.1371/journal.pgen.1003801

Takagaki, N., Ohta, A., Ohnishi, K., Kawanabe A., Minakuchi, Y., Toyoda, A., Fujiwara, Y., & Kuhara, A. (2020). The mechanoreceptor DEG-1 regulates cold tolerance in *Caenorhabditis elegans. EMBO reports*, e48671, 1–14.

Takeishi, A., Yu, Y.V., Hapiak, V.M., Bell, H.W., O'Leary, T., & Sengupta, P. (2016). Receptor-type guanylyl cyclases confer thermosensory responses in *C. elegans. Neuron, 90*, 235–244. doi:10.1016/j.neuron.2016.03.002

Tanizawa, Y., Kuhara, A., Inada, H., Kodama, E., Mizuno, T., & Mori, I. (2006). Inositol monophosphatase regulates localization of synaptic components and behavior in the mature nervous system of *C. elegans. Genes and Development, 20*, 3296–3310. doi:10.1101/gad.1497806

Tobin, D.M., Madsen, D.M., Kahn-Kirby, A., Peckol, E.L., Moulder, G., Barstead, R., ... Bargmann, C.I. (2002). Combinatorial expression of TRPV channel proteins defines their sensory functions and subcellular localization in *C. elegans* neurons. *Neuron, 35*, 307–318. doi:10.1016/S0896-6273(02)00757-2

Tsukada, Y., Yamao, M., Naoki, H., Shimowada, T., Ohnishi, N., Kuhara, A., … Mori, I. (2016). Reconstruction of spatial thermal gradient encoded in thermosensory neuron AFD in *Caenorhabditis elegans*. *The Journal of Neuroscience, 36*, 2571–2581. doi:10.1523/JNEUROSCI.2837-15.2016

Ujisawa, T., Ohta, A., Ii, T., Minakuchi, Y., Toyoda, A., Ii, M., & Kuhara, A. (2018). Endoribonuclease ENDU-2 regulates multiple traits including cold tolerance via cell autonomous and nonautonomous controls in *Caenorhabditis elegans*. *Proceedings of the National Academy of Sciences of the United States of America, 115*, 8823–8828. doi:10.1073/pnas.1808634115

Ujisawa T., Ohta A., Okahata M., Sonoda S., Kuhara A. (2014). Cold tolerance assay for studying cultivation-temperature-dependent cold habituation in *C. elegans*. *Protocol Exchange*, doi:10.1038/protex.2014.032

Ujisawa, T., Ohta, A., Uda-Yagi, M., & Kuhara, A. (2016). Diverse regulation of temperature sensation by trimeric G-protein signaling in *Caenorhabditis elegans*. *PLOS One, 11*(10), e0165518, 1–20. doi:10.1371/journal.pone.0165518

Venkatachalam, V., Ji, N., Wang, X., Clark, C., Mitchell, J.K., Klein, M., … Samuel, A.D.T. (2016). Pan-neuronal imaging in roaming *Caenorhabditis elegans*. *Proceedings of the National Academy of Sciences of the United States of America, 113*, E1082–E1088. doi:10.1073/pnas.1507109113

Voets, T. (2014). TRP channels and thermosensation. *Handbook of Experimental Pharmacology, 223*, 729–741. doi:10.1007/978-3-319-05161-1_1

Wang, D., O'Halloran, D., & Goodman, M.B. (2013). GCY-8, PDE-2, and NCS-1 are critical elements of the cGMP-dependent thermotransduction cascade in the AFD neurons responsible for *C. elegans* thermotaxis. *The Journal of General Physiology, 142*, 437–449. doi:10.1085/jgp.201310959

Wasserman, S.M., Beverly, M., Bell, H.W., & Sengupta, P. (2011). Regulation of response properties and operating range of the AFD thermosensory neurons by cGMP signaling. *Current Biology, 21*, 353–362. doi:10.1016/j.cub.2011.01.053

Wharton, D., & Ferns, D. (1995). Survival of intracellular freezing by the Antarctic nematode *Panagrolaimus davidi*. *The Journal of Experimental Biology, 198*, 1381–1387.

Wharton, D.A., & Brown, I.M. (1991). Cold-tolerance mechanisms of the Antarctic nematode *Panagrolaimus davidi*. *Journal of Experimental Biology, 155*(1), 629–641. doi:10.1242/jeb.00083

White, J.G., Southgate, E., Thomson, J.N., & Brenner, S. (1986). The structure of the nervous system of the nematode *Caenorhabditis elegans*. *Philosophical Transactions of the Royal Society of London. Series B, Biological Sciences, 314*, 1–340. doi:10.1098/rstb.1986.0056

Wittenburg, N., & Baumeister, R. (1999). Thermal avoidance in *Caenorhabditis elegans*: An approach to the study of nociception. *Proceedings of the National Academy of Sciences of the United States of America, 96*, 10477–10482. doi:10.1073/pnas.96.18.10477

Woldemariam, S., Nagpal, J., Hill, T., Li, J., Schneider, M.W., Shankar, R., … L'Etoile, N. (2019). Using a robust and sensitive GFP-based cGMP sensor for real-time imaging in intact *Caenorhabditis elegans*. *Genetics, 213*(1), 59–77. doi:10.1534/genetics.119.302392

Wu, X., Zhu, X., Wang, Y., Liu, X., Chen, L., & Duan, Y. (2018). The cold tolerance of the northern root-knot nematode, *Meloidogyne hapla*. *PLOS One, 13*(1), e0190531. doi:10.1371/journal.pone.0190531

Xiao, R., Zhang, B., Dong, Y., Gong, J., Xu, T., Liu, J., & Xu, X.Z. (2013). A genetic program promotes *C. elegans* longevity at cold temperatures via a thermosensitive TRP channel. *Cell, 152*, 806–817. doi:10.1016/j.cell.2013.01.020

Yamada, Y., & Ohshima, Y. (2003). Distribution and movement of *Caenorhabditis elegans* on a thermal gradient. *Journal of Experimental Biology, 206*, 2581–2593. doi:10.1242/jeb.00477

Yoshida, A., Nakano, S., Suzuki, T., Ihara, K., Higashiyama, T., & Mori, I. (2016). A glial K(+)/Cl(-) cotransporter modifies temperature-evoked dynamics in *Caenorhabditis elegans* sensory neurons. *Genes, Brain and Behavior, 15*, 429–440. doi:10.1111/gbb.12260

Yu, S., Avery, L., Baude, E., & Garbers, D.L. (1997). Guanylyl cyclase expression in specific sensory neurons: A new family of chemosensory receptors. *Proceedings of the National Academy of Sciences of the United States of America, 94*, 3384–3387. doi:10.1073/pnas.94.7.3384

Yu, Y.V., Bell, H.W., Glauser, D., Van Hooser, S.D., Goodman, M.B., & Sengupta, P. (2014). CaMKI-dependent regulation of sensory gene expression mediates experience-dependent plasticity in the operating range of a thermosensory neuron. *Neuron, 84*, 919–926. doi:10.1016/j.neuron.2014.10.046

Zariwala, H.A., Miller, A.C., Faumont, S., & Lockery, S.R. (2003). Step response analysis of thermotaxis in *Caenorhabditis elegans*. *The Journal of Neuroscience, 23*, 4369–4377. doi:10.1523/JNEUROSCI.23-10-04369.2003

Mechano-gated channels in *C. elegans*

Umar Al-Sheikh and Lijun Kang

ABSTRACT
Mechanosensation such as touch, hearing and proprioception, is functionally regulated by mechano-gated ion channels through the process of transduction. Mechano-gated channels are a subtype of gated ion channels engaged in converting mechanical stimuli to chemical or electrical signals thereby modulating sensation. To date, a few families of mechano-gated channels (DEG/ENaC, TRPN, K_2P, TMC and Piezo) have been identified in eukaryotes. Using a tractable genetic model organism *Caenorhabditis elegans*, the molecular mechanism of mechanosensation have been the focus of much research to comprehend the process of mechanotransduction. Comprising of almost all metazoans classes of ion channels, transporters and receptors, *C. elegans* is a powerful genetic model to explore mechanosensitive behaviors such as touch sensation and proprioception. The nematode relies primarily on its sensory abilities to survive in its natural environment. Genetic screening, calcium imaging and electrophysiological analysis have established that ENaC proteins and TRPN channel (TRP-4 protein) can characterize mechano-gated channels in *C. elegans*. A recent study reported that TMCs are likely the pore-forming subunit of a mechano-gated channel in *C. elegans*. Nevertheless, it still remains unclear whether Piezo as well as other candidate proteins can form mechano-gated channels in *C. elegans*.

Introduction

Mechano-gated channels are evolutionarily conserved mechanical gates regulating mechanosensation like touch, hearing and proprioception (Kung, 2005). Through mechanotransduction process, gated ion channels convert mechanical stimuli into electrochemical signals thereby triggering mechanosensation. When a mechanical stimulus is applied, membrane tension or force spring leads to structural deformation of the gated protein. As such, the gated channel opens in order to allow the flow of ions to generate graded receptor potentials which triggers mechanosensation (Marshall & Lumpkin, 2012). Bacterial mechanosensitive channels (MscL, MscS and MscM) in *E.coli* are gated by changes of membrane tension forming non-selective pores through which hydrated ions and solutes can flow, and act as osmosensors for turgor control (Rasmussen & Rasmussen, 2018). In eukaryotes, a few cation-selective channels — degenerin and epithelial sodium channels (DEG/ENaC), N-type Transient receptor potential (TRPN), two-pore potassium channels (K_2P), transmembrane-like proteins (TMC) and Piezo have been classified as *bona fide* mechano-gated channels (Delmas & Coste, 2013; Jin, Jan, & Jan, 2020). It is still enigmatic whether any anion channel such as chlorides can characterize mechano-gated channels.

In the mammalian nervous system, mechano-gated channels are often located around subcutaneous regions and inner ear hair cells to mediate touch sensation, hearing and proprioception (Wu *et al.*, 2017). Previous studies have demonstrated that Piezo2 required for cutaneous Merkel cell mechanotransduction, plays a major role in light-touch mechanosensation (Woo *et al.*, 2014). Strikingly, latest reports reveal that TMC1/2, TMIE, PCDH15/CDH23, CIB, ankyrin and PIP_2 serve as key gating components of hair cell mechanotransduction channel, in charge of mammalian hearing and balance (Al-Sheikh & Kang, 2020). However, most channel proteins engaged in mechanosensation may not be the primary mechanoreceptor but function indirectly as a secondary messenger in response to a mechanical stimulus (Christensen & Corey, 2007). Yet, it is quite difficult to investigate primary and secondary mechanoreceptors in mammals because (1) mammalian mechanosensory cells are sparsely abundant; (2) the expression levels of mechano-gated channel proteins are usually very low; (3) mechanotransduction apparatus is typically formed as a protein complex, which makes it extremely difficult to be functionally reconstituted in a heterologous system; (4) even though new genetic tools such as CRISPR-Cas9 system ease mammalian genetic manipulations, it is arduous to set up a whole-genome wide genetic screening to identify new candidates of mechano-gated channels in a mammalian system.

Therefore the invertebrate *Caenorhabditis elegans* fits in as a perfect model organism to uncover novel channel candidates or complexes in order to decipher the molecular mechanism of mechanotransduction. *C. elegans* depends predominantly on its sensory abilities to survive and reproduce in its environment (Takeishi, Takagaki, & Kuhara, 2020). The worms are mainly hermaphrodites which can be differentiated from their male counterparts based on behavioral responses and anatomical features. The eyeless nematodes can detect at least two distinct types of touch stimuli—gentle touch and harsh touch. They respond to mechanical stimulation with a bundle of touch behaviors namely nose touch response, gentle touch response, harsh touch response, basal slowing response, proprioception, head withdrawal and head swing response (Ardiel & Rankin, 2008; Goodman, 2006; Kang, Gao, Schafer, Xie, & Xu, 2010; Li, Kang, Piggott, Feng, & Xu, 2011). Furthermore, out of their 302 neurons, hermaphrodites have more than 40 sensory neurons that might be recruited to detect mechanical stimulation (Goodman, 2006; Li *et al.*, 2011) (Figure 1). Males possess at least 52 additional putative mechanoreceptor neurons which are responsible for mating (Goodman, 2006). Notably, hermaphrodites have 56 glial cells, some of which may be also directly involved in sensory transduction (Ding *et al.*, 2015). In addition, easily tractable *C. elegans* behaviors with fast genetic screening enabled us to explore mechanically activated sensation. Lastly, powerful functional assays such as *in vivo* patch-clamp recordings in live worms contribute to determine whether a candidate protein is a primary gated

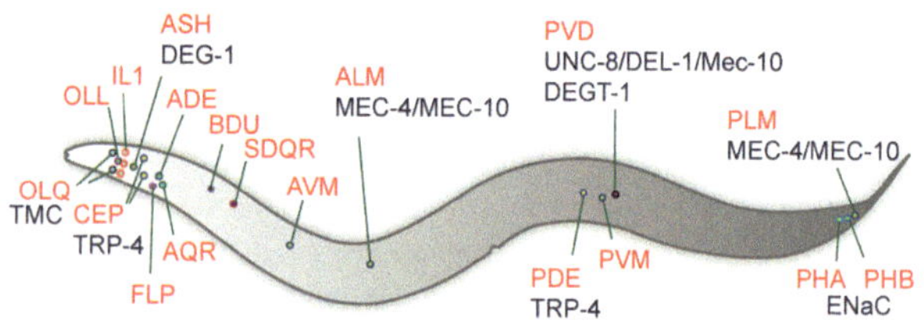

Figure 1. Putative mechanosensory neurons in hermaphrodites. More than 40 putative sensory neurons are recruited to detect mechanical stimulation in hermaphrodites. Note that the processes of the neurons are not shown and only the mechano-gated channels in *C. elegans* are also shown. (Please refer to: (Geffeney *et al.*, 2011; Kang *et al.*, 2010; Katta, Krieg, & Goodman, 2015; Li *et al.*, 2011; O'Hagan *et al.*, 2005; Tang *et al.*, 2020; Tao *et al.*, 2019; Zou *et al.*, 2017).

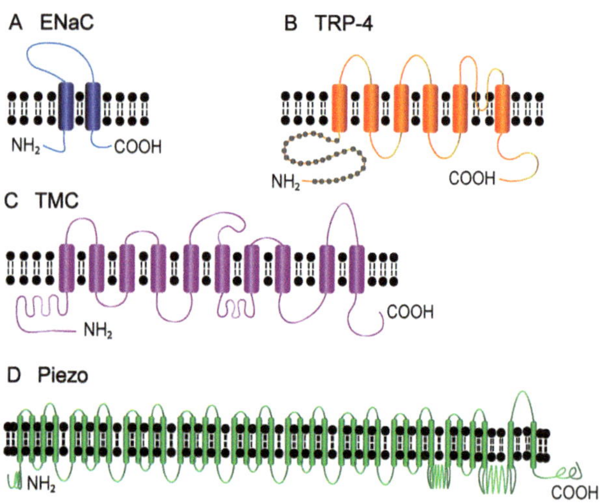

Figure 2. A schematic model denoting subunits of ENaC, TRP-4, TMC and Piezo. Some ENaCs (MEC-4, MEC-10, DEG-1, DEL-1, UNC-8, DEGT-1), TRP (TRP-4) and TMC-1 can form mechano-gated channels in *C. elegans*. It is yet unknown whether Piezo/PEZO-1 can form mechano-gated channel in *C. elegans*.

channel modulator or indirectly activated by a secondary messenger in response to a mechanical stimulus (Goodman, 2006; Kang *et al.*, 2010; Li *et al.*, 2011).

DEG/ENaC channels

The degenerin (DEG) and epithelial sodium channels (ENaC) are a metazoan superfamily of cation channels, with homologous sequences and protein structures, expressed in epithelial cells to the nervous system (Lumpkin, Marshall, & Nelson, 2010) (Figure 2). DEG/ENaC ion channels have been involved in numerous biophysiological functions particularly sensory responses in nematodes, flies, and mammals (Hill & Ben-Shahar, 2018). The epithelial sodium channels (ENaC) family is a voltage independent and amiloride-sensitive Na^+ channel which transport ions from epithelial cells and act as acid sensing ion channels that may contribute to pain perception and mechanosensation (Chalfie, 2009). In contrast, degenerins (DEG) are *C. elegans* specific ion channels characterized for the unusual gain-of-function mutations which induce degenerations like swelling, vacuolation and eventually cell death (Hall *et al.*, 1997). So far, about 30 genes encoding DEG/ENaC superfamily have been discovered in the *C. elegans* genome (Fechner *et al.*, 2020). Studies in mechanoreceptor currents of DEG/ENaC channels revealed similarities to mechanoelectrical transduction of gentle and nociceptive mechanical stimuli. Classical genetic, molecular and electrophysiological approaches enabled us to identify several DEG/ENaC family members implicated in gentle or harsh touch-transduction mechanisms (Chalfie, 2009; Geffeney *et al.*, 2011).

MEC-4/MEC-10 is the first metazoan touch-mediated complex discovered which disrupts gentle touch sensitivity of the worms (O'Hagan, Chalfie, & Goodman, 2005; Suzuki *et al.*, 2003). MEC-4 and MEC-10 subunits interact to form a heteromeric mechanotransduction gated channel along with two structural components, stomatin-like protein, MEC-2 and paraoxonase-like protein, MEC-6 (Brown, Liao, & Goodman, 2008; O'Hagan *et al.*, 2005). While mutations in the MEC-4/MEC-10 complex decrease neuronal intracellular calcium levels in response to gentle touch, the worms turn out to be touch insensitive—suggesting that MEC-4/MEC-10 complex is crucial to touch mechanistic processes (Chalfie & Sulston, 1981; Chatzigeorgiou *et al.*, 2010). Using *in vivo* whole-cell patch-clamp technique, the MEC-4/MEC-10 complex evoked mechanically activated currents which are responsible for transducing mechanical stimulus of light touch sensitivity in gentle touch neurons like ALM and PLM (O'Hagan *et al.*, 2005).

Another DEG/ENaC family member, DEG-1 is required for nose-touch response which is mediated by ASH nociceptive neurons. Mutations of *deg-1* gene dramatically reduce the magnitude of ASH mechanotransduction currents evoked by nose touch responses (Ding *et al.*, 2015; Geffeney *et al.*, 2011). Similarly, another degenerin gene, *unc-8* mutations lead to swelling and degeneration of interneurons along with uncoordinated body movement. An interesting study reported that distinct DEG/ENaC channels expressed in the multidendritic PVD neurons sense two kinds of mechanical stimuli—external touch and proprioceptive body movement using parallel signaling pathways (Chatzigeorgiou *et al.*, 2010; Tao *et al.*, 2019). In short, a heterotrimeric complex comprising of UNC-8/DEL-1/MEC-10 forms a mechanotransduction channel which controls proprioception by the release of NLP-12 neuropeptides from DVA interneurons at neuromuscular junctions while DEGT-1 evoked calcium responses in PVD neurons facilitating harsh touch sensation (Tao *et al.*, 2019).

Mammalian somatosensory neurons express six subtypes of acid-sensing ion channels (ASICs), which are proton-gated ion channels belonging to the DEG/ENaC family. It has been suggested that ASICs may play dual functions in acid nociception and neurosensory mechanotransduction (Lee & Chen, 2018; Lin, Sun, & Chen, 2015). In *C. elegans*, two ASIC channels—DEL-7 and DEL-3 mediate food responses in the serotonergic neuron NSM that controls foraging behaviors (Rhoades *et al.*, 2019). However, it remains unknown whether ASICs directly form mechano-gated channels in either mammals or *C. elegans* (Lin *et al.*, 2015; Rhoades *et al.*, 2019).

TRP channels: TRP-4

Transient receptor potential (TRP) channels are from a family of cation channels involved in a range of sensory processes counting chemosensation, thermosensation, mechanosensation and pain sensation (Arnadottir & Chalfie, 2010; Christensen & Corey, 2007) (Figure 2). TRP channels are classified into seven subfamilies (TRPA, TRPC, TRPML, TRPM, TRPN, TRPP and TRPV), which are tetrameric cation channels that can link to other molecular complexes required in various functions (Christensen & Corey, 2007; Montell, 2005). The importance of TRP ion channels is highlighted by a plethora of diseases and channelopathies in all major organs subjected to dysfunctions or mutations (Christensen & Corey, 2007; Nilius, Voets, & Peters, 2005). Interestingly, TRP channels are also present in single-celled organisms (protozoa) like yeast but limited to only TRPL, TRPM and TRPP channels (Venkatachalam, Luo, & Montell, 2014).

TRPN channels are also known as NOMPC—no mechanoreceptor potential C (Li, Feng, Sternberg, & Xu, 2006). The *C. elegans* TRP family, TRP-4 is homologous to the *Drosophila* NOMPC and zebrafish TRPN, but there is no mammalian TRPN homologue (Walker, Willingham, & Zuker, 2000). The nematodes move at a fast speed in the absence of bacteria, constantly moving its nose tip around (foraging) to detect food across different surfaces (Liu, Qin, *et al.*, 2019). Prior to feeding, *C. elegans* exhibit basal slowing response reducing locomotion in the presence of food (Rivard *et al.*, 2010). TRP-4 channel mediates basal slowing response in dopaminergic neurons (CEP, PDE and ADE), proprioception in two interneurons (DVA and DVC), and nose tip touch sensation in CEP neurons (Kang *et al.*, 2010; Li *et al.*, 2011). Also, TRP-4 channel is directly activated by mechanical stimuli, and mutations in its pore forming domain alters ion selectivity of touch-evoked conductance (Kang *et al.*, 2010; Li *et al.*, 2011). 29 ankyrin repeats of *Drosophila* NOMPC protein were reported to tether microtubules of the cytoskeleton to form a gate spring in response to mechanical stimulation (Zhang *et al.*, 2018).

Further, *C. elegans* mating is among the most complex behaviors comprising of orchestrated physiological processes from both males and hermaphrodites. The male depends on mechanosensation for contact response to locate and penetrate the hermaphrodite vulva with its ciliated tail. Two mammalian TRPP (polycystin) genes, *PKD1* and *PKD2*, homologous to the *C. elegans lov-1* and *pkd-2*, respectively have been implicated in the mating behavior (Barr *et al.*, 2001). *pkd-2* mutant worms did not exhibit any defects in touch-evoked calcium responses of the male tail ray neurons even though PKD-2 is required for male mating behavior; suggesting that PKD-2 might not be the primary modulator of the mechano-gated channel in these neurons (Zou *et al.*, 2017). In addition, two *C. elegans* TRPV (vanilloid) channels—OSM-9 and OCR-2 are instrumental in ASH neurons arbitrating osmosensation, nociception and mechanosensation (Tobin *et al.*, 2002). OSM-9 modulates calcium kinetics of touch sensation in male tail ray neurons (Zhang *et al.*, 2018). However, the collective evidence from a number of studies suggest that OSM-9 is perhaps not a mechano-sensor but a modulator in touch receptor neurons (Chatzigeorgiou & Schafer, 2011; Geffeney *et al.*, 2011; Tang *et al.*, 2020; Zhang *et al.*, 2018).

Two-pore potassium (K_2P) channels

Evolutionary genetics revealed that potassium channels are widely expressed archaic ion channels across species. Potassium channels control the influx and efflux of K^+ ions through cell membranes (Douguet & Honore, 2019). The opposing polarization and depolarization of potassium versus calcium and sodium channels promote membrane potential/cell excitability for numerous vital cellular mechanics as well as survival. To date, four main classes of potassium channels are known—Calcium-activated (K_{ca}), Inward rectifying (K_{ir}), Tandem/two pore domain (K_2P) and voltage-gated potassium channels (K_V). Previously known as K^+ background (leak) channels, K_2P channel subunits are encoded by 15 KCNK mammalian genes, 11 *Drosophila* genes and 50 putative *C. elegans* genes. Out of the four potassium ion channels, only two K_2P subfamilies—Tandem pore domain in weak rectifying K^+ channel (TWIK) and TWIK-related K^+ channel (TREK) have been divulged as mechano-gated channels. Thus far, TREK1 (KCNK2), TREK2 (KCNK10) and TRAAK (TWIK-related arachidonic acid-stimulated K^+) channels have been found to be mechano-gated channels in mammals (Chalfie, 2009) but there is yet no mechanically activated K_2P channel exposed in *C. elegans*.

TMC protein family

Transmembrane channel-like (TMC) protein is a conserved non-selective ion channel-like family present from *C. elegans* to mammals (Yue *et al.*, 2018) (Figure 2). TMC proteins are implicated in numerous functions comprising of chemosensation and mechanosensation. The TMC proteins are essential for the proper functioning and survival of inner ear hair cells where mutations cause deafness in mice and humans (Cunningham *et al.*, 2020; Kawashima, Kurima, Pan, Griffith, & Holt, 2015; Pan *et al.*, 2013). Mammalian TMC1 and TMC2 (TMC1/2) are evolutionarily related to the *C. elegans* TMC-1 and TMC-2 proteins, respectively (Wang, Li, Liu, Liu, & Xu, 2016). Unlike other species, the nematode does not possess an advanced hearing organ but relies predominantly on its sensory neurons for mechano-perception (Kang *et al.*, 2010). Interestingly, TMC1/2 form sodium-leak channels in *C. elegans* neurons and muscles while modulating membrane potentials, which are essential for calcium activities of neurons and muscle cells (Yue *et al.*, 2018). Consistent with its role in *C. elegans*, TMC1 was lately reported to form a leak channel that modulates tonotopy and excitability of auditory hair cells in mice (Liu, Qin, *et al.*, 2019). In addition, TMC-1 evokes an inward current required for alkaline-sensation in ASH neurons corresponding to the requirements of *in vivo* nociceptive chemosensation (Wang *et al.*, 2016). Notably, it is reported that TMC-1 encodes a sodium-sensitive channel and mediates salt chemosensation in ASH neurons (Chatzigeorgiou, Bang, Hwang, & Schafer, 2013), however, other studies show that TMC-1 may not be required for sodium induced avoidance behaviors (Dao, Lee, Drecksel, Bittlingmaier, & Nelson, 2020; Wang *et al.*, 2016). Not long ago, TMC1/2 have been reported as pore-forming subunits for the mechanosensitive channels of auditory perception (Jia *et al.*, 2020), but still not yet clear how the mouse TMC1/2 are gated by mechanical stimulation. Recently uncovered, ankyrin binds to the actin cytoskeleton and TMC-1/CIB complex, acting as an intracellular tether to confer mechanosensation in both OLQ neurons and body wall muscles in *C. elegans* (Tang *et al.*, 2020). It should be noted that a recent study reported that TMC proteins do not contribute to mechanically-gated currents in body-wall muscles in *C. elegans* (Yan, Su, Cheng, & Liu, 2020), raising the possibility that TMCs may play an indirect modulative role in these cells. Further genetic and electrophysiological studies will be required to identify the roles and mechanisms of TMCs underlying *C. elegans* mechanosensation, which may provide new cues to uncover the gating mechanisms of mechanotransduction.

Piezo ion channels

Piezo protein family is a polymodal non-selective mechanically activated channel regulating transmembrane cation influx, most preferably calcium ions (Xiao & Xu, 2010) (Figure 2). Piezo channels participate in fundamental mechanosensitive roles in touch, hearing, proprioception, cellular homeostasis, blood pressure regulation, shear stress, and alveolar epithelium and endothelium instigating the Hering–Breuer inflation reflex to prevent over-inflation of the lungs (Zhong, Komarova, Rehman, & Malik, 2018). Piezo proteins are orthologous in countless eukaryotes such as plants, vertebrates, insects, nematodes, and pathogenic protozoans (Prole & Taylor, 2013). The sole *C. elegans* orthologue, *pezo-1* is expressed throughout the nematode development with strong expression in reproductive tissues to promote appropriate ovulation and fertilization (Bai *et al.*, 2019). Piezos (Piezo1 & Piezo2) are revealed as touch-sensors in fly, zebrafish, mouse and humans; however it seems like the *C. elegans* Piezo protein is absent from the nervous system and it is a challenge to elucidate whether it can function as a direct mechano-gated channel in the worm.

Gating mechanism

Channel gating regulates the kinetics of ions across cell membranes, required in vital cellular processes. A few mechano-gating mechanisms are known such as membrane tension and two primary models—single-tether and dual-tether (Chalfie, 2009; Douguet & Honore, 2019; Ranade, Syeda, & Patapoutian, 2015). Membrane tension gating is mediated by forces (osmotic pressures or mechanical stimuli) stretching the lipid bilayer to provide a pathway for ions to flow across the membrane. Membrane tension gating or "force from lipid" have been observed in eukaryotes and prokaryotes (Chalfie, 2009; Douguet & Honore, 2019). The dual tether model was proposed for vertebrate hair cell mechanotransduction where tip links proteins (PCDH15/CDH23) would act as an extracellular tether and an intracellular adaptation motor chained to the actin cytoskeleton and transduction channel would open and close as operating as a gate spring accordingly (Chalfie, 2009). The single-tether model is simple with an intracellular tether and membrane forces congregate to regulate the passage of ions across the channel (Chalfie, 2009). The gating mechanism of hair cell mechanotransduction might be very complex incorporating features of known tethering models and more (Al-Sheikh & Kang, 2020). For instance, ankyrin recently reported to tether the cytoskeleton and the TMC/CIB complex, conforms to the dual-tether model alongside the tip link proteins (PCDH15/CDH23) (Tang *et al.*, 2020). Whereas, another latest study about TMIE and PIP_2 (Phosphatidylinositol 4, 5-bisphosphate) revealed to promote membrane tension and mechanosensation (Cunningham *et al.*, 2020). As such, the mammalian hair cell mechanotransduction machinery is still enigmatic even after 62 years since the first report of hereditary deafness in a mouse model (Al-Sheikh & Kang, 2020).

Conclusion and perspective

C. elegans established a crucial role in the process to decipher mechanosensation and identify candidates of mechano-gated channels. So far, only ENaC, TRP (TRP-4) and TMC channels have been implicated to be mechano-gated channels in *C. elegans* while Piezo/PEZO-1 protein is under extensive scrutiny. TMC proteins form a sodium leak channel in both neurons and muscles and function as an alkaline-sensor in ASH neurons and probably operate as *C. elegans* mechano-sensors though the electrophysiological evidence is still lacking. Since, PEZO-1 was first reported in the reproductive functionalities, it is not yet confirmed as mechano-gated in *C. elegans*. It would be fascinating to decode the functions of PEZO-1 and other candidate proteins such as K_2P and ASIC channels in mechanosensation of nematodes. If they do form mechano-gated channels, it would be an exciting paragon to investigate the physiological function and gating mechanisms. If they do not sense mechanical stimulation, what makes them different from their homologues in higher organisms? In mammals, only a few classes of mechano-gated channels (Piezos, TMCs, and K_2Ps) have been recognized to mediate physiological events such as touch sensation, hearing, balance, cell volume regulation, shear force sensation, and even cancer cells migration. There may be some other mechano-gated channels not yet characterized whether protein complexes-forming gated channel activated by mechanical forces in both vertebrates and invertebrates. Further studies using the model organism *C. elegans* will bring forth deeper understanding of the molecular modalities and characteristics of mechano-gated channels.

Disclosure statement

No potential conflict of interest was reported by the author(s).

ORCID

Umar Al-Sheikh http://orcid.org/0000-0003-4977-822X
Lijun Kang http://orcid.org/0000-0001-9939-5134

References

Al-Sheikh, U., & Kang, L. (2020). Molecular crux of hair cell mechanotransduction machinery. *Neuron*, *107*(3), 404–406. doi:10.1016/j.neuron.2020.07.007

Ardiel, E.L., & Rankin, C.H. (2008). Behavioral plasticity in the *C. elegans* mechanosensory circuit. *Journal of Neurogenetics*, *22*(3), 239–255. doi:10.1080/01677060802298509

Arnadottir, J., & Chalfie, M. (2010). Eukaryotic mechanosensitive channels. *Annual Review of Biophysics*, *39*, 111–137. doi:10.1146/annurev.biophys.37.032807.125836

Bai, X., Bouffard, J., Lord, A., Brugman, K., Sternberg, P.W., Cram, E.J., & Golden, A. (2019). Caenorhabditis elegans PIEZO channel coordinates multiple reproductive tissues to govern ovulation. 9, *bioRxiv*, 847392. doi:10.1101/847392

Barr, M.M., DeModena, J., Braun, D., Nguyen, C.Q., Hall, D.H., & Sternberg, P.W. (2001). The *Caenorhabditis elegans* autosomal dominant polycystic kidney disease gene homologs lov-1 and pkd-2 act in the same pathway. *Current Biology : CB*, *11*(17), 1341–1346. doi:10.1016/S0960-9822(01)00423-7

Brown, A.L., Liao, Z., & Goodman, M.B. (2008). MEC-2 and MEC-6 in the *Caenorhabditis elegans* sensory mechanotransduction complex: Auxiliary subunits that enable channel activity. *The Journal of General Physiology*, *131*(6), 605–616. doi:10.1085/jgp.200709910

Chalfie, M. (2009). Neurosensory mechanotransduction. *Nature Reviews. Molecular Cell Biology*, *10*(1), 44–52. doi:10.1038/nrm2595

Chalfie, M., & Sulston, J. (1981). Developmental genetics of the mechanosensory neurons of *Caenorhabditis elegans*. *Developmental Biology*, *82*(2), 358–370. doi:10.1016/0012-1606(81)90459-0

Chatzigeorgiou, M., Bang, S., Hwang, S.W., & Schafer, W.R. (2013). tmc-1 encodes a sodium-sensitive channel required for salt chemosensation in *C. elegans*. *Nature*, *494*(7435), 95–99. doi:10.1038/nature11845

Chatzigeorgiou, M., Grundy, L., Kindt, K.S., Lee, W.H., Driscoll, M., & Schafer, W.R. (2010). Spatial asymmetry in the mechanosensory phenotypes of the *C. elegans* DEG/ENaC gene mec-10. *Journal of Neurophysiology*, *104*(6), 3334–3344. doi:10.1152/jn.00330.2010

Chatzigeorgiou, M., & Schafer, W.R. (2011). Lateral facilitation between primary mechanosensory neurons controls nose touch perception in *C. elegans*. *Neuron*, *70*(2), 299–309. doi:10.1016/j.neuron.2011.02.046

Christensen, A.P., & Corey, D.P. (2007). TRP channels in mechanosensation: Direct or indirect activation? *Nature Reviews. Neuroscience*, *8*(7), 510–521. doi:10.1038/nrn2149

Cunningham, C.L., Qiu, X., Wu, Z., Zhao, B., Peng, G., Kim, Y.-H., … Müller, U. (2020). TMIE defines pore and gating properties of the mechanotransduction channel of mammalian cochlear hair cells. *Neuron*, *107*(1), 126–143.e8. doi:10.1016/j.neuron.2020.03.033

Dao, J., Lee, A., Drecksel, D.K., Bittlingmaier, N.M., & Nelson, T.M. (2020). Characterization of TMC-1 in *C. elegans* sodium chemotaxis and sodium conditioned aversion. *BMC Genetics*, *21*(1), 37. doi:10.1186/s12863-020-00844-4

Delmas, P., & Coste, B. (2013). Mechano-gated ion channels in sensory systems. *Cell*, *155*(2), 278–284. doi:10.1016/j.cell.2013.09.026

Ding, G., Zou, W., Zhang, H., Xue, Y., Cai, Y., Huang, G., … Kang, L. (2015). In vivo tactile stimulation-evoked responses in *Caenorhabditis elegans* amphid sheath glia. *PLoS One*, *10*(2), e0117114. doi:10.1371/journal.pone.0117114

Douguet, D., & Honore, E. (2019). Mammalian mechanoelectrical transduction: Structure and function of force-gated ion channels. *Cell*, *179*(2), 340–354. doi:10.1016/j.cell.2019.08.049

Fechner, S., D'Alessandro, I., Wang, L., Tower, C., Tao, L., & Goodman, M.B. (2020). Functional and pharmacological characterization of *C. elegans* DEG/ENaC/ASIC channels. *Biophysical Journal*, *118*(3), 114a. doi:10.1016/j.bpj.2019.11.768

Geffeney, S.L., Cueva, J.G., Glauser, D.A., Doll, J.C., Lee, T.H.-C., Montoya, M., … Goodman, M.B. (2011). DEG/ENaC but not TRP channels are the major mechanoelectrical transduction channels in a C. elegans nociceptor. *Neuron*, *71*(5), 845–857. doi:10.1016/j.neuron.2011.06.038

Goodman, M.B. (2006). Mechanosensation. *WormBook*, 1–14. doi:10.1895/wormbook.1.62.1

Hall, D.H., Gu, G., García-Añoveros, J., Gong, L., Chalfie, M., &., & Driscoll, M. (1997). Neuropathology of degenerative cell death in *Caenorhabditis elegans. The Journal of Neuroscience*, *17*(3), 1033–1045. doi:10.1523/JNEUROSCI.17-03-01033.1997

Hill, A.S., & Ben-Shahar, Y. (2018). The synaptic action of Degenerin/Epithelial sodium channels. *Channels (Austin, Tex.)*, *12*(1), 262–275. doi:10.1080/19336950.2018.1495006

Jia, Y., Zhao, Y., Kusakizako, T., Wang, Y., Pan, C., Zhang, Y., … Yan, Z. (2020). TMC1 and TMC2 proteins are pore-forming subunits of mechanosensitive ion channels. *Neuron*, *105*(2), 310–321.e313. doi:10.1016/j.neuron.2019.10.017

Jin, P., Jan, L.Y., & Jan, Y.N. (2020). Mechanosensitive ion channels: Structural features relevant to mechanotransduction mechanisms. *Annual Review of Neuroscience*, *43*, 207–229. doi:10.1146/annurev-neuro-070918-050509

Kang, L., Gao, J., Schafer, W.R., Xie, Z., & Xu, X.Z. (2010). *C. elegans* TRP family protein TRP-4 is a pore-forming subunit of a native mechanotransduction channel. *Neuron*, *67*(3), 381–391. doi:10.1016/j.neuron.2010.06.032

Katta, S., Krieg, M., & Goodman, M.B. (2015). Feeling force: Physical and physiological principles enabling sensory mechanotransduction. *Annual Review of Cell and Developmental Biology*, *31*, 347–371. doi:10.1146/annurev-cellbio-100913-013426

Kawashima, Y., Kurima, K., Pan, B., Griffith, A.J., & Holt, J.R. (2015). Transmembrane channel-like (TMC) genes are required for auditory and vestibular mechanosensation. *Pflugers Archiv : European Journal of Physiology*, *467*(1), 85–94. doi:10.1007/s00424-014-1582-3

Kung, C. (2005). A possible unifying principle for mechanosensation. *Nature, 436*(7051), 647–654. doi:10.1038/nature03896

Lee, C.H., & Chen, C.C. (2018). Roles of ASICs in nociception and proprioception. *Advances in Experimental Medicine and Biology, 1099*, 37–47. doi:10.1007/978-981-13-1756-9_4

Li, W., Feng, Z., Sternberg, P.W., & Xu, X.Z. (2006). A *C. elegans* stretch receptor neuron revealed by a mechanosensitive TRP channel homologue. *Nature, 440*(7084), 684–687. doi:10.1038/nature04538

Li, W., Kang, L., Piggott, B.J., Feng, Z., & Xu, X.Z. (2011). The neural circuits and sensory channels mediating harsh touch sensation in *Caenorhabditis elegans. Nature Communications, 2*, 315. doi:10.1038/ncomms1308

Lin, S.H., Sun, W.H., & Chen, C.C. (2015). Genetic exploration of the role of acid-sensing ion channels. *Neuropharmacology, 94*, 99–118. doi:10.1016/j.neuropharm.2014.12.011

Liu, H., Qin, L.W., Li, R., Zhang, C., Al-Sheikh, U., & Wu, Z.X. (2019). Reciprocal modulation of 5-HT and octopamine regulates pumping via feedforward and feedback circuits in *C. elegans. Proceedings of the National Academy of Sciences of the United States of America, 116*(14), 7107–7112. doi:10.1073/pnas.1819261116

Liu, S., Wang, S., Zou, L., Li, J., Song, C., Chen, J., … Xiong, W. (2019). TMC1 is an essential component of a leak channel that modulates tonotopy and excitability of auditory hair cells in mice. *eLife, 8*. doi:10.7554/eLife.47441

Lumpkin, E.A., Marshall, K.L., & Nelson, A.M. (2010). The cell biology of touch. *The Journal of Cell Biology, 191*(2), 237–248. doi:10.1083/jcb.201006074

Marshall, K.L., & Lumpkin, E.A. (2012). The molecular basis of mechanosensory transduction. *Advances in Experimental Medicine and Biology, 739*, 142–155. doi:10.1007/978-1-4614-1704-0_9

Montell, C. (2005). The TRP superfamily of cation channels. *Science Signaling, 2005*(272), re3. doi:10.1126/stke.2722005re3

Nilius, B., Voets, T., & Peters, J. (2005). TRP channels in disease. *Science's STKE : Signal Transduction Knowledge Environment, 2005*(295), re8. doi:10.1126/stke.2952005re8

O'Hagan, R., Chalfie, M., & Goodman, M.B. (2005). The MEC-4 DEG/ENaC channel of *Caenorhabditis elegans* touch receptor neurons transduces mechanical signals. *Nature Neuroscience, 8*(1), 43–50. doi:10.1038/nn1362

Pan, B., Géléoc, G.S., Asai, Y., Horwitz, G.C., Kurima, K., Ishikawa, K., … Holt, J.R. (2013). TMC1 and TMC2 are components of the mechanotransduction channel in hair cells of the mammalian inner ear. *Neuron, 79*(3), 504–515. doi:10.1016/j.neuron.2013.06.019

Prole, D.L., & Taylor, C.W. (2013). Identification and analysis of putative homologues of mechanosensitive channels in pathogenic protozoa. *PLoS One, 8*(6), e66068. doi:10.1371/journal.pone.0066068

Ranade, S.S., Syeda, R., & Patapoutian, A. (2015). Mechanically activated ion channels. *Neuron, 87*(6), 1162–1179. doi:10.1016/j.neuron.2015.08.032

Rasmussen, T., & Rasmussen, A. (2018). Bacterial mechanosensitive channels. *Sub-cellular Biochemistry, 87*, 83–116. doi:10.1007/978-981-10-7757-9_4

Rhoades, J.L., Nelson, J.C., Nwabudike, I., Yu, S.K., McLachlan, I.G., Madan, G.K., … Flavell, S.W. (2019). ASICs mediate food responses in an enteric serotonergic neuron that controls foraging behaviors. *Cell, 176*(1–2), 85–97.e14. doi:10.1016/j.cell.2018.11.023

Rivard, L., Srinivasan, J., Stone, A., Ochoa, S., Sternberg, P.W., & Loer, C.M. (2010). A comparison of experience-dependent locomotory behaviors and biogenic amine neurons in nematode relatives of *Caenorhabditis elegans. BMC Neuroscience, 11*, 22. doi:10.1186/1471-2202-11-22

Suzuki, H., Kerr, R., Bianchi, L., Frøkjaer-Jensen, C., Slone, D., Xue, J., … Schafer, W.R. (2003). In vivo imaging of *C. elegans* mechanosensory neurons demonstrates a specific role for the MEC-4 channel in the process of gentle touch sensation. *Neuron, 39*(6), 1005–1017. doi:10.1016/j.neuron.2003.08.015

Takeishi, A., Takagaki, N., & Kuhara, A. (2020). Temperature signaling underlying thermotaxis and cold tolerance in *Caenorhabditis elegans. J Neurogenet*, 1–12. doi:10.1080/01677063.2020.1734001

Tang, Y.Q., Lee, S.A., Rahman, M., Vanapalli, S.A., Lu, H., & Schafer, W.R. (2020). Ankyrin is an intracellular tether for TMC mechanotransduction channels. *Neuron, 107*(4), 759–761. doi:10.1016/j.neuron.2020.03.026

Tao, L., Porto, D., Li, Z., Fechner, S., Lee, S.A., Goodman, M.B., … Shen, K. (2019). Parallel processing of two mechanosensory modalities by a single neuron in *C. elegans. Developmental Cell, 51*(5), 617–631.e3. doi:10.1016/j.devcel.2019.10.008

Tobin, D.M., Madsen, D.M., Kahn-Kirby, A., Peckol, E.L., Moulder, G., Barstead, R., … Bargmann, C.I. (2002). Combinatorial expression of TRPV channel proteins defines their sensory functions and subcellular localization in *C. elegans* neurons. *Neuron, 35*(2), 307–318. doi:10.1016/S0896-6273(02)00757-2

Venkatachalam, K., Luo, J., & Montell, C. (2014). Evolutionarily conserved, multitasking TRP channels: Lessons from worms and flies. *Handbook of Experimental Pharmacology, 223*, 937–962. doi:10.1007/978-3-319-05161-1_9

Walker, R.G., Willingham, A.T., & Zuker, C.S. (2000). A Drosophila mechanosensory transduction channel. *Science (New York, N.Y.), 287*(5461), 2229–2234. doi:10.1126/science.287.5461.2229

Wang, X., Li, G., Liu, J., Liu, J., & Xu, X.Z. (2016). TMC-1 mediates alkaline sensation in *C. elegans* through nociceptive neurons. *Neuron, 91*(1), 146–154. doi:10.1016/j.neuron.2016.05.023

Woo, S.-H., Ranade, S., Weyer, A.D., Dubin, A.E., Baba, Y., Qiu, Z., … Patapoutian, A. (2014). Piezo2 is required for Merkel-cell mechanotransduction. *Nature, 509*(7502), 622–626. doi:10.1038/nature13251

Wu, Z., Grillet, N., Zhao, B., Cunningham, C., Harkins-Perry, S., Coste, B., … Mueller, U. (2017). Mechanosensory hair cells express two molecularly distinct mechanotransduction channels. *Nature Neuroscience, 20*(1), 24–33. doi:10.1038/nn.4449

Xiao, R., & Xu, X.Z. (2010). Mechanosensitive channels: In touch with Piezo. *Current Biology : CB, 20*(21), R936–R938. doi:10.1016/j.cub.2010.09.053

Yan, Z., Su, Z., Cheng, X., & Liu, J. (2020). *Caenorhabditis elegans* body wall muscles sense mechanical signals with an amiloride-sensitive cation channel. *Biochemical and Biophysical Research Communications, 527*(2), 581–587. doi:10.1016/j.bbrc.2020.04.130

Yue, X., Zhao, J., Li, X., Fan, Y., Duan, D., Zhang, X., … Kang, L. (2018). TMC proteins modulate egg laying and membrane excitability through a background leak conductance in *C. elegans. Neuron, 97*(3), 571–585.e575. doi:10.1016/j.neuron.2017.12.041

Zhang, H., Yue, X., Cheng, H., Zhang, X., Cai, Y., Zou, W., … Kang, L. (2018). OSM-9 and an amiloride-sensitive channel, but not PKD-2, are involved in mechanosensation in *C. elegans* male ray neurons. *Scientific Reports, 8*(1), 7192. doi:10.1038/s41598-018-25542-1

Zhong, M., Komarova, Y., Rehman, J., & Malik, A.B. (2018). Mechanosensing Piezo channels in tissue homeostasis including their role in lungs. *Pulmonary Circulation, 8*(2), 2045894018767393. doi:10.1177/2045894018767393

Zou, W., Cheng, H., Li, S., Yue, X., Xue, Y., Chen, S., & Kang, L. (2017). Polymodal responses in *C. elegans* phasmid neurons rely on multiple intracellular and intercellular signaling pathways. *Scientific Reports, 7*, 42295. doi:10.1038/srep42295

OPEN ACCESS

What can a worm learn in a bacteria-rich habitat?

He Liu and Yun Zhang

ABSTRACT

With a nervous system that has only a few hundred neurons, *Caenorhabditis elegans* was initially not regarded as a model for studies on learning. However, the collective effort of the *C. elegans* field in the past several decades has shown that the worm displays plasticity in its behavioral response to a wide range of sensory cues in the environment. As a bacteria-feeding worm, *C. elegans* is highly adaptive to the bacteria enriched in its habitat, especially those that are pathogenic and pose a threat to survival. It uses several common forms of behavioral plasticity that last for different amounts of time, including imprinting and adult-stage associative learning, to modulate its interactions with pathogenic bacteria. Probing the molecular, cellular and circuit mechanisms underlying these forms of experience-dependent plasticity has identified signaling pathways and regulatory insights that are conserved in more complex animals.

Caenorhabditis elegans senses and responds to diverse environmental cues

Animals live in different ecological niches that are characteristic of different chemical, physical and biological cues and have likely evolved sensorimotor systems that are able to detect and respond to the environmental conditions of their habitats. *C. elegans* feeds on bacteria and is often found in decaying fruits or other organic matters that are rich in bacteria (Felix & Duveau, 2012; Frezal & Felix, 2015; Samuel, Rowedder, Braendle, Felix, & Ruvkun, 2016). It navigates its environment by detecting and responding to various chemical cues, including odorants and salts, temperature, pheromones, gases, as well as mechanical stimuli [(Figure 1) and (Aoki & Mori, 2015; Bargmann, 2006; Brandt *et al.*, 2012; Bretscher, Busch, & de Bono, 2008; Butcher, Fujita, Schroeder, & Clardy, 2007; Chalfie, 2009; Cheung, Cohen, Rogers, Albayram, & de Bono, 2005; de Bono & Maricq, 2005; Goodman *et al.*, 2014; Goodman & Sengupta, 2019; Gray *et al.*, 2004; Hallem *et al.*, 2011; Hao *et al.*, 2018; Jeong *et al.*, 2005; Kaplan & Horvitz, 1993; Kim *et al.*, 2009; Macosko *et al.*, 2009; Pierce-Shimomura, Faumont, Gaston, Pearson, & Lockery, 2001; Reddy, Hunter, Bhatla, Newman, & Kim, 2011; Schafer, 2015; Srinivasan *et al.*, 2008; 2012; White *et al.*, 2007; White & Jorgensen, 2012)]. Some of the odorants that are attractive to *C. elegans* can be produced by plants and may serve as cues representing an environment that is abundant in bacteria. In addition, *C. elegans* is known to navigate within a thermal gradient and the ambient temperature significantly regulates the development and life span of a worm. The sensorimotor response to chemical cues and temperature have been extensively studied in *C. elegans*. For example, a few ciliated sensory neurons use G-protein coupled seven-transmembrane receptors and cyclic nucleotide-gated channels (CNGs), to detect and mediate responses to odorants. The calcium-permeable CNGs transform odorant information into intracellular signals, which produce intercellular signals to engage postsynaptic interneurons and downstream motor neurons to generate movement towards or away from the odorants (Bargmann, 2006; de Bono & Maricq, 2005). Similarly, the major sensory neurons, as well as their intracellular signaling pathways, that perceive and respond to external salt concentration, ambient temperature gradient, pheromones, gases and mechanical cues have been identified and characterized (Aoki & Mori, 2015; Bargmann, 2006; Brandt *et al.*, 2012; Bretscher *et al.*, 2008; Butcher *et al.*, 2007; Chalfie, 2009; Cheung *et al.*, 2005; de Bono & Maricq, 2005; Goodman *et al.*, 2014; Goodman & Sengupta, 2019; Gray *et al.*, 2004; Hallem *et al.*, 2011; Hao *et al.*, 2018; Jeong *et al.*, 2005; Kaplan & Horvitz, 1993; Kim *et al.*, 2009; Macosko *et al.*, 2009; Pierce-Shimomura *et al.*, 2001; Reddy *et al.*, 2011; Schafer, 2015; Srinivasan *et al.*, 2008; 2012; White *et al.*, 2007; White & Jorgensen, 2012). The behavioral strategies and the underlying neural circuits through which *C. elegans* navigates a sensory environment are also intensively investigated (Aprison & Ruvinsky, 2019; Bargmann, 2006; Chalasani *et al.*, 2007; de Bono & Maricq, 2005; Donnelly *et al.*, 2013; Goodman & Sengupta, 2019; Gordus, Pokala, Levy, Flavell, & Bargmann, 2015; Gray, Hill, & Bargmann, 2005; Iino & Yoshida, 2009; Ikeda *et al.*, 2020; Jang *et al.*, 2012; Kaplan, Salazar Thula,

Khoss, & Zimmer, 2020; Kato *et al.*, 2015; Kunitomo *et al.*, 2013; Li, Liu, Zheng, & Xu, 2014; Liu *et al.*, 2018; Luo *et al.*, 2014; Macosko *et al.*, 2009; Mori & Ohshima, 1995; Pierce-Shimomura, Morse, & Lockery, 1999; Schafer, 2015; Tsalik & Hobert, 2003; Venkatachalam *et al.*, 2016; Wen, Gao, & Zhen, 2018; White *et al.*, 2007). The molecular, cellular and circuit bases for these sensorimotor responses provide the substrates for experience-dependent regulation. The studies that investigate various forms of learning in *C. elegans* have been reviewed elsewhere (Alcedo & Zhang, 2013; de Bono & Maricq, 2005; McDiarmid, Yu, & Rankin, 2019; Sasakura & Mori, 2013). Here, we will focus on several learning paradigms that regulate the interaction between *C. elegans* and pathogenic bacteria.

Environmental cues induce plasticity across different timescales

C. elegans displays both adaptation and habituation, two common forms of non-associative learning (Figure 1). *C. elegans* is attracted to several chemical odorants, such as isoamyl alcohol and benzaldehyde; however, prolonged exposure to these volatile chemicals reduces the sensory response to the odorants and generates adaptation that lasts for a couple hours (Colbert & Bargmann, 1995; Inoue *et al.*, 2013; Kaye, Rose, Goldsworthy, Goga, & L'Etoile, 2009). It is shown that during adaptation the endogenous small RNA (endo-siRNA)-mediated regulation of gene expression in the sensory neuron that detects isoamyl alcohol and benzaldehyde downregulates a guanylyl cyclase that is critical for the G-protein coupled signaling pathway underlying the sensing of the odorants (Juang *et al.*, 2013). These results reveal a novel function of endo-siRNA pathways in regulating gene expression in response to olfactory experience. In addition, the worm reverses from a mild mechanical stimulus that is delivered to its body or nose and senses the stimulus using receptor neurons, several of which contain distinct morphological features (Chalfie, 2009; Kaplan & Horvitz, 1993; Schafer, 2015). Tapping the cultivating plate also generates mechanical stimuli that trigger reversals. However, tapping for multiple times reduces the amplitude of the reversals (Rankin, Beck, & Chiba, 1990). This type of behavioral changes is analogous to habituation previously characterized in *Aplysia* and cats, where multiple stimulations with a benign mechanical stimulus lead to a reduction in response (Bailey & Chen, 1983; Spencer, Thompson, & Neilson, 1966). Repeated habituation training under certain conditions can generate memory that lasts for 24 h (Rose, Kaun, & Rankin, 2002).

In addition to non-associative learning, previous studies have shown that olfactory responses can be respectively enhanced or weakened by paring odorant exposure with the presence or absence of food, which presumably represents an appetitive or aversive environment (Figure 1). Various neuronal circuits and molecular pathways have been characterized in regulating these associative learning behaviors [(Alcedo & Zhang, 2013; de Bono & Maricq, 2005) and the references therein]. *C. elegans* also remembers the salt

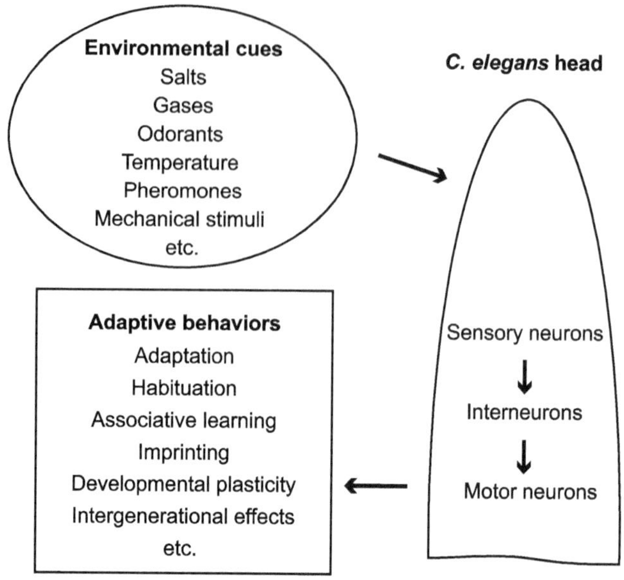

Figure 1. Diverse adaptive behaviors in response to environmental cues in *C. elegans*.

concentration under its cultivation condition and seeks this concentration when tested in a salt gradient after the training. However, if the worm is kept at a salt concentration in the absence of food, it avoids the concentration during the post-training rest (Kunitomo *et al.*, 2013; Luo *et al.*, 2014; Saeki, Yamamoto, & Iino, 2001; Tomioka *et al.*, 2006). As a critical condition, the cultivation temperature significantly modulates the navigation of the worm in a temperature gradient (Aoki & Mori, 2015; Biron *et al.*, 2006; Goodman *et al.*, 2014; Goodman & Sengupta, 2019; Hedgecock & Russell, 1975; Mori & Ohshima, 1995). Some of these forms of behavioral plasticity resemble associative learning identified in vertebrate animals and in fruit flies. While a one-time massed training in these paradigms often generates a memory for a couple hours, spaced training can generate a long-term memory that lasts for 16 h (Kauffman, Ashraf, Corces-Zimmerman, Landis, & Murphy, 2010).

Experimental power of *C. elegans* facilitates dissection of plasticity mechanisms

The ease of using forward and reverse genetic approaches to characterize gene function in *C. elegans* and the knowledge of genetic identities and synaptic connections of the worm neurons facilitate studies on learning and behavioral plasticity in *C. elegans* in several important ways:

1. Because the worm neurons are defined in their genetic making and synaptic connection, we are able to identify the neurons where the gene products implicated in learning are generated and act, as well as their pre-synaptic and post-synaptic neurons. These analyses provide us with knowledge on neuronal circuits underlying various forms of learning behaviors.
2. By applying *in vivo* imaging and genetic manipulations, we can identify experience-dependent changes in the activity and the connection of the learning circuit that

are correlated with behavioral changes and characterize the causality of these changes in generating learned behavior.

3. Once the key neurons underlying learning are identified, we can also analyze gene expression in these neurons in naive and trained animals in order to identify genes that display training-correlated changes in their expression and address the function of these molecules in learning.
4. Meanwhile, the ease of performing genetic analyses in the *C. elegans* nervous system also makes it feasible to conduct genetic screens in order to identify new functions of characterized genes and pathways in learning, as well as identify new genes with previously unknown functions.

Interactions with pathogenic bacteria that modulate behavior

C. elegans feeds on bacteria in the wild and laboratories. A wide range of different bacteria strains, including many in the *Pseudomonas* genus, are found to be associated with *C. elegans* isolated from its natural habitats (Felix & Duveau, 2012; Frezal & Felix, 2015; Samuel *et al.*, 2016; Schulenburg & Felix, 2017). While some of these bacteria serve as food sources, others are pathogenic and kill *C. elegans* through infections or with secreted toxins [(Hoffman & Aballay, 2019; Irazoqui, Urbach, & Ausubel, 2010; Kim & Ewbank, 2018) and the references therein]. Because bacteria play a vital role in the development and survival of *C. elegans*, it is conceivable that *C. elegans* has evolved diverse strategies to mediate its interactions with the environmental bacteria.

Bacteria produce multiple types of sensory cues that can be used by the worm to detect and respond to the microbes. In addition to odorants, bacteria also produce water-soluble metabolites, generate or alter concentration of gases. The border and the texture of a bacterial lawn may also generate mechanical stimulation to moving worms. These bacteria-derived sensory cues act in a combinatorial manner to elicit behavioral responses in *C. elegans* [(Bargmann, Hartwieg, & Horvitz, 1993; Brandt & Ringstad, 2015; Bretscher *et al.*, 2008; Calhoun *et al.*, 2015; Cheung *et al.*, 2005; Flavell *et al.*, 2013; Kim & Flavell, 2020; Gramstrup Petersen *et al.*, 2013; Gray *et al.*, 2004; Ha *et al.*, 2010; Hallem *et al.*, 2011; Hao *et al.*, 2018; Harris *et al.*, 2019; Meisel, Panda, Mahanti, Schroeder, & Kim, 2014; Ooi & Prahlad, 2017; Pradel *et al.*, 2007; Reddy *et al.*, 2011; Rhoades *et al.*, 2019; Sawin, Ranganathan, & Horvitz, 2000; Tran *et al.*, 2017) and the references therein]. The diversity of the sensory cues is consistent with multiple signaling pathways that are identified to mediate bacteria-worm interactions.

Adult-stage learning of pathogenic bacteria

Some pathogenic bacteria, such as the *Pseudomonas aeruginosa* strain PA14, infect *C. elegans* after being ingested, which leads to a slow death of the worm over several days (Tan, Mahajan-Miklos, & Ausubel, 1999). Thus, pathogenic bacteria likely signal both food and danger to the worm. The odorants of several pathogenic bacteria, including PA14, are attractive to the worms that are cultivated standard conditions by feeding on *E. coli* OP50 at $20-22\,^{\circ}C$ (Ha *et al.*, 2010; Jin, Pokala, & Bargmann, 2016; Zhang, Lu, & Bargmann, 2005). When newly transferred to a lawn of PA14, worms feed on the lawn. However, in the next few hours worms start to leave the lawn [Figure 2 and (Chang, Paek, & Kim, 2011)]. The virulence of the bacteria, as well as several bacteria-derived chemicals act together to repel the worms from the lawn. The mechanisms underlying the changes in behavior and physiology of the worms over this process are separately reviewed (Hoffman & Aballay, 2019; Kim & Flavell, 2020; Kim & Ewbank, 2018; Meisel & Kim, 2014).

Do worms learn to associate the aversiveness of PA14 with sensory cues produced by the pathogenic bacteria to generate retrievable memory of the bacterium? This question can be addressed by testing the response to PA14-derived sensory cues in the naive, i.e. *E. coli*-raised, worms and the trained, i.e. PA14-fed, worms [(Ha *et al.*, 2010; Jin *et al.*, 2016; Liu *et al.*, 2018; Zhang *et al.*, 2005) and Figure 2(A–F)]. Previously, by probing the worms with an assay that resembles the chemotaxis assay on olfactory responses or with an assay that uses airstreams saturated with the odorants of tested bacteria, it is shown that after feeding on PA14 for 4–6 h, adult worms learn to reduce their preference for the odorants of the bacterium (Figure 2(A–D)) (Ha *et al.*, 2010; Jin *et al.*, 2016; Liu *et al.*, 2018; Zhang *et al.*, 2005). This type of learning in the adult *C. elegans* is contingent on the pathogenicity of the training bacteria and a serotonin signal. The training-dependent change in the olfactory response is specific for the odorants of the training bacteria (Figure 2(A–C)) (Choi, Liu, Wu, Yang, & Zhang, 2020; Ha *et al.*, 2010; Jin *et al.*, 2016; Liu *et al.*, 2018; Zhang *et al.*, 2005). Together, these results indicate a learned association between pathogenicity and the odorants of the training bacterium. Since pathogenic bacteria represent a critical constraint to the survival of *C. elegans*, it is conceivable that *C. elegans* has evolved the ability to associate the olfactory cues of some pathogenic bacteria with the virulence, which regulates subsequent interactions with the pathogens. The learned behavioral response is reversible, which suggests a temporary modulation of the nervous system (Ha *et al.*, 2010; Jin *et al.*, 2016; Liu *et al.*, 2018; Zhang *et al.*, 2005).

Several different methods have been used to analyze the changes in behavioral responses to pathogenic bacterium PA14 after feeding on PA14 [Figure 2 and (Ha *et al.*, 2010; Hao *et al.*, 2018; Horspool & Chang, 2017; Jin *et al.*, 2016; Lee & Mylonakis, 2017; Liu *et al.*, 2018; Ma, Zhang, Dai, Khan, & Zou, 2017; Meisel *et al.*, 2014; Miller, Grandi, Giannini, Robinson, & Powell, 2015; Moore, Kaletsky, & Murphy, 2019; Ooi & Prahlad, 2017; Singh & Aballay, 2019; Wolfe *et al.*, 2019; Zhang *et al.*, 2005)]. Some of these methods differ in the types of the sensory cues that they examine, the spatial and temporal patterns of the cues, and the behavioral strategies that they measure (Figure 2). For example, the behavioral strategies used to distinguish between

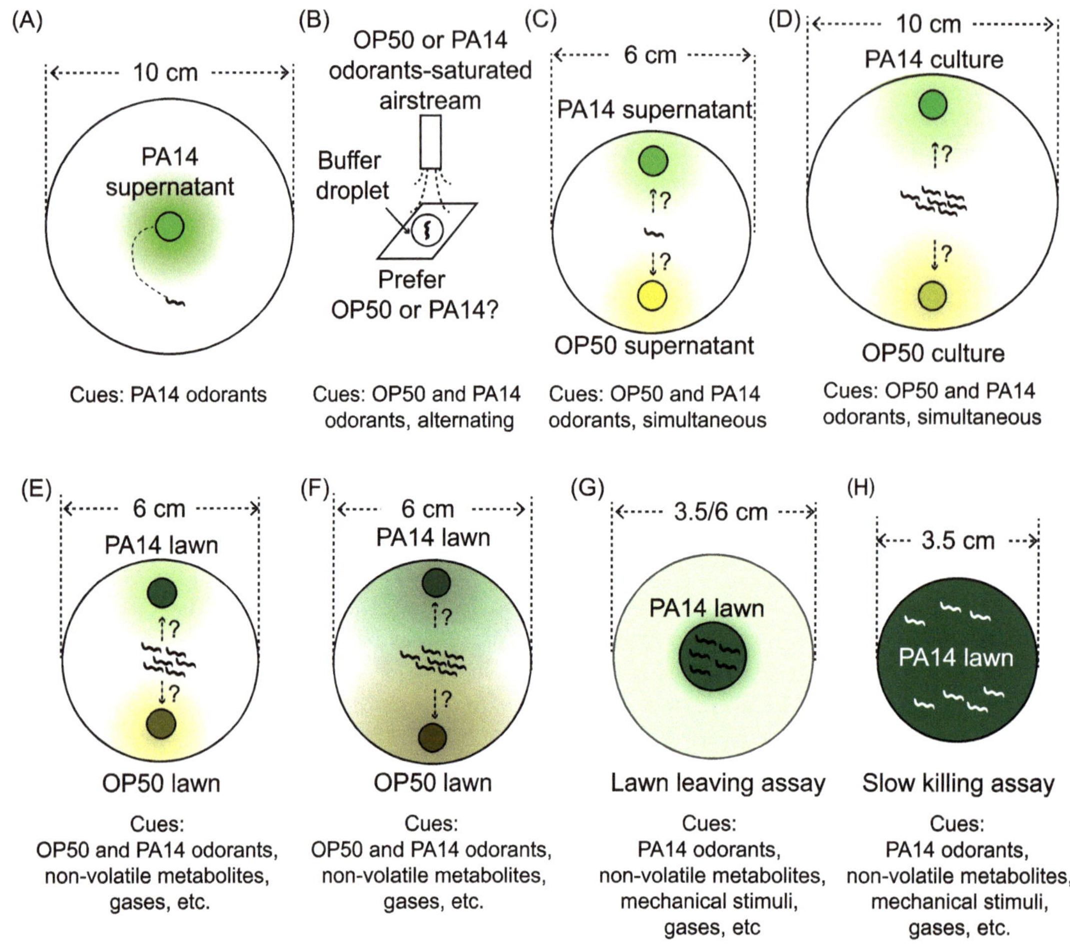

Figure 2. Schematic diagrams showing assays for behavioral responses to PA14. (A) A small drop of supernatant of PA14 culture is used as the source of odorants to examine attractive steering movements, when a worm starts from a position relatively close to the odorant source. (B) To test the relative preference between the odorants of PA14 and the odorants of *E. coli* OP50, two airstreams saturated with the odorants of OP50 or the odorants of PA14 are used to deliver alternating stimuli to individual worms swimming in a small drop of buffer in an airtight chamber. (C, D) Two small drops of supernatant of bacterial culture are put on a plate immediately before the assay (C) or two small drops of bacterial culture are quickly air-dried before the assay (D) to measure the preference between the two odorant mixtures in a single worm (C) or a population of worms (D). In D, the plate does not contain peptone and therefore does not support growth of the bacteria. (E, F) A small plate with two bacteria lawns grown on the plate for a few hours (E) or for 24–48 hours (F) to be used as odorant sources. During cultivation, the bacterial lawns may produce cues diffused into the medium, produce or alter the concentration of gases in the lawn areas. (G, H) A bacterial lawn centered on a small plate (G) or completely covering a small plate (H) prepared by fully growing first at 37 °C and then at 25 °C to examine the lawn avoidance/occupancy or survival of the worms over time.

simultaneously present odorants from two bacterial strains (Figure 2(C–F)) are likely different from those used to respond to two alternating odorant stimuli (Figure 2(B)). The odorant gradient established by a point source is not linear (Tanimoto *et al.*, 2017). Therefore, the size of the testing plate is important for the intensity and the spatial pattern of the bacterial odorants sensed by the worms (Figure 2(C and D)). The duration for which the testing bacteria were placed or grown on the testing plate or the temperature used to cultivate the testing bacteria generates different mixtures of the sensory cues (Figure 2(E and F)). In addition, separating the training process from the testing assay makes it possible to examine whether a retrievable memory is formed. Furthermore, the assay that measures the occupancy of a bacterial lawn grown on an assay plate likely measures sensory responses elicited by metabolites that are not volatile and mechanical cues, in addition to olfactory responses (Figure 2). While worms feeding on a lawn of pathogenic bacterium PA14 gradually leave the lawn over time (Figure 2(G and H)), it usually takes longer for the worms to significantly leave the lawn than to learn to reduce the preference for the odorants of PA14. Because different molecular and neuronal apparatus are employed to detect and generate behavioral responses to these various types of sensory information, studies using these different assay conditions highlight the robustness of the behavioral responses and potentially allow us to examine different pathways through which the worm interacts with pathogenic bacteria.

Imprinting

Interestingly, training the worms by feeding on pathogenic bacteria during the first larval stage (L1) for 12 h forms the aversive memory of the odorants of the pathogens that can be retrieved during the adult stage [Figure 1 and (Jin *et al.*, 2016)]. This form of memory is comparable to the imprinted memory characterized in various vertebrate animals (Lorenz, 1935; Nevitt, Dittman, Quinn, & Moody, 1994; Wilson &

Sullivan, 1994). The worms imprint not only the odorants associated with the aversive experience, but also the odorants associated with food sources to form a long-lasting appetitive memory [Figure 1 and (Remy & Hobert, 2005)]. Mapping the neural circuits for the imprinting of pathogen odorants and the retrieval of the aversive memory, which take place two days apart, show that different circuits subserve learning and retrieval (Jin *et al.*, 2016). In addition to the odorants that often represent food, pheromones also signal significant environmental conditions, such as the density of the conspecifics, to the worm (Butcher *et al.*, 2007; Jeong *et al.*, 2005; Macosko *et al.*, 2009; Srinivasan *et al.*, 2008; 2012; White *et al.*, 2007). Exposing the worms during the L1 stage to a repulsive pheromone enhances the avoidance of the pheromone during the adult stage by strengthening the synaptic connection between a pheromone-sensing neuron and its downstream motor neurons [Figure 1 and (Hong *et al.*, 2017)]. Starvation during the L1 stage also profoundly alters the wiring of the nervous system by regulating neurotransmitters that respond to food availability (Bayer & Hobert, 2018). It is conceivable that during the early larval development when the nervous system is being formed, strong neuronal activities in response to environmental conditions reprogram the developmental process to generate persistent changes. A couple of studies show that harsh conditions during development systematically regulate gene expression and modulate the anatomy and activity of the nervous system, which produce behavioral changes that last into the adult stage. For example, when the worm density is high and food is relatively sparse, the larval worms can enter a diapause state, dauer, that halts the development for days until the conditions improve (Golden & Riddle, 1984). Adult animals that have experienced the dauer stage exhibit distinct behavioral traits, including those important for food seeking. The molecules that regulate chromatin structures and endogenous RNAi pathways mediate dauer formation, which potentially modulate the expression of genes underlying dauer-inducing changes in anatomy and behavior (Bharadwaj & Hall, 2017; Hall, Beverly, Russ, Nusbaum, & Sengupta, 2010; Ow, Borziak, Nichitean, Dorus, & Hall, 2018; Pradhan, Quilez, Homer, & Hendricks, 2019).

Intergenerational effects

Because pathogenic bacteria serve as food sources and critical survival constraints to the worm, it is plausible that exposure to the pathogens modulates the nervous system and the behavior of the progenies. A recent study shows that training adult hermaphrodites by feeding on PA14 for 4 h, which is known to produce a robust aversive memory that associates PA14 odorants with virulence in the adult mothers, increases the progeny's preference for the PA14 odorants (Pereira, Gracida, Kagias, & Zhang, 2020). Many animals prefer the food that they are exposed to *in utero* (Liu & Urban, 2017; Nehring, Kostka, von Kries, & Rehfuess, 2015; Todrank, Heth, & Restrepo, 2011). These results suggest that with 4-h exposure the food response elicited by PA14 is significant in the hermaphrodite mothers and that the resulting signals modulate the progeny developing in the uterus. Increasing the duration of PA14 exposure to 8 h enhances the infection to the mothers (Troemel *et al.*, 2006) and reduces the preference for PA14 in the progeny (Pereira *et al.*, 2020). These findings suggest that longer exposure to PA14 induces a stronger response to the pathogenicity of PA14, which changes the response of the progeny to PA14 from attraction to avoidance. While 4 to 8-h parental training with PA14 produces robust aversive memory in the hermaphrodite mothers, their modulatory effects on the progeny are limited to the first generation of the offspring (Pereira *et al.*, 2020).

Further increasing the duration of PA14 exposure to 24 h that starts at the L4 larval stage not only generates avoidance of PA14 in the exposed mothers, but also produces PA14 avoidance in the progeny for 4 generations through a separate mechanism (Moore *et al.*, 2019). However, both the effect of 4-h training on the progeny and the multi-generational effect require the endo-RNAi pathway and the piRNA pathway (Moore *et al.*, 2019; Pereira *et al.*, 2020). Small RNA pathways also mediate olfactory adaptation by regulating the expression of genes critical for odorant sensation (Juang *et al.*, 2013). However, these behavioral changes differ in their durations and modulatory effects, which suggest distinct mechanisms through which the underlying small RNA pathways alter the nervous system and behavior.

In addition to food-seeking related behavior, exposing the worms to pathogenic bacteria generates intergenerational effects on other physiological traits critical for survival. For example, exposure to *Pseudomonas vranovensis*, another bacterium pathogenic to *C. elegans*, generates multigenerational effects and enhances immune resistance to the pathogen in the offspring (Burton *et al.*, 2019). Meanwhile, exposure to certain pathogenic bacteria for two consecutive generations induces formation of dauers, a dormant development stage that is highly resistant to environmental stresses (Palominos *et al.*, 2017). These studies reveal multiple ways that the worm has evolved to adapt its development and function to the bacteria in its environment. Interestingly, pathogenic bacteria are not the only environmental conditions that impact the worm for multiple generations. It has been shown that parental experiences, including dietary restriction, osmotic stress, temperature changes, olfactory imprinting, and prolonged starvation, can regulate the physiology of the offspring, some of which last for several generations and are mediated by small RNA pathways (Burton *et al.*, 2017; 2019; Das *et al.*, 2020; Demoinet, Li, & Roy, 2017; Greer *et al.*, 2011; Hibshman, Hung, & Baugh, 2016; Jobson *et al.*, 2015; Klosin, Casas, Hidalgo-Carcedo, Vavouri, & Lehner, 2017; Ni *et al.*, 2016; Palominos *et al.*, 2017; Posner *et al.*, 2019; Rechavi *et al.*, 2014; Remy, 2010; Schott, Yanai, & Hunter, 2014). These studies together show that *C. elegans* has evolved diverse adaptive strategies to generate long-term plasticity that lasts for multiple generations. The short lifespan of *C. elegans* and the ease to conduct genetic experiments on it makes this line of research productive. It will be informative to compare the mechanistic insights identified in studies on intergenerational effects in different animals to understand the difference and similarity in the logic of these regulations.

Outlook

C. elegans lives in a bacteria-rich environment that represents a vast amount of opportunities and constraints to its survival, reproduction and evolution. We are at the beginning stage to understand its impressive adaptive responses encoded in a compact genome. In addition to its responses to pathogenic bacteria, several recent studies also reveal interesting interactions between *C. elegans* and its commensal bacteria species. These studies show that *C. elegans* utilizes neurotransmitters or vitamins produced by the environmental bacteria (O'Donnell, Fox, Chao, Schroeder, & Sengupta, 2020; Urrutia *et al.*, 2020; Wei & Ruvkun, 2020) to maintain or modulate various physiological events and neural functions. These findings together with those investigating the interactions of *C. elegans* and pathogenic bacteria have established *C. elegans* as a promising system to probe mechanisms underlying gut-brain interactions. Together, these studies allow us to leverage the experimental powers provided by a model organism to investigate the function of the nervous system in an ethological and evolutionary context.

Disclosure statement

No potential conflict of interest was reported by the author(s).

Funding

The work in the Zhang laboratory is supported by The National Institutes of Health [DC009852, R21MH117386, NS115484].

References

Alcedo, J., & Zhang, Y. (2013). Molecular and cellular circuits underlying *Caenorhabditis elegans* olfactory plasticity. In R. Menzel, and P. Benjamine, eds., *Invertebrate learning and memory* (pp. 112–123). San Diego, CA: Elsevier Academic Press Inc.

Aoki, I., & Mori, I. (2015). Molecular biology of thermosensory transduction in *C. elegans*. *Current Opinion in Neurobiology, 34*, 117–124. doi:10.1016/j.conb.2015.03.011

Aprison, E.Z., & Ruvinsky, I. (2019). Coordinated behavioral and physiological responses to a social signal are regulated by a shared neuronal circuit. *Current Biology* , *29* (23), 4108–4115 e4104. doi:10.1016/j.cub.2019.10.012

Bailey, C.H., & Chen, M. (1983). Morphological basis of long-term habituation and sensitization in Aplysia. *Science, 220* (4592), 91–93. doi:10.1126/science.6828885

Bargmann, C.I. (2006). Chemosensation in *C. elegans*. WormBook, p. 1–29.

Bargmann, C.I., Hartwieg, E., & Horvitz, H.R. (1993). Odorant-selective genes and neurons mediate olfaction in *C. elegans*. *Cell, 74* (3), 515–527. doi:10.1016/0092-8674(93)80053-H

Bayer, E.A., & Hobert, O. (2018). Past experience shapes sexually dimorphic neuronal wiring through monoaminergic signalling. *Nature, 561* (7721), 117–121. doi:10.1038/s41586-018-0452-0

Bharadwaj, P.S., & Hall, S.E. (2017). Endogenous RNAi pathways are required in neurons for Dauer formation in *Caenorhabditis elegans*. *Genetics, 205* (4), 1503–1516. doi:10.1534/genetics.116.195438

Biron, D., Shibuya, M., Gabel, C., Wasserman, S.M., Clark, D.A., Brown, A., … Samuel, A.D. (2006). A diacylglycerol kinase modulates long-term thermotactic behavioral plasticity in *C. elegans*. *Nature Neuroscience, 9*, 1499–1505. doi:10.1038/nn1796

Brandt, J.P., Aziz-Zaman, S., Juozaityte, V., Martinez-Velazquez, L.A., Petersen, J.G., Pocock, R., & Ringstad, N. (2012). A single gene target of an ETS-family transcription factor determines neuronal CO_2-chemosensitivity. *PLoS One, 7* (3), e34014. doi:10.1371/journal.pone.0034014

Brandt, J.P., & Ringstad, N. (2015). Toll-like receptor signaling promotes development and function of sensory neurons required for a *C. elegans* pathogen-avoidance behavior. *Current Biology, 25* (17), 2228–2237. doi:10.1016/j.cub.2015.07.037

Bretscher, A.J., Busch, K.E., & de Bono, M. (2008). A carbon dioxide avoidance behavior is integrated with responses to ambient oxygen and food in *Caenorhabditis elegans*. *Proceedings of the National Academy of Sciences, 105* (23), 8044–8049. doi:10.1073/pnas.0707607105

Burton, N.O., Furuta, T., Webster, A.K., Kaplan, R.E., Baugh, L.R., Arur, S., & Horvitz, H.R. (2017). Insulin-like signalling to the maternal germline controls progeny response to osmotic stress. *Nature Cell Biology, 19* (3), 252–257. doi:10.1038/ncb3470

Burton, N.O., Riccio, C., Dallaire, A., Price, J., Jenkins, B., Koulman, A., & Miska, E.A. (2019). *C. elegans* heritably adapts to *P. vranovensis* infection via a mechanism that requires the cysteine synthases cysl-1 and cysl-2. *Nature Communications*. doi:10.1038/s41467-020-15555-8

Butcher, R.A., Fujita, M., Schroeder, F.C., & Clardy, J. (2007). Small-molecule pheromones that control dauer development in *Caenorhabditis elegans*. *Nature Chemical Biology, 3* (7), 420–422. doi:10.1038/nchembio.2007.3

Calhoun, A.J., Tong, A., Pokala, N., Fitzpatrick, J.A., Sharpee, T.O., & Chalasani, S.H. (2015). Neural mechanisms for evaluating environmental variability in *Caenorhabditis elegans*. *Neuron, 86* (2), 428–441. doi:10.1016/j.neuron.2015.03.026

Chalasani, S.H., Chronis, N., Tsunozaki, M., Gray, J.M., Ramot, D., Goodman, M.B., & Bargmann, C.I. (2007). Dissecting a circuit for olfactory behaviour in *Caenorhabditis elegans*. *Nature, 450* (7166), 63–70. doi:10.1038/nature06292

Chalfie, M. (2009). Neurosensory mechanotransduction. *Nature Reviews. Molecular Cell Biology, 10* (1), 44–52. doi:10.1038/nrm2595

Chang, H.C., Paek, J., & Kim, D.H. (2011). Natural polymorphisms in *C. elegans* HECW-1 E3 ligase affect pathogen avoidance behaviour. *Nature, 480* (7378), 525–529. doi:10.1038/nature10643

Cheung, B.H., Cohen, M., Rogers, C., Albayram, O., & de Bono, M. (2005). Experience-dependent modulation of *C. elegans* behavior by ambient oxygen. *Current Biology, 15* (10), 905–917. doi:10.1016/j.cub.2005.04.017

Choi, M.-K., Liu, H., Wu, T., Yang, W., & Zhang, Y. (2020). NMDAR-mediated modulation of gap junction circuit regulates olfactory learning in *C. elegans*. *Nature Communications, 11* (1), 1–16. doi:10.1038/s41467-020-17218-0

Colbert, H.A., & Bargmann, C.I. (1995). Odorant-specific adaptation pathways generate olfactory plasticity in *C. elegans*. *Neuron, 14* (4), 803–812. doi:10.1016/0896-6273(95)90224-4

Das, S., Ooi, F.K., Cruz Corchado, J., Fuller, L.C., Weiner, J.A., & Prahlad, V. (2020). Serotonin signaling by maternal neurons upon stress ensures progeny survival. *eLife, 9*, e55246. doi:10.7554/eLife.55246

de Bono, M., & Maricq, A.V. (2005). Neuronal substrates of complex behaviors in *C. elegans*. *Annual Review of Neuroscience, 28*, 451–501. doi:10.1146/annurev.neuro.27.070203.144259

Demoinet, E., Li, S., & Roy, R. (2017). AMPK blocks starvation-inducible transgenerational defects in *Caenorhabditis elegans*. *Proceedings of the National Academy of Sciences of the United States of America, 114* (13), E2689–E2698. doi:10.1073/pnas.1616171114

Donnelly, J.L., Clark, C.M., Leifer, A.M., Pirri, J.K., Haburcak, M., Francis, M.M., … Alkema, M.J. (2013). Monoaminergic orchestration of motor programs in a complex *C. elegans* behavior. *PLoS Biology, 11* (4), e1001529. doi:10.1371/journal.pbio.1001529

Felix, M.A., & Duveau, F. (2012). Population dynamics and habitat sharing of natural populations of *Caenorhabditis elegans* and *C. briggsae. BMC Biology, 10*, 59. doi:10.1186/1741-7007-10-59

Flavell, S.W., Pokala, N., Macosko, E.Z., Albrecht, D.R., Larsch, J., & Bargmann, C.I. (2013). Serotonin and the neuropeptide PDF initiate and extend opposing behavioral states in *C. elegans. Cell, 154* (5), 1023–1035. doi:10.1016/j.cell.2013.08.001

Frezal, L., & Felix, M.A. (2015). *C. elegans* outside the Petri dish. *Elife, 4*, e05849. doi:10.7554/eLife.05849.

Golden, J.W., & Riddle, D.L. (1984). The *Caenorhabditis elegans* dauer larva: developmental effects of pheromone, food, and temperature. *Developmental Biology, 102* (2), 368–378. doi:10.1016/0012-1606(84)90201-X

Goodman, M.B., Klein, M., Lasse, S., Luo, L., Mori, I., Samuel, A., … Wang, D. (2014). Thermotaxis navigation behavior. WormBook, ed. The *C. elegans* Research Community, WormBook, doi:10.1895/wormbook.1.168.1, http://www.wormbook.org.

Goodman, M.B., & Sengupta, P. (2019). How *Caenorhabditis elegans* senses mechanical stress, temperature, and other physical stimuli. *Genetics, 212* (1), 25–51. doi:10.1534/genetics.118.300241

Gordus, A., Pokala, N., Levy, S., Flavell, S.W., & Bargmann, C.I. (2015). Feedback from network states generates variability in a probabilistic olfactory circuit. *Cell, 161* (2), 215–227. doi:10.1016/j.cell.2015.02.018

Gramstrup Petersen, J., Rojo Romanos, T., Juozaityte, V., Redo Riveiro, A., Hums, I., Traunmuller, L., … Pocock, R. (2013). EGL-13/SoxD specifies distinct O_2 and CO_2 sensory neuron fates in *Caenorhabditis elegans. PLoS Genetics, 9* (5), e1003511. doi:10.1371/journal.pgen.1003511

Gray, J.M., Hill, J.J., & Bargmann, C.I. (2005). A circuit for navigation in *Caenorhabditis elegans. Proceedings of the National Academy of Sciences of the United States of America, 102* (9), 3184–3191. doi:10.1073/pnas.0409009101

Gray, J.M., Karow, D.S., Lu, H., Chang, A.J., Chang, J.S., Ellis, R.E., … Bargmann, C.I. (2004). Oxygen sensation and social feeding mediated by a *C. elegans* guanylate cyclase homologue. *Nature, 430* (6997), 317–322. doi:10.1038/nature02714

Greer, E.L., Maures, T.J., Ucar, D., Hauswirth, A.G., Mancini, E., Lim, J.P., … Brunet, A. (2011). Transgenerational epigenetic inheritance of longevity in *Caenorhabditis elegans. Nature, 479* (7373), 365–371. doi:10.1038/nature10572

Ha, H.I., Hendricks, M., Shen, Y., Gabel, C.V., Fang-Yen, C., Qin, Y., … Zhang, Y. (2010). Functional organization of a neural network for aversive olfactory learning in *Caenorhabditis elegans. Neuron, 68* (6), 1173–1186. doi:10.1016/j.neuron.2010.11.025

Hall, S.E., Beverly, M., Russ, C., Nusbaum, C., & Sengupta, P. (2010). A cellular memory of developmental history generates phenotypic diversity in *C. elegans. Current Biology, 20* (2), 149–155. doi:10.1016/j.cub.2009.11.035

Hallem, E.A., Spencer, W.C., McWhirter, R.D., Zeller, G., Henz, S.R., Ratsch, G., … Ringstad, N. (2011). Receptor-type guanylate cyclase is required for carbon dioxide sensation by *Caenorhabditis elegans. Proceedings of the National Academy of Sciences of the United States of America, 108* (1), 254–259. doi:10.1073/pnas.1017354108

Hao, Y., Yang, W., Ren, J., Hall, Q., Zhang, Y., & Kaplan, J.M. (2018). Thioredoxin shapes the *C. elegans* sensory response to Pseudomonas produced nitric oxide. *eLife, 7*, e36833. doi:10.7554/eLife.36833

Harris, G., Wu, T., Linfield, G., Choi, M.K., Liu, H., & Zhang, Y. (2019). Molecular and cellular modulators for multisensory integration in *C. elegans. PLoS Genetics, 15* (3), e1007706. doi:10.1371/journal.pgen.1007706

Hedgecock, E.M., & Russell, R.L. (1975). Normal and mutant thermotaxis in the nematode *Caenorhabditis elegans. Proceedings of the National Academy of Sciences of the United States of America, 72* (10), 4061–4065. doi:10.1073/pnas.72.10.4061

Hibshman, J.D., Hung, A., & Baugh, L.R. (2016). Maternal diet and insulin-like signaling control intergenerational plasticity of progeny size and starvation resistance. *PLoS Genetics, 12* (10), e1006396. doi:10.1371/journal.pgen.1006396

Hoffman, C., & Aballay, A. (2019). Role of neurons in the control of immune defense. *Current Opinion in Immunology, 60*, 30–36. doi:10.1016/j.coi.2019.04.005

Hong H, , M., Ryu, L., Ow, M.C., Kim, J., Je, A.R., Chinta, S., … Choi, H. (2017). Early pheromone experience modifies a synaptic activity to influence adult pheromone responses of *C. elegans. Current Biology, 27* (20), 3168–3177. e3163. doi:10.1016/j.cub.2017.08.068

Horspool, A.M., & Chang, H.C. (2017). Superoxide dismutase SOD-1 modulates *C. elegans* pathogen avoidance behavior. *Scientific Reports, 7*, 45128. doi:10.1038/srep45128

Iino, Y., & Yoshida, K. (2009). Parallel use of two behavioral mechanisms for chemotaxis in *Caenorhabditis elegans. The Journal of Neuroscience, 29* (17), 5370–5380. doi:10.1523/JNEUROSCI.3633-08.2009

Ikeda, M., Nakano, S., Giles, A.C., Xu, L., Costa, W.S., Gottschalk, A., & Mori, I. (2020). Context-dependent operation of neural circuits underlies a navigation behavior in *Caenorhabditis elegans. Proceedings of the National Academy of Sciences of the United States of America, 117* (11), 6178–6188. doi:10.1073/pnas.1918528117

Inoue, A., Sawatari, E., Hisamoto, N., Kitazono, T., Teramoto, T., Fujiwara, M., … Ishihara, T. (2013). Forgetting in *C. elegans* is accelerated by neuronal communication via the TIR-1/JNK-1 pathway. *Cell Reports, 3* (3), 808–819. doi:10.1016/j.celrep.2013.02.019

Irazoqui, J.E., Urbach, J.M., & Ausubel, F.M. (2010). Evolution of host innate defence: Insights from *Caenorhabditis elegans* and primitive invertebrates. *Nature Reviews. Immunology, 10* (1), 47–58. doi:10.1038/nri2689

Jang, H., Kim, K., Neal, S.J., Macosko, E., Kim, D., Butcher, R.A., … Sengupta, P. (2012). Neuromodulatory state and sex specify alternative behaviors through antagonistic synaptic pathways in *C. elegans. Neuron, 75* (4), 585–592. doi:10.1016/j.neuron.2012.06.034

Jeong, P.Y., Jung, M., Yim, Y.H., Kim, H., Park, M., Hong, E., … Paik, Y.K. (2005). Chemical structure and biological activity of the *Caenorhabditis elegans* dauer-inducing pheromone. *Nature, 433* (7025), 541–545. doi:10.1038/nature03201

Jin, X., Pokala, N., & Bargmann, C.I. (2016). Distinct circuits for the formation and retrieval of an imprinted olfactory memory. *Cell, 164* (4), 632–643. doi:10.1016/j.cell.2016.01.007

Juang, B.T., Gu, C., Starnes, L., Palladino, F., Goga, A., Kennedy, S., & L'Etoile, N.D. (2013). Endogenous nuclear RNAi mediates behavioral adaptation to odor. *Cell, 154* (5), 1010–1022. doi:10.1016/j.cell.2013.08.006

Jobson, M.A., Jordan, J.M., Sandrof, M.A., Hibshman, J.D., Lennox, A.L., & Baugh, L.R. (2015). Transgenerational effects of early life starvation on growth, reproduction, and stress resistance in *Caenorhabditis elegans. Genetics, 201* (1), 201–212. doi:10.1534/genetics.115.178699

Kaplan, H.S., Salazar Thula, O., Khoss, N., & Zimmer, M. (2020). Nested neuronal dynamics orchestrate a behavioral hierarchy across timescales. *Neuron, 105* (3), 562–576. e569. doi:10.1016/j.neuron.2019.10.037

Kaplan, J.M., & Horvitz, H.R. (1993). A dual mechanosensory and chemosensory neuron in *Caenorhabditis elegans. Proceedings of the National Academy of Sciences of the United States of America, 90* (6), 2227–2231. doi:10.1073/pnas.90.6.2227

Kato, S., Kaplan, H.S., Schrodel, T., Skora, S., Lindsay, T.H., Yemini, E., … Zimmer, M. (2015). Global brain dynamics embed the motor command sequence of *Caenorhabditis elegans. Cell, 163* (3), 656–669. doi:10.1016/j.cell.2015.09.034

Kauffman, A.L., Ashraf, J.M., Corces-Zimmerman, M.R., Landis, J.N., & Murphy, C.T. (2010). Insulin signaling and dietary restriction differentially influence the decline of learning and memory with age. *PLoS Biology, 8* (5), e1000372. doi:10.1371/journal.pbio.1000372

Kaye, J.A., Rose, N.C., Goldsworthy, B., Goga, A., & L'Etoile, N.D. (2009). A 3'UTR pumilio-binding element directs translational activation in olfactory sensory neurons. *Neuron, 61* (1), 57–70. doi:10.1016/j.neuron.2008.11.012

Klosin, A., Casas, E., Hidalgo-Carcedo, C., Vavouri, T., & Lehner, B. (2017). Transgenerational transmission of environmental information in *C. elegans*. *Science, 356* (6335), 320–323. doi:10.1126/science.aah6412

Kim, D.H., & Ewbank, J.J. (2018). Signaling in the innate immune response. *WormBook: The Online Review of C. elegans Biology, 2018*, 1–35. doi:10.1895/wormbook.1.83.2

Kim, D.H. & Flavell, S.W. (2020). Host-microbe interactions and the behavior of *Caenorhabditis elegans*. *J Neurogenet.* doi:10.1080/01677063.2020.1802724

Kim, K., Sato, K., Shibuya, M., Zeiger, D.M., Butcher, R.A., Ragains, J.R., … Sengupta, P. (2009). Two chemoreceptors mediate developmental effects of dauer pheromone in *C. elegans*. *Science, 326* (5955), 994–998. doi:10.1126/science.1176331

Kunitomo, H., Sato, H., Iwata, R., Satoh, Y., Ohno, H., Yamada, K., & Iino, Y. (2013). Concentration memory-dependent synaptic plasticity of a taste circuit regulates salt concentration chemotaxis in *Caenorhabditis elegans*. *Nature Communications, 4*, 2210. doi:10.1038/ncomms3210

Lee, K., & Mylonakis, E. (2017). An intestine-derived neuropeptide controls avoidance behavior in *Caenorhabditis elegans*. *Cell Reports, 20* (10), 2501–2512. doi:10.1016/j.celrep.2017.08.053

Li, Z., Liu, J., Zheng, M., & Xu, X.Z. (2014). Encoding of both analog- and digital-like behavioral outputs by one *C. elegans* interneuron. *Cell , 159* (4), 751–765. doi:10.1016/j.cell.2014.09.056

Liu, A., & Urban, N.N. (2017). Prenatal and early postnatal odorant exposure heightens odor-evoked mitral cell responses in the mouse olfactory bulb. *eNeuro, 4*(5), ENEURO.0129-17.2017. doi:10.1523/ENEURO.0129-17.2017

Liu, H., Yang, W., Wu, T., Duan, F., Soucy, E., Jin, X., & Zhang, Y. (2018). Cholinergic sensorimotor integration regulates olfactory steering. *Neuron, 97* (2), 390–405 e393. doi:10.1016/j.neuron.2017.12.003

Lorenz, K. (1935). Der kumpan in der umwelt des vogels. *Journal of Ornithology, 83* (2), 137–213. doi:10.1007/BF01905355

Luo, L., Wen, Q., Ren, J., Hendricks, M., Gershow, M., Qin, Y., … Smith-Parker, H.K. (2014). Dynamic encoding of perception, memory, and movement in a *C. elegans* chemotaxis circuit. *Neuron, 82* (5), 1115–1128. doi:10.1016/j.neuron.2014.05.010

Ma, Y.C., Zhang, L., Dai, L.L., Khan, R.U., & Zou, C.G. (2017). mir-67 regulates *P. aeruginosa* avoidance behavior in *C. elegans*. *Biochemical and Biophysical Research Communications, 494* (1–2), 120–125. doi:10.1016/j.bbrc.2017.10.069

Macosko, E.Z., Pokala, N., Feinberg, E.H., Chalasani, S.H., Butcher, R.A., Clardy, J., & Bargmann, C.I. (2009). A hub-and-spoke circuit drives pheromone attraction and social behaviour in *C. elegans*. *Nature, 458* (7242), 1171–1175. doi:10.1038/nature07886

McDiarmid, T.A., Yu, A.J., & Rankin, C.H. (2019). Habituation is more than learning to ignore: Multiple Mechanisms serve to facilitate shifts in behavioral strategy. *BioEssays: News and Reviews in Molecular, Cellular and Developmental Biology, 41* (9), e1900077 doi:10.1002/bies.201900077

Meisel, J.D., & Kim, D.H. (2014). Behavioral avoidance of pathogenic bacteria by *Caenorhabditis elegans*. *Trends in Immunology, 35* (10), 465–470. doi:10.1016/j.it.2014.08.008

Meisel, J.D., Panda, O., Mahanti, P., Schroeder, F.C., & Kim, D.H. (2014). Chemosensation of bacterial secondary metabolites modulates neuroendocrine signaling and behavior of *C. elegans*. *Cell, 159* (2), 267–280. doi:10.1016/j.cell.2014.09.011

Miller, E.V., Grandi, L.N., Giannini, J.A., Robinson, J.D., & Powell, J.R. (2015). The conserved g-protein coupled receptor FSHR-1 regulates protective host responses to infection and oxidative stress. *PLoS One, 10* (9), e0137403. doi:10.1371/journal.pone.0137403

Moore, R.S., Kaletsky, R., & Murphy, C.T. (2019). Piwi/PRG-1 argonaute and TGF-β mediate transgenerational learned pathogenic avoidance. *Cell, 177* (7), 1827–1841 e1812. doi:10.1016/j.cell.2019.05.024

Mori, I., & Ohshima, Y. (1995). Neural regulation of thermotaxis in *Caenorhabditis elegans*. *Nature, 376* (6538), 344–348. doi:10.1038/376344a0

Nehring, I., Kostka, T., von Kries, R., & Rehfuess, E.A. (2015). Impacts of in utero and early infant taste experiences on later taste acceptance: A systematic review. *The Journal of Nutrition, 145* (6), 1271–1279. doi:10.3945/jn.114.203976

Nevitt, G.A., Dittman, A.H., Quinn, T.P., & Moody, W.J. Jr. (1994). Evidence for a peripheral olfactory memory in imprinted salmon. *Proceedings of the National Academy of Sciences of the United States of America, 91*(10), 4288–4292. doi:10.1073/pnas.91.10.4288

Ni, J.Z., Kalinava, N., Chen, E., Huang, A., Trinh, T., & Gu, S.G. (2016). A transgenerational role of the germline nuclear RNAi pathway in repressing heat stress-induced transcriptional activation in *C. elegans*. *Epigenetics and Chromatin, 9*(1), 1–15.

O'Donnell, M.P., Fox, B.W., Chao, P.H., Schroeder, F.C., & Sengupta, P. (2020). A neurotransmitter produced by gut bacteria modulates host sensory behaviour. *Nature, 583*(7816), 415–420. doi:10.1038/s41586-020-2395-5

Ooi, F.K., & Prahlad, V. (2017). Olfactory experience primes the heat shock transcription factor HSF-1 to enhance the expression of molecular chaperones in *C. elegans*. *Science Signaling, 10*(501), eaan4893. doi:10.1126/scisignal.aan4893

Ow, M.C., Borziak, K., Nichitean, A.M., Dorus, S., & Hall, S.E. (2018). Early experiences mediate distinct adult gene expression and reproductive programs in *Caenorhabditis elegans*. *PLoS Genetics, 14* (2), e1007219. doi:10.1371/journal.pgen.1007219

Palominos, M.F., Verdugo, L., Gabaldon, C., Pollak, B., Ortiz-Severin, J., Varas, M.A., … Calixto, A. (2017). Transgenerational diapause as an avoidance strategy against bacterial pathogens in *Caenorhabditis elegans*. *mBio, 8*(5), e01234. doi:10.1128/mBio.01234-17

Pereira, A.G., Gracida, X., Kagias, K., & Zhang, Y. (2020). *C. elegans* aversive olfactory learning generates diverse intergenerational effects. J Neurogenet. doi: 10.1080/01677063.2020.1819265.

Pierce-Shimomura, J.T., Faumont, S., Gaston, M.R., Pearson, B.J., & Lockery, S.R. (2001). The homeobox gene lim-6 is required for distinct chemosensory representations in *C. elegans*. *Nature, 410* (6829), 694–698. doi:10.1038/35070575

Pierce-Shimomura, J.T., Morse, T.M., & Lockery, S.R. (1999). The fundamental role of pirouettes in *Caenorhabditis elegans* chemotaxis. *The Journal of Neuroscience, 19* (21), 9557–9569. doi:10.1523/JNEUROSCI.19-21-09557.1999

Posner, R., Toker, I.A., Antonova, O., Star, E., Anava, S., Azmon, E., … Rechavi, O. (2019). Neuronal small RNAs control behavior transgenerationally. *Cell, 177* (7), 1814–1826 e1815. doi:10.1016/j.cell.2019.04.029

Pradel, E., Zhang, Y., Pujol, N., Matsuyama, T., Bargmann, C.I., & Ewbank, J.J. (2007). Detection and avoidance of a natural product from the pathogenic bacterium *Serratia marcescens* by *Caenorhabditis elegans*. *Proceedings of the National Academy of Sciences of the United States of America, 104* (7), 2295–2300. doi:10.1073/pnas.0610281104

Pradhan, S., Quilez, S., Homer, K., & Hendricks, M. (2019). Environmental programming of adult foraging behavior in *C. elegans*. *Current Biology, 29* (17), 2867–2879 e2864. doi:10.1016/j.cub.2019.07.045

Rankin, C.H., Beck, C.D., & Chiba, C.M. (1990). *Caenorhabditis elegans*: A new model system for the study of learning and memory. *Behavioural Brain Research, 37* (1), 89–92. doi:10.1016/0166-4328(90)90074-O

Rechavi, O., Houri-Ze'evi, L., Anava, S., Goh, W.S.S., Kerk, S.Y., Hannon, G.J., & Hobert, O. (2014). Starvation-induced transgenerational inheritance of small RNAs in *C. elegans*. *Cell, 158* (2), 277–287. doi:10.1016/j.cell.2014.06.020

Reddy, K.C., Hunter, R.C., Bhatla, N., Newman, D.K., & Kim, D.H. (2011). *Caenorhabditis elegans* NPR-1-mediated behaviors are suppressed in the presence of mucoid bacteria. *Proceedings of the National Academy of Sciences of the United States of America, 108* (31), 12887–12892. doi:10.1073/pnas.1108265108

Remy, J.J. (2010). Stable inheritance of an acquired behavior in *Caenorhabditis elegans*. *Current Biology, 20* (20), R877–878. doi:10.1016/j.cub.2010.08.013

Remy, J.J., & Hobert, O. (2005). An interneuronal chemoreceptor required for olfactory imprinting in *C. elegans*. *Science*, *309* (5735), 787–790. doi:10.1126/science.1114209

Rhoades, J.L., Nelson, J.C., Nwabudike, I., Yu, S.K., McLachlan, I.G., Madan, G.K., … Flavell, S.W. (2019). ASICs mediate food responses in an enteric serotonergic neuron that controls foraging behaviors. *Cell*, *176* (1–2), 85–97 e14. doi:10.1016/j.cell.2018.11.023

Rose, J.K., Kaun, K.R., & Rankin, C.H. (2002). A new group-training procedure for habituation demonstrates that presynaptic glutamate release contributes to long-term memory in *Caenorhabditis elegans*. *Learning & Memory*, *9* (3), 130–137. doi:10.1101/lm.46802

Saeki, S., Yamamoto, M., & Iino, Y. (2001). Plasticity of chemotaxis revealed by paired presentation of a chemoattractant and starvation in the nematode *Caenorhabditis elegans*. *The Journal of Experimental Biology*, *204* (Pt 10), 1757–1764.

Samuel, B.S., Rowedder, H., Braendle, C., Felix, M.A., & Ruvkun, G. (2016). *Caenorhabditis elegans* responses to bacteria from its natural habitats. *Proceedings of the National Academy of Sciences of the United States of America*, *113* (27), E3941–3949. doi:10.1073/pnas.1607183113

Sasakura, H., & Mori, I. (2013). Behavioral plasticity, learning, and memory in *C. elegans*. *Current Opinion in Neurobiology*, *23* (1), 92–99. doi:10.1016/j.conb.2012.09.005

Sawin, E.R., Ranganathan, R., & Horvitz, H.R. (2000). *C. elegans* locomotory rate is modulated by the environment through a dopaminergic pathway and by experience through a serotonergic pathway. *Neuron*, *26* (3), 619–631. doi:10.1016/S0896-6273(00)81199-X

Schafer, W.R. (2015). Mechanosensory molecules and circuits in *C. elegans*. *Pflugers Archiv: European Journal of Physiology*, *467* (1), 39–48. doi:10.1007/s00424-014-1574-3

Schott, D., Yanai, I., & Hunter, C.P. (2014). Natural RNA interference directs a heritable response to the environment. *Scientific Reports*, *4*, 7387. doi:10.1038/srep07387

Schulenburg, H., & Felix, M.A. (2017). The natural biotic environment of *Caenorhabditis elegans*. *Genetics*, *206* (1), 55–86. doi:10.1534/genetics.116.195511

Singh, J., & Aballay, A. (2019). Intestinal infection regulates behavior and learning via neuroendocrine signaling. *eLife*, *8*, e50033. doi:10.7554/eLife.50033

Spencer, W.A., Thompson, R.F., & Neilson, D.R. Jr. (1966). Response decrement of the flexion reflex in the acute spinal cat and transient restoration by strong stimuli. *Journal of Neurophysiology*, *29*(2), 221–239. doi:10.1152/jn.1966.29.2.221

Srinivasan, J., Kaplan, F., Ajredini, R., Zachariah, C., Alborn, H.T., Teal, P.E., … Schroeder, F.C. (2008). A blend of small molecules regulates both mating and development in *Caenorhabditis elegans*. *Nature*, *454* (7208), 1115–1118. doi:10.1038/nature07168

Srinivasan, J., von Reuss, S.H., Bose, N., Zaslaver, A., Mahanti, P., Ho, M.C., … Schroeder, F.C. (2012). A modular library of small molecule signals regulates social behaviors in *Caenorhabditis elegans*. *PLoS Biology*, *10* (1), e1001237. doi:10.1371/journal.pbio.1001237

Tan, M.W., Mahajan-Miklos, S., & Ausubel, F.M. (1999). Killing of *Caenorhabditis elegans* by *Pseudomonas aeruginosa* used to model mammalian bacterial pathogenesis. *Proceedings of the National Academy of Sciences of the United States of America*, *96* (2), 715–720. doi:10.1073/pnas.96.2.715

Tanimoto, Y., Yamazoe-Umemoto, A., Fujita, K., Kawazoe, Y., Miyanishi, Y., Yamazaki, S.J., … Nakai, J. (2017). Calcium dynamics regulating the timing of decision-making in *C. elegans*. *eLife*, *6*, e21629. doi:10.7554/eLife.21629

Todrank, J., Heth, G., & Restrepo, D. (2011). Effects of in utero odorant exposure on neuroanatomical development of the olfactory bulb and odour preferences. *Proceedings Biological Sciences*, *278* (1714), 1949–1955. doi:10.1098/rspb.2010.2314

Tomioka, M., Adachi, T., Suzuki, H., Kunitomo, H., Schafer, W.R., & Iino, Y. (2006). The insulin/PI 3-kinase pathway regulates salt chemotaxis learning in *Caenorhabditis elegans*. *Neuron*, *51* (5), 613–625. doi:10.1016/j.neuron.2006.07.024

Tran, A., Tang, A., O'Loughlin, C.T., Balistreri, A., Chang, E., Coto Villa, D., … Pyle, J., *et al.* (2017). *C. elegans* avoids toxin-producing Streptomyces using a seven transmembrane domain chemosensory receptor. *eLife*, *6*, e23770. doi:10.7554/eLife.23770

Troemel, E.R., Chu, S.W., Reinke, V., Lee, S.S., Ausubel, F.M., & Kim, D.H. (2006). p38 MAPK regulates expression of immune response genes and contributes to longevity in *C. elegans*. *PLoS Genetics*, *2* (11), e183. doi:10.1371/journal.pgen.0020183

Tsalik, E.L., & Hobert, O. (2003). Functional mapping of neurons that control locomotory behavior in *Caenorhabditis elegans*. *Journal of Neurobiology*, *56* (2), 178–197. doi:10.1002/neu.10245

Urrutia, A., Garcia-Angulo, V.A., Fuentes, A., Caneo, M., Legue, M., Urquiza, S., … Calixto, A. (2020). Bacterially produced metabolites protect *C. elegans* neurons from degeneration. *PLoS Biology*, *18* (3), e3000638. doi:10.1371/journal.pbio.3000638

Venkatachalam, V., Ji, N., Wang, X., Clark, C., Mitchell, J.K., Klein, M., … Greenwood, J. (2016). Pan-neuronal imaging in roaming *Caenorhabditis elegans*. *Proceedings of the National Academy of Sciences of the United States of America*, *113* (8), E1082–E1088. doi:10.1073/pnas.1507109113

Wei, W., & Ruvkun, G. (2020). Lysosomal activity regulates *Caenorhabditis elegans* mitochondrial dynamics through vitamin B12 metabolism. Proceedings of the National Academy of Sciences of the United States of America, 117(33):19970–19981.

Wen, Q., Gao, S., & Zhen, M. (2018). *Caenorhabditis elegans* excitatory ventral cord motor neurons derive rhythm for body undulation. *Philosophical Transactions of the Royal Society of London. Series B, Biological Sciences*, 373(1758):20170370. doi:10.1098/rstb.2017.0370

White, J.Q., & Jorgensen, E.M. (2012). Sensation in a single neuron pair represses male behavior in hermaphrodites. *Neuron*, *75* (4), 593–600. doi:10.1016/j.neuron.2012.03.044

White, J.Q., Nicholas, T.J., Gritton, J., Truong, L., Davidson, E.R., & Jorgensen, E.M. (2007). The sensory circuitry for sexual attraction in *C. elegans* males. *Current Biology*, *17* (21), 1847–1857. doi:10.1016/j.cub.2007.09.011

Wilson, D.A., & Sullivan, R.M. (1994). Neurobiology of associative learning in the neonate: Early olfactory learning. *Behavioral and Neural Biology*, *61* (1), 1–18. doi:10.1016/S0163-1047(05)80039-1

Wolfe, G.S., Tong, V.W., Povse, E., Merritt, D.M., Stegeman, G.W., Flibotte, S., & van der Kooy, D. (2019). A receptor tyrosine kinase plays separate roles in sensory integration and *eneuro*, *6*(4), ENEURO.0244-18.2019. doi: 10.1523/ENEURO.0244-18.2019.

Zhang, Y., Lu, H., & Bargmann, C.I. (2005). Pathogenic bacteria induce aversive olfactory learning in *Caenorhabditis elegans*. *Nature*, *438*(7065), 179–184. doi:10.1038/nature04216

C. elegans aversive olfactory learning generates diverse intergenerational effects

Ana Goncalves Pereira, Xicotencatl Gracida, Konstantinos Kagias and Yun Zhang

ABSTRACT

Parental experience can modulate the behavior of their progeny. While the molecular mechanisms underlying parental effects or inheritance of behavioral traits have been studied under several environmental conditions, it remains largely unexplored how the nature of parental experience affects the information transferred to the next generation. To address this question, we used *C. elegans*, a nematode that feeds on bacteria in its habitat. Some of these bacteria are pathogenic and the worm learns to avoid them after a brief exposure. We found, unexpectedly, that a short parental experience increased the preference for the pathogen in the progeny. Furthermore, increasing the duration of parental exposure switched the response of the progeny from attraction to avoidance. To characterize the underlying molecular mechanisms, we found that the RNA-dependent RNA Polymerase (RdRP) RRF-3, required for the biogenesis of 26G endo-siRNAs, regulated both types of intergenerational effects. Together, we show that different parental experiences with the same environmental stimulus generate different effects on the behavior of the progeny through small RNA-mediated regulation of gene expression.

Introduction

In many organisms, the behaviour and physiology of progeny are modulated by parental experience, ultimately generating traits that often resemble the acquired phenotype of the parents (Bohacek & Mansuy, 2015; Dias, Maddox, Klengel, & Ressler, 2015; Horsthemke, 2018; Miska & Ferguson-Smith, 2016; Weigel & Colot, 2012). A growing body of evidence documents the intergenerational transfer of acquired traits, such as stress response or olfactory sensitivity in rodents (Dias & Ressler, 2014; Gapp *et al.*, 2014). This form of plasticity "anticipates" that the next generation will experience a similar environment to their parents, in which case it may prove to be adaptive. However, the same parental experience can affect progeny differentially depending on their sex, genotype, further ancestral history and life experience (Bohacek & Mansuy, 2015; Deas, Blondel, & Extavour, 2019; Kundakovic *et al.*, 2013; Palominos *et al.*, 2017). These observations suggest that inter- and transgenerational regulations are influenced by many different factors and do not always copy the parentally acquired traits.

Despite our increased knowledge on transgenerational effects, the impact of duration, intensity or frequency of parental experience on modulating certain traits in their progeny remains largely unexplored. Here, we examine this question using *C. elegans*, a nematode that feeds on bacteria of different genuses. The chemotactic behaviour of *C. elegans* towards this food source is highly dependent on the olfactory system, which is critical for detecting food and avoiding dangers. Bacteria from the *Pseudomonas* genus are abundant in the natural habit of the nematode (Félix & Duveau, 2012; Samuel, Rowedder, Braendle, Félix, & Ruvkun, 2016), but certain *Pseudomonas* strains, such as *Pseudomonas aeruginosa* PA14, infect the worm after being ingested, which results in a slow death over several days (Tan, Mahajan-Miklos, & Ausubel, 1999). Therefore, *Pseudomonas aeruginosa* PA14 is both food and pathogen to *C. elegans* and plays a critical role in the survival of the worm.

We have previously shown that adult *C. elegans* robustly learns to reduce their preference for the odorants of PA14 after transiently feeding on the pathogen for 4 h (Ha *et al.*, 2010; Zhang, Lu, & Bargmann, 2005). This form of learning is contingent on the pathogenesis of the bacteria and resembles the Garcia effect, a form of conditioned aversion that allows animals to learn to avoid the smell or taste of a food that makes them ill (Garcia, Hankins, & Rusiniak, 1974; Ha *et al.*, 2010; Zhang *et al.*, 2005). Because continuously ingesting PA14 can be lethal to the worm, intergenerational transfer of information regarding the valence of the cues that signal this food source could be adaptive for the next generation. Therefore, we asked whether learning to avoid PA14 in *C. elegans* mothers regulated the olfactory response to

PA14 in progeny. Furthermore, we asked whether parental learning experience of different durations differentially affected the behavioral response of the progeny.

To elucidate the molecular pathways underlying parental experience-induced behavioral changes in progeny, we characterized the small non-coding RNA pathways previously implicated in epigenetic inheritance in different species (Bohacek & Mansuy, 2015; Gapp *et al.*, 2014; Miska & Ferguson-Smith, 2016; Moore *et al.*, 2019; Okamura & Lai, 2008; Palominos *et al.*, 2017; Posner *et al.*, 2019; Rechavi *et al.*, 2014). Three major classes of small RNAs have been described in *C. elegans*: small interfering RNAs (endo-siRNAs and exo-siRNAs for endogenously and exogenously produced small interfering RNAs, respectively), microRNAs and PIWI-interacting RNAs (piRNAs). Endo-siRNAs are encoded in the genome and target worm transcripts; the RNA-dependent RNA Polymerase (RdRP) RRF-3 uses endogenous mRNAs as templates to synthesize the complementary strand. The resulting double-stranded RNAs (dsRNAs) are cleaved into 26 G endo-siRNAs that target genes expressed in the soma and germline. Mutating RRF-3 downregulates 26 G endo-siRNAs in the sperm, oocytes and embryos and upregulates the expression of many genes (Gent *et al.*, 2010; Han *et al.*, 2009; Lee, Hammell, & Ambros, 2006; Simmer *et al.*, 2002). The piRNAs are 21 to 22-nucleotide RNAs enriched in the germline where they protect the integrity of the genome and regulate germ cell development and fertility. The piRNAs form complexes with the PIWI-clade Argonautes, such as PRG-1, to target both endogenous and exogenous sequences (Ashe *et al.*, 2012; Batista *et al.*, 2008; Das *et al.*, 2008; Lee *et al.*, 2012; Wang & Reinke, 2008)) . Given the documented role of these two pathways in epigenetic inheritance, we characterized the function of RRF-3 and PRG-2 in aversive learning of PA14 in *C. elegans* adults and the effects of the parental learning experience on the olfactory response of the progeny. Our results show that parental learning experience of different durations differentially modulate the behavior of the progeny and that the endo-siRNA pathways regulate these intergenerational effects.

Materials and methods

Experimental models and subject details

C. elegans hermaphrodites were used in this study and cultivated at 20 °C under standard conditions (Brenner, 1974). The strains used in this study include: N2 (Bristol), ZC2834 *rrf-3(pk1426)* II, YY13 *rrf-3(mg373)* II, ZC2987 *prg-2(n4358)* IV, WM162 *prg-2(tm1094)* IV, CX4998 kyIs140 I; *nsy-1(ky397)* II.

Bacterial strains

For this study we used the non-pathogenic bacteria *Escherichia coli* OP50 and the pathogenic bacteria *Pseudomonas aeruginosa* PA14.

Method details

Aversive olfactory training of P0s with PA14 and cultivation of F1s

Training of P0s (mothers) with PA14 was performed as previously described (Ha *et al.*, 2010; Zhang *et al.*, 2005) with minor modifications. Individual OP50 or PA14 colonies were inoculated in 50 mL nematode growth medium (NGM) and cultivated with shaking at 27 °C overnight. 800 µL (for 4-h training) or 400 µL (for 8-h training) of OP50 or PA14 culture was spread onto 10 cm NGM plate and incubated at 27 °C for 2 days to prepare the naive control plates and the training plates, respectively. Tests performed prior to the beginning of the experiments showed that 800 µL-seeded plates generated a very stable learning after 4-h exposure. However, the same amount of bacteria seeded in training plates used for 8-h training caused severe locomotor impairment in P0s that could not be tested in the droplet assay. Therefore, we used 400 µL PA14 culture to seed the training plates for 8-h training.

P0 eggs were first extracted from 1-day adult hermaphrodites cultivated under standard conditions (Brenner, 1974) and grown until they reached adulthood. Synchronized young adult P0s were then transferred to the training (PA14 seeded) or the control (OP50 seeded) plate and kept at 20 °C for 4h or 8 h depending on the experimental procedure. By the end of the training, some naive control and trained P0 worms were randomly picked from the plates and tested in parallel for their olfactory choice and learning in the droplet assay (see next section for details). The rest of the P0 worms were collected with S-basal medium and passed through a cell strainer 40-micrometer nylon filter (Falcon) to remove laid eggs. The P0 worms were then treated with a bleach solution to isolate F1 embryos, which were hatched and cultivated on plates seeded with an OP50 culture grown overnight in LB medium and kept at 20 °C until the adult stage.

Olfactory preference assay using the automated olfactory assay

The automated assay (Droplet assay) that quantifies olfactory preference in individual animals was performed as described [(Ha *et al.*, 2010) and Figure 1] with some modifications. Briefly, after training, 6 naive and 6 trained P0s per assay were randomly picked, washed, and individually placed in the droplets of 2 µL NGM buffer in an enclosed chamber. Worms were exposed to alternating airstreams odorized with OP50 or PA14 by blowing clean air through the supernatant of freshly generated bacterial cultures. Each olfactory stimulation lasted for 30 s and each assay contained 12 cycles of stimulation. Naive and trained worms were tested in parallel. The locomotion of the worms was recorded and the large body bends (Ω bends, i.e. body bends in the shape of Ω) of the tested worms were quantified with computational analysis performed in Matlab. Because Ω bends are followed by reorientation, a higher rate of Ω bends indicates a lower preference for the tested airstream (Ha *et al.*, 2010). The P0 Choice Index (CI) of each worm was defined as the turning

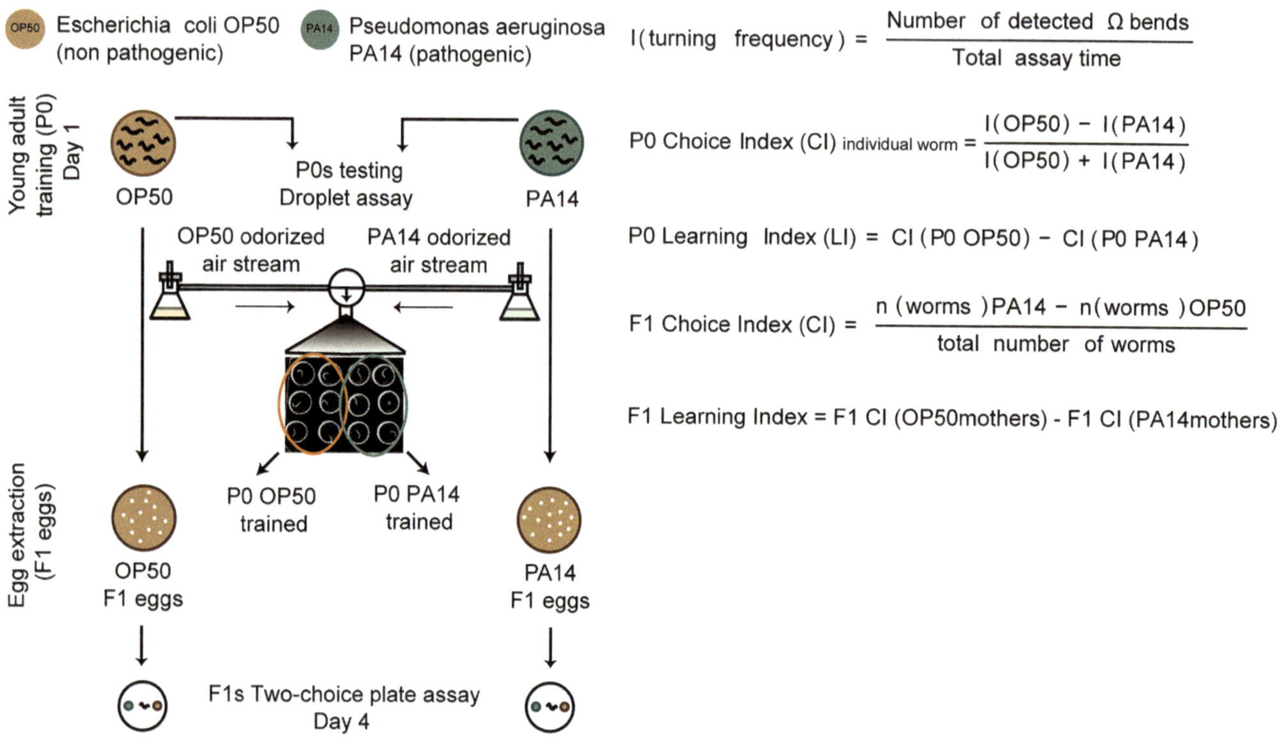

Figure 1. Experiment design to examine effect of parental experience with PA14 on progeny. Schematics for the training procedure and the learning assays.

rate evoked by OP50 smell minus the turning rate evoked by PA14 smell and normalized by the sum of the rates (Figure 1). The CI of each assay was quantified as the median value of the CIs of the individual P0 worms in that assay. Each test session consisted of two assays (with a total of 12 naive and 12 trained worms being tested). Individual data points in the graphs represent the average CI of the two assays of the test session. If technical problems occurred during testing, the affected assay was excluded from analysis. Any worm that was injured during the process of transferring to the testing chamber was also excluded. The P0 Learning Index (LI) of each session was defined as the CI of the naive worms (P0 OP50) minus the CI of the trained worms (P0 PA14) (Figure 1). Individual data points in the graphs represent the average LIs of the two assays of individual test sessions. A positive LI indicates learned avoidance of PA14. The total number of worms per experiment is specified in Supplemental Table 1.

Importantly, the droplet assay allowed us to test several P0s simultaneously, with minimal disturbance of the other P0s left on the plate, whose embryos (F1s) were collected for testing.

Olfactory preference assay using the two-choice assay

The two-choice plate assay is similar to the one previously described (Zhang *et al.*, 2005), except for several modifications. To measure the olfactory preference for bacteria odorants, a drop of 5 µL supernatant of OP50 and a drop of 5 µL supernatant of PA14 freshly generated bacterial cultures were put 2 cm apart on a 6 cm NGM plate. In each assay, one worm was placed on the plate equidistant to the two drops of the culture supernatant right after the drops were put on the plate and allowed to crawl to the preferred stimulus.

For each assay 8 worms were tested per condition. The worms that did not make a choice after 10 min were counted in the total number. The worms that disappeared during the assay by moving outside of the plate were not included. The Choice Index in the two-choice assay was defined as the number of worms that chose PA14 minus the number of worms that chose OP50 normalized by the total number of worms tested (Figure 1). When testing the progeny, the F1 Learning Index for the two-choice assay was the Choice Index of F1 progeny from naive P0s ($OP50_{mothers}$) minus the Choice Index of F1 progeny from trained P0s ($PA14_{mothers}$) (Figure 1). A positive F1 Learning Index indicates an increased avoidance of PA14 in F1s induced by the parental experience with PA14. Although we name this variable Learning Index, no F1 worm was exposed to PA14 training plates at any time. The total number of worms per experiment is shown in Supplemental Table 1.

The behavior of the F1 worms was recorded during the two-choice assay and specific behavioral features were quantified (Liu *et al.*, 2018), which allowed us to examine the odorant-guided movements of the progeny in detail. The analysis was performed with the Wormlab tracker (https://www.mbfbioscience.com/wormlab) and further analysed in Matlab. While the two-choice plate assay allowed us to analyse odorant-guided chemotactic movements, it tested one worm each time. In comparison, the droplet assay tested several worms simultaneously within a few minutes, allowing us to quickly sample the learning of the hermaphrodite mothers before isolating F1 embryos.

Slow killing assay

The plates for the slow killing assay were prepared as previously described (Tan *et al.*, 1999). Twenty young adult hermaphrodites were transferred onto each slow killing plate and kept at 25 °C. The dead and the live worms were counted at the specific time points as shown in the figure.

The analysis of parental effects induced by 8-h training

To analyse the parental effects induced by 8-h training, we plotted the values of F1 Learning Index as a function of the values of the Learning Index of their respective P0s and tested for a significant linear correlation between the variables using a Permutation test where the P0 and F1 learning indexes were randomly paired for 5000 times. We then calculated the average P0 Learning Index and separated the experiments into two groups—the experiments in which the value of the P0 Learning Index was higher than the average Learning Index, and the experiments in which the P0 Learning Indexes were lower than the average Learning Index (Figure 3(E)).

Quantification and statistical analysis

All statistical data analyses were conducted using Matlab_R2015b. In graphical representations each data point represents a biological replicate (n = number of biological replicates) unless stated otherwise. Data from different groups presented in the same graph (such as trained and control groups, WT and mutants) were collected in parallel. When data were normally distributed, we used parametric tests for comparison between groups and data were represented as average ± sem. When data were not normally distributed, non-parametric tests were used and data were graphically represented as median with first and third quartiles. The statistical methods, sample size and number of the replicates are shown in Supplemental Table 1.

Results

Parental exposure to PA14 for 4 h increases preference for pathogen in progeny

To examine how parental experience modulates offspring behavior, we exposed *C. elegans* to the pathogenic bacteria *Pseudomonas aeruginosa* PA14 and asked whether maternal aversive learning experience altered the olfactory response to PA14 in progeny.

We trained adult hermaphrodite mothers (P0s) on PA14 for 4 h, while in parallel the naive controls were fed with OP50 (Figure 1, Materials and methods). Naive *C. elegans* worms prefer the odorants of PA14 in comparison with OP50, and previous results in our lab have shown that 4-h exposure to PA14 reduces this olfactory preference (Ha *et al.*, 2010). Therefore, we employed this experimental model to study how an experience that changed the behavior in mothers affected the behavior of the next generation.

First, we quantified the preference between the smell of OP50 and the smell of PA14 in naive and trained mothers using the droplet assay, and consistent with our previous findings (Ha *et al.*, 2010; Zhang *et al.*, 2005), mothers trained with PA14 (P0 WT PA14) showed a reduced preference for the odorants of the pathogen in comparison with naive mothers (P0 WT OP50). This difference is reflected in a positive P0 Learning index, indicating that training with PA14 in P0s reduces the preference for PA14 in comparison with naive controls (Figure 2(A,B), data on the left side of the graphs depicted by empty circles). We also analyzed the preference of P0s using the two-choice assay on plate (see Material and methods) and found similar training effects (Supplemental Figure 1(A)), indicating that both the droplet assay and the two-choice assay can be used to measure olfactory preference between two bacterial food sources.

After sampling the olfactory preference in the naive and trained mothers, we harvested the embryos from the hermaphrodite mothers that remained on the control and training plates using a bleach solution, which dissolved P0 bodies and the associated bacteria and prevented the progeny from a direct contact with PA14. We cultivated F1s on OP50 under standard conditions and examined whether the learning experience of their P0 mothers modulated their olfactory behavior when they reached adulthood. To this end we tested F1 worms in the two-choice assay on plate and measured the difference between the choice index of the progeny of the naive mothers (F1 WT $OP50_{mothers}$) and the choice index of the progeny of the trained mothers (F1 WT $PA14_{mothers}$) (Figure 1, Materials and methods). While the fast speed of the droplet assay allows us to sample learning in P0s before isolating F1 embryos carried by the trained hermaphrodites, the two-choice assay allows us to record and analyze chemotactic movements in F1s. Similar to the droplet assay, a positive Choice Index in the two-choice assay indicates a preference for PA14. In contrast with P0s, we found that the progeny of the trained mothers (F1 WT $PA14_{mothers}$) showed a stronger preference for PA14 in comparison with the progeny of the naive mothers (F1 WT $OP50_{mothers}$). This is reflected in a negative F1 Learning Index, which results from an average preference for PA14 in the progeny of the trained mothers and an average preference for OP50 in the progeny of the naive mothers (Figure 2(C,D), data on the left side of the graphs depicted by empty triangles). Because the F1 progenies are isolated from trained hermaphrodites and cultivated on non-pathogenic OP50 without a direct exposure to PA14, this surprising result suggests that the food choices of the progeny are modulated by maternal exposure. A short exposure to the pathogen PA14 reduces the preference for the odorants of PA14 in mothers, but increases this preference in the progeny.

The endo-siRNA and piRNA pathways regulate parental experience modulation of behavior

We hypothesized that parental experience modulated the behaviour of the next generation through epigenetic mechanisms that regulate gene expression in progeny. In *C. elegans*, the endogenous RNA interference pathway can respond to environmental conditions to regulate

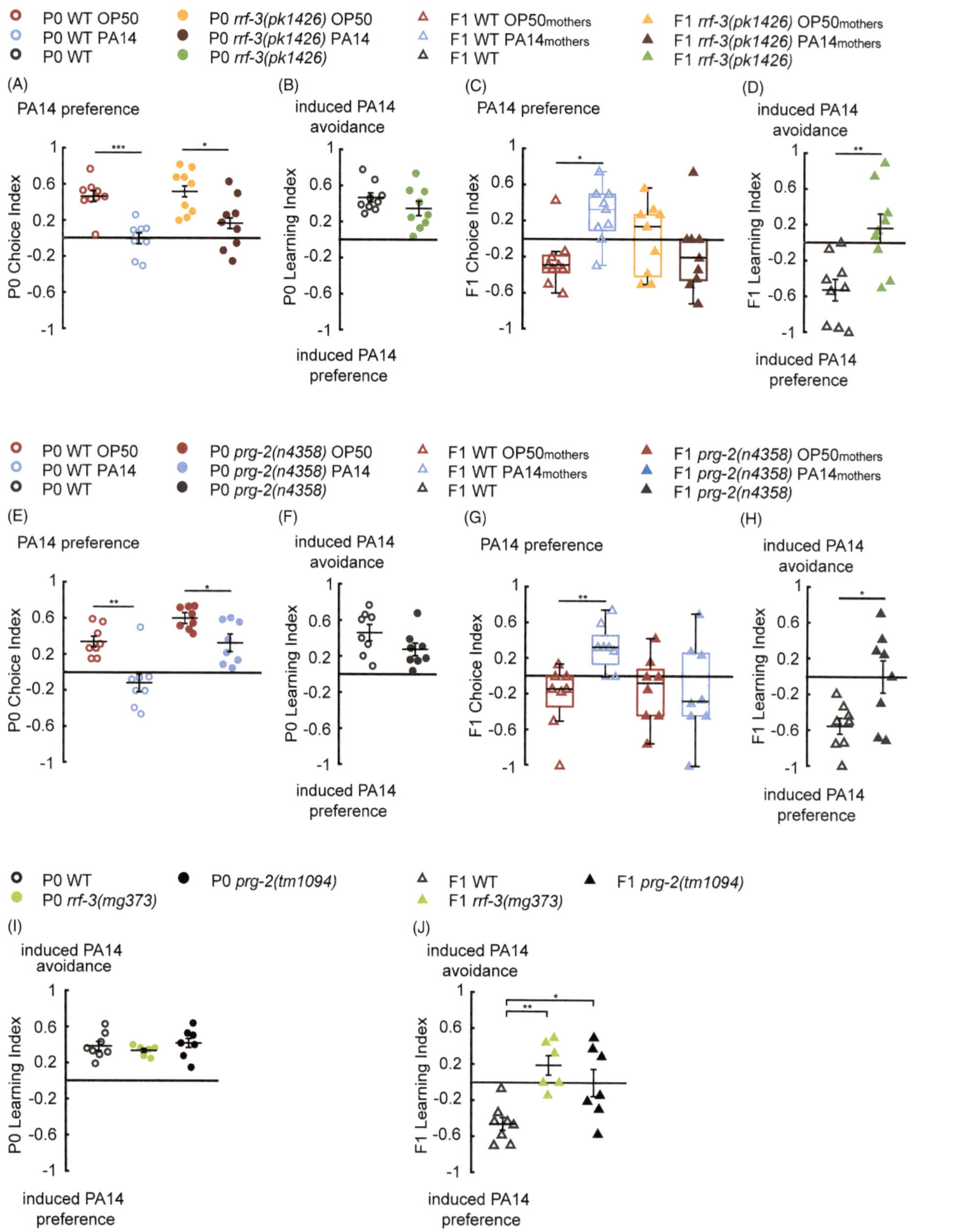

Figure 2. Training with pathogenic PA14 for 4 h increases offspring olfactory preference for PA14, in a small RNA pathway dependent manner. (A,B) Naive wild-type (WT) animals fed on *E. coli* OP50 (P0 WT OP50) prefer the odorants of PA14 (A), and training with PA14 (P0 WT PA14) decreases the preference (A,B). The deletion mutation *rrf-3(pk1426)* does not alter PA14 aversive learning in P0 mothers (A,B). $n = 9$ biological replicates each, droplet assay. (C,D) Progeny of WT trained mothers (F1 WT $PA14_{mothers}$) show an increased preference for PA14 compared with progeny of WT naive mothers (F1 WT $OP50_{mothers}$). The *rrf-3(pk1426)* deletion abolishes this increase (C), which produces a F1 Learning Index different from WT (D). $n = 9$ biological replicates each, two-choice assay. (E,F) The *prg-2(n4358)* mothers (P0s) learn to avoid PA14 similarly to WT. Further comparisons also show that CI of P0 WT OP50 is different from CI of P0 *prg-2(n4358)* OP50 ($p < 0.01$) and CI of P0 WT PA14 is different from CI of P0 *prg-2(n4358)* PA14 ($p < 0.01$). $n = 8$ biological replicates each, droplet assay. (G,H) The learning experience of P0s does not alter the olfactory preference in *prg-2(n4358)* F1s (G), which display defective F1 Learning Index in comparison with WT (H), $n = 8$ biological replicates each, two-choice assay. (I,J) The *rrf-3(mg373)* and *prg-2(tm1094)* mutant mothers learn similarly to WT (I, droplet assay) but show a defective F1 Learning Index (J, two-choice assay). $n = 8$, 6 and 7 biological replicates for WT, *rrf-3(mg373)* and *prg-2(tm1094)*, respectively. For all, $^{***}p < 0.001$, $^{**}p < 0.01$, $^{*}p < 0.05$. Data are normally distributed in A, B, D–F, H–J and presented with Mean ± SEM; data are not normally distributed in C, G and presented with box plots (showing median, first and third quartile, whiskers extending to values within 2.7 standard deviations). More details in statistical analyses are in Supplemental Table 1.

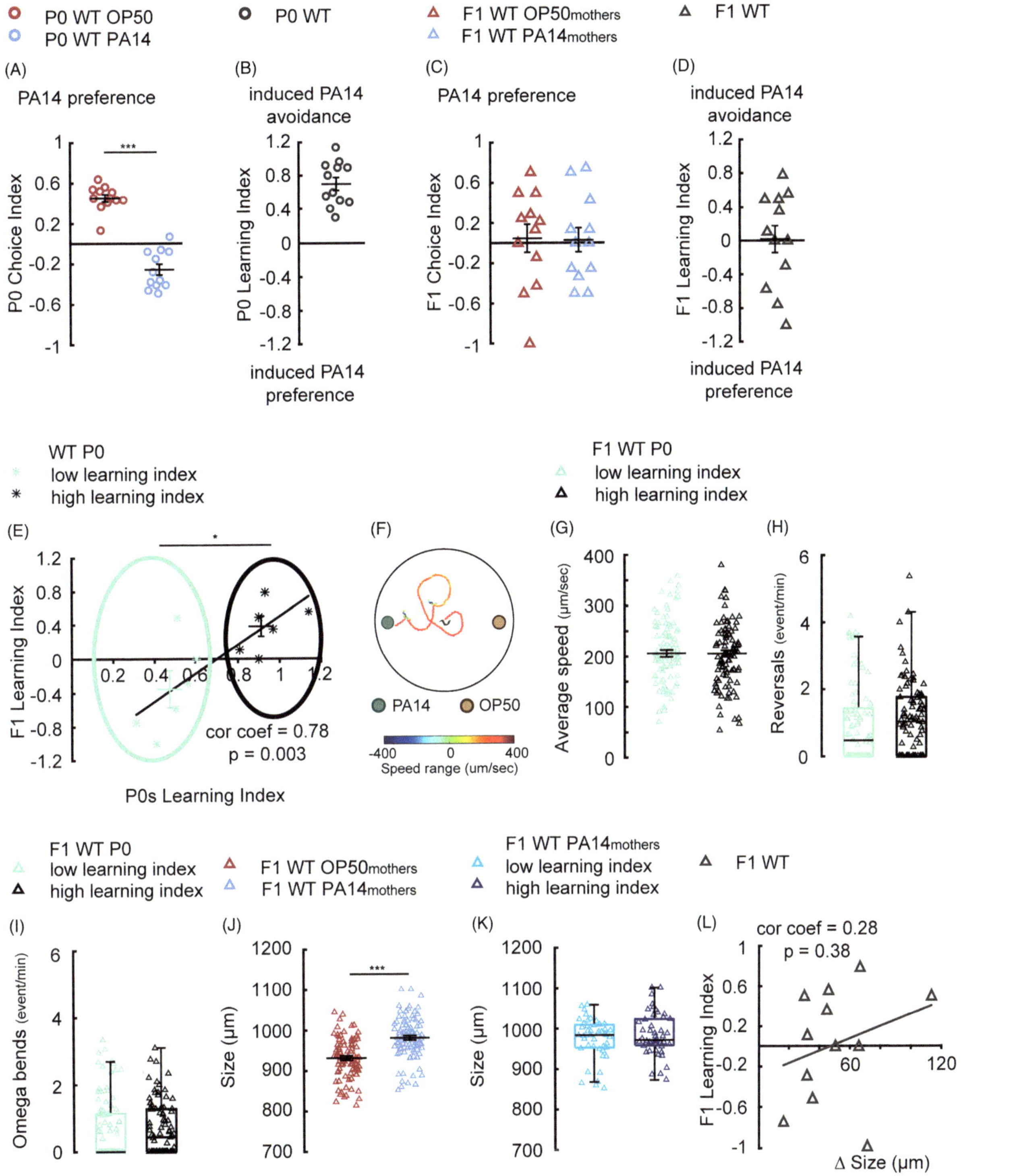

Figure 3. Extending parental training to 8 h switches offspring response from attraction to aversion. (A,B) Training with PA14 for 8 h decreases the preference for PA14 odorants in wild-type (WT) mothers (A), resulting in learned avoidance of PA14 (B), $n = 12$ biological replicates each, droplet assay. (C,D) The progeny of naive (F1 WT $OP50_{mothers}$) and trained (F1 WT $PA14_{mothers}$) mothers show similar preference for the pathogen, $n = 12$ biological replicates each, two-choice assay. (E) The Learning Index in F1s is linearly correlated with the Learning Index of their mothers after 8-h training. The progeny from the "WT P0 low learning index" group (light green stars) respond to PA14 differently from the progeny from the "WT P0 high learning index" group (dark grey stars). $N = 12$ biological replicates. (F) Schematics showing the trajectory of a F1 worm in the two-choice plate assay. (G–I) The movement speed (G), reversal rate (H) and the rate of omega bends (I) of F1s from mothers from the "WT P0 low learning index" group ($n = 90$ worms) and "WT P0 high learning index" group ($n = 93$ worms). (J) The effect of parental learning on body size of F1 WT $OP50_{mothers}$ ($n = 90$ worms) in comparison with F1 WT $PA14_{mothers}$ ($n = 93$ worms). (K) F1s of the trained mothers have similar body size. F1 WT $PA14_{mothers}$ low learning index ($n = 45$ worms) compared with F1 WT $PA14_{mothers}$ high learning index ($n = 48$ worms). (L) The F1 learning index is not correlated with the difference in F1 body size, $n = 12$ biological replicates. For all, $^{***}p < 0.001$, $^{*}p < 0.05$. Data are normally distributed in A–E, G, J and presented with Mean ± SEM; data are not normally distributed in H, I, K and presented with box plots (showing median, first and third quartile, whiskers extending to values within 2.7 standard deviations). More details in statistical analyses are in Supplemental Table 1.

physiological events in the offspring (Moore *et al.*, 2019; Ni *et al.*, 2016; Palominos *et al.*, 2017; Posner *et al.*, 2019; Rechavi *et al.*, 2014; Schott, Yanai, & Hunter, 2014). Moreover, gene expression in the germline and oocytes is regulated by RRF-3, a RNA-dependent RNA polymerase that is needed for the biogenesis of endo-siRNAs (Gent *et al.*, 2010; Han *et al.*, 2009; Lee *et al.*, 2006). The role of this pathway in olfactory behaviours has been previously demonstrated (Juang *et al.*, 2013; Sims *et al.*, 2016), which prompted us to ask how the disruption of the biogenesis of 26 G endo-siRNAs affected the modulation of olfactory choices in F1s. To this end we tested animals with a loss of function mutation in *rrf-3(pk1426)* (Simmer *et al.*, 2002) and found that the mutant mothers displayed normal learning after 4-h training with PA14 (Figure 2(A,B), data on the right side of the graphs depicted by filled circles). However, this learning experience did not alter the olfactory preference of their progeny, demonstrated by the similar choice indexes in *rrf-3(pk1426)* F1s from naive and trained mothers and a F1 Learning Index significantly different from that of wild type (WT) (Figure 2(C,D), data on the right side of the graphs depicted by filled symbols). To further confirm the role of *rrf-3,* we tested another independently generated mutant allele, *mg373,* that harbours a missense mutation for a conserved catalytic residue (Pavelec, Lachowiec, Duchaine, Smith, & Kennedy, 2009). We found that *rrf-3(mg373)* mutant animals showed a similar phenotype to the *rrf-3(pk1426)* mutants in both P0s and F1s (Figure 2(I,J); Supplemental Figure 2(A,B,D,E)). Together, these results indicate that the *rrf-3*-mediated endo-siRNA pathway regulates the modulation of offspring olfactory preference by parental experience.

We next examined the piRNA pathway that mainly regulates gene silencing and germline integrity (Ashe *et al.*, 2012; Batista *et al.*, 2008; Cox *et al.*, 1998; Das *et al.*, 2008; Lee *et al.*, 2012; Weick & Miska, 2014). The piRNA population has been shown to be altered by exposure to environmental stressors, including pathogenic *Pseudomonas aeruginosa* strain PA14 (Belicard, Jareosettasin, & Sarkies, 2018). While the PIWI-clade Argonaute *prg-1* has been previously implicated in regulating a form of pathogen avoidance that lasts for several generations (Moore *et al.*, 2019), little is known about the function of its homolog *prg-2* (Ashe *et al.*, 2012; Batista *et al.*, 2008; Cox *et al.*, 1998; Das *et al.*, 2008; Kasper, Gardner, & Reinke, 2014; Lee *et al.*, 2012). We tested a deletion mutation, *n4358,* in the *prg-2* gene. At least three alternatively spliced transcripts are predicted from the genomic locus of *prg-2* and the deletion in *n4358* affects all the transcripts. Two of the predicted transcripts encode pseudogenes and the third transcript encodes a homologous protein of PRG-1. We detected the third transcript in a wild-type cDNA library (Supplemental Figure 2(G,H)). PRG-1 regulates 21 U-RNAs (piRNAs) and mediates gene silencing, germline integrity and fertility (Ashe *et al.*, 2012; Batista *et al.*, 2008; Cox *et al.*, 1998; Das *et al.*, 2008; Lee *et al.*, 2012; Weick & Miska, 2014). Although the putative PRG-2 protein is highly similar to PRG-1 and the *prg-2* transcripts are found primarily in the germline (Hashimshony, Feder, Levin, Hall, & Yanai, 2015), the function of PRG-2 is not well characterized. We found that *prg-2(n4358)* P0s exhibited a significantly decreased preference for PA14 after training and generated learning indexes comparable to wild type (Figure 2(E,F), data on the right side of the graphs depicted by filled circles). However, the F1s of the naive and trained *prg-2(n4358)* mothers displayed similar choice indexes for PA14, resulting in a F1 Learning Index significantly different from WT (Figure 2(G,H), data on the right side of the graphs depicted by filled triangles). Furthermore, an independently generated deletion mutation, *tm1094,* in *prg-2* similarly disrupted F1 Learning Index (Figure 2(I,J), Supplemental Figure 2(C,F)). Taken together, these findings suggest the function of the piRNA pathway in transducing parental learning experience to the progeny and modulating offspring behavior.

Extending duration of parental exposure to PA14 switches attraction to aversion in progeny

We were intrigued by the increased olfactory preference for the pathogenic bacteria PA14 in the progeny of PA14-trained mothers. We asked whether stabilizing the parental environment by extending parental training with PA14 would generate a different response in the offspring. We found that training mothers with PA14 for 8 h also induced aversive learning of PA14 in the mothers (Figure 3(A,B)). However, the preference for PA14 in the progeny of the trained mothers was similar to that in the progeny of the naive mothers, which is reflected in a F1 Learning Index that has a mean value close to zero (Figure 3(C,D)). One feature in both the choice indexes and learning indexes of these progenies is their widespread distribution centred on 0. We noticed that the number of biological replicates in which the progeny of PA14-trained mothers preferred PA14 (negative Learning Index in F1) was similar to the number of biological replicates in which the progeny of trained mothers preferred OP50 (positive Learning Index in F1). This observation prompted us to examine whether there was a correlation between the learning of the mothers and the behavior of their progeny.

We plotted the Learning Index of F1 (*y*-axis) as a function of the Learning Index of their mothers in each independent experiment and found that they were linearly correlated—the more the mothers learned to avoid PA14, the more their offspring avoided the pathogen (Figure 3(E)). We calculated the average Learning Index of the P0s and divided the experiments in two groups—"P0 low learning index" and "P0 high learning index" (Figure 3(E)). Comparing these two groups we found that the F1s in experiments where mothers learned to avoid PA14 at a lower level, i.e. "P0 low learning index", had a negative average Learning Index (Figure 3(E)). This indicates that in these experiments the progeny of trained mothers preferred PA14 more than the progeny of naive mothers, similarly to the progeny of 4-h trained mothers. In contrast, the F1s in the experiments where mothers learned to strongly avoid PA14, i.e. "P0 high learning index", had a positive average F1

Learning Index (Figure 3(E)). This value suggests that the progeny of trained mothers avoided PA14 more than the progeny of naive mothers. Consistently, the F1s of the trained mothers from the "P0 high learning index" group avoided the pathogen significantly more than the F1s of the trained mothers from the "P0 low learning index" group (Supplemental Figure 3(A)). Together, these results show that the wide-spread behavioral responses to PA14 in the progeny of 8-h trained mothers result from the difference in the learning of their mothers.

We then asked how the progeny of the "P0 low learning index" group and the progeny of the "P0 high learning index" group differed in their chemotactic movements in the two-choice assay. We video tracked the worms and quantified several behavioral parameters in order to address if any specific features were modulated by parental experience and may underlie the different food odorant choices (Figure 3(F), Material and methods). We found no difference in locomotor speed, reversal rate and frequency of omega bends (i.e. the big body bends that have the shape of the letter Omega) between the F1s from the "P0 low learning index" and the F1s from the "P0 high learning index" groups (Figure 3(G - I)). These results suggest that parental learning experience with PA14 regulates the choice of the offspring without affecting these specific locomotor parameters, and likely modulates offspring behavior by regulating a higher level of sensorimotor functions.

Meanwhile, we also noticed that F1s of the trained mothers were larger in body size than F1s of the naive mothers (Figure 3(J)). This intergenerational effect is consistent with previous findings showing that animals infected with PA14 retain their eggs more than worms raised on OP50 (Tan *et al.*, 1999) and thus their progeny are kept inside the mother until a more advanced stage. Therefore, we examined if the difference in body size, which could be due to their difference in the developmental stage, could be related with the difference in the food choice. We found no differences when comparing the body size between F1s of the trained mothers in the "P0 low learning index" group and F1s of the trained mothers in the "P0 high learning index" group (Figure 3(K)), as well as no correlation between F1 Learning Index and the increase in body size in the progeny of the trained mothers compared with the progeny of the naive mothers (Figure 3(L)). In addition, we found that 8-h exposure to PA14, while inducing robust avoidance of PA14 in P0 mothers, did not alter the innate resistance to PA14 in F1s (Supplemental Figure 3(B)). Together, these results show that the body size and the innate immune resistance do not play a significant role in regulating parental experience-induced difference in offspring olfactory preference.

Next, we asked whether the learning experience in the P0 mothers trained with 8-h exposure to PA14 regulated F2s. We isolated F2 embryos from F1s and cultivated F2s under standard conditions on *E. coli* OP50. Similarly, we quantified the learning index in F2s in each experiment and found no correlation with the learning index of their F1 mothers (Supplemental Figure 4). These results indicate that the parental experience with PA14 only regulates the olfactory response of their first-generation progeny, potentially leaving the flexibility for F2s to respond to further changes in the environment.

Furthermore, we found that mutating *rrf-3* also disrupted the parental experience-induced behavioral change in the 8-h training experiments. Despite the wild-type learning ability in the *rrf-3(pk1426)* P0 mothers (Figure 4(A,B)), the learning index of *rrf-3(pk1426)* F1s is no longer correlated with the learning of their mothers (Figure 4(C)). Together, these results show that the *rrf-3*-mediated endo-siRNA pathway also regulates intergenerational effects generated by training P0 mothers for 8 h. In addition, we found that during the two-choice assay, while *rrf-3* F1s had a locomotor speed and a rate of reversals comparable to those in wild-type F1s, their rate of omega bends was lower (Figure 4(D–F)). These results identify the regulation of omega bends as the potential downstream effector of the *rrf-3*-mediated pathway in regulating intergenerational effects.

Discussion

Previous studies addressing the impact of parental experience on offspring show that intergenerational and transgenerational regulations are more complex than a unidirectional transmission of phenotypic traits acquired by the progenitors (Bohacek & Mansuy, 2015; Gapp *et al.*, 2014; Palominos *et al.*, 2017; Remy, 2010). Here, we find that a brief experience (4 h) with the pathogenic strain of *Pseudomonas aeruginosa*, PA14, while reducing the preference for the odorants of the pathogen in the trained mothers, increases the preference in the progeny. Furthermore, extending the maternal exposure time to 8 h switches the response of the progeny from attraction to avoidance in a way that is dependent on the experience of the mother. This intergenerational effect is limited to the first generation. Because the tested F1 progenies are isolated from the trained hermaphrodites at the end of the 4-h or 8-h training and because under standard conditions fertilized eggs spend 2–3 h to develop *in utero* before egg-laying (Hall, Herndon, & Altun, 2017), the behavioral effects in F1s that never directly interact with PA14 likely result from maternal exposure. Recently it has been shown that an even longer exposure to PA14 (24 h, starting at L4 larval stage) induces avoidance in mothers and their offspring for several generations through a separate mechanism involving a bacterial small RNA (Moore et al., 2019; Kaletsky, Moore, Parsons, & Murphy, 2019). Together, these results suggest that different maternal experience with the same type of bacteria can lead to diverse responses in progeny. They also suggest that the valence of the parental experience that is transmitted to the offspring is likely mediated by the physiological, but not behavioral, response of the parents.

After 4-h exposure to the pathogen, *C. elegans* adults shift their preference towards food sources that are non-pathogenic (Zhang *et al.*, 2005), a behavioral defense against increased infection. However, when presented with the option between PA14 and no food, PA14-trained worms still prefer PA14 (Ha *et al.*, 2010). In addition, in a lawn-leaving

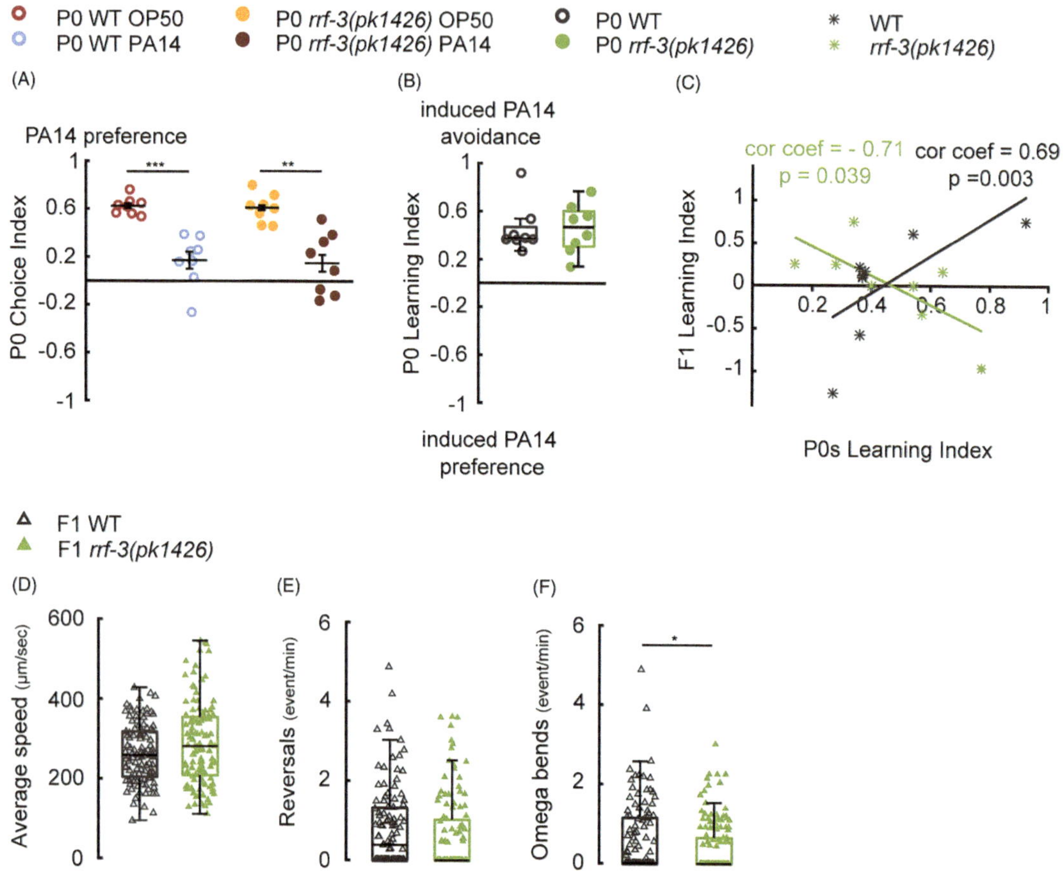

Figure 4. The RRF-3-pathway regulates intergenerational effects generated by 8-h parental training with PA14. (A–C) The *rrf-3(pk1426)* mutant P0s display wild-type choice indexes (A) and learning (B) after 8-h training with PA14; however, the learning indexes in *rrf-3(pk1426)* F1s do not positively correlate with their mothers' learning (C), $n=8$ biological replicates for each genotype. P0s and F1s are respectively tested in droplet assay and two-choice assay. (D–F) The movement speed (D), reversal rate (E) and the rate of omega bends (F) in wild-type (WT) F1s ($n=107$ worms) and *rrf-3(pk1426)* F1s ($n=115$ worms). For all, $^{***}p<0.001$, $^{**}p<0.01$, $^{*}p<0.05$. Data are normally distributed in A (Mean ± SEM) and not normally distributed in B, D–F (box plots showing median, first and third quartile, whiskers extending to values within 2.7 standard deviations). More details in statistical analyses are in Supplemental Table 1.

assay where worms feed on PA14 but can leave over time, almost all worms are still inside the lawn after 4 h and start to significantly leave only after 8 h [(Singh & Aballay, 2019) and unpublished results from our lab]. Under such conditions, it is conceivable that after a 4-h exposure PA14 is still perceived as a food source, albeit pathogenic, and positively modulates the preference of the progeny as the food consumed by their mothers. These findings are consistent with several studies in mice and humans showing that consuming foods containing certain taste or smell by mothers during gestation increases the preference of the progeny for the same taste or the odorant after birth (Hepper, Wells, Dornan, & Lynch, 2013; Todrank, Heth, & Restrepo, 2011). This modulation is likely to be adaptive since the food consumed by the mother is a potential source of safe food in the future. Extending the adult exposure to 8 h enhances the innate immune response to PA14 (Troemel *et al.*, 2006), which likely represents a more stable presence of virulence, resulting in the switch of offspring behavior from attraction to avoidance. A longer exposure of 24 h to a pathogenic lawn grown under conditions that enhance bacterial pathogenicity (Moore *et al.*, 2019) probably signals a constant presence of virulence that generates the transmission of aversive information for several generations. Together, these results on parental exposure to pathogenic bacteria are in agreement with observation in other species in which the intergenerational and transgenerational modulations of physiological and behavioral traits are influenced by the components of the diet of previous generations (Deas *et al.*, 2019; Öst *et al.*, 2014).

rrf-3 mutants are depleted in the 26G RNA population, which affects endogenous gene expression during spermatogenesis, oogenesis and zygotic development (Gent *et al.*, 2010; Han *et al.*, 2009; Lee *et al.*, 2006). Thus, our results suggest that these *rrf-3*-mediated gene expression programs can be modulated by parental experience in a way that affects the behavior of the progeny. Furthermore, one of the *prg-2* transcripts encodes a homolog of a *C. elegans* Argonaute/Piwi-related protein PRG-1 that regulates piRNAs. Despite the high level of similarity between *prg-1* and *prg-2*, the function of *prg-2* is not well understood (Ashe *et al.*, 2012; Batista *et al.*, 2008; Cox *et al.*, 1998; Das *et al.*, 2008; Lee *et al.*, 2012; Wang & Reinke, 2008). Our study reveals a role of *prg-2* in regulating parental learning experience-induced modulation of olfactory preference. Thus, we show that both the endo-siRNA and piRNA pathways play a role in the modulation of olfactory preference in progeny after 4-h maternal exposure. Whether they affect the expression of the same genes is however an open question. Behaviorally, while both mutations abolish the increased preference for PA14 in F1s as a result of training their mothers for 4 h, their phenotypes are slightly different. This observation suggests that they may act in different ways to regulate parental experience-induced behavioral changes.

A major role of the endo-siRNA pathways is to modulate endogenous gene expression (Han *et al.*, 2009; Lee *et al.*, 2006; Mello & Conte, 2004; Okamura & Lai, 2008)—in its absence the basal level and the condition-dependent expression of these genes may be compromised. If parental experience-induced changes in the next generation depends on the modulation of gene expression in the progeny, the lack of the endo-siRNA pathway may compromise the effects of parental experience. Interestingly, a recent paper reports that the absence of a dsRNA-binding protein RDE-4 in neurons affects the pool of small RNAs in the germline in a heritable manner (Posner *et al.*, 2019). Previous studies show that exogenously generated dsRNAs or bacterially derived small RNAs can also regulate gene expression or behavior in offspring and some of the processes affect the germ line (Ashe *et al.*, 2012; Buckley *et al.*, 2012; Fire *et al.*, 1998; Kaletsky, Moore, Parsons, & Murphy, 2019; Palominos *et al.*, 2017; Vastenhouw *et al.*, 2006). Together, the results from these and our studies suggest that the normal function of the gene expression program regulated by small RNA pathways is important for parental experience-modulated behaviors. The findings showing that different parental experiences employ the small RNA pathways to generate modulatory effects in the offspring reveal the flexibility of the small RNA-mediated gene expression programs to encode various environmental conditions to regulate animal physiology across generations.

Disclosure statement

No potential conflict of interest was reported by the author(s).

Author contributions

A.P., X.G., K.K. and Y.Z. designed the experiments, interpreted the results and wrote the manuscript. A.P., X.G. and K.K. performed the experiments and analyzed the data.

Funding

We thank *Caenorhabditis Genetics Center*, funded by NIH Office of Research Infrastructure Programs [P40 OD010440], for strains. Ana Pereira was funded by the Dean's Competitive Fund for Promising Scholarship, Harvard University. Y.Z. is funded by NIH [DC009852].

Data availability statement

The data and codes supporting this study are accessible in the Harvard Dataverse repository, https://doi.org/10.7910/DVN/UQAYAK

References

Ashe, A., Sapetschnig, A., Weick, E.M., Mitchell, J., Bagijn, M.P., Cording, A.C., … Miska, E.A. (2012). PiRNAs can trigger a multigenerational epigenetic memory in the germline of *C. elegans*. *Cell*, *150*(1), 88–99. doi:10.1016/j.cell.2012.06.018

Batista, P.J., Ruby, J.G., Claycomb, J.M., Chiang, R., Fahlgren, N., Kasschau, K.D., … Mello, C.C. (2008). PRG-1 and 21U-RNAs Interact to Form the piRNA Complex Required for Fertility in C. elegans. *Molecular Cell*, *31*(1), 67–78. doi:10.1016/j.molcel.2008.06.002

Belicard, T., Jareosettasin, P., & Sarkies, P. (2018). The piRNA pathway responds to environmental signals to establish intergenerational adaptation to stress. *BMC Biology*, *16*(1), 1–14. doi:10.1186/s12915-018-0571-y

Bohacek, J., & Mansuy, I.M. (2015). Molecular insights into transgenerational non-genetic inheritance of acquired behaviours. *Nature Reviews. Genetics*, *16*(11), 641–652. doi:10.1038/nrg3964

Brenner, S. (1974). The genetics of *Caenorhabditis elegans*. *Genetics*, *77*(1), 71–94.

Buckley, B.A., Burkhart, K.B., Gu, S.G., Spracklin, G., Kershner, A., Fritz, H., … Kennedy, S. (2012). A nuclear Argonaute promotes multigenerational epigenetic inheritance and germline immortality. *Nature*, *489*(7416), 447–451. doi:10.1038/nature11352

Cox, D.N., Chao, A., Baker, J., Chang, L., Qiao, D., & Lin, H. (1998). A novel class of evolutionarily conserved genes defined by piwi are essential for stem cell self-renewal. *Genes & Development*, *12*(23), 3715–3727. doi:10.1101/gad.12.23.3715

Das, P.P., Bagijn, M.P., Goldstein, L.D., Woolford, J.R., Lehrbach, N.J., Sapetschnig, A., … Miska, E.A. (2008). Piwi and piRNAs Act upstream of an endogenous siRNA pathway to suppress Tc3 transposon mobility in the *Caenorhabditis elegans* germline. *Molecular Cell*, *31*(1), 79–90. doi:10.1016/j.molcel.2008.06.003

Deas, J.B., Blondel, L., & Extavour, C.G. (2019). Ancestral and offspring nutrition interact to affect life-history traits in *Drosophila melanogaster*. *Proceedings of the Royal Society B: Biological Sciences*, *286*(1897), 20182778. doi:10.1098/rspb.2018.2778

Dias, B.G., Maddox, S.A., Klengel, T., & Ressler, K.J. (2015). Epigenetic mechanisms underlying learning and the inheritance of learned behaviors. *Trends in Neurosciences*, *38*(2), 96–107. doi:10.1016/j.tins.2014.12.003

Dias, B.G., & Ressler, K.J. (2014). Parental olfactory experience influences behavior and neural structure in subsequent generations. *Nature Neuroscience*, *17*(1), 89–96. doi:10.1038/nn.3594

Félix, M.A., & Duveau, F. (2012). Population dynamics and habitat sharing of natural populations of *Caenorhabditis elegans* and *C. briggsae*. *BMC Biology*, *10*(1), 59. doi:10.1186/1741-7007-10-59

Fire, A., Xu, S., Montgomery, M.K., Kostas, S.A., Driver, S.E., & Mello, C.C. (1998). Potent and specific genetic interference by double-stranded RNA in *Caenorhabditis elegans*. *Nature*, *391* (6669), 806–811. doi:10.1038/35888

Gapp, K., Jawaid, A., Sarkies, P., Bohacek, J., Pelczar, P., Prados, J., … Mansuy, I.M. (2014). Implication of sperm RNAs in transgenerational inheritance of the effects of early trauma in mice. *Nature Neuroscience*, *17*(5), 667–669. doi:10.1038/nn.3695

Garcia, J., Hankins, W.G., & Rusiniak, K.W. (1974). Behavioral regulation of the milieu interne in man and rat. *Science (New York, N.Y.)*, *185*(4154), 824–831. doi:10.1126/science.185.4154.824

Gent, J.I., Lamm, A.T., Pavelec, D.M., Maniar, J.M., Parameswaran, P., Tao, L., … Fire, A.Z. (2010). Distinct phases of siRNA synthesis in an endogenous RNAi pathway in *C. elegans* Soma. *Molecular Cell*, *37*(5), 679–689. doi:10.1016/j.molcel.2010.01.012

Ha, H. i., Hendricks, M., Shen, Y., Gabel, C.V., Fang-Yen, C., Qin, Y., … Zhang, Y. (2010). Functional organization of a neural network for aversive olfactory learning in *Caenorhabditis elegans. Neuron, 68*(6), 1173–1186. doi:10.1016/j.neuron.2010.11.025

Han, T., Manoharan, A.P., Harkins, T.T., Bouffard, P., Fitzpatrick, C., Chu, D.S., ... Kim, J.K. (2009). 26G endo-siRNAs regulate spermatogenic and zygotic gene expression in *Caenorhabditis elegans. Proceedings of the National Academy of Sciences USA, 106*(44), 18674–18676. doi:10.1073/pnas.0906378106

Hashimshony, T., Feder, M., Levin, M., Hall, B.K., & Yanai, I. (2015). Spatiotemporal transcriptomics reveals the evolutionary history of the endoderm germ layer. *Nature, 519*(7542), 219–222. doi:10.1038/nature13996

Hepper, P.G., Wells, D.L., Dornan, J.C., & Lynch, C. (2013). Long-term flavor recognition in humans with prenatal garlic experience. *Developmental Psychobiology, 55*(5), 568–574. doi:10.1002/dev.21059

Hall, D.H., Herndon, L.A., & Altun, Z. (2017). Introduction to *C.elegans* embryo anatomy. In *Worm Atlas*. doi:10.3908/wormatlas.4.1

Horsthemke, B. (2018). A critical view on transgenerational epigenetic inheritance in humans. *Nature Communications, 9*(1):2973. doi:10.1038/s41467-018-05445-5

Juang, B.-T., Gu, C., Starnes, L., Palladino, F., Goga, A., Kennedy, S., & L'Etoile, N.D. (2013). Endogenous nuclear RNAi mediates behavioral adaptation to odor. *Cell, 154*(5), 1010–1022. doi:10.1016/j.cell.2013.08.006

Kaletsky, R., Moore, R.S., Parsons, L.L., & Murphy, C.T. (2019). Cross-kingdom recognition of bacterial small RNAs induces transgenerational pathogenic avoidance. biorXiv preprint. doi:10.1101/697888

Kasper, D.M., Gardner, K.E., & Reinke, V. (2014). Homeland security in the *C. elegans* germ line: Insights into the biogenesis and function of pirnas. *Epigenetics, 9*(1), 62–74. doi:10.4161/epi.26647

Kundakovic, M., Gudsnuk, K., Franks, B., Madrid, J., Miller, R.L., Perera, F.P., & Champagne, F.A. (2013). Sex-specific epigenetic disruption and behavioral changes following low-dose in utero bisphenol a exposure. *Proceedings of the National Academy of Sciences of the United States of America, 110*(24), 9956–9961. doi:10.1073/pnas.1214056110

Lee, H.C., Gu, W., Shirayama, M., Youngman, E., Conte, D., & Mello, C.C. (2012). *C. elegans* piRNAs mediate the genome-wide surveillance of germline transcripts. *Cell, 150*(1), 78–87. doi:10.1016/j.cell.2012.06.016

Lee, R.C., Hammell, C.M., & Ambros, V. (2006). Interacting endogenous and exogenous RNAi pathways in *Caenorhabditis elegans. RNA (New York, N.Y.), 12*(4), 589–597. doi:10.1261/rna.2231506

Liu, H., Yang, W., Wu, T., Duan, F., Soucy, E., Jin, X., & Zhang, Y. (2018). Cholinergic sensorimotor integration regulates olfactory steering. *Neuron, 97*(2), 390–405.e3. doi:10.1016/j.neuron.2017.12.003

Mello, C.C., & Conte, D.J.. (2004). Revealing the world of RNA interference. *Nature, 431*(7006), 338–342. doi:10.1038/nature02872

Miska, E.A., & Ferguson-Smith, A.C. (2016). Transgenerational inheritance: Models and mechanisms of non – DNA sequence – based inheritance. *Science (New York, N.Y.), 354*(6308), 59–782. doi:10.1126/science.aaf4945

Moore, R.S., Kaletsky, R., & Murphy, C.T. (2019). Piwi/PRG-1 argonaute and TGF-β mediate transgenerational learned pathogenic avoidance. *Cell, 177*(7), 1827–1841. doi:10.1016/j.cell.2019.05.024

Ni, J.Z., Kalinava, N., Chen, E., Huang, A., Trinh, T., & Gu, S.G. (2016). A transgenerational role of the germline nuclear RNAi pathway in repressing heat stress-induced transcriptional activation in *C. elegans. Epigenetics & Chromatin, 9*(1), 3–15. doi:10.1186/s13072-016-0052-x

Okamura, K., & Lai, E.C. (2008). Endogenous small interfering RNAs in animals. *Nature Reviews Molecular Cell Biology, 9*(9), 673–678. doi:10.1038/nrm2479

Öst, A., Lempradl, A., Casas, E., Weigert, M., Tiko, T., Deniz, M., ... Pospisilik, J.A. (2014). Paternal diet defines offspring chromatin state and intergenerational obesity. *Cell, 159*(6), 1352–1364. doi:10.1016/j.cell.2014.11.005

Palominos, M.F., Verdugo, L., Gabaldon, C., Pollak, B., Ortíz-Severín, J., Varas, M.A., ... Calixto, A. (2017). Transgenerational diapause as an avoidance strategy against bacterial pathogens in *Caenorhabditis elegans. mBio, 8*(5), 1–18. doi:10.1128/mBio.01234-17

Pavelec, D.M., Lachowiec, J., Duchaine, T.F., Smith, H.E., & Kennedy, S. (2009). Requirement for the ERI/DICER complex in endogenous RNA interference and sperm development in Caenorhabditis elegans. *Genetics, 183*(4), 1283–1295. doi:10.1534/genetics.109.108134

Posner, R., Toker, I.A., Antonova, O., Star, E., Anava, S., Azmon, E., ... Rechavi, O. (2019). Neuronal small RNAs control behavior transgenerationally. *Cell, 177*(7), 1814–1826.e15. doi:10.1016/j.cell.2019.04.029

Rechavi, O., Houri-Ze'evi, L., Anava, S., Goh, W.S.S., Kerk, S.Y., Hannon, G.J., & Hobert, O. (2014). Starvation-induced transgenerational inheritance of small RNAs in *C. elegans. Cell, 158*(2), 277–287. doi:10.1016/j.cell.2014.06.020

Remy, J. (2010). Stable inheritance of an acquired behavior in *Caenorhabditis elegans. Current Biology : CB, 20*(20), R877–R875. https://doi.org/10.1016/j.cub.2010.08.013.Acknowledgements. doi:10.1016/j.cub.2010.08.013

Samuel, B.S., Rowedder, H., Braendle, C., Félix, M.A., & Ruvkun, G. (2016). Caenorhabditis elegans responses to bacteria from its natural habitats. *Proceedings of the National Academy of Sciences of the United States of America, 113*(27), E3941–E3949. doi:10.1073/pnas.1607183113

Schott, D., Yanai, I., & Hunter, C.P. (2014). Natural RNA interference directs a heritable response to the environment. *Scientific Reports, 4*, 7387. doi:10.1038/srep07387

Simmer, F., Tijsterman, M., Parrish, S., Koushika, S.P., Nonet, M.L., Fire, A., ... Plasterk, R.H.A. (2002). Loss of the putative RNA-directed RNA polymerase RRF-3 makes C. Elegans hypersensitive to RNAi. *Current Biology : CB, 12*(15), 1317–1319. doi:10.1016/S0960-9822(02)01041-2

Sims, J.R., Ow, M.C., Nishiguchi, M.A., Kim, K., Sengupta, P., & Hall, S.E. (2016). Developmental programming modulates olfactory behavior in *C. elegans* via endogenous RNAi pathways. *eLife, 5*, 1–26. doi:10.7554/eLife.11642

Singh, J., & Aballay, A. (2019). Microbial colonization activates an immune fight-and-flight response via neuroendocrine signaling. *Developmental Cell, 49*(1), 89–99.e4. doi:10.1016/j.devcel.2019.02.001

Tan, M.W., Mahajan-Miklos, S., & Ausubel, F.M. (1999). Killing of *Caenorhabditis elegans* by *Pseudomonas aeruginosa* used to model mammalian bacterial pathogenesis. *Proceedings of the National Academy of Sciences of the United States of America, 96*(2), 715–720. doi:10.1073/pnas.96.2.715

Todrank, J., Heth, G., & Restrepo, D. (2011). Effects of in utero odorant exposure on neuroanatomical development of the olfactory bulb and odour preferences. *Proceedings. Biological Sciences, 278*(1714), 1949–1955. doi:10.1098/rspb.2010.2314

Troemel, E.R., Chu, S.W., Reinke, V., Lee, S.S., Ausubel, F.M., & Kim, D.H. (2006). p38 MAPK regulates expression of immune response genes and contributes to longevity in *C. elegans. PLoS Genetics, 2*(11), e183. doi:10.1371/journal.pgen.0020183

Vastenhouw, N., Brunschwig, K., Okihara, K.L., Muller, F., Tijsterman, M., & Plasterk, R.H.A. (2006). Long-term gene silencing by RNAi. *Nature, 442*(7105), 882. doi:10.1038/442882a

Wang, G., & Reinke, V. (2008). A *C. elegans* Piwi, PRG-1, regulates 21U-RNAs during spermatogenesis. *Current Biology : CB, 18*(12), 861–867. doi:10.1016/j.cub.2008.05.009

Weick, E.M., & Miska, E.A. (2014). piRNAs: From biogenesis to function. *Development (Cambridge, England), 141*(18), 3458–3471. doi:10.1242/dev.094037

Weigel, D., & Colot, V. (2012). Epialleles in plant evolution. *Genome Biology, 13*(10), 249–246. doi:10.1186/gb-2012-13-10-249

Zhang, Y., Lu, H., & Bargmann, C.I. (2005). Pathogenic bacteria induce aversive olfactory learning in *Caenorhabditis elegans. Nature, 438*(7065), 179–184. doi:10.1038/nature04216

Part IV

Social and sexual behaviors

Social and sexual behaviors in *C. elegans*: the first fifty years

Douglas S. Portman

ABSTRACT

For the first 25 years after the landmark 1974 paper that launched the field, most *C. elegans* biologists were content to think of their subjects as solitary creatures. *C. elegans* presented no shortage of fascinating biological problems, but some of the features that led Brenner to settle on this species—in particular, its free-living, self-fertilizing lifestyle—also seemed to reduce its potential for interesting social behavior. That perspective soon changed, with the last two decades bringing remarkable progress in identifying and understanding the complex interactions between worms. The growing appreciation that *C. elegans* behavior can only be meaningfully understood in the context of its ecology and evolution ensures that the coming years will see similarly exciting progress.

In January 1996, when I made my first worm pick, the idea that worms might interact with each other in interesting ways was just beginning to be appreciated. Of course, signaling between worms wasn't unknown; on the contrary, it was well-established that soluble, worm-derived population density cues could trigger entry into the long-lived, stress-resistant dauer stage. Classic studies by Don Riddle demonstrated the involvement of a pheromonal cue in this process (Golden & Riddle, 1982, 1984, 1985) and studies by Riddle, Jim Thomas, Cori Bargmann, and others had explored the genetic control of dauer entry and its underlying neural basis (Albert, Brown, & Riddle, 1981; Bargmann & Horvitz, 1991; Riddle, Swanson, & Albert, 1981; Thomas, Birnby, & Vowels, 1993; Vowels & Thomas, 1992). The possibility that worms might use pheromones for other purposes, though, hadn't received much attention.

At the time, meaningful behavioral interactions between worms were thought to be limited to copulation. Katherine Liu and Paul Sternberg's landmark 1995 paper on the sensory control of male mating marked the beginning of careful studies of the neural and genetic mechanisms underlying worm sexual behavior (Liu & Sternberg, 1995) (Figure 1). However, this behavior was thought to be driven largely by mechanosensory cues; at that point, there had been no reports that males could find mates via diffusible signals. Further, the sex-specificity of these behaviors was thought to be driven by sex-specific circuits, with little attention to the possibility that sex-specific modulation of shared circuits might also have a role. (Progress in understanding male copulatory behavior itself will not be discussed here; for recent reviews of this topic, see (Barr, Garcia, & Portman, 2018; Emmons, 2018)).

Three's Company: Social feeding in *C. elegans*

Over the following decade or so, several key discoveries overturned the idea that copulation was the only interesting behavioral interaction between worms. The first of these came in 1998, with Mario de Bono and Cori Bargmann's finding that many wild isolates of *C. elegans*, including the highly divergent Hawaiian strain CB4856, exhibited 'social feeding' behavior: animals aggregate at the edges of the bacterial lawn, where the food tends to be a bit thicker, and feed in groups that can range in size from several to several hundred (de Bono & Bargmann, 1998) (Figure 2). This behavior, also called 'bordering' or 'clumping,' had been noted before, both in wild strains and in some dauer-constitutive mutants (Hodgkin & Doniach, 1997; Thomas *et al.*, 1993). de Bono and Bargmann's contribution, a seminal one, was to identify a molecular basis for this behavior: they demonstrated that the non-social, solitary behavior seen in the laboratory strain N2 resulted from a gain-of-function variant in the neuropeptide receptor *npr-1*.

The discovery of the regulation of social behavior by *npr-1* launched a long series of papers from the Bargmann and de Bono groups that provided a paradigm for understanding the relationships between genes, circuits, evolution, and behavior in *C. elegans*. An important early clue to the neural mechanisms underlying aggregation was that it depended on sensory information from both amphid chemosensory neurons and internal gas-sensing neurons (Coates & de Bono, 2002; de Bono, Tobin, Davis, Avery, & Bargmann, 2002). Subsequent work showed that the N2 allele of *npr-1* blunts O_2 avoidance; in wild strains carrying the ancestral *npr-1* allele, strong O_2 aversion causes worms to clump when cultured on plates, as animals seek relief from high ambient O_2

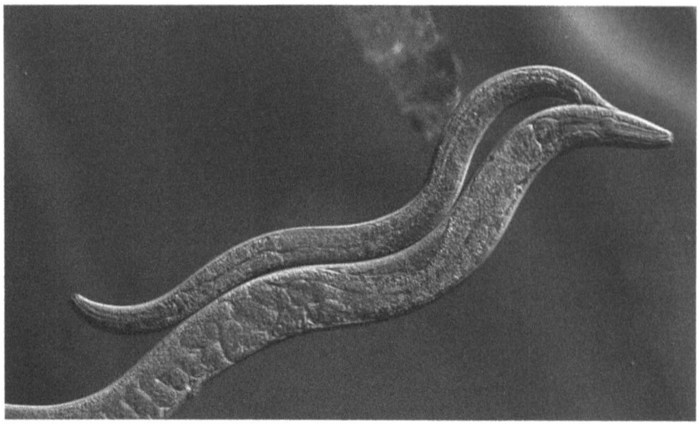

Figure 1. *C. elegans* mating behavior. The male (above) is engaged in 'scanning' behavior, in which it moves backwards while holding the ventral side of its body against the hermaphrodite (below), searching for the vulval opening.

in the low-oxygen environment of the aggregate (Cheung, Arellano-Carbajal, Rybicki, & de Bono, 2004; Gray *et al.*, 2004). *npr-1* influences O_2 avoidance, as well as other aggregation-related behaviors, by modulating a hub-and-spoke circuit in which the interneuron RMG regulates and integrates signals from the many sensory neurons to which it is connected, including the O_2-sensing URX neurons (Busch *et al.*, 2012; Fenk & de Bono, 2017; Jang *et al.*, 2012, 2017; Laurent *et al.*, 2015; Macosko *et al.*, 2009). However, variation in *npr-1* is not the only source of the difference in aggregation between N2 and wild strains: N2 also harbors a loss-of-function allele of the neuroglobin *glb-5*, which acts in O_2-sensing neurons to tune their sensory responses (Bendesky *et al.*, 2012; McGrath *et al.*, 2009; Oda, Toyoshima, & de Bono, 2017; Persson *et al.*, 2009). Interestingly, the N2 alleles of *npr-1* and *glb-5* were found to be laboratory-derived variants; it was thought that they became fixed as a result of artificial selection against aggregation during routine culture (McGrath *et al.*, 2009). However, recent work has shown that it is more likely that *npr-1(gf)* and *glb-5(lf)* increase fitness in the laboratory by modulating food intake and reproductive timing (Zhao *et al.*, 2018). Even though the difference in social behavior between N2 and wild strains does not represent true natural variation, there is no doubt that this series of studies has provided numerous foundational insights into *C. elegans* neurobiology and neurogenetics.

A modular chemical language

Another series of important advances during this time concerned the nature of *C. elegans* chemical communication. We now know that this species produces a remarkably complex set of signaling molecules, but in the early 2000s, it was apparent only that worms produced one or more soluble compounds that could trigger young larvae to commit to the dauer developmental program. One of the first indications that there was more to the story was a 2002 report from Paul Sternberg's lab demonstrating that adult hermaphrodites produced a diffusible cue that elicited a male-specific behavioral response (Simon & Sternberg, 2002). This was somewhat controversial; others had been unable to detect such a cue in *C. elegans* and had proposed that ancestral pheromones produced by *Caenorhabditis* females had been lost in *C. elegans* with the transition to self-fertility (Chasnov, So, Chan, & Chow, 2007; Chasnov & Chow, 2002). Though this impression persisted in the literature (Frezal & Felix, 2015), it is now clear that *C. elegans* hermaphrodites produce at least two distinct chemical classes of sex pheromone that can attract or retain males, and that males produce compounds that have behavioral and physiological effects on hermaphrodites.

The first insights into the molecular nature of *C. elegans* pheromones came in 2005, with the identification by Young-Ki Paik's group of 'daumone,' purified as a dauer-inducing activity from 300 L (!) of *C. elegans* conditioned media (Jeong *et al.*, 2005). Daumone, now called ascr#1, is a fatty acid derivative of the dideoxy sugar ascarylose and a member of a larger class of compounds called ascarosides, first isolated from parasitic nematodes in the early twentieth century (Ludewig & Schroeder, 2013; Park, Joo, Park, & Paik, 2019). Subsequent activity-guided fractionation in John Clardy's lab identified additional ascarosides—ascr#2, ascr#3, and ascr#5—whose dauer-inducing activity is significantly

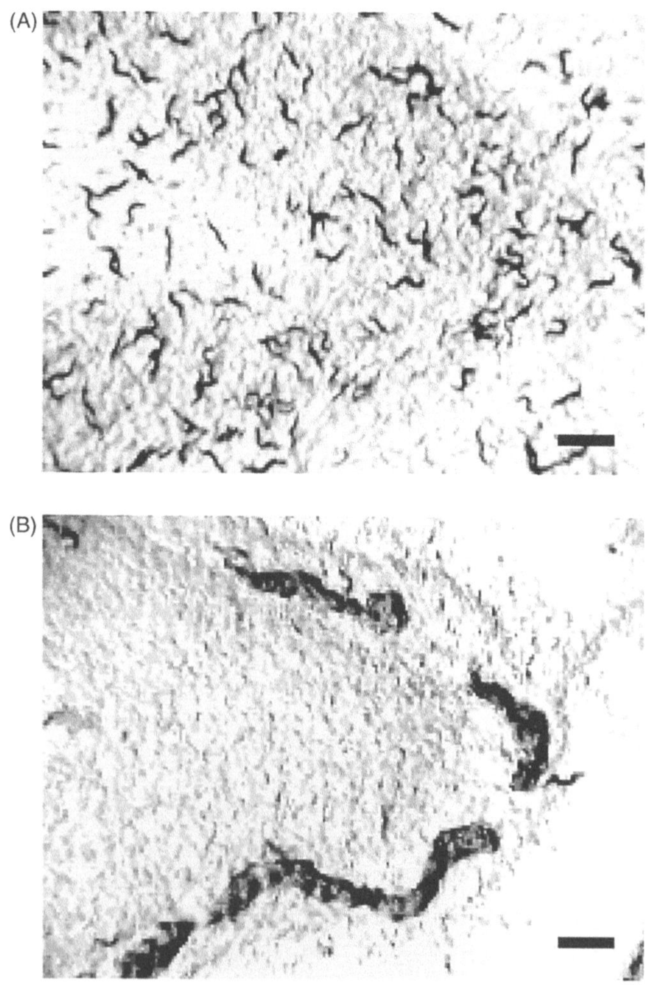

Figure 2. *C. elegans* social feeding behavior. (A) Solitary N2 hermaphrodites disperse on a bacterial lawn. (B) Social *npr-1* hermaphrodites form aggregates, particularly at the border of the bacterial lawn. Reproduced with permission from (de Bono & Bargmann, 1998).

more potent than that of ascr#1 (Butcher, Fujita, Schroeder, & Clardy, 2007; Butcher, Ragains, Kim, & Clardy, 2008).

Remarkably, ascarosides also affect the behavior and physiology of *C. elegans* adults. This was first shown by Jagan Srinivasan, working in Paul Sternberg's lab, who purified hermaphrodite-derived signals capable of attracting males. Strikingly, in a collaboration with Frank Schroeder, it was found that the active components were ascarosides: a mixture of ascr#2, ascr#3, and ascr#4 potently retained males but had little effect on hermaphrodites (Srinivasan *et al.*, 2008). This important 2008 paper also showed that the CEM and ASK neurons, the former male-specific but the latter sex-shared, are important for this effect.

Since then, a great deal of progress has been made by many groups in exploring the vast chemical space of ascarosides produced by *C. elegans* and characterizing their biological activities. Multiple recent reviews, including one in this issue, have covered this exciting work (Butcher, 2017; McGrath & Ruvinsky, 2019; Muirhead & Srinivasan, 2020; Park *et al.*, 2019; Schroeder, 2015; von Reuss, 2018). Among the most important advances have been the demonstration that ascaroside production varies according to many aspects of worm physiology, including sex, developmental stage, and nutritional status, as well as the identification of mechanisms underlying this variation (Artyukhin *et al.*, 2013; Faghih *et al.*, 2020; Izrayelit *et al.*, 2012; Joo *et al.*, 2010; 2016; Kaplan *et al.*, 2011; Panda *et al.*, 2017; Zhang, Li, Jones, Bruner, & Butcher, 2016; Zhang, Wang, Perez, Jones Lipinski, & Butcher, 2018; Zhang *et al.*, 2015; Zhou, Zhang, & Butcher, 2019; Zhou *et al.*, 2018). Ascarosides can have multiple behavioral effects on worms, inducing aggregation, attraction/retention, aversion, and promotion of foraging, depending on their chemical structures, concentrations, and interactions; moreover, the response of the receiver depends on its sex, stage, previous experience, genetic background, and other factors (Aprison & Ruvinsky, 2019; Borne, Kasimatis, & Phillips, 2017; Dong, Dolke, & von Reuss, 2016; Fagan *et al.*, 2018; Greene *et al.*, 2016; Hong *et al.*, 2017; Izrayelit *et al.*, 2012; Jang *et al.*, 2012; Lee *et al.*, 2019; Macosko *et al.*, 2009; Pungaliya *et al.*, 2009; Ryu *et al.*, 2018; Scott *et al.*, 2017; Sims *et al.*, 2016; Srinivasan *et al.*, 2008; 2012; von Reuss *et al.*, 2012; Zhang, Sanchez-Ayala, Sternberg, Srinivasan, & Schroeder, 2017). (Plasticity in the aversive responses to ascarosides is the subject of another review in this issue (Cheon, Hwang, & Kim, 2020)). Furthermore, exposure to ascarosides modulates other worm sensory behaviors (Wu *et al.*, 2019; Yamada *et al.*, 2010; Yoshimizu, Shidara, Ashida, Hotta, & Oka, 2018) and can also cause physiological changes, including altered germline proliferation and germ cell function, increased stress resistance and lifespan, and changes in lipid metabolism (Aprison & Ruvinsky, 2015, 2016; Hussey *et al.*, 2017; Ludewig *et al.*, 2013; McKnight *et al.*, 2014).

Ascarosides are not the only pheromones that mediate social behaviors. Using conditioned media, several groups have described hermaphrodite-derived factors that attract males but are likely not ascarosides. One of these papers, from Jamie White and Erik Jorgensen in 2007, showed that male-specific features of the shared chemosensory neurons AWA and AWC were important for the attraction of males to non-ascaroside pheromones produced by hermaphrodites; together with a companion paper from my lab, this was the first demonstration that genetic sex could modulate the function of sex-shared neurons (Lee & Portman, 2007; White *et al.*, 2007). Others have shown that hermaphrodites produce volatile, non-ascaroside pheromones as a function of their germline status (Leighton, Choe, Wu, & Sternberg, 2014) and that the chemoreceptor *srd-1* acts in the AWA neurons to generate male-specific attraction to non-ascaroside pheromones (Wan *et al.*, 2019).

How else might chemical signals influence worm social behaviors? Fascinatingly, some species of nematophagous fungi can lure unsuspecting *C. elegans* victims by producing compounds that mimic sex pheromones (Hsueh *et al.*, 2017; Yang *et al.*, 2020). *C. elegans* might also use pheromones to avoid predators, as it can also respond to alarm pheromones produced by injured nematodes (Zhou *et al.*, 2017) and to sulfolipids produced by *Pristionchus* (Liu *et al.*, 2018). Indeed, many nematodes produce ascarosides, raising the possibility that *C. elegans* has evolved complex responses to heterospecific competitors or predators (Choe *et al.*, 2012). Another review in this issue explores the interesting possibility that interactions between *Pristionchus*, *C. elegans*, and their bacterial food sources might represent aggressive behavior, a social interaction that has received little attention in nematodes (Quach & Chalasani, 2020).

For years, the relevance of inter-worm behaviors observed in the lab was uncertain, as precious little was known about *C. elegans* ecology. Thankfully, this is changing: our understanding of *C. elegans* ecology and evolution is advancing rapidly, and there is a growing appreciation of the complexity of its lifestyle in the wild (Frezal & Felix, 2015; Viney & Harvey, 2017). Viewing social behaviors through this lens is essential for achieving a nuanced understanding of their proximate and ultimate causes. Over the coming decades, progress toward this goal is sure to benefit from the interdisciplinary, integrative style of investigation that has been a hallmark of the *C. elegans* community since its inception.

Acknowledgements

I am grateful to Jagan Srinivasan, as well as the anonymous reviewers, for feedback that improved this Perspective.

Disclosure statement

No potential conflict of interest was reported by the author(s).

Funding

Research in the author's laboratory is financially supported by National Institute of General Medical Sciences the NIH [R01 GM108885, R01 GM130136].

References

Albert, P.S., Brown, S.J., & Riddle, D.L. (1981). Sensory control of dauer larva formation in Caenorhabditis elegans. *The Journal of Comparative Neurology, 198*(3), 435–451. doi:10.1002/cne.901980305

Aprison, E.Z., & Ruvinsky, I. (2015). Sex pheromones of *C. elegans* males prime the female reproductive system and ameliorate the effects of heat stress. *PLoS Genetics, 11*(12), e1005729. doi:10.1371/journal.pgen.1005729

Aprison, E.Z., & Ruvinsky, I. (2016). Sexually antagonistic male signals manipulate germline and soma of *C. elegans* Hermaphrodites. *Curr Biol, 26*(20), 2827–2833. doi:10.1016/j.cub.2016.08.024

Aprison, E.Z., & Ruvinsky, I. (2019). Coordinated behavioral and physiological responses to a social signal are regulated by a shared neuronal circuit. *Current Biology, 29*(23), 4108–4115 e4104. doi:10.1016/j.cub.2019.10.012

Artyukhin, A.B., Yim, J.J., Srinivasan, J., Izrayelit, Y., Bose, N., von Reuss, S.H., … Cheong, M. (2013). Succinylated octopamine ascarosides and a new pathway of biogenic amine metabolism in *Caenorhabditis elegans. The Journal of Biological Chemistry, 288*(26), 18778–18783. doi:10.1074/jbc.C113.477000

Bargmann, C.I., & Horvitz, H.R. (1991). Control of larval development by chemosensory neurons in Caenorhabditis elegans. *Science, 251*(4998), 1243–1246. doi:10.1126/science.2006412

Barr, M.M., Garcia, L.R., & Portman, D.S. (2018). Sexual dimorphism and sex differences in caenorhabditis elegans neuronal development and behavior. *Genetics, 208*(3), 909–935. doi:10.1534/genetics.117.300294

Bendesky, A., Pitts, J., Rockman, M.V., Chen, W.C., Tan, M.W., Kruglyak, L., & Bargmann, C.I. (2012). Long-range regulatory polymorphisms affecting a GABA receptor constitute a quantitative trait locus (QTL) for social behavior in Caenorhabditis elegans. *PLoS Genetics, 8*(12), e1003157. doi:10.1371/journal.pgen.1003157

Borne, F., Kasimatis, K.R., & Phillips, P.C. (2017). Quantifying male and female pheromone-based mate choice in Caenorhabditis nematodes using a novel microfluidic technique. *PLoS One, 12*(12), e0189679. doi:10.1371/journal.pone.0189679

Busch, K.E., Laurent, P., Soltesz, Z., Murphy, R.J., Faivre, O., Hedwig, B., … de Bono, M. (2012). Tonic signaling from O_2 sensors sets neural circuit activity and behavioral state. *Nature Neuroscience, 15*(4), 581–591. doi:10.1038/nn.3061

Butcher, R.A. (2017). Decoding chemical communication in nematodes. *Natural Product Reports, 34*(5), 472–477. doi:10.1039/c7np00007c

Butcher, R.A., Fujita, M., Schroeder, F.C., & Clardy, J. (2007). Small-molecule pheromones that control dauer development in *Caenorhabditis elegans. Nature Chemical Biology, 3*(7), 420–422. doi:10.1038/nchembio.2007.3

Butcher, R.A., Ragains, J.R., Kim, E., & Clardy, J. (2008). A potent dauer pheromone component in Caenorhabditis elegans that acts synergistically with other components. *Proceedings of the National Academy of Sciences of the United States of America, 105*(38), 14288–14292. doi:10.1073/pnas.0806676105

Chasnov, J., & Chow, K. (2002). Why are there males in the hermaphroditic species *Caenorhabditis elegans*? *Genetics S, 160*(3), 983–994.

Chasnov, J., So, W., Chan, C., & Chow, K. (2007). The species, sex, and stage specificity of a Caenorhabditis sex pheromone. *Proceedings of the National Academy of Sciences of the United States of America, 104*(16), 6730–6735. doi:10.1073/pnas.0608050104

Cheon, Y., Hwang, H., & Kim, K. (2020). Plasticity of pheromone-mediated avoidance behavior in *C. elegans. J Neurogenet*, 1–7. doi:10.1080/01677063.2020.1802723

Cheung, B.H., Arellano-Carbajal, F., Rybicki, I., & de Bono, M. (2004). Soluble guanylate cyclases act in neurons exposed to the body fluid to promote *C. elegans* aggregation behavior. *Current Biology, 14*(12), 1105–1111. doi:10.1016/j.cub.2004.06.027

Choe, A., von Reuss, S.H., Kogan, D., Gasser, R.B., Platzer, E.G., Schroeder, F.C., & Sternberg, P.W. (2012). Ascaroside signaling is widely conserved among nematodes. *Current Biology, 22*(9), 772–780. doi:10.1016/j.cub.2012.03.024

Coates, J.C., & de Bono, M. (2002). Antagonistic pathways in neurons exposed to body fluid regulate social feeding in Caenorhabditis elegans. *Nature, 419*(6910), 925–929. doi:10.1038/nature01170

de Bono, M., & Bargmann, C.I. (1998). Natural variation in a neuropeptide Y receptor homolog modifies social behavior and food response in *C. elegans. Cell, 94*(5), 679–689. doi:10.1016/S0092-8674(00)81609-8

de Bono, M., Tobin, D.M., Davis, M.W., Avery, L., & Bargmann, C.I. (2002). Social feeding in *Caenorhabditis elegans* is induced by neurons that detect aversive stimuli. *Nature, 419*(6910), 899–903. doi:10.1038/nature01169

Dong, C., Dolke, F., & von Reuss, S.H. (2016). Selective MS screening reveals a sex pheromone in Caenorhabditis briggsae and species-specificity in indole ascaroside signalling. *Organic & Biomolecular Chemistry, 14*(30), 7217–7225. doi:10.1039/c6ob01230b

Emmons, S.W. (2018). Neural circuits of sexual behavior in *Caenorhabditis elegans. Annual Review of Neuroscience, 41*, 349–369. doi:10.1146/annurev-neuro-070815-014056

Fagan, K.A., Luo, J., Lagoy, R.C., Schroeder, F.C., Albrecht, D.R., & Portman, D.S. (2018). A single-neuron chemosensory switch determines the valence of a sexually dimorphic sensory behavior. *Current Biology, 28*(6), 902–914 e905. doi:10.1016/j.cub.2018.02.029

Faghih, N., Bhar, S., Zhou, Y., Dar, A.R., Mai, K., Bailey, L.S., … Butcher, R.A. (2020). A large family of enzymes responsible for the modular architecture of nematode pheromones. *Journal of the American Chemical Society, 142*(32), 13645–13650. doi:10.1021/jacs.0c04223

Fenk, L.A., & de Bono, M. (2017). Memory of recent oxygen experience switches pheromone valence in *Caenorhabditis elegans. Proceedings of the National Academy of Sciences of the United States of America, 114*(16), 4195–4200. doi:10.1073/pnas.1618934114

Frezal, L., & Felix, M.A. (2015). *C. elegans* outside the Petri dish. *Elife*, 4:e05849. doi:10.7554/eLife.05849

Golden, J.W., & Riddle, D.L. (1982). A pheromone influences larval development in the nematode Caenorhabditis elegans. *Science, 218*(4572), 578–580. doi:10.1126/science.6896933

Golden, J.W., & Riddle, D.L. (1984). ACaenorhabditis elegans dauer-inducing pheromone and an antagonistic component of the food supply. *Journal of Chemical Ecology, 10*(8), 1265–1280. doi:10.1007/BF00988553

Golden, J.W., & Riddle, D.L. (1985). A gene affecting production of the Caenorhabditis elegans dauer-inducing pheromone. *Molecular & General Genetics , 198*(3), 534–536. doi:10.1007/BF00332953

Gray, J.M., Karow, D.S., Lu, H., Chang, A.J., Chang, J.S., Ellis, R.E., … Bargmann, C.I. (2004). Oxygen sensation and social feeding mediated by a *C. elegans* guanylate cyclase homologue. *Nature, 430*(6997), 317–322. doi:10.1038/nature02714

Greene, J.S., Brown, M., Dobosiewicz, M., Ishida, I.G., Macosko, E.Z., Zhang, X., … Bargmann, C.I. (2016). Balancing selection shapes density-dependent foraging behaviour. *Nature, 539*(7628), 254–258. doi:10.1038/nature19848

Hodgkin, J., & Doniach, T. (1997). Natural variation and copulatory plug formation in Caenorhabditis elegans. *Genetics, 146*(1), 149–164.

Hong, M., Ryu, L., Ow, M.C., Kim, J., Je, A.R., Chinta, S., … Choi, H. (2017). Early pheromone experience modifies a synaptic activity to influence adult pheromone responses of *C. elegans. Current Biology, 27*(20), 3168–3177 e3163. doi:10.1016/j.cub.2017.08.068

Hsueh, Y.P., Gronquist, M.R., Schwarz, E.M., Nath, R.D., Lee, C.H., Gharib, S., … Sternberg, P.W. (2017). Nematophagous fungus *Arthrobotrys oligospora* mimics olfactory cues of sex and food to lure its nematode prey. *eLife, 6*, 20023.

Hussey, R., Stieglitz, J., Mesgarzadeh, J., Locke, T.T., Zhang, Y.K., Schroeder, F.C., & Srinivasan, S. (2017). Pheromone-sensing neurons regulate peripheral lipid metabolism in *Caenorhabditis elegans. PLoS Genetics, 13*(5), e1006806. doi:10.1371/journal.pgen.1006806

Izrayelit, Y., Srinivasan, J., Campbell, S.L., Jo, Y., von Reuss, S.H., Genoff, M.C., … Schroeder, F.C. (2012). Targeted metabolomics reveals a male pheromone and sex-specific ascaroside biosynthesis in Caenorhabditis elegans. *ACS Chem Biol, 7*(8), 1321–1325. doi:10.1021/cb300169c

Jang, H., Kim, K., Neal, S.J., Macosko, E., Kim, D., Butcher, R.A., … Sengupta, P. (2012). Neuromodulatory state and sex specify alternative behaviors through antagonistic synaptic pathways in *C. elegans*. *Neuron*, *75*(4), 585–592. doi:10.1016/j.neuron.2012.06.034

Jang, H., Levy, S., Flavell, S.W., Mende, F., Latham, R., Zimmer, M., & Bargmann, C.I. (2017). Dissection of neuronal gap junction circuits that regulate social behavior in *Caenorhabditis elegans*. *Proceedings of the National Academy of Sciences of the United States of America*, *114*(7), E1263–E1272. doi:10.1073/pnas.1621274114

Jeong, P.Y., Jung, M., Yim, Y.H., Kim, H., Park, M., Hong, E., … Paik, Y.K. (2005). Chemical structure and biological activity of the *Caenorhabditis elegans* dauer-inducing pheromone. *Nature*, *433*(7025), 541–545. doi:10.1038/nature03201

Joo, H.J., Kim, K.Y., Yim, Y.H., Jin, Y.X., Kim, H., Kim, M.Y., & Paik, Y.K. (2010). Contribution of the peroxisomal acox gene to the dynamic balance of daumone production in Caenorhabditis elegans. *The Journal of Biological Chemistry*, *285*(38), 29319–29325. doi:10.1074/jbc.M110.122663

Joo, H.J., Park, S., Kim, K.Y., Kim, M.Y., Kim, H., Park, D., & Paik, Y.K. (2016). HSF-1 is involved in regulation of ascaroside pheromone biosynthesis by heat stress in Caenorhabditis elegans. *The Biochemical Journal*, *473*(6), 789–796. doi:10.1042/BJ20150938

Kaplan, F., Srinivasan, J., Mahanti, P., Ajredini, R., Durak, O., Nimalendran, R., … Edison, A.S. (2011). Ascaroside expression in *Caenorhabditis elegans* is strongly dependent on diet and developmental stage. *PLoS One*, *6*(3), e17804. doi:10.1371/journal.pone.0017804

Laurent, P., Soltesz, Z., Nelson, G.M., Chen, C., Arellano-Carbajal, F., Levy, E., & de Bono, M. (2015). Decoding a neural circuit controlling global animal state in *C. elegans*. *eLife*, *4*, e04241. doi:10.7554/eLife.04241

Lee, D., Zdraljevic, S., Cook, D.E., Frezal, L., Hsu, J.C., Sterken, M.G., … Braendle, C. (2019). Selection and gene flow shape niche-associated variation in pheromone response. *Nat Ecol Evol*, *3*(10), 1455–1463. doi:10.1038/s41559-019-0982-3

Lee, K., & Portman, D. (2007). Neural sex modifies the function of a *C. elegans* sensory circuit. *Current Biology*, *17*(21), 1858–1863. doi:10.1016/j.cub.2007.10.015

Leighton, D.H., Choe, A., Wu, S.Y., & Sternberg, P.W. (2014). Communication between oocytes and somatic cells regulates volatile pheromone production in Caenorhabditis elegans. *Proceedings of the National Academy of Sciences of the United States of America*, *111*(50), 17905–17910. doi:10.1073/pnas.1420439111

Liu, K.S., & Sternberg, P.W. (1995). Sensory regulation of male mating behavior in Caenorhabditis elegans. *Neuron*, *14*(1), 79–89. doi:10.1016/0896-6273(95)90242-2

Liu, Z., Kariya, M.J., Chute, C.D., Pribadi, A.K., Leinwand, S.G., Tong, A., … Srinivasan, J. (2018). Predator-secreted sulfolipids induce defensive responses in *C. elegans*. *Nature Communications*, *9*(1), 1128. doi:10.1038/s41467-018-03333-6

Ludewig, A.H., Izrayelit, Y., Park, D., Malik, R.U., Zimmermann, A., Mahanti, P., … Riddle, D.L. (2013). Pheromone sensing regulates *Caenorhabditis elegans* lifespan and stress resistance via the deacetylase SIR-2.1. *Proceedings of the National Academy of Sciences of the United States of America*, *110*(14), 5522–5527. doi:10.1073/pnas.1214467110

Ludewig AH. and Schroeder FC. Ascaroside signaling in C. elegans (January 18, 2013), WormBook, ed. The C. elegans Research Community, WormBook, doi/10.1895/wormbook.1.155.1, http://www.wormbook.org.

Macosko, E.Z., Pokala, N., Feinberg, E.H., Chalasani, S.H., Butcher, R.A., Clardy, J., & Bargmann, C.I. (2009). A hub-and-spoke circuit drives pheromone attraction and social behaviour in *C. elegans*. *Nature*, *458*(7242), 1171–1175. doi:10.1038/nature07886

McGrath, P.T., Rockman, M.V., Zimmer, M., Jang, H., Macosko, E.Z., Kruglyak, L., & Bargmann, C.I. (2009). Quantitative mapping of a digenic behavioral trait implicates globin variation in *C. elegans* sensory behaviors. *Neuron*, *61*(5), 692–699. doi:10.1016/j.neuron.2009.02.012

McGrath, P.T., & Ruvinsky, I. (2019). A primer on pheromone signaling in is for systems biologists. *Curr Opin Syst Biol*, *13*, 23–30. doi:10.1016/j.coisb.2018.08.012

McKnight, K., Hoang, H.D., Prasain, J.K., Brown, N., Vibbert, J., Hollister, K.A., … Miller, M.A. (2014). Neurosensory perception of environmental cues modulates sperm motility critical for fertilization. *Science*, *344*(6185), 754–757. doi:10.1126/science.1250598

Muirhead, C.S., & Srinivasan, J. (2020). Small molecule signals mediate social behaviors in *C. elegans*. *Journal of Neurogenetics*, 1–9. doi:10.1080/01677063.2020.1808634

Oda, S., Toyoshima, Y., & de Bono, M. (2017). Modulation of sensory information processing by a neuroglobin in *Caenorhabditis elegans*. *Proceedings of the National Academy of Sciences of the United States of America*, *114*(23), E4658–E4665. doi:10.1073/pnas.1614596114

Panda, O., Akagi, A.E., Artyukhin, A.B., Judkins, J.C., Le, H.H., Mahanti, P., … Schroeder, F.C. (2017). Biosynthesis of Modular Ascarosides in *C. elegans*. *Angewandte Chemie*, *56*(17), 4729–4733. doi:10.1002/anie.201700103

Park, J.Y., Joo, H.J., Park, S., & Paik, Y.K. (2019). Ascaroside pheromones: Chemical biology and pleiotropic neuronal functions. *International Journal of Molecular Sciences.*, *20*(16), 3898. doi:10.3390/ijms20163898

Persson, A., Gross, E., Laurent, P., Busch, K.E., Bretes, H., & de Bono, M. (2009). Natural variation in a neural globin tunes oxygen sensing in wild Caenorhabditis elegans. *Nature*, *458*(7241), 1030–1033. doi:10.1038/nature07820

Pungaliya, C., Srinivasan, J., Fox, B.W., Malik, R.U., Ludewig, A.H., Sternberg, P.W., & Schroeder, F.C. (2009). A shortcut to identifying small molecule signals that regulate behavior and development in *Caenorhabditis elegans*. *Proceedings of the National Academy of Sciences of the United States of America*, *106*(19), 7708–7713. doi:10.1073/pnas.0811918106

Quach, K.T., & Chalasani, S.H. (2020). Intraguild predation between Pristionchus pacificus, Caenorhabditis elegans, and bacteria: A complex interspecific interaction with the potential for aggressive behavior. *J Neurogenet*, *(115)*, 54404. doi:10.1080/01677063.2020.1833004

Riddle, D.L., Swanson, M.M., & Albert, P.S. (1981). Interacting genes in nematode dauer larva formation. *Nature*, *290*(5808), 668–671. doi:10.1038/290668a0

Ryu, L., Cheon, Y., Huh, Y.H., Pyo, S., Chinta, S., Choi, H., … Kim, K. (2018). Feeding state regulates pheromone-mediated avoidance behavior via the insulin signaling pathway in Caenorhabditis elegans. *The EMBO Journal*, *37*(15), e98402.

Schroeder, F.C. (2015). Modular assembly of primary metabolic building blocks: A chemical language in *C. elegans*. *Chemistry & Biology*, *22*(1), 7–16. doi:10.1016/j.chembiol.2014.10.012

Scott, E., Hudson, A., Feist, E., Calahorro, F., Dillon, J., de Freitas, R., … Holden-Dye, L. (2017). An oxytocin-dependent social interaction between larvae and adult *C. elegans*. *Scientific Reports*, *7*(1), 10122. doi:10.1038/s41598-017-09350-7

Simon, J.M., & Sternberg, P.W. (2002). Evidence of a mate-finding cue in the hermaphrodite nematode Caenorhabditis elegans. *Proceedings of the National Academy of Sciences of the United States of America*, *99*(3), 1598–1603. doi:10.1073/pnas.032225799

Sims, J.R., Ow, M.C., Nishiguchi, M.A., Kim, K., Sengupta, P., & Hall, S.E. (2016). Developmental programming modulates olfactory behavior in *C. elegans* via endogenous RNAi pathways. *eLife*, *5*, 11642.

Srinivasan, J., Kaplan, F., Ajredini, R., Zachariah, C., Alborn, H., Teal, P., … Schroeder, F. (2008). A blend of small molecules regulates both mating and development in *Caenorhabditis elegans*. *Nature*, *454*(7208), 1115–1118. doi:10.1038/nature07168

Srinivasan, J., von Reuss, S.H., Bose, N., Zaslaver, A., Mahanti, P., Ho, M.C., … Schroeder, F.C. (2012). A modular library of small molecule signals regulates social behaviors in Caenorhabditis elegans. *PLoS Biology*, *10*(1), e1001237. doi:10.1371/journal.pbio.1001237

Thomas, J.H., Birnby, D.A., & Vowels, J.J. (1993). Evidence for parallel processing of sensory information controlling dauer formation in *Caenorhabditis elegans*. *Genetics*, *134*(4), 1105–1117.

Viney, M., & Harvey, S. (2017). Reimagining pheromone signalling in the model nematode Caenorhabditis elegans. *PLoS Genetics, 13*(11), e1007046. doi:10.1371/journal.pgen.1007046

von Reuss, S.H. (2018). Exploring modular glycolipids involved in nematode chemical communication. *Chimia, 72*(5), 297–303. doi:10.2533/chimia.2018.297

von Reuss, S.H., Bose, N., Srinivasan, J., Yim, J.J., Judkins, J.C., Sternberg, P.W., & Schroeder, F.C. (2012). Comparative metabolomics reveals biogenesis of ascarosides, a modular library of small-molecule signals in *C. elegans*. *Journal of the American Chemical Society, 134*(3), 1817–1824. doi:10.1021/ja210202y

Vowels, J.J., & Thomas, J.H. (1992). Genetic analysis of chemosensory control of dauer formation in Caenorhabditis elegans. *Genetics, 130*(1), 105–123.

Wan, X., Zhou, Y., Chan, C.M., Yang, H., Yeung, C., & Chow, K.L. (2019). SRD-1 in AWA neurons is the receptor for female volatile sex pheromones in *C. elegans* males. *EMBO Reports, 20*(3), e46288. doi:10.15252/embr.201846288

White, J., Nicholas, T., Gritton, J., Truong, L., Davidson, E., & Jorgensen, E. (2007). The sensory circuitry for sexual attraction in *C. elegans* males. *Current Biology, 17*(21), 1847–1857. doi:10.1016/j.cub.2007.09.011

Wu, T., Duan, F., Yang, W., Liu, H., Caballero, A., Fernandes de Abreu, D.A., … Butcher, R.A. (2019). Pheromones modulate learning by regulating the balanced signals of two insulin-like peptides. *Neuron, 104*(6), 1095–1109 e1095. doi:10.1016/j.neuron.2019.09.006

Yamada, K., Hirotsu, T., Matsuki, M., Butcher, R.A., Tomioka, M., Ishihara, T., … Iino, Y. (2010). Olfactory plasticity is regulated by pheromonal signaling in Caenorhabditis elegans. *Science, 329*(5999), 1647–1650. doi:10.1126/science.1192020

Yang, C.T., Vidal-Diez de Ulzurrun, G., Goncalves, A.P., Lin, H.C., Chang, C.W., Huang, T.Y., … Schroeder, F.C. (2020). Natural diversity in the predatory behavior facilitates the establishment of a robust model strain for nematode-trapping fungi. *Proceedings of the National Academy of Sciences of the United States of America, 117*(12), 6762–6770. doi:10.1073/pnas.1919726117

Yoshimizu, T., Shidara, H., Ashida, K., Hotta, K., & Oka, K. (2018). Effect of interactions among individuals on the chemotaxis behaviours of *Caenorhabditis elegans*. *The Journal of Experimental Biology, 221*(11), jeb182790. doi:10.1242/jeb.182790

Zhang, X., Feng, L., Chinta, S., Singh, P., Wang, Y., Nunnery, J.K., & Butcher, R.A. (2015). Acyl-CoA oxidase complexes control the chemical message produced by *Caenorhabditis elegans*. *Proceedings of the National Academy of Sciences of the United States of America, 112*(13), 3955–3960. doi:10.1073/pnas.1423951112

Zhang, X., Li, K., Jones, R.A., Bruner, S.D., & Butcher, R.A. (2016). Structural characterization of acyl-CoA oxidases reveals a direct link between pheromone biosynthesis and metabolic state in *Caenorhabditis elegans*. *Proceedings of the National Academy of Sciences of the United States of America, 113*(36), 10055–10060. doi:10.1073/pnas.1608262113

Zhang, X., Wang, Y., Perez, D.H., Jones Lipinski, R.A., & Butcher, R.A. (2018). Acyl-CoA Oxidases Fine-tune the production of ascaroside pheromones with specific side chain lengths. *ACS Chemical Biology, 13*(4), 1048–1056. doi:10.1021/acschembio.7b01021

Zhang, Y.K., Sanchez-Ayala, M.A., Sternberg, P.W., Srinivasan, J., & Schroeder, F.C. (2017). Improved synthesis for modular ascarosides uncovers biological activity. *Organic Letters, 19*(11), 2837–2840. doi:10.1021/acs.orglett.7b01009

Zhao, Y., Long, L., Xu, W., Campbell, R.F., Large, E.E., Greene, J.S., & McGrath, P.T. (2018). Changes to social feeding behaviors are not sufficient for fitness gains of the *Caenorhabditis elegans* N2 reference strain. *eLife, 7*, 38675.

Zhou, Y., Loeza-Cabrera, M., Liu, Z., Aleman-Meza, B., Nguyen, J.K., Jung, S.K., … Zhong, W. (2017). Potential nematode alarm pheromone induces acute avoidance in *Caenorhabditis elegans*. *Genetics, 206*(3), 1469–1478. doi:10.1534/genetics.116.197293

Zhou, Y., Wang, Y., Zhang, X., Bhar, S., Jones Lipinski, R.A., Han, J., … Butcher, R.A. (2018). Biosynthetic tailoring of existing ascaroside pheromones alters their biological function in *C. elegans*. *eLife, 7*, 33286.

Zhou, Y., Zhang, X., & Butcher, R.A. (2019). Tryptophan metabolism in *Caenorhabditis elegans* links aggregation behavior to nutritional status. *ACS Chemical Biology, 14*(1), 50–57. doi:10.1021/acschembio.8b00872

Small molecule signals mediate social behaviors in *C. elegans*

Caroline S. Muirhead and Jagan Srinivasan

ABSTRACT
The last few decades have seen the structural and functional elucidation of small-molecule chemical signals called ascarosides in *C. elegans*. Ascarosides mediate several biological processes in worms, ranging from development, to behavior. These signals are modular in their design architecture, with their building blocks derived from metabolic pathways. Behavioral responses are not only concentration dependent, but also are influenced by the current physiological state of the animal. Cellular and circuit-level analyses suggest that these signals constitute a complex communication system, employing both synergistic molecular elements and sex-specific neuronal circuits governing the response. In this review, we discuss research from multiple laboratories, including our own, that detail how these chemical signals govern several different social behaviors in *C. elegans*. We propose that the ascaroside repertoire represents a link between diverse metabolic and neurobiological life-history traits and governs the survival of *C. elegans* in its natural environment.

Structural diversity of ascaroside signals

Conspecific communication allows animals to locate mates, alert each other of danger, and find food more easily. These communiques can be visual, auditory, or most commonly: chemical in their nature (Yohe & Brand, 2018). The roundworm, *Caenorhabditis elegans*, offers a unique model for studying chemical communication, as they rely solely on chemosensation, as reviewed by Bargmann (Bargmann, 2006). *C. elegans* communicate using ascarosides (small molecule pheromones), a large class of small molecule signals that regulate both behavioral and developmental traits as extensively reviewed by several articles (Butcher, 2017; Chute & Srinivasan, 2014; Park, Joo, Park, & Paik, 2019). These molecules are sensed through G protein-coupled receptors (GPCRs) (Butcher, 2017; Kim et al., 2009; McGrath et al., 2011; Park et al., 2012). Ascarosides grant worms the ability to communicate living conditions and the proximity of other worms (Golden & Riddle, 1982; Golden & Riddle, 1984; McGrath & Ruvinsky, 2019; Srinivasan et al., 2008). Multiple studies state that *C. elegans* produce and secrete multiple ascarosides in different concentrations, depending on their current physiological state and the information being communicated (Aprison & Ruvinsky, 2017; Butcher, 2017; Chute & Srinivasan, 2014; Kaplan et al., 2011; Schroeder, 2015; Srinivasan et al., 2008; Srinivasan et al., 2012). Various wild isolates of *C. elegans* have been shown to produce different cocktails of ascarosides in response to the same stimuli (Diaz et al., 2014).

Since the discovery of the ascarosides, the chemical biology of these molecules has been investigated extensively by various research groups as reviewed extensively in the following references (Butcher, 2019; Park et al., 2019). These studies suggest that social signaling in *C. elegans* is based on a *modular language* of small molecules, derived from combinatorial assembly of several structurally distinct building blocks (Figure 1). Different combinations of the building blocks lead to distinct functions for these molecules. Similarly, changing one or more of the building blocks can change the behavior they elicit. For instance, changing the head group of the ascaroside from an indole moiety in icas#3 to a succinyl octopamine moiety in osas#9, causes the behavior to change from aggregation to avoidance (Figure 1). Ascarosides are generally derived from three basic metabolic pathways: carbohydrate metabolism, peroxisomal fatty-acid β-oxidation, and amino acid metabolism and the modularity of these molecules suggests an intricate biochemical framework used to derive the diversity (Butcher, 2019). This raises the possibility that social signaling via small molecules transduces input from the overall metabolic state of the organism. Several reviews confirm that both physiological state together with other environmental factors control the biosynthesic pathways of ascarosides to generate specific pheromone blends that differentially regulate dauer development, aggregation, mate attraction, and developmental timing (Chute & Srinivasan, 2014; Ludewig & Schroeder, 2013). Functional roles of ascarosides in regulating developmental and behavioral phenotypes have been reviewed extensively by Edison, Ludewig and Schroeder, Chute and Srinivasan, and Butcher (Butcher, 2017; Chute & Srinivasan, 2014; Edison, 2009; Ludewig & Schroeder, 2013). In the following sections of the review, we will specifically discuss the role that ascarosides play in different social behaviors.

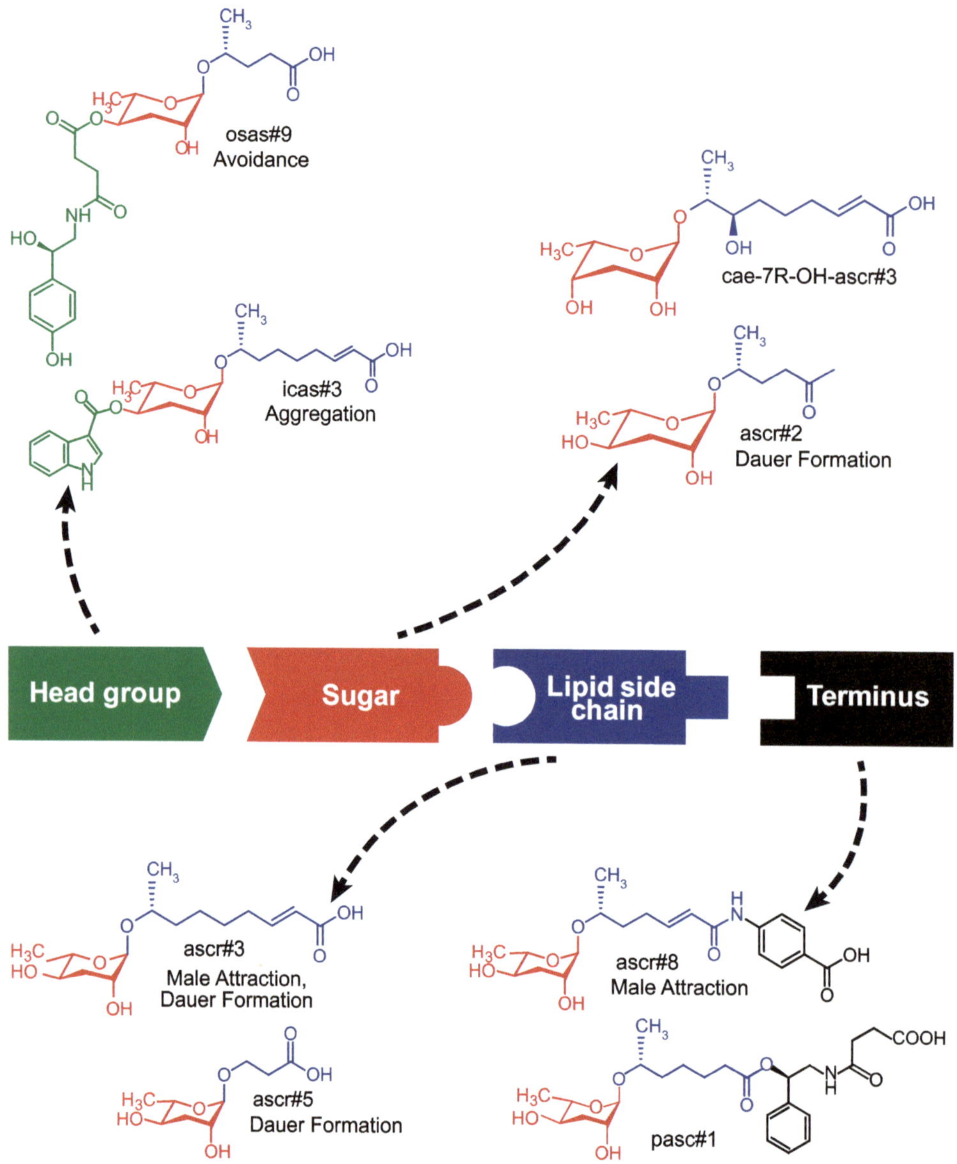

Figure 1. Social signaling in *C. elegans* is regulated by a modular language of small molecules called ascarosides. Ascarosides are alkyl glycosides that carry a fatty acid-derived lipophilic side chain. The diversity of these molecules is a result of combinatorial regulation of building blocks. These building blocks are derived from different metabolic pathways. Given the diverse metabolic origin of these molecules, the structure of the ascaroside dictates the behavior it elicits. The figure does not represent the vast diversity of ascarosides both within *C. elegans* and in other nematode species.

Ascarosides across the phylum Nematoda

In 1996, six different classes of ascarosides from *Ascaris suum*, a parasitic nematode, were structurally characterized using nuclear magnetic resonance (NMR) spectroscopy and electrospray mass spectrometry (Bartley, Bennett, & Darben, 1996). It has since been discovered that ascarosides are secreted and used for communication all across the Nematoda phylum (Choe, von Reuss, et al., 2012; von Reuss, 2018)

Choe et al examined the ascaroside profiles in both parasitic and free-living nematode species and discovered that despite the large variation in ascaroside profiles across species, the pathways responsible for ascaroside biosynthesis are conserved (Choe, von Reuss, et al., 2012). For instance, ascr#1 is produced by both nematodes *Panagrellus redivivus* and *C. elegans*. *P. redivivus* is a free-living nematode that is a distant relative of *C. elegans* (Srinivasan et al., 2013). Their abilities to sense ascr#1, a male attractant for *P. redivivus* and a component of dauer pheromone for *C. elegans*, suggests that these nematodes share a common biosynthetic pathway. However, the functional differences in response to ascr#1 demonstrate that the ascaroside response is not necessarily conserved (Choe, Chuman, et al., 2012). Ascaroside secretions are known to vary between different isolates of the same species. For example, *P. pacificus* isolates from different locations secrete varying ascarosides cocktails and enter dauer in response to different ascaroside conditions (Bose et al., 2014). Recently, it has been shown, that along with attachment of additional building blocks from primary metabolic pathways (Figure 1), dimerization of monomeric units represents a highly effective mechanism to generate species-specific ascarosides, which form a complex chemical language in nematode communication (Dong, Dolke, Bandi, Paetz, & von Reuß, 2020)

Typically, niche-sharing and phylogenetically close worms have more similar ascaroside secretions (Choe, von Reuss, et al., 2012). And while there is variation in ascaroside response and production across species, there are some interesting instances where ascarosides induce the same response in different nematode species. *P. redivivus* females attract *C. elegans* and *P. redivivus* males with their secretions (Choe, von Reuss, et al., 2012). Additionally, *C. elegans* dispersal blend of pheromones induces dispersal in *S. feltiae* (Kaplan et al., 2012). Given their role in nematode communication, ascarosides offer a way to study the intricacies of chemical communication and decision-making.

Ascarosides and dauer development

Pheromones have the ability induce the dauer state in *C. elegans* (Golden & Riddle, 1982; Golden & Riddle, 1984). Dauer state is characterized by a thicker, more robust cuticle and tolerance to starvation, which enables worms to survive in poor conditions. When food is re-introduced to worms, they have the ability to exit dauer state and resume development (Cassada & Russell, 1975). Worms exposed to daumone (a dauer-inducing ascaroside cocktail) or unfavorable conditions, like high population density or low levels of food (Butcher, Fujita, Schroeder, & Clardy, 2007), enter the dauer state (Golden & Riddle, 1984).

The dauer pheromone is a cocktail of multiple different ascarosides: including, at least, ascr#1, ascr#2, and ascr#3, with ascr#2 and ascr#3 having the greatest dauer-inducing strength (Butcher et al., 2007) (Figure 1). Under starvation conditions, worms produce increased amounts of ascr#2 compared to fed worms (Kaplan et al., 2011). However, it should be noted that isolates of wild type worms induce dauer state in response to differing levels of environmental stress and communicate with varying amounts of ascr#2 and ascr#3 (Diaz et al., 2014).

Multiple neurons have been found to be involved in dauer formation. ADF, ASG, ASI, and ASJ neuron-ablated worms enter dauer state regardless of environmental conditions, suggesting that these neurons are involved in either dauer pheromone or food sensation (Bargmann & Horvitz, 1991). The ASI neuron is unique in that it expresses the *daf-7* gene, encoding a constitutively expressed TGF-β ligand (Bargmann & Horvitz, 1991). Dauer pheromone inhibits *daf-7* and thereby TGF-β action, limiting growth and inducing dauer state. As such, *daf-7* mRNA is more abundant in well-fed worms (Ren et al., 1996).

Interestingly, animals that are unable to produce daumone are still able to sense it. DAF-22 is required for synthesis of ascarosides with short fatty acid-like side chains (Butcher et al., 2009). Jeong et al. found that while *daf-22* mutant animals could not produce ascr#1, they could still respond to it, indicating that sensation and production do not share identical pathways (Jeong et al., 2005). Additionally, *daf-22* mutants also lack olfactory plasticity: an animal's ability to stop sensing a constant stimulus, such as food. When *daf-22* mutants were exposed to daumone, they regained olfactory plasticity. Plasticity is further regulated by population density, an environmental factor conveyed through daumone signaling, suggesting that dauer pheromone also controls olfactory plasticity (Yamada et al., 2010).

G protein-coupled receptors (GPCRs) are involved in daumone sensation; without the G proteins encoded by *gpa-2* and *gpa-3*, worms cannot induce dauer state in response to dauer-inducing pheromone (Zwaal, Mendel, Sternberg, & Plasterk, 1997). SRBC-64 and SRBC-66 are GPCRs expressed in the ASK neuron; mutations in either GPCR gene disrupts ascr#1, ascr#2, and ascr#3 sensation, preventing entry into the dauer state (Kim et al., 2009). When the ascr#2-sensing GPCR, DAF-37, is expressed in the ASI neuron it mediates dauer formation, and when it is expressed in the ASK neuron, it mediates hermaphrodite avoidance. DAF-38 and DAF-37 heterodimerize and inhibit cAMP in response to ascr#2 (Park et al., 2012). *srg-36* and *srg-37* encode GPCRs expressed on the ASI neuron, and mutations in either eliminate dauer formation in response to ascr#5 (McGrath et al., 2011). Despite these advances in our understanding of the molecular players, it is still largely unclear whether different ascarosides involved in dauer development are associated with different pathways (and neurons), or if all ascarosides signal via the same pathways.

Ascaroside influenced mate attraction

Unfortunately, there are no male-specific studies of dauer inducing-ascarosides because dauer formation happens in the L2 larval stage (Hu, 2005–2018), when male worms are indistinguishable from hermaphrodites (Emmons, 2005). Despite this challenge, studying differences in ascaroside signaling and response between the sexes is valuable because it allows researchers to understand sex-specific neural differences. Mating-related ascaroside signaling offers a more readily available avenue for studying sex-specific neural differences.

C. elegans are capable of self-fertilizing, and only a small portion of the *C. elegans* population are male. However, both males and hermaphrodites attract each other using ascaroside signals. Both ascr#3 and ascr#8, produced primarily by hermaphrodites, attract males (Izrayelit et al., 2012; Narayan et al., 2016). Ablation experiments reveal that male-specific CEM neurons are solely responsible for sensation of ascr#8 and mediation of a worm's ability to detect different concentrations of ascr#8 (Narayan et al., 2016). In male worms, the ASK and CEM neurons are responsible for sensing ascr#3 (Narayan et al., 2016; Srinivasan et al., 2008). Additional recent research has also shown that the ADF neuron is responsible for ascr#3 sensation which is implicated in male attraction, but only when the ADF neuron has undergone masculine differentiation (Fagan et al., 2018). Hermaphrodites also sense ascr#3 via the ADL neurons, although they are not attracted to ascr#3 (Fagan et al., 2018; Jang et al., 2012),

ascr#10, produced by male worms, attracts hermaphrodites, and the sensory ADL neurons play a role in mediating the germline response to ascr#10 (Aprison & Ruvinsky, 2017; Izrayelit et al., 2012). Despite hermaphrodite's attraction to ascr#10, it causes them to age faster and reduce exploration (Erin Z. Aprison & Ruvinsky, 2016; Aprison & Ruvinsky, 2019a). However, the faster aging does not come without survival benefits at the species level; hermaphrodites exposed to male ascarosides lay more eggs and have a slower loss of germline precursor cells (Aprison & Ruvinsky, 2016).

The impact of attractive ascarosides can be diminished by the presence of other ascarosides. ascr#3 limits the effects of ascr#10, and therefore decreases hermaphroditic response to ascr#10. The antagonistic relationship between ascr#10 and ascr#3 may serve the purpose of allowing hermaphrodites to recognize when males are present by detecting the ratio of ascr#10 to ascr#3 (Aprison & Ruvinsky, 2017).

Both OSM-9 and OCR-2 are transient receptor potential vanilloid channel required for ascr#10 response (Aprison & Ruvinsky, 2017). On the other hand, mutations in the cyclic nucleotide gated channels, TAX-2 and TAX-4, abolish the ascr#3 behavioral response in worms. These findings show that ascr#10 and ascr#3 sensation use different downstream signaling. Surprisingly, TAX-2 and TAX-4, are expressed in the ascr#10-sensing ASI neurons (Aprison & Ruvinsky, 2017). The reason for this is likely that the ASI neuron is implicated in dauer formation in response to ascr#3-containing dauer pheromone (Bargmann & Horvitz, 1991; Butcher et al., 2007). DAF-7, secreted by the ASI neuron, regulates dauer formation in response to ascr#3 and regulates production of germline precursor cell proliferation in response to ascr#10 (Aprison & Ruvinsky, 2016; Aprison & Ruvinsky, 2017; Dalfó, Michaelson, Hubbard, & Jane, 2012). The dual role of the ASI neuron may explain why it contains both transient receptor potential vanilloid channels and cyclic nucleotide gated channels (Aprison & Ruvinsky, 2017).

ascr#3 and ascr#8 are not the only ascarosides known to generate male attraction. A blend of ascr#2, ascr#3, and ascr#4 can act synergistically at low concentrations (as low as fmol and pmol concentrations) to induce male attraction (Srinivasan et al., 2008). The CEM, AWA, and AWC neurons all express OSM-9 and are involved in male attractive behavior (White et al., 2007). Genetic and laser ablation experiments show that L3 worms or younger can compensate for removal of a single class of these attractive neurons as long at the other attractive neurons remain intact. L4 and adult worms do not retain this ability; removing a single class of attractive neurons impairs their attraction indefinitely (White et al., 2007).

A slight variation of ascaroside structure can alter the behavior elicited from the ascaroside. For example, while ascr#3 and ascr#10 have opposite effects on male and hermaphrodite behavior, they are very structually similar; a lone double bond differentiates ascr#3 and ascr#10 (Izrayelit et al., 2012). Additionally, icas#3, sensed by the ASK neuron, is identical to ascr#3 but for the addition of an indole group attached at the fourth position of the ascarylose sugar (Srinivasan et al., 2012; Figure 1). While ascr#3 is a hermaphrodite repellant (Izrayelit et al., 2012), icas#3 is a strong hermaphrodite attractant (Srinivasan et al., 2012). Unlike the effects of ascr#3, attraction to icas#3 is not sex-specific, although hermaphrodites are more strongly attracted than male worms (Srinivasan et al., 2012).

Ascaroside production and sensation play a large role in *C. elegans* reproduction. Sex-specific ascaroside production allows worms to identify members of the opposite sex (Aprison & Ruvinsky, 2017) and in addition, sensing these ascarosides can change a worm's behavior (Park et al., 2019).

Social, roaming, and aggregation behaviors

While many attractive and aversive behaviors elicited by ascarosides are considered sex-specific due to their implication in reproduction, not all are. Non-sex-specific avoidance inducing pheromones were identified through high-performance liquid chromatography-mass spectrometry, including an ascr#9 derivative wherein the ascarylose is connected to an *N*-succinylated octopamine, osas#9 (Figure 1). osas#9 is produced by starved L1-arrest worms and only causes avoidance in starved worms (Artyukhin et al., 2013). TYRA-2, a GPCR, senses osas#9 within the sensory cilia of the ASH neurons to induce avoidance (Chute et al., 2019). The presence of osas#9 prompts worms to turn away from the starvation signal and continue their exploration foraging elsewhere.

While starvation signals like osas#9 influence where worms forage, other ascarosides influence the amount of time worms spend roaming. Differences in the *roam-1* locus were identified via quantitative trace locus analyses (QTL) using different recombinant inbred worms created from various wild isolates with varying icas#9 sensitivities. The *roam-1* locus encodes five GPCRs, including *srx-43*, which is expressed in ASI neuron (Greene, Brown, et al., 2016). icas#9 elicits aggregation in hermaphrodites and males (Srinivasan et al., 2012), though wild type isolates that produce the same level of icas#9 have different sensitivities to it (Greene, Brown, et al., 2016). The varying sensitivities to icas#9 rely on the *srx-43* promoter within the *roam-1* locus. *daf-7* has decreased expression in the presence of icas#9, suggesting that icas#9 plays a role in regulating exploratory behavior by targeting DAF-7, as DAF-7 regulates exploratory behavior in response to ascr#10 (Greene, Brown, et al., 2016).

However, *srx-43* is not the only gene responsible for icas#9 sensitivity; *srx-44*, also found within the *roam-1* locus, affects icas#9 response. While the *srx-43* promoter sequence affects the level at which it is expressed, the promoter sequence of *srx-44* controls where it is expressed. When *srx-44* is expressed in ASJ, it inhibits icas#9 response, while ADL expression increases the icas#9 aggregation response (Greene, Dobosiewicz, Butcher, McGrath, & Bargmann, 2016). This research stresses the importance of investigating protomers for genes in addition to the gene itself when examining gene function.

Wild isolates of *C. elegans* exhibit both social and solitary behaviors due to variation in the *npr-1* gene. A loss of function *npr-1* mutation causes solitary strains of worms to exhibit social behavior (de Bono & Bargmann, 1998), due to changes in oxygen preferences (Gray et al., 2004). Solitary wild type hermaphrodites are repelled by ascarosides that attract social and *npr-1* loss of function hermaphrodites. Through calcium imaging, it was discovered that the ASK neuron responds to ascarosides in worms with low activity NPR-1 or without NPR-1 (Macosko et al., 2009). The ASK neuron is connected by a gap junction to the RMG neuron and the RMG neuron is required for social behavior. NPR-1 modifies social behavior by altering gap junctions between the RMG neuron and the ASK neuron (Macosko et al., 2009). The dual roles of the ASK neuron in ascaroside sensation and social behavior could to explain the difference in ascaroside attraction between social and solitary strains of worms (Macosko et al., 2009). Ascarosides not only influence the social behavior of worms (Srinivasan et al., 2012), but also can have different effects on worms depending on their social state (Macosko et al., 2009).

Sex-Specific neural circuits

The circuit that senses ascr#10 in hermaphrodites is closely related to the reproductive system (Aprison & Ruvinsky, 2019b). When HSN neurons, which control egg-laying, are removed, ascr#10 does not result in decreased exploration (Aprison & Ruvinsky, 2019b; Schafer, 2006). HSN signals to vm2 vulva muscles, and vm2 mutants have increased egg laying, but not reduced exploration in response to ascr#10, suggesting the vm2 has a role in the ascr#10 circuit (Aprison & Ruvinsky, 2019b).

TPH-1 is the biosynthetic enzyme yielding serotonin, and animals lacking this enzyme are unable to respond to ascr#10 (Aprison & Ruvinsky, 2019a; Sze, Victor, Loer, Shi, & Ruvkun, 2000). Additionally, both exogenous serotonin and ascr#10 increase dwell time in hermaphrodites (Aprison & Ruvinsky, 2019a; Flavell et al., 2013). TPH-1 function is needed in NSM and HSN neurons, and expression of the serotonin-gated chloride channel, MOD-1, is required in the AIY and RIF interneurons for ascr#10 response (Aprison & Ruvinsky, 2019a; Flavell et al., 2013; Figure 2(A)). The reason ascr#10 may reduce exploration is because decreased hermaphrodite movement allows for more successful copulation (Aprison & Ruvinsky, 2019a)

Male worms are capable of ignoring adverse cues in the presence of hermaphroditic ascarosides. When a single-sex plate of worms is subjected to starvation conditions and salt, the animals learn to avoid salt because they associate salt with starvation. However, when conditioning takes place on a mixed plate, male worms do not learn to avoid the salt because they associate salt with ascarosides, and additional signals produced by hermaphrodites (Sakai et al., 2013). The response of male worms requires a distinctly male neural circuit (Sakai et al., 2013). Males with a non-specific gain of function mutation in the transcription factor that controls sexual differentiation by suppressing male differentiation, *tra-1*, do not respond to sexual conditioning with hermaphrodites (Lee & Portman, 2007; Sakai et al., 2013). However, when only the salt-sensing ASE neuron is feminized, males still respond to conditioning (Adachi et al., 2010; Sakai et al., 2013).

Surprisingly, while the CEM neuron is involved in ascr#3 sensation, male worms lacking CEM neurons are still attracted to ascr#3 (Fagan et al., 2018; Narayan et al., 2016; Srinivasan et al., 2008). Instead, when the ADF neuron is ablated, the attraction of males to ascr#3 is reversed, but when it is ablated in hermaphrodites, there is no visible change in their behavior. Feminizing male ADF neurons using a TRA-1 mutation similarly abolishes male attraction to hermaphroditic ascarosides; masculinizing the ADF neuron in hermaphrodites caused hermaphrodites to gain attraction to ascr#3 (Fagan et al., 2018). Through calcium imaging, it was shown that the ADL neuron responds to ascr#3 in both sexes of worm; however, the ADL response is stronger in hermaphrodites (Jang et al., 2012). The ADL is primarily responsible for ascr#3 avoidance in hermaphrodites; the masculinized ADF neuron encourages ascr#3 attraction in males while the ASK neuron inhibits avoidance caused by the ADL neuron (Fagan et al., 2018; Jang et al., 2012; Figure 2(B)).

Hermaphrodite *npr-1* mutants exhibit decreased aversion to ascr#3, and expressing NPR-1 in the RMG neuron, not in the ADL neuron, rescued ascr#3 response. Additionally, a gap junction subunit mutation increased the ADL neuronal response to ascr#3 in *npr-1* mutants, indicating that a gap junction connects the RMG and ADL neurons. *npr-1* males have a greater ASK response than wild type worms (Jang et al., 2012; Figure 2(B)). Unsurprisingly, there are notable differences between male and hermaphrodite neural circuits. Therefore, neural circuits not only govern different behavioral responses for the same chemical, but also recruit different sensory neurons.

Conclusions

The rapid developments in the ascaroside field in the past decade highlight the importance of understanding ascaroside receptor-mediated neuronal signaling pathways and the relevant biosynthetic pathways in enacting different social behaviors. Areas that may see continued investigations are the following:

Ascaroside pheromone recognition and processing: As the number of newly discovered ascaroside pheromones increases, how they are sensed and transduced within the nervous system remains unresolved. Are common signaling pathways recruited for multiple ascarosides and what are the mechanisms? Furthermore, several behavioral effects induced by these pheromones are synergistic, i.e., single pheromones do not always act alone (Butcher et al., 2007; Srinivasan et al., 2008). Thus, ascaroside sensation and signaling are likely complex and elaborately intertwined and untangling these pathways could provide important clues for understanding neuronal signaling in other species.

Ascaroside receptor research: Fully understanding the chemical biology of ascarosides requires additional research on receptors and a cross-species comparison of these receptors. To date, there are more than 200 ascarosides identified in the Nematoda phylum (Choe, Millar, & Rust, 2009). Despite this large diversity of ascarosides, the number of

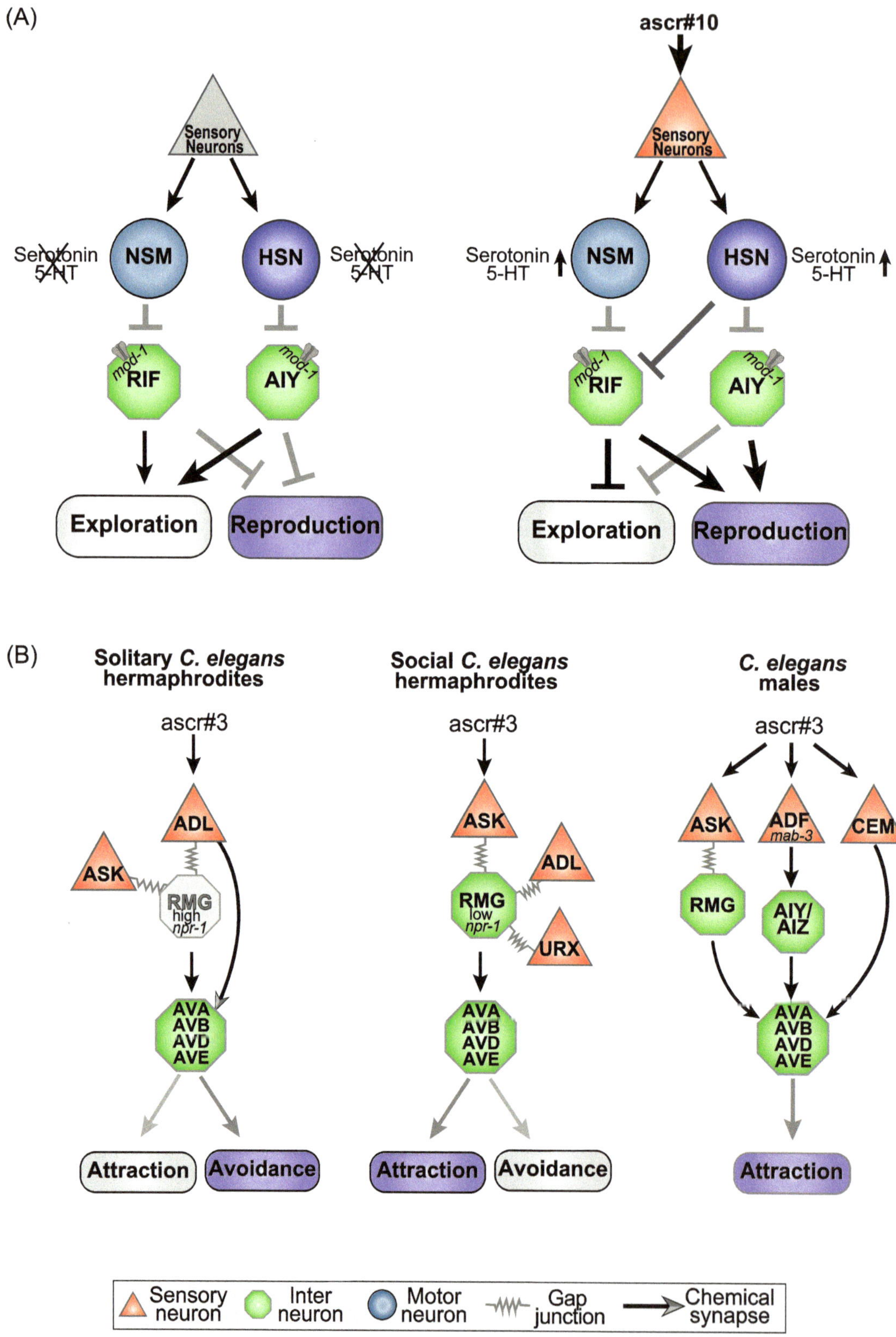

Figure 2. Neural circuits mediating responses to different ascarosides. (A) *ascr#10 decreases exploration and increases reproduction in hermaphrodites.* In the absence of ascr#10, biosynthesis of serotonin in either NSM or HSN neurons is absent in hermaphrodites. This allows both RIF and AIY interneurons to remain active, resulting in exploratory behavior and also slowing of germline development. However, upon sensing ascr#10, serotonin biosynthesis is activated leading to increased levels of serotonin in the HSN neurons. This resulting serotonin (5HT) release causes inhibition of the RIF and AIY neurons decreasing exploratory behavior decreases and increasing the rate of germ cell production. (B) *Differential behavioral responses to ascaroside ascr#3 is regulated by complex neural circuits.* ascr#3 exposure results in avoidance behavior in solitary strains of *C. elegans* hermaphrodites, which has high levels of the neuropeptide receptor gene *npr-1*. The sensory neuron ADL senses the ascaroside ascr#3, results in avoidance via electrical synapses. The social strain of *C. elegans*, which has low *npr-1* levels does not cause avoidance to ascr#3, instead it shows attraction to the chemical. This change the valence of response to ascr#3 is regulated by the hub-and-spoke gap junction circuit, specifically ASK and RMG neurons. In addition, sex of the worm regulates the response to ascr#3. In males, the sex-specific sensory neuron CEM, and the core sensory neuron, ASK is involved in detecting ascr#3. Both these neurons likely engage in the shared interneuron circuitry downstream to promote attraction in males. In addition, changing the genetic sex of individual neurons enables them to detect ascr#3. Males expressing *mab-3* in ADF sensory neurons are attracted to ascr#3. It is likely that this attractive response to the chemical is mediated by ADF neurons opposing the ADL- promoted avoidance responses either via the first layer amphid interneurons or direct synaptic input to command interneurons.

ascaroside receptors characterized remains as low as ten (Ludewig & Schroeder, 2013). The slow pace of identification of ascaroside receptors hampers our insights into the biological implications of ascarosides in development and behavior. The potential large number of GPCRs in *C. elegans* likely exceeds 1000, many of which are uncharacterized, though not all of these are chemosensory (Bargmann, 1998). In order to identify ascaroside receptors, previous studies have employed reverse genetics screens, (Chute et al., 2019; Kim et al., 2009) as well as quantitative trait locus analyses (Greene, Brown, et al., 2016; Greene, Dobosiewicz, et al., 2016; McGrath et al., 2011). However, identifying receptors is easier with the development of photoaffinity labeling for generating targeted probes for individual ascarosides (Park et al., 2012; Zhang, Reilly, Yu, Srinivasan, & Schroeder, 2019).

Ascaroside signaling and behavior: Interestingly, the effects associated with ascarosides are almost exclusively influenced by external environmental cues, many of which involve stress (e.g., poor nutrition, overcrowding, and heat). Therefore, it will be interesting to decipher the links between ascaroside function and stress responses.

Biosynthetic pathways involving ascaroside production: While certain aspects of ascarosides biosynthesis, like the shortening of side chains, are well-studied (Butcher, 2017, 2019; Ludewig & Schroeder, 2013; Park et al., 2019), not all aspects of biosynethic pathways are fully understood. Models that propose biosynthetic pathways are still subject to change as more components of pathways need to be ascertained (Butcher, 2017; Ludewig & Schroeder, 2013; Park et al., 2019). Investigations into these biosynthetic pathways will create a better understanding of *C. elegans'* productional regulation of ascarosides and could help reveal more about the specific function of ascarosides (Butcher, 2017).

Neuronal Imaging strategies: Ascaroside signaling can have both short-term and long-term influences on behavior; whole-brain imaging of ascaroside exposed worms will enable characterization of these short-term and long-term changes in behaviors. This technology will allow for integration of sensory neurons with motor neurons to be quantified, elucidating the role of multiple neurons within the connectome simultaneously (Nguyen et al., 2016; Venkatachalam et al., 2016).

Insights from these research areas will enable an interdisciplinary understanding of chemosensation, highlighting a synergy between social behavior and chemical biology within living organisms.

Acknowledgements

The authors thank the anonymous reviewers and members of the worm community. Additionally, the authors thank all (current and alumni) members of the Srinivasan Laboratory for their dedication and sincere efforts to uncover diverse aspects of neurobiology of ascaroside signaling using the *C. elegans* model system.

Disclosure statement

The authors declare no other competing interests.

References

Adachi, T., Kunitomo, H., Tomioka, M., Ohno, H., Okochi, Y., Mori, I., & Iino, Y. (2010). Reversal of salt preference is directed by the insulin/PI3K and Gq/PKC signaling in Caenorhabditis elegans. *Genetics*, *186*(4), 1309–1319. doi:10.1534/genetics.110.119768

Aprison, E.Z., & Ruvinsky, I. (2016). Sexually antagonistic male signals manipulate germline and soma of C. elegans hermaphrodites. *Current Biology*, *26*(20), 2827–2833. doi:10.1016/j.cub.2016.08.024

Aprison, E.Z., & Ruvinsky, I. (2017). Counteracting ascarosides act through distinct neurons to determine the sexual identity of C. elegans pheromones. *Current Biology*, *27*(17), 2589–2599.e2583. doi:10.1016/j.cub.2017.07.034

Aprison, E.Z., & Ruvinsky, I. (2019a). Coordinated behavioral and physiological responses to a social signal are regulated by a shared neuronal circuit. *Current Biology*, *29*(23), 4108–4115.e4104. doi:10.1016/j.cub.2019.10.012

Aprison, E.Z., & Ruvinsky, I. (2019b). Dynamic regulation of adult-specific functions of the nervous system by signaling from the reproductive system. *Current Biology*, *29*(23), 4116–4123.e4113. doi:10.1016/j.cub.2019.10.011

Artyukhin, A.B., Yim, J.J., Srinivasan, J., Izrayelit, Y., Bose, N., von Reuss, S.H., … Schroeder, F.C. (2013). Succinylated octopamine ascarosides and a new pathway of biogenic amine metabolism in Caenorhabditis elegans. *The Journal of Biological Chemistry*, *288*(26), 18778–18783. doi:10.1074/jbc.C113.477000

Bargmann, C.I. (1998). Neurobiology of the caenorhabditis elegans genome. *Science*, *282*(5396), 2028–2033. doi:10.1126/science.282.5396.2028

Bargmann, C.I. (2006). Chemosensation in C. elegans. *WormBook: the online review of C. elegans biology*. p. 1–29. doi:10.1895/wormbook.1.123.1

Bargmann, C.I., & Horvitz, H.R. (1991). Control of larval development by chemosensory neurons in Caenorhabditis elegans. *Science*, *251*(4998), 1243–1246. doi:10.1126/science.2006412

Bartley, J.P., Bennett, E.A., & Darben, P.A. (1996). Structure of the ascarosides from ascaris suum. *Journal of Natural Products*, *59*(10), 921–926. doi:10.1021/np960236 + doi:10.1021/np960236+

Bose, N., Meyer, J.M., Yim, J.J., Mayer, M.G., Markov, G.V., Ogawa, A., … Sommer, R.J. (2014). Natural variation in dauer pheromone production and sensing supports intraspecific competition in nematodes. *Current Biology*, *24*(13), 1536–1541. doi:10.1016/j.cub.2014.05.045

Butcher, R.A. (2017). Small-molecule pheromones and hormones controlling nematode development. *Nature Chemical Biology*, *13*(6), 577–586. doi:10.1038/nchembio.2356

Butcher, R.A. (2019). Natural products as chemical tools to dissect complex biology in C. elegans. *Current Opinion in Chemical Biology*, *50*, 138–144. doi:10.1016/j.cbpa.2019.03.005

Butcher, R.A., Fujita, M., Schroeder, F.C., & Clardy, J. (2007). Small-molecule pheromones that control dauer development in Caenorhabditis elegans. *Nature Chemical Biology*, *3*(7), 420–422. doi:10.1038/nchembio.2007.3

Butcher, R.A., Ragains, J.R., Li, W., Ruvkun, G., Clardy, J., & Mak, H.Y. (2009). Biosynthesis of the Caenorhabditis elegans dauer pheromone. *Proceedings of the National Academy of Sciences of the United States of America*, *106*(6), 1875–1879. doi:10.1073/pnas.0810338106

Cassada, R.C., & Russell, R.L. (1975). The dauerlarva, a post-embryonic developmental variant of the nematode Caenorhabditis elegans. *Developmental Biology*, *46*(2), 326–342. doi:10.1016/0012-1606(75)90109-8

Choe, A., Chuman, T., von Reuss, S.H., Dossey, A.T., Yim, J.J., Ajredini, R., … Edison, A.S. (2012). Sex-specific mating pheromones in the nematode Panagrellus redivivus. *Proceedings of the National Academy of Sciences of the United States of America*, *109*(51), 20949–20954. doi:10.1073/pnas.1218302109

Choe, A., von Reuss, S.H., Kogan, D., Gasser, R.B., Platzer, E.G., Schroeder, F.C., & Sternberg, P.W. (2012). Ascaroside signaling is widely conserved among nematodes. *Current Biology*, *22*(9), 772–780. doi:10.1016/j.cub.2012.03.024

Choe, D.-H., Millar, J.G., & Rust, M.K. (2009). Chemical signals associated with life inhibit necrophoresis in Argentine ants. *Proceedings of the National Academy of Sciences*, *106*(20), 8251–8255. doi:10.1073/pnas.0901270106

Chute, C.D., DiLoreto, E.M., Zhang, Y.K., Reilly, D.K., Rayes, D., Coyle, V.L., … Srinivasan, J. (2019). Co-option of neurotransmitter signaling for inter-organismal communication in C. elegans. *Nature Communications*, *10*(1), 3186–3186. doi:10.1038/s41467-019-11240-7

Chute, C.D., & Srinivasan, J. (2014). Chemical mating cues in C. elegans. *Seminars in Cell & Developmental Biology*, *33*, 18–24. doi:10.1016/j.semcdb.2014.06.002

Dalfó, D., Michaelson, D., Hubbard, E., & Jane, A. (2012). Sensory regulation of the C. elegans germline through TGF-β-dependent signaling in the niche. *Current Biology*, *22*(8), 712–719. doi:10.1016/j.cub.2012.02.064

de Bono, M., & Bargmann, C.I. (1998). Natural variation in a neuropeptide Y receptor homolog modifies social behavior and food response in C. elegans. *Cell*, *94*(5), 679–689. doi:10.1016/S0092-8674(00)81609-8

Diaz, S.A., Brunet, V., Lloyd-Jones, G.C., Spinner, W., Wharam, B., & Viney, M. (2014). Diverse and potentially manipulative signalling with ascarosides in the model nematode C. elegans. *BMC Evolutionary Biology*, *14*(1), 46–46. doi:10.1186/1471-2148-14-46

Dong, C., Dolke, F., Bandi, S., Paetz, C., & von Reuß, S.H. (2020). Dimerization of conserved ascaroside building blocks generates species-specific male attractants in Caenorhabditis nematodes. *Organic & Biomolecular Chemistry*, *18*(27), 5253–5263. doi:10.1039/D0OB00799D

Edison, A.S. (2009). Caenorhabditis elegans pheromones regulate multiple complex behaviors. *Current Opinion in Neurobiology*, *19*(4), 378–388. doi:10.1016/j.conb.2009.07.007

Emmons SW. Male development. WormBook. 2005;1–22. Published 2005 Nov 10. doi:10.1895/wormbook.1.33.1

Fagan, K.A., Luo, J., Lagoy, R.C., Schroeder, F.C., Albrecht, D.R., & Portman, D.S. (2018). A single-neuron chemosensory switch determines the valence of a sexually dimorphic sensory behavior. *Current Biology*, *28*(6), 902–914.e905. doi:10.1016/j.cub.2018.02.029

Flavell, S.W., Pokala, N., Macosko, E.Z., Albrecht, D.R., Larsch, J., & Bargmann, C.I. (2013). Serotonin and the neuropeptide PDF initiate and extend opposing behavioral states in C. elegans. *Cell*, *154*(5), 1023–1035. doi:10.1016/j.cell.2013.08.001

Golden, J.W., & Riddle, D.L. (1982). A pheromone influences larval development in the nematode Caenorhabditis elegans. *Science (New York, N.Y.).)*, *218*(4572), 578–580. doi:10.1126/science.6896933

Golden, J.W., & Riddle, D.L. (1984). The Caenorhabditis elegans dauer larva: Developmental effects of pheromone, food, and temperature. *Developmental Biology*, *102*(2), 368–378. doi:10.1016/0012-1606(84)90201-X

Gray, J.M., Karow, D.S., Lu, H., Chang, A.J., Chang, J.S., Ellis, R.E., … Bargmann, C.I. (2004). Oxygen sensation and social feeding mediated by a C. elegans guanylate cyclase homologue. *Nature*, *430*(6997), 317–322. doi:10.1038/nature02714

Greene, J.S., Brown, M., Dobosiewicz, M., Ishida, I.G., Macosko, E.Z., Zhang, X., … Bargmann, C.I. (2016). Balancing selection shapes density-dependent foraging behaviour. *Nature*, *539*(7628), 254–258. doi:10.1038/nature19848

Greene, J.S., Dobosiewicz, M., Butcher, R.A., McGrath, P.T., & Bargmann, C.I. (2016). Regulatory changes in two chemoreceptor genes contribute to a Caenorhabditis elegans QTL for foraging behavior. *eLife*, *5*, e21454. doi:10.7554/eLife.21454

Hu, P.J., Dauer (August 08, 2007), *WormBook*, ed. The *C. elegans* Research Community, WormBook, doi/10.1895/wormbook.1.144.1, http://www.wormbook.org.

Izrayelit, Y., Srinivasan, J., Campbell, S.L., Jo, Y., von Reuss, S.H., Genoff, M.C., … Schroeder, F.C. (2012). Targeted metabolomics reveals a male pheromone and sex-specific ascaroside biosynthesis in Caenorhabditis elegans. *ACS Chemical Biology*, *7*(8), 1321–1325. doi:10.1021/cb300169c

Jang, H., Kim, K., Neal, S.J., Macosko, E., Kim, D., Butcher, R.A., … Sengupta, P. (2012). Neuromodulatory state and sex specify alternative behaviors through antagonistic synaptic pathways in C. elegans. *Neuron*, *75*(4), 585–592. doi:10.1016/j.neuron.2012.06.034

Jeong, P.-Y., Jung, M., Yim, Y.-H., Kim, H., Park, M., Hong, E., … Paik, Y.-K. (2005). Chemical structure and biological activity of the Caenorhabditis elegans dauer-inducing pheromone. *Nature*, *433*(7025), 541–545. doi:10.1038/nature03201

Kaplan, F., Alborn, H.T., von Reuss, S.H., Ajredini, R., Ali, J.G., Akyazi, F., … Teal, P.E. (2012). Interspecific nematode signals regulate dispersal behavior. *PloS One*, *7*(6), e38735. doi:10.1371/journal.pone.0038735

Kaplan, F., Srinivasan, J., Mahanti, P., Ajredini, R., Durak, O., Nimalendran, R., … Alborn, H.T. (2011). Ascaroside expression in Caenorhabditis elegans is strongly dependent on diet and developmental stage. *PloS One*, *6*(3), e17804. doi:10.1371/journal.pone.0017804

Kim, K., Sato, K., Shibuya, M., Zeiger, D.M., Butcher, R.A., Ragains, J.R., … Sengupta, P. (2009). Two chemoreceptors mediate developmental effects of dauer pheromone in *C. elegans*. *Science (New York, N.Y.).)*, *326*(5955), 994–998. doi:10.1126/science.1176331

Lee, K., & Portman, D.S. (2007). Neural sex modifies the function of a C. elegans sensory circuit. *Current Biology*, *17*(21), 1858–1863. doi:10.1016/j.cub.2007.10.015

Ludewig, A.H., & Schroeder, F.C. (2013). Ascaroside signaling in C. elegans. In WormBook: the online review of C. elegans biology, 1–22. doi:10.1895/wormbook.1.155.1

Macosko, E.Z., Pokala, N., Feinberg, E.H., Chalasani, S.H., Butcher, R.A., Clardy, J., & Bargmann, C.I. (2009). A hub-and-spoke circuit drives pheromone attraction and social behaviour in C. elegans. *Nature*, *458*(7242), 1171–1175. doi:10.1038/nature07886

McGrath, P.T., & Ruvinsky, I. (2019). A primer on pheromone signaling in Caenorhabditis elegans for systems biologists. *Current Opinion in Systems Biology*, *13*, 23–30. doi:10.1016/j.coisb.2018.08.012

McGrath, P.T., Xu, Y., Ailion, M., Garrison, J.L., Butcher, R.A., & Bargmann, C.I. (2011). Parallel evolution of domesticated Caenorhabditis species targets pheromone receptor genes. *Nature*, *477*(7364), 321–325. doi:10.1038/nature10378

Narayan, A., Venkatachalam, V., Durak, O., Reilly, D.K., Bose, N., Schroeder, F.C., … Sternberg, P.W. (2016). Contrasting responses within a single neuron class enable sex-specific attraction in Caenorhabditis elegans. *Proceedings of the National Academy of Sciences of the United States of America*, *113*(10), E1392–E1401. doi:10.1073/pnas.1600786113

Nguyen, J.P., Shipley, F.B., Linder, A.N., Plummer, G.S., Liu, M., Setru, S.U., … Leifer, A.M. (2016). Whole-brain calcium imaging with cellular resolution in freely behaving Caenorhabditis elegans. *Proceedings of the National Academy of Sciences of the United States of America*, *113*(8), E1074–E1081. doi:10.1073/pnas.1507110112

Park, D., O'Doherty, I., Somvanshi, R.K., Bethke, A., Schroeder, F.C., Kumar, U., & Riddle, D.L. (2012). Interaction of structure-specific and promiscuous G-protein-coupled receptors mediates small-molecule signaling in Caenorhabditis elegans. *Proceedings of the National Academy of Sciences of the United States of America*, *109*(25), 9917–9922. doi:10.1073/pnas.1202216109

Park, Y.J., Joo, H.-J., Park, S., & Paik, Y.-K. (2019). Ascaroside pheromones: chemical biology and pleiotropic neuronal functions. *International Journal of Molecular Sciences*, *20*(16), 3898. doi:10.3390/ijms20163898

Ren, P., Lim, C.-S., Johnsen, R., Albert, P.S., Pilgrim, D., & Riddle, D.L. (1996). Control of C. elegans larval development by neuronal expression of a TGF-beta homolog . *Science (New York, N.Y.).)*, *274*(5291), 1389–1391. doi:10.1126/science.274.5291.1389

Sakai, N., Iwata, R., Yokoi, S., Butcher, R.A., Clardy, J., Tomioka, M., & Iino, Y. (2013). A sexually conditioned switch of chemosensory behavior in C. elegans. *PloS One*, *8*(7), e68676doi:10.1371/journal.pone.0068676

Schafer, W.R. (2006). Genetics of egg-laying in worms. *Annual Reviews Genetics*, *40*(1), 487–509. doi:10.1146/annurev.genet.40.110405.090527

Schroeder, F.C. (2015). Modular assembly of primary metabolic building blocks: a chemical language in C. elegans. *Chemistry & Biology*, *22*(1), 7–16. doi:10.1016/j.chembiol.2014.10.012

Srinivasan, J., Dillman, A.R., Macchietto, M.G., Heikkinen, L., Lakso, M., Fracchia, K.M., ... Sternberg, P.W. (2013). The draft genome and transcriptome of panagrellus redivivus are shaped by the harsh demands of a free-living lifestyle. *Genetics*, *193*(4), 1279–1295. doi:10.1534/genetics.112.148809

Srinivasan, J., Kaplan, F., Ajredini, R., Zachariah, C., Alborn, H.T., Teal, P.E.A., ... Schroeder, F.C. (2008). A blend of small molecules regulates both mating and development in Caenorhabditis elegans. *Nature*, *454*(7208), 1115–1118. doi:10.1038/nature07168

Srinivasan, J., von Reuss, S.H., Bose, N., Zaslaver, A., Mahanti, P., Ho, M.C., ... Schroeder, F.C. (2012). A modular library of small molecule signals regulates social behaviors in Caenorhabditis elegans. *PLoS Biology*, *10*(1), e1001237. doi:10.1371/journal.pbio.1001237

Sze, J.Y., Victor, M., Loer, C., Shi, Y., & Ruvkun, G. (2000). Food and metabolic signalling defects in a Caenorhabditis elegans serotonin-synthesis mutant. *Nature*, *403*(6769), 560–564. doi:10.1038/35000609

Venkatachalam, V., Ji, N., Wang, X., Clark, C., Mitchell, J.K., Klein, M., ... Samuel, A.D.T. (2016). Pan-neuronal imaging in roaming Caenorhabditis elegans . *Proceedings of the National Academy of Sciences of the United States of America*, *113*(8), E1082–E1088. doi:10.1073/pnas.1507109113

von Reuss, S.H. (2018). Exploring modular glycolipids involved in nematode chemical communication. *Chimia*, *72*(5), 297–303. doi:10.2533/chimia.2018.297

White, J.Q., Nicholas, T.J., Gritton, J., Truong, L., Davidson, E.R., & Jorgensen, E.M. (2007). The sensory circuitry for sexual attraction in C. elegans males. *Current Biology*, *17*(21), 1847–1857. doi:10.1016/j.cub.2007.09.011

Yamada, K., Hirotsu, T., Matsuki, M., Butcher, R.A., Tomioka, M., Ishihara, T., ... Iino, Y. (2010). Olfactory plasticity is regulated by pheromonal signaling in Caenorhabditis elegans. *Science (New York, N.Y.).)*, *329*(5999), 1647–1650. doi:10.1126/science.1192020

Yohe, L.R., & Brand, P.; Handling editor: Rebecca Fuller. (2018). Evolutionary ecology of chemosensation and its role in sensory drive. *Current Zoology*, *64*(4), 525–533. doi:10.1093/cz/zoy048

Zhang, Y.K., Reilly, D.K., Yu, J., Srinivasan, J., & Schroeder, F.C. (2019). Photoaffinity probes for nematode pheromone receptor identification. *Organic & Biomolecular Chemistry*, *18*(1), 36–40. doi:10.1039/C9OB02099C

Zwaal, R.R., Mendel, J.E., Sternberg, P.W., & Plasterk, R.H. (1997). Two neuronal G proteins are involved in chemosensation of the *Caenorhabditis* elegans Dauer-inducing pheromone. *Genetics*, *145*(3), 715–727. Retrieved from https://www.ncbi.nlm.nih.gov/pubmed/9055081

OPEN ACCESS

Intraguild predation between *Pristionchus pacificus* and *Caenorhabditis elegans*: a complex interaction with the potential for aggressive behaviour

Kathleen T. Quach and Sreekanth H. Chalasani

ABSTRACT

The related nematodes *Pristionchus pacificus* and *Caenorhabditis elegans* both eat bacteria for nutrition and are therefore competitors when they exploit the same bacterial resource. In addition to competing with each other, *P. pacificus* is a predator of *C. elegans* larval prey. These two relationships together form intraguild predation, which is the killing and sometimes eating of potential competitors. In killing *C. elegans*, the intraguild predator *P. pacificus* may achieve dual benefits of immediate nutrition and reduced competition for bacteria. Recent studies of *P. pacificus* have characterized many aspects of its predatory biting behaviour as well as underlying molecular and genetic mechanisms. However, little has been explored regarding the potentially competitive aspect of *P. pacificus* biting *C. elegans*. Moreover, aggression may also be implicated if *P. pacificus* intentionally bites *C. elegans* with the goal of reducing competition for bacteria. The aim of this review is to broadly outline how aggression, predation, and intraguild predation relate to each other, as well as how these concepts may be applied to future studies of *P. pacificus* in its interactions with *C. elegans*.

Introduction

The nematode *Pristionchus pacificus* was first introduced by Sommer, Carta, Kim, and Sternberg (1996) to serve as a counterpoint species to *Caenorhabditis elegans* in comparative studies (Sommer, 2015). Since then, numerous studies have characterized the similarities, differences, and interactions between *P. pacificus* and *C. elegans*. *Pristionchus pacificus* and *C. elegans* are separated by an order of 100 million years of evolution (Dieterich *et al.*, 2008), and share a remarkable level of similarity. On a gross morphological level, *P. pacificus* and *C. elegans* are both vermiform in shape and roughly the same size, approximately 1 mm long as young adults (Figure 1). *Pristionchus pacificus*, like most nematodes, are also conveniently eutelic and have a fixed number of developmentally determined somatic cells (Hong & Sommer, 2006b; Sommer, 2015). While number, neuroanatomical positions, and processes of homologous neurons are highly conserved between the two nematodes, subtle changes in neuroanatomical features of amphid neurons (Hong *et al.*, 2019; Sommer, 2015; Srinivasan, Durak, & Sternberg, 2008) and massive wiring of the pharyngeal motor system have been reported (Bumbarger, Riebesell, Rödelsperger, & Sommer, 2013). Despite having similar life cycle length, early *P. pacificus* development differs from that of *C. elegans* in that *P. pacificus* eggs hatch at the J2 stage, one full larval stage later than the corresponding *C. elegans* L1 stage (von Lieven, 2005). Although dauer formation in both nematode species share conserved endocrine signalling (Ogawa, Streit, Antebi, & Sommer, 2009), exit from dauer in *P. pacificus* strongly biases development of a non-predatory mouth form (Bento, Ogawa, & Sommer, 2010).

The most striking difference between *P. pacificus* and *C. elegans* relates to how they feed. While both species eat bacteria, *P. pacificus*, but not *C. elegans*, can also kill and consume non-self nematode larvae with the use of teeth-like denticles (Figure 1). *Pristionchus pacificus*, as do most Diplogastrids, possesses a dorsal tooth and lacks the pharyngeal grinder (Figure 1) that *C. elegans* uses to grind bacteria (von Lieven & Sudhaus, 2000). This dramatic restructuring of the buccal cavity is accompanied by drastic rewiring of the *P. pacificus* pharyngeal motor system relative to that of *C. elegans* (Bumbarger *et al.*, 2013). *Pristionchus pacificus* exhibits a developmental dimorphism in which a proportion of individuals known as stenostomatous develop only a dorsal tooth, while eurystomatous individuals develop a larger dorsal tooth and an additional ventral tooth (Figure 1). The relative proportions of eurystomatous and stenostomatous individuals in a population are affected by starvation, crowding, and the sulfatase EUD-1, all of which promote the eurystomatous mouth form (Bento *et al.*, 2010; Namdeo *et al.*, 2018; Ragsdale, Müller, Rödelsperger, & Sommer, 2013). The eurystomatous mouth form is adaptive for predating on nematode larvae, while the stenostomatous mouth form is ineffective for killing prey and is restricted to bacteriovory

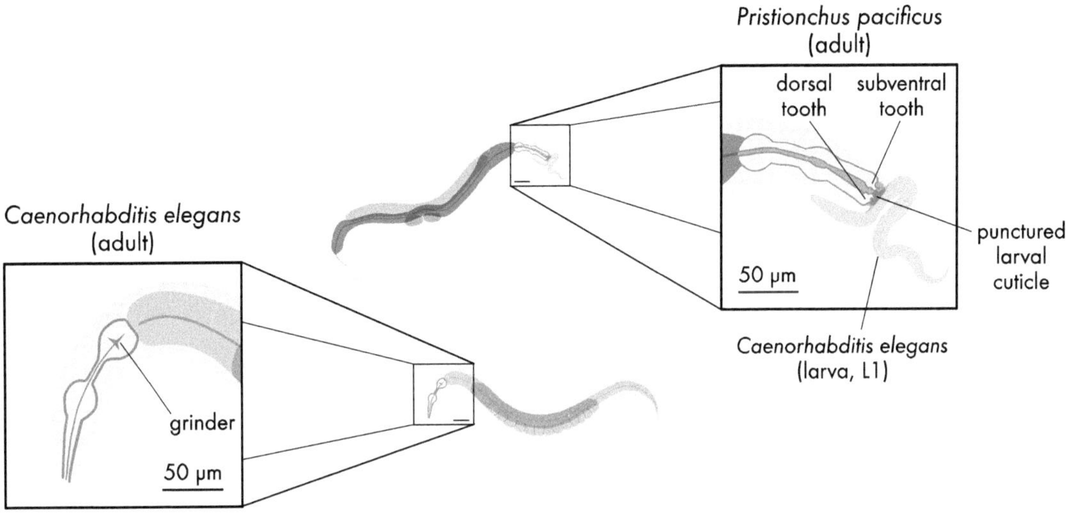

Figure 1. *Pristionchus pacificus* and *Caenorhabditis elegans* are similar in size and body form at the young adult stage. *Caenorhabditis elegans* possesses a grinder that it uses to lyse bacteria for consumption. Instead of a grinder, *P. pacificus* instead has one or two teeth that it uses to puncture the cuticle of larval *C. elegans* prey. The non-predatory stenostomatous dimorph of *P. pacificus* has only dorsal tooth, while the predation-enabled eurystomatous dimorph possesses a larger claw-like dorsal tooth and an additional subventral tooth.

and prey scavenging (Serobyan, Ragsdale, & Sommer, 2014; Wilecki, Lightfoot, Susoy, & Sommer, 2015). The remainder of this perspective will only discuss eurystomatous individuals since they are able to inflict harm on other nematodes and therefore have the potential to be aggressive.

Some indirect evidence exists to suggest that *P. pacificus* and *C. elegans* may compete with each other. Both nematodes have overlapping bacterial diets in the wild (Akduman, Rödelsperger, & Sommer, 2018; Samuel, Rowedder, Braendle, Félix, & Ruvkun, 2016) and have been found to co-occur in nature on bacteria-rich rotting plant material (Félix, Ailion, Hsu, Richaud, & Wang, 2018). In exploiting the same bacterial resources, *P. pacificus* and *C. elegans* likely compete with each other in an indirect manner. Direct competition may also occur if *P. pacificus* interferes with *C. elegans* access to bacteria. *Pristionchus pacificus* may achieve this by using its teeth, which are the only implements of direct physical harm that *P. pacificus* possesses. However, *P. pacificus* teeth have traditionally been attributed to predatory function, so further research must be done before a competitive function can be ascribed to biting.

The killing and sometimes feeding on an interspecific potential competitor is called intraguild predation (Polis, Myers, & Holt, 1989). When *P. pacificus* kills and feeds on *C. elegans*, it can simultaneously achieve both a prey meal and decreased competition for bacterial resources. However, it is unclear whether this competitive benefit is intentional or just a side effect of predation and the motivation of an intraguild predator is notoriously difficult to dissect. An animal's goal is obscured when a single behaviour produces multiple simultaneous benefits. Motivation further eludes simple inference when the eliciting stimuli and behavioural expression of killing appear similar regardless of whether killing is motivated by hunger, competition for a shared resource, or a combination of both. If *P. pacificus* is motivated by the goal of killing and eating prey, then the killing of *C. elegans* is predation. On the other hand, if *P. pacificus* is motivated by the goal of reducing competition for bacteria, then the killing of *C. elegans* is interspecific aggression. Although both involve intentional harm of others, aggression has been traditionally distinguished from predation in their respective competitive and nutritional goals for harm (Archer, 1988; Nelson, 2005).

While many studies have explored the ecological ramifications of intraguild predation on a community level, little is known about the motivation that drives attack behaviour on the individual intraguild predator level. Intraguild predation is widespread throughout the animal kingdom and is a key trophic module in many food webs (Arim & Marquet, 2004). After introducing intraguild predation as a concept (Polis & Holt, 1992), Holt and Polis (1997) articulated a theoretical framework of intraguild predation that predicted immense impacts on biodiversity and community structure. Since then, most studies of intraguild predation have focused on validating or invalidating those predictions by measuring population patterns and dynamics. Field studies are well-suited for these macroecological investigations of intraguild predation: with access to the full complexity of an open ecosystem, field studies of intraguild predation have unsurpassable ecological validity. However, open ecosystems preclude fine control and manipulation of environmental elements that may instigate and influence the predator to attack. This makes it is difficult to control the experiences of any single animal. A deeper understanding of the individual intraguild predator's internal state will enrich understanding of observed behaviour in the field as well as provide more accurate predictions of the ecological effects of intraguild predation. For example, prey avoidance of intraguild predators has been shown to be a critical constraint on species coexistence (Pringle *et al.*, 2019; Sommers & Chesson, 2019). However, little is known about how the intraguild predator's motivation influences its proclivity to attack intraguild prey, which in turn may influence the level of prey avoidance.

We suggest that the laboratory study of a simple tripartite community module consisting of *P. pacificus*, *C. elegans*, and bacteria is ideal for elucidating the context-dependent motivations underlying intraguild predation. In contrast to vertebrate, the use of invertebrate prey circumvents ethical qualms of purposefully subjecting vertebrates to being painfully killed and eaten as prey. Additionally, *P. pacificus* and *C. elegans* have large brood sizes and short life cycles of only 3–4 days in optimal conditions (Byerly, Cassada, & Russell, 1976; Félix *et al.*, 1999), allowing for fast quantification of fitness consequences. Both nematodes are cultivated in the laboratory using the same standard bacterial strain *Escherichia coli* OP50 (Brenner, 1974; Sommer *et al.*, 1996), although other bacterial strains can be fed to explore effects on diet and competition. Perhaps the most powerful advantage of studying the proposed tripartite system is the relative ease of applying genetic tools to *P. pacificus*, bacteria, and especially *C. elegans*. All three organisms conveniently produce genetically identical progeny: *C. elegans* and *P. pacificus* are self-fertilizing species (Brenner, 1974; Sommer *et al.*, 1996), while bacteria reproduce asexually. Genetic modification methods such as RNAi, DNA-mediated transformation, and genome editing have been established for *P. pacificus* (Cinkornpumin & Hong, 2011; Schlager, Wang, Braach, & Sommer, 2009; Witte *et al.*, 2015) and *C. elegans* (Dickinson & Goldstein, 2016; Nance & Frøkjaer-Jensen, 2019). The genomes of the laboratory *E. coli* OP50 strain (May *et al.*, 2009), as well as wild microbiomes from *P. pacificus* (Akduman *et al.*, 2018; Koneru, Salinas, Flores, & Hong, 2016; Rae *et al.*, 2008) and *C. elegans* (Dirksen *et al.*, 2016; Samuel *et al.*, 2016; Schulenburg & Félix, 2017), will allow for correlation of bacterial genetic components with resource-dependent perturbations of nematode behaviour. Furthermore, bacterial transformation methods (Sheth, Cabral, Chen, & Wang, 2016) can be used to engineer bacteria in order to causally identify which bacterial signals trigger nematode competitive responses.

This review is unconventional in that it is intended to provide a broad conceptual foundation for catalysing future laboratory experiments of nematode intraguild predation, which are currently non-existent in the published corpus of nematode literature. To begin to unravel aggressive and predatory motivational components of intraguild predation between *P. pacificus* and *C. elegans*, this review considers relevant key concepts, identifies guiding principles and highlights approaches in aggression, predation, and intraguild predation. First, we establish definitions of aggression that are broadly applicable and discuss interspecific aggression. Second, predation is reviewed to explore which predatory behaviours allow the possibility for the predatory attack to be intentionally harmful. Third, field observations and theoretical predictions of intraguild predation are outlined as a conceptual framework for future work. Finally, *P. pacificus*, *C. elegans*, and their trophic relationships with each other and bacteria are characterized as the focal intraguild predation community module of this review.

Aggression

'Aggression' is an unbound term used to refer to a subset of complex social interactions. Although numerous definitions of aggression have been proposed, none concisely encapsulate the behavioural diversity of aggression. Furthermore, many of these definitions are fraught with stipulations about motivations that are not readily observable. Despite a lack of consensus, it is generally accepted that a hallmark feature of aggression is intentional harm or injury to others (Berkowitz, 1981). From this, a minimal definition of aggression can be framed as any behaviour that is intended to inflict harm to another individual (Berkowitz, 1993; Buss, 1961; Gendreau & Archer, 2005; Olivier & Young, 2002). This minimal definition inherently possesses little value for discriminating between aggressive behaviours and does not capture the multifaceted complexity of aggression. Several taxonomies have been developed to meaningfully characterize differences between aggressive behaviours and sort them into discrete subtypes. These classification systems vary in which dimensions of aggression they use to compare aggressive behaviours. These dimensions include behavioural expression, eliciting stimulus, motivation, functional value, and underlying neurophysiological mechanisms (Gendreau & Archer, 2005). Of these classification dimensions, motivation is the most difficult to evaluate because it must be inferred from the others.

In all aggression taxonomies, competition is the most representative and often defining function of aggression (Archer, 1988; Nelson, 2005). We will therefore introduce a more stringent definition of aggression that we will refer to as the **competitive definition of aggression**, which we define as any behaviour that is intended to (1) inflict harm to another individual and (2) deal with competition. It is important to note that this definition requires that both harm and competition be intentional. Aggression that conforms to this is the competitive definition of aggression include some of the most distinctive aggressive behaviours. For example, aggression associated with male-male competition for mates is often marked by conspicuous behavioural expression (ritual combat) that is specifically elicited (by male targets) for a singular observable function (access to mates) (Chen, Lee, Bowens, Huber, & Kravitz, 2002; Crane, 1966; Darwin, 1896; Huxley, 1966; Issa & Edwards, 2006; Kravitz & Huber, 2003; Moynihan & Moynihan, 1998). It has been suggested that ritualized aggression evolved as a way for social species to settle intraspecific contests without killing conspecifics (De Waal, 2000; Nelson, 2000). In the case of ritualized aggression, one-to-one mapping between behavioural expression, eliciting stimulus, and function provide unambiguous support that mate competition is the driving motivation of aggression.

Interspecific aggression

In contrast to mate competition that is necessarily intraspecific, territoriality is the most commonly studied form of agonistic interactions between species (Grether, Losin, Anderson, & Okamoto, 2009; Peiman & Robinson, 2010). Although first described in birds (Howard, 1920), territoriality evolved in many animals such as fish (Gerking, 1959), mammals (Burt, 1943), reptiles (Brattstrom, 1974), and insects (Baker, 1983). 'Territory' is any defended area in which a dominant individual or group has a priority of access to resources (Kaufmann, 1983; Nice, 1941). Notably, this dominance must be achieved through social interaction, often with aggressive attacks and threats. Territorial aggression is adaptive only when the resource benefits outweigh the energetic costs of defending territory (Brown & Orians, 1970; MacLean & Seastedt, 1979). In general, territorial aggression serves to reduce intruder trespass by driving out intruders and inducing avoidance, which ensures a future supply of resources for the aggressor.

Interspecific aggression as exerted by a focal species is frequently evaluated by comparing it to intraspecific aggression that occurs in that species. In the case of interspecific territoriality involving phylogenetically related species, this kind of comparison is particularly useful for determining whether interspecific aggression is a by-product of misidentification of heterospecifics as conspecifics due to apparent similarity (Murray, 1981), or if it is a case of alpha-selection, in which interspecific territorial aggression is an adaptive response to resource overlap with another species and is selected for separately from intraspecific aggression (Gill, 1974). For example, a study of two species of reciprocally aggressive salamanders showed that one species likely misidentifies since it is equally aggressive to conspecifics and heterospecifics across levels of sympatry and interspecific competition, while the other species were equally aggressive to both heterospecifics and conspecifics only when the interspecific competition was strong (Nishikawa, 1987). Comparison between intraspecific and interspecific territoriality is also useful for understanding the evolution of associated phenotypes, also known as agonistic character displacement (Grether *et al.*, 2009, 2013). A species that benefits from dealing with both conspecific and heterospecific intruders could do so most efficiently if competitor recognition cues were similar in both species, driving the convergence of characteristics in both species (Cody, 1969). Conversely, divergence of characteristics may occur when interspecific aggression is maladaptive (Lorenz, 1966; Tynkkynen, Rantala, & Suhonen, 2004).

Predation

Little is known about how interspecific aggression evolved in animals that do not already possess intraspecific aggression. In cases like this, interspecific aggression can be compared to predator-prey encounters, since both are agonistic behaviours that often have similar motor or action patterns despite having different functions (King, 1973). In order to accurately relate predation to aggression, predation must first be explicitly defined. Predation at its broadest refers to **an organism killing another organism for nutritional purposes** (Taylor, 1984). This definition differs from the previously described minimal definition of aggression in three important ways: (1) harm is ideally lethal, (2) harm does not need to be intentional, and (3) nutrition is the function for behaviour. This last point is the main cause of contention regarding whether predatory killing should be included as a subtype of aggression. Moyer (1968) was first to outline a stimulus-based taxonomy of aggression, in which predatory aggression was defined as behaviour that is elicited by and targeted at prey. However, a subsequent classification scheme, based on function rather than eliciting stimulus, rejected predation as a valid form of aggression because it did not fulfil any competitive, protective, or parental purpose (Archer, 1988). In discussing predation, we will not yet impose these exclusionary criteria based on function, though they should be acknowledged for their classification value. Instead, this section will explore how predation may overlap with aggression, based on the broad definition of predation and minimal definition of aggression. Specifically, this section will focus on the **intentionality of the harm** inflicted during predation, as well as how it fulfils a requirement of aggression.

Nonaggressive predatory behaviours

Predation likely first evolved when the first unicellular life forms appeared and have since evolved independently many times across all domains and many kingdoms of life (Bengtson, 2002). In contrast, aggression is typically considered to only occur between animals. A key factor for disqualifying simpler predators from aggression is whether predatory attack and feeding are simultaneous or occur in separate phases. Unicellular organisms, especially protozoa, can predate on each other by using phagocytosis to engulf a whole prey (Lancaster, Ho, Hipolito, Botelho, & Terebiznik, 2019). Predatory phagocytosis is strongly implicated in the origin of mitochondria and chloroplasts as resident prokaryotes that survived engulfment (McFadden, Gilson, Hofmann, Adcock, & Maier, 1994; Roger, 1999), leading to the origin of eukaryotes (Cavalier-Smith, 2009; Davidov & Jurkevitch, 2009). Engulfment is a simple and compressed form of predation in which killing and feeding are achieved simultaneously – there is no separate attack phase. In this case, the killing of the engulfed prey is incidental to feeding on the prey and is generally not considered intentional harm, a requirement for aggression. A similar logic can also be applied to exclude multicellular suspension/filter feeders from being considered aggressive.

Other instances in which predator-prey interactions are not deemed aggressive concern the prey's response. For example, herbivores that kill plant or algae in the process of grazing are not considered predators. Unlike engulfers and suspension/filter feeders, grazer-type herbivores can kill and feed in separate steps. For example, sea urchins can use its rasping teeth to incrementally carve away and feed on portions of kelp without necessarily killing it first (Harrold & Reed, 1985). In other words, sea urchins do not have to subjugate the kelp first to reap nutritional rewards. The kelp only dies when it receives a critical amount of damage, and once again, killing is a side effect of feeding, albeit delayed. Feeding without killing is possible when the prey is too large for engulfment and does not physically evade harm. Plants certainly can suffer from harm inflicted by herbivores and have accordingly evolved anti-herbivore defences, including chemical defences and tolerance to herbivory (Agrawal, 2011). However, these plant defences are largely passive or invisible to the herbivore, and therefore the predatory grazer lacks discernible cues for associating its own harmful actions with a correlated harm response from the prey. From an epistemological perspective, the predatory grazer cannot intend harm if it does not 'know' that its grazing is harmful to the prey. From an evolutionary perspective, the predatory grazer cannot intend harm if evolution did not select for it, particularly when the predator has no additional adaptive benefit from inflicting harm separately from feeding.

Potentially aggressive predatory behaviours

A predator receives feedback that its actions are harmful to prey when prey must be sufficiently maimed or killed before consumption. The potential for predation to be aggressive arises as prey become more difficult to kill and predation transforms from a simple process into a complex sequence of steps in which killing must precede feeding (Figure 2). Predation exerts a stronger selective pressure on prey than on predators. Referred to as the 'life-dinner principle', failure costs the prey its life, whereas it only costs the predator a meal (Dawkins & Krebs, 1979). Mutations that are disadvantageous for predation survive longer in the predator gene pool than in the prey gene pool. This suggests that prey can quickly evolve antipredatory adaptations and accelerate co-evolution between predator and prey. Such antipredatory adaptations, such as increased size and speed, make prey more resistant to harm and ingestion and more able to escape. As prey become too big to swallow and motile instead of sessile, engulfment and grazing cease to be adequate predatory strategies. Instead of achieving harm and feeding in the same step, predation now requires considerably more effort to capture the prey before feeding can even commence. The predatory process leading to capture can be subdivided into a sequence of escalating steps: encounter, detection, pursuit, attack, and capture (Lima & Dill, 1990). The prey has the opportunity to escape at any of these points of escalation, placing selective pressure on the predator to develop efficient hunting skills. In this elongated predation process, the harm is temporally separated from feeding.

The particular temporal order of harm and feeding affects the degree to which intentionality of harm can be inferred. As previously described, it is difficult to disprove that harm is incidental to feeding when killing coincides with or follows feeding. In contrast, when killing precedes feeding, a causal relationship between the two becomes available as a possibility. More specifically, the predatory attack may be vitally instrumental in capturing prey and contribute directly to the predator's ability to feed on prey (Figure 2, 1st question). In order to argue a case for predatory aggression, it must be demonstrated that harm inflicted by the predatory attack is intentionally perpetrated. However, the close sequential proximity between killing and feeding insinuates that killing may be directly associated with feeding as part of a programmed feeding behavioural sequence, which would rule out aggressive intent. The predatory attack can fulfil the intentionality requirement of aggression only if it can operate separately from feeding. Therefore, studies that argue for an aggressive quality to predation have outlined ways in which predatory attack is a deliberate and separate behaviour that can operate in an uncoordinated way from feeding.

Behavioural evidence for an incongruous relationship between the tendency to kill, the tendency to feed, and hunger have existed for some time (Polsky, 1975). The most prominent indication comes from widespread observations that predators often kill prey in excess of what they need to fulfil their nutritional requirements, with numerous instances in which killed prey is abandoned without being consumed. Surplus killing behaviour has been readily observed in the wild for a variety of predators, including mammalian carnivores (Jedrzejewska & Jedrzejewski, 1989; Kruuk, 2009; Lincoln & Quinn, 2019; Rasa, 1973; Schaller, 2009; Zimmermann, Sand, Wabakken, Liberg, & Andreassen, 2015), rodents, (Boice & Schmeck, 1968; Desisto & Huston, 1970), birds (Nunn, Klem, Kimmel, & Merriman, 1976; Solheim, 1984), and insects (Lounibos, Makhni, Alto, & Kesavaraju, 2008). Experimental efforts to differentially influence killing and feeding behaviour largely come from studies of muricide by rats. Rats are known to predate on mice in the wild and in the laboratory (Karli, 1956; O'Boyle, 1974). When presented with mice, a small proportion of laboratory rats kill mice (Karli, 1956). Notably, rats that kill will only eat a portion of killed prey and with variable latency after killing. These 'killers' attack regardless of whether they are hungry or fully satiated. Further exploration into water deprivation, food deprivation, and time of testing relative to regularly scheduled feeding time failed to show any significant effect on the tendency of killers to attack mice (Paul, 1972; Paul, Miley, & Baenninger, 1971). Conversely, 'nonkillers' cannot be coerced into killing mice with extreme food deprivation - some rats were reported to have even starved to death in the presence of prey (Karli, 1956). Studies have also shown that the respective tendencies to kill and eat are not mutually reinforcing and do not follow each other as one is selectively repressed or promoted. For example, killing does not decrease when the rat is prevented from feeding on its prey (Myer, 1967, 1969, 1971), and killing experience is sufficient to promote killing tendency (Leyhausen, 1973). However, killing does not potentiate subsequent feeding. Rats presented with pre-killed prey were just as likely to feed as rats who were allowed to kill their own prey (Paul & Posner, 1973). Therefore, promoting killing does not always enhance feeding, nor is the inverse true. Altogether, this body of evidence suggests that predation does not always proceed as a unitary behavioural chain of killing and feeding. Rather, predatory attack can be influenced by factors other than those that influence feeding. The predatory attack may be more aptly described as an aggressive behavioural module that is intentionally, though not necessarily, deployed as a means to acquire prey.

Mouse killing is peculiarly situated in between two other rat behaviours that involve harming others: predation of

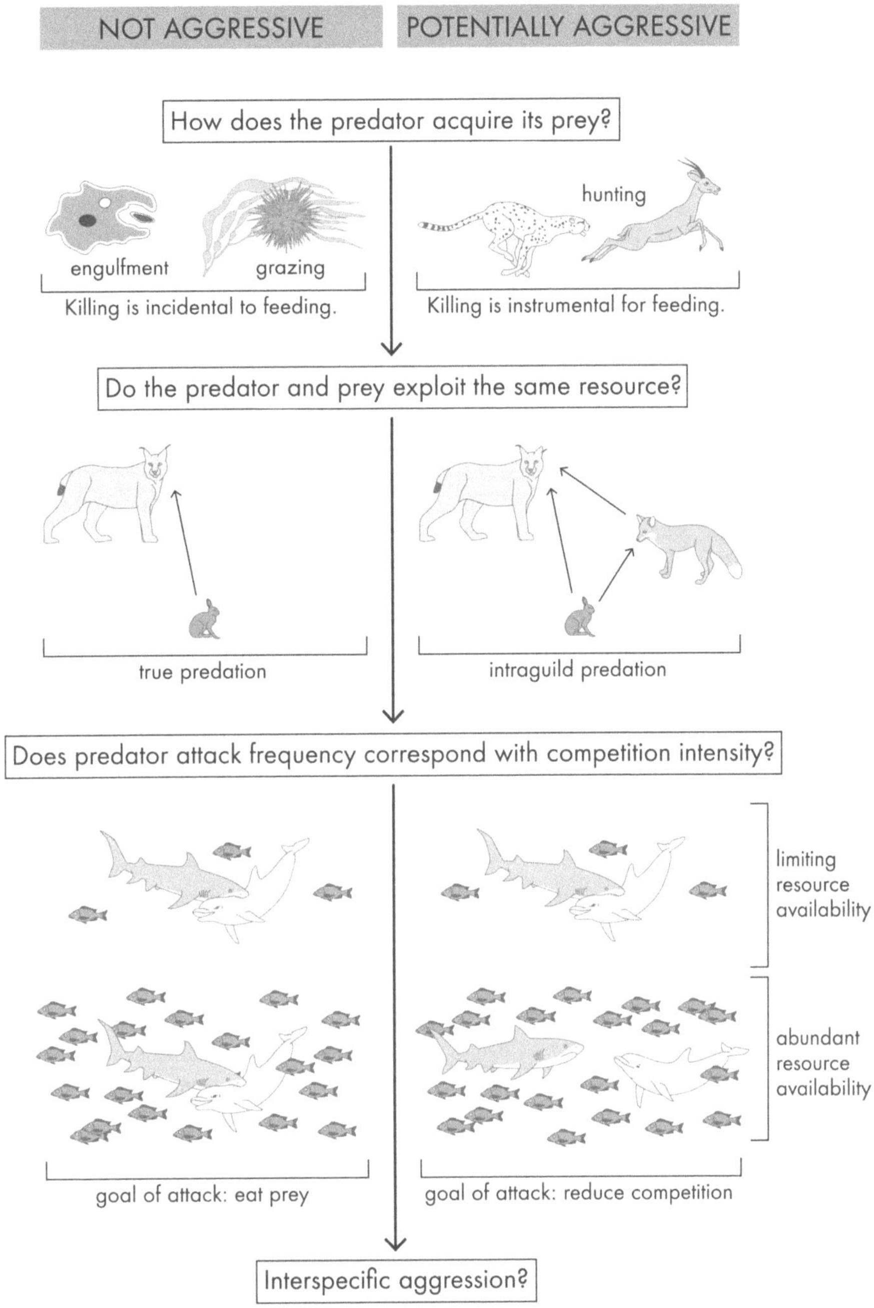

Figure 2. Three key questions are critical for determining whether a particular predatory behaviour has the potential to be aggressive. The first question establishes whether harm in predation is intentionally inflicted. The second question identifies whether the prey can be a potential competitor. Finally, the third question explores whether the predator intentionally harms the prey for interference competition.

other less-related species and aggression against conspecific intruders. Unlike mice prey, predation of phylogenetically distant species is characterized by much higher and more consistent rates of attack and subsequent feeding of prey such as frogs, turtles, chicks (Bandler, 1970; Desisto & Huston, 1970), and insects (Kemble & Davies, 1981). In one study, nearly all tested rats attacked frogs or turtles placed in the same cage, while only 17% of rats attacked mice in the same cage (Bandler, 1970). The killing of frog and turtle prey was almost always accompanied by eating of the corpse (Landry, 1970). Mouse killing, therefore, differs starkly from predation by rats' most common prey food, and brings into question whether mouse killing possesses some non-predatory component. Since mice are phylogenetically close to rats, Blanchard, Takahashi, and Blanchard (1977) surmised that mouse killing shares aspects of conspecific aggression between rats. Rat colonies are known to attack strange intruder rats, and experience with intruders leads to increased attack behaviour (Blanchard *et al.*, 1977). If mouse killing resembles intraspecific aggression against conspecific intruders, then the increase in aggression induced by exposure to conspecific rat intruders should also lead to an increase in aggression against heterospecific mice intruders. However, previous aggressive exposure to conspecifics failed to induce any change in the readiness of rats to attack either mouse targets or roach controls (Kemble & Davies, 1981). If a predatory attack is indeed aggressive, it is not influenced by the same factors that govern intraspecific aggression.

Thus, the behavioral signatures of intraspecific aggression cannot be referenced for the identification and validation of a predatory attack as a form of aggression.

In addition to behavioural evidence, hypothalamic stimulation studies in cats have shown that that feeding and killing are separable on the neuroanatomical level. While some hypothalamic sites can elicit both predatory attack and eating (Hutchinson & Renfrew, 1966), stimulation of a particular site in the lateral hypothalamus in cats has been shown to selectively elicit predatory attack (Siegel & Brutus, 1990; Siegel & Pott, 1988; Siegel & Shaikh, 1997). In order to ascertain that this lateral hypothalamic site is indeed specifically dedicated to the attack aspect of predation, Flynn and associates conducted an exhaustive set of behavioural experiments in which they attempted to coax eating behaviour out of cats while they were stimulated (Flynn, 1967; Flynn, Vanegas, Foote, & Edwards, 1970; Polsky, 1975). First, researchers increased stimulation to the hypothalamic site that reliably induces a cat to attack a rat, finding that even the highest intensities could not induce most tested cats to eat their captured rat prey. Similarly, persistent stimulation duration past the point of attack did not lead to consummatory feeding after the predatory attack of a rat had already been evoked. Second, the researchers presented easily attainable non-prey food to reduce the effort needed to eat. When a dish of non-prey food was presented during stimulation, most cats attacked the dish but never consumed the food (Wasman & Flynn, 1962). When horsemeat is placed closer than an anaesthetized rat prey in relation to a cat, stimulation-induced most cats to pass over the horsemeat and attack the rat. Finally, the researchers increased motivation eat by starving cats for three days. The starved cats were then fed non-prey food and stimulated while eating. Amazingly, most of the cats halted eating of the non-prey food and proceeded to attack a nearby rat. Altogether, these cat studies indicate that a predatory attack site of the lateral hypothalamus exists that is functionally selective in influencing the attack component of predation and is neuroanatomically distinct from other neighbouring sites that influence eating or the predatory process as a whole. Combined with previously described behavioural experiments of muricide by rats, a strong body of evidence suggests that predatory attack is dissociable from feeding, thus opening up the possibility for the predatory attack to be applied for other functions, such as reducing competition.

Intraguild predation

While predatory attack as described above has been labelled as predatory aggression by a relatively small cohort of aggression researchers, the consensus remains far out of reach. One explanation for this hesitancy is that it is unsatisfactory to only show that predatory attack can be dissociated from feeding – something else must replace feeding as the motivation for and function of the attack. For many, the most convincing motivation and function is competition (Archer, 1988; Nelson, 2005). We will henceforth adopt the competitive definition of aggression, which requires not only intentional harm but also competition as the goal of that harm.

One class of interspecific interaction that can potentially satisfy both competitive motivation and function of a predatory attack, and thus aggression in a more widely accepted sense, is intraguild predation (Figure 2, 2nd question). In intraguild predation, a predator kills and sometimes eats a potential interspecific competitor (Polis *et al.*, 1989). A guild consists of a group of species that exploit the same resource in a similar way (Simberloff & Dayan, 1991). From a food chain perspective, intraguild predation is the set of relationships between three trophic levels: the intraguild predator, the intraguild prey, and the shared resource. A basic model of intraguild predation has the following trophic structure: (1) Both the intraguild predator and intraguild prey exploit the same shared resource, and (2) the intraguild predator is facultative and can also eat the intraguild prey (Holt & Huxel, 2007; Holt & Polis, 1997; Polis *et al.*, 1989). This type of intraguild predation is asymmetric because only one of the guild species consistently predates on the other.

Two types of interspecific competition

Two general forms of competition, exploitation and interference, are involved in this basic form of intraguild predation. First is exploitative competition, in which two species indirectly negatively affect each other by consuming the same resource and thereby reducing resource abundance (Case & Gilpin, 1974; Tilman, 1982; Vance, 1984). If two species have the exact same resource needs and only engage in exploitative competition, the species that is more efficient at consuming the shared resource should theoretically emerge as the winner, while the less efficient consumer is driven to extinction or a different niche (Vance, 1984). In order for intraguild predation to be robust and its participating species to coexist, it must include a second form of competition, interference competition (Amarasekare, 2002; Hsu, 1982; Vance, 1984). In interference competition, one species reduces the ability of the other to exploit the shared resource (Case & Gilpin, 1974; Hsu, 1982; Vance, 1984). Intraguild predation involves a severe form of interference competition in which the competitor is killed. With these two forms of competition in mind, there are three key predictions of a simple model of stable intraguild predation (Holt & Huxel, 2007; Holt & Polis, 1997):

1. The intraguild prey is superior in exploiting the shared resource.
2. The intraguild predator should have greater fitness from predating on the intraguild prey than from competing on a purely exploitative level.
3. The intraguild predator, by reducing the population of the more efficient consumer species, indirectly increases the abundance of the shared resource at equilibrium.

Interference competition is the component of intraguild predation that is most relevant to demonstrating that predatory attack can be aggressive. By definition, predation of the intraguild prey eliminates competitors for a shared resource and thus fulfils a competitive function for the intraguild predator. Competitive motivation, on the other hand, is difficult to prove in intraguild predation. The set of interactions that comprise intraguild predation are difficult to disentangle. Predation and interference competition are especially difficult to delineate because they usually occur simultaneously, which add another dimension to intentionality: in addition to harm being intentionally inflicted, is competition also intentional? Or is it an accidental benefit that emerges from facultative generalists that consume multiple trophic levels? Unfortunately, most intraguild predation research focuses on the ecological effects on intraguild predation on community structure, rather than on the individual scale. Specifically, much of the interest in intraguild predation lies in understanding if and how intraguild predation promotes species coexistence and biodiversity, often with complex variations of the intraguild predation community module.

Uneaten killed prey

Meanwhile, little research has been done to dissect the motivations of an intraguild predator, even when field examples seem to conform to the simplest form of intraguild predation. When the intraguild predator successfully kills and eats the intraguild prey, nutrition and competition benefits are simultaneously achieved and thus the corresponding motivations are difficult to distinguish. However, when the intraguild predator does not consume a proportion of intraguild prey that it kills, an opportunity arises to use the percentage of uneaten intraguild prey as a proxy indicator of non-predatory motivation.

This idea is reminiscent of the aforementioned studies of mouse-killing by rats, in which some mouse prey are left uneaten after being killed (Karli, 1956). Since both rats and mice and are phylogenetically related, it was previously hypothesized and then rejected that perhaps the killing of mice mimicked intraspecific competition against invader rats (Kemble & Davies, 1981). Instead of intraspecific competition, phylogenetic relatedness may more strongly suggest that rats and mice have overlapping resource niches. Indeed, field studies indicate that rats and mice compete intensely for the same food resources and reciprocally affect each other's population numbers (King *et al.*, 1996; Ruscoe & Murphy, 2005). Rats have also been previously described as intraguild predators of competing mice (O'Boyle, 1975). Field studies of poisoned or trapped rats have shown that mice dramatically increase in abundance when rats are removed, even if mice were also being eradicated at the same rate (Brown, Moller, Innes, & Alterio, 1996; Innes, Warburton, Williams, Speed, & Bradfield, 1995; Miller & Miller, 1995). In what is sometimes referred to as 'competitor release', the increase in mouse population from rat removal is much higher than expected from exploitation competition alone and strongly implicates interference competition through predation (Brown *et al.*, 1996; Caut *et al.*, 2007; Stapp, 1997). In order to validate whether this interference competition against mice is intentional, or just simple predatory behaviour with incidental competitive benefits, Bridgman *et al.* (2013) looked for (1) threat and display features associated with intraspecific aggression, and (2) uneaten prey. Results taken from wild rats indicated a lack of threat and display features towards mice, and all well-fed and starved rats ate at least a portion of euthanized mice. These findings led the researchers to conclude that interference competition, in this case, was predatory behaviour and not intentionally competitive.

There are two important caveats to this conclusion. First, it is important to note that here, just as in the aforementioned mouse-killing studies, the researchers used similarity to intraspecific competition as an indirect metric for whether interference competition is intentional. The similarity to the intraspecific competition does not address competition in a definitional sense that directly accounts for resource motivations. Additionally, intraspecific competition, especially for mates, likely evolved display postures and ritualized fighting as a way to establish dominance without killing of conspecifics (De Waal, 2000; Nelson, 2000). These social methods of communicating a threat and determining the winner may serve as species-preserving restrictions on the severity of harm, and as such may not be applicable to competition between recognizably different species. Second, it is known that wild rats consume most killed mice, while laboratory rats consume only a small portion of killed mice (Karli, 1956). Laboratory rats were used to demonstrate that feeding and killing were behaviourally dissociable components of predation. While wild rats are more pertinent for the ecologically valid representation of an actual ecosystem, laboratory rats may have been more valuable for extricating competitive and predatory motivations for eating or not eating prey.

In contrast to the aforementioned studies of intraguild predation in wild rats, Sunde, Overskaug, and Kvam (1999) investigated the motivation of the intraguild predator by comparing intraguild predation to conventional predation, rather than to intraspecific competition. In this study, lynxes are the intraguild predator and foxes are the intraguild prey. Lynxes and foxes both predate on smaller animals such as roe deer and mountain hares. Since they do not compete with lynxes, roe deer and mountain hares are referred to as 'true' prey species. Predation of true prey species is considered 'true' foraging because it only serves nutritional purposes and does not confer competitive benefits. If the nutritional need is the only factor motivating the killing of foxes, then the proportion of uneaten fox corpses should closely match the proportions of uneaten roe deer and hares. On the other hand, if something other than nutrition also motivates killing of foxes, then killed foxes should be left uneaten more often than roe deer and hare. The latter prediction was vindicated: 37% of foxes killed by lynxes are uneaten, while 2% of roe deer and 0% of hares were uneaten. This finding is similar to the previously mentioned behaviour of lab rats that attack and eat almost all frog, turtle, or insect prey but only a small percentage of mice (Bandler, 1970; Desisto & Huston, 1970), and insects

(Kemble & Davies, 1981). The notable difference between these rat-mouse-true prey studies and the lynx-fox-true prey study is that intraguild predation relationships were only explicitly described in the latter. This opens a line of questioning about interference competition, rather than intraspecific competition, as a potential 'other' factor for driving killing of the intraguild prey.

While it may be tempting to conclude that competition is the putative other factor that motivates lynxes to kill but only sometimes eat foxes, the field study was unable to account for the relative abundance of foxes, roe deer, and hares. Specifically, they could not account for how often lynxes encountered foxes or true prey by coincidence. Even if the absolute population counts of true prey were large, they may be effectively scarce to lynxes if true prey is good at evading lynx detection. On the other hand, foxes may be effectively abundant if they were poor at evading lynx detection and lynxes encountered them more often by chance. In the latter case, lynxes may find that the extra immediate energy required to subdue a fox prey may be worthwhile if they do not require as much time and energy for prey search. In short, scarcity of true prey species should increase uneaten fox corpses, while the abundance of foxes should increase eating of foxes. Without full control and understanding of the relative abundances of intraguild prey and shared resources, it is difficult to concretely attribute uneaten intraguild prey to the competition. Firm evidence of competition must be acquired before competitive aggression can be argued for.

The use of a percentage of uneaten killed prey does not clearly delineate between predatory and competitive motivations for attacking. Killing prey without immediately feeding can have advantages that indirectly promote predation, such as caching uneaten prey for possible later consumption, benefiting other members of a same social unit, or gaining experience that may facilitate later kills (Kruuk, 2009). Therefore, some have narrowed the definition of 'surplus killing' to refer to cases in which the predator makes no use of the kills whatsoever (Mueller & Hastings, 1977). It has been suggested that selective consumption and discarding of killed prey is an optimal foraging strategy when the focal prey is larger than can be consumed in one feeding or there is a high density of prey (Cook & Cockrell, 1978; Formanowicz, 1984; Sih, 1980; Zong *et al.*, 2012). For example, bears discard killed salmon during high prey abundance, and when prey is low in nutritional quality, which is consistent with a strategy to maximize energy intake (Lincoln & Quinn, 2019). Therefore, it is critical to consider the energetic costs, density, and nutritional differences between true prey and competing prey before the predation of both can be compared.

To get around the problems of interpreting uneaten killed competing prey, we suggest supplementing measure of uneaten killed prey with an attack-based metric to allow for more balanced and direct measurement of predatory and competitive motivations. Harm is integral and instrumental to both predation and aggression, but feeding on prey is only relevant to predation. Therefore, uneaten killed prey can only tell us that something other than immediately feeding on that prey is motivating the predator, but does not point to what that other motivation may be. Without a positive indicator of competitive motivation for attacking, it is difficult to rule out some distally predatory function for uneaten killed competing prey. To facilitate the equal detection of both predatory and competitive intent for attacking in intraguild predation, we suggest measuring how the frequency of attack changes across resource contexts (Figure 2, 3rd question). In the language of motivation, the frequency of attack indicates the intensity of pursuit, while how the frequency of attack changes across resource conditions indicates whether predation or aggression is the goal of attacking.

A similar approach has been applied to determine the motivation for interspecific territorial aggression between phylogenetically related species. For example, Nishikawa (1987) measured how the frequency of aggressive behaviour between two species of salamanders varied across different levels of sympatry and interspecific competition in order to answer whether interspecific aggression was due to the misidentification of heterospecifics as conspecifics (Murray, 1981), or whether interspecific territoriality is adaptive interference (Gill, 1974). If the latter were true, the frequency of aggressive behaviour should increase as interspecific competition increases. This concept also applies to interspecific aggression in intraguild predation. Specifically, an intraguild predator motivated by interference competition should attack the competing prey more frequently when the shared resource is more scarce or valuable. In contrast, an intraguild predator motivated by predation should attack the competing prey most when the shared resource is absent and the competing is the only available food option. Motivation directs behaviour by specifying a goal and setting the intensity with which to pursue that goal (Simpson & Balsam, 2016).

Nematode intraguild predation

While field studies can provide insight into the true mix of selective pressures that an animal faces in its natural life, the laboratory setting potentially offers greater control over the many variables that can affect the intraguild predator's motivation for attacking a competing prey. To study intraguild predation in the lab in an efficient manner, we recommend the nematodes *P. pacificus* and *C. elegans* as intraguild predator and prey, respectively, with bacteria as the shared resource (Figure 3). In this section, we will review the literature about *P. pacificus* and *C. elegans* with relation to each other and to bacteria. The goal of this section is to outline what is known about the participants and interactions that constitute this proposed nematode model of intraguild predation.

Intraguild predator: Pristionchus pacificus

Intraguild predators, including *P. pacificus*, are omnivores by definition. As a facultative predator, adult *P. pacificus* can

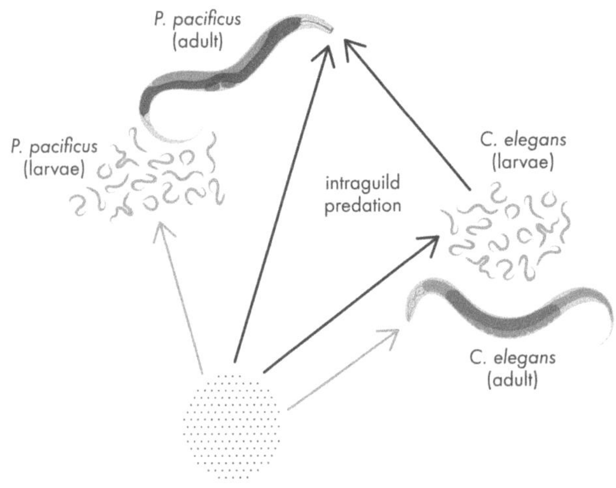

Figure 3. This food web diagram shows the directions in which different types of food travel between *P. pacificus, C. elegans*, and a bacterial food that both species exploit. Arrows originate from a food source and point to the organism that eats that food. Black arrows lead between direct participants in intraguild predation, while grey arrows indicate feeding interactions that are indirectly involved. The intraguild predator is adult *P. pacificus*, which predates on larval *C. elegans* as its intraguild prey. Adult and larval stages of *P. pacificus* and *C. elegans* consume bacteria.

derive nutrition from grazing on bacterial food and predating on nematode larva. The consumption of each of these food types flexibly engages different feeding rhythms that vary in the rate of pharyngeal pumping and dorsal tooth movement (Wilecki *et al.*, 2015). While eating bacteria, pharyngeal pumping is high and tooth movements are rare. When switching to predatory feeding, pharyngeal pumping decreases to about 66% of the bacterial rate and tooth movement increases dramatically until it matches pharyngeal pumping in a 1:1 ratio. Exogenous treatment of serotonin triggers predatory rhythms in the absence of prey, while interruption of serotonin synthesis and ablation of serotonergic neurons result in uncoordinated rhythms (Okumura, Wilecki, & Sommer, 2017; Wilecki *et al.*, 2015).

Pristionchus pacificus seems to prefer bacterial food over nematode prey. When *P. pacificus* is presented with an excess of both larval *C. elegans* and bacteria, *P. pacificus* bite larval prey less often than when bacteria are absent (Wilecki *et al.*, 2015). Reduced biting of larvae on bacteria suggests that predatory drive decreases when bacteria become available as an alternate food. Consistently, *P. pacificus* chemotaxes toward a source of *E. coli* OP50 bacteria when presented on the same plate as a source of larval *C. elegans* prey (Wilecki *et al.*, 2015). Despite preference for naturally co-occurring bacteria over *E. coli* OP50, *P. pacificus* fecundity and survival is as high or better on a diet of *E. coli* (Akduman *et al.*, 2018; Rae *et al.*, 2008). In fact, sometimes this preference is displayed for pathogenic bacteria, such as those of the *Serratia* genus (Akduman *et al.*, 2018). Overall, bacterial preference does not strongly correlate with the suitability of the food source (Akduman *et al.*, 2018). It may also be that this discordance between nutrition and food preference may also extend to prey food that vary in species and life stage. Whether *P. pacificus* is more motivated to predate or compete for bacteria will likely depend on the relative valuation of the bacterial and prey foods selected for a particular intraguild predation experiment. In the previously mentioned study by (Wilecki *et al.*, 2015) in which *P. pacificus* reduces biting of larval *C. elegans*, one could imagine that switching out *E. coli* OP50 to as undesirable bacteria may attenuate the reduction in biting and perhaps even elevate larval prey as the preferred food relative to the undesirable bacterial option.

It is important to note that the convention of feeding *E. coli* OP50 to *P. pacificus* in the laboratory setting was established out of convenience and a desire to ease the adoption of *P. pacificus* into existing *C. elegans* laboratories. Several studies have surveyed the microbiomes of the *P. pacificus* collected from natural settings (Akduman *et al.*, 2018; Koneru *et al.*, 2016; Meyer *et al.*, 2017; Rae *et al.*, 2008). Although *P. pacificus* can also be found in rotting plant material (Félix *et al.*, 2018), these microbiome studies focused on the bacteria present alongside *P. pacificus* in scarab beetles. *Pristionchus pacificus* can have a necromenic association with scarab beetles, whereby they reside exclusively as dauer larvae inside the living beetle and resume development once the beetle starts to decay (Herrmann *et al.*, 2007; Meyer *et al.*, 2017; Ragsdale, Kanzaki, & Herrmann, 2015). Enterobaceriaceae was found in many of this studies to be the most abundant family of bacteria present *P. pacificus* harvested from beetles, although many other types of bacteria were also isolated (Koneru *et al.*, 2016; Meyer *et al.*, 2017). While *E. coli* is part of the Enterobaceriaceae family, *Escherichia* species were rarely encountered (Koneru *et al.*, 2016). *Pristionchus pacificus* grown on *E. coli* OP50 preferred many of the bacteria isolated from beetles and soil, as measured by chemotaxis assays (Akduman *et al.*, 2018; Koneru *et al.*, 2016; Rae *et al.*, 2008).

In additional to being bacterial generalists, *P. pacificus* are generalist predators of larvae of many nematode species (Lightfoot *et al.*, 2019). *Pristionchus pacificus* uses highly specific small peptide-mediated self-recognition that allows them to discriminate between their own larvae and those of other species as well as different geographical isolates of *P. pacificus* (Lightfoot *et al.*, 2019). This provides strong evidence that, if interspecific aggression indeed exists between *P. pacificus* and *C. elegans*, it is highly unlikely to be caused by misidentification of heterospecifics as conspecifics, especially since intraspecific aggression has yet to be seen between members of the same *P. pacificus* isolate strain.

Presented with an excess of larval *C. elegans* in the absence of bacteria, the standard *P. pacificus* laboratory strain PS312 readily bites larval prey, with about 34\% of bites resulting in killed corpses (Wilecki *et al.*, 2015). However, the same study noted that only about half of larvae corpses were eaten and surmised that surplus killing by *P. pacificus* may serve to eliminate competition. Although intraguild predation was not explicitly mentioned in (Wilecki *et al.*, 2015), the metric of uneaten killed prey once again raises the question of whether a non-predatory component is behind the motivation for killing prey. It still remains to be demonstrated whether or not competition is in fact the non-predatory motivation in question. Recent findings reveal that the level of surplus killing by *P. pacificus* is influenced by the nutrient composition of its bacterial diet (Akduman *et al.*, 2019). Specifically, B12 derived from the bacterial strain *Novosphingobium* L76 was found to double the killing efficiency of *P. pacificus* without co-ordinately increasing feeding rate. *Pristionchus pacificus* raised on a *Novosphingobium* L76 diet versus an *E. coli* OP50 diet exhibited differential expression of genes involved in fatty acid metabolism. Thus, in addition to bacterial preference, physiological changes induced by bacterial diet also affect predatory behaviour. In order to design contexts that may potentially discourage predation for eating prey and instead promote competition against *C. elegans* for bacteria, multiple bacterial variables such as abundance, preference, caloric value, and nutrient composition can be individually manipulated. How killing efficiency and surplus killing changes across bacterial conditions may provide answers to the question of whether competition can motivate *P. pacificus* to bite *C. elegans* in conditions that exacerbate competition for bacterial food.

Intraguild prey: Caenorhabditis elegans

Successful interspecific aggression often depends on the response of the target, which is *C. elegans* in the proposed nematode intraguild predation model. The *C. elegans* response to *P. pacificus* has not been studied in-depth, which is likely due to the fact that the smallest larval stage (L1) of *C. elegans* are most often used to assay P. pacificus predatory behaviour. Often, the larval *C. elegans* is killed immediately upon contact with *P. pacificus* nose, thereby precluding any subsequent *C. elegans* response. A recent study removed the danger of live *P. pacificus* by instead using an extract of excretions collected from live *P. pacificus* animals (Liu *et al.*, 2018). Interestingly, adult *C. elegans* immediately avoided this 'predator cue' when it was collected from starved *P. pacificus*, but not when the cue was collected from well-fed *P. pacificus*. This suggests that *P. pacificus* may be a more serious threat to *C. elegans* when bacteria are absent as a preferred food source for *P. pacificus*.

The concept of intentionality in motivated behaviour is useful not only for exploring aggression in *P. pacificus*, but also for designing experiments to characterize risk-taking and fear in *C. elegans* responses in intraguild predation interactions. By first understanding how environmental and internal conditions modulate *P. pacificus* motivation to attack *C. elegans*, *P. pacificus* can then be deliberately manipulated to pose particular levels of risk toward *C. elegans*. With additional manipulation of bacterial variables, *C. elegans* responses in intraguild predation interactions can be measured as a reflection of internal balancing of appetitive bacterial factors and aversive *P. pacificus* factors, the latter acquiring more weight with induced fear. More broadly, the construction of behavioural experiments to probe *C. elegans* intentionality opens the way for analysing more complex computations and cognitive processes underlying motivation and decision-making.

In contrast to *C. elegans* behavioural responses to *P. pacificus*, much more is known about the relationship between *C. elegans* and bacteria. Importantly, this knowledge may inform homology-based hypotheses about how *P. pacificus* senses and responds to bacteria. It is known that *P. pacificus* and *C. elegans* have disparate responses to the same set of odorants, with some odorants that are attractive to one and repulsive to the other (Hong, 2015; Hong & Sommer, 2006a). Therefore, any discussion of potential conserved bacteria responses and underlying mechanisms will have to involve direct sensation of bacteria and not of proxy odorants, such as benzaldehyde and diacetyl, that putatively represent bacteria in *C. elegans*. The first notable change in behaviour that *C. elegans* exhibits upon finding a bacterial lawn is to decrease its locomotory rate (Sawin, Ranganathan, & Horvitz, 2000). This basal slowing response requires dopamine, as dopamine synthesis mutants continue moving through bacteria at the same rate as when bacteria is absent (Sawin *et al.*, 2000).

In addition to binary detection of the presence or absence of bacteria, *C. elegans* is also able to distinguish and seek out the boundary of a bacterial lawn from its circumscribed region. Some social wild strains of *C. elegans* and *npr-1* mutants that lack the neuropeptide Y receptor naturally migrate to and aggregate at the border of a bacterial lawn, where bacteria is thickest (De Bono & Bargmann, 1998). This bordering tendency involves oxygen sensing by guanylate cyclase, which promotes aerotaxing away from regions of higher oxygen levels towards areas of lower oxygen levels in both wildtype and *npr-1* mutants (Gray *et al.*, 2004). Thick *E. coli* OP50 bacterial lawns consume oxygen more quickly than can be replenished by ambient diffusion, and borders with the highest concentration of bacteria were observed to have lower effective oxygen concentrations (Gray *et al.*, 2004). Acute reduction of ambient oxygen levels abolished bordering behaviour in *C. elegans* (Gray *et al.*, 2004), as well as in *P. pacificus* (Moreno, McGaughran, Rödelsperger, Zimmer, & Sommer, 2016). Therefore, *C. elegans*, as well as *P. pacificus*, may use relative lower oxygen concentrations to find and demarcate the lawn edge. The ability to detect the edge of a lawn opens up the possibility of estimating the size of the lawn. Indeed, guanylate cyclase *C. elegans* mutants were unable to distinguish between small and large lawns of bacteria (Calhoun *et al.*, 2015). The mechanism for computing lawn size experience depends on the variability in bacteria levels that *C. elegans* senses during its exploration of the lawn. The thick edge relative to the thinner interior of the lawn means that *C. elegans* will experience changing bacteria levels more often in a small lawn, where the animal will encounter the edge at a higher rate. Large bacterial variability is sensed by ASI and ASK neurons and result in downstream dopamine release.

C. elegans has been cultivated in the laboratory setting with *E. coli*, OP50 since its debut as a model organism (Brenner, 1974). However, like *P. pacificus*, *C. elegans* is found in nature with a variety of other bacteria species, with *Enterobacteriaceae* and *Acetobacteraceae* species associated with high proliferation (Dirksen *et al.*, 2016; Samuel *et al.*,

2016; Schulenburg & Félix, 2017). *Caenorhabditis elegans* also displays a preference for bacterial species other than *E. coli* OP50, particularly if the other bacteria is higher quality food, as measured by growth rate (Shtonda & Avery, 2006). Furthermore, *C. elegans* raised on higher-quality bacteria leave mediocre bacteria more often (Shtonda & Avery, 2006). One such high-quality bacterial strain is *Comamonas sp.*, which was isolated from a soil environment (Avery & Shtonda, 2003). Interestingly, the list of bacteria naturally found with and preferred by *P. pacificus* also includes the *Comamonadaceae* family (Akduman *et al.*, 2018; Koneru *et al.*, 2016). Additionally, a *Comamonas sp.* DA1877 diet has been shown to increase surplus killing in *P. pacificus* via increased the same B12 mechanism as in a *Novosphingobium* L76 diet (Akduman *et al.*, 2019). Therefore, *Comamonas sp.* may be useful in mutually exacerbating competition between *P. pacificus* and *C. elegans.*

Concluding remarks

The aim of this perspective was to outline key concepts about interspecific interactions and specifically identify feeding-related nematode literature that are relevant to answering whether *P. pacificus* biting of *C. elegans* may be a form of interspecific aggression derived from intraguild predation. In particular, establishing whether the goal of biting is to kill prey for consumption or to defend bacterial resources will be critical to answering this question.

Interspecific aggression between nematodes has been previously observed between *Steinernema* species that compete for host resources (O'Callaghan, Zenner, Hartley, & Griffin, 2014), but these parasitic nematodes also exhibit intraspecific aggression between males (Zenner, O'Callaghan, & Griffin, 2014). If interspecific aggression exists in *P. pacificus*, it likely arose *de novo* as a modification of some non-aggressive behaviour. Without intraspecific aggression as a point of comparison, we suggest contrasting potential interspecific aggression between *P. pacificus* and *C. elegans* to an agonistic interaction that already exists between the two species, predation. Second, while inter- and intraspecific aggression necessarily involve different targets of different species, *C. elegans* is the same target regardless of whether *P. pacificus* is motivated by predatory or competitive goals. To obtain interspecific aggression from intraspecific aggression, an animal only needs to change how they recognize competitors to include both conspecific and heterospecific targets. In contrast, to achieve interspecific aggression from intraguild predation, *P. pacificus* must be able to flexibly change which goals motivates it to harm *C. elegans*, either to eat or compete with it for bacteria. We hope that our proposed nematode intraguild predation model may provide insight into how a behaviour as complex as interspecific aggression can arise in a simple nematode without having intraspecific aggression as a convenient behavioural substrate.

We have presented a series of relevant concepts in aggression, predation, and intraguild predation that together provide one possible approach for determining the motivation driving the intraguild predator *P. pacificus* when it kills its intraguild prey *C. elegans*. This approach begins with establishing core criteria for aggression, which we distil into two components: intentional harm and a competitive goal for harm. More criteria can be added to achieve more face validity with aggressive behaviour as it is typically considered in the field. Since members of different species do not compete for mates, territorial aggression for the defence of overlapping resources is the most probable form of interspecific aggression. The next step is to assess whether predatory behaviour is potentially aggressive, which we take to mean intentionally harmful. We disqualify engulfing and grazing because harm is incidental to feeding actions. Instead, we suggest that harm that precedes feeding and directly contributes to the capture and killing of prey can be intentional. Once intentional harm is established in predatory behaviour, the major task at hand is to demonstrate that a competitive goal for harm can increase *P. pacificus* attack frequency in conditions in which competition for bacteria in intensified. For intraguild predation, this requires an assessment of both exploitative and interference competition. Careful consideration of multiple bacterial and prey factors will be crucial for designing conditions and experiments that are informative about how *P. pacificus* food experience and relative valuation of bacterial and prey food factor into its motivation for attacking *C. elegans*.

Disclosure statement

No potential conflict of interest was reported by the author(s).

Funding

This work was supported by the National Institutes of Health (R01MH113905).

ORCID

Sreekanth H. Chalasani http://orcid.org/0000-0003-2522-8338

References

Agrawal, A.A. (2011). Current trends in the evolutionary ecology of plant defence. *Functional Ecology*, *25*(2), 420–432. doi:10.1111/j.1365-2435.2010.01796.x

Akduman, N., Lightfoot, J. W., Röseler, W., Witte, H., Lo, W. S., Rödelsperger, C., & Sommer, R. J. (2020). Bacterial vitamin B 12 production enhances nematode predatory behavior. The ISME journal, 14(6), 1494–1507. doi:10.1038/s41396-020-0626-2.

Akduman, N., Rödelsperger, C., & Sommer, R.J. (2018). Culture-based analysis of *Pristionchus*-associated microbiota from beetles and figs for studying nematode-bacterial interactions. *PLoS One*, *13*(6), e0198018. doi:10.1371/journal.pone.0198018

Amarasekare, P. (2002). Interference competition and species coexistence. *Proceedings. Biological Sciences*, *269*(1509), 2541–2550. doi:10.1098/rspb.2002.2181

Archer, J. (1988). *The behavioral biology of aggression*. Cambridge, UK: Cambridge University Press.

Arim, M., & Marquet, P.A. (2004). Intraguild predation: A widespread interaction related to species biology. *Ecology Letters*, *7*(7), 557–564. doi:10.1111/j.1461-0248.2004.00613.x

Avery, L., & Shtonda, B.B. (2003). Food transport in the C. elegans pharynx. *Journal of Experimental Biology, 206*(14), 2441–2457. doi: 10.1242/jeb.00433

Baker, R.R. (1983). Insect territoriality. *Annual Review of Entomology, 28*(1), 65–89. doi:10.1146/annurev.en.28.010183.000433

Bandler, R.J. Jr, (1970). Animals spontaneously attacked by rats. *Communications Behavior and Biology, 5*, 177–182.

Bengtson, S. (2002). Origins and early evolution of predation. *The Paleontological Society Papers, 8*, 289–318. doi:10.1017/S1089332600001133

Bento, G., Ogawa, A., & Sommer, R.J. (2010). Co-option of the hormone-signalling module dafachronic acid-DAF-12 in nematode evolution. *Nature, 466*(7305), 494–497. doi:10.1038/nature09164

Berkowitz, L. (1981). The concept of aggression. In P.E. Brain & D. Benton (Eds.), *Multidisciplinary approaches to aggression research* (pp. 3–15). Amsterdam, Netherlands: Elsevier.

Berkowitz, L. (1993). *Aggression: Its causes, consequences, and control.* New York, NY: McGraw-Hill.

Blanchard, R.J., Takahashi, L.K., & Blanchard, D.C. (1977). The development of intruder attack in colonies of laboratory rats. *Animal Learning & Behavior, 5*(4), 365–369.

Boice, R., & Schmeck, R.R. (1968). Predatory behaviors of grasshopper mice (*Onychomys leucogaster*). *American Zoologist, 8*(4), 751.

Brattstrom, B.H. (1974). The evolution of reptilian social behavior. *American Zoologist, 14*(1), 35–49. doi:10.1093/icb/14.1.35

Brenner, S. (1974). The genetics of *Caenorhabditis elegans. Genetics, 77*(1), 71–94.

Bridgman, L.J., Innes, J., Gillies, C., Fitzgerald, N.B., Miller, S., & King, C.M. (2013). Do ship rats display predatory behaviour towards house mice? *Animal Behaviour, 86*(2), 257–268. doi:10.1016/j.anbehav.2013.05.013

Brown, J.L., & Orians, G.H. (1970). Spacing patterns in mobile animals. *Annual Review of Ecology and Systematics, 1*(1), 239–262. doi:10.1146/annurev.es.01.110170.001323

Brown, K.P., Moller, H., Innes, J., & Alterio, N. (1996). Calibration of tunnel tracking rates to estimate relative abundance of ship rats (*Rattus rattus*) and mice (*Mus musculus*) in a New Zealand forest. *New Zealand Journal of Ecology, 20*(2), 271–275.

Bumbarger, D.J., Riebesell, M., Rödelsperger, C., & Sommer, R.J. (2013). System-wide rewiring underlies behavioral differences in predatory and bacterial-feeding nematodes. *Cell, 152*(1–2), 109–119. doi:10.1016/j.cell.2012.12.013

Burt, W.H. (1943). Territoriality and home range concepts as applied to mammals. *Journal of Mammalogy, 24*(3), 346–352. doi:10.2307/1374834

Buss, A.H. (1961). *The psychology of aggression*. New York, NY: Wiley.

Byerly, L., Cassada, R.C., & Russell, R.L. (1976). The life cycle of the nematode *Caenorhabditis elegans*. I. Wild-type growth and reproduction. *Developmental Biology, 51*(1), 23–33. doi:10.1016/0012-1606(76)90119-6

Calhoun, A.J., Tong, A., Pokala, N., Fitzpatrick, J.A., Sharpee, T.O., & Chalasani, S.H. (2015). Neural mechanisms for evaluating environmental variability in *Caenorhabditis elegans. Neuron, 86*(2), 428–441. doi:10.1016/j.neuron.2015.03.026

Case, T.J., & Gilpin, M.E. (1974). Interference competition and niche theory. *Proceedings of the National Academy of Sciences, 71*(8), 3073–3077. doi:10.1073/pnas.71.8.3073

Caut, S., Casanovas, J.G., Virgos, E., Lozano, J., Witmer, G.W., & Courchamp, F. (2007). Rats dying for mice: Modelling the competitor release effect. *Austral Ecology, 32*(8), 858–868. doi:10.1111/j.1442-9993.2007.01770.x

Cavalier-Smith, T. (2009). Predation and eukaryote cell origins: A coevolutionary perspective. *The International Journal of Biochemistry & Cell Biology, 41*(2), 307–322. doi:10.1016/j.biocel.2008.10.002

Chen, S., Lee, A.Y., Bowens, N.M., Huber, R., & Kravitz, E.A. (2002). Fighting fruit flies: A model system for the study of aggression. *Proceedings of the National Academy of Sciences, 99*(8), 5664–5668. doi:10.1073/pnas.082102599

Cinkornpumin, J.K., & Hong, R.L. (2011). RNAi mediated gene knockdown and transgenesis by microinjection in the necromenic nematode *Pristionchus pacificus. JoVE (Journal of Visualized Experiments), 56*, e3270.

Cody, M.L. (1969). Convergent characteristics in sympatric species: A possible relation to interspecific competition and aggression. *The Condor, 71*(3), 223–239. doi:10.2307/1366300

Cook, R.M., & Cockrell, B.J. (1978). Predator ingestion rate and its bearing on feeding time and the theory of optimal diets. *The Journal of Animal Ecology, 47*(2), 529–547. doi:10.2307/3799

Crane, J. (1966). Combat, display and ritualization in fiddler crabs (Ocypodidae, genus *Uca*). *Philosophical Transactions of the Royal Society of London. Series B, Biological Sciences, 251*(772), 459–472.

Darwin, C. (1896). *Charles Darwin's works: The descent of man and selection in relation to sex* (Vol. 9). New York, NY: D. Appleton.

Davidov, Y., & Jurkevitch, E. (2009). Predation between prokaryotes and the origin of eukaryotes. *BioEssays: News and Reviews in Molecular, Cellular and Developmental Biology, 31*(7), 748–757. doi:10.1002/bies.200900018

Dawkins, R., & Krebs, J.R. (1979). Arms races between and within species. *Proceedings of the Royal Society of London. Series B. Biological Sciences, 205*(1161), 489–511.

De Bono, M., & Bargmann, C.I. (1998). Natural variation in a neuropeptide Y receptor homolog modifies social behavior and food response in *C. elegans. Cell, 94*(5), 679–689. doi:10.1016/S0092-8674(00)81609-8

De Waal, F.B. (2000). Primates-a natural heritage of conflict resolution. *Science, 289*(5479), 586–590. doi:10.1126/science.289.5479.586

Desisto, M.J., & Huston, J.P. (1970). Effect of territory on frog-killing by rats. *The Journal of General Psychology, 83*, 179–184. doi:10.1080/00221309.1970.9710800

Dickinson, D.J., & Goldstein, B. (2016). CRISPR-based methods for *Caenorhabditis elegans* genome engineering. *Genetics, 202*(3), 885–901. doi:10.1534/genetics.115.182162

Dieterich, C., Clifton, S.W., Schuster, L.N., Chinwalla, A., Delehaunty, K., Dinkelacker, I., … Sommer, R.J. (2008). The *Pristionchus pacificus* genome provides a unique perspective on nematode lifestyle and parasitism. *Nature Genetics, 40*(10), 1193–1198. doi:10.1038/ng.227

Dirksen, P., Marsh, S.A., Braker, I., Heitland, N., Wagner, S., Nakad, R., … Schulenburg, H. (2016). The native microbiome of the nematode *Caenorhabditis elegans*: Gateway to a new host-microbiome model. *BMC Biology, 14*(1), 38. doi:10.1186/s12915-016-0258-1

Félix, M.A., Hill, R.J., Schwarz, H., Sternberg, P.W., Sudhaus, W., & Sommer, R.J. ((1999). *Pristionchus pacificus*, a nematode with only three juvenile stages, displays major heterochronic changes relative to *Caenorhabditis elegans. Proceedings of the Royal Society of London. Series B: Biological Sciences, 266*(1429), 1617–1621. 1999). doi:10.1098/rspb.1999.0823

Félix, M.-A., Ailion, M., Hsu, J.C., Richaud, A., & Wang, J. (2018). *Pristionchus nematodes* occur frequently in diverse rotting vegetal substrates and are not exclusively necromenic, while *Panagrellus redivivoides* is found specifically in rotting fruits. *PLoS One, 13*(8), e0200851. doi:10.1371/journal.pone.0200851

Flynn, J.P. (1967). The neural basis of aggression in cats. In D.C. Glass (Ed.), *Neurophysiology and emotion* (pp. 40–60). New York, NY: Rockefeller University Press and Russell Sage Foundation.

Flynn, J.P., Vanegas, H., Foote, W., & Edwards, S. (1970). Neural mechanisms involved in a cat's attack on a rat. In *The neural control of behavior* (pp. 135–173). Waltham, MA: Academic Press.

Formanowicz, D.R. Jr, (1984). Foraging tactics of an aquatic insect: Partial consumption of prey. *Animal Behaviour, 32*(3), 774–781. doi:10.1016/S0003-3472(84)80153-0

Gendreau, P.L., & Archer, J. (2005). Subtypes of Aggression in Humans and Animals. In R.E. Tremblay, W.W. Hartup, & J. Archer (Eds.), *Developmental origins of aggression* (pp. 25–46). New York, NY: Guilford Press.

Gerking, S.D. (1959). The restricted movement of fish populations. *Biological Reviews, 34*(2), 221–242. doi:10.1111/j.1469-185X.1959.tb01289.x

Gill, D.E. (1974). Intrinsic rate of increase, saturation density, and competitive ability. II. The evolution of competitive ability. *The American Naturalist, 108*(959), 103–116. doi:10.1086/282888

Gray, J.M., Karow, D.S., Lu, H., Chang, A.J., Chang, J.S., Ellis, R.E., … Bargmann, C.I. (2004). Oxygen sensation and social feeding mediated by a *C. elegans* guanylate cyclase homologue. *Nature, 430*(6997), 317–322. doi:10.1038/nature02714

Grether, G.F., Anderson, C.N., Drury, J.P., Kirschel, A.N., Losin, N., Okamoto, K., & Peiman, K.S. (2013). The evolutionary consequences of interspecific aggression. *Annals of the New York Academy of Sciences, 1289*(1), 48–68. doi:10.1111/nyas.12082

Grether, G.F., Losin, N., Anderson, C.N., & Okamoto, K. (2009). The role of interspecific interference competition in character displacement and the evolution of competitor recognition. *Biological Reviews of the Cambridge Philosophical Society, 84*(4), 617–635. doi:10.1111/j.1469-185X.2009.00089.x

Harrold, C., & Reed, D.C. (1985). Food availability, sea urchin grazing, and kelp forest community structure. *Ecology, 66*(4), 1160–1169. doi:10.2307/1939168

Herrmann, M., Mayer, W.E., Hong, R.L., Kienle, S., Minasaki, R., & Sommer, R.J. (2007). The nematode *Pristionchus pacificus* (Nematoda: Diplogastridae) is associated with the oriental beetle Exomala orientalis (Coleoptera: Scarabaeidae) in Japan. *Zoological Science, 24*(9), 883–889. doi:10.2108/zsj.24.883

Holt, R.D., & Huxel, G.R. (2007). Alternative prey and the dynamics of intraguild predation: Theoretical perspectives. *Ecology, 88*(11), 2706–2712. doi:10.1890/06-1525.1

Holt, R.D., & Polis, G.A. (1997). A theoretical framework for intraguild predation. *The American Naturalist, 149*(4), 745–764. doi:10.1086/286018

Hong, R.L. (2015). *Pristionchus pacificus* olfaction. In R.J. Sommer (Ed.), *Pristionchus pacificus: a nematode model for comparative and evolutionary biology.*, (pp. 331–352. Leiden, Netherlands: Brill.

Hong, R.L., & Sommer, R.J. (2006a). Chemoattraction in *Pristionchus nematodes* and implications for insect recognition. *Current Biology, 16*(23), 2359–2365. doi:10.1016/j.cub.2006.10.031

Hong, R.L., & Sommer, R.J. (2006b). *Pristionchus pacificus*: A well-rounded nematode. *BioEssays: News and Reviews in Molecular, Cellular and Developmental Biology, 28*(6), 651–659. doi:10.1002/bies.20404

Hong, R.L., Riebesell, M., Bumbarger, D.J., Cook, S.J., Carstensen, H.R., Sarpolaki, T., … Hobert, O. (2019). Evolution of neuronal anatomy and circuitry in two highly divergent nematode species. *Elife, 8*, e47155.

Howard, H.E. (1920). *Territory in bird life*. New York, NY: EP. Dutton & Company.

Hsu, S.B. (1982). On a resource based ecological competition model with interference. *Journal of Mathematical Biology, 12*(1), 45–52. doi:10.1007/BF00275202

Hutchinson, R.R., & Renfrew, J.W. (1966). Stalking attack and eating behaviors elicited from the same sites in the hypothalamus. *Journal of Comparative and Physiological Psychology, 61*(3), 360–367. doi:10.1037/h0023250

Huxley, J. (1966). Introduction: A discussion on ritualization of behaviour in animals and man. *Philosophical Transactions of the Royal Society of London, Series B}, 251*, 249–271.

Innes, J., Warburton, B., Williams, D., Speed, H., & Bradfield, P. (1995). Large-scale poisoning of ship rats (*Rattus rattus*) in indigenous forests of the North Island, New Zealand. *New Zealand Journal of Ecology, 19*(1), 5–17.

Issa, F.A., & Edwards, D.H. (2006). Ritualized submission and the reduction of aggression in an invertebrate. *Current Biology, 16*(22), 2217–2221. doi:10.1016/j.cub.2006.08.065

Jedrzejewska, B., & Jedrzejewski, W. (1989). Seasonal surplus killing as hunting strategy of the weasel Mustela nivalis-test of a hypothesis. *Acta Theriologica, 34*(12–28), 347–360.

Karli, P. (1956). The Norway rat's killing response to the white mouse: An experimental analysis. *Behaviour, 10*(1), 81–103. doi:10.1163/156853956X00110

Kaufmann, J.H. (1983). On the definitions and functions of dominance and territoriality. *Biological Reviews, 58*(1), 1–20. doi:10.1111/j.1469-185X.1983.tb00379.x

Kemble, E.D., & Davies, V.A. (1981). Effects of prior environmental enrichment and amygdaloid lesions on consumatory behavior, activity, predation, and shuttlebox avoidance in male and female rats. *Physiological Psychology, 9*(4), 340–346. doi:10.3758/BF03326991

King, C.M., Innes, J.G., Flux, M., Kimberley, M.O., Leathwick, J.R., & Williams, D.S. (1996). Distribution and abundance of small mammals in relation to habitat in Pureora Forest Park. *New Zealand Journal of Ecology, 20*(2), 215–240.

King, J.A. (1973). The ecology of aggressive behavior. *Annual Review of Ecology and Systematics, 4*(1), 117–138. doi:10.1146/annurev.es.04.110173.001001

Koneru, S.L., Salinas, H., Flores, G.E., & Hong, R.L. (2016). The bacterial community of entomophilic nematodes and host beetles. *Molecular Ecology, 25*(10), 2312–2324. doi:10.1111/mec.13614

Kravitz, E.A., & Huber, R. (2003). Aggression in invertebrates. *Current Opinion in Neurobiology, 13*(6), 736–743. doi:10.1016/j.conb.2003.10.003

Kruuk, H. (2009). Surplus killing by carnivores. *Journal of Zoology, 166*(2), 233–244. doi:10.1111/j.1469-7998.1972.tb04087.x

Lancaster, C.E., Ho, C.Y., Hipolito, V.E., Botelho, R.J., & Terebiznik, M.R. (2019). Phagocytosis: What's on the menu? *Biochemistry and Cell Biology = Biochimie et Biologie Cellulaire, 97*(1), 21–29. doi:10.1139/bcb-2018-0008

Landry, S.O. Jr, (1970). The Rodentia as omnivores. *The Quarterly Review of Biology, 45*(4), 351–372. doi:10.1086/406647

Leyhausen, P. (1973). On the function of the relative hierarchy of moods (as exemplified by the phylogenetic and ontogenetic development of prey-catching in carnivores). In K. Lorenz & P. Leyhausen (Eds.), *Motivation of human and animal behavior* (pp. 144–247). London, UK: Van Nostrand Reinhold.

Lightfoot, J.W., Wilecki, M., Rödelsperger, C., Moreno, E., Susoy, V., Witte, H., & Sommer, R.J. (2019). Small peptide-mediated self-recognition prevents cannibalism in predatory nematodes. *Science, 364*(6435), 86–89. doi:10.1126/science.aav9856

Lima, S.L., & Dill, L.M. (1990). Behavioral decisions made under the risk of predation: A review and prospectus. *Canadian Journal of Zoology, 68*(4), 619–640. doi:10.1139/z90-092

Lincoln, A.E., & Quinn, T.P. (2019). Optimal foraging or surplus killing: Selective consumption and discarding of salmon by brown bears. *Behavioral Ecology, 30*(1), 202–212. doi:10.1093/beheco/ary139

Liu, Z., Kariya, M.J., Chute, C.D., Pribadi, A.K., Leinwand, S.G., Tong, A., … Chalasani, S.H. (2018). Predator-secreted sulfolipids induce defensive responses in C. elegans. *Nature Communications, 9*(1), 1128. doi:10.1038/s41467-018-03333-6

Lorenz, K. (1966). *On aggression*. New York, NY: Harcourt, Brace and World.

Lounibos, L.P., Makhni, S., Alto, B.W., & Kesavaraju, B. (2008). Surplus killing by predatory larvae of *Corethrella appendiculata*: Prepupal timing and site-specific attack on mosquito prey. *Journal of Insect Behavior, 21*(2), 47–54. doi:10.1007/s10905-007-9103-2

MacLean, S.F., Jr., & Seastedt, T.R. (1979). Avian territoriality: Sufficient resources or interference competition. *The American Naturalist, 114*(2), 308–312.

May, R.C., Loman, N.J., Haines, A.S., Pallen, M.J., Boehnisch, C., Penn, C.W., … Kim, J. (2009). The genome sequence of *E. coli* OP50. *Worm Breeding Gazette, 18*, 24.

McFadden, G.I., Gilson, P.R., Hofmann, C., Adcock, G.J., & Maier, U.-G. (1994). Evidence that an amoeba acquired a chloroplast by retaining part of an engulfed eukaryotic alga. *Proceedings of the National Academy of Sciences, 91*(9), 3690–3694. doi:10.1073/pnas.91.9.3690

Meyer, J.M., Baskaran, P., Quast, C., Susoy, V., Rödelsperger, C., Glöckner, F.O., & Sommer, R.J. (2017). Succession and dynamics of *Pristionchus nematodes* and their microbiome during decomposition of Oryctes borbonicus on La Réunion Island. *Environmental Microbiology, 19*(4), 1476–1489. doi:10.1111/1462-2920.13697

Miller, C.J., & Miller, T.K. (1995). Population dynamics and diet of rodents on Rangitoto Island, New Zealand, including the effect of a

1080 poison operation. *New Zealand Journal of Ecology, 19*(1), 19–27.
Moreno, E., McGaughran, A., Rödelsperger, C., Zimmer, M., & Sommer, R.J. (2016). Oxygen-induced social behaviours in *Pristionchus pacificus* have a distinct evolutionary history and genetic regulation from *Caenorhabditis elegans. Proceedings. Biological Sciences, 283*(1825), 20152263. doi:10.1098/rspb.2015.2263
Moyer, K.E. (1968). Kinds of aggression and their physiological basis. *Communications in Behavioral Biology, 2*(2), 65–87.
Moynihan, M., & Moynihan, M. (1998). *The social regulation of competition and aggression in animals.* Washington, DC: Smithsonian Institution Press.
Mueller, D.L., & Hastings, B.C. (1977). A clarification of "surplus killing". *Animal Behaviour, 25*, 1065. doi:10.1016/0003-3472(77)90059-8
Murray, B.G. (1981). The origins of adaptive interspecific territorialism. *Biological Reviews, 56*(1), 1–22. doi:10.1111/j.1469-185X.1981.tb00341.x
Myer, J.S. (1967). Prior killing experience and the suppressive effects of punishment on the killing of mice by rats. *Animal Behaviour, 15*(1), 59–61. doi:10.1016/S0003-3472(67)80011-3
Myer, J.S. (1969). Early experience and the development of mouse killing by rats. *Journal of Comparative and Physiological Psychology, 67*(1), 46–49. doi:10.1037/h0026657
Myer, J.S. (1971). Experience and the stability of mouse killing by rats. *Journal of Comparative and Physiological Psychology, 75*(2), 264–268. doi:10.1037/h0030819
Namdeo, S., Moreno, E., Rödelsperger, C., Baskaran, P., Witte, H., & Sommer, R.J. (2018). Two independent sulfation processes regulate mouth-form plasticity in the nematode *Pristionchus pacificus. Development, 145*(13), dev166272. doi:10.1242/dev.166272
Nance, J., & Frøkjaer-Jensen, C. (2019). The *Caenorhabditis elegans* transgenic toolbox. *Genetics, 212*(4), 959–990. doi:10.1534/genetics.119.301506
Nelson, R.J. (2000). Affiliative and aggressive behavior. In *An introduction to behavioral endocrinology* (3rd ed., pp. 395–445). Sunderland, MA: Sinauer.
Nelson, R.J. (Ed.). (2005). *Biology of aggression.* Oxford, UK: Oxford University Press.
Nice, M.M. (1941). The role of territory in bird life. *American Midland Naturalist, 26*(3), 441–487. doi:10.2307/2420732
Nishikawa, K.C. (1987). Interspecific aggressive behaviour in salamanders: Species-specific interference or misidentification? *Animal Behaviour, 35*(1), 263–270. doi:10.1016/S0003-3472(87)80232-4
Nunn, G.L., Klem, D., Jr, Kimmel, T., & Merriman, T. (1976). Surplus killing and caching by American kestrels (*Falco sparverius*). *Animal Behaviour, 24*(4), 759–763. doi:10.1016/S0003-3472(76)80005-X
O'Callaghan, K.M., Zenner, A.N., Hartley, C.J., & Griffin, C.T. (2014). Interference competition in entomopathogenic nematodes: Male Steinernema kill members of their own and other species. *International Journal for Parasitology, 44*(13), 1009–1017.
O'Boyle, M. (1974). Rats and mice together: The predatory nature of the rat's mouse-killing response. *Psychological Bulletin, 81*(4), 261–269. doi:10.1037/h0036175
O'Boyle, M. (1975). The rat as a predator. *Psychological Bulletin, 82* (3), 460–462. doi:10.1037/0033-2909.82.3.460
Ogawa, A., Streit, A., Antebi, A., & Sommer, R.J. (2009). A conserved endocrine mechanism controls the formation of dauer and infective larvae in nematodes. *Current Biology, 19*(1), 67–71. doi:10.1016/j.cub.2008.11.063
Okumura, M., Wilecki, M., & Sommer, R.J. (2017). Serotonin drives predatory feeding behavior via synchronous feeding rhythms in the nematode *Pristionchus pacificus. G3, 7*(11), 3745–3755. doi:10.1534/g3.117.300263
Olivier, B., & Young, L.J. (2002). Animal models of aggression. In K.L. Davis, D. Charney, J.T. Coyle, & C. Nemeroff, *Neuropsychopharmacology: The Fifth Generation of Progress* (pp. 1699–1706). Philadelphia, PA: Lippincott.
Paul, L. (1972). Predatory attack by rats: Its relationship to feeding and type of prey. *Journal of Comparative and Physiological Psychology, 78*(1), 69–76. doi:10.1037/h0032187
Paul, L., & Posner, I. (1973). Predation and feeding: Comparisons of feeding behavior of killer and nonkiller rats. *Journal of Comparative and Physiological Psychology, 84*(2), 258–264. doi:10.1037/h0035321
Paul, L., Miley, W.M., & Baenninger, R. (1971). Mouse killing by rats: Roles of hunger and thirst in its initiation and maintenance. *Journal of Comparative and Physiological Psychology, 76*(2), 242–249. doi:10.1037/h0031394
Peiman, K., & Robinson, B. (2010). Ecology and evolution of resource-related heterospecific aggression. *The Quarterly Review of Biology, 85*(2), 133–158. doi:10.1086/652374
Polis, G.A., Myers, C.A., & Holt, R.D. (1989). The ecology and evolution of intraguild predation: Potential competitors that eat each other. *Annual Review of Ecology and Systematics, 20*(1), 297–330. doi:10.1146/annurev.es.20.110189.001501
Polis, G. A., & Holt, R. D. (1992). Intraguild predation: the dynamics of complex trophic interactions. Trends in Ecology & Evolution, 7(5), 151–154.
Polsky, R.H. (1975). Hunger, prey feeding, and predatory aggression. *Behavioral Biology, 13*(1), 81–93. doi:10.1016/S0091-6773(75)90823-8
Pringle, R. M., Kartzinel, T. R., Palmer, T. M., Thurman, T. J., Fox-Dobbs, K., Xu, C. C., … & Gotanda, K. M. (2019). Predator-induced collapse of niche structure and species coexistence. Nature, 570(7759), 58–64. doi: 10.1038/s41586-019-1264-6.
Rae, R., Riebesell, M., Dinkelacker, I., Wang, Q., Herrmann, M., Weller, A.M., … Sommer, R.J. (2008). Isolation of naturally associated bacteria of necromenic *Pristionchus nematodes* and fitness consequences. *The Journal of Experimental Biology, 211*(12), 1927–1936. doi:10.1242/jeb.014944
Ragsdale, E.J., Kanzaki, N., & Herrmann, M. (2015). Taxonomy and natural history: the genus *Pristionchus.* In R.J. Sommer (Ed.), *Pristionchus pacificus: a nematode model for comparative and evolutionary biology* (pp. 77–120). Leiden, The Netherlands: Brill.
Ragsdale, E.J., Müller, M.R., Rödelsperger, C., & Sommer, R.J. (2013). A developmental switch coupled to the evolution of plasticity acts through a sulfatase. *Cell, 155*(4), 922–933. doi:10.1016/j.cell.2013.09.054
Rasa, O.A.E. (1973). Prey capture, feeding techniques, and their ontogeny in the African dwarf mongoose, *Helogale undulata rufula. Zeitschrift Fur Tierpsychologie, 32*(5), 449–488. doi:10.1111/j.1439-0310.1973.tb01117.x
Roger, A.J. (1999). Reconstructing early events in eukaryotic evolution. *The American Naturalist, 154*(S4), S146–S163. doi:10.1086/303290
Ruscoe, W.A., & Murphy, E.C. (2005). House mouse. In C.M. King (Ed.), *The handbook of New Zealand mammals* (pp. 204–221). Oxford, UK: Oxford University Press.
Samuel, B.S., Rowedder, H., Braendle, C., Félix, M.A., & Ruvkun, G. (2016). *Caenorhabditis elegans* responses to bacteria from its natural habitats. *Proceedings of the National Academy of Sciences, 113*(27), E3941–E3949. doi:10.1073/pnas.1607183113
Sawin, E.R., Ranganathan, R., & Horvitz, H.R. (2000). C. elegans locomotory rate is modulated by the environment through a dopaminergic pathway and by experience through a serotonergic pathway. *Neuron, 26*(3), 619–631. doi:10.1016/S0896-6273(00)81199-X
Schaller, G.B. (2009). *The Serengeti lion: a study of predator-prey relations.* Chicago, IL: University of Chicago Press.
Schlager, B., Wang, X., Braach, G., & Sommer, R.J. (2009). Molecular cloning of a dominant roller mutant and establishment of DNA-mediated transformation in the nematode *Pristionchus pacificus. Genesis, 47*(5), 300–304. doi:10.1002/dvg.20499
Schulenburg, H., & Félix, M.A. (2017). The natural biotic environment of *Caenorhabditis elegans. Genetics, 206*(1), 55–86. doi:10.1534/genetics.116.195511
Serobyan, V., Ragsdale, E.J., & Sommer, R.J. (2014). Adaptive value of a predatory mouth-form in a dimorphic nematode. *Proceedings of the Royal Society. Biological Sciences, 281*(1791), 20141334. doi:10.1098/rspb.2014.1334
Sheth, R.U., Cabral, V., Chen, S.P., & Wang, H.H. (2016). Manipulating bacterial communities by in situ microbiome engineering. *Trends in Genetics, 32*(4), 189–200. doi:10.1016/j.tig.2016.01.005

Shtonda, B.B., & Avery, L. (2006). Dietary choice behavior in *Caenorhabditis elegans. The Journal of Experimental Biology, 209*(1), 89–102. doi:10.1242/jeb.01955

Siegel, A., & Brutus, M. (1990). Neural substrates of aggression and rage in the cat. *Progress in Psychobiology and Physiological Psychology, 14,* 135–233.

Siegel, A., & Pott, C.B. (1988). Neural substrates of aggression and flight in the cat. *Progress in Neurobiology, 31*(4), 261–283. doi:10.1016/0301-0082(88)90015-9

Siegel, A., & Shaikh, M.B. (1997). The neural bases of aggression and rage in the cat. *Aggression and Violent Behavior, 2*(3), 241–271. doi:10.1016/S1359-1789(96)00010-9

Sih, A. (1980). Optimal foraging: Partial consumption of prey. *The American Naturalist, 116*(2), 281–290. doi:10.1086/283626

Simberloff, D., & Dayan, T. (1991). The guild concept and the structure of ecological communities. *Annual Review of Ecology and Systematics, 22*(1), 115–143. doi:10.1146/annurev.es.22.110191.000555

Simpson, E.H., & Balsam, P.D. (Eds.). (2016). *Behavioral neuroscience of motivation* (pp. 1–12). New York, NY: Springer.

Solheim, R. (1984). Caching behaviour, prey choice and surplus killing by pygmy owls *Glaucidium passerinum* during winter, a functional response of a generalist predator. *Annales Zoologici Fennici, 21*(3), 301–308.

Sommer, R.J. (Ed). (2015). *Pristionchus pacificus: A nematode model for comparative and evolutionary biology.* Leiden, The Netherlands: Brill.

Sommer, R.J., Carta, L.K., Kim, S-y., & Sternberg, P.W. (1996). Morphological, genetic and molecular description of *Pristionchus pacificus* sp. n.(Nematoda: Neodiplogasteridae). *Fundamental and Applied Nematology, 19,* 511–522.

Sommers, P., & Chesson, P. (2019). Effects of predator avoidance behavior on the coexistence of competing prey. The American Naturalist, 193(5), E132–E148. doi:10.1086/701780

Srinivasan, J., Durak, O., & Sternberg, P.W. (2008). Evolution of a polymodal sensory response network. *BMC Biology, 6*(1), 52. doi:10.1186/1741-7007-6-52

Stapp, P. (1997). Community structure of shortgrass-prairie rodents: Competition or risk of intraguild predation? *Ecology, 78*(5), 1519–1530.

Sunde, P., Overskaug, K., & Kvam, T. (1999). Intraguild predation of lynxes on foxes: Evidence of interference competition? *Ecography, 22*(5), 521–523. doi:10.1111/j.1600-0587.1999.tb01281.x

Taylor, R.J. (1984). *Predation.* London, UK: Chapman and Hall.

Tilman, D. (1982). *Resource competition and community structure.* Princeton, NJ: Princeton University Press.

Tynkkynen, K., Rantala, M.J., & Suhonen, J. (2004). Interspecific aggression and character displacement in the damselfly *Calopteryx splendens. Journal of Evolutionary Biology, 7*(4), 759–767.

Vance, R.R. (1984). Interference competition and the coexistence of two competitors on a single limiting resource. *Ecology, 65*(5), 1349–1357. doi:10.2307/1939115

von Lieven, A.F., & Sudhaus, W. (2000). Comparative and functional morphology of the buccal cavity of Diplogastrina (Nematoda) and a first outline of the phylogeny of this taxon. *Journal of Zoological Systematics and Evolutionary Research, 38*(1), 37–63. doi:10.1046/j.1439-0469.2000.381125.x

von Lieven, A.F. (2005). The embryonic moult in diplogastrids (Nematoda)–homology of developmental stages and heterochrony as a prerequisite for morphological diversity. *Zoologischer Anzeiger-A Journal of Comparative Zoology, 244*(1), 79–91.

Wasman, M., & Flynn, J.P. (1962). Directed attack elicited from hypothalamus. *Archives of Neurology, 6*(3), 220–227. doi:10.1001/archneur.1962.00450210048005

Wilecki, M., Lightfoot, J.W., Susoy, V., & Sommer, R.J. (2015). Predatory feeding behaviour in *Pristionchus nematodes* is dependent on phenotypic plasticity and induced by serotonin. *The Journal of Experimental Biology, 218*(Pt 9), 1306–1313. doi:10.1242/jeb.118620

Witte, H., Moreno, E., Rödelsperger, C., Kim, J., Kim, J. S., Streit, A., & Sommer, R. J. (2015). Gene inactivation using the CRISPR/Cas9 system in the nematode Pristionchus pacificus. Development Genes and Evolution, *225*(1), 55–62. doi:10.1007/s00427-014-0486-8

Zenner, A.N., O'Callaghan, K.M., & Griffin, C.T. (2014). Lethal fighting in nematodes is dependent on developmental pathway: Male-male fighting in the entomopathogenic nematode Steinernema longicaudum. *PLoS One, 9*(2), e89385. doi:10.1371/journal.pone.0089385

Zimmermann, B., Sand, H., Wabakken, P., Liberg, O., & Andreassen, H.P. (2015). Predator-dependent functional response in wolves: From food limitation to surplus killing . *The Journal of Animal Ecology, 84*(1), 102–112. doi:10.1111/1365-2656.12280

Zong, C., Wauters, L.A., Rong, K., Martinoli, A., Preatoni, D., & Tosi, G. (2012). Nutcrackers become choosy seed harvesters in a mast-crop year. *Ethology Ecology & Evolution, 24*(1), 54–61.

OPEN ACCESS

Plasticity of pheromone-mediated avoidance behavior in *C. elegans*

YongJin Cheon, Hyeonjeong Hwang and Kyuhyung Kim

ABSTRACT

Caenorhabditis elegans secretes a complex cocktail of small chemicals collectively called ascaroside pheromones which serves as a chemical language for intra-species communication. Subsets of ascarosides have been shown to mediate a broad spectrum of *C. elegans* behavior and development, such as gender-specific attraction, repulsion, aggregation, olfactory plasticity, and dauer formation. Recent studies show that specific components of ascarosides elicit a rapid avoidance response that allows animals to avoid predators and escape from unfavorable conditions. Moreover, this avoidance behavior is modulated by external conditions, internal states, and previous experience, indicating that pheromone avoidance behavior is highly plastic. In this review, we describe molecular and circuit mechanisms underlying plasticity in pheromone avoidance behavior which pave a way to better understanding circuit mechanisms underlying behavioral plasticity in higher animals, including humans.

Introduction

Animals consistently modulate their behavior in response to the ever-changing environmental conditions and internal states. Thus, a major goal in neuroscience is to understand how animals sense environmental signals and internal metabolic status and how neuronal circuits process and translate this information into the appropriate behavioral programs. The correct sensation, integration, and processing of the external and internal cues is critical for the maintenance of physiology and behavior, and dysregulation of these pathways leads to multiple psychiatric and/or other neurological disorders. However, it is tremendously challenging to identify the underlying neuronal circuits and molecular and cellular mechanisms in higher animals. The nervous system of these animals is complex, and it is difficult to correlate specific behavioral changes with disruption of function of particular neurons and molecules.

The nematode *C. elegans* provides an excellent model system in which to explore the neural and molecular basis of environmentally influenced behavior. *C. elegans* has a relatively simple nervous system with only 302 neurons and over 7,000 synapses in hermaphrodites and 385 neurons and over 8,000 synapses in males, and their wiring diagrams have been completely reconstructed (Bhattacharya, Aghayeva, Berghoff, & Hobert, 2019; Cook *et al.*, 2019; Hall & Russell, 1991; Jarrell *et al.*, 2012; White, Southgate, Thomson, & Brenner, 1986). However, *C. elegans* displays a broad spectrum of behaviors, such as locomotion, chemosensation, nociception, foraging, and feeding. In addition, *C. elegans* exhibits more complex behaviors, including social and sleep-like behaviors as well as learning and memory [See review (Ardiel & Rankin, 2010; Byrne *et al.*, 2019; Rengarajan & Hallem, 2016)]. Moreover, most, if not all, of these behaviors are plastic and can be modulated by external and/or internal conditions [See review (Ardiel & Rankin, 2008; Garcia & Portman, 2016; Hobert, 2003; Sasakura & Mori, 2013; Sengupta, 2013)]. Here, we review recent findings regarding particular chemosensory behaviors, avoidance or attraction elicited by pheromone cues, as well as discuss their modulation by circuit state, sex, previous experience and the feeding state in *C. elegans*.

Ascaroside pheromones

Pheromones are blends of released chemicals that play major roles in intraspecies chemical communication (Karlson & Luscher, 1959). *C. elegans* secretes a complex cocktail of small chemicals that are collectively called ascaroside pheromones; these affect many aspects of *C. elegans* biology [See review (Butcher, 2019; Edison, 2009; Ludewig & Schroeder, 2013; McGrath & Ruvinsky, 2019; J. Park, Choi, Dar, Butcher, & Kim, 2019; Schroeder, 2015)]. Since *C. elegans* ascaroside pheromones were discovered as a dauer-inducing metabolite in 1982 (Golden & Riddle, 1982), the chemical components of ascaroside pheromones have been identified as hundreds of structurally related compounds (Artyukhin *et al.*, 2013; Butcher, Fujita, Schroeder, & Clardy, 2007;

Butcher, Ragains, Kim, & Clardy, 2008; Butcher *et al.*, 2009; Jeong *et al.*, 2005; Pungaliya *et al.*, 2009; Srinivasan *et al.*, 2008; Srinivasan *et al.*, 2012).

As a dauer-inducing cue, ascaroside pheromones act as a population density indicator to regulate growth in the early larval developmental stage: high pheromone concentrations cause animals to enter into an alternative non-aging, stress-resistant, and developmentally arrested dauer larval stage (Butcher *et al.*, 2007; Butcher *et al.*, 2009; Golden & Riddle, 1982, 1984; Jeong *et al.*, 2005; Pungaliya *et al.*, 2009; Srinivasan *et al.*, 2008) (Table 1). Subsequently, over a dozen research groups have shown that ascaroside pheromones play additional roles in several behaviors, including gender-specific attraction and repulsion, aggregation, and foraging (Artyukhin *et al.*, 2013; Chute *et al.*, 2019; Fagan *et al.*, 2018; Greene, Dobosiewicz, Butcher, McGrath, & Bargmann,

Table 1 Structure and function of ascaroside pheromones in *C. elegans*.

Name	Chemical structure	Function						Reference
		Avoidance	Dauer formation	Male attraction	Hermaphrodite attraction	Foraging	Aggregation	
ascr#1 (daumone-1, C7)	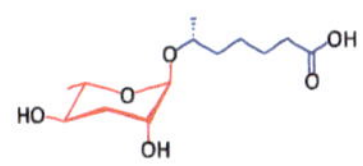	O	O					1, 12, 17
ascr#2 (daumone-2, asc-C6-MK, C6)	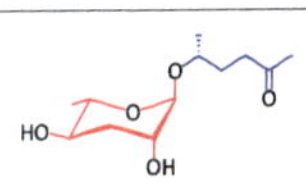	O	O	O		O		3, 7, 10, 11, 12, 14, 17, 21
ascr#3 (daumone-3, asc-ΔC9, C9)	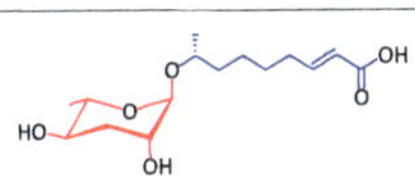	O	O	O		O		3, 7, 10, 11, 12, 15, 16, 17, 18, 19, 21
ascr#4	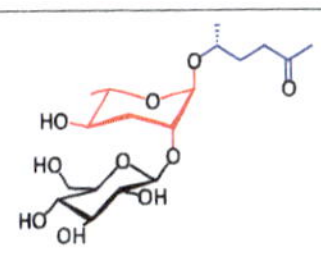		O	O				7
ascr#5 (asc ωC3, C3)	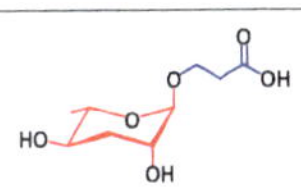	O	O					4, 11, 13, 21
ascr#6.1	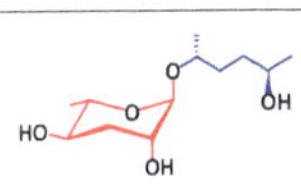		O					6
ascr#8	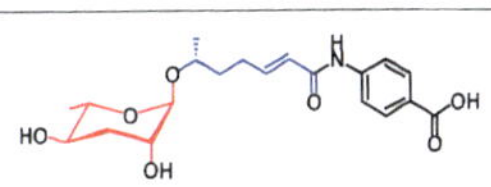		O	O		O		6, 10, 15
icas#3	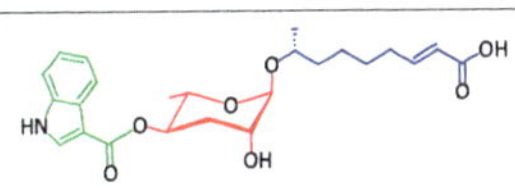		O	O		O		8
icas#9 (C5)	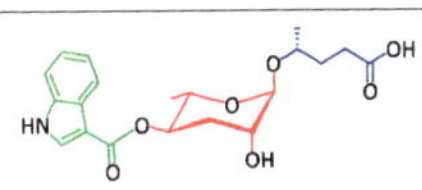			O			O	5, 8, 10, 22
hbas#3	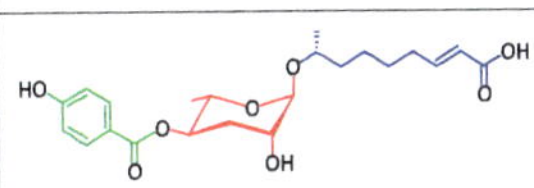		O	O		O	O	2
mbas#3	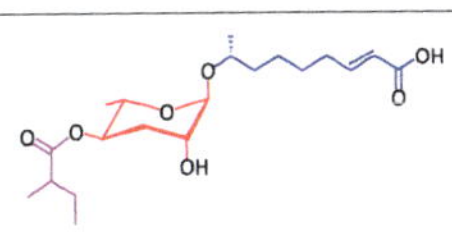				O			2, 20
osas#9	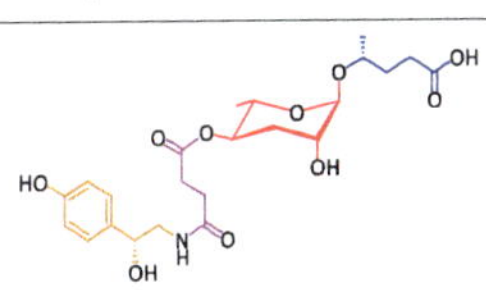	O						9

1. Jeong *et al.*, 2005; 2. von Reuss *et al.*, 2012; 3. Butcher *et al.*, 2007; 4. Butcher *et al.*, 2008; 5. Butcher *et al.*, 2009; 6. Pungaliya *et al.*, 2009; 7. Srinivasan *et al.*, 2008; 8. Srinivasan *et al.*, 2012; 9. Artyukhin *et al.*, 2013; 10. Greene *et al.*, 2016; 11. Macosko *et al.*, 2009; 12. Kim K *et al.*, 2009; 13. McGrath *et al.*, 2011; 14. Park *et al.*, 2012; 15. Narayan *et al.*, 2016; 16. Fagan *et al.*, 2018; 17. Park *et al.*, 2017; 18. Hong *et al.*, 2017; 19. Ryu *et al.*, 2018; 20. Zhang *et al.*,2017; 21. Fenk *et al.*, 2017; 22. Greene *et al.*, 2016.

2016; Jang *et al.*, 2012; Macosko *et al.*, 2009; Narayan *et al.*, 2016; D. Park *et al.*, 2017; Pungaliya *et al.*, 2009; Srinivasan *et al.*, 2008; Srinivasan *et al.*, 2012; von Reuss *et al.*, 2012; Zhang, Sanchez-Ayala, Sternberg, Srinivasan, & Schroeder, 2017) (Table 1). Moreover, complexity of ascaroside signaling further modulates these biological processes. Thus, *C. elegans* ascaroside pheromones elicit long-term changes in development and physiology as well as short-term behavioral changes, making it an outstanding model organism to explore the mechanisms by which biologically relevant cues are sensed and integrated to direct critically important developmental and behavioral programs.

Ascr#3 avoidance behavior and its plasticity

A few pheromones that elicit chemosensory responses have been well-characterized in *C. elegans*. Srinivasan *et al.* showed that adult wild-type males were attractive to a blend of femto- or pico-molar concentrations of three ascarosides, namely ascr#2 (asc-C6-MK, C6), ascr#3(asc-ΔC9, C9), and ascr#4, whereas adult wild-type hermaphrodites avoided higher concentrations of ascr#2 or ascr#3 (Srinivasan *et al.*, 2008). This group also showed that in males, ascr#3 appeared to be sensed by two chemosensory neuron-types, the ASK amphid neurons and the male-specific CEM neurons (Srinivasan *et al.*, 2008). Macosko *et al.* further showed that hermaphrodites from social feeding strains with low *npr-1* neuropeptide receptor gene acitivity (215-valine) or *npr-1* loss-of-function mutant (*ad609lf*) strains exhibited attraction behavior to a 10 nM mixture of ascr#2, ascr#3, and ascr#5 (asc ωC3, C3) but not to any one of these pheromone components individually (Macosko *et al.*, 2009). However, wild-type solitary feeding hermaphrodites with high *npr-1* activity (215-phenylalanine) avoided the ascaroside mixture (Macosko *et al.*, 2009). The ASK pheromone-sensing neurons elicited acute Ca^{2+} transient upon exposure to this pheromone mixture: pheromone-evoked Ca^{2+} response was much stronger in *npr-1(lf)* mutant animals than in wild-type animals (Macosko *et al.*, 2009). This enhanced Ca^{2+} response in the ASK neurons upon pheromone exposure may contribute to attraction behavior toward the pheromone in *npr-1(lf)* mutant animals. These two studies indicate that pheromone responses in *C. elegans* vary depending on sex and a neuropeptide circuit that controls solitary or social feeding behavior.

Jang *et al.* further dissected molecular and circuit mechanisms by which worms modulate pheromone-mediated chemosensory behaviors depending on circuit state and sex (Jang *et al.*, 2012). Using the pheromone drop test that detects acute avoidance responses upon pheromone exposure (Hilliard, Bargmann, & Bazzicalupo, 2002; Jang & Bargmann, 2013), the group showed that adult wild-type hermaphrodites specifically avoided nano-molar concentrations of ascr#3 but not of ascr#2 or ascr#5 (Jang *et al.*, 2012). Calcium imaging, genetic analysis, behavioral analysis, and tools to manipulate neuronal activity supported that the ADL amphid chemosensory neurons sense ascr#3 and mediate ascr#3 avoidance in adult wild-type hermaphrodites (Jang *et al.*, 2012). Subsequently, the ADL ascr#3 Ca^{2+} activities were significantly decreased in *npr-1(lf)* hermaphrodites and wild-type males, indicating that reduced ADL ascr#3 responses are required for ascr#3 attraction behavior in *npr-1(lf)* hermaphrodites and wild-type males (Jang *et al.*, 2012). Strikingly, the ADL neurons drove ascr#3 avoidance in wild-type hermaphrodites via ADL chemical synapses likely onto the AVA and AVD command interneurons for backward locomotion and drove attraction through gap junctions via the RMG neurons in *npr-1(lf)* hermaphrodites (Jang *et al.*, 2012). In males, the ADL neurons exhibited decreased ascr#3 responses which were independent of *npr-1* activities and were possibly due to differences in their own sexual state and/or inputs from other sexually dimorphic neurons (Cook *et al.*, 2019; Jang *et al.*, 2012). These results suggest that the ADL neurons are capable of contributing to either attraction to or avoidance from ascr#3 and that their sensitivity to ascr#3 is modified in worms based on the function of the *npr-1* gene and animal's sex (Figure 1(A,B)). Furthermore, since the *npr-1* genotype seems to reflect a stress-related behavioral state (de Bono, Tobin, Davis, Avery, & Bargmann, 2002; Rogers, Persson, Cheung, & de Bono, 2006), ascr#3 avoidance behavior could be modulated by stressful conditions. More recently, Fagan et al showed that the ADF neurons of males sensed and promoted attraction to ascr#3, while ADF-ablated hermaphrodites were repelled to ascr#3. In this sexual dimorphic process, the sexual regulator *tra-1* and its direct target *mab-3* cell-autonomously regulated the male state of ADF to promote ascr#3 attraction and the hermaphrodite state of ADF to permit ascr#3 repulsion (Fagan *et al.*, 2018).

The ascr#3 avoidance behavior is further modulated by the feeding state. Ryu *et al.* found that starved animals showed increased ascr#3 avoidance behavior compared to well-fed animals and that this effect was dependent on the *daf-2* insulin-like receptor gene (Ryu *et al.*, 2018). Although *daf-2* mutant animals exhibited normal ADL Ca^{2+} activities upon ascr#3 exposure, they exhibited reduced synaptic transmission from the ADL ascr#3 sensing neurons to first-order AVA command interneurons (Ryu *et al.*, 2018). The group further showed that *daf-2* acted through the canonical IGF pathway in the ADL neurons to downregulate the *snb-1* presynaptic protein (Ryu *et al.*, 2018). Remarkably, an intestinal insulin-like peptide, INS-18, acted upstream of DAF-2 in an inhibitory manner and modulated ascr#3 avoidance, and the INS-18 release was reduced in starved worms (Ryu *et al.*, 2018). This study significantly advances our understanding of how the feeding state can fundamentally change the behavioral output via hormonal signaling (Figure 1(C)).

Hong *et al.* investigated whether ascr#3 avoidance behavior is modulated by previous experience. The group demonstrated that in hermaphrodites, exposure to the ascr#3 pheromone during the critical period of the L1 larval stage led to enhanced ascr#3-specific avoidance behavior as an adult (Hong *et al.*, 2017). Interestingly, this increased ascr#3 avoidance was not attributed to increase in the synaptic transmission via ADL–AVA synapses but to the recruitment of the SMB motor neurons into the circuit activated by the

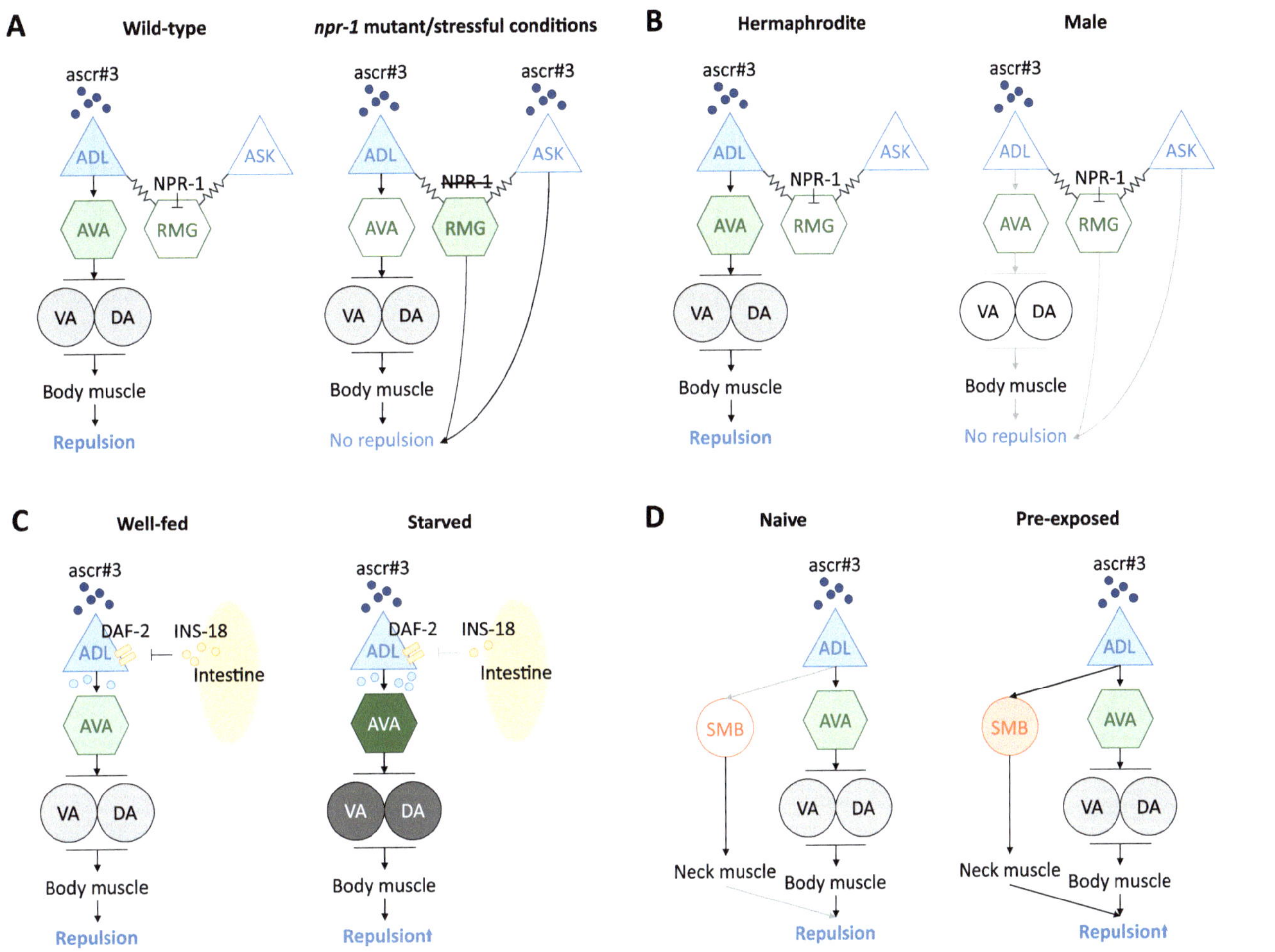

Figure 1. Models for behavioral plasticity of ascr#3 avoidance. (A) In *npr-1(lf)* hermaphrodite, ASK and RMG antagonize ADL chemical synapses and decrease ascr#3 avoidance. (B) In wild-type male, ADL ascr#3 response is decreased due to sexual dimorphism. ASK and RMG circuit antagonizes ADL output to further reduce ascr#3 avoidance. (C) In starved conditions, secretion of INS-18 from the intestine is decreased, which activates the DAF-2 signaling of ADL, resulting in increase in synaptic release from ADL to downstream neurons, and promotes enhanced ascr#3 avoidance. (D) In pre-exposed condition, ADL-SMB synaptic activities that are inactive in naïve animals is altered to promote enhanced ascr#3 avoidance.

ascr#3-sensing ADL neurons (Hong *et al.*, 2017). The SMB motor neurons, which are one of over twenty post-synaptic neurons (Cook *et al.*, 2019; White *et al.*, 1986), exhibited Ca^{2+} activities upon ascr#3 exposure only in ascr#3 pre-exposed animals (Hong *et al.*, 2017). These results indicated that ascr#3-experienced animals form a long-lasting memory or imprint for ascr#3 experience (Figure 1(D)). Thus, this work provides fundamental circuit mechanisms underlying how individual synapses are functionally or anatomically altered upon sensory imprinting.

Avoidance behavior to other ascarosides

In addition to ascr#3, osas#9 ascaroside was characterized as a single pheromone component which causes avoidance behavior in wild-type hermaphrodites (Artyukhin *et al.*, 2013; Chute *et al.*, 2019). osas#9 is an interesting ascaroside since its side chain is derived from the neurotransmitter octopamine and it is produced mainly by starved young L1 larvae (Artyukhin *et al.*, 2013). Chute *et al.* found that starved but not well-fed, young adult hermaphrodites avoided micro-molar concentrations of osas#9 (Chute *et al.*, 2019). This osas#9 avoidance was mediated by not the ADL neurons but the nociceptive ASH neurons via the osas#9-sensing *tyra-2* tyramine/octopamine G protein-coupled receptor (GPCR) (Chute *et al.*, 2019). Interestingly, the group found that *trya-2* expression was increased in starved worms, suggesting that changes in the expression level of the osas#9-chemoreceptor gene depending on the feeding state are responsible for starvation-dependent osas#9 avoidance. The ways in which the feeding state regulates the expression of the *trya-2* gene remain to be explored. Moreover, since the osas#9-sensing ASH neurons also participate in the RMG gap junction circuit (Cook *et al.*, 2019; White *et al.*, 1986), synergistic avoidance with ascr#3 and osas#9 needs to be examined in starved animals, which will provide further molecular and circuit mechanisms underlying starvation-dependent ascaroside avoidance behavior.

Park *et al.* found that micromolar concentrations of ascr#1 (daumone, C7) elicited avoidance behavior in young adult hermaphrodites (D. Park *et al.*, 2017). The group also suggested that the ascr#1 avoidance behavior is mediated by the *daf-16*/*FoxO* gene and glutamatergic transmission via the

AWB chemosensory neurons but not either by ADL or ASH neurons (D. Park *et al.*, 2017). However, roles of the AWB neurons in ascr#1 avoidance need to be further characterized for better understanding the ascr#1 avoidance behavior.

Avoidance behavior to other pheromones

Worms appear to avoid other putative pheromone components. Zhou *et al.* found that the internal fluid from injured worms elicited repulsive behavior in adult hermaphrodites via the ASI and ASK chemosensory neurons (Zhou *et al.*, 2017). The group also found that this internal fluid did not affect the lifespan of worms, suggesting that this fluid acts as not a harmful repellent chemical but as an alarm pheromone (Zhou *et al.*, 2017). This potential pheromone was none of the known ascarosides but comprised at least three nonvolatile components (Zhou *et al.*, 2017). Notably, this avoidance was decreased in starved animals, indicating that this behavior is also dependent on the feeding state (Zhou *et al.*, 2017). The chemical identity of these pheromones needs to be examined to further unravel the molecular and circuit mechanisms underlying this distinct pheromone avoidance behavior.

Pristionchus pacificus is a necromantic insect-dwelling nematode and a facultative predator of *C. elegans* (Serobyan, Ragsdale, & Sommer, 2014). It was reported that *P. pacificus* secreted a few ascarosides, indicating that *C. elegans* avoids this predator nematode (Choe *et al.*, 2012). Interestingly, Liu *et al.*, found that starved, but not well-fed, *P. pacificus* secreted additional non-volatile chemicals which elicited a strong avoidance behavior in *C. elegans* (Liu *et al.*, 2018). They identified these chemical signals as a mixture of sulfolipids and found them to be similar to a known *C. elegans* repellent, sodium dodecyl sulfate (SDS) (Liu *et al.*, 2018). These sulfolipids appeared to be detected redundantly by multiple chemosensory neurons, including the ADL, ASH, ASI, and ASJ neurons (Liu *et al.*, 2018). Taken together, these results suggest that *C. elegans* ensures rapid and life-saving avoidance behavior against the predator by detecting predator-secreted ascaroside as a pheromone as well as sulfolipid as a kairomone (Brown, Eisner, & Whittaker, 1970; Stowe, Turlings, Loughrin, Lewis, & Tumlinson, 1995).

Conclusions and future outlook

Connectomics, the description of comprehensive maps of anatomical connections between every neuron in an organism, is a rapidly evolving field in neuroscience. These large-scale wiring diagrams are expected to offer a unique opportunity to understand not only the structural details of neural connectivity but also neural functions. However, previous studies demonstrated that the wiring diagrams are insufficient to decipher circuit and neural mechanisms of behavior. Studies on the plasticity of *C. elegans* pheromone avoidance further support that behavior cannot be predicted by anatomy alone and eventually requires delicate physiological and behavioral analysis at high resolutions. Studies on *C. elegans*' ascr#3 pheromone avoidance behavior and its functional circuits have provided plentiful information on anatomical and functional logics of the circuit mechanisms including the ASK–RMG–ADL "hub and spoke" and "push and pull" circuit models which are conserved in higher organisms (Bargmann & Marder, 2013). It is intriguing to examine whether ascr#3 avoidance behavior is further modulated by temperature, mating experience, maternal age, and other internal or external conditions (Parida, Neogi, & Padmanabhan, 2014; Perez, Francesconi, Hidalgo-Carcedo, & Lehner, 2017; Shi & Murphy, 2014), that could lead to findings of unexpected but crucial circuit mechanisms. Moreover, it will help us more deeply understand how the nervous system is functionally organized to generate behavioral outcomes.

Understanding how individual synapses are functionally or anatomically altered in response to changes in environmental conditions and internal states is the essential first step toward being able to dissect out the molecular and neuronal mechanisms underlying many forms of behavioral plasticity. Given that the structures and functions of neural circuits are evolutionarily conserved, these works will lead to the development of a general framework for understanding how circuits are modulated in higher animals, including humans, and in disease states (Aitman *et al.*, 2011; Aryal & Lee, 2019; Wangler *et al.*, 2017). Moreover, these works will not only pave the way to identify molecular mechanisms underlying behavioral plasticity but also provide a useful platform to develop tools to modulate behaviors.

Disclosure statement

No potential conflict of interest was reported by the author(s).

Funding

This study was supported by the National Research Foundation of Korea [NRF-2020R1A4A1019436, NRF-2018R1A2A3074987].

ORCID

YongJin Cheon http://orcid.org/0000-0003-4033-2918
Hyeonjeong Hwang http://orcid.org/0000-0002-3327-6199
Kyuhyung Kim http://orcid.org/0000-0002-9943-5092

References

Aitman, T.J., Boone, C., Churchill, G.A., Hengartner, M.O., Mackay, T.F., & Stemple, D.L. (2011). The future of model organisms in human disease research. *Nature Reviews. Genetics*, *12*(8), 575–582. doi:10.1038/nrg3047

Ardiel, E.L., & Rankin, C.H. (2008). Behavioral plasticity in the *C. elegans* mechanosensory circuit. *Journal of Neurogenetics*, *22*(3), 239–255. doi:10.1080/01677060802298509

Ardiel, E.L., & Rankin, C.H. (2010). An elegant mind: learning and memory in *Caenorhabditis elegans*. *Learning & Memory (Cold Spring Harbor, N.Y.).)*, *17*(4), 191–201. doi:10.1101/lm.960510

Artyukhin, A.B., Yim, J.J., Srinivasan, J., Izrayelit, Y., Bose, N., von Reuss, S.H., … Schroeder, F.C. (2013). Succinylated octopamine ascarosides and a new pathway of biogenic amine metabolism in

Caenorhabditis elegans. *The Journal of Biological Chemistry*, *288*(26), 18778–18783. doi:10.1074/jbc.C113.477000

Aryal, B., & Lee, Y. (2019). Disease model organism for Parkinson disease: Drosophila melanogaster. *BMB Reports*, *52*(4), 250–258. Retrieved from https://www.ncbi.nlm.nih.gov/pubmed/30545438. doi:10.5483/BMBRep.2019.52.4.204

Bargmann, C.I., & Marder, E. (2013). From the connectome to brain function. *Nature Methods*, *10*(6), 483–490. doi:10.1038/nmeth.2451

Bhattacharya, A., Aghayeva, U., Berghoff, E.G., & Hobert, O. (2019). Plasticity of the electrical connectome of *C. elegans*. *Cell*, *176*(5), 1174–1189. e1116. doi:10.1016/j.cell.2018.12.024

Brown, W.L., Jr., Eisner, T., & Whittaker, R.H. (1970). Allomones and kairomones: Transspecific chemical messengers. *BioScience*, *20*(1), 21–21. doi:10.2307/1294753

Butcher, R.A. (2019). Natural products as chemical tools to dissect complex biology in *C. elegans*. *Current Opinion in Chemical Biology*, *50*, 138–144. doi:10.1016/j.cbpa.2019.03.005

Butcher, R.A., Fujita, M., Schroeder, F.C., & Clardy, J. (2007). Small-molecule pheromones that control dauer development in *Caenorhabditis elegans*. *Nature Chemical Biology*, *3*(7), 420–422. doi:10.1038/nchembio.2007.3

Butcher, R.A., Ragains, J.R., Kim, E., & Clardy, J. (2008). A potent dauer pheromone component in *Caenorhabditis elegans* that acts synergistically with other components. *Proceedings of the National Academy of Sciences of the United States of America*, *105*(38), 14288–14292. doi:10.1073/pnas.0806676105

Butcher, R.A., Ragains, J.R., Li, W., Ruvkun, G., Clardy, J., & Mak, H.Y. (2009). Biosynthesis of the *Caenorhabditis elegans* dauer pheromone. *Proceedings of the National Academy of Sciences of the United States of America*, *106*(6), 1875–1879. doi:10.1073/pnas.0810338106

Byrne, J.J., Soh, M.S., Chandhok, G., Vijayaraghavan, T., Teoh, J.-S., Crawford, S., … Neumann, B. (2019). Disruption of mitochondrial dynamics affects behaviour and lifespan in *Caenorhabditis elegans*. *Cellular and Molecular Life Sciences : Cmls*, *76*(10), 1967–1985. doi:10.1007/s00018-019-03024-5

Choe, A., von Reuss, S.H., Kogan, D., Gasser, R.B., Platzer, E.G., Schroeder, F.C., & Sternberg, P.W. (2012). Ascaroside signaling is widely conserved among nematodes. *Current Biology : CB*, *22*(9), 772–780. doi:10.1016/j.cub.2012.03.024

Chute, C.D., DiLoreto, E.M., Zhang, Y.K., Reilly, D.K., Rayes, D., Coyle, V.L., … Srinivasan, J. (2019). Co-option of neurotransmitter signaling for inter-organismal communication in *C. elegans*. *Nature Communications*, *10*(1), 3186. doi:10.1038/s41467-019-11240-7

Cook, S.J., Jarrell, T.A., Brittin, C.A., Wang, Y., Bloniarz, A.E., Yakovlev, M.A., … Emmons, S.W. (2019). Whole-animal connectomes of both *Caenorhabditis elegans* sexes. *Nature*, *571*(7763), 63–71. doi:10.1038/s41586-019-1352-7

de Bono, M., Tobin, D.M., Davis, M.W., Avery, L., & Bargmann, C.I. (2002). Social feeding in *Caenorhabditis elegans* is induced by neurons that detect aversive stimuli. *Nature*, *419*(6910), 899–903. doi:10.1038/nature01169

Edison, A.S. (2009). *Caenorhabditis elegans* pheromones regulate multiple complex behaviors. *Current Opinion in Neurobiology*, *19*(4), 378–388. doi:10.1016/j.conb.2009.07.007

Fagan, K.A., Luo, J.T., Lagoy, R.C., Schroeder, F.C., Albrecht, D.R., & Portman, D.S. (2018). A single-neuron chemosensory switch determines the valence of a sexually dimorphic sensory behavior. *Current Biology: CB*, *28*(6), 902–914.e5. +. doi:10.1016/j.cub.2018.02.029

Fenk, L. A., & de Bono, M.. (2017). Memory of recent oxygen experience switches pheromone valence in Caenorhabditis elegans. *Proceedings of the National Academy of Sciences of the United States of America*, 114(16), 4195–4200. doi:10.1073/pnas.1618934114

Garcia, L.R., & Portman, D.S. (2016). Neural circuits for sexually dimorphic and sexually divergent behaviors in *Caenorhabditis elegans*. *Current Opinion in Neurobiology*, *38*, 46–52. doi:10.1016/j.conb.2016.02.002

Golden, J.W., & Riddle, D.L. (1982). A pheromone influences larval development in the nematode *Caenorhabditis elegans*. *Science (New York, N.Y.).)*, *218*(4572), 578–580. doi:10.1126/science.6896933

Golden, J.W., & Riddle, D.L. (1984). The *Caenorhabditis elegans* dauer larva: Developmental effects of pheromone, food, and temperature. *Developmental Biology*, *102*(2), 368–378. doi:10.1016/0012-1606(84)90201-X

Greene, J.S., Dobosiewicz, M., Butcher, R.A., McGrath, P.T., & Bargmann, C.I. (2016). Regulatory changes in two chemoreceptor genes contribute to a *Caenorhabditis elegans* QTL for foraging behavior. *eLife*, *5*, e21454. doi:10.7554/eLife.21454

Hall, D.H., & Russell, R.L. (1991). The posterior nervous system of the nematode *Caenorhabditis elegans*: Serial reconstruction of identified neurons and complete pattern of synaptic interactions. *The Journal of Neuroscience : The Official Journal of the Society for Neuroscience*, *11*(1), 1–22. Retrieved from https://www.ncbi.nlm.nih.gov/pubmed/1986064. doi:10.1523/JNEUROSCI.11-01-00001.1991

Hilliard, M.A., Bargmann, C.I., & Bazzicalupo, P. (2002). *C. elegans* responds to chemical repellents by integrating sensory inputs from the head and the tail. *Current Biology: CB*, *12*(9), 730–734. doi:10.1016/S0960-9822(02)00813-8

Hobert, O. (2003). Behavioral plasticity in *C. elegans*: paradigms, circuits, genes. *Journal of Neurobiology*, *54*(1), 203–223. doi:10.1002/neu.10168

Hong, M., Ryu, L., Ow, M.C., Kim, J., Je, A.R., Chinta, S., … Kim, K. (2017). Early pheromone experience modifies a synaptic activity to influence adult pheromone responses of *C. elegans*. *Current Biology: CB*, *27*(20), 3168.e3–3177.e3. doi:10.1016/j.cub.2017.08.068

Jang, H., & Bargmann, C.I. (2013). Acute behavioral responses to pheromones in *C. elegans* (adult behaviors: attraction, repulsion). *Methods in Molecular Biology (Clifton, N.J.).)*, *1068*, 285–292. doi:10.1007/978-1-62703-619-1_21

Jang, H., Kim, K., Neal, S.J., Macosko, E., Kim, D., Butcher, R.A., … Sengupta, P. (2012). Neuromodulatory state and sex specify alternative behaviors through antagonistic synaptic pathways in *C. elegans*. *Neuron*, *75*(4), 585–592. doi:10.1016/j.neuron.2012.06.034

Jarrell, T.A., Wang, Y., Bloniarz, A.E., Brittin, C.A., Xu, M., Thomson, J.N., … Emmons, S.W. (2012). The connectome of a decision-making neural network. *Science (New York, N.Y.).)*, *337*(6093), 437–444. doi:10.1126/science.1221762

Jeong, P.-Y., Jung, M., Yim, Y.-H., Kim, H., Park, M., Hong, E., … Paik, Y.-K. (2005). Chemical structure and biological activity of the *Caenorhabditis elegans* dauer-inducing pheromone. *Nature*, *433*(7025), 541–545. doi:10.1038/nature03201

Karlson, P., & Luscher, M. (1959). Pheromones': a new term for a class of biologically active substances. *Nature*, *183*(4653), 55–56. doi:10.1038/183055a0

Liu, Z., Kariya, M.J., Chute, C.D., Pribadi, A.K., Leinwand, S.G., Tong, A., … Chalasani, S.H. (2018). Predator-secreted sulfolipids induce defensive responses in *C. elegans*. *Nature Communications*, *9*(1), 1128. doi:10.1038/s41467-018-03333-6

Ludewig, A.H., & Schroeder, F.C. (2013). Ascaroside signaling in *C. elegans*. *WormBook*, 1–22. doi:10.1895/wormbook.1.155.1

Macosko, E.Z., Pokala, N., Feinberg, E.H., Chalasani, S.H., Butcher, R.A., Clardy, J., & Bargmann, C.I. (2009). A hub-and-spoke circuit drives pheromone attraction and social behaviour in *C. elegans*. *Nature*, *458*(7242), 1171–1175. doi:10.1038/nature07886

McGrath, P.T., & Ruvinsky, I. (2019). A primer on pheromone signaling in *Caenorhabditis elegans* for systems biologists. *Current Opinion in Systems Biology*, *13*, 23–30. doi:10.1016/j.coisb.2018.08.012

McGrath, P. T., Xu, Y., Ailion, M., Garrison, J. L., Butcher, R. A., & Bargmann, C. I. (2011). Parallel evolution of domesticated Caenorhabditis species targets pheromone receptor genes. Nature, 477(7364), 321–325. doi:10.1038/nature10378

Narayan, A., Venkatachalam, V., Durak, O., Reilly, D.K., Bose, N., Schroeder, F.C., … Sternberg, P.W. (2016). Contrasting responses within a single neuron class enable sex-specific attraction in *Caenorhabditis elegans*. *Proceedings of the National Academy of Sciences of the United States of America*, *113*(10), E1392–1401. doi:10.1073/pnas.1600786113

Parida, L., Neogi, S., & Padmanabhan, V. (2014). Effect of temperature pre-exposure on the locomotion and chemotaxis of *C. elegans*. *PLoS One*, *9*(10), e111342. doi:10.1371/journal.pone.0111342

Park, J., Choi, W., Dar, A.R., Butcher, R.A., & Kim, K. (2019). Neuropeptide signaling regulates pheromone-mediated gene expression of a chemoreceptor gene in *C. elegans*. *Molecules and Cells*, *42*(1), 28–35. doi:10.14348/molcells.2018.0380

Park, D., O'Doherty, I., Somvanshi, R. K., Bethke, A., Schroeder, F. C., Kumar, U., & Riddle, D. L. (2012). Interaction of structure-specific and promiscuous G-protein-coupled receptors mediates small-molecule signaling in Caenorhabditis elegans. *Proceedings of the National Academy of Sciences of the United States of America*, *109*(25), 9917–9922. doi:10.1073/pnas.1202216109

Park, D., Hahm, J.-H., Park, S., Ha, G., Chang, G.-E., Jeong, H., ... Paik, Y.-K. (2017). A conserved neuronal DAF-16/FoxO plays an important role in conveying pheromone signals to elicit repulsion behavior in *Caenorhabditis elegans*. *Scientific Reports*, *7*(1), 7260. doi:10.1038/s41598-017-07313-6

Perez, M.F., Francesconi, M., Hidalgo-Carcedo, C., & Lehner, B. (2017). Maternal age generates phenotypic variation in *Caenorhabditis elegans*. *Nature*, *552*(7683), 106–109. doi:10.1038/nature25012

Pungaliya, C., Srinivasan, J., Fox, B.W., Malik, R.U., Ludewig, A.H., Sternberg, P.W., & Schroeder, F.C. (2009). A shortcut to identifying small molecule signals that regulate behavior and development in *Caenorhabditis elegans*. *Proceedings of the National Academy of Sciences of the United States of America*, *106*(19), 7708–7713. doi:10.1073/pnas.0811918106

Rengarajan, S., & Hallem, E.A. (2016). Olfactory circuits and behaviors of nematodes. *Current Opinion in Neurobiology*, *41*, 136–148. doi:10.1016/j.conb.2016.09.002

Rogers, C., Persson, A., Cheung, B., & de Bono, M. (2006). Behavioral motifs and neural pathways coordinating O_2 responses and aggregation in *C. elegans*. *Current Biology: CB*, *16*(7), 649–659. doi:10.1016/j.cub.2006.03.023

Ryu, L., Cheon, Y., Huh, Y.H., Pyo, S., Chinta, S., Choi, H., ... Kim, K. (2018). Feeding state regulates pheromone-mediated avoidance behavior via the insulin signaling pathway in *Caenorhabditis elegans*. *The EMBO Journal*, *37*(15), e98402. doi:10.15252/embj.201798402

Sasakura, H., & Mori, I. (2013). Behavioral plasticity, learning, and memory in *C. elegans*. *Current Opinion in Neurobiology*, *23*(1), 92–99. doi:10.1016/j.conb.2012.09.005

Schroeder, F.C. (2015). Modular assembly of primary metabolic building blocks: a chemical language in *C. elegans*. *Chemistry & Biology*, *22*(1), 7–16. doi:10.1016/j.chembiol.2014.10.012

Sengupta, P. (2013). The belly rules the nose: feeding state-dependent modulation of peripheral chemosensory responses. *Current Opinion in Neurobiology*, *23*(1), 68–75. doi:10.1016/j.conb.2012.08.001

Serobyan, V., Ragsdale, E.J., & Sommer, R.J. (2014). Adaptive value of a predatory mouth-form in a dimorphic nematode. *Proceedings of the Royal Society B: Biological Sciences*, *281*(1791), 20141334. doi:10.1098/rspb.2014.1334

Shi, C., & Murphy, C.T. (2014). Mating induces shrinking and death in Caenorhabditis mothers. *Science (New York, N.Y.).)*, *343*(6170), 536–540. doi:10.1126/science.1242958

Srinivasan, J., Kaplan, F., Ajredini, R., Zachariah, C., Alborn, H.T., Teal, P.E.A., ... Schroeder, F.C. (2008). A blend of small molecules regulates both mating and development in *Caenorhabditis elegans*. *Nature*, *454*(7208), 1115–1118. doi:10.1038/nature07168

Srinivasan, J., von Reuss, S.H., Bose, N., Zaslaver, A., Mahanti, P., Ho, M.C., ... Schroeder, F.C. (2012). A modular library of small molecule signals regulates social behaviors in *Caenorhabditis elegans*. *PLoS Biology*, *10*(1), e1001237. doi:10.1371/journal.pbio.1001237

Stowe, M.K., Turlings, T.C., Loughrin, J.H., Lewis, W.J., & Tumlinson, J.H. (1995). The chemistry of eavesdropping, alarm, and deceit. *Proceedings of the National Academy of Sciences of the United States of America*, *92*(1), 23–28. doi:10.1073/pnas.92.1.23

von Reuss, S.H., Bose, N., Srinivasan, J., Yim, J.J., Judkins, J.C., Sternberg, P.W., & Schroeder, F.C. (2012). Comparative metabolomics reveals biogenesis of ascarosides, a modular library of small-molecule signals in *C. elegans*. *Journal of the American Chemical Society*, *134*(3), 1817–1824. doi:10.1021/ja210202y

Wangler, M.F., Yamamoto, S., Chao, H.-T., Posey, J.E., Westerfield, M., Postlethwait, J., ... Bellen, H.J, Members of the Undiagnosed Diseases Network (UDN). (2017). Model organisms facilitate rare disease diagnosis and therapeutic research. *Genetics*, *207*(1), 9–27. doi:10.1534/genetics.117.203067

White, J.G., Southgate, E., Thomson, J.N., & Brenner, S. (1986). The structure of the nervous system of the nematode *Caenorhabditis elegans*. *Philosophical Transactions of the Royal Society of London. B, Biological Sciences* , *314*(1165), 1–340. doi:10.1098/rstb.1986.0056

Zhang, Y.K., Sanchez-Ayala, M.A., Sternberg, P.W., Srinivasan, J., & Schroeder, F.C. (2017). Improved synthesis for modular ascarosides uncovers biological activity. *Organic Letters*, *19*(11), 2837–2840. doi:10.1021/acs.orglett.7b01009

Zhou, Y., Loeza-Cabrera, M., Liu, Z., Aleman-Meza, B., Nguyen, J.K., Jung, S.-K., ... Zhong, W. (2017). Potential nematode alarm pheromone induces acute avoidance in *Caenorhabditis elegans*. *Genetics*, *206*(3), 1469–1478. doi:10.1534/genetics.116.197293

Part V

Quiescence and sleep

Worms sleep: a perspective

David Raizen

ABSTRACT
I review the history of sleep research in *Caenorhabditis elegans*, briefly introduce the four articles in this issue focused on worm sleep and propose future directions our field might take.

During sleep, we are vulnerable and unproductive. Yet, we spend one-third of our lives asleep. When scientists search for sleep in other animals, including in those distantly related to us, they find it. The neuroscientist Allan Rechtschaffen famously said, 'If sleep doesn't serve an absolutely vital function, it is the greatest mistake evolution ever made' (Rechtschaffen, 1971).

I became interested in worm sleep when, in the 90 s doing thesis work in Leon Avery's lab, I noticed that RC301 males stopped moving and feeding after mating. This behavior was noted by others (LeBoeuf *et al.*, 2014). When I returned to lab after completing clinical training, I initially dabbled in fruit fly sleep research, but ultimately returned to worms due to their faster life cycle and simpler, more transparent nervous system.

Worms stop moving and eating during lethargus, a 2–3 h period at the transition between larval stages (Cassada & Russell, 1975; Singh & Sulston, 1978). Insightfully, Sir John Sulston noted that movement quiescence was broken up by brief bouts of motion (Singh & Sulston, 1978), suggesting that lethargus was a behavioral state. My imagination was captured by a 1999 Rougvie lab paper, showing that expression of *lin-42*, the *C. elegans* ortholog of the circadian clock gene *period*, oscillates during development (Jeon *et al.*, 1999). There was a fixed phase relationship between expression of *lin-42* and worm quiescence, just as there was between expression of *period* and circadian sleep in other animals.

Several worm biologists I chatted with (Meera Sundaram, Anne Hart, Victor Ambros, Piali Sengupta, Young-jai You, and others) had similar ideas and encouraged me to pursue the idea that worms sleep during lethargus. I presented our preliminary behavioral results on worm sleep at the 2005 LA Worm Meeting. They were well-received but with a healthy dose of skepticism. I am grateful for both the encouragement and the critiques. At that same worm meeting, I was asked to chair a parallel slide session on worm behavior. It was my first time chairing a session at a scientific meeting. I think I was more nervous than any of the speakers and had trouble focusing on the talks. But in a moment of scientific serendipity, one of the speakers was Cheryl van Buskirk, then a post-doc in the Sternberg lab. Cheryl showed a video of a worm after over-expression of the gene *lin-3*. There was absolutely no movement in the video! I panicked thinking this was a malfunction of AV equipment, which I had no idea how to fix. Only on returning to Philadelphia and discussing meeting highlights with Meera Sundaram and Gautam Kao did it occur to me that Cheryl's video was in fact working but that the worm was completely quiescent. The field of 'worm sleep' was born.

In the early days, we and other worm sleep groups were asked by editors and reviewers to use the term 'sleep-like' instead of 'sleep'. We pushed back because 'sleep-like' implies it is not really sleep but a state similar to sleep. I also did not like the term because sleep itself can be very different across species. Is the behavior like sleep in brown bats, who are quiescent 18 h each day? Is it like sleep in dolphins, who alternate sleep in the two sides of their brains? Or is it like sleep in fruit flies, whose sleep is strongly circadian?

The sociology of science can be funny. Sleep in flies was first described in 2000 (Hendricks *et al.*, 2000; Shaw *et al.*, 2000), eight years before worm sleep was proposed. Yet, the term 'sleep' did not fully stick in the fly literature *until* worm sleep was described. The collective reaction of the sleep field, which was made up of mostly mammalian sleep researchers, was, '… ok ok, we can now accept that a fruit fly sleeps… but worms? no way!'. Similarly, the term 'worm sleep' seemed to stick only after three Cal. Tech labs collaborated to show that jellyfish sleep (Nath *et al.*, 2017). Maybe it will require a description of sleep in sponges in order to fully accept that jellyfish sleep? The first appearance of sleep on this planet keeps getting pushed back.

There remains some healthy scientific skepticism, but 'worm sleep' is pretty well accepted now. It has also become an important model for sleep in other organisms. This is powerfully illustrated by a paper describing the first murine forward genetic screen for sleep mutants. In that screen, Funato and colleagues identified the salt-induced kinase SIK3 as a sleep regulator (Funato *et al.*, 2016). Remarkably, to make the point of evolutionary conservation of this gene function, they showed that the worm ortholog of SIK3, called KIN-29, regulated worm sleep. (As an aside, a role for *kin-29* in quiescence regulation was actually reported almost a decade earlier, by Young-jai You (van der Linden *et al.*, 2008). She just did not call it 'sleep' back then.)

The focus of the 'worm sleep' field in its first 5–10 years was on developmentally timed sleep (DTS) during lethargus. But beginning with the van Buskirk's lab's identification of sleep in adult animals recovering from exposures such as high heat that cause cellular stress (Hill *et al.*, 2014), the breadth of 'worm sleep' has been expanding. We often refer to this latter type of sleep, which is relevant to sleepiness and fatigue during human illness, as 'stress- or sickness-induced sleep' (SIS). Sleep behavior has also been described during prolonged fasts in both larvae and adults (Skora *et al.*, 2018; Wu *et al.*, 2018), and quiescence has been described in the setting of metabolic satiation (You *et al.*, 2008).

This worm sleep field is still small—for every worm sleep lab, there are probably >50 fly sleep labs. Yet our field has already made tremendous progress, which, at some levels, is at least on par with progress in understanding fly and mouse sleep. This is particularly true for studies of sleep circuitry, where I would argue our understanding in worms exceeds that in other systems. These studies have converged on two second-order interneurons called RIS and ALA, identified by the Bringmann (Turek *et al.*, 2013) and van Buskirk (Hill *et al.*, 2014) labs, respectively. Nichols and colleagues showed that among nearly all neurons, only RIS is strongly activated during lethargus quiescent bouts (Nichols *et al.*, 2017). While there are some differences between the roles of RIS and ALA (Konietzka *et al.*, 2020; Robinson *et al.*, 2019), it is likely that these are the main two neurons executing the sleep state. Both of these neurons are peptidergic: ALA secretes a cocktail of neuropeptides that include FLP-13, FLP-24, and NLP-8 to promote quiescence (Nath *et al.*, 2016; Nelson *et al.*, 2014) whereas RIS secretes FLP-11 (and perhaps also NLP-8) for its quiescent output (Turek *et al.*, 2016). A current challenge in the field is to understand the mechanism of RIS and ALA activation as well as the downstream mechanisms. Several other neurons, as well as non-neural cells, have been implicated in sleep regulation (Choi *et al.*, 2013, 2015; Grubbs *et al.*, 2019; Singh *et al.*, 2011; Skora *et al.*, 2018) in some fashion or another but details regarding how they connect to RIS and/or ALA are still being worked out.

Each of the four worm sleep papers published in this edition of the Journal of Neurogenetics carries an important message. The van Buskirk lab paper (Goetting, *et al*, J. Neurogenet., in press) contributes to our understanding of the mechanism of SIS. They also describe a new trigger for SIS: skin injury, which is relevant to the human complaint of severe fatigue after an operation. The Bringmann lab paper (Busack *et al*, J. Neurogenet., in press) describes a method for long-term optogenetic manipulation of worms. Developing such methods is important because prior worm tools have been optimized for much shorter durations of manipulation and observations. The Nelson lab describes the role of orcokinins, neuropeptides conserved among molting animals, in regulating sleep (Honer *et al*, J. Neurogenet., in press). This study again emphasizes the important, but complex roles of neuropeptides in behavioral state modulation. Finally, the Hart lab paper reminds us that not all that stops moving is sleep and that we must remain self-skeptical as a field. They suggest that cessation of swimming is better explained by neuromuscular fatigue than by sleep (Schuch *et al*, J. Neurogenet., in press).

What does the future hold for worm sleep? Thus far, the field has primarily studied genes and neurons with large effect sizes. This has taken us far and has led to the identification of key neurons and genes. But sleep regulation is complex and there are likely numerous other genes and neurons with smaller, quantitative roles in sleep regulation. Understanding the sleep/wake circuit is a solvable problem in *C. elegans* using currently available optogenetic tools for manipulating and recording physiological activity. A lovely example of such a circuit-interrogation was recently published (Maluck *et al.*, 2020). Finding new genes will require higher throughput forward genetic screens and increased reliance on quantitative measurements of quiescence.

Given the conservation of sleep, we should be able to use the worm to model human sleep disorders. A step in that direction has been reported (Huang *et al.*, 2017). Finally, by comparing what happens during sleep and sleep curtailment in worms to other animals, we can gain insight into the ancient function of sleep and maybe understand why we cannot live without it (Anafi *et al.*, 2019; Bennett *et al.*, 2018; Driver *et al.*, 2013; Fry *et al.*, 2016; Hill *et al.*, 2014; Wu *et al.*, 2018).

Acknowledgements

The author thanks Ron Anafi for editing this perspective.

Disclosure statement

No potential conflict of interest was reported by the author(s).

Funding

This work is supported by the National Institutes of Health [R01NS107969, R01NS088432, and R21CA224267].

ORCID

David Raizen http://orcid.org/0000-0001-5935-0476

References

Anafi, R.C., Kayser, M.S., & Raizen, D.M. (2019). Exploring phylogeny to find the function of sleep. *Nature Reviews. Neuroscience, 20* (2), 109–116. doi:10.1038/s41583-018-0098-9

Bennett, H.L., Khoruzhik, Y., Hayden, D., Huang, H., Sanders, J., Walsh, M.B., Biron, D., Hart, A.C. (2018). Normal sleep bouts are not essential for C. elegans survival and FoxO is important for compensatory changes in sleep. *BMC Neuroscience, 19* (1), 10. doi:10.1186/s12868-018-0408-1

Cassada, R.C., & Russell, R.L. (1975). The dauerlarva, a post-embryonic developmental variant of the nematode Caenorhabditis elegans. *Developmental Biology, 46* (2), 326–342. doi:10.1016/0012-1606(75)90109-8

Choi, S., Chatzigeorgiou, M., Taylor, K.P., Schafer, W.R., & Kaplan, J.M. (2013). Analysis of NPR-1 reveals a circuit mechanism for behavioral quiescence in C. elegans. *Neuron, 78* (5), 869–880. doi:10.1016/j.neuron.2013.04.002

Choi, S., Taylor, K.P., Chatzigeorgiou, M., Hu, Z., Schafer, W.R., & Kaplan, J.M. (2015). Sensory neurons arouse C. elegans locomotion via both glutamate and neuropeptide release. *PLoS Genetics, 11* (7), e1005359 doi:10.1371/journal.pgen.1005359

Driver, R.J., Lamb, A.L., Wyner, A.J., & Raizen, D.M. (2013). DAF-16/FOXO regulates homeostasis of essential sleep-like behavior during larval transitions in C. elegans. Curr. Biol, *23* (6), 501–506. doi:10.1016/j.cub.2013.02.009

Fry, A.L., Laboy, J.T., Huang, H., Hart, A.C., & Norman, K.R. (2016). A Conserved GEF for Rho-Family GTPases Acts in an EGF Signaling Pathway to Promote Sleep-like Quiescence in Caenorhabditis elegans. *Genetics, 202* (3), 1153–1166. doi:10.1534/genetics.115.183038

Funato, H., Miyoshi, C., Fujiyama, T., Kanda, T., Sato, M., Wang, Z., ... Yonezawa, T., (2016). Forward-genetics analysis of sleep in randomly mutagenized mice. *Nature, 539* (7629), 378–383. doi:10.1038/nature20142

Grubbs, J.J., Lopes, L.E., der Linden, AMv., & Raizen, D.M. (2019). A salt-induced kinase (SIK) is required for the metabolic regulation of sleep. bioRxiv, 586701.

Hendricks, J.C., Finn, S.M., Panckeri, K.A., Chavkin, J., Williams, J.A., Sehgal, A., & Pack, A.I. (2000). Rest in Drosophila is a sleep-like state. *Neuron, 25* (1), 129–138. doi:10.1016/S0896-6273(00)80877-6

Hill, A.J., Mansfield, R., Lopez, J.M., Raizen, D.M., & Van Buskirk, C. (2014). Cellular stress induces a protective sleep-like state in C. elegans. *Current Biology : Cb, 24* (20), 2399–2405. doi:10.1016/j.cub.2014.08.040

Huang, H., Zhu, Y., Eliot, M.N., Knopik, V.S., McGeary, J.E., Carskadon, M.A., & Hart, A.C. (2017). Combining human epigenetics and sleep studies in Caenorhabditis elegans: a cross-species approach for finding conserved genes regulating sleep. *Sleep, 40* (6), zsx063. doi:10.1093/sleep/zsx063

Jeon, M., Gardner, H.F., Miller, E.A., Deshler, J., & Rougvie, A.E. (1999). Similarity of the C. elegans developmental timing protein LIN-42 to circadian rhythm proteins. *Science (New York, N.Y.).), 286* (5442), 1141–1146. doi:10.1126/science.286.5442.1141

Konietzka, J., Fritz, M., Spiri, S., McWhirter, R., Leha, A., Palumbos, S., Miller, D.M., 3rd., et al. (2020). Epidermal growth factor signaling promotes sleep through a combined series and parallel neural circuit. *Current Biology : Cb, 30*(1), 1–16 e13. doi:10.1016/j.cub.2019.10.048

LeBoeuf, B., Correa, P., Jee, C., & Garcia, L.R. (2014). Caenorhabditis elegans male sensory-motor neurons and dopaminergic support cells couple ejaculation and post-ejaculatory behaviors. *eLife, 3*, e02938. doi:10.7554/eLife.02938

Maluck, E., Busack, I., Besseling, J., Masurat, F., Turek, M., Busch, K.E., & Bringmann, H. (2020). A wake-active locomotion circuit depolarizes a sleep-active neuron to switch on sleep. *PLoS Biology, 18* (2), e3000361 doi:10.1371/journal.pbio.3000361

Nath, R.D., Bedbrook, C.N., Abrams, M.J., Basinger, T., Bois, J.S., Prober, D.A., ... Goentoro, L. (2017). The jellyfish cassiopea exhibits a sleep-like state. *Current Biology, 27* (19), 2984–2990 e2983. doi:10.1016/j.cub.2017.08.014

Nath, R.D., Chow, E.S., Wang, H., Schwarz, E.M., & Sternberg, P.W. (2016). C. elegans stress-induced sleep emerges from the collective action of multiple neuropeptides. *Current Biology : Cb, 26* (18), 2446–2455. doi:10.1016/j.cub.2016.07.048

Nelson, M.D., Lee, K.H., Churgin, M.A., Hill, A.J., Van Buskirk, C., Fang-Yen, C., & Raizen, D.M. (2014). FMRFamide-like FLP-13 neuropeptides promote quiescence following heat stress in Caenorhabditis elegans. *Current Biology : Cb, 24* (20), 2406–2410. doi:10.1016/j.cub.2014.08.037

Nichols, A.L.A., Eichler, T., Latham, R., & Zimmer, M. (2017). A global brain state underlies C. elegans sleep behavior. *Science, 356*(6344), eaam6851. doi:10.1126/science.aam6851

Rechtschaffen, A. (1971). The Control of Sleep. In W. A. H. ed.*Human Behavior and its Control.*, Shenkman Publishing Company, Inc.

Robinson, B., Goetting, D.L., Cisneros Desir, J., & Van Buskirk, C. (2019). aptf-1 mutants are primarily defective in head movement quiescence during C. elegans sleep.

Shaw, P.J., Cirelli, C., Greenspan, R.J., & Tononi, G. (2000). Correlates of sleep and waking in Drosophila melanogaster. *Science (New York, N.Y.).), 287* (5459), 1834–1837. doi:10.1126/science.287.5459.1834

Singh, K., Chao, M.Y., Somers, G.A., Komatsu, H., Corkins, M.E., Larkins-Ford, J., ... Hart, A.C. (2011). C. elegans Notch signaling regulates adult chemosensory response and larval molting quiescence. *Current Biology : Cb, 21* (10), 825–834. doi:10.1016/j.cub.2011.04.010

Singh, R.N., & Sulston, J.E. (1978). Some observations on moulting in Caenorhabditis Elegans. *Nematologica, 24* (1), 63–71. doi:10.1163/187529278X00074

Skora, S., Mende, F., & Zimmer, M. (2018). Energy scarcity promotes a brain-wide sleep state modulated by insulin signaling in c. elegans. *Cell Reports, 22* (4), 953–966. doi:10.1016/j.celrep.2017.12.091

Turek, M., Besseling, J., Spies, J.P., Konig, S., & Bringmann, H. (2016). Sleep-active neuron specification and sleep induction require FLP-11 neuropeptides to systemically induce sleep. *eLife, 5* doi:10.7554/eLife.12499

Turek, M., Lewandrowski, I., & Bringmann, H. (2013). An AP2 transcription factor is required for a sleep-active neuron to induce sleep-like quiescence in C. elegans. *Current Biology : Cb, 23* (22), 2215–2223. doi:10.1016/j.cub.2013.09.028

van der Linden, A.M., Wiener, S., You, Y.J., Kim, K., Avery, L., & Sengupta, P. (2008). The EGL-4 PKG acts with KIN-29 salt-inducible kinase and protein kinase A to regulate chemoreceptor gene expression and sensory behaviors in Caenorhabditis elegans. *Genetics, 180* (3), 1475–1491. doi:10.1534/genetics.108.094771

Wu, Y., Masurat, F., Preis, J., & Bringmann, H. (2018). Sleep counteracts aging phenotypes to survive starvation-induced developmental arrest in C. elegans. *Current Biology, 28*(22), 3610–3624. e3618. doi:10.1016/j.cub.2018.10.009

You, Y.J, Kim, J., Raizen, D.M. Avery, L.,(2008). Insulin, cGMP, and TGF-beta signals regulate food intake and quiescence in C. elegans: a model for satiety. *Cell metabolism, 7*, 249–257. doi:10.1016/j.cmet.2008.01.005

Cellular damage, including wounding, drives *C. elegans* stress-induced sleep

Desiree L. Goetting, Richard Mansfield, Rony Soto and Cheryl Van Buskirk

ABSTRACT

Across animal phyla, sleep is associated with increased cellular repair, suggesting that cellular damage may be a core component of sleep pressure. In support of this notion, sleep in the nematode *Caenorhabditis elegans* can be triggered by damaging conditions, including noxious heat, high salt, and ultraviolet light exposure. It is not clear, however, whether this stress-induced sleep (SIS) is a direct consequence of cellular damage, or of a resulting energy deficit, or whether it is triggered simply by the sensation of noxious conditions. Here, we show that thermosensation is dispensable for heat-induced sleep, that osmosensation is dispensable for salt-induced sleep, and that wounding is also a sleep trigger, together indicating that SIS is not triggered by sensation of noxious environments. We present evidence that genetic variation in cellular repair pathways impacts sleep amount, and that SIS involves systemic monitoring of cellular damage. We show that the low-energy sensor AMP-activated protein kinase (AMPK) is not required for SIS, suggesting that energy deficit is not the primary sleep trigger. Instead, AMPK-deficient animals display enhanced SIS responses, and pharmacological activation of AMPK reduces SIS, suggesting that ATP-dependent repair of cellular damage mitigates sleep pressure.

Introduction

Though the cellular function of sleep is of debate, its importance is evident in the range of deleterious consequences associated with sleep loss as well as in its conservation across phyla (reviewed in Cirelli & Tononi, 2008; Anafi, Kayser, & Raizen, 2019). Insight into the function of sleep at the cellular level has come in recent years from the study of model organisms. Molecular correlates of sleep and wakefulness across species have been identified that implicate sleep in synaptic pruning, allocation of energy resources and clearance of metabolic waste (reviewed in Krueger, Frank, Wisor, & Roy, 2016). Other studies point to a link between sleep and restoration of cellular homeostasis. For example, in all species studied, extended wakefulness correlates with an up-regulation of components of the unfolded protein response (reviewed in Naidoo, 2009), and in zebrafish, DNA lesions accumulate during wakefulness whereas chromosome dynamics that stimulate DNA repair processes are observed during sleep (Zada, Bronshtein, Lerer-Goldshtein, Garini, & Appelbaum, 2019). These and other studies indicate that cellular homeostasis and repair are facilitated by sleep, but whether perturbations of cellular homeostasis constitute the basis of sleep need is not known.

While sleep is influenced by circadian cues in most animals, sleep in the nematode *Caenorhabditis elegans* does not appear to fall under circadian control and has thus emerged as a model for investigation of the cellular basis of sleep need. Interestingly, this nematode engages in sleep following exposure to damaging conditions, a phenomenon known as stress-induced sleep (SIS) (Hill, Mansfield, Lopez, Raizen, & Van Buskirk, 2014; Nelson *et al.*, 2014). SIS is dependent on Epidermal Growth Factor Receptor (EGFR) signaling in the peptidergic ALA interneuron (Hill *et al.*, 2014; Nelson *et al.*, 2014), the specification of which is dependent on the homeobox transcription factors CEH-10, CEH-14, and CEH-17 (Van Buskirk & Sternberg, 2010). ALA promotes a coordinated quiescent state through the collective action of neuropeptides that impact an array of sleep sub-behaviors (Nath, Chow, Wang, Schwarz, & Sternberg, 2016; Iannacone *et al.*, 2017). SIS can be triggered by heat, noxious cold, osmotic stress, ethanol, ingestion of pore-forming toxin, and UV irradiation (Hill *et al.*, 2014; DeBardeleben, Lopes, Nessel, & Raizen, 2017) and appears to enhance cellular repair, as sleepless ALA-defective mutant animals are impaired for survival after noxious heat exposure (Hill *et al.*, 2014; Fry, Laboy, Huang, Hart, & Norman, 2016). The observation that *C. elegans* sleep can be triggered by a variety of damaging conditions suggests that cellular damage, or a molecular consequence of it, constitutes the basis of sleep need. It is also possible that it is not the damage itself, but rather the sensory perception of noxious conditions, that promotes sleep. In this study, we aim to distinguish between these possibilities.

Here, we show that thermosensory components are dispensable for heat-SIS and that osmosensory components are dispensable for salt-SIS, indicating that sensory perception of noxious conditions is not the sleep trigger. We find that chaperone-defective mutants have an enhanced heat-SIS response, pointing to a disruption of proteostasis as a determinant of heat-induced sleep. Similarly, we find that DNA repair-defective mutants have an enhanced SIS response following UV exposure. We show that ALA-dependent sleep can be triggered by wounding, and we present evidence that cellular damage to any tissue or body area can promote SIS. Last, we find that SIS does not rely on activation of the energy status sensor AMP kinase and that forced AMP-activated protein kinase (AMPK) activation suppresses, rather than promotes, SIS. Together, these findings indicate that systemic monitoring of cellular damage, and not the sensation of noxious conditions or an energetic crisis, triggers SIS in *C. elegans*.

Methods

Strains & worm cultivation

Worms were grown and maintained on nematode growth media (NGM) plates and fed OP50 *Escherichia coli* bacteria as a food source (Brenner, 1974). All strains were raised at 20 °C.

Standard for all assays

SIS was examined under a stereomicroscope as previously described (Hill *et al.*, 2014). Briefly, pre-fertile young adults animals were examined for normal locomotion and feeding behavior prior to each experiment. During SIS, plates were gently slid into the field of view 45 s prior to examination and plate movement was minimized during scoring. Unless otherwise noted in the figure legend, individual trials consisted of 25 animals each, and behavioral quiescence was defined as a quickly reversible cessation of both locomotor and feeding activity. Cessation of locomotion was defined as complete immobility (no body or head movement) and cessation of feeding was defined as the absence of contractions of the posterior pharyngeal bulb. This was achieved by examining each animal for 5 s at 250× magnification. An animal showing any movement or feeding was categorized as non-quiescent. Where possible, the experimenter was blind to genotype and condition. Statistical tests, indicated in figure legends, were performed using Prism 8 software (San Diego, USA).

Heat-induced sleep

All heat shocks were delivered by placing young adult animals onto NGM plates seeded with OP50 that were sealed with parafilm prior to being placed upright (agar side down) in a heated water bath. After heat shock, plates were cooled to room temperature by placing them on ice for 3 min, and then moved to a stereomicroscope for examination of SIS.

Protocol 1 (Figures 1 and 4): Animals were transferred to 35 × 10 mm plates (5 ml NGM) and placed at 37 °C for 11 min (Goetting, Soto, & Van Buskirk, 2018).

Protocol 2 (Figures 2 and 3): Animals were transferred to 60 × 15 mm plates (12 ml NGM) and placed at 37 °C for 30 min (Hill *et al.*, 2014).

Salt-induced sleep

High salt exposure was performed by moving young adult animals to unseeded NGM plates with a sodium chloride concentration of 300 mM for 30 min. After exposure, animals were transferred to regular NGM plates (50 mM NaCl) seeded with OP50 and examined for SIS behavior.

UV-induced sleep

Young adult animals were transferred to NGM plates seeded with OP50 *E. coli*. Plates were placed lid-down (agar side up) on a UVP M-10E mini benchtop 302 nm (UVB) 60 mW/cm^2 transilluminator gel box for 30 s.

Wounding-induced sleep

Young adult animals were transferred individually to small (35 × 10 mm) NGM plates seeded with OP50 *E. coli* and allowed to recover from the stimulation of locomotion due to transfer. Under a Leica M165 continuous zoom stereomicroscope, a 26.5 gauge needle tip (attached to a 100 ul pipet tip for easier manipulation) was used to puncture the cuticle on either the head, midbody, or tail of the animal. Successful puncture was indicated by a deformation of cuticle followed by a release in tension. After gentle removal of the needle, SIS was examined as described above. Animals were excluded from the experiment if they showed extrusion of tissue following wounding, which we found to be associated with a failure to recover.

Cry-5B-induced sleep

Young adult animals were placed for 10 min onto NGM plates containing 60 μg/ml carbenicillin and 1 mM IPTG that had been seeded with JM103 bacteria harboring an IPTG-inducible Cry5B toxin (Wei *et al.*, 2003). At the end of the ten minute exposure, animals were transferred to NGM plates seeded with OP50 *E. coli* and assayed for SIS as described above.

AICAR exposure

Young adult animals were exposed to the adenosine analog and selective AMPK activator AICAR for 6 h by placing them onto NGM plates containing 2 mM AICAR that had been seeded with OP50 *E. coli*. At the end of the exposure, animals were transferred to standard NGM–OP50 plates and assayed for SIS as described above.

Strains

The following strains were obtained from the *Caenorhabditis* Genetics Center (CGC) unless otherwise noted, and are listed in order of appearance in the figures: N2 wild type, IB16 *ceh-17(np1) I*, TB528 *ceh-14(ch3) X*, PR767 *ttx-1(p767) V*, PY1283 *ttx-1(oy29) V*, GN112 *pgIs2[gcy-8p:TU#813 + gcy-8p:TU#814 + unc-122p:GFP + gcy-8p:mCherry + gcy-8p:GFP + ttx-3p:GFP]*, FK129 *tax-4(ks11) III*, IK597 *gcy-23(nj37) gcy-*

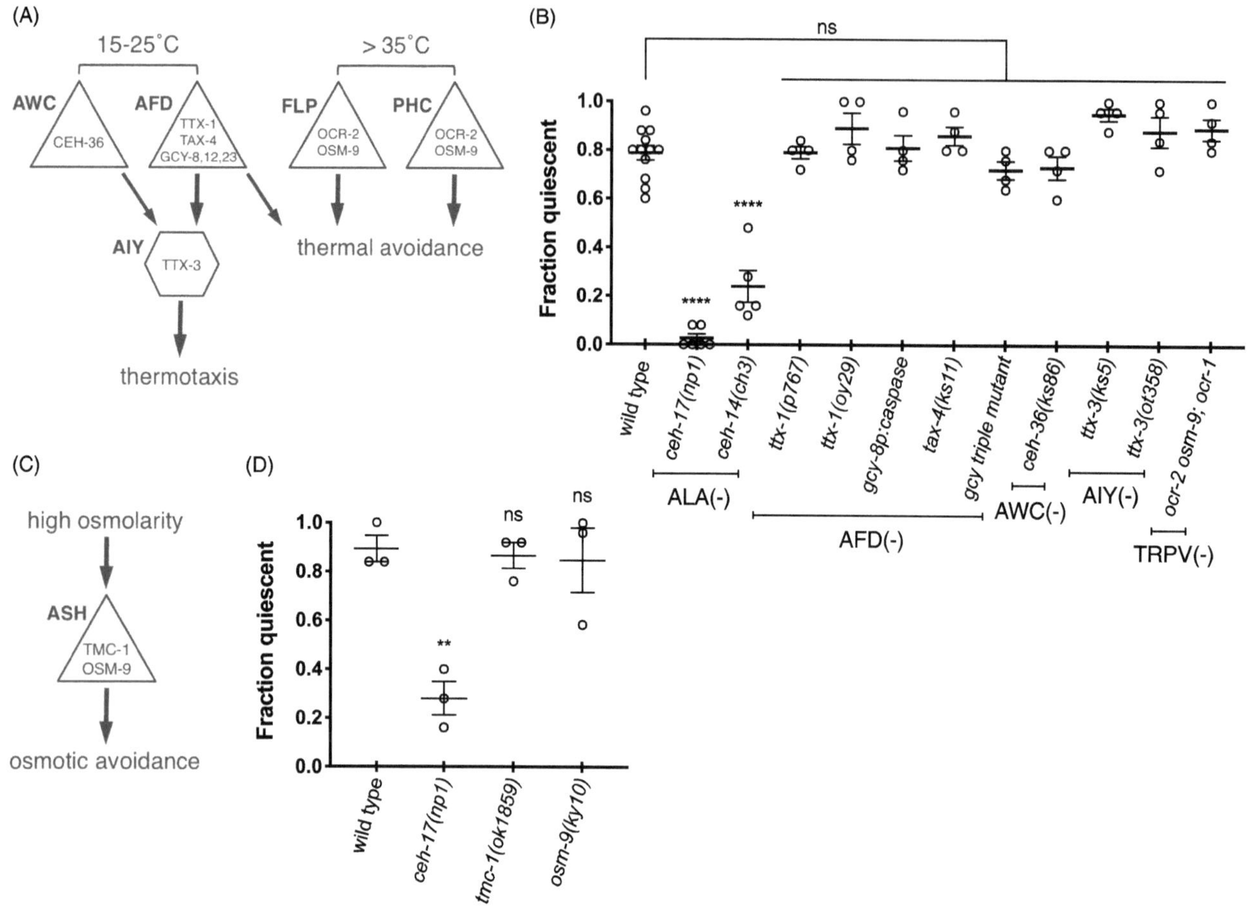

Figure 1. Sensory components are not required for stress-induced sleep. (A) Schematic of neural circuits mediating known behavioral responses to heat. Thermotaxis is mediated by the AFD and AWC sensory neurons as well as the AIY interneurons, and avoidance of noxious heat is mediated by the FLP and PHC sensory neurons with some input from AFD. Genetic components contributing to thermal responses are indicated within respective neurons. (B) Animals defective in the differentiation or function of thermosensory neurons do not show defects in heat stress-induced sleep. Quiescence was scored 8 min after heat shock (heat protocol 1 in 'Methods' section). (C) Schematic of known components of osmosensation. Avoidance of high osmolarity is mediated by the ASH neuron and is dependent on OSM-9 and TMC-1 activity. (D) Components of osmosensation are not required for osmotic stress-induced sleep. *tmc-1(rf)* and *osm-9(rf)* mutant animals exhibit wild-type sleep 20 min after a 30 min exposure to 300 mM salt. (B, D) $^{****}p < .0001$, $^{**}p < .01$, ns: not significant vs. wild type, one-way ANOVA with Dunnett's multiple comparisons test. ALA neuron-defective *ceh-17(np1)* and *ceh-14(ch3)* animals are included as SIS-impaired controls. Individual data points represent the fraction of quiescent animals in one trial of at least 20 young adult animals. Mean and SEM of multiple trials are shown.

8(oy44) gcy-18(nj38) IV, FK311 *ceh-36(ks86) X*, FK134 *ttx-3(ks5) X*, OH9331 *ttx-3(ot358) X*, FG125 *ocr-2(ak47) osm-9(ky10) IV; ocr-1(ak46) V*, RB1546 *tmc-1(ok1859) X*, CX10 *osm-9(ky10) IV*, JT6130 *daf-21(p673) V*, RB864 *xpa-1(ok698) I*, CVB24 *xpa-1(ok698) I; ceh-14(ch3) X* (this study), PS3551 *hsf-1(sy441) I*, OS3062 *hsf-1(sy441) I; nsEx1730[myo-2p:hsf-1 + hsp-16–2::GFP + hsp-16–41::GFP + rol-6(su1006)]*, IK0982 *hsf-1(sy441); njEx393[myo-3p:hsf-1cDNA, ges-1p:nls-gfp]* (Ikue Mori), IK0983 *hsf-1(sy441); njEx394[ges-1p:hsf-1cDNA, ges-1p:nls-gfp]* (Ikue Mori), IK0984 *hsf-1(sy441); njEx395 [unc-14p:hsf-1cDNA, ges-1p:nls-gfp]* (Ikue Mori), *AGD397 aak-1(tm1944);aak-2(ok524); uthEx202[crtc-1p::crtc-1cDNA::tdTomato::unc-54 3'UTR + rol-6(su1006)]*.

Results

Sensory perception of noxious conditions is not required to initiate SIS

SIS in *C. elegans* ensues after exposure to noxious conditions (Hill *et al.*, 2014; Nelson *et al.*, 2014; DeBardeleben *et al.*, 2017). To determine whether the sensation of these conditions, or the resulting cellular damage, is the primary sleep trigger, we examined whether the genetic and neural components that sense environmental conditions are required to initiate sleep. We first investigated the SIS response that is triggered by heat. In *C. elegans*, the neural circuits that mediate heat-responsive behaviors, such as thermotaxis and noxious heat avoidance, have been well characterized (Figure 1(A)). Thermotaxis – the ability to move within a temperature gradient toward a past cultivation temperature – is mediated by the AIY interneurons, with major and minor input from the AFD and AWC sensory neurons respectively (reviewed in Kimata, Sasakura, Ohnishi, Nishio, & Mori, 2012). Noxious heat avoidance (above 35 °C) depends on AFD as well as the polymodal FLP and PHC neurons (Liu, Schulze, & Baumeister, 2012). To investigate the requirement for the AFD/AWC/AIY thermosensory circuit in heat-induced sleep, we examined SIS in animals with partial reduction-of-function (rf) mutations in TTX-1, an Otd/Otx homeodomain transcription factor required for all differentiated characteristics of the AFD sensory neuron (Satterlee *et al.*, 2001). We found that *ttx-1(rf)* mutants exhibit wild-type SIS (Figure 1(B)),

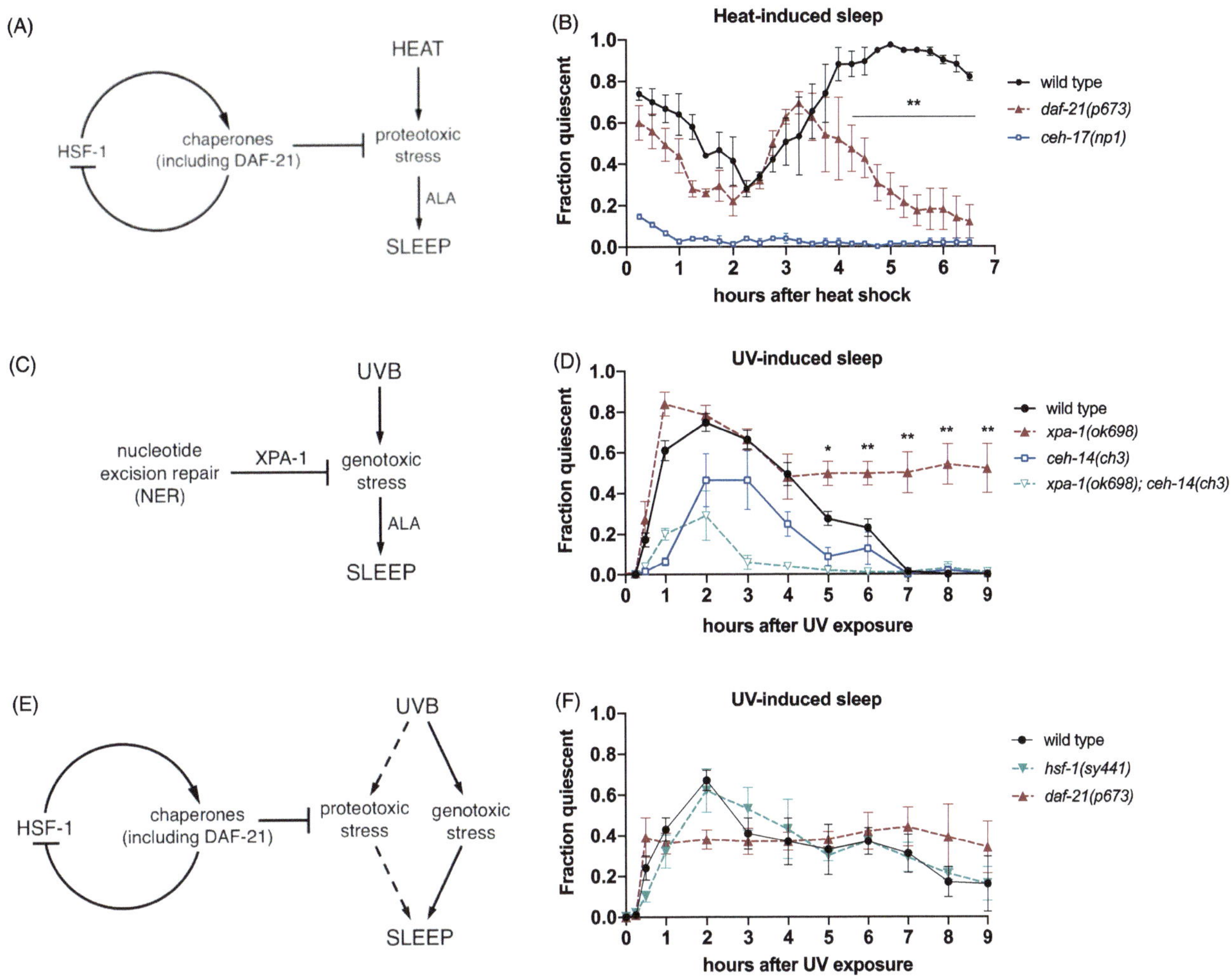

Figure 2. Cellular repair mechanisms modulate SIS. (A) Proposed relationship between the heat shock response (HSR) and heat-induced sleep. HSF-1 initiates a broad transcriptional response to proteotoxic stress, elevating the expression of chaperones including DAF-21/HSP90, which attenuates the HSR. Our findings indicate that proteotoxic stress is the major trigger of heat-induced sleep. (B) *daf-21(rf)* mutants exhibit reduced SIS during the second bout of sleep following a stringent heat shock (heat protocol 2 in 'Methods' section). $**p < .01$ *daf-21(rf)* vs. wild type, multiple Student's *t*-tests with Holm–Sidak correction. (C) Proposed relationship between DNA repair and UV-SIS. Loss of XPA-1 impairs nucleotide excision repair following UV light exposure. Our findings indicate that impairment of DNA repair increases the duration of UV-induced sleep. (D) *xpa-1* null animals exhibit prolonged SIS following a 30 s UVB light exposure. The ALA-defective *ceh-14(ch3)* mutation is sufficient to suppress the enhanced sleep phenotype of *xpa-1* mutant animals. $*p < .05$, $**p < .01$, *xpa-1(ok698)* vs. wild type, multiple Student's *t*-tests with Holm–Sidak correction. (B, D) Mean and SEM of a minimum of four trials are shown ($n = 25$ each). (E) Model for relationship between proteotoxic stress, genotoxic stress, and UV-induced sleep. Our findings indicate that UV-SIS is triggered largely by DNA damage. (F) *daf-21(rf)* and *hsf-1(rf)* mutants exhibit wild-type sleep following a 30 s UVB light exposure. No significant difference vs. wild type, multiple Student's *t*-tests with Holm–Sidak correction. The area under the curve (AUC) was also calculated to approximate the total amount of sleep for each genotype, and compared using a two-tailed Student's *t*-test; no significant difference from wild type was found. Mean and SEM of four trials are shown ($n = 25$ each).

suggesting that the AFD neuron is dispensable for heat-SIS. To further test AFD involvement, we examined animals in which the AFD neuron had been killed by expression of a constitutively active caspase under the control of the AFD-specific *gcy-8* promoter (Glauser *et al.*, 2011). These AFD-ablated animals show robust SIS following heat shock (Figure 1(B)), indicating that heat-induced sleep does not require thermosensory input from AFD. Consistent with this, we found that animals lacking AFD-expressed genetic components of thermosensation including the cyclic nucleotide-gated channel TAX-4 (Komatsu, Mori, Rhee, Akaike, & Ohshima, 1996) and guanylyl cyclases (Inada *et al.*, 2006) show wild-type sleep (Figure 1(B)). We also evaluated potential contribution by the AWC sensory neurons by examining animals with reduced activity of the CEH-36 OTX homeodomain transcription factor, which is required for AWC differentiation (Lanjuin, VanHoven, Bargmann, Thompson, & Sengupta, 2003). We found that *ceh-36(rf)* mutants display wild-type heat-induced sleep (Figure 1(B)), suggesting that heat sensation by AWC is also dispensable for SIS. We then examined the role of the AIY interneurons, the development of which is dependent on the LIM homeodomain transcription factor TTX-3 (Hobert *et al.*, 1997). We found that *ttx-3(rf)* mutants also display wild-type heat-induced sleep (Figure 1(B)). Together these data suggest that the AFD/AWC/AIY thermosensory circuit is not required to promote sleep in response to heat. We next examined the requirement for the nociceptive FLP/PHC circuit in heat-induced sleep. Noxious heat sensation by these neurons utilizes the heat/capsaicin-sensitive Transient Receptor Potential Vanilloid (TRPV) channels encoded by *ocr-2* and *osm-9* (Liu *et al.*,

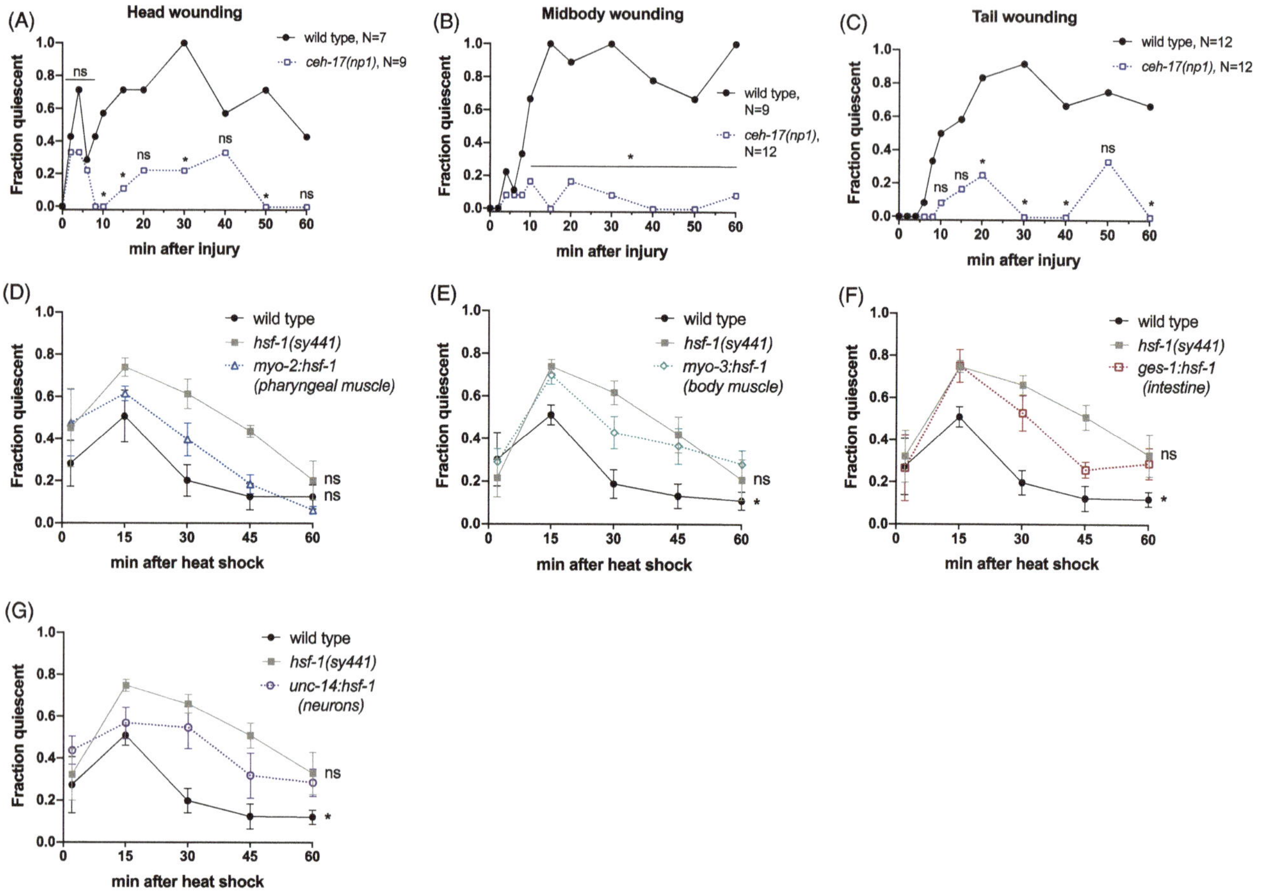

Figure 3. Cellular damage in multiple body regions and tissues contributes to SIS. Animals engage in SIS following wounding to the head (A), midbody (B), or tail (C). Wild-type animals show significantly more sleep than ALA-defective *ceh-17(np1)* animals. $*p < .05$, ns: not significant, Fisher's exact tests. The total number of animals examined for each genotype is indicated by *N* in each panel. (D–G) The enhanced SIS phenotype of *hsf-1(sy441)* is partially suppressed by expression of HSF-1 in pharyngeal muscle. Heat-SIS (heat protocol 2 in 'Methods' section) was examined following tissue-specific rescue of HSF-1 under the pharyngeal muscle-specific *myo-2* promoter (D), body muscle-specific *myo-3* promoter (E), intestine-specific *ges-1* promoter (F), or pan-neuronal *unc-14* promoter (G). While SIS in each HSF-1 rescue line is reduced relative to *hsf-1(rf)*, this reduction is not significant. However, the pharyngeal muscle rescue line is also not significantly different from wild type, pointing to partial rescue of the *hsf-1(rf)* enhanced SIS phenotype. Mean and SEM of four trials ($n = 20$ each) are shown. $*p < .05$, ns: not significant, tissue-specific rescue strain vs. *hsf-1(sy441)* or vs. wild type as indicated, one-way ANOVA comparing area under the curve (AUC) with Sidak's multiple comparisons test.

2012). We found that animals defective in TRPV channel activity also show wild-type heat-induced sleep (Figure 1(B)), suggesting that the FLP/PHC circuit is not required for SIS. Thus, heat-induced sleep in *C. elegans* does not appear to be dependent on thermosensation by known heat-sensing neurons.

To investigate the potential role of sensory perception in the SIS response to a noxious stimulus other than heat, we examined osmotic SIS (Hill *et al.*, 2014). The detection of high osmolarity relies on the polymodal ASH sensory neurons (reviewed by Kim & Jin, 2015; Figure 1(C)). The TRPV channel OSM-9 is required for several ASH-mediated avoidance behaviors, including salt avoidance (Colbert, Smith, & Bargmann, 1997) and the multipass transmembrane protein TMC-1 is required specifically for salt chemosensation (Chatzigeorgiou, Bang, Hwang, & Schafer, 2013). We found neither *tmc-1(rf)* nor *osm-9(rf)* mutants to be impaired for SIS following exposure to high salt (Figure 1(D)), indicating that known osmosensory mechanisms are dispensable for osmotic SIS. Together, these data support the notion that cellular damage caused by noxious conditions, rather than sensory perception of noxious stimuli, promotes SIS.

Cellular repair mechanisms modulate SIS

Our observation that noxious stimuli can trigger sleep in the absence of functional sensory circuits suggests that SIS may be induced by the cellular damage resulting from exposure to noxious conditions. Two previous observations support this notion. First, in response to a longer heat shock (protocol 2 in 'Methods' section), two bouts of sleep are observed: one immediately following the heat shock, and another that peaks several hours later (Hill *et al.*, 2014), the timing of which makes it unlikely to be a consequence of thermosensation. Second, the duration of this additional SIS bout is increased in animals with reduced activity of the chaperone master regulator HSF-1 (Hill *et al.*, 2014). To further test the notion that proteotoxic stress exacerbates sleep need, we wished to determine whether heightened HSF-1 activity might reduce SIS. To this end we examined heat-induced sleep in animals with reduced function of DAF-21, a member of the HSP90 family of heat shock proteins that negatively regulates the activity of HSF-1 (Guisbert, Czyz, Richter, McMullen, & Morimoto, 2013; Van Oosten-Hawle, Porter, & Morimoto, 2013; Kijima *et al.*, 2018; Figure 2(A)). We exposed *daf-21(rf)* animals to a heat shock and found

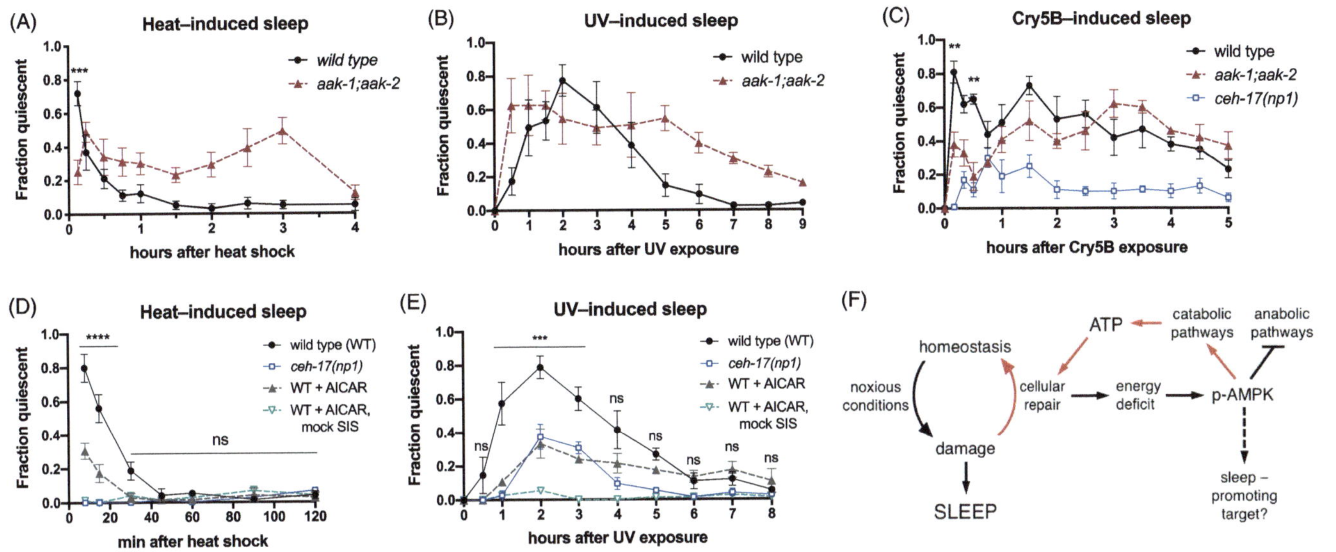

Figure 4. AMPK activity limits the need for sleep. (A) AMPK null mutants show enhanced heat-SIS. While impaired for SIS at the initial time point (***$p < .001$, Student's *t*-test), *aak-1(tm1944);aak-2(ok524)* mutants exhibit a greater amount of sleep than wild-type animals ($p < .0001$, two-tailed Student's *t*-test comparing area under the curve (AUC)). Mean and SEM of four trials ($n = 25$ each) are indicated. (B) Following a 30 s UVB light exposure, *aak-1;aak-2* null mutants sleep more throughout the SIS time course ($p = .0054$, two-tailed Student's *t*-test comparing area under the curve). The peak of quiescence is reduced, but not significantly ($p = .13$, Student's *t*-test, 2 h time point). Mean and SEM of three trials ($n = 25$ each) are indicated. (C) Following a 15 min Cry5B toxin exposure, *aak-1;aak-2* null mutants are SIS-defective at certain time points (**$p < .01$, ***$p < .001$ vs. wild-type control, multiple Student's *t*-tests with Holm–Sidak correction) and show a non-significant trend toward increased sleep at latter time points, with a similar amount of sleep overall (ns, two-tailed Student's *t*-test comparing AUC). SIS-defective *ceh-17(lf)* animals are included to control for non-SIS-related effects of the Cry5B toxin. Mean and SEM of four trials ($n = 25$) each are indicated. (D–E) Activation of AMPK in wild-type animals by the AMP mimetic AICAR significantly suppresses sleep induced by heat (D) and UV light (E). Mean and SEM of three trials ($n = 25$ each) are indicated. ****$p < .0001$, ***$p < .001$, ns: not significant, WT + AICAR vs. WT control, multiple Student's *t*-tests with Holm–Sidak correction. (F) Model proposing mechanism of dual action of AMPK activation on sleep. The reduced SIS in AMPK(–) at earlier time points may point to an AMPK target involved in promoting sleep (dotted arrow), and the increased SIS observed in these animals at later time points may reflect an indirect role for AMPK, limiting sleep need via stimulation of ATP-dependent cellular repair (red arrows). p-AMPK = phosphorylated (active) AMPK.

that *daf-21(rf)* mutants show reduced SIS compared to wild type, with a strikingly limited second sleep bout (Figure 2(B)), suggesting that heightened HSF-1 activity limits SIS. As DAF-21 interacts with many client proteins other than HSF-1, other interpretations are possible. However, together with the previously observed exaggerated sleep phenotype of *hsf-1(rf)*, these data suggest that restoration of protein homeostasis mitigates sleep need in *C. elegans*.

We next investigated the cellular repair component of a different SIS trigger, ultraviolet light (DeBardeleben *et al.*, 2017). We examined the role of DNA damage caused by UVB radiation, which is normally ameliorated by the nucleotide excision repair (NER) pathway (reviewed in Budden & Bowden, 2013). *C. elegans* XPA-1 is a core NER component orthologous to human XPA, mutated in Xeroderma Pigmentosum (reviewed in Lans & Vermeulen, 2011; Figure 2(C)). We reasoned that if DNA damage can promote SIS, animals with impaired XPA-1 function will show increased sleep following UV exposure. To test this prediction, we examined SIS in *xpa-1(ok698)* loss-of-function (lf) animals following exposure to 302 nm UVB radiation, and found that they exhibit significantly enhanced quiescence (Figure 2(D)). To determine whether this increased quiescence represents ALA-dependent sleep, rather than general malaise due to increased genotoxic stress, we examined *xpa-1(ok698);ceh-14(ch3)* double mutants, which are expected to be severely impaired in ALA neuron differentiation (Van Buskirk & Sternberg, 2010). We found the enhanced sleep of *xpa-1(lf)* to be completely suppressed by the *ceh-14* mutation (Figure 2(D)), supporting the notion that defective repair of DNA lesions increases the drive for ALA-dependent sleep.

Along with induction of genotoxic stress, UVB irradiation is known to induce up-regulation of chaperones and proteasome components (Perluigi *et al.*, 2010). We therefore wished to examine whether a component of UV-SIS is attributable to proteotoxic stress (Figure 2(E)). To this end, we examined whether UV-induced sleep is impacted by mutations affecting protein homeostasis. We found that both *hsf-1(rf)* and *daf-21(rf)* animals have a wild-type UV-SIS response (Figure 2(F)). In the case of *daf-21(rf)*, we observed a trend toward a unique sleep profile, but this was not statistically different from wild type (Figure 2(F)). These data suggest that proteostatic disruption is not a major contributor to UVB-SIS, consistent with previous findings on UVC-SIS (DeBardeleben *et al.*, 2017.) Taken together with our observation that DNA repair-defective *xpa-1(lf)* mutants show enhanced UV-SIS, these data indicate that sleep following UV exposure is largely attributable to DNA damage.

SIS can be triggered by wounding the head, body, or tail

Previous studies have shown that in addition to heat and UV, a variety of stimuli can trigger SIS, including exposure to high salt, ethanol, and bacterial toxin (Hill *et al.*, 2014). We wished to determine whether physical injury could also trigger ALA-dependent sleep. To address this, we compared the effects of wounding on wild-type animals and ALA-defective *ceh-17(np1)* mutants. Incision with a needle causes local disruption of the cuticle, epidermis and associated extracellular matrix, and may also damage internal tissues (Pujol *et al.*, 2008). We used a fine gauge needle to target a small incision to the head, midbody, or tail of the animals, and examined their behavior. We excluded animals in which incision led to extrusion of tissue; in all other cases, animals recovered from the injury. We found that wounding to any area led to behavioral quiescence, the majority of which was ALA-dependent (Figure 3(A–C)). Incisions at the midbody produced the most robust SIS response, but incisions at the head and tail also triggered SIS, with a peak of ALA-dependent behavioral quiescence 30 min following wounding. These data indicate that wounding can trigger SIS, and suggest that damage to any region can result in the production of a sleep-promoting signal.

The region-independence of wounding-induced sleep suggests that cellular damage may be monitored systemically for the assessment of sleep need. To investigate this possibility, we returned to an examination of heat-induced sleep. We used tissue-specific expression of HSF-1 in a *hsf-1(rf)* background (Sugi, Nishida, & Mori, 2011) to restore chaperone activity to one of four tissues: pharyngeal muscle, body muscle, the intestine, or the nervous system. If one of these tissues is the source of a large fraction of the damage signal, we expect HSF-1 expression in that tissue to rescue the majority of the *hsf-1(rf)* enhanced SIS phenotype. Alternatively, if proteotoxic stress in multiple tissues contributes to the damage signal, we expect a partial suppression, at most, of the *hsf-1(rf)* SIS phenotype. We found that HSF-1 rescue in pharyngeal muscle produces a partial suppression of the enhanced SIS phenotype of *hsf-1(rf)* mutants (Figure 3(D–G)), with the other lines showing a non-significant reduction in SIS. We interpret these results to indicate that proteotoxic stress within multiple tissues contributes to SIS, and that among the tissues examined here, SIS is most sensitive to proteostatic disruptions within pharyngeal muscle. Our data do not address the contribution of proteotoxic stress within the epidermis nor the germline.

AMPK modulates SIS in a complex manner

Our data indicate that cellular damage, or a consequence of it, contributes to sleep pressure in *C. elegans*. We reasoned that the initiation of energy-intensive cellular repair processes may lead to a drop in ATP levels that in turn promotes SIS, a model supported by observed decreases in ATP levels in *C. elegans* following exposure to genotoxic stress (Grubbs, Lopes, van der Linden, & Raizen, 2019). To investigate the role of energy status in driving sleep, we examined AMPK, the major cellular energy sensor across species including *C. elegans* (Apfeld, O'Connor, McDonagh, DiStefano, & Curtis, 2004). AMPK is activated during energy decline by phosphorylation of its alpha subunit, and phospho-AMPK in turn promotes ATP production and energy conservation *via* activation of catabolic processes and suppression of anabolic ones (reviewed in Hardie, Ross, & Hawley, 2012). In *C. elegans, aak-1* and *aak-2* encode the two alternate alpha subunits of AMPK. To determine whether AMPK-mediated monitoring of cellular energy is required for SIS, we examined SIS in *aak-1;aak-2* loss-of-function mutant animals. We first examined heat-SIS and found that while AMPK(lf) animals show reduced peak quiescence, they sleep significantly more overall than their wild-type counterparts (Figure 4(A)). We then examined UV-SIS, and observed an increase in total sleep amount in AMPK(lf) animals (Figure 4(B)). We also observed a trend toward decreased peak quiescence, but this was not significant. These observations indicate that AMPK activation may be a component of the sleep-promoting signal following damage, but that the major role of AMPK in SIS is to limit sleep. We examined this further using a different type of SIS trigger, ingestion of the pore-forming toxin Cry5B (Hill *et al.*, 2014). In this case, we observed reduced peak quiescence in AMPK(lf) animals, with a total sleep amount similar to wild type (Figure 4(C)). Together these findings suggest that the relative impact of sleep-promoting and sleep-limiting roles of AMPK may differ depending on the source of cellular damage.

We continued to explore the function of AMPK by examining the effect of forced AMPK activation on SIS. We employed pharmacological activation of AMPK by AICAR, an adenosine analog that is phosphorylated to form an AMP mimetic that allosterically activates AMPK (reviewed by Kim, Yang, Kim, Kim, & Ha, 2016), and we examined both heat-SIS and UV-SIS. Consistent with the notion that AMPK activation alone is not a major component of sleep drive, we found that pharmacological activation of AMPK does not trigger sleep (Figure 4(D,E)). Instead, AICAR exposure significantly reduces SIS following exposure to noxious heat and UV light (Figure 4(D,E)), supporting our findings that the major role for AMPK in certain types of SIS is to limit sleep. We speculate that AMPK activity limits sleep indirectly, by enhancing ATP production that in turn promotes efficient cellular repair (Figure 4(F)).

Discussion

C. elegans SIS is a programmed sleep state that follows exposure to noxious stimuli (Hill *et al.*, 2014). Here, we investigated whether this SIS is triggered by sensation of noxious conditions, or by the cellular damage that ensues. While the former would provide the benefit of conserving resources preemptively for repair of cellular damage, the latter would maximize opportunity for escape. Data presented here indicate that SIS in *C. elegans* is not a consequence of sensory perception of noxious conditions, but instead is triggered by cellular damage. First, heat-SIS and salt-SIS do not require known components of thermosensation and osmosensation, respectively. Second, SIS responses are exaggerated in animals defective in cellular repair, and reduced in animals with heightened repair mechanisms. Last, these findings build on prior work showing that in some cases, sleep bouts can be initiated after removal from the noxious condition (Hill *et al.*, 2014). Together these data indicate that cellular damage, or a downstream consequence of it, drives *C. elegans* SIS.

Using a wounding-induced sleep assay, we have shown that damage to the head, mid-body, or tail can trigger ALA-dependent sleep, suggesting that damage may be monitored systemically in the assessment of sleep need. In support of this, our tissue-specific HSF-1 rescue experiments indicate that at least during heat-SIS, proteostatic disruption within pharyngeal muscle contributes a fraction of the damage signal. HSF-1 rescue in other tissues (body muscle, intestine, neurons) produced mild effects on SIS that did not reach significance. Together with our wounding data, these findings point to a systemic monitoring of cellular damage that promotes ALA-dependent sleep. Interestingly, it has been shown that the *C. elegans* heat shock response (HSR) is also systemic, with proteostatic disruption in one tissue activating chaperone expression in other tissues (Van Oosten-Hawle *et al.*, 2013). A model emerges in which localized proteotoxic stress activates an organismal HSR as well as ALA-dependent sleep, coordinating an organismal chaperone response with a state of behavioral quiescence that enhances restoration of proteostasis. It will be of interest to determine whether this paradigm extends to forms of SIS that do not involve proteotoxic stress, such as UV-SIS, which we have found to be triggered largely by DNA damage. Of potential relevance is the observation that among cancer patients undergoing radiation therapy (RT), fatigue is commonly experienced even when the RT is delivered to peripheral organs (Monga, Kerrigan, Thornby, & Monga, 1999; Janda *et al.*, 2000), pointing to a systemic monitoring of genotoxic stress that impacts sleep drive.

Work here and elsewhere reveals that the conditions that trigger SIS are quite varied, including noxious heat, cold, osmotic stress, UV light, bacterial toxin, and wounding. Here, we investigated the possibility that the sleep-promoting signal across these damaging conditions may be a drop in cellular energy. Specifically, we examined the role of AMPK, a potent energy sensor activated by elevated AMP/ATP ratio and reported to play a role in sleep homeostasis (reviewed in Chikahisa & Séi, 2011). Our data indicate that while AMPK activation may contribute to SIS, it is not the central sleep-promoting signal, as animals lacking both catalytic alpha subunits engage in all forms of SIS examined. Instead, for certain types of SIS, AMPK activity appears to limit sleep, as the total sleep amount is increased by loss of AMPK, and reduced by pharmacological activation of AMPK. One interpretation of these data is that AMPK-mediated energy mobilization facilitates ATP-dependent cellular repair, thus limiting the drive for sleep. Alternatively, these data may point to an AMPK target with a role in sleep suppression.

Our finding that AMPK activity modulates SIS predicts that cellular damage should elicit an increase in the active, phosphorylated form of AMPK. However, whole-animal levels of phospho-AMPK appear to decrease, rather than increase, during *C. elegans* UV-induced sleep (Grubbs *et al.*, 2019). These observations may be reconciled by positing that localized, rather than whole-animal, AMPK activation is critical in modulating SIS. Our work adds to a growing body of knowledge surrounding the complex interplay between metabolism and sleep (reviewed in Anafi *et al.*, 2019). For example, *C. elegans* that have been food deprived for several hours (Goetting *et al.*, 2018), or have genetic defects in energy mobilization (Grubbs *et al.*, 2019), are SIS-defective, indicating that low energy, at least under certain conditions, can suppress sleep. Indeed, sleep can be suppressed by starvation across species (MacFadyen, Oswald, & Lewis, 1973; Keene *et al.*, 2010), potentially for prioritization of foraging until there is sufficient energy for repair processes. Taken with previous work, our data support a model in which energy deficit, rather than constituting the basis of sleep need, acts as a potent context-dependent modulator of sleep.

If a drop in cellular energy is not the primary sleep-promoting signal upstream of ALA activation, then what is? It is plausible that the sleep signal should depend on activation of stress responses, which in *C. elegans* are governed largely by HSF-1 and DAF-16/FOXO (reviewed in Rodriguez, Snoek, De Bono, & Kammenga, 2013). However, similar to AMPK, mutations in these genes are associated with reduced quiescence at certain time points, but with significantly enhanced SIS overall (Hill *et al.*, 2014). This indicates that activation of these stress response pathways may contribute to sleep drive, but the major effect of these factors is to limit SIS, presumably by promoting restoration of cellular homeostasis. This raises the possibility that SIS is initiated largely in parallel to, rather than downstream of, the activation of stress responses. The relationship between cellular damage and SIS is likely to be complex, though it is worth investigating given that SIS occurs across species (Lenz, Xiong, Nelson, Raizen, & Williams, 2015; Zada *et al.*, 2019) and systemic monitoring of cellular damage may be a deeply conserved component of sleep drive.

Conclusions

We report that *C. elegans* SIS is driven by cellular damage rather than by sensation of noxious conditions, and we present evidence that this damage is monitored systemically in the assessment of sleep need. How the varied forms of cellular stress (proteotoxic, genotoxic, wounding) lead to EGFR-dependent activation of the sleep-inducing ALA neuron remains unknown. As expression of EGF is highly restricted within the animal (Saffer, Kim, van Oudenaarden, & Horvitz, 2011), we posit that a cell non-autonomous damage signal that triggers EGF signaling is yet to be identified. Our data indicate that the energy sensor AMPK modulates SIS in a complex manner, with a major role in limiting sleep. Further investigation in this and other systems is required to identify conserved mechanisms linking energy status, cellular damage, sleep, and repair.

Acknowledgements

The authors thank the *Caenorhabditis* Genetics Center (CGC) and Ikue Mori for strains. Special thanks to Alex Hernandez of BIOL447/L for observing that wounding could trigger ALA-dependent sleep. This article is dedicated to Sydney Brenner.

Disclosure statement

No potential conflict of interest was reported by the author(s).

Funding

This work was supported by an NSF-CAREER award [IOS 1553673] to CVB. The Caenorhabditis Genetics Center is funded by the NIH Office of Research Infrastructure Programs [P40 OD010440].

ORCID

Cheryl Van Buskirk http://orcid.org/0000-0003-1929-5948

References

Anafi, R.C., Kayser, M.S., & Raizen, D.M. (2019). Exploring phylogeny to find the function of sleep. *Nature Reviews Neuroscience*, *20*, 109–116. doi:10.1038/s41583-018-0098-9

Apfeld, J., O'Connor, G., McDonagh, T., DiStefano, P.S., & Curtis, R. (2004). The AMP-activated protein kinase AAK-2 links energy levels and insulin-like signals to lifespan in *C. elegans*. *Genes & Development*, *18*, 3004–3009. doi:10.1101/gad.1255404

Brenner, S. (1974). The genetics of *Caenorhabditis elegans*. *Genetics*, *77*, 71–94.

Budden, T., & Bowden, N.A. (2013). The role of altered nucleotide excision repair and UVB-induced DNA damage in melanomagenesis. *International Journal of Molecular Sciences*, *14*, 1132–1151. doi: 10.3390/ijms14011132

Chatzigeorgiou, M., Bang, S., Hwang, S.W., & Schafer, W.R. (2013). Tmc-1 encodes a sodium-sensitive channel required for salt chemosensation in *C. elegans*. *Nature*, *494*, 95–99. doi:10.1038/nature11845

Chikahisa, S., & Séi, H. (2011). The role of ATP in sleep regulation. *Frontiers in Neurology*, *2*, 87. doi: 10.3389/fneur.2011.00087

Cirelli, C., & Tononi, G. (2008). Is sleep essential? *PLoS Biology*, *6*, e216. doi:10.1371/journal.pbio.0060216

Colbert, H.A., Smith, T.L., & Bargmann, C.I. (1997). OSM-9, a novel protein with structural similarity to channels, is required for olfaction, mechanosensation, and olfactory adaptation in *Caenorhabditis elegans*. *The Journal of Neuroscience*, *17*, 8259–8269. doi:10.1523/JNEUROSCI.17-21-08259.1997

DeBardeleben, H.K., Lopes, L.E., Nessel, M.P., & Raizen, D.M. (2017). Stress-induced sleep after exposure to ultraviolet light is promoted by p53 in *Caenorhabditis elegans*. *Genetics*, *207*, 571–582. doi: 10.1534/genetics.117.300070

Fry, A.L., Laboy, J.T., Huang, H., Hart, A.C., & Norman, K.R. (2016). A conserved GEF for rho-family GTPases acts in an EGF signaling pathway to promote sleep-like quiescence in *Caenorhabditis elegans*. *Genetics*, *202*, 1153–1166. doi:10.1534/genetics.115.183038

Glauser, D.A., Chen, W.C., Agin, R., Macinnis, B.L., Hellman, A.B., Garrity, P.A., … Goodman, M.B. (2011). Heat avoidance is regulated by transient receptor potential (TRP) channels and a neuropeptide signaling pathway in *Caenorhabditis elegans*. *Genetics*, *188*, 91–103. doi:10.1534/genetics.111.127100

Goetting, D.L., Soto, R., & Van Buskirk, C. (2018). Food-dependent plasticity in caenorhabditis elegans stress-induced sleep is mediated by TOR-FOXA and TGF-β signaling. *Genetics*, *209*, 1183–1195. doi: 10.1534/genetics.118.301204

Grubbs, J.J., Lopes, L.E., van der Linden, A.M., & Raizen, D.M. (2019). A salt-induced kinase (SIK) is required for the metabolic regulation of sleep. *BioRxiv*, 586701. doi:10.1101/586701

Guisbert, E., Czyz, D.M., Richter, K., McMullen, P.D., & Morimoto, R.I. (2013). Identification of a Tissue-Selective Heat Shock Response Regulatory Network. *PLOS Genet.*, *9*, e1003466. doi:10.1371/journal.pgen.1003466

Hardie, D.G., Ross, F.A., & Hawley, S.A. (2012). AMPK: A nutrient and energy sensor that maintains energy homeostasis. *Nature Reviews Molecular Cell Biology*, *13*, 251–262. doi:10.1038/nrm3311

Hill, A.J., Mansfield, R., Lopez, J.M.N.G., Raizen, D.M., & Van Buskirk, C. (2014). Cellular stress induces a protective sleep-like state in *C. elegans*. *Current Biology*, *24*, 2399–2405. doi:10.1016/j.cub.2014.08.040

Hobert, O., Mori, I., Yamashita, Y., Honda, H., Ohshima, Y., Liu, Y., & Ruvkun, G. (1997). Regulation of interneuron function in the *C. elegans* thermoregulatory pathway by the ttx-3 LIM homeobox gene. *Neuron*, *19*, 345–357. doi:10.1016/S0896-6273(00)80944-7

Iannacone, M.J., Beets, I., Lopes, L.E., Churgin, M.A., Fang-Yen, C., Nelson, M.D., … Raizen, D.M. (2017). The RFamide receptor DMSR-1 regulates stress-induced sleep in. *eLife*, *6*, e19837. doi:10.7554/eLife.19837

Inada, H., Ito, H., Satterlee, J., Sengupta, P., Matsumoto, K., & Mori, I. (2006). Identification of guanylyl cyclases that function in thermosensory neurons of *Caenorhabditis elegans*. *Genetics*, *172*, 2239–2252. doi:10.1534/genetics.105.050013

Janda, M., Gerstner, N., Obermair, A., Fuerst, A., Wachter, S., Dieckmann, K., & Pötter, R. (2000). Quality of life changes during conformal radiation therapy for prostate carcinoma. *Cancer*, *89*, 1322–1328. doi:10.1002/1097-0142(20000915)89:6 < 1322::AID-CNCR18 > 3.0.CO;2-D

Keene, A.C., Duboué, E.R., McDonald, D.M., Dus, M., Suh, G.S.B., Waddell, S., & Blau, J. (2010). *Clock* and *cycle* limit starvation-induced sleep loss in *Drosophila*. *Current Biology*, *20*, 1209–1215. doi:10.1016/j.cub.2010.05.029

Kijima, T., Prince, T.L., Tigue, M.L., Yim, K.H., Schwartz, H., Beebe, K., … Neckers, L. (2018). HSP90 inhibitors disrupt a transient HSP90-HSF1 interaction and identify a noncanonical model of HSP90-mediated HSF1 regulation. *Scientific Reports*, *8*, 6976. doi: 10.1038/s41598-018-25404-w

Kim, J., Yang, G., Kim, Y., Kim, J., & Ha, J. (2016). AMPK activators: Mechanisms of action and physiological activities. *Experimental & Molecular Medicine*, *48*, e224–e224. doi:10.1038/emm.2016.16

Kim, K.W., & Jin, Y. (2015). Neuronal responses to stress and injury in *C. elegans*. *FEBS Letters*, *589*, 1644–1652. doi:10.1016/j.febslet.2015.05.005

Kimata, T., Sasakura, H., Ohnishi, N., Nishio, N., & Mori, I. (2012). Thermotaxis of *C. elegans* as a model for temperature perception, neural information processing and neural plasticity. *Worm*, *1*, 31–41. doi:10.4161/worm.19504

Komatsu, H., Mori, I., Rhee, J.S., Akaike, N., & Ohshima, Y. (1996). Mutations in a cyclic nucleotide-gated channel lead to abnormal thermosensation and chemosensation in *C. elegans*. *Neuron*, *17*, 707–718. doi:10.1016/S0896-6273(00)80202-0

Krueger, J.M., Frank, M.G., Wisor, J.P., & Roy, S. (2016). Sleep function: Toward elucidating an enigma. *Sleep Medicine Reviews*, *28*, 46–54. doi:10.1016/j.smrv.2015.08.005

Lanjuin, A., VanHoven, M.K., Bargmann, C.I., Thompson, J.K., & Sengupta, P. (2003). Otx/otd homeobox genes specify distinct sensory neuron identities in *C. elegans*. *Developmental Cell*, *5* (4), 621–633. doi:10.1016/S1534-5807(03)00293-4

Lans, H., & Vermeulen, W. (2011). Nucleotide excision repair in *Caenorhabditis elegans*. *Molecular Biology International*, *2011*, 1–12. doi:10.4061/2011/542795

Lenz, O., Xiong, J., Nelson, M.D., Raizen, D.M., & Williams, J.A. (2015). FMRFamide signaling promotes stress-induced sleep in *Drosophila*. *Brain, Behavior, and Immunity*, *47*, 141–148. doi:10.1016/j.bbi.2014.12.028

Liu, S., Schulze, E., & Baumeister, R. (2012). Temperature- and touch-sensitive neurons couple CNG and TRPV channel activities to control heat avoidance in *Caenorhabditis elegans*. *PLoS One*, *7*, e32360. doi:10.1371/journal.pone.0032360

MacFadyen, U.M., Oswald, I., & Lewis, S.A. (1973). Starvation and human slow wave sleep. *Journal of Applied Physiology*, *35*, 391–394. doi:10.1152/jappl.1973.35.3.391

Monga, U., Kerrigan, A.J., Thornby, J., & Monga, T.N. (1999). Prospective study of fatigue in localized prostate cancer patients undergoing radiotherapy. *Radiation Oncology Investigations, 7,* 178–185. doi:10.1002/(SICI)1520-6823(1999)7:3 < 178::AID-ROI7 > 3.0.CO;2-0

Naidoo, N. (2009). Cellular stress/the unfolded protein response: Relevance to sleep and sleep disorders. *Sleep Medicine Reviews, 13,* 195–204. doi:10.1016/j.smrv.2009.01.001

Nath, R.D., Chow, E.S., Wang, H., Schwarz, E.M., & Sternberg, P.W. (2016). *C. elegans* stress-induced sleep emerges from the collective action of multiple neuropeptides. *Current Biology, 26,* 2446–2455. doi:10.1016/j.cub.2016.07.048

Nelson, M.D., Lee, K.H., Churgin, M.A., Hill, A.J., Van Buskirk, C., Fang-Yen, C., & Raizen, D.M. (2014). FMRFamide-like FLP-13 neuropeptides promote quiescence following heat stress in *Caenorhabditis elegans. Current Biology, 24,* 2406–2410. doi:10.1016/j.cub.2014.08.037

Perluigi, M., Di Domenico, F., Blarzino, C., Foppoli, C., Cini, C., Giorgi, A., ... Coccia, R. (2010). Effects of UVB-induced oxidative stress on protein expression and specific protein oxidation in normal human epithelial keratinocytes: A proteomic approach. *Proteome Science, 8,* 13. doi:10.1186/1477-5956-8-13

Pujol, N., Cypowyj, S., Ziegler, K., Millet, A., Astrain, A., Goncharov, A., ... Ewbank, J.J. (2008). Distinct innate immune responses to infection and wounding in the *C. elegans* epidermis. *Current Biology, 18,* 481–489. doi:10.1016/j.cub.2008.02.079

Rodriguez, L., Snoek, L.B., De Bono, M., & Kammenga, J.E. (2013). Worms under stress: *C. elegans* stress response and its relevance to complex human disease and aging. *Trends in Genetics, 29,* 367–374. doi:10.1016/j.tig.2013.01.010

Saffer, A.M., Kim, D.H., van Oudenaarden, A., & Horvitz, H.R. (2011). The *Caenorhabditis elegans* synthetic multivulva genes prevent RAS pathway activation by tightly repressing global ectopic expression of lin-3 EGF. *PLoS Genetics, 7,* e1002418. doi:10.1371/journal.pgen.1002418

Satterlee, J.S., Sasakura, H., Kuhara, A., Berkeley, M., Mori, I., & Sengupta, P. (2001). Specification of thermosensory neuron fate in *C. elegans* requires ttx-1, a homolog of otd/Otx. *Neuron, 31,* 943–956. doi:10.1016/S0896-6273(01)00431-7

Sugi, T., Nishida, Y., & Mori, I. (2011). Regulation of behavioral plasticity by systemic temperature signaling in *Caenorhabditis elegans. Nature Neuroscience, 14,* 984–992. doi:10.1038/nn.2854

Van Buskirk, C., & Sternberg, P.W. (2010). Paired and LIM class homeodomain proteins coordinate differentiation of the *C. elegans* ALA neuron. *Development, 137,* 2065–2074. doi:10.1242/dev.040881

Van Oosten-Hawle, P., Porter, R.S., & Morimoto, R.I. (2013). Regulation of organismal proteostasis by transcellular chaperone signaling. *Cell, 153,* 1366–1378. doi:10.1016/j.cell.2013.05.015

Wei, J.-Z., Hale, K., Carta, L., Platzer, E., Wong, C., Fang, S.-C., & Aroian, R.V. (2003). *Bacillus thuringiensis* crystal proteins that target nematodes. *Proceedings of the National Academy of Sciences of the United States of America, 100,* 2760–2765. doi:10.1073/pnas0538072100

Zada, D., Bronshtein, I., Lerer-Goldshtein, T., Garini, Y., & Appelbaum, L. (2019). Sleep increases chromosome dynamics to enable reduction of accumulating DNA damage in single neurons. *Nature Communications, 10,* 895. doi: 10.1038/s41467-019-08806-w

Orcokinin neuropeptides regulate sleep in *Caenorhabditis elegans*

Madison Honer, Kristen Buscemi, Natalie Barrett, Niknaz Riazati, Gerald Orlando and Matthew D. Nelson

ABSTRACT

Orcokinin neuropeptides are conserved among ecdysozoans, but their functions are incompletely understood. Here, we report a role for orcokinin neuropeptides in the regulation of sleep in the nematode *Caenorhabditis elegans*. The *C. elegans* orcokinin peptides, which are encoded by the *nlp-14* and *nlp-15* genes, are necessary and sufficient for quiescent behaviors during developmentally timed sleep (DTS) as well as during stress-induced sleep (SIS). The five orcokinin neuropeptides encoded by *nlp-14* have distinct but overlapping functions in the regulation of movement and defecation quiescence during SIS. We suggest that orcokinins may regulate behavioral components of sleep-like states in nematodes and other ecdysozoans.

Introduction

Ecdysozoa is comprised of the most diverse group of animals on earth. This clade includes arthropods and nematodes, as well as other smaller phyla, which are united by having a molting cycle (Aguinaldo *et al.*, 1997). Molts occur periodically during growth and are accompanied by elaborate and specific molting behaviors. In the nematode *C. elegans*, the molting cycle is similar to the circadian cycle in other animals (Hendriks, Gaidatzis, Aeschimann, & Großhans, 2014). Molt timing is regulated by LIN-42, a worm homolog of the circadian protein PERIOD, and behavior during the molt resembles sleep controlled by circadian timing in other animals (Raizen *et al.*, 2008).

Orcokinin neuropeptides are strikingly conserved across Ecdysozoa. They have been described in nematodes (Nathoo, Moeller, Westlund, & Hart, 2001), in several arthropods including cockroaches (Hofer, Dircksen, Tollback, & Homberg, 2005; Hofer & Homberg 2006), kissing bugs (Wulff *et al.*, 2017), fruit flies (Chen *et al.*, 2015), crayfish (Yasuda-Kamatani & Yasuda, 2000), lobsters (Dickinson *et al.*, 2009), and in tardigrades (Koziol, 2018). Orcokinins are related to pedal peptides (Kim, Go, Oh, Elphick, & Park, 2018), identified in mollusks (Lloyd & Connolly, 1989) and to smooth muscle relaxant peptides (SMPs), identified in echinoderms (Kim *et al.*, 2016; Rowe & Elphick, 2012), suggesting that an ancestor to ecdysozoan orcokinins was present in early bilaterians (Jekely, 2013; Semmens & Elphick, 2017).

Orcokinins regulate insect ecdysis and circadian activity. In kissing bugs, disruption of orcokinin signaling causes molting defects (Wulff *et al.*, 2017), perhaps partially due to a role for these peptides in the biosynthesis of the ecdysteroids (Yamanaka *et al.*, 2011; Zitnan *et al.*, 1999). In cockroaches, orcokinin peptides injected into the brain induce a phase shift in circadian sleep/wake behavior (Hofer & Homberg 2006). A role in the regulation of both molting and circadian rhythms suggests the intriguing hypothesis that an ancestral role for orcokinins is in the regulation of behavioral rhythms, specifically sleep/wake behavior. We pursued this hypothesis in *C. elegans*, given the similarity of its molting cycle to circadian rhythms in other organisms.

C. elegans orcokinins are encoded by the genes *nlp-14* and *nlp-15* (Nathoo *et al.*, 2001). NLP-14 peptides modulate cholinergic signaling during male mating (Sherlekar *et al.*, 2013) and mediate decision-making during nociceptive behaviors (Hapiak *et al.*, 2013). *nlp-14* transcripts are upregulated during the 2-h period prior to ecdysis (George-Raizen, Shockley, Trojanowski, Lamb, & Raizen, 2014), when the animals display sleep-like quiescent behavior (Raizen *et al.*, 2008). Single-cell transcriptomic data suggests that *nlp-14* is expressed in the sleep-promoting ALA neuron (Nath, Chow, Wang, Schwarz, & Sternberg, 2016; Taylor *et al.*, 2019), while its paralog *nlp-15* is expressed in both the ALA neuron and in another sleep promoting neuron called RIS (Taylor *et al.*, 2019). Thus, we sought to test the hypothesis that orcokinin neuropeptides regulate sleep.

C. elegans sleep is regulated by neuropeptides. Developmentally timed sleep (DTS) occurs during larval transitions, coincident with the molt (Raizen *et al.*, 2008; Singh & Sulston, 1978). Movement quiescence during DTS is controlled primarily by the RIS neuron which releases FLP-11 neuropeptides (Turek, Besseling, Spies, Konig, & Bringmann, 2016; Turek, Lewandrowski, & Bringmann, 2013) with a minor role for NLP-22 peptides released from the RIA neurons (Nelson *et al.*, 2013). Arousal during DTS is mediated by pigment dispersing factor (PDF) neuropeptides, which also mediate arousal in insects (Choi, Chatzigeorgiou, Taylor, Schafer, & Kaplan, 2013; Renn, Park, Rosbash, Hall, & Taghert, 1999). SIS is controlled by both the ALA and RIS interneurons, via the release of a collection of neuropeptides (Lenz, Xiong, Nelson, Raizen, & Williams, 2015; Nelson *et al.*, 2014; Turek *et al.*, 2016). Based on their spatial and temporal expression patterns, and on their roles in regulating behavioral rhythms in other ecdysozoans, we hypothesized that NLP-14 and NLP-15 play a role in sleep regulation in *C. elegans*.

We combined the analysis of loss-of-function with over-expression to characterize the function of the *C. elegans* orcokinins and find that NLP-14 and NLP-15 are required for movement and defecation quiescence that occur during sleep; NLP-14 peptides play a larger role. This work expands our knowledge of the function of orcokinins and suggests a previously unappreciated role in sleep regulation.

Methods

Worm maintenance and strains

Animals were maintained at 20 °C on agar plates containing nematode growth medium and fed the OP50 derivative bacterial strain DA837 (Davis *et al.*, 1995). The following strains were used in this study:

- N2 (Bristol - *wild type*)
- KG532=*kin-2(ce179)* X
- VC1063=*nlp-15(ok1512)* I
- PS5009=*pha-1(e2132ts)*; *syEx723[hsp-16.2p::lin-3C;myo-2p:gfp; pha-1(+)]*
- SJU6=*stjEx3[hsp-16.2p::nlp-14; myo-2p::mCherry]*
- SJU27=*stjIs2[hsp-16.2p::nlp-14; myo-2p::mCherry]*
- SJU44=*stjEx32[hsp-16.2p::nlp-14*(1-3); *myo-2p::gfp]*
- SJU47=*stjEx36[hsp-16.2p::nlp-14*(1-2); *myo-2p::gfp]*
- SJU95=*stjEx76[ida-1p::mCherry; nlp-14p::gfp]*
- SJU96=*stjEx77[ida-1p::mCherry; nlp-14p::gfp]*
- SJU102=*kin-2(ce179)* X; *stjIs2*
- SJU109=*sjtEx123[hsp-16.2p::nlp-14*(1); *myo-2p::mCherry]*
- SJU110=*sjtEx123[hsp-16.2p::nlp-14*(1); *myo-2p::mCherry]*
- SJU121=*stjEx91[hsp-16.2p::nlp-15; myo-3p::mCherry]*
- SJU122=*stjEx92[hsp-16.2p::nlp-15; myo-3::mCherry]*
- SJU154=*nlp-14(tm1880)* X
- SJU178=*nlp-14(stj10)* X
- SJU207=*nlp-14(tm1880)* X; *stjEx146[ida-1p::nlp-14; myo-3p::mCherry]* (Line#1)
- SJU208=*nlp-14(tm1880)* X; *stjEx147[ida-1p::nlp-14; myo-3p::mCherry]* (Line#2)
- SJU209=*nlp-14(tm1880)* X; *stjEx148[ida-1p::nlp-14;myo-3p::mCherry]* (Line#3)
- SJU232=*stjEx163[hsp-16.2p::nlp-14*(3); *myo-2p::mCherry]*
- SJU233=*nlp-14(stj18)* X
- SJU241=*stjIs160[hsp-16.2p::nlp-14; myo-2p::mCherry]*
- SJU244=*stjEx167[hsp-16.2p::nlp-14*(3); *myo-2p::mCherry]*
- SJU245=*stjEx168[hsp-16.2p::nlp-14*(3); *myo-2p::mCherry]*
- SJU246=*stjEx169[hsp-16.2p::nlp-14*(3); *myo-2p::mCherry]*
- SJU247=*stjEx170[hsp-16.2p::nlp-14*(3); *myo-2p::mCherry]*
- SJU254=*stjEx171[hsp-16.2p::nlp-14*(1-4); *myo-2p::mCherry]*
- SJU255=*stjEx172[hsp-16.2p::nlp-14*(1-4); *myo-2p::mCherry]*
- SJU256=*stjEx173[hsp-16.2p::nlp-14*(1-4); *myo-2p::mCherry]*
- SJU257=*stjEx174[hsp-16.2p::nlp-14*(1-4); *myo-2p::mCherry]*
- SJU258=*stjEx175[hsp-16.2p::nlp-14*(1-4); *myo-2p::mCherry]*
- SJU260=*nlp-14(stj19)* X
- SJU262=*nlp-15(ok1512)* I; *nlp-14(stj18)* X
- SJU272=*nlp-14(stj18)* X; *pha-1(e2132ts)* III; *syEx723[hsp16.2p::lin-3C; myo-2p::gfp; pha-1(+)]*
- SJU273=*nlp-14(tm1880)* X; *pha-1(e2132ts)* III; *syEx723[hsp16.2p::lin-3C; myo-2p::gfp; pha-1(+)]*
- SJU281=*nlp-14(stj13)* X; *nlp-15(stj25)* I
- SJU282=*stjEx184[nlp-15p::gfp;ida-1p::mCherry; myo-3p::mCherry]*
- SJU312=*nlp-15(stj25)* I

Molecular biology and transgenesis

DNA for transgenesis was constructed using overlap extension-polymerase chain reaction (OE-PCR) (Nelson & Fitch, 2011). The promoter of the gene *hsp-16.2* and the coding sequences of *nlp-14* and *nlp-15* were amplified from genomic DNA by PCR. The amplicons were fused together by OE-PCR. To over-express subsets of NLP-14 peptides, the *hsp-16.2* promoter and a portion of the *nlp-14* gene coding for the N-terminal signal peptide and NLP-14(1), (1–2), (1–3) or (1–4) peptide(s), followed by a stop codon, were amplified from genomic DNA. Next, the operon sequence from the genes *gpd-2* and *gpd-3* and the coding sequence for the red fluorescent protein RFP were amplified from the plasmid pLR304. The three amplicons were fused by OE-PCR. To over-express the NLP-14(3) peptide, the plasmid pSJU8 was commercially engineered to contain sequence for the *hsp-16.2* promoter, the coding sequence for the N-terminal signal peptide and NLP-14(3) peptide of the *nlp-14* gene and the 3′untranslated region of the gene *unc-54* (GeneScript ©). The *nlp-14*, *nlp-15* and *ida-1* fluorescent reporters were constructed by amplifying 5′ regulatory DNA for each gene from genomic DNA and the green fluorescent protein (*gfp*) or mCherry coding sequence from the plasmids pPD95.75 or pCFJ90 (Addgene, Watertown, MA). The promoter and *gfp* amplicons were fused for each gene using OE-PCR. To express *nlp-14* in the ALA neuron, 5′ regulatory DNA of the gene *ida-1* was fused by OE-PCR to the *nlp-14* coding sequence. Transgenesis was performed by microinjection, as described (Stinchcomb, Shaw, Carr, & Hirsh, 1985). The strains SJU27, SJU241 and SJU242 were integrated using UV irradiation, as described (Mello & Fire, 1995).

Reverse transcription-PCR (RT-PCR) of *nlp-14(tm1880)* was accomplished by isolating total RNA using an RNeasy mini kit (Qiagen ©, Hilden, Germany), followed by cDNA synthesis using SuperScript™ One-Step RT-PCR System

(Thermo Fisher ©, Waltham, MA). Oligonucleotides used are in Table S1. Extrachromosomal arrays and DNA concentrations are listed in Table S2.

Construction of mutants

SJU178, SJU233, SJU262, SJU260, SJU262 and SJU281 were constructed by CRISPR/Cas9 gene editing. Using a published protocol (Arribere *et al.*, 2014), insertions were made in the *nlp-14* or *nlp-15* gene at defined sites. Simultaneously, an edit of the *dpy-10* gene was made which resulted in an easily identifiable dumpy (Dpy) or roller (Rol) phenotype, to allow for screening. Specifically, a mixture of guide RNA (gRNA) duplexed with Alt-R® CRISPR-Cas9 tracrRNA (IDT ©), Alt-R® S.p. Cas9 Nuclease V3 (IDT) and oligonucleotide repair templates were injected into day-1 adult wild-type or SJU233 animals. Dpy or Rol progeny of the injected animals was transferred to individual plates and maintained to the next generation. The genomic DNA of 10–15 progeny was used as templates for PCR to amplify a portion of the *nlp-14* or *nlp-15* gene. The amplicon was treated with NheI restriction enzyme and analyzed by agarose gel electrophoresis. Fifteen to twenty non-Dpy non-Rol animals from plates with the desired edit were transferred individually to fresh plates and grown to the next generation. These worms were again screened by PCR combined with restriction digest, and the alleles were confirmed by sequencing (Genewiz ©). The sequences of reagents are listed in Table S3.

The strain SJU154 was generated by crossing the nlp-14 (*tm1880*) strain obtained from the National BioResource Project (PI, Shohei Mitani), to male N2 animals, and then crossing resultant males back to tm1880. This procedure was repeated three times to reduce the number of unlinked mutations on the five autosomal chromosomes.

WorMotel behavioral assays

Movement quiescence was quantified during both DTS and stress-induced sleep (SIS), using the WorMotel, as previously described (Churgin *et al.*, 2017). For DTS, we monitored active L4 animals (pre-lethargus) of each genotype for 12-h. Due to day-to-day and chip-to-chip variability in sleep, we statistically compared strains housed in different wells of the same WorMotel. A combination of 24 wild-type, mutant, and/or transgenic active L4 animals were picked onto the agar surfaces of individual wells of the WorMotel polydimethylsiloxane (PDMS) chip. Images were captured every 10-s for 12 h, and quiescence was quantified and DTS was manually measured based on a definable peak of quiescence, as previously described (Raizen *et al.*, 2008). For SIS, a combination of 24 wild-type, mutant, and/or transgenic day-1 adults were picked individually onto the agar surfaces of a welled PDMS microchip. The chip was placed into a UV-cross linker (Ultraviolet, 254 UVP) and exposed to 1500 J/m^2 of UV light to induce SIS (DeBardeleben, Lopes, Nessel, & Raizen, 2017). For over-expression experiments, day-1 adults were heat-shocked on standard growth plates, by submerging them, wrapped in para-film, in a 33 °C water bath for 30 min. The heat-shocked animals were individually transferred to the agar surfaces of a welled PDMS microchip. For SIS and over-expression, images were captured every 10-s for 8 or 4 h, respectively, and total minutes of quiescence was determined.

Body bending analysis following over-expression

Day-1 adults were heat-shocked by submerging standard growth plates, wrapped in para-film, in a 33 °C water bath for 30 min. Body bends were counted manually using a stereomicroscope for 60-s, 2 h after heat exposure. A body bend was defined as the movement of the body just posterior to the pharynx to the opposite position from the previous maximal bend.

Defecation analysis during SIS and following overexpression

The defecation cycle was measured manually for 5–6 min by visual inspection using a stereomicroscope, using described criteria (Thomas, 1990). For over-expression, day-1 adults, on standard growth plates, were submerged, wrapped in para-film, in a 33 °C water bath for 30 min and analyzed for 5 min between 2 and 2.5 h after heat exposure. For SIS, day-1 adults were exposed to 1500 J/m^2 of UV light in a UV-cross linker (Ultraviolet, 254 UVP) on growth plates and the defecation cycle was measured for 5–6 min, 85–95 min post-UV. For temporal SIS analyses, a single animal was examined for 5-min every 30 min for 4-h, post-UV.

Microscopy

Fluorescence microscopy was conducted using an Olympus BX63 wide-field fluorescence microscope equipped with a Hamamatsu FLASH 4.0V3 digital camera and CellSens Dimension Version 2 software. Day-1 adult transgenic animals were immobilized on glass slides containing a 5% agar pad supplemented with 25 mM sodium azide.

Alignments

Peptide sequences were obtained from the National Center for Biotechnology Information (NCBI) and aligned using online T-coffee software (Notredame, Higgins, & Heringa, 2000). Peptide alignments were annotated using Boxshade (https://embnet.vital-it.ch/software/BOX_doc.html).

Results

NLP-14 and NLP-15 are orcokinin homologs expressed in the ALA and RIS neurons

Neuropeptides coded by the genes *nlp-14* and *nlp-15* are classified as orcokinins based on sequence similarity to insect and crustacean peptides (Nathoo *et al.*, 2001). All five NLP-14 peptides are conserved at the C-terminus, while the NLP-15 peptides show greater conservation at the N-terminus (Figure 1(A)). *nlp-14* expression has been

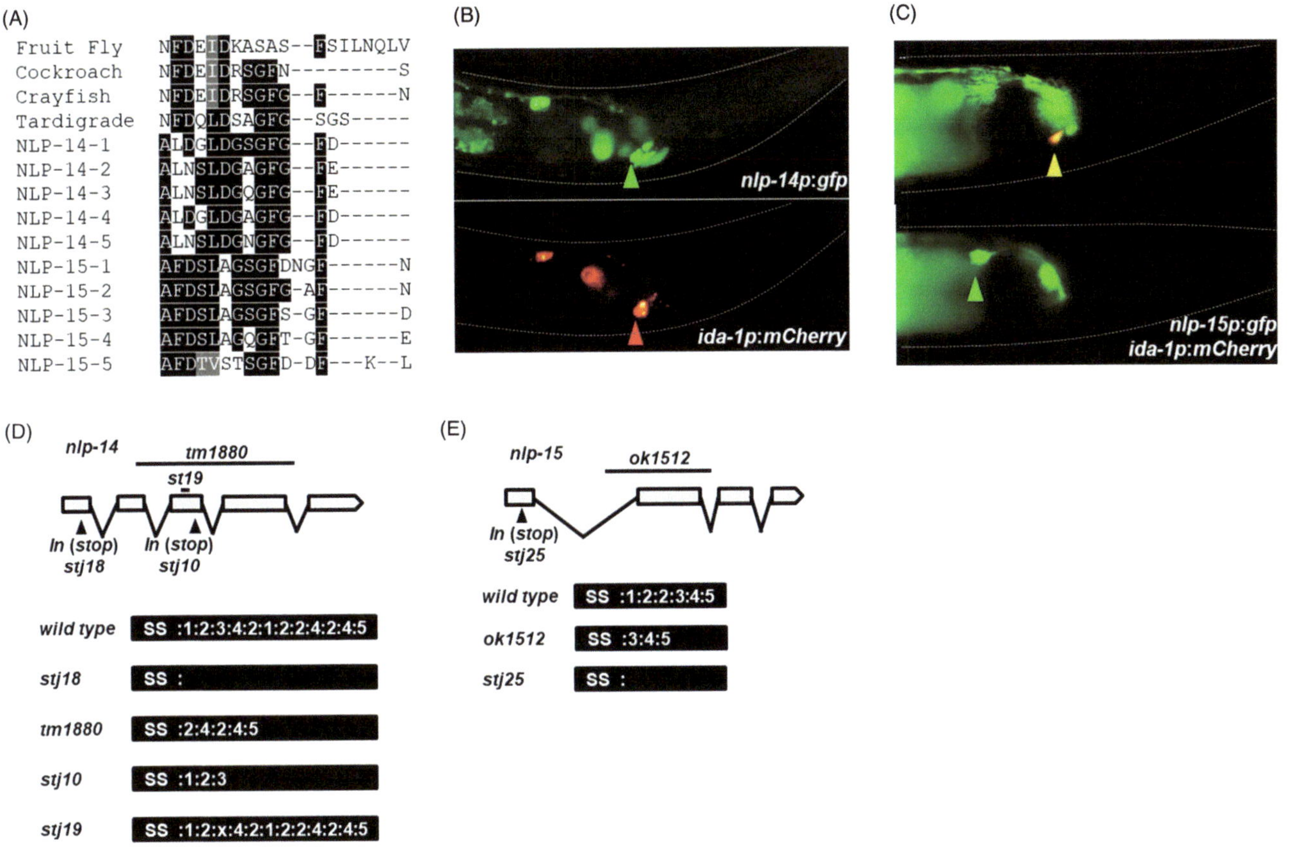

Figure 1. NLP-14 and NLP-15 neuropeptides are related to arthropod orcokinins and are expressed in sleep-promoting cells. (A) Peptide alignment of orcokinins from: *Drosophila melanogaster* (fruit fly), *Blattella germanica* (cockroach), *Orconectes limosus* (crayfish), *Hypsibius dujardini* (Tardigrade) and *C. elegans* (Nematode). (B) Representative images of an animal expressing *gfp* from the promoter of *nlp-14* (top image) and mCherry from the promoter of *ida-1* (bottom image). ALA expression is denoted by arrowheads. Anterior – right; dorsal – bottom, ventral – top. (C) Representative images of an animal expressing *gfp* from the promoter of *nlp-15* and the same image superimposed with mCherry from the promoter of *ida-1*. Expression in the ALA (top - gfp and mCherry) and RIS (bottom - gfp only) are denoted by arrowheads. Anterior – right; dorsal – bottom, ventral – top. (D) Gene and protein structure for NLP-14, highlighting the location of deletion and insertion alleles. (E) Gene and protein diagram for NLP-15, highlighting the location of a deletion and insertion allele. SS denotes signal sequence.

demonstrated in the ventral cord and some sensory and interneurons (Nathoo *et al.*, 2001), as well as in male-specific neurons (Sherlekar *et al.*, 2013). Single-cell gene expression studies revealed enrichment of both *nlp-14* and *nlp-15* transcripts in the ALA neuron and enrichment of *nlp-15* (but not *nlp-14*) in the RIS neuron (Nath *et al.*, 2016; Taylor *et al.*, 2019). ALA and RIS are central sleep-promoting neurons (Hill, Mansfield, Lopez, Raizen, & Van Buskirk, 2014; Konietzka *et al.*, 2020; Turek *et al.*, 2013). In support of the single-cell transcriptomic data, we found that a GFP transcriptional reporter for *nlp-14* colocalizes with an mCherry transcriptional reporter for *ida-1*, which is strongly expressed in ALA (Zahn, Macmorris, Dong, Day, & Hutton, 2001) (Figure 1(B)). Similarly, a GFP transcriptional reporter for *nlp-15* showed expression in both ALA and RIS neurons (Figure 1(C)). The combination of expression during the molt, a *C. elegans* sleep state, with expression in sleep-regulating neurons led us to hypothesize that *nlp-14* and *nlp-15* regulate sleep.

To test this hypothesis, we obtained mutant strains for *nlp-14* and *nlp-15*, which carry the deletion alleles, *tm1880* and *ok1512*, respectively. *tm1880* is predicted to cause an in-frame deletion that preserved the signal sequence as well as peptides 2, 4, and 5 of *nlp-14*. We confirmed this *in silico* prediction using RT-PCR (Figure 1(D)). We refer to *tm1880* from hereon as *nlp-14*(2,4,5). *ok1512* too is predicted to cause an in-frame deletion, which preserves the signal sequence as well as peptides 3–5 of *nlp-15* (Figure 1(E)). We refer to *ok1512* as *nlp-15*(3–5).

To construct complete loss-of-function mutants as well as other mutants, we used CRISPR/Cas9 gene editing technology (Paix, Folkmann, & Seydoux, 2017). The *nlp-14*(*stj18*) strain contains a stop codon in the first exon, after the signal sequence but prior to the sequence encoding all NLP-14 peptides (Figure 1(D)). Similarly, the *nlp-15*(*stj25*) strain contains an insertion of a stop codon 5′ of all five NLP-15 peptides (Figure 1(E)). Hence, *stj18* and *stj25* are predicted to make null alleles of *nlp-14* and *nlp-15*, respectively, so we will refer to them as *nlp-14*(*null*) and *nlp-15*(*null*). The *nlp-14*(*stj10*) strain carries an insertion of a stop codon at the 3′-end of the sequence encoding peptide 3 so we will refer to it as *nlp-14*(*1-3*). The *nlp-14*(*stj19*) strain contains an in-frame deletion that removes only peptide 3, so we will refer to it from hereon as *nlp-14*(*1, 2, 4, 5*).

The C. elegans orcokinins are required for DTS

DTS occurs prior to ecdysis in *C. elegans* (Raizen *et al.*, 2008; Singh & Sulston, 1978). Orcokinins regulate ecdysis

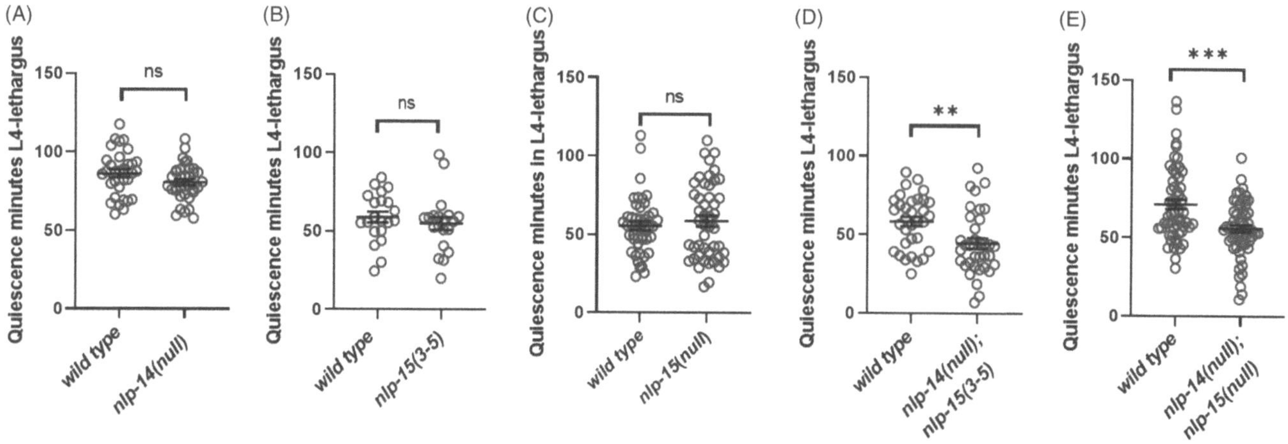

Figure 2. The *C. elegans* orcokinins play a small role during DTS. (A) Movement quiescence during L4 lethargus in wild-type and *nlp-14(stj18)* animals ($N \geq 32$). (B) Movement quiescence during L4 lethargus in wild-type and *nlp-15(ok1512)* animals ($N \geq 20$). (C) Movement quiescence during L4 lethargus in wild-type and *nlp-15(stj25)* animals ($N \geq 46$). (D) Movement quiescence during L4 lethargus in wild-type and *nlp-14(stj18); nlp-15(ok1512)* animals ($N \geq 33$, $^{**}p<.01$). (E) Movement quiescence during L4 lethargus in wild-type and *nlp-14(stj18); nlp-15(stj25)* animals ($N \geq 58$, $^{***}p<.001$). Statistical significance was calculated using Student's *t*-test. All error bars represent mean ± SEM.

and ecdysteroid biosynthesis in insects (Wulff *et al.*, 2017; Yamanaka *et al.*, 2011). We hypothesized that DTS and/or the molt would be disrupted in *nlp-14* and/or *nlp-15* mutants. Using the WorMotel (Churgin *et al.*, 2017), we found that DTS was unaltered in *nlp-14(null)*, *nlp-15(3-5)* or *nlp-15(null)* single mutant animals (Figure 2(A–C)). As we suspected there might be functional redundancy between the two genes, we tested animals that were mutant for both *nlp-14(null)* and *nlp-15(3-5)* and *nlp-14(null)* and *nlp-15(null)*. DTS was modestly reduced in both double mutants (Figure 2(D,E)). We found no molting defects by carefully scanning double mutant strains via stereomicroscopy or by inspecting select animals at 1000× using wide-field differential interference contrast (DIC) microscopy. Based on these results, we conclude that NLP-14 and NLP-15 are required for movement quiescence during DTS but play a minor role in this behavior perhaps due to compensatory action of other neuropeptides (Nelson *et al.*, 2013; Turek *et al.*, 2016). They are not required for the successful completion of the molt.

NLP-14 peptides are required for movement quiescence during SIS

Since *nlp-14* and *nlp-15* are expressed in the RIS and/or ALA neurons, which are central regulators of SIS (Hill *et al.*, 2014; Konietzka *et al.*, 2020), we tested their necessity for movement quiescence. SIS was induced by UV irradiation (DeBardeleben *et al.*, 2017) and animals were monitored on a WorMotel (Churgin *et al.*, 2017). The *nlp-14(null)*, *nlp-14(2,4,5)* and *nlp-14(1-3)* animals all displayed reductions in movement quiescence (Figure 3(A–C,H,I); Figure S1), while *nlp-14(1,2,4,5)*, *nlp-15(3-5)* and *nlp-15(null)* animals did not (Figure 3(D,E); Figure S1). *nlp-14(null)*; *nlp-15(3-5)* and *nlp-14(null)*; *nlp-15(null)* double mutants displayed reduced total movement quiescence similar to that observed in *nlp-14(null)* single mutants; i.e. *nlp-15* mutations did not enhance the phenotype caused by *nlp-14* mutations (Figure 3(F,G,J); Figure S1). However, the *nlp-15* mutations increased the variance of the *nlp-14(null)* phenotype (p=.03, *nlp-14(null);nlp-15(null)*; p=.003, *nlp-14(null);nlp-15(3-5)*; Levene's test). Based on these results, we conclude that one or more of the NLP-14 peptides (but not peptide 3) are required for movement quiescence during SIS and that NLP-15 peptides are likely dispensable for SIS total quiescence, but modulate the *nlp-14* phenotype in some way.

The timing of movement quiescence is regulated by NLP-14 and NLP-15 peptides

We noted that the *nlp-14(null)*, *nlp-14(2,4,5)* and *nlp-14(1-3)* mutants all displayed movement quiescence earlier than wild-type controls (Figure 3(H,I)). In the first hour post UV-stress, each mutant displayed significantly more quiescence than wild-type animals (Figure S2). *nlp-14(2,4,5)* mutants had the most severe defect. We expected that *nlp-15* mutants may enhance this phenotype but, to our surprise, the early quiescence phenotype of *nlp-14* mutants was suppressed rather than enhanced by *nlp-15* mutations. *nlp-14(null)*; *nlp-15(3-5)* and *nlp-14(null)*; *nlp-15(null)* double mutants did not show these timing defects (Figure 3(J); Figure S2). These data suggest that removal of *nlp-14* alters the timing of SIS through *nlp-15*-dependent mechanisms.

NLP-14 peptides are required for defecation quiescence during SIS

Insect and crustacean orcokinins regulate rhythmic intestinal muscle contractions (Chen *et al.*, 2015; Hofer & Homberg 2006; Stangier, Hilbich, Burdzik, & Keller, 1992). In *C. elegans*, the rhythmic defecation cycle is inhibited during SIS. Defecation is precisely timed by NLP-40 neuropeptides released from the posterior intestines (Wang *et al.*, 2013), and during sleep it is partially inhibited by peptides released from the ALA (Nath *et al.*, 2016). Defecation consists of three behaviors, which occur every 50–60 s in the following order: a posterior body contraction (pBoc), an anterior body contraction (aBoc) and an expulsion (Exp) (Thomas, 1990). The defecation rates of unstressed wild-type and *nlp-14(2, 4, 5)*

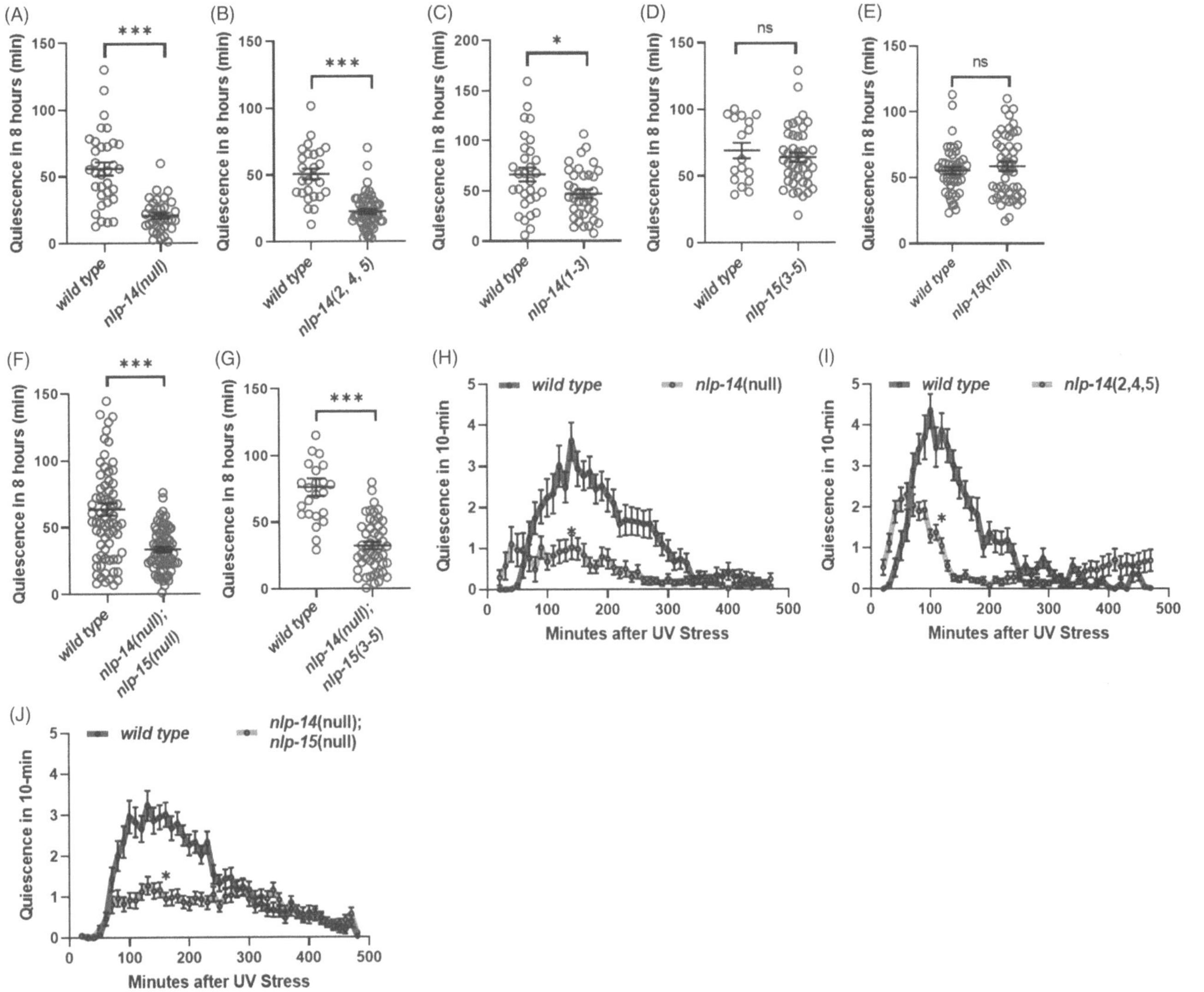

Figure 3. NLP-14 neuropeptides are required for movement quiescence during SIS. (A) Movement quiescence during UV-induced SIS in wild-type and *nlp-14(stj18)* animals ($N \geq 33$, ***$p<.001$). (B) Movement quiescence during UV-induced SIS in wild-type and *nlp-14(tm1880)* animals ($N \geq 26$, ***$p<.001$). (C) Movement quiescence during UV-induced SIS in wild-type and *nlp-14(stj10)* animals ($N \geq 30$, *$p<.05$). (D) Movement quiescence during UV-induced SIS in wild-type and *nlp-15(ok1512)* animals ($N \geq 17$). (E) Movement quiescence during UV-induced SIS in wild-type and *nlp-15(stj25)* animals ($N \geq 33$). (F) Movement quiescence during UV-induced SIS in wild-type and *nlp-14(stj18); nlp-15(stj25)* animals ($N \geq 68$, ***$p<.001$). (G) Movement quiescence during UV-induced SIS in wild-type and *nlp-14(stj18); nlp-15(ok1512)* animals ($N \geq 24$, ***$p<.001$). (A–G) Statistical significance was calculated using Student's *t*-test. (H) Average quiescence in 10-min windows over 8-h during UV-induced SIS of wild-type and *nlp-14(stj18)* animals ($N \geq 33$, *$p<.01$ at 70, 90–270 min). (I) Average quiescence in 10-min windows over 8-h during UV-induced SIS of wild-type and *nlp-14(tm1880)* animals ($N \geq 26$, *$p<.01$ at 40, 50, 80–190 min). (J) Average quiescence in 10-min windows over 8-h during UV-induced SIS of wild-type and *nlp-14(stj18); nlp-15(stj25)* animals ($N \geq 68$, *$p<.01$ at 80–230 min). (H–J) Statistical significance was calculated using two-way ANOVA followed by Sidak's multiple comparisons test. All error bars represent mean ± SEM.

animals were similar (Figure 4(A)); however, in the 10-min period beginning 85 min after UV exposure, the number of expulsions was significantly increased in the *nlp-14(2,4,5)* and *nlp-14(null)* animals (Figure 4(B,D)). We also observed that *nlp-14(2,4,5)* animals performed more pBoc and aBoc events without an Exp (Figure 4(C)). In contrast, *nlp-14(1-3)* mutant animals did not display defects in defecation quiescence (Figure 4(E)), suggesting that NLP-14 peptides 4 and 5 are not needed for this behavior. Based on defects observed in the various mutants, we conclude that the NLP-14 peptides 1 and/or peptide 3 are required for the quiescence of defecation during SIS.

We performed a temporal analysis of defecation quiescence by counting expulsions for 5-min every 30 min after UV irradiation (Table S4). As expected, wild-type animals slowed their defecation rate throughout the first 4 h of UV-induced SIS. Also as expected, *nlp-14(2,4,5)* and *nlp-14(null)* animals displayed more frequent events, but there was high variation between animals (Figure 4(F,G), Table S4). Both the *nlp-14(1-3)* and *nlp-14(1,2,4,5)* animals showed defecation temporal profiles similar to wild type (Figure 4(H); Table S4). However, the *nlp-14(1-3)* mutant animals showed reduced expulsions throughout SIS, which is significantly lower than wild-type animals at later time points. The *nlp-15(3-5)* single mutants showed no defect in defecation quiescence (Table S4); surprisingly, both *nlp-14(null); nlp-15(3-5)* and *nlp-14(null); nlp-15(null)* double mutants also were similar to wild-type animals during these 4 h (Table S4). Taken together, these data suggest that NLP-14 peptides 1 and 3 are required for defecation quiescence. However,

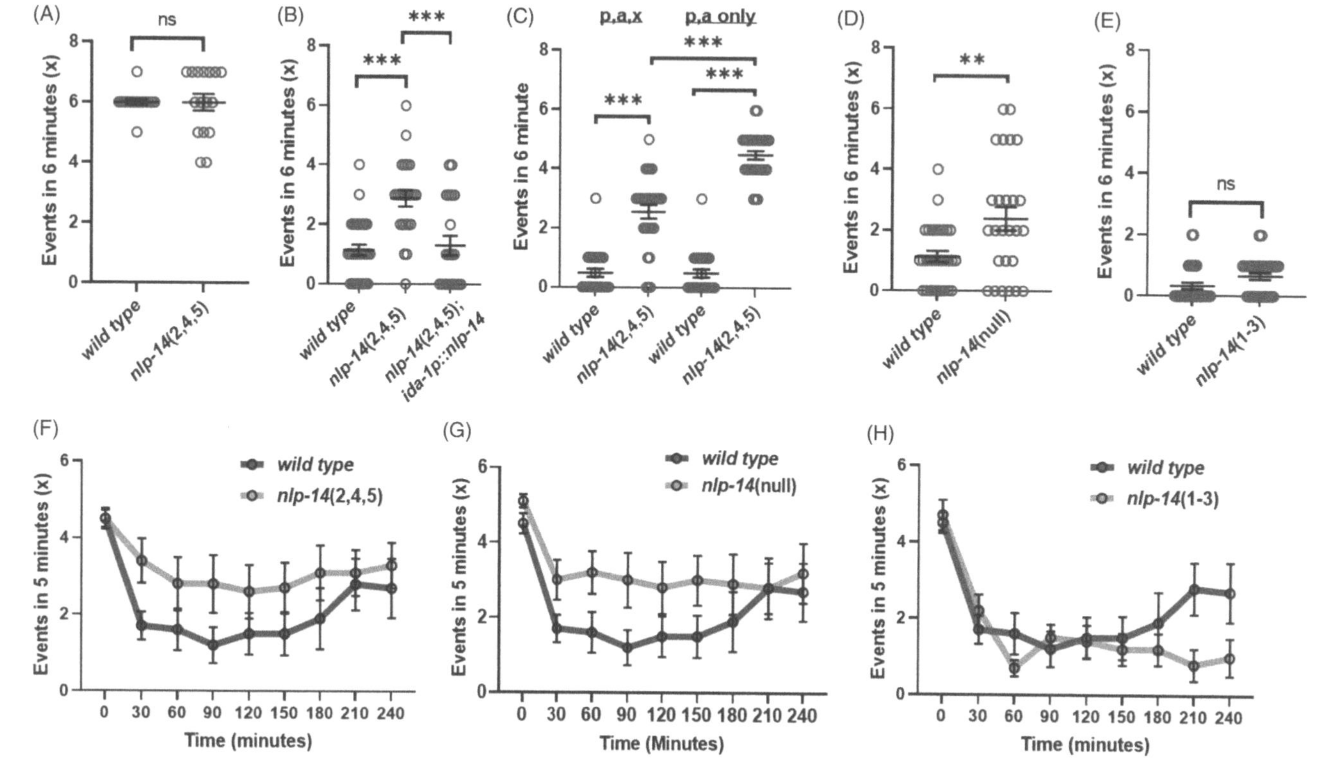

Figure 4. NLP-14 neuropeptides are indispensable for defecation quiescence during SIS. (A) Expulsions (x) performed in 6-min, of wild-type and *nlp-14(tm1880)* animals, who had not been exposed to any external stress. (B) Wild-type, *nlp-14(tm1880)* and *nlp-14(tm1880); ida-1p::nlp-14* animals 85–95 min following UV-stress ($N \geq 20$, ***$p<.001$). (C) p- (posterior body contraction or pBoc) and a- (anterior body contraction or aBoc) events or p-, a- and x-events in 6-min of wild-type and *nlp-14(tm1880)* animals, 85–95 min post-UV stress ($N \geq 26$, ***$p<.001$). (D) x-events performed in 6-min, of wild-type and *nlp-14(stj18)* animals ($N \geq 26$, **$p<.01$) and (E) wild-type and *nlp-14(stj10)* animals ($N=30$). (F) Defecation quiescence profile of *nlp-14(tm1880)*, (G) *nlp-14(stj18)*, and (H) *nlp-14(stj10)* animals ($N=10$, *$p<.05$, 210 min). (A, D, E) Statistical significance was calculated using Student's *t*-test. (B, C) Statistical significance was calculated using one-way ANOVA followed by Tukey's multiple comparisons test. (H) Statistical significance was calculated using two-way ANOVA followed by Sidak's multiple comparisons test. All error bars represent mean ± SEM.

similar to what was observed with the timing defects of *nlp-14(null)* and *nlp-14(2,4,5)* animals, the defecation defects may be dependent upon the presence of *nlp-15*.

NLP-14 peptides are secreted from the ALA during SIS

Since *nlp-14* is required for quiescence during SIS and is expressed in the ALA neuron (Nath *et al.*, 2016; Taylor *et al.*, 2019), we predicted that quiescence induced by strong activation of this neuron would be blunted in *nlp-14* mutants. Epidermal growth factor (EGF) induces sleep in mammals (Kramer *et al.*, 2001; Kushikata, Fang, Chen, Wang, & Krueger, 1998), *Drosophila* (Foltenyi, Greenspan, & Newport, 2007) and *C. elegans* (Van Buskirk & Sternberg, 2007), and acts during SIS by stimulating neuropeptide release from both the ALA and RIS neurons (Konietzka *et al.*, 2020; Nath *et al.*, 2016; Nelson *et al.*, 2014). Overexpression of *lin-3*, coding for EGF, induces movement quiescence (i.e. EGF-induced sleep) (Van Buskirk & Sternberg, 2007). Using the WorMotel, we found that *lin-3* overexpression caused prolonged movement quiescence in otherwise wild-type animals (Figure 5(A,B)), but this quiescence was significantly attenuated in *nlp-14(null)*, *nlp-14(2,4,5)* and *nlp-14(1-3)* mutant animals (Figure 5(A,B)). These results suggest that one or more NLP-14 peptides are required for EGF-induced sleep.

In addition to quiescence of body and feeding movements, *lin-3* over-expression also induced quiescence of defecation in wild-type animals, where not a single animal performed a pBoc, aBoc, or Exp (Figure 5(C–E)). EGF-induced defecation quiescence was variably attenuated (i.e. defecation events were observed) in *nlp-14(null)* and in *nlp-14(2,4,5)* mutants (Figure 5(C–E)). We conclude that NLP-14 peptides are functioning downstream of EGF signaling to promote quiescence of both movement and defecation.

To test whether activity of NLP-14 peptides in ALA is sufficient to restore quiescent behavior, we made transgenic animals in which *nlp-14* expression was controlled by the *ida-1* promoter; *ida-1* is expressed strongly in ALA, but is also expressed in a few other neurosecretory neurons (Zahn *et al.*, 2001). In three independent transgenic lines, movement quiescence during SIS was significantly increased relative to *nlp-14(2,4,5)* mutants (Figure 5(F)). Also, the timing defects we observed in *nlp-14(2,4,5)* animals were corrected by expressing *nlp-14* from the *ida-1* promoter (Figure S2). Additionally, defecation quiescence during SIS was much more prevalent in these *ida-1p::nlp-14* transgenic animals (Figure 4(B)) and, in fact, it was reduced below that of wild-type control levels (Table S4). Based on expression pattern, the requirement of *nlp-14* for EGF-induced sleep and rescue from the *ida-1* promoter, our data suggest that the NLP-14 peptides are released from the ALA neuron to regulate both

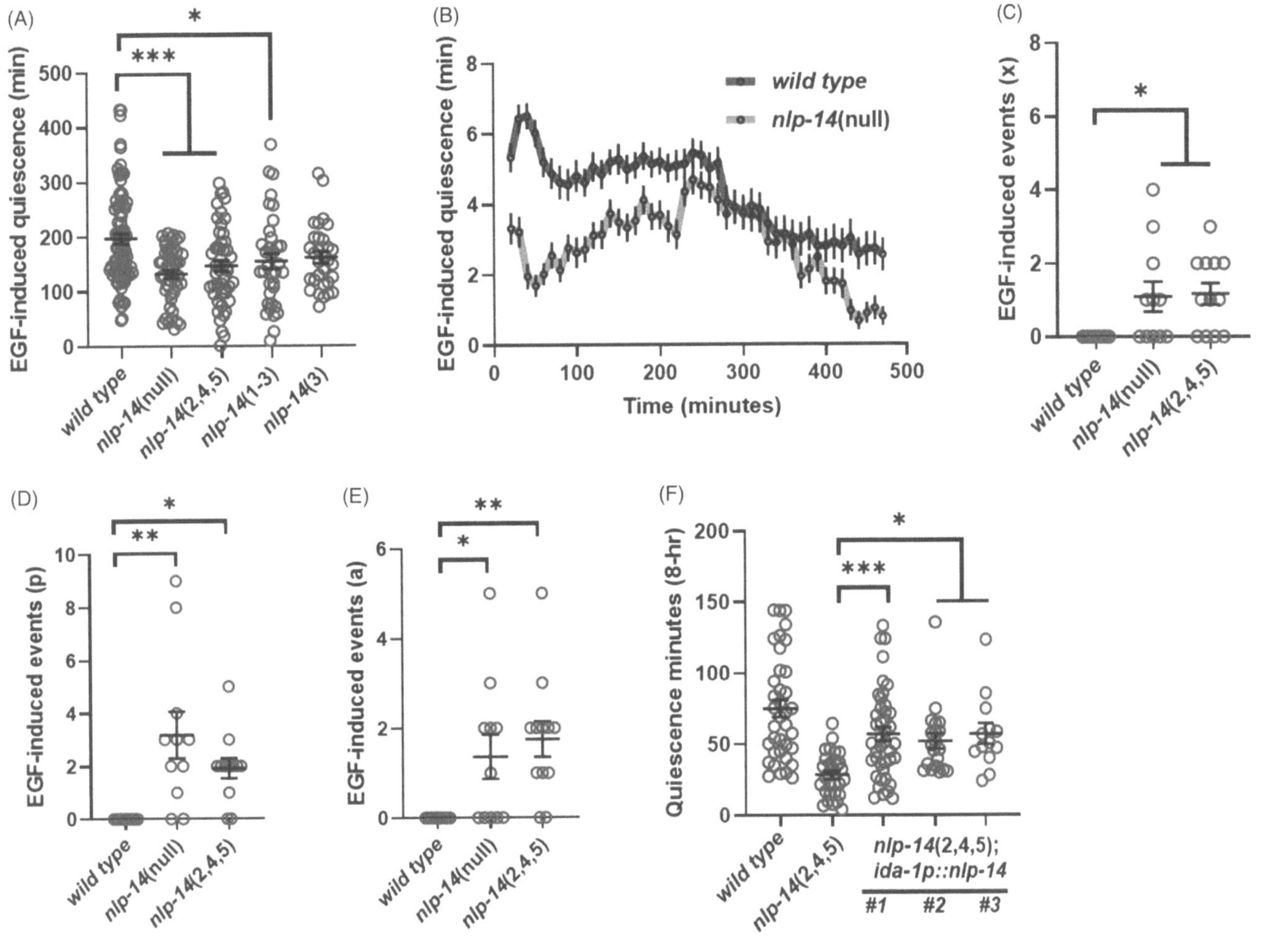

Figure 5. NLP-14 peptides function downstream of EGF from the ALA neuron. (A) Total minutes of quiescence in 4-h, following overexpression of *lin-3*, in the following genetic backgrounds: wild type, *nlp-14(stj18)*, *nlp-14(tm1880)*, *nlp-14(stj10)*, and *nlp-14(stj19)* ($N \geq 31$, $^*p<.05$, $^{***}p<.001$). (B) Average quiescence in 10-min windows over 4-h following *lin-3* over-expression in wild-type and *nlp-14(stj18)* animals ($N = 48$, $p<.05$, 100–120 min, $p<.001$, 20–80 min). Statistical significance was calculated using two-way ANOVA followed by Sidak's multiple comparisons test. (C) x-, (D) p- and (E) a-events performed in 5 min, 2-h after the overexpression of *lin-3* in wild-type, *nlp-14(stj18)* and *nlp-14(tm1880)* animals ($N \geq 10$, $^*p<.05$, $^{**}p<.01$). (F) Movement quiescence during UV-induced SIS in wild-type, *nlp-14(tm1880)* and *nlp-14(tm1880); ida-1p::nlp-14* animals ($N \geq 11$, $^*p<.05$, $^{***}p<.001$). Numbers below *nlp-14(tm1880); ida-1p::nlp-14* indicate distinct transgenic lines. (A, C–F) Statistical significance was calculated using one-way ANOVA followed by Tukey's multiple comparisons test. All error bars represent mean ± SEM.

movement and defecation quiescence during SIS. However, our data do not rule out the possibility that NLP-14 peptides are released from other cells as well.

Overexpression of nlp-14 induces movement and defecation quiescence

We predicted that overexpression of *nlp-14* would induce quiescence of movement and defecation in active animals, like that observed for other somnogenic neuropeptides such as *flp-11*, *flp-13*, *flp-24*, *nlp-8*, and *nlp-22* (Nath *et al.*, 2016; Nelson *et al.*, 2013, 2014; Turek *et al.*, 2016). We constructed multiple transgenic lines in which *nlp-14* expression is controlled by a heat-inducible promoter. To induce strong pan-somatic expression of the gene, we subjected *hsp-16p::nlp-14* animals to a 30-min heat pulse and then waited 2 h before analysis of behavior. At this 2-h time point, any direct effect of heat on behavior, which is minor at temperatures less than 35°, had fully dissipated. Wild-type control animals were exposed to the same conditions.

Overexpression of *nlp-14* strongly suppressed body movement, which we measured by counting body bends (Figure 6(A); Table S5) and by using machine vision, the WorMotel (Figure 6(B,C)). Overexpression of *nlp-14* also caused a significant reduction in defecation events (Figure 6(D); Table S6). Thus, NLP-14 peptides are both required for and capable of inducing quiescence of movement and defecation. In addition to the movement and defecation phenotypes we were focused on, we incidentally noted that many *hsp-16.2p::nlp-14* transgenic animals, even before induced overexpression, displayed a kinked body posture phenotype, where their body resembled a question mark (Video S1).

Many neuropeptides signal through GPCRs, which increase or decrease signaling of second messenger pathways. Movement quiescence is antagonized by signaling through the cyclic adenosine monophosphate/protein kinase A pathway (Cianciulli *et al.*, 2019). In *C. elegans*, PKA activity can be experimentally increased by genetic impairment in the gene *kin-2*, which encodes a regulatory subunit of PKA (Charlie, Thomure, Schade, & Miller, 2006). We found that the increased PKA activity of *kin-2(ce179)* mutants (Charlie *et al.*, 2006), stimulated movement but not

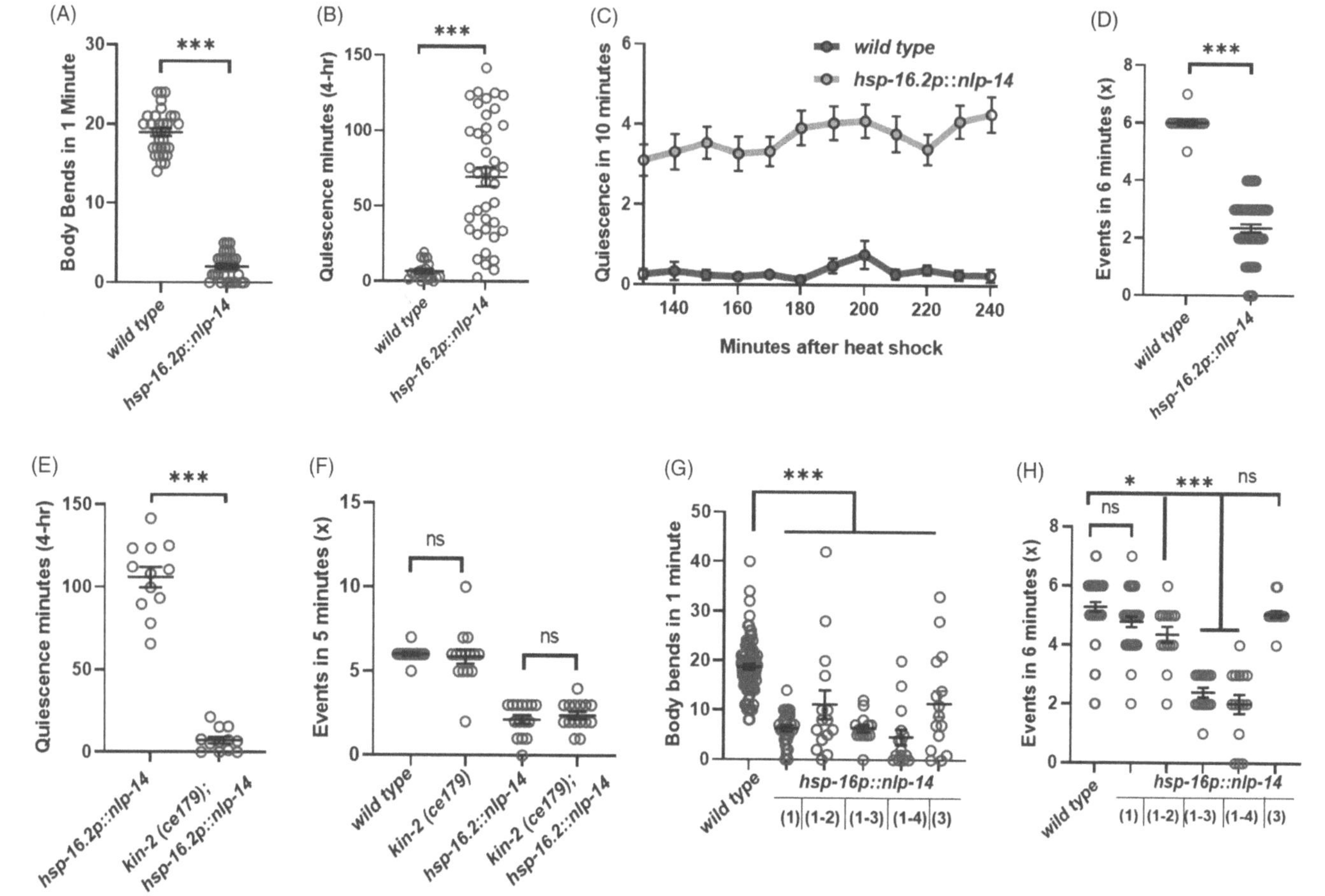

Figure 6. NLP-14 peptides are capable of inducing sleep-like behaviors. (A) Body bends performed in 1 min, 2–2.5-h after a 30-min, 33 °C heat shock, by wild-type and *hsp-16.2p::nlp-14* (strain SJU27) animals ($N = 32$, ***$p < .001$). (B) Movement quiescence during 4-h following a 30-min, 33 °C heat shock, by wild-type and *hsp-16.2p::nlp-14* (strain SJU27) animals ($N \geq 19$, ***$p < .001$). (C) Average quiescence in 10-min windows over 4-h following a 30-min, 33 °C heat shock, wild-type and *hsp-16.2p::nlp-14* (strain SJU27) animals ($N \geq 19$, ***$p < .001$ at all time points displayed). (D) x-events performed in 6-min, 2–2.5-h after a 30-min, 33 °C heat shock, by wild-type and *hsp-16.2p::nlp-14* (strain SJU27) animals ($N = 16$, ***$p < .001$). (E) Movement quiescence during 4-h following a 30-min, 33 °C heat shock, *hsp-16.2p::nlp-14* (strain SJU27) and *kin-2(ce179); hsp-16.2p::nlp-14* animals ($N = 12$, *$p < .001$). (F) x-events performed in 5-min by wild-type and *kin-2(ce179)* animals without exposure to heat shock and by *hsp-16.2p::nlp-14* (strain SJU27) and *kin-2(ce179); hsp-16.2p::nlp-14* animals 2–2.5-h after a 30-min, 33 °C heat shock ($N = 15$). (G) Body bends performed in 1-min, 2–2.5-h after a 30-min, 33 °C heat shock, by wild type and animals over-expressing *nlp-14*(1), (1–2), (1–3), (1–4) or (3) ($N \geq 15$, ***$p < .001$). (H) x-events performed in 5-min, 2–2.5-h after a 30-min, 33 °C heat shock, by wild type and animals overexpressing *nlp-14*(1), (1–2), (1–3), (1–4) or (3) ($N \geq 15$, *$p < .05$, ***$p < .001$). (A, B, D, E) Statistical significance was calculated using Student's *t*-test. (C) Statistical significance was calculated using two-way ANOVA followed by Sidak's multiple comparisons test. (F–H) Statistical significance was calculated using one-way ANOVA followed by Tukey's multiple comparisons test. All error bars represent mean ± SEM.

defecation following *nlp-14* overexpression (Figure 6(E,F)). These data suggest that NLP-14 peptides inhibit cAMP/PKA in cells regulating movement quiescence.

Based on our analysis of different *nlp-14* loss-of-function alleles, suggesting that removal of subsets of NLP-14 peptides affected behavioral quiescence in unique ways, we over expressed different combinations of the peptides. Overexpression of NLP-14-1 or NLP-14-3 induced quiescence of movement but not defecation, while overexpression of NLP-14-1 and NLP-14-2 strongly induced quiescence of movement and weakly of defecation. Overexpression of NLP-14-(1-3) or NLP-14-(1-4) strongly induced quiescence of both movement and defecation (Figure 6(G,H)). These data, together with the loss-of-function analyses, suggest that all five NLP-14 peptides regulate movement quiescence; defecation quiescence, however, is prominently regulated by peptides 1 and 3, while the other peptides may play more subtle modulating roles.

Orcokinin receptors are unknown for all ecdysozoans

Orcokinin receptors have not been identified in any animal, despite screening attempts using heterologous expression systems (Yamanaka *et al.*, 2010, 2011). In *C. elegans*, the receptor NPR-10 has been proposed as an NLP-14 receptor, based on genetic interactions and anatomical connectivity (Hapiak *et al.*, 2013). We did not detect changes in movement quiescence during SIS in the presumptive *npr-10(ok1442)* null mutants. *ok1442* mutants have a 788 bp deletion that is predicted to result in a frameshift and premature stop in exon 5 and therefore a truncated protein composed of only five transmembrane domains (Figure S1). Our data suggest that NPR-10 is not the receptor for NLP-14 during sleep regulation or that other receptors function redundantly together with NPR-10. To date, no orcokinin receptor for any animal has been convincingly identified.

Discussion

Orcokinin neuropeptides are conserved in Ecdysozoa, which consists of organisms that undergo molting (Aguinaldo *et al.*, 1997). Millions of years of evolution separate these animals, yet orcokinin peptide sequences are highly similar (Chen *et al.*, 2015; Dickinson *et al.*, 2009; Hofer *et al.*, 2005; Hofer & Homberg 2006; Koziol, 2018; Nathoo *et al.*, 2001; Wulff *et al.*, 2017; Yasuda-Kamatani & Yasuda, 2000). Functional studies have demonstrated that they regulate insect circadian rhythms and molting (Hofer & Homberg 2006; Wulff *et al.*, 2017), rhythmic smooth muscle contractions of insects and crustaceans (Li *et al.*, 2002; Skiebe, Dreger, Meseke, Evers, & Hucho, 2002; Stangier *et al.*, 1992) and decision making behaviors and male mating in *C. elegans* (Hapiak *et al.*, 2013; Sherlekar *et al.*, 2013). Here, we described a novel function for the orcokinins encoded by *nlp-14* and *nlp-15* during the regulation of sleep.

Using a combination of loss-of-function and overexpression studies we find that NLP-14 and NLP-15 peptides regulate two sleep states, DTS, which resembles sleep in animals that are strongly circadian (Trojanowski & Raizen, 2016), and SIS, a behavior required for recovery following exposure to damaging stress (Hill *et al.*, 2014). The five NLP-14 peptides play a larger role in the regulation of SIS than DTS. They promote movement and defecation quiescence, the latter of which is largely regulated by NLP-14 peptides 1 and 3. We propose that these sleep-regulatory roles are more conserved in Ecdysozoa.

Developmentally timed sleep

Both *nlp-14* and *nlp-15* are expressed in the sleep regulating neurons ALA and RIS (Turek *et al.*, 2013; Van Buskirk & Sternberg, 2007). Individually, they are dispensable for DTS but removal of both genes reduces movement quiescence without causing molting defects. This is in contrast to studies done with the kissing bug *Rhodnius prolixus* where disruption of orcokinins by RNA interference (RNAi) caused molting defects (Wulff *et al.*, 2017). We also did not observe molting difficulties when *nlp-14* was over-expressed, suggesting that either the role of these peptides is strictly behavioral or that there is degeneracy in the control of molting (Choi *et al.*, 2013; Nelson *et al.*, 2013; Turek *et al.*, 2016).

Stress-induced sleep

In contrast to its relatively minor roles in DTS, NLP-14 is more important during SIS, where the removal of all or subsets of peptides causes strong defects in movement and defecation quiescence. Numerous neuropeptides regulate movement quiescence, including FLP-11, secreted from the RIS (Konietzka *et al.*, 2020) and FLP-13, FLP-24, and NLP-8 (Nath *et al.*, 2016; Nelson *et al.*, 2015), released from ALA. These molecules signal through many GPCRs (Iannacone *et al.*, 2017; Nelson *et al.*, 2015), reducing cAMP/PKA signaling in different cells (Cianciulli *et al.*, 2019). The orcokinins can be added to this expanding list of somnogenic neuropeptides. This observation in *C. elegans* that multiple peptides can induce quiescence when over-expressed is consistent with studies in fish, which have identified several somnogenic neuropeptides using an over-expression approach (Chiu *et al.*, 2016; Lee *et al.*, 2017). Therefore, this complexity to sleep regulation appears to be phylogenetically conserved and demonstrates the importance of sleep to all animals.

Our data, however, suggest that the orcokinins in *C. elegans* are not acting strictly as somnogens. Removal of *nlp-14* shifts the timing of SIS, such that it occurs earlier. We propose that NLP-14 peptides may be functioning to promote aversive behaviors associated with nociception, a previously described role (Hapiak *et al.*, 2013), and act as a somnogen only at later stages, to facilitate recovery from the stressful exposure.

Surprisingly, this early increased quiescence in *nlp-14* mutants is dependent on the presence of *nlp-15*. An interpretation of this could be that NLP-15 and NLP-14 peptides are antagonizing one another during the injurious response to UV, promoting both quiescence and arousal, respectively. At early time points after UV exposure, NLP-14 peptides may promote behavioral arousal, perhaps to allow for an escape response, whereas NLP-15 may promote quiescence at all time points.

At later time points, both promote sleep. After exposures to injurious conditions such as UV light or high heat, animals must balance the benefits of aversion and escape with those of recovery, which are linked to sleep. More work needs to be done to test this idea.

The defecation motor program is both stimulated (Wang *et al.*, 2013) and inhibited by neuropeptides (Nath *et al.*, 2016). When awake, NLP-40 peptides are released from the posterior intestines following rhythmic calcium fluxes, bind their receptor AEX-2, stimulating cAMP/PKA and calcium signaling in the AVL and DVB neurons. This stimulates GABA release, which excites the enteric muscles and initiates an expulsion (Wang *et al.*, 2013). During SIS, the ALA neuron releases NLP-8 peptides to inhibit defecation (Nath *et al.*, 2016). We find that ALA also releases NLP-14 peptides. Our data indicate that NLP-14's effects on defecation are cAMP/PKA-independent. PKA functions in the AVL and DVB motor neurons during defecation to increase their activity leading to enteric muscle contractions that drive the expulsion events (Wang & Sieburth, 2013). Based on this, our data suggest that NLP-14 peptides are functioning either directly on the enteric muscles or downstream of PKA in AVL and DVB. The notion that orcokinins act directly on GI motility would be consistent with observations of crustacean orcokinins, which directly regulate smooth muscle of the gut (Li *et al.*, 2002) and deuterostome starfish myorelaxant peptides (SMPs), which promote the relaxation of stomach muscle (Lin, Egertova, Zampronio, Jones, & Elphick, 2018). There is a sequence similarity between the SMPs and NLP-14 peptides (Kim *et al.*, 2016). Therefore, the role of orcokinin/SMPs in smooth muscle regulation may be conserved.

A conserved sleep-regulating role for orcokinins neuropeptides

Are these sleep functions of NLP-14 and NLP-15 more broadly conserved in Ecdysozoa? Though prior studies have not reported the requirement of orcokinins for sleep, some observations point towards a sleep-regulating role in insects too. Elegant work by Hofer and Homber showed that orcokinin injections result in a circadian phase shift, measured by wheel-running activity in cockroaches. Interestingly, while the authors do not emphasize this point, their actographic data indicate strong inhibition of activity 24–48 h after the orcokinin injection (Hofer & Homberg 2006). Hence, they observed both a change in sleep timing and in sleep/activity in response to orcokinin injections, much as we observe a change in timing and in sleep following *nlp-14* overexpression in *C. elegans.*

DTS in *C. elegans* occurs coincident with the molt (Raizen *et al.*, 2008). Insect larvae can sleep between molts (Szuperak *et al.*, 2018) and also become quiescent during the molt, a behavior called molt-sleep (Reinecke, Buckner, & Grugel, 1980). Gene-expression analysis suggests that molt-sleep is regulated by neuropeptide signaling (MacWilliam, Arensburger, Higa, Cui, & Adams, 2015). Removal of orcokinins in the kissing bug causes molting failure and death (Wulff *et al.*, 2017). It is possible that inhibition of either the behavioral or physiological aspects of molt-sleep is the cause of this lethality. To test for a conserved sleep-regulating role during the molt, it will be important to measure sleep following orcokinin manipulation during inter-molt and molt-sleep in insects and sleep in crustaceans.

In contrast to effects during DTS, the NLP-14 peptides play a more important role during SIS regulation. Heat-induced recovery sleep occurs in *Drosophila melanogaster* and is regulated by the same family of neuropeptides controlling SIS in *C. elegans* (Lenz *et al.*, 2015). In an effort to test the generalizability of our findings, we propose that an initial approach would be to test the necessity of orcokinins in *Drosophila.* It would be particularly interesting to test for an orcokinin role in crayfish, which display slow-wave brain activity, similar to mammals (Ramon, Hernandez-Falcon, Nguyen, & Bullock, 2004; Ramon, Mendoza-Angeles, & Hernandez-Falcon, 2012). Considering that sickness and injury increase sleep in mammals (Imeri & Opp, 2009), SIS may exist in crustaceans as well.

What about tardigrades? These amazingly hardy animals can survive some of the harshest conditions, like desiccation and extreme heat and osmotic pressure. They do so by entering a state of extended quiescence referred to as cryptobiosis (Crowe, 1975). This can promote survival for years, but is reversible, at which point their bodies can be remarkably repaired (Wright, Westh, & Ramlov, 2010). Cryptobiosis may represent an extreme version of SIS. Is this protective behavioral state regulated by orcokinins? If so, orcokinins may be an evolutionarily ancient mechanism controlling protective behavioral quiescence in Ecdysozoa.

Acknowledgements

The authors thank the National BioResource Project (PI, Shohei Mitani) for strains.

Disclosure statement

No potential conflict of interest was reported by the author(s).

Funding

Some strains were provided by the CGC, which is funded by NIH Office of Research Infrastructure Programs (P40 OD010440). We would also like to thank the SJU Summer Scholars Program, the John P. McNulty Fellows Program and the Peter and Dorothy Kowey Fellowship for student funding. MDN was supported by the National Institute of General Medical Sciences of the National Institutes of Health Grant R15GM122058 and the National Science Foundation Grants IOS-CAREER-1845020 and DBI-MRI-1919847.

ORCID

Madison Honer http://orcid.org/0000-0002-4279-0894
Natalie Barrett http://orcid.org/0000-0001-6732-418X
Matthew D. Nelson http://orcid.org/0000-0002-2085-8974

References

Aguinaldo, A.M., Turbeville, J.M., Linford, L.S., Rivera, M.C., Garey, J.R., Raff, R.A., & Lake, J.A. (1997). Evidence for a clade of nematodes, arthropods and other moulting animals. *Nature, 387*(6632), 489–493. doi:10.1038/387489a0

Arribere, J.A., Bell, R.T., Fu, B.X., Artiles, K.L., Hartman, P.S., & Fire, A.Z. (2014). Efficient marker-free recovery of custom genetic modifications with CRISPR/Cas9 in *Caenorhabditis elegans. Genetics, 198*(3), 837–846. doi:10.1534/genetics.114.169730

Charlie, N.K., Thomure, A.M., Schade, M.A., & Miller, K.G. (2006). The Dunce cAMP phosphodiesterase PDE-4 negatively regulates G alpha(s)-dependent and G alpha(s)-independent cAMP pools in the *Caenorhabditis elegans* synaptic signaling network. *Genetics, 173*(1), 111–130. doi:10.1534/genetics.105.054007

Chen, J., Choi, M.S., Mizoguchi, A., Veenstra, J.A., Kang, K., Kim, Y.J., & Kwon, J.Y. (2015). Isoform-specific expression of the neuropeptide orcokinin in *Drosophila melanogaster. Peptides, 68*, 50–57. doi:10.1016/j.peptides.2015.01.002

Chiu, C.N., Rihel, J., Lee, D.A., Singh, C., Mosser, E.A., Chen, S., … Prober, D.A. (2016). A zebrafish genetic screen identifies neuromedin U as a regulator of sleep/wake states. *Neuron, 89*(4), 842–856. doi:10.1016/j.neuron.2016.01.007

Choi, S., Chatzigeorgiou, M., Taylor, K.P., Schafer, W.R., & Kaplan, J.M. (2013). Analysis of NPR-1 reveals a circuit mechanism for behavioral quiescence in *C. elegans. Neuron, 78*(5), 869–880. doi:10.1016/j.neuron.2013.04.002

Churgin, M.A., Jung, S.K., Yu, C.C., Chen, X., Raizen, D.M., & Fang-Yen, C. (2017). Longitudinal imaging of *Caenorhabditis elegans* in a microfabricated device reveals variation in behavioral decline during aging. *eLife, 6*, e26652. doi:10.7554/eLife.26652

Cianciulli, A., Yoslov, L., Buscemi, K., Sullivan, N., Vance, R.T., Janton, F., … Nelson, M.D. (2019). Interneurons regulate locomotion quiescence via cyclic adenosine monophosphate signaling during stress-induced sleep in *Caenorhabditis elegans. Genetics, 213*(1), 267–279. doi:10.1534/genetics.119.302293

Crowe, J. (1975). The physiology of cryptobiosis in tardigrades. *Memorie dell'Istituto italiano di idrobiologia 32*, 37–59.

Davis, M.W., Somerville, D., Lee, R.Y., Lockery, S., Avery, L., & Fambrough, D.M. (1995). Mutations in the *Caenorhabditis elegans* Na,K-ATPase alpha-subunit gene, eat-6, disrupt excitable cell function. *The Journal of Neuroscience, 15*(12), 8408–8418. doi:10.1523/JNEUROSCI.15-12-08408.1995

DeBardeleben, H.K., Lopes, L.E., Nessel, M.P., & Raizen, D.M. (2017). Stress-induced sleep after exposure to ultraviolet light is promoted by p53 in *Caenorhabditis elegans*. *Genetics*, *207*(2), 571–582. doi:10.1534/genetics.117.300070

Dickinson, P.S., Stemmler, E.A., Barton, E.E., Cashman, C.R., Gardner, N.P., Rus, S., … Christie, A.E. (2009). Molecular, mass spectral, and physiological analyses of orcokinins and orcokinin precursor-related peptides in the lobster Homarus americanus and the crayfish *Procambarus clarkii*. *Peptides*, *30*(2), 297–317. doi:10.1016/j.peptides.2008.10.009

Foltenyi, K., Greenspan, R.J., & Newport, J.W. (2007). Activation of EGFR and ERK by rhomboid signaling regulates the consolidation and maintenance of sleep in *Drosophila*. *Nature Neuroscience*, *10*(9), 1160–1167. doi:10.1038/nn1957

George-Raizen, J.B., Shockley, K.R., Trojanowski, N.F., Lamb, A.L., & Raizen, D.M. (2014). Dynamically-expressed prion-like proteins form a cuticle in the pharynx of *Caenorhabditis elegans*. *Biology Open*, *3*(11), 1139–1149. doi:10.1242/bio.20147500

Hapiak, V., Summers, P., Ortega, A., Law, W.J., Stein, A., & Komuniecki, R. (2013). Neuropeptides amplify and focus the monoaminergic inhibition of nociception in *Caenorhabditis elegans*. *The Journal of Neuroscience*, *33*(35), 14107–14116. doi:10.1523/JNEUROSCI.1324-13.2013

Hendriks, G.-J., Gaidatzis, D., Aeschimann, F., & Großhans, H. (2014). Extensive oscillatory gene expression during *C. elegans* larval development. *Molecular Cell*, *53*(3), 380–392. doi:10.1016/j.molcel.2013.12.013

Hill, A.J., Mansfield, R., Lopez, J.M., Raizen, D.M., & Van Buskirk, C. (2014). Cellular stress induces a protective sleep-like state in *C. elegans*. *Current Biology: CB*, *24*(20), 2399–2405. doi:10.1016/j.cub.2014.08.040

Hofer, S., Dircksen, H., Tollback, P., & Homberg, U. (2005). Novel insect orcokinins: Characterization and neuronal distribution in the brains of selected dicondylian insects. *The Journal of Comparative Neurology*, *490*(1), 57–71. doi:10.1002/cne.20650

Hofer, S., & Homberg, U. (2006). Evidence for a role of orcokinin-related peptides in the circadian clock controlling locomotor activity of the cockroach *Leucophaea maderae*. *The Journal of Experimental Biology*, *209*(Pt 14), 2794–2803. doi:10.1242/jeb.02307

Hofer, S., & Homberg, U. (2006). Orcokinin immunoreactivity in the accessory medulla of the cockroach *Leucophaea maderae*. *Cell and Tissue Research*, *325*(3), 589–600. doi:10.1007/s00441-006-0155-y

Iannacone, M.J., Beets, I., Lopes, L.E., Churgin, M.A., Fang-Yen, C., Nelson, M.D., … Raizen, D.M. (2017). The RFamide receptor DMSR-1 regulates stress-induced sleep in *C. elegans*. *eLife*, *6*, e19837. doi:10.7554/eLife.19837

Imeri, L., & Opp, M.R. (2009). How (and why) the immune system makes us sleep. *Nature Reviews. Neuroscience*, *10*(3), 199–210. doi:10.1038/nrn2576

Jekely, G. (2013). Global view of the evolution and diversity of metazoan neuropeptide signaling. *Proceedings of the National Academy of Sciences of the United States of America*, *110*(21), 8702–8707. doi:10.1073/pnas.1221833110

Kim, C.H., Go, H.J., Oh, H.Y., Elphick, M.R., & Park, N.G. (2018). Identification of evolutionarily conserved residues required for the bioactivity of a pedal peptide/orcokinin-type neuropeptide. *Peptides*, *103*, 10–18. doi:10.1016/j.peptides.2018.03.007

Kim, C.H., Kim, E.J., Go, H.J., Oh, H.Y., Lin, M., Elphick, M.R., & Park, N.G. (2016). Identification of a novel starfish neuropeptide that acts as a muscle relaxant. *Journal of Neurochemistry*, *137*(1), 33–45. doi:10.1111/jnc.13543

Konietzka, J., Fritz, M., Spiri, S., McWhirter, R., Leha, A., Palumbos, S., … Bringmann, H. (2020). Epidermal growth factor signaling promotes sleep through a combined series and parallel neural circuit. *Current Biology: CB*, *30*(1), 1–16e13. doi:10.1016/j.cub.2019.10.048

Koziol, U. (2018). Precursors of neuropeptides and peptide hormones in the genomes of tardigrades. *General and Comparative Endocrinology*, *267*, 116–127. doi:10.1016/j.ygcen.2018.06.012

Kramer, A., Yang, F.C., Snodgrass, P., Li, X., Scammell, T.E., Davis, F.C., & Weitz, C.J. (2001). Regulation of daily locomotor activity and sleep by hypothalamic EGF receptor signaling. *Science (New York, N.Y.)*, *294*(5551), 2511–2515. doi:10.1126/science.1067716

Kushikata, T., Fang, J., Chen, Z., Wang, Y., & Krueger, J.M. (1998). Epidermal growth factor enhances spontaneous sleep in rabbits. *The American Journal of Physiology*, *275*(2), R509–R514. doi:10.1152/ajpregu.1998.275.2.R509

Lee, D.A., Andreev, A., Truong, T.V., Chen, A., Hill, A.J., Oikonomou, G., … Prober, D.A. (2017). Genetic and neuronal regulation of sleep by neuropeptide VF. *eLife*, *6*, e25727. doi:10.7554/eLife.25727

Lenz, O., Xiong, J., Nelson, M.D., Raizen, D.M., & Williams, J.A. (2015). FMRFamide signaling promotes stress-induced sleep in *Drosophila*. *Brain, Behavior, and Immunity*, *47*, 141–148. doi:10.1016/j.bbi.2014.12.028

Li, L., Pulver, S.R., Kelley, W.P., Thirumalai, V., Sweedler, J.V., & Marder, E. (2002). Orcokinin peptides in developing and adult crustacean stomatogastric nervous systems and pericardial organs. *The Journal of Comparative Neurology*, *444*(3), 227–244. doi:10.1002/cne.10139

Lin, M., Egertova, M., Zampronio, C.G., Jones, A.M., & Elphick, M.R. (2018). Functional characterization of a second pedal peptide/orcokinin-type neuropeptide signaling system in the starfish *Asterias rubens*. *The Journal of Comparative Neurology*, *526*(5), 858–876. doi:10.1002/cne.24371

Lloyd, P.E., & Connolly, C.M. (1989). Sequence of pedal peptide: A novel neuropeptide from the central nervous system of Aplysia. *The Journal of Neuroscience*, *9*(1), 312–317. doi:10.1523/JNEUROSCI.09-01-00312.1989

MacWilliam, D., Arensburger, P., Higa, J., Cui, X., & Adams, M.E. (2015). Behavioral and genomic characterization of molt-sleep in the tobacco hornworm, *Manduca sexta*. *Insect Biochemistry and Molecular Biology*, *62*, 154–167. doi:10.1016/j.ibmb.2015.01.012

Mello, C., & Fire, A. (1995). DNA transformation. *Methods in Cell Biology*, *48*, 451–482. doi:10.1016/S0091-679X(08)61399-0

Nath, R.D., Chow, E.S., Wang, H., Schwarz, E.M., & Sternberg, P.W. (2016). *C. elegans* stress-induced sleep emerges from the collective action of multiple neuropeptides. *Current Biology: CB*, *26*(18), 2446–2455. doi:10.1016/j.cub.2016.07.048

Nathoo, A.N., Moeller, R.A., Westlund, B.A., & Hart, A.C. (2001). Identification of neuropeptide-like protein gene families in *Caenorhabditis elegans* and other species. *Proceedings of the National Academy of Sciences of the United States of America*, *98*(24), 14000–14005. doi:10.1073/pnas.241231298

Nelson, M.D., & Fitch, D.H. (2011). Overlap extension PCR: An efficient method for transgene construction. *Methods in Molecular Biology (Clifton, N.J.)*, *772*, 459–470. doi:10.1007/978-1-61779-228-1_27

Nelson, M.D., Janssen, T., York, N., Lee, K.H., Schoofs, L., & Raizen, D.M. (2015). FRPR-4 is a G-protein coupled neuropeptide receptor that regulates behavioral quiescence and posture in *Caenorhabditis elegans*. *PLoS One*, *10*(11), e0142938. doi:10.1371/journal.pone.0142938

Nelson, M.D., Lee, K.H., Churgin, M.A., Hill, A.J., Van Buskirk, C., Fang-Yen, C., & Raizen, D.M. (2014). FMRFamide-like FLP-13 neuropeptides promote quiescence following heat stress in *Caenorhabditis elegans*. *Current Biology*, *24*(20), 2406–2410. doi:10.1016/j.cub.2014.08.037

Nelson, M.D., Trojanowski, N.F., George-Raizen, J.B., Smith, C.J., Yu, C.C., Fang-Yen, C., & Raizen, D.M. (2013). The neuropeptide NLP-22 regulates a sleep-like state in *Caenorhabditis elegans*. *Nature Communications*, *4*, 2846. doi:10.1038/ncomms3846

Notredame, C., Higgins, D.G., & Heringa, J. (2000). T-Coffee: A novel method for fast and accurate multiple sequence alignment. *Journal of Molecular Biology*, *302*(1), 205–217. doi:10.1006/jmbi.2000.4042

Paix, A., Folkmann, A., & Seydoux, G. (2017). Precision genome editing using CRISPR-Cas9 and linear repair templates in *C. elegans*. *Methods (San Diego, California)*, *121–122*, 86–93. doi:10.1016/j.ymeth.2017.03.023

Raizen, D.M., Zimmerman, J.E., Maycock, M.H., Ta, U.D., You, Y.J., Sundaram, M.V., & Pack, A.I. (2008). Lethargus is a *Caenorhabditis elegans* sleep-like state. *Nature*, *451*(7178), 569–572. doi:10.1038/nature06535

Ramon, F., Hernandez-Falcon, J., Nguyen, B., & Bullock, T.H. (2004). Slow wave sleep in crayfish. *Proceedings of the National Academy of Sciences of the United States of America, 101*(32), 11857–11861. doi: 10.1073/pnas.0402015101

Ramon, F., Mendoza-Angeles, K., & Hernandez-Falcon, J. (2012). Sleep in invertebrates: Crayfish. *Frontiers in Bioscience (Scholar Edition), 4,* 1190–1200. doi:10.2741/s325

Reinecke, J., Buckner, J.S., & Grugel, S.R. (1980). Life cycle of laboratory-reared tobacco hornworms, *Manduca sexta*, a study of development and behavior, using time-lapse cinematography. *The Biological Bulletin, 158*(1), 129–140. doi:10.2307/1540764

Renn, S.C., Park, J.H., Rosbash, M., Hall, J.C., & Taghert, P.H. (1999). A pdf neuropeptide gene mutation and ablation of PDF neurons each cause severe abnormalities of behavioral circadian rhythms in *Drosophila. Cell, 99*(7), 791–802. doi:10.1016/S0092-8674(00)81676-1

Rowe, M.L., & Elphick, M.R. (2012). The neuropeptide transcriptome of a model echinoderm, the sea urchin *Strongylocentrotus purpuratus. General and Comparative Endocrinology, 179*(3), 331–344. doi: 10.1016/j.ygcen.2012.09.009

Semmens, D.C., & Elphick, M.R. (2017). The evolution of neuropeptide signalling: Insights from echinoderms. *Briefings in Functional Genomics, 16*(5), 288–298. doi:10.1093/bfgp/elx005

Sherlekar, A.L., Janssen, A., Siehr, M.S., Koo, P.K., Caflisch, L., Boggess, M., & Lints, R. (2013). The *C. elegans* male exercises directional control during mating through cholinergic regulation of sex-shared command interneurons. *PLoS One, 8*(4), e60597. doi:10.1371/journal.pone.0060597

Singh, R.N., & Sulston, J.E. (1978). Some observations on moulting in *Caenorhabditis elegans. Nematologica, 24*(1), 63–71. doi:10.1163/187529278X00074

Skiebe, P., Dreger, M., Meseke, M., Evers, J.F., & Hucho, F. (2002). Identification of orcokinins in single neurons in the stomatogastric nervous system of the crayfish, Cherax destructor. *The Journal of Comparative Neurology, 444*(3), 245–259. doi:10.1002/cne.10145

Stangier, J., Hilbich, C., Burdzik, S., & Keller, R. (1992). Orcokinin: A novel myotropic peptide from the nervous system of the crayfish, *Orconectes limosus. Peptides, 13*(5), 859–864. doi:10.1016/0196-9781(92)90041-Z

Stinchcomb, D.T., Shaw, J.E., Carr, S.H., & Hirsh, D. (1985). Extrachromosomal DNA transformation of *Caenorhabditis elegans. Molecular and Cellular Biology, 5*(12), 3484–3496. doi:10.1128/mcb.5.12.3484

Szuperak, M., Churgin, M.A., Borja, A.J., Raizen, D.M., Fang-Yen, C., & Kayser, M.S. (2018). A sleep state in *Drosophila* larvae required for neural stem cell proliferation. *eLife, 7,* e33220. doi:10.7554/eLife.33220

Taylor, S.R., Santpere, G., Reilly, M., Glenwinkel, L., Poff, A., McWhirter, R., ... Miller, D.M. (2019). Expression profiling of the mature *C. elegans* nervous system by single-cell RNA-sequencing. bioRxiv: 737577. doi:10.1101/737577

Thomas, J.H. (1990). Genetic analysis of defecation in *Caenorhabditis elegans. Genetics, 124*(4), 855–872.

Trojanowski, N.F., & Raizen, D.M. (2016). Call it worm sleep. *Trends in Neurosciences, 39*(2), 54–62. doi:10.1016/j.tins.2015.12.005

Turek, M., Besseling, J., Spies, J.P., Konig, S., & Bringmann, H. (2016). Sleep-active neuron specification and sleep induction require FLP-11 neuropeptides to systemically induce sleep. *eLife, 5,* e12499. doi:10.7554/eLife.12499

Turek, M., Lewandrowski, I., & Bringmann, H. (2013). An AP2 transcription factor is required for a sleep-active neuron to induce sleep-like quiescence in *C. elegans. Current Biology: CB, 23*(22), 2215–2223. doi:10.1016/j.cub.2013.09.028

Van Buskirk, C., & Sternberg, P.W. (2007). Epidermal growth factor signaling induces behavioral quiescence in *Caenorhabditis elegans. Nature Neuroscience, 10*(10), 1300–1307. doi:10.1038/nn1981

Wang, H., Girskis, K., Janssen, T., Chan, J.P., Dasgupta, K., Knowles, J.A., ... Sieburth, D. (2013). Neuropeptide secreted from a pacemaker activates neurons to control a rhythmic behavior. *Current Biology: CB, 23*(9), 746–754. doi:10.1016/j.cub.2013.03.049

Wang, H., & Sieburth, D. (2013). PKA controls calcium influx into motor neurons during a rhythmic behavior. *PLoS Genetics, 9*(9), e1003831. doi:10.1371/journal.pgen.1003831

Wright, J.C., Westh, P., & Ramlov, H. (2010). Cryptobiosis in tardigrada. *Biological Reviews, 67* (1), 1–29. doi:10.1111/j.1469-185X.1992.tb01657.x

Wulff, J.P., Sierra, I., Sterkel, M., Holtof, M., Van Wielendaele, P., Francini, F., ... Ons, S. (2017). Orcokinin neuropeptides regulate ecdysis in the hemimetabolous insect *Rhodnius prolixus. Insect Biochemistry and Molecular Biology, 81,* 91–102. doi:10.1016/j.ibmb.2017.01.003

Yamanaka, N., Hua, Y.J., Roller, L., Spalovska-Valachova, I., Mizoguchi, A., Kataoka, H., & Tanaka, Y. (2010). Bombyx prothoracicostatic peptides activate the sex peptide receptor to regulate ecdysteroid biosynthesis. *Proceedings of the National Academy of Sciences of the United States of America, 107*(5), 2060–2065. doi:10.1073/pnas.0907471107

Yamanaka, N., Roller, L., Zitňan, D., Satake, H., Mizoguchi, A., Kataoka, H, Tanaka, Y. (2011). Bombyx orcokinins are brain-gut peptides involved in the neuronal regulation of ecdysteroidogenesis. *The Journal of Comparative Neurology, 519*(2), 238–246. doi:10.1002/cne.22517

Yasuda-Kamatani, Y., & Yasuda, A. (2000). Identification of orcokinin gene-related peptides in the brain of the crayfish *Procambarus clarkii* by the combination of MALDI-TOF and on-line capillary HPLC/Q-Tof mass spectrometries and molecular cloning. *General and Comparative Endocrinology, 118*(1), 161–172. doi:10.1006/gcen.1999.7453

Zahn, T.R., Macmorris, M.A., Dong, W., Day, R., & Hutton, J.C. (2001). IDA-1, a *Caenorhabditis elegans* homolog of the diabetic autoantigens IA-2 and phogrin, is expressed in peptidergic neurons in the worm. *The Journal of Comparative Neurology, 429*(1), 127–143. doi:10.1002/1096-9861(20000101)429:1<127::AID-CNE10>3.0.CO;2-H

Zitnan, D., Ross, L.S., Zitnanova, I., Hermesman, J.L., Gill, S. S., & Adams, M. E., (1999). Steroid induction of a peptide hormone gene leads to orchestration of a defined behavioral sequence. *Neuron, 23*(3), 523–535. doi:10.1016/S0896-6273(00)80805-3

Discriminating between sleep and exercise-induced fatigue using computer vision and behavioral genetics

Kelsey N. Schuch, Lakshmi Narasimhan Govindarajan, Yuliang Guo, Saba N. Baskoylu, Sarah Kim, Benjamin Kimia, Thomas Serre and Anne C. Hart

ABSTRACT

Following prolonged swimming, *Caenorhabditis elegans* cycle between active swimming bouts and inactive quiescent bouts. Swimming is exercise for *C. elegans* and here we suggest that inactive bouts are a recovery state akin to fatigue. It is known that cGMP-dependent kinase (PKG) activity plays a conserved role in sleep, rest, and arousal. Using *C. elegans* EGL-4 PKG, we first validate a novel learning-based computer vision approach to automatically analyze *C. elegans* locomotory behavior and an edge detection program that is able to distinguish between activity and inactivity during swimming for long periods of time. We find that *C. elegans* EGL-4 PKG function impacts timing of exercise-induced quiescent (EIQ) bout onset, fractional quiescence, bout number, and bout duration, suggesting that previously described pathways are engaged during EIQ bouts. However, EIQ bouts are likely not sleep as animals are feeding during the majority of EIQ bouts. We find that genetic perturbation of neurons required for other *C. elegans* sleep states also does not alter EIQ dynamics. Additionally, we find that EIQ onset is sensitive to age and DAF-16 FOXO function. In summary, we have validated behavioral analysis software that enables a quantitative and detailed assessment of swimming behavior, including EIQ. We found novel EIQ defects in aged animals and animals with mutations in a gene involved in stress tolerance. We anticipate that further use of this software will facilitate the analysis of genes and pathways critical for fatigue and other *C. elegans* behaviors.

Introduction

Fatigue is a commonly experienced phenomenon that is typically defined by a feeling of exhaustion combined with decreased muscle output (Wan, Qin, Wang, Sun, & Liu, 2017). Feelings of fatigue are common after vigorous physical activity or exercise, but fatigue is also a hallmark symptom of a variety of health disorders and diseases, including cancer, mood disorders, neurodegenerative disorders, and chronic fatigue syndrome. Fatigue is not limited to vertebrates; exercise eventually drives decreased spontaneous locomotion in invertebrates as well. The molecular pathways and mechanisms involved in fatigue have not been fully delineated in any animal species.

The nematode *Caenorhabditis elegans* provides a potentially powerful model system to interrogate genetic mechanisms and cellular pathways underlying fatigue. Several methods have been developed to test the neuromuscular output of *C. elegans* during locomotion, including burrowing assays through media of varying densities (Beron *et al.*, 2015) and pillar deflection strength measuring assays (Rahman *et al.*, 2018). *Caenorhabditis elegans* are typically grown on solid media, but swimming exercise in liquid media has been shown to be energetically costly (Laranjeiro, Harinath, Burke, Braeckman, & Driscoll, 2017). Following prolonged swimming, *C. elegans* begin to spontaneously cycle between periods of active swimming with vigorous body undulations (active bouts) and periods of immobility that lack body undulations (quiescent bouts) (Ghosh & Emmons, 2008). Because swimming is exercise for *C. elegans*, these quiescent bouts may be fatigue, as they occur after the exertion of swimming and represent a decline in muscle output. The initial vigorous swimming activity could also be a result of introduction to a new environment; however, previous studies show that *C. elegans* have a reduction in recovery crawl distance with longer swimming duration, indicating that the animals are likely fatigued (Laranjeiro *et al.*, 2017). Definition and dissection of exercise-induced quiescence (EIQ) pave the way to study conserved mechanisms fundamental to fatigue in all animals.

However, using *C. elegans* to study EIQ and fatigue requires analysis of swimming behavior over long periods of time, and therefore presents logistical and computational challenges. Several methods have been specifically developed for automatically estimating the pose of small laboratory animals, including *C. elegans* (Gomez-Marin, Partoune, Stephens, & Louis, 2012; Jung, Aleman-Meza, Riepe, & Zhong, 2014; Patel *et al.*, 2014; Restif *et al.*, 2014). For the most part, methods developed for automatically estimating the pose of small laboratory animals rely on simple image processing (e.g. background subtraction) to extract the silhouette of a body before computing a medial axis transform. A major drawback of such methods, with non-parameterized poses, is their inability to discriminate between the front and rear ends of the body, forcing researchers to rely on simple heuristics instead (Jung *et al.*, 2014; Restif *et al.*, 2014) (e.g. by computing the direction of movement and assuming that the animal moves forward). Moreover, the performance of these methods relied heavily on certain hyperparameter choices. For example, methods which perform background subtraction are dictated by factors including the granularity of temporal sampling and thresholds for change detection. Though these are in principle automated methods, some choices for method hyperparameters can yield erroneous pose estimates, making the end user's workflow tedious. Additionally, in the context of biological research, these failures need to be detected—either automatically (Restif *et al.*, 2014) or manually (Jung *et al.*, 2014; Stephens, Johnson-Kerner, Bialek, & Ryu, 2008) to exclude the corresponding frames from further behavioral analysis. These human-in-the-loop systems suffer from two drawbacks: (i) the effective throughput of the system is conditioned on the required frequency of human intervention, and (ii) opportunistic pose tracking may lead to biases in behavioral analyses if those system failures co-occur more frequently with certain behaviors (e.g. for those behaviors that yield significant self-contact and/or self-occlusion such as omega turns and coiling). Other high-throughput computer vision systems have attempted to screen the body postures of up to 120 individual animals at a time (Swierczek, Giles, Rankin, & Kerr, 2011). However, with an increasing number of animals, the pixel resolution on each of these individuals decreases, which can yield either coarse or unreliable pose metrics.

In this study, we extended previous computer-vision work (Yang & Ramanan, 2013) with a learning-based approach [see Methods and Guo, Govindarajan, Kimia, and Serre (2018) for details] which outperforms competing solutions including a representative deep neural network that exhibits state-of-the-art accuracy for human tracking. Unlike humans, *C. elegans* lack distinctive body parts, presenting a significant challenge for deep neural networks and related approaches that rely heavily on appearance alone. Moreover, the employment of machine learning alleviates the need for manual parameter tuning, and instead yields a model that best describes an animal's posture directly from the data. This system can efficiently distinguish between periods of activity and inactivity in freely swimming *C. elegans* (Guo *et al.*, 2018)—addressing the need for high-resolution analysis that can handle both extended periods of swimming and quiescent behaviors in *C. elegans*. Unlike automated tracking systems that limit body 'pose' to just a center of mass, our system permits fine-grained analysis of body movements and assessment of locomotion changes in *C. elegans* swimming over time and additionally handles complex postures more accurately. Here, we refer to this computer vision system as poseEIQ.

Using poseEIQ, we examined *C. elegans* locomotory behaviors and EIQ after prolonged swimming. To more efficiently determine only whether animals are active or inactive, we also developed a new behavioural analysis system called edgeEIQ, which we validated using mutant strains known to have altered EIQ (Ghosh & Emmons, 2008). With this, we describe how EIQ changes over time with extended free swimming and report previously undescribed changes in EIQ as animals age. We also determined that most EIQ bouts are not a sleep state. These new computer-vision analysis systems should be valuable for *C. elegans* researchers in any field that requires accurate assessment of locomotory behavior over extended time intervals.

Methods

Strains and maintenance

Wild type N2 Bristol, MT1072 *egl-4(n477)*, DA521 *egl-4(ad450)*, HBR227 *aptf-1(gk794)*, HBR232 *aptf-1(tm3287)*, IB16 *ceh-17(np1)*, GR1307 *daf-16(mgDf50)*, and CF1038 *daf-16(mu86)* strains were used. *C. elegans* were grown on NGM agar plates with *E. coli* OP50 bacterial food at 20 °C (Brenner, 1974). Animals were obtained by selecting L4 larval stage animals; after 24 h, animals were used in assays as day 1 adult animals. For aging assays, 5-fluoro-2′-deoxyuridine (FUDR) was not used to suppress progeny production. Animals were gently serially passaged using bacteria on a pick to avoid overcrowding of plates with progeny.

Microfluidic chip preparation

PDMS microfluidic chips with 24 wells (1.6 mm wide, 0.07 mm deep, 0.4 mm gap between wells) organized in four rows and six columns were created using a custom mold. A Sylgard 184 silicone elastomer kit (Dow Chemical) was used to make PDMS, which was poured onto the mold to a thickness of 4 mm. Freshly poured PDMS was degassed in a dessiccator using a vacuum until air bubbles had dissipated, then placed in a 55 °C oven for 18 h to cure. PDMS was removed from the mold and cut into chips using a razor. To decrease hydrophobicity before use in swimming assays, these chips were then soaked in *E. coli* OP50 culture overnight, washed with water and ethanol, then left to dry for a week.

Kanamycin-treated E. coli OP50 preparation

Kanamycin-treated *E. coli* OP50 food solution was prepared as previously described (Huang, Singh, & Hart, 2017). In brief, *E. coli* OP50 was streaked onto LB agar plates and cultured overnight at 37 °C. A single colony was used to inoculate 100 ml of liquid LB and cultured at 30 °C shaking at 220 rpm for approximately 12 h. The culture was grown until it reached an optical density of 2–2.5 at 550 nm and concentrated to a final OD_{550} of 10 (OD determinations made using diluted cultures to stay within the linear range of the spectrophotometer). 0.2 mg/mL Kanamycin was added and the culture was placed at 4 °C for 1 week to yield a static bacterial culture. Kanamycin-treated OP50 was discarded after 6 weeks of antibiotic treatment. Immediately prior to assays, 200 microliters of kanamycin-treated OP50 was pelleted and resuspended in 300 microliters of liquid NGM.

Swimming behavioral assays

Chips were cleaned of dust and debris using laboratory tape. Microfluidic chips were then placed into 35 × 10 mm Petri dishes. Each well was loaded with kanamycin-treated OP50 in liquid NGM until a dome of the liquid droplet was visible, but no dark shadows were visible (approximately 0.5 microliters). Water was added to the Petri dish until just level with the chip surface and paraffin oil was layered over droplets and surrounding water to prevent evaporation. To conserve oil, water was used underneath paraffin oil. Before loading in individual wells, animals were picked to an unseeded plate to avoid contaminating wells with additional food. In assays with multiple genotypes, loading order always changed between trials. Animals were recorded swimming at 30 frames per second for 6 hours with a Grasshopper3 4.1MP Mono USB3 Vision camera (GS3-U3-41C6M) mounted on a Zeiss Discovery V20 microscope with a 1.25× objective providing 14.8× magnification. For image capture, we used FlyCap2 version 2.12.3.2 or SpinView version 1.13.0.33. Image resolution was 2048 × 1600 pixels, 270 × 270 pixels per chamber with approximately 1,800 pixels per animal. Representative video available at DOI: 10.5281/zenodo.3604455.

For pumping analysis, videos were recorded using the same setup but at 72.0× magnification to allow for visualization of the pharynx. For quiescent bout analysis, these videos were scaled to match the resolution of all other videos to ensure that quiescent bouts were called in a similar manner. After identification of quiescent bouts, the original videos were manually analyzed for pumping status. If the pharynx was for any reason not visible during a bout (self-occlusion or debris), those bouts were not counted (five instances).

Beat rate was manually collected by analyzing the first minute of swimming. One beat was defined as a full-body bend by the animal in both directions. The experimenter was blinded during this analysis.

Automated well detection

As a pre-processing stage, wells were automatically segmented using MATLAB's connected component algorithm (function *bwboundaries*) on the output of an edge detector (Kimia, Li, Guo, & Tamrakar, 2019). The top 24 connected components were then selected—each corresponding to a different well.

edgeEIQ and activity level analysis

The behavior of each animal was analyzed for each well independently. We implemented a simple 'active' vs. 'quiescent' classifier by considering the binary output of an edge detector (Kimia *et al.*, 2019) and computing the edge difference between consecutive frames using the Jaccard index defined as: $IoU = Area\ of\ Overlap/Area\ of\ Union$, where *Area of Overlap* is the number of edge pixels present in two consecutive frames and *Area of Union* is the sum of the total number of edge pixels across consecutive frames. If the *IoU* was above a threshold $\theta = 0.9$, the behavior was set to quiescent. Otherwise, the behavior was set to active. Using edges rather than pixels yielded robustness to noise compared to simpler systems based on pixels (Restif *et al.*, 2014). A final class label was computed by voting between the two behaviors over a 1 s (30 frames) time window. The activity threshold θ was treated as a hyperparameter and a grid search strategy was employed to identify the optimal value. By systematically varying θ from 0.1 to 1 (with a step size of 0.05) and computing the Jaccard index between the estimations and manually annotated labels, we were able to automatically determine optimal θ^* that yielded the highest agreement. The annotated dataset comprised active/quiescent labels from video recordings of 8 held out animals, each video lasting 20 min at 30 frames per second. The optimal threshold value θ^* was identified to be 0.9, and yielded an agreement of 92%.

Analysis of binary quiescence data

For each hour, binary quiescence data for each animal were analyzed using MATLAB to determine fractional quiescence, time to first bout, bout number, and bout duration. First, data for each animal was iterated through to determine the start and end times for every quiescent bout across the 6-h assay. Arbitrarily, we defined the minimum duration of a quiescent bout as three seconds. Fractional quiescence for each hour was determined by summing the duration of quiescent bouts observed that hour and dividing this by time in that hour. To determine bout number per hour, the number of bouts observed in an hour was tallied. For this metric, if a bout crossed over multiple hours, this bout was counted as a fractional bout. Bout duration was determined by dividing the total time spent in quiescent bouts by the number of bouts that occurred in that hour. For bout duration, if a bout crossed over multiple hours it was counted in each hour for its full duration to more accurately portray the demographics of quiescent bouts observed during each hour. This rarely resulted in bouts being counted more than once in bout duration calculations. For fractional quiescence, if a bout crossed over multiple hours, only the portion of the bout occurring in the given hour contributed to the fractional quiescence for that hour. The time to first bout was calculated by determining the second in which the first quiescent bout for an animal occurred.

poseEIQ and kinematic analysis

The approach is based on a pose tracking algorithm developed in-house (Guo *et al.,* 2018). Briefly, nine body points were chosen along the medial axis of the animal body from head to tail. We collected 10 representative video sequences (each corresponding to a different animal) with a total length of 1200 frames (30 frames per second). Ground-truth poses were manually annotated every 10 frames by marking the location of the nine body points. To train and test our computer vision system, we used a leave-one-video out procedure. Detailed evaluation can be found in (Guo *et al.,* 2018). On average, 83% of the points were correctly detected (i.e. within a 5px radius around the ground-truth landmark). This strategy outperformed competing systems for human pose estimation, specifically the original deformable part model (Yang & Ramanan, 2013) and a leading neural network architecture for human pose estimation, the convolutional pose machine (Wei, Ramakrishna, Kanade, & Sheikh, 2016). These positional points were then used to compute metrics informative of body movements, as defined by Restif *et al.* (2014). Since these measures rely heavily on canonical reorientation of the worm with respect to its head, differentiating the head from the tail becomes particularly important. To alleviate this concern, we specifically ran a second round of tracking, using the Hungarian algorithm (Kuhn, 1955), with the initial head assignment manually checked by a trained expert. Representative video of tracking available at DOI: 10.5281/zenodo.3606717.

Statistical analysis

Statistical analysis was performed using GraphPad Prism 7.0 software (La Jolla, CA). Statistical significance in fractional quiescence, bout number, and bout duration was determined using a two-way ANOVA and Dunnett's multiple comparisons testing. Statistical significance for time to first quiescent bout was determined using Kruskal–Wallis and Dunn's multiple-comparison tests. A value of $p < 0.05$ was used to determine statistical significance ($^{*}p < 0.05$, $^{**}p < 0.01$, $^{***}p < 0.001$). Error bars in figures represent the standard error of the mean.

Results

Behavioral analysis of swimming C. elegans

Analysis of *C. elegans* movements during swimming is critical for the analysis of diverse circuits and behavior. To further characterize *C. elegans* locomotion in liquid media, we took advantage of a new computer-vision system for analysis of swimming animals briefly described by Guo *et al.* (2018), referred to here as poseEIQ (Figure 1(A)), and assessed changes in *C. elegans* locomotory behaviors during prolonged swimming. This method addresses several problems common in current image analysis methods. With respect to background subtraction methods, the temporal granularity of sampling can lead to scenarios where the animal is construed as part of the background if it is not moving. Image

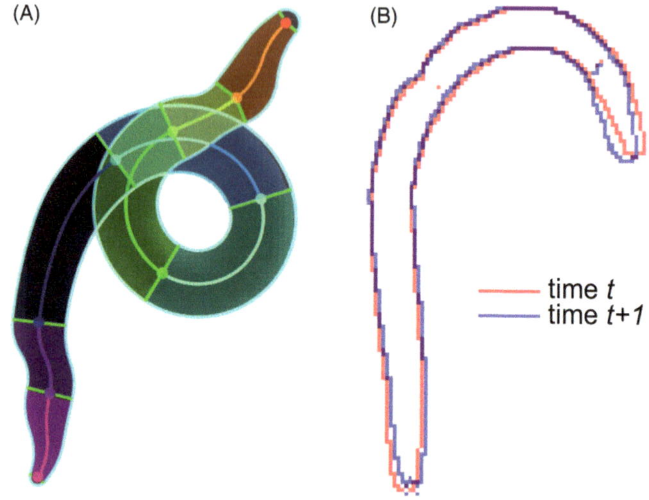

Figure 1. Visual representations of poseEIQ (A) and edgeEIQ (B). (A) poseEIQ uses a shape-consistent flexible mixture of parts model to track *C. elegans* locomotion. Image adapted from Guo *et al.* (2018). (B) edgeEIQ compares edge overlap between consecutive frames to determine activity level.

analysis tracking algorithms are also sensitive to other hyperparameter choices that can vary drastically across imaging sessions because of changes in illumination, background contrast, and imaging resolution. The poseEIQ method is a supervised machine learning algorithm, which learns appropriate configurations of body pose model parameters by optimizing an objective function that factors in both the local visual appearance of body parts and their global geometry. This alleviates the need for explicit background selection and the hand selection of hyperparameters. poseEIQ is more robust by design as long as representative image samples are included in the training phase of this model.

For this analysis, individual young adult animals were placed in small liquid drops with static bacteria as food and high-frame rate, high-resolution videos were obtained over a 6-h period. We examined both control animals and *egl-4* mutant animals, which are known to have altered locomotion. The cGMP-dependent protein kinase encoded by the *egl-4* gene is critical for a wide range of behaviors, including arousal, locomotion, and sleep (Raizen *et al.,* 2008). *egl-4(n477)* loss of function animals and *egl-4(ad450)* gain of function animals were expected to show opposing differences in locomotory behaviors, as increased EGL-4 function promotes quiescent behaviour in swimming animals (Ghosh & Emmons, 2008; McCloskey, Fouad, Churgin, & Fang-Yen, 2017).

Accurate assessment of locomotory behaviors during swimming

To determine if poseEIQ detected changes in locomotion, we analyzed recordings of 12 wild type animals and 12 animals for each of the *egl-4* mutant alleles. We used poseEIQ for two time intervals within the 6-h recording window: the first 30 min of swimming (early), when quiescent bouts are unlikely to occur were compared to 30 min of swimming (late) about 4 h (228.25 ± 4.28 min) into the videos.

Behavioral outputs are for both active and quiescent animals, but outputs can easily be changed to analysis of only active or quiescent bouts. For ease of comparison to previous work, we used locomotion parameters that were previously defined in Restif *et al.* (2014) (Supplementary Table 1). To undertake analysis for these locomotion parameters, it was necessary to assign the head and tail of each animal. Initial head and tail assignments were made computationally and verified manually. As the poseEIQ system performs shape-based pose tracking, these associations were maintained over the course of the video in most scenarios, alleviating the need for further human intervention. The resulting description of locomotion parameters is shown in Figure 2 (statistics in Table 1). Swimming beat rate was manually counted to validate poseEIQ derived wave initiation rate, a similar metric. One beat was defined as a full body movement in both directions. This yielded comparable values (Supplementary Figure 1), and values obtained running the swimming tracking program CeleST on our videos were also comparable (Restif *et al.*, 2014) (Supplementary Table 2).

As expected, the locomotory behavior of *egl-4* animals differed from wild-type animals; loss and gain of function animals are expected to have a diametrically opposed impact on behaviors. For example, at both early and late time points, average wave initiation rate and activity index were decreased in *egl-4(ad450*gf*)* animals and increased in *egl-4(n477*lf*)* animals, compared to wild-type animals (Figure 2(A,B), Table 1). We also found increased curling activity in *egl-4(ad450*gf*)* animals and decreased curling activity in *egl-4(n477*lf*)* animals compared to wild type at both early and late time points (Figure 2(C); Table 1). At the early time point, differences in stretch were observed, as curvature range was found to be increased in *egl-4(ad450*gf*)* animals and decreased in *egl-4(n477*lf*)* animals when compared to wild type (Figure 2(D); Table 1). Differences from wild type were also observed in attenuation during the early time point, with increased body wave attenuation in *egl-4(ad450*gf*)* animals and decreased body wave attenuation in *egl-4(n477*lf*)* animals (Figure 2(E); Table 1). At the later time point, *egl-4(n477*lf*)* animals showed increased brush stroke and *egl-4(ad450*gf*)* animals showed decreased brush stroke compared to wild type (Figure 2(F); Table 1). We also observed a significant increase in body wave number of *egl-4(ad450*gf*)* animals compared to wild type (Figure 2(G); Table 1). *egl-4(n477*lf*)* animals at the late time point showed increased curvature range and body wave attenuation compared to wild type (Figure 2(D,E); Table 1). The large number of diametrically opposed differences observed in *egl-4* animals suggests that poseEIQ can accurately discriminate between normal and mutant locomotion.

Prolonged swimming changes locomotory behaviors

We compared locomotion parameters across time, comparing early *versus* late time points after prolonged swimming within genotypes. Wave initiation rate and activity index decreased in wild type, *egl-4(n477*lf*)*, and *egl-4(ad450*gf*)* animals over time (Figure 2(A, B); Table 1). Body wave number increased and brush stroke decreased in wild type and *egl-4(ad450*gf*)* animals between early and late time points (Figure 2(F,G); Table 1). Finally, curling activity and stretch increased in wild type animals over time (Figure 2(C,D); Table 1). The differences observed in each genotype show that not only can this approach measure the effects of prolonged exercise on locomotion using poseEIQ and these parameters, but also that changes in locomotion after prolonged swimming differ based on genotype. Overall, the changes observed at the late time point are consistent with less vigorous locomotion after 4 h of swimming.

Swimming behavioral states can be represented and analyzed using binary data or unsupervised hidden Markov models (HMM)

One drawback of detecting locomotion changes using poseEIQ is that the analysis is computationally intensive and takes a substantial amount of time to run (approximately 10 s/frame for pose estimation). Therefore, we focused on quiescent behavior during prolonged swimming and developed an edge-detection program, edgeEIQ (Figure 1(B)), that more efficiently identifies EIQ bouts in swimming animals. Knowing that diminished EGL-4 activity decreases quiescent behavior after prolonged swimming (Ghosh & Emmons, 2008), we used edgeEIQ to detect quiescent bouts on the video used above. Ethograms constructed after analysis of wild type and *egl-4* mutant animals showed clear differences in EIQ bouts for each genotype. The downside of this approach was that the activity threshold was a manually selected parameter (manual optimization of the threshold for behavior to be determined quiescent and EIQ set at a minimum duration of 3 s). As an alternative, we explored using an unsupervised Hidden Markov Model (HMM) to extract latent underlying state temporal sequences of activity. This circumvented the need for an explicit selection of EIQ minimum duration or other model parameters. Modeling was done using the open-source package *hmmlearn* in Python. The two-state HMM yielded a latent state sequence that was qualitatively similar to the ethogram constructed from the manually thresholded edgeEIQ binary data (Figure 3(A)). In a three-state HMM, the state that seems to correspond best to the inactive state from the manually thresholded binary ethogram seems to be atomic in nature, and thus cannot be further decomposed. The state corresponding best to the active state from the manually thresholded binary ethogram decomposed into two states (Supplementary Figure 2(A)). We found that HMMs with more than three states resulted in degenerate latent states, i.e. states for which more than one observation is very rare. We can identify such states from analyzing the transition matrices (Supplementary Figure 2(B,C)) and locating states for which all outgoing transition probabilities are close to zero.

To explore transitions between states predicted by HMM modeling, we computed the log-transformed transition count matrices (Supplementary Figure 2(B,C)). Prior to this, the per-frame latent state sequences were clumped into bouts of length 30 s. The equivalent latent state of the bout

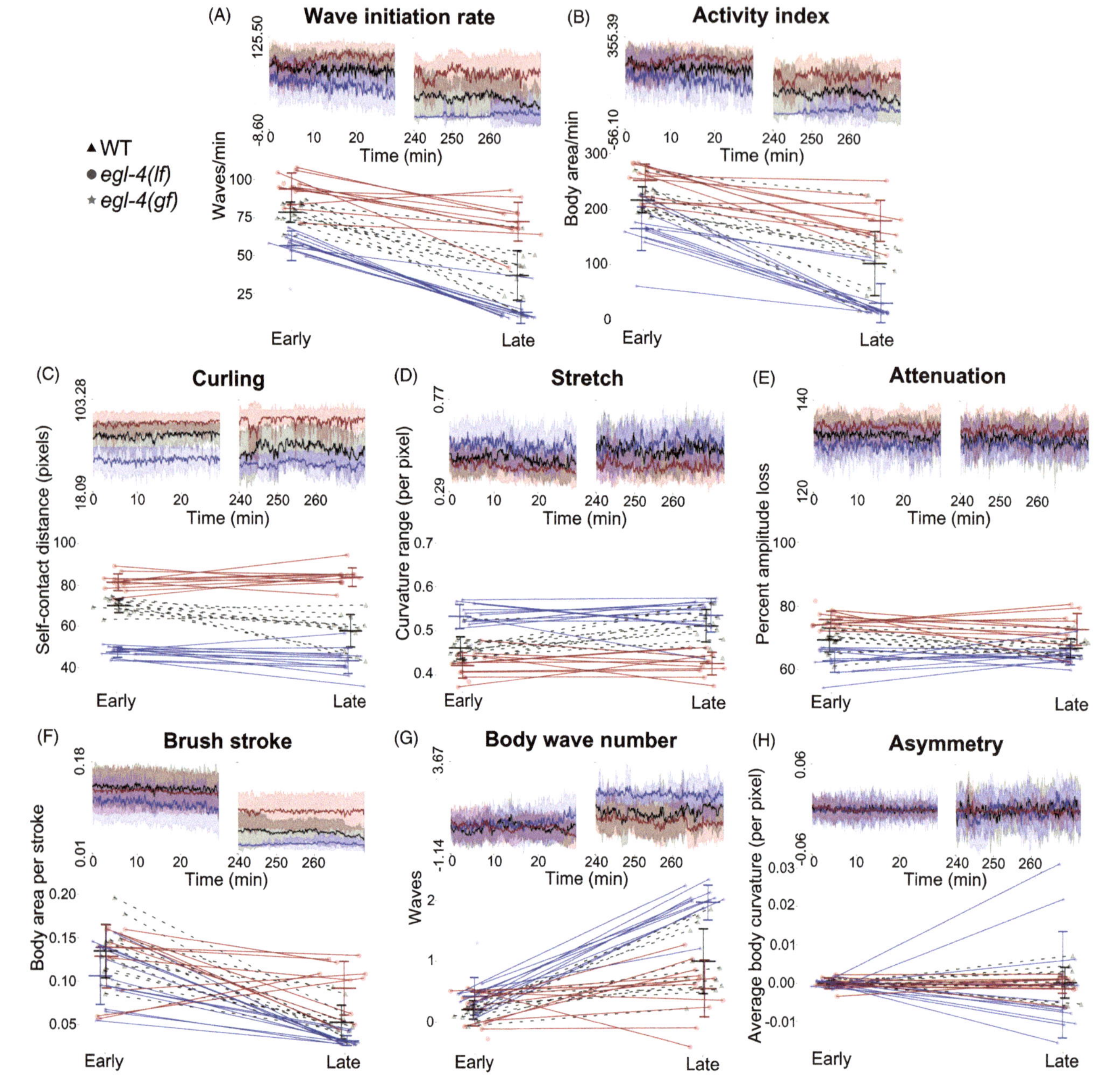

Figure 2. Locomotion analysis of *egl-4* mutant animals during prolonged swimming. Evaluation of parameters (A) Wave Initiation Rate, (B) Activity Index, (C) Curling, (D) Stretch, (E) Attenuation, (F) Brush Stroke, (G) Body Wave Number, (H) Asymmetry; each indicative of different aspects of locomotion over two windows of 30 min each. For a detailed explanation of these parameters, please refer to Restif *et al.* (2014). The 'early' time point is at the very beginning of the 6-h long behavioral assay while the 'late' time point is approximately at the 4 h mark (228.25 ± 4.28 min). The choice of these time points was motivated by the ethograms of Figure 3(A). Behavioral parameters for 12 animals from each of the three genotypes: wild type (WT), *egl-4(n477lf)* (*egl-4(lf)*), and *egl-4(ad450gf)* (*egl-4(gf)*) are shown here. The within-group mean temporal course of each parameter is shown as insets in the respective panel, with the early time point on the top left and the late time point on the top right. The average parameter value (over the 30 min window) for each individual animal is shown in the respective panel; corresponding early/late points are connected by dashed/straight lines. Activity index and brush stroke were normalized to body size (calculated in pixels). Statistical analysis available in Table 1.

Table 1. Statistical analysis of *egl-4* mutant animal locomotion parameters during prolonged swimming.

	Early		Late		Early vs. late		
	WT vs. *e.g.,l-4(lf)*	WT vs. *e.g.,l-4(gf)*	WT vs. *e.g.,l-4(lf)*	WT vs. *e.g.,l-4(gf)*	WT	*egl-4(lf)*	*egl-4(gf)*
A. Wave initiation rate	*	***	**	**	***	**	***
B. Activity index	n.s.	**	*	**	**	**	***
C. Curling	***	***	**	***	**	n.s.	n.s.
D. Stretch	*	**	**	n.s.	*	n.s.	n.s.
E. Attenuation	**	*	*	n.s.	n.s.	n.s.	n.s.
F. Brush stroke	n.s.	n.s.	*	*	***	n.s.	***
G. Body wave number	n.s.	n.s.	n.s.	**	***	n.s.	***
H. Asymmetry	n.s.	n.s.	n.s.	n.s.	n.s.	n.s.	n.s.

The non-parametric Kruskal–Wallis test (with Bonferroni correction for multiple comparisons) was used for testing significance of inter-group differences within and across time points, as well as within-group differences across time points. Error bars indicate ± SEM. *$p < 0.05$; **$p < 0.01$; ***$p < 0.001$. WT = wild type; *egl-4(lf)* = *egl-4(n477lf)*; *egl-4(gf)* = *egl-4(ad450gf)*.

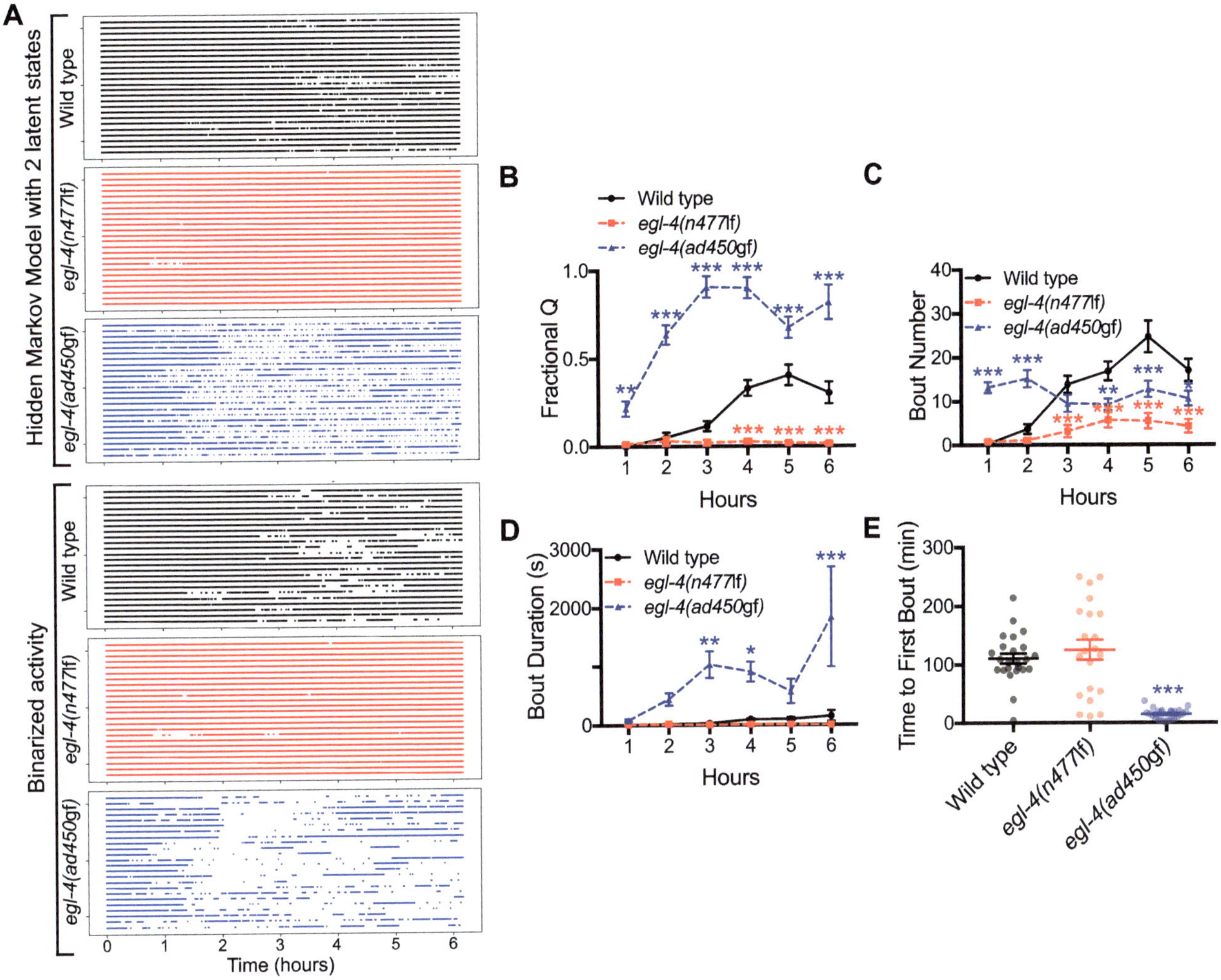

Figure 3. Analysis of exercise-induced quiescence in *egl-4* mutant animals. 24 animals per genotype. (A) Left panel: Ethograms generated using an unsupervised Hidden Markov Model for wild type, *egl-4(n477*lf), and *egl-4(ad450*gf) animals. Each row represents the latent states of a single animal over the course of the 6 h experiment. Right panel: ethograms constructed from manually thresholded binary activity data. Filled (or empty) regions can be interpreted as an 'active' (or 'inactive') state, respectively. (B) On average, the *egl-4(n477*lf) animals showed decreased fractional quiescence (fraction of each hour spent in quiescent bouts) compared to wild type animals during hours 4 through 6. *egl-4(ad450*gf) animals showed increased average fractional quiescence at all time points compared to wild type animals. (C) *egl-4(n477*lf) animals showed decreased average number of bouts (per hour) in hours 3 through 6, compared to wild type animals. *egl-4(ad450*gf) animals showed increased average number of bouts for hours 1 and 2, and decreased average number of bouts in hours 4 through 6, compared to wild type. (D) Average duration of quiescent bouts did not differ between wild type and the *egl-4(n477*lf) animals. *egl-4(ad450*gf) animals showed increased average bout duration during hours 3, 4, and 6. Animals from three independent biological replicates (3 different days). Two-way ANOVA and Dunnett's multiple comparisons test. (E) *egl-4(ad450*gf) animals initiated quiescent bouts earlier than wild type animals, while *egl-4(n477*lf) animals were not different. Kruskal–Wallis test and Dunn's multiple comparisons test. Error bars indicate ± SEM. $*p < 0.05$; $**p < 0.01$; $***p < 0.001$.

was assigned as the statistical mode of the latent states of the constituent frames. The complete absence of state 2 in *egl-4(ad450)* gain of function animals, coupled with the relative infrequency of transitions between states 1 and 2 in the wild-type animals lends support to the aforementioned atomicity of the latent states, i.e. whether or not a given state can be decomposed further into unique latent states (Supplementary Figure 2(B,C)). The most straightforward conclusion from HMM modeling of locomotory behavior is that swimming *C. elegans* have a single inactivity state, based on activity. Additionally, swimming *C. elegans* likely have two active states, which is consistent with previous work (McCloskey *et al.*, 2017).

Using our two-state HMM, we randomly sampled body poses of animals in states that likely represent activity and quiescence (Supplementary Figure 3(A)). Similar to previous studies, eigen decomposition revealed three modes that account for >95% of the total variance in body postures (Brown, Yemini, Grundy, Jucikas, & Schafer, 2013; Stephens *et al.*, 2008), and the statistical mean body posture of active and quiescent animals appears quite similar (Supplementary Figure 3(B,C)). Although quiescent animals sometimes assume a straight posture (Supplementary Figure 3(A)), this similarity between active and quiescent postures contrasts with previous observations reporting that quiescent animals in liquid gradually assume a rod-like posture (Ghosh & Emmons, 2008). The posture discrepancy is likely explained by differences in the definition of quiescent bouts and sampling rates. Here, short quiescent bouts are included; these were not detected or excluded in the previous study. The lack of a straight posture in a quiescent animal can be observed in the representative video available at DOI:

10.5281/zenodo.3606717. *C. elegans* display stereotypical body postures during sleep (Iwanir *et al.*, 2013; Schwarz, Spies, & Bringmann, 2012; Tramm, Oppenheimer, Nagy, Efrati, & Biron, 2014); the lack of clear difference between active and quiescent body postures here indicates that EIQ is likely distinct from sleep.

We also examined whether the duration of an active bout was related to the duration of the following quiescent bout and found that poor correlation was seen for bout durations (Spearman's $r = -0.29$) (Supplementary Figure 4(A)). This differs from the relationship between active and subsequent quiescent bouts observed during developmentally timed sleep, as a positive correlation was found between active and quiescent bout durations (Iwanir *et al.*, 2013). To explore this further, we used the raw edge-overlap data output by edgeEIQ (which is indicative of how much an animal is moving from frame to frame) to determine whether the intensity of an active bout affects the following quiescent bout. To look at animals with differing intensities of activity before entering a quiescent bout, we split animals into two groups, more active and less active, by the median activity prior to entering quiescence. We saw no difference in quiescent bout duration between the two groups (Supplementary Figure 4(B)). We also found that the less active animals returned to their initial activity level after leaving a quiescent bout—this was also true of the more active animals (Supplementary Figure 4(B)). We conclude that the relationship between active and quiescent bouts is more complicated than a linear relationship where increased activity leads to increased rest.

Accurate assessment of quiescent behavior during prolonged swimming

To confirm that our edgeEIQ program could detect previously described differences in EIQ, we compared wild-type animals swimming in the presence of food to animals carrying previously described *egl-4* mutant alleles. *egl-4(n477)* loss of function animals were predicted to show decreased EIQ and *egl-4(ad450)* gain of function animals were expected to show increased EIQ. 6-h videos of swimming animals were recorded and analyzed with the program. The fraction of time quiescent (fractional quiescence) was determined for each hour in individual animals across the 6-h experiment (Figure 3(B)). At every time point, *egl-4(ad450)* gain of function animals showed increased fractional quiescence compared to wild type, while *egl-4(n477)* loss of function animals showed decreased fractional quiescence for hours 4, 5, and 6, which is consistent with our prediction and previous work (Ghosh & Emmons, 2008). Next, we examined the average number of quiescent bouts in each hour and the average quiescence bout duration for each hour. *egl-4(n477)* animals had decreased bout numbers, starting in hour three and onward (Figure 3(C)). *egl-4(ad450)* gain of function animals showed increased bout numbers per hour during the first 2 h of swimming and decreased bout numbers for hours 4 through 6 (Figure 3(C)) and had increased bout durations at almost all time points, with the exception of hour one (Figure 3(D)). Finally, we examined when wild type and *egl-4* mutant animals first entered a quiescent bout after prolonged swimming. On average, *egl-4(ad450)* gain of function animals showed quiescence at an earlier time than wild type animals (first quiescent bout of $\geq$3 s, Figure 3(E)). The loss of *egl-4* function did not alter quiescent bout onset. Overall, these results obtained are entirely consistent with previous work and suggest that the edgeEIQ program can robustly identify differences in EIQ and other quiescent behaviors of swimming animals.

EIQ does not require pathways necessary for developmentally timed sleep or stress-induced sleep

A well-characterized quiescent state in *C. elegans* is sleep (Hill, Mansfield, Lopez, Raizen, & Van Buskirk, 2014; Raizen *et al.*, 2008). To determine whether sleep was occurring during EIQ bouts, we manually examined quiescent bouts in wild type animals to determine if feeding was occurring. In *C. elegans*, pharyngeal pumping can be used as a metric for food intake and feeding. During each hour, pharyngeal pumping was observed in the majority of quiescent bouts; very few bouts were observed where no pharyngeal pumping occurred (Figure 4(A)). However, during several quiescent bouts, animals did not pump in the first part of the quiescent bout, then resumed pumping for the remainder of the bout. We called these 'mixed bouts' and found that the percentage of mixed bouts increased as quiescent bout duration increased (right panel, Figure 4(A)). However, in the majority of quiescent bouts animals were pumping, suggesting that most *C. elegans* EIQ bouts are not sleep.

Examination of mutant strains confirms that there is little mechanistic overlap between EIQ and previously defined *C. elegans* sleep states. Changes in EGL-4 kinase activity impact all known types of *C. elegans* sleep (Hill *et al.*, 2014; Raizen *et al.*, 2008; You, Kim, Raizen, & Avery, 2008). Perturbation of EGL-4 also alters sensory response and changes locomotion in waking animals (Figure 2). The AP2 transcription factor APTF-1 is specifically required for locomotion quiescence during *C. elegans* developmentally timed sleep (Turek, Lewandrowski, & Bringmann, 2013) and the paired homeodomain transcription factor CEH-17 is required for locomotion quiescence in stress-induced sleep (Hill *et al.*, 2014). We tested *aptf-1(gk794)* and *aptf-1(tm3287)* loss of function mutants for defects in EIQ and found that loss of APTF-1 does not alter fractional quiescence, bout number, bout duration, or time to first bout, when compared to wild type animals (Figure 4(B–E)). Although *aptf-1(tm3287)* differed from wild type in time to first bout as well as fractional quiescence and bout duration at hour five, similar changes were not observed in *aptf-1(gk794)* animals, decreasing confidence that these changes can be attributed to decreased *aptf-1* function (Figure 4(B,D,E)). We found that *ceh-17(np1)* loss of function animals were not different than wild type animals in fractional quiescence, bout number, bout duration, or time to first bout (Figure 4(F–I)). Combined, these results

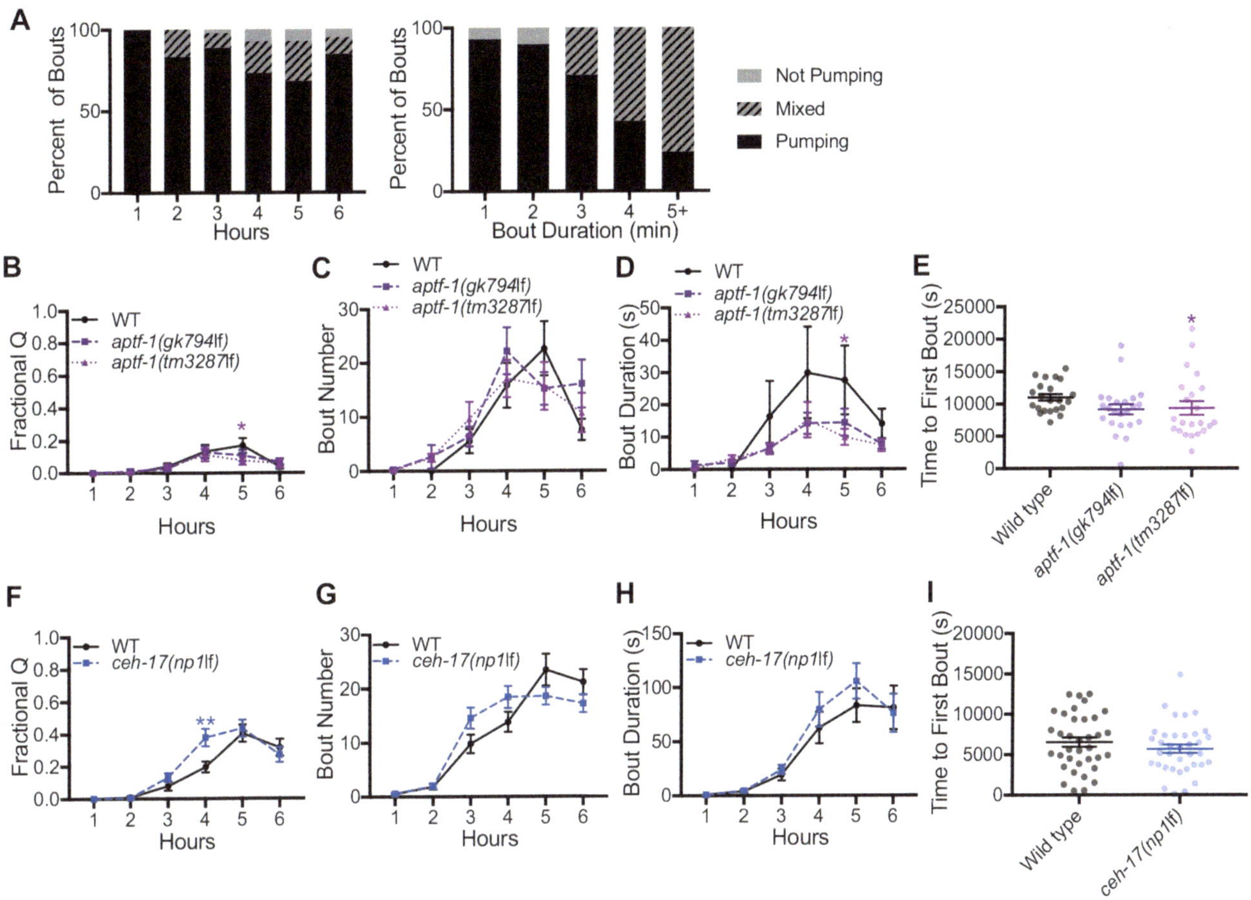

Figure 4. Exercise-induced quiescent bouts are not sleep. (A) Behavior of wild type animals during quiescent bouts was classified as 'pumping' (exhibited pharyngeal pumping throughout a quiescent bout), 'mixed' (began a quiescent bout without pumping, and resumed pumping midway through the bout), and 'not pumping' (no pharyngeal pumping). Left panel: The majority of quiescent bouts were classified as pumping, regardless of when they occurred. Right panel: When classified based on bout duration, most bouts lasting 3 min or less were classified as not pumping, while bouts longer than 4 min were usually classified as mixed bouts. 199 total quiescent bouts classified drawn from 5 animals. Loss of function *aptf-1(gk794lf)* and *aptf-1(tm3287lf)* animals did not differ from wild type in fractional quiescence (A), bout number (B), or bout duration (C), with the exception of fractional quiescence (B) and bout duration (B) of *aptf-1(tm3287lf)* animals at hour 5. Two-way ANOVA and Dunnett's multiple comparisons test. (E) Average time to first bout was slightly sooner in *aptf-1(tm3287lf)*, but not *aptf-1(gk794lf)*, animals *versus* wild type. Kruskal-Wallis test and Dunn's multiple comparisons test. $n = 24$ per genotype. Loss of function *ceh-17(np1lf)* showed no difference in fractional quiescence (F), bout number (G), and bout duration (H), with the exception of increased fractional quiescence at hour 5 (F). 2-way ANOVA and Dunnett's multiple comparisons test. (I) No difference in time to first bout was observed between *ceh-17(np1lf)* animals and wild type. Kruskal-Wallis test and Dunn's multiple comparisons test. $n = 24$ per genotype. Error bars indicate ± SEM. *$p < 0.05$; **$p < 0.01$; ***$p < 0.001$.

suggest that different molecular mechanisms underlie EIQ *versus C. elegans* locomotion quiescence during sleep.

DAF-16/FOXO function delays initiation of quiescent bout cycling after prolonged swimming

The DAF-16/FOXO transcription factor plays a critical role in response to multiple stressors, including sleep restriction (Driver, Lamb, Wyner, & Raizen, 2013; Henderson & Johnson, 2001). In fasting conditions, *daf-16(mu86)* loss of function animals have total quiescent activity equivalent to wild type animals (McCloskey *et al.*, 2017). We examined both *daf-16(mgDf50)* and *daf-16(mu86)* loss of function mutant animals for changes in EIQ timing. *daf-16(mgDf50)* animals showed increased fractional quiescence at hours three, four, and six (Figure 5(A)), as well as increased bout number and bout duration at hours three and four (Figure 5(B,C)). However, these differences were not observed in *daf-16(mu86)* animals. Both loss of function strains showed decreased time to first quiescent bout (Figure 5(D)). Because this last defect was seen in both mutant *daf-16* strains, we suggest that DAF-16 plays a role in the response to stress caused by prolonged swimming that is important to determining EIQ onset.

Aged animals enter quiescence at an earlier time and show increased quiescence during prolonged swimming

With age, organisms experience loss of muscle mass, known as sarcopenia, which is believed to contribute to increased frailty, decreased muscle strength, and fatigue in aged populations (Marty, Liu, Samuel, Or, & Lane, 2017). Age-related muscle deterioration has previously been observed in *C. elegans* body wall muscle (Herndon *et al.*, 2002), and aged *C. elegans* have deficits in various locomotion assays, including swim rate (Mulcahy, Holden-Dye, & O'Connor, 2013; Restif

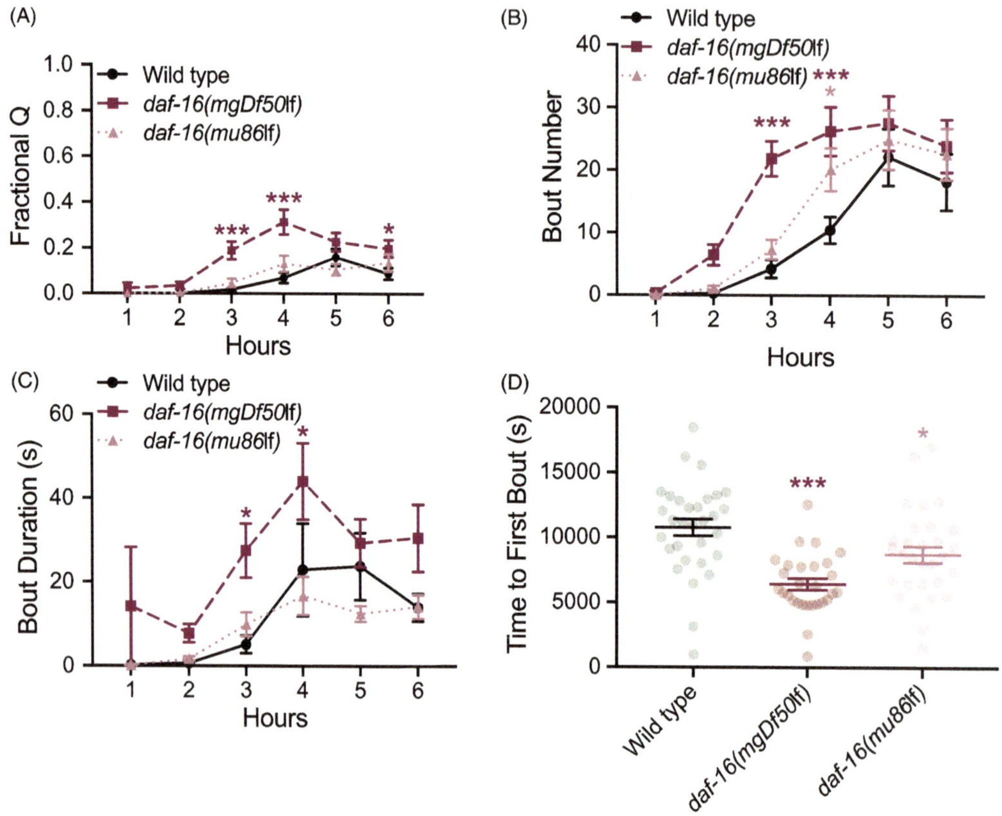

Figure 5. Analysis of exercise-induced quiescence in *daf-16* mutant animals. (A) Loss of function *daf-16(mgDf50lf)* animals showed increased average fractional quiescence, compared to wild type at hours 3, 4, and 6. This difference was not repeated in *daf-16(mu86lf)* animals. (B) *daf-16(mu86lf)* animals showed increased average number of bouts per hour, compared to wild type in hour 4, and *daf-16(mgDf50lf)* animals showed an increase during hours 3 and 4. (C) *daf-16(mgDf50lf)* animals showed increased average bout duration, compared to wild type during hours 3 and 4, while *daf-16(mu86lf)* animals showed no difference. Two-way ANOVA and Dunnett's multiple comparisons test. (D) The *daf-16(mgDf50lf)* and *daf-16(mu86lf)* animals both showed decreased average time to first bout, compared to wild type. Kruskal–Wallis test and Dunn's multiple comparisons test. $n = 32$ per genotype. Error bars indicate ± SEM. *$p < 0.05$; **$p < 0.01$; ***$p < 0.001$.

et al., 2014). To test whether aged animals show defective EIQ, we aged wild type animals for 1, 2, 3, 4, 7, and 10 days into adulthood. Then, we quantified differences in EIQ timing. Day 7 and 10 adult animals showed increased fractional quiescence compared to day 1 adults, while days 2, 3, and 4 adults were generally indistinguishable from day 1 adults (Figure 6(A)). No dramatic differences in bout duration were seen between different aged animals. However, bout number was increased from day 7 at hours 2, 3, and 6 and increased in day 10 adults at hours one, two, three, five, and six (compared to day 1; Figure 6(B,C)). On days 3, 4, 7, and 10, animals began cycling between activity and inactivity more quickly than day 1 adults (Figure 6(D)). A similar decrease in beat rate per minute was also observed at days 2, 3, 4, 7, and 10 (Supplementary Figure 5). These results suggest that locomotory output declines in aged animals, consistent with diminished muscle function in aging animals. Additionally, we noted that different aspects of EIQ metrics decline with age at different rates; time to first EIQ bout decreased with age more rapidly than fractional EIQ increased with age.

Discussion

Here, we work with the computer vision programs poseEIQ and edgeEIQ, which reveal in finer detail changes in *C. elegans* locomotion after prolonged periods of swimming. Increased inactivity was observed after extended swimming, as were differences in swimming locomotory behaviors between wild-type and mutant animals. Loss of the EGL-4 cGMP-dependent kinase and DAF-16 FOXO function altered EIQ after prolonged swimming. However, loss of the proteins APTF-1 and CEH-17, which are required for developmentally timed sleep and stress-induced sleep in *C. elegans*, respectively, did not affect EIQ. Aged animals showed increased EIQ. Based on examining pharyngeal pumping, animals are actively feeding during the majority of EIQ bouts, indicating that EIQ is usually not a sleep state. Computer-vision programs used here enabled in-depth analysis of EIQ and locomotory behaviors across time and revealed previously undescribed defects. This work and development of these automated analysis strategies enables future work that will interrogate the molecular pathways underlying behaviors associated with exercise and fatigue.

C. elegans locomotory behavior after prolonged swimming has not been thoroughly studied. Previous studies were limited by reliance on manual annotation, which hinders research depth, or by reliance on constrictive microfluidic devices, which may induce mechanical stress (Ghosh & Emmons, 2008; Gonzales, Zhou, Fan, & Robinson, 2019). Using both of our new systems, we can provide a detailed analysis of how *C. elegans* locomotory behaviors change after prolonged swimming. When comparing wild type and *egl-4* mutant animals, differences were found in multiple

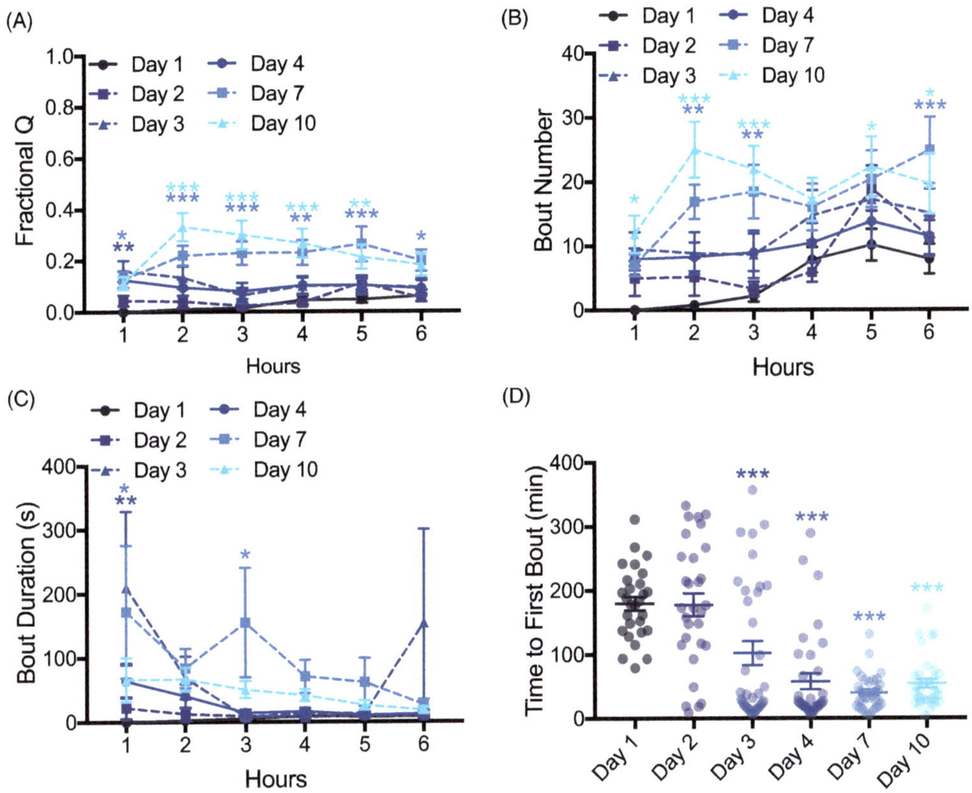

Figure 6. Analysis of exercise-induced quiescence in aged animals. Wild-type animals were aged 1, 2, 3, 4, 7, and 10 days into adulthood. (A) Day 3 adult animals showed increased average fractional quiescence, compared to day 1 adult animals at hour 1. Compared to day 1 adult animals, day 7 adult animals showed increased average fractional quiescence at all time points, while day 10 animals showed increased average fractional quiescence at hours 2 through 5. (B) An increase in average number of bouts per hour, compared to day 1 adult animals, was observed at day 7 adult animals during hours 2, 3, and 6. Day 10 adult animals also showed increased average number of bouts (per hour) during hours 1, 2, 3, 5, and 6, compared to day 1 adult animals. (C) No difference in average bout duration was found amongst the different age groups, with the exception of increased average bout duration in day 3 adult animals at hour 1 and at day 7 adult animals at hours 1 and 3. Two-way ANOVA and Dunnett's multiple comparisons test. (D) Day 3, 4, 7, and 10 adult animals all showed decreased average time to first bout, compared to day 1 adult animals. Kruskal–Wallis test and Dunn's multiple comparisons test. $n = 36$ per group. Error bars indicate ± SEM. $^{*}p < 0.05$; $^{**}p < 0.01$; $^{***}p < 0.001$.

parameters, including wave initiation rate at early and late time points. Interestingly, there were also differences in which locomotion parameters changed over time between wild type and mutant animals. For example, curling activity and stretch were found to change over time in wild type animals, but not in *egl-4* mutant animals. In future studies of *C. elegans* fatigue, parameters like wave initiation rate and brush stroke will likely provide information about how vigorously an animal swims and can be used to track decreased muscle output after prolonged swimming exercise. We note that quiescence levels can vary across experiments, even in wild type animals. This may be caused by differences in ambient conditions during rearing (e.g. humidity levels, vibration). Here, all experimental groups in a trial were reared in tandem to control for these differences.

It is important to note that poseEIQ, used herein for analysis of swimming locomotory behaviour, is computationally expensive and requires substantial time for processing. To increase efficiency, we developed edgeEIQ which is faster, but less comprehensive, as it only distinguishes active and inactive states during swimming. Loss of function of the gene *egl-4* has previously been associated with decreased quiescence after prolonged swimming (Ghosh & Emmons, 2008, 2010). To test whether edgeEIQ could detect the impact of *egl-4* mutations on EIQ, we analyzed the behavior of gain and loss of function mutants of *egl-4*. As expected, gain and loss of function mutations in *egl-4* were associated with increased and decreased EIQ activity, respectively. cGMP-dependent protein kinase EGL-4 is also known to promote quiescent activity during all known forms of *C. elegans* sleep (Hill *et al.*, 2014; Raizen *et al.*, 2008). We originally hypothesized that the quiescent behavior after prolonged swimming might also be a sleep state. Usually sleeping animals will stop both feeding and locomotion. But, we found that pharyngeal pumping was usually observed in animals during EIQ bouts, suggesting that EIQ bouts are usually not sleep. We also used genetic strategies to explore the relationship between sleep and EIQ bouts. The RIS neuron is critical for developmentally timed sleep, and function of the APTF-1 transcription factor is required for RIS-mediated sleep induction (Turek *et al.*, 2013). Likewise, the ALA sensory neuron is required for quiescent behavior during stress-induced sleep (Hill *et al.*, 2014). CEH-17 loss alters gene expression in the ALA neuron, and loss of this protein leads to shortened ALA axons and inability to enter a sleep state following cellular stress (Hill *et al.*, 2014; Pujol, Torregrossa, Ewbank, & Brunet, 2000; Van Buskirk & Sternberg, 2010). We tested animals lacking *aptf-1* and *ceh-17* function for defects in EIQ to determine whether exercise-induced locomotion quiescence was mediated by pathways involved in mediating locomotion quiescence in developmentally timed or stress-induced sleep. We found that these mutant animals had normal locomotion quiescence. Therefore, EIQ and sleep are not identical and are likely mediated by overlapping, but distinct molecular and cellular pathways.

Prolonged exercise is stressful. In *C. elegans*, the transcription factor DAF-16/FOXO localizes to the nucleus during cellular stress, where it upregulates genes involved with stress response, including oxidative stress response genes (Henderson & Johnson, 2001; Murphy *et al.*, 2003). As prolonged swimming by *C. elegans* leads to transcriptional changes in oxidative stress response genes (Laranjeiro *et al.*, 2017), we reasoned that DAF-16 might play a role in EIQ and stress response. Loss of function *daf-16* mutant animals initiated EIQ earlier than wild type animals, suggesting that DAF-16 normally promotes expression of genes that allow animals to endure the stress of exercise for a longer initial swim period of time.

As they age, animals experience deterioration that can result in fatigue, weakness, and sarcopenia. We examined EIQ as *C. elegans* age and, as predicted, older animals showed increased EIQ. Surprisingly, a dramatic decrease in initial swim period time was seen in animals three or more days into adulthood. This deterioration is not mirrored in fractional EIQ, as animals only show an increase in fractional EIQ at seven or more days into adulthood. Further analysis of EIQ should provide insight into the effects of the aging process on muscle function over time. Changes in EIQ in aged *C. elegans* are likely a marker of healthspan and the mechanisms underlying these changes may be conserved mechanisms relevant to fatigue in aging humans.

Here, we developed systems for automatic analysis of *C. elegans* swimming locomotory behavior with a long-range goal of understanding locomotion quiescience and associated mechanisms. The cellular and molecular pathways that communicate fatigue from muscles to the nervous system during exercise remain obscure. In complex animals, afferent neuronal pathways are thought to carry information from the periphery to the central nervous system, which eventually results in a 'feeling of exhaustion' that results in decreased locomotion. Energy depletion in peripheral organs could also result in decreased locomotion. Results presented here and other works (Beron *et al.*, 2015; Lesanpezeshki *et al.*, 2019; Rahman *et al.*, 2018) have demonstrated that invertebrates show aspects of fatigue, even though they lack extensive afferent neuronal pathways. Further dissection of these behaviors in invertebrates should provide insight into conserved cellular and molecular pathways involved in fatigue, as well as other aspects of endurance and exercise.

Conclusions

Computer-vision systems have been developed for accurate analysis of *C. elegans* locomotory behavior during prolonged swimming exercise (over 6 h). This system is complemented by a faster edge detection-based system that can detect pauses in locomotion. We find that most prolonged swimming results in exercise-induced quiescence (EIQ) bouts that are dependent on EGL-4/PKG function, but not on the function of genes that are specifically required for *C. elegans* sleep. The timing of EIQ bout initiation is dependent on DAF-16/FOXO function. As *C. elegans* age, the timing, duration, and frequency of EIQ bouts change, consistent with diminished healthspan.

Acknowledgements

The authors acknowledge the Cloud TPU hardware resources that Google made available via the TensorFlow Research Cloud (TFRC) program. Some strains were provided by the CGC, which is funded by the NIH Office of Research Infrastructure Programs (P40 OD010440).

Disclosure statement

No potential conflict of interest was reported by the author(s).

Funding

This research was supported by funds from Brown's Office for the Vice-President for Research (A.C.H. and T.S.). Additional support provided by the Carney Institute for Brain Science, the Center for Vision Research (CVR) and the Center for Computation and Visualization (CCV) at Brown University and a Karen T. Romer Undergraduate Teaching and Research Awards (S.K.).

ORCID

Anne C. Hart http://orcid.org/0000-0001-7239-4350

References

Beron, C., Vidal-Gadea, A.G., Cohn, J., Parikh, A., Hwang, G., & Pierce-Shimomura, J.T. (2015). The burrowing behavior of the nematode *Caenorhabditis elegans*: a new assay for the study of neuromuscular disorders. *Genes, Brain, and Behavior*, *14*(4), 357–368. doi:10.1111/gbb.12217

Brenner, S. (1974). The genetics of *Caenorhabditis elegans*. *Genetics*, *77*(1), 71–94.

Brown, A.E.X., Yemini, E.I., Grundy, L.J., Jucikas, T., & Schafer, W.R. (2013). A dictionary of behavioural motifs reveals clusters of genes affecting *Caenorhabditis elegans* locomotion. *Proceedings of the National Academy of Sciences of the United States of America*, *110*(2), 791–796. doi:10.1073/pnas.1211447110

Driver, R.J., Lamb, A.L., Wyner, A.J., & Raizen, D.M. (2013). DAF-16/FOXO regulates homeostasis of essential sleep-like behavior during larval transitions in *C. elegans*. *Current Biology*, *23*(6), 501–506. doi:10.1016/j.cub.2013.02.009

Ghosh, R., & Emmons, S.W. (2008). Episodic swimming behavior in the nematode *C. elegans*. *The Journal of Experimental Biology*, *211*(Pt 23), 3703–3711. doi:10.1242/jeb.023606

Ghosh, R., & Emmons, S.W. (2010). Calcineurin and protein kinase G *regulate C. elegans* behavioral quiescence during locomotion in liquid. *BMC Genetics*, *11*, 7. doi:10.1186/1471-2156-11-7

Gomez-Marin, A., Partoune, N., Stephens, G.J., & Louis, M. (2012). Automated tracking of animal posture and movement during exploration and sensory orientation behaviors. *PloS One*, *7*(8), e41642. doi:10.1371/journal.pone.0041642

Gonzales, D.L., Zhou, J., Fan, B., & Robinson, J.T. (2019). A microfluidic-*induced C. elegans* sleep state. *Nature Communications*, *10*(1), 5035. doi:10.1038/s41467-019-13008-5

Guo, Y., Govindarajan, L., Kimia, B., & Serre, T. (2018). Robust pose tracking with a joint model of appearance and shape. arXiv, 1-11. arXiv:1806.1.

Henderson, S.T., & Johnson, T.E. (2001). *daf-16* integrates developmental and environmental inputs to mediate aging in the nematode *Caenorhabditis elegans*. *Current Biology*, *11*(24), 1975–1980. doi:10.1016/S0960-9822(01)00594-2

Herndon, L.A., Schmeissner, P.J., Dudaronek, J.M., Brown, P.A., Listner, K.M., Sakano, Y., ... Driscoll, M. (2002). Stochastic and genetic factors influence tissue-specific decline in ageing *C. elegans*. *Nature*, *419*(6909), 808–814. doi:10.1038/nature01135

Hill, A.J., Mansfield, R., Lopez, J.M.N.G., Raizen, D.M., & Van Buskirk, C. (2014). Cellular stress induces a protective sleep-like state in *C. elegans*. *Current Biology*, *24*(20), 2399–2405. doi:10.1016/j.cub.2014.08.040

Huang, H., Singh, K., & Hart, A.C. (2017). Measuring *Caenorhabditis elegans* sleep during the transition to adulthood using a microfluidics-based system. *Bio-protocol*, *7*(6), e2174. doi:10.21769/BioProtoc.2174

Iwanir, S., Tramm, N., Nagy, S., Wright, C., Ish, D., & Biron, D. (2013). The microarchitecture of *C. elegans* behavior during lethargus: homeostatic bout dynamics, a typical body posture, and regulation by a central neuron. *Sleep*, *36*(3), 385–395. doi:10.5665/sleep.2456

Jung, S.-K., Aleman-Meza, B., Riepe, C., & Zhong, W. (2014). QuantWorm: a comprehensive software package for *Caenorhabditis elegans* phenotypic assays. *PloS One*, *9*(1), e84830. doi:10.1371/journal.pone.0084830

Kimia, B.B., Li, X., Guo, Y., & Tamrakar, A. (2019). Differential geometry in edge detection: accurate estimation of position, orientation and curvature. *IEEE Transactions on Pattern Analysis and Machine Intelligence*, *41*(7), 1573–1586. doi:10.1109/TPAMI.2018.2846268

Kuhn, H.W. (1955). The Hungarian method for the assignment problem. *Naval Research Logistics Quarterly*, *2*(1-2), 83–97. doi:10.1002/nav.3800020109

Laranjeiro, R., Harinath, G., Burke, D., Braeckman, B.P., & Driscoll, M. (2017). Single swim sessions *in C. elegans* induce key features of mammalian exercise. *BMC Biology*, *15*(1), 30. doi:10.1186/s12915-017-0368-4

Lesanpezeshki, L., Hewitt, J.E., Laranjeiro, R., Antebi, A., Driscoll, M., Szewczyk, N.J., ... Vanapalli, S.A. (2019). Pluronic gel-based burrowing assay for rapid assessment of neuromuscular health in C. elegans. *Scientific Reports*, *9*(1), 15246. doi:10.1038/s41598-019-51608-9

Marty, E., Liu, Y., Samuel, A., Or, O., & Lane, J. (2017). A review of sarcopenia: enhancing awareness of an increasingly prevalent disease. *Bone*, *105*, 276–286. doi:10.1016/j.bone.2017.09.008

McCloskey, R.J., Fouad, A.D., Churgin, M.A., & Fang-Yen, C. (2017). Food responsiveness regulates episodic behavioral states in *Caenorhabditis elegans*. *Journal of Neurophysiology*, *117*(5), 1911–1934. doi:10.1152/jn.00555.2016

Mulcahy, B., Holden-Dye, L., & O'Connor, V. (2013). Pharmacological assays reveal age-related changes in synaptic transmission at the *Caenorhabditis elegans* neuromuscular junction that are modified by reduced insulin signalling. *The Journal of Experimental Biology*, *216*(Pt 3), 492–501. doi:10.1242/jeb.068734

Murphy, C.T., McCarroll, S.A., Bargmann, C.I., Fraser, A., Kamath, R.S., Ahringer, J., ... Kenyon, C. (2003). Genes that act downstream of DAF-16 to influence the lifespan of *Caenorhabditis elegans*. *Nature*, *424*(6946), 277–283. doi:10.1038/nature01789

Patel, T.P., Gullotti, D.M., Hernandez, P., O'Brien, W.T., Capehart, B.P., Morrison, B., ... Meaney, D.F. (2014). An open-source toolbox for automated phenotyping of mice in behavioral tasks. *Frontiers in Behavioral Neuroscience*, *8*, 349. doi:10.3389/fnbeh.2014.00349

Pujol, N., Torregrossa, P., Ewbank, J.J., & Brunet, J.F. (2000). The homeodomain protein CePHOX2/CEH-17 controls antero-posterior axonal growth in *C. elegans*. *Development*, *127*(15), 3361–3371.

Rahman, M., Hewitt, J.E., Van-Bussel, F., Edwards, H., Blawzdziewicz, J., Szewczyk, N.J., ... Vanapalli, S.A. (2018). NemaFlex: a microfluidics-based technology for standardized measurement of muscular strength of *C. elegans*. *Lab on a Chip*, *18*(15), 2187–2201. doi:10.1039/C8LC00103K

Raizen, D.M., Zimmerman, J.E., Maycock, M.H., Ta, U.D., You, Y.-J., Sundaram, M.V., & Pack, A.I. (2008). Lethargus is a *Caenorhabditis elegans* sleep-like state. *Nature*, *451*(7178), 569–572. doi:10.1038/nature06535

Restif, C., Ibáñez-Ventoso, C., Vora, M.M., Guo, S., Metaxas, D., & Driscoll, M. (2014). CeleST: computer vision software for quantitative analysis of *C. elegans* swim behavior reveals novel features of locomotion. *PLoS Computational Biology*, *10*(7), e1003702. doi:10.1371/journal.pcbi.1003702

Schwarz, J., Spies, J.-P., & Bringmann, H. (2012). Reduced muscle contraction and a relaxed posture during sleep-like lethargus. *Worm*, *1*(1), 12–14. doi:10.4161/worm.19499

Stephens, G.J., Johnson-Kerner, B., Bialek, W., & Ryu, W.S. (2008). Dimensionality and dynamics in the behavior of *C. elegans*. *PLoS Computational Biology*, *4*(4), e1000028. doi:10.1371/journal.pcbi.1000028

Swierczek, N.A., Giles, A.C., Rankin, C.H., & Kerr, R.A. (2011). High-throughput behavioural analysis in *C. elegans*. *Nature Methods*, *8*(7), 592–598. doi:10.1038/nmeth.1625

Tramm, N., Oppenheimer, N., Nagy, S., Efrati, E., & Biron, D. (2014). Why do sleeping nematodes adopt a hockey-stick-like posture?. *PLoS One*, *9*(7), e101162. doi:10.1371/journal.pone.0101162

Turek, M., Lewandrowski, I., & Bringmann, H. (2013). An AP2 transcription factor is required for a sleep-active neuron to induce sleep-like quiescence in *C. elegans*. *Current Biology*, *23*(22), 2215–2223. doi:10.1016/j.cub.2013.09.028

Van Buskirk, C., & Sternberg, P.W. (2010). Paired and LIM class homeodomain proteins coordinate differentiation of the *C. elegans* ALA neuron. *Development*, *137*(12), 2065–2074. doi:10.1242/dev.040881

Wan, J.-J., Qin, Z., Wang, P.-Y., Sun, Y., & Liu, X. (2017). Muscle fatigue: general understanding and treatment. *Experimental and Molecular Medicine*, *49*(10), e384.

Wei, S.-E., Ramakrishna, V., Kanade, T., & Sheikh, Y. (2016). Convolutional pose machines. arXiv, 1-9. arXiv:1602.00134.

Yang, Y., & Ramanan, D. (2013). Articulated human detection with flexible mixtures of parts. *IEEE Transactions on Pattern Analysis and Machine Intelligence*, *35*(12), 2878–2890. doi:10.1109/TPAMI.2012.261

You, Y.-J., Kim, J., Raizen, D.M., & Avery, L. (2008). Insulin, cGMP, and TGF-beta signals regulate food intake and quiescence in *C. elegans*: a model for satiety. *Cell Metabolism*, *7*(3), 249–257. doi:10.1016/j.cmet.2008.01.005

The OptoGenBox – a device for long-term optogenetics in *C. elegans*

Inka Busack, Florian Jordan, Peleg Sapir and Henrik Bringmann

ABSTRACT

Optogenetics controls neural activity and behavior in living organisms through genetically targetable actuators and light. This method has revolutionized biology and medicine as it allows controlling cells with high temporal and spatial precision. Optogenetics is typically applied only at short time scales, for instance to study specific behaviors. Optogenetically manipulating behavior also gives insights into physiology, as behavior controls systemic physiological processes. For example, arousal and sleep affect aging and health span. To study how behavior controls key physiological processes, behavioral manipulations need to occur at extended time scales. However, methods for long-term optogenetics are scarce and typically require expensive compound microscope setups. Optogenetic experiments can be conducted in many species. Small model animals such as the nematode *C. elegans* have been instrumental in solving the mechanistic basis of medically important biological processes. We developed the OptoGenBox, an affordable stand-alone and simple-to-use device for long-term optogenetic manipulation of *C. elegans*. The OptoGenBox provides a controlled environment and is programmable to allow the execution of complex optogenetic manipulations over long experimental times of many days to weeks. To test our device, we investigated how optogenetically increased arousal and optogenetic sleep deprivation affect survival of arrested first larval stage *C. elegans*. We optogenetically activated the nociceptive ASH sensory neurons using ReaChR, thus triggering an escape response and increase in arousal. In addition, we optogenetically inhibited the sleep neuron RIS using ArchT, a condition known to impair sleep. Both optogenetic manipulations reduced survival. Thus, the OptoGenBox presents an affordable system to study the long-term consequences of optogenetic manipulations of key biological processes in *C. elegans* and perhaps other small animals.

Introduction

Optogenetics can control many physiological processes by actively influencing biochemical reactions and manipulating neuronal activity (Fenno, Yizhar, & Deisseroth, 2011). A light-sensitive actuator can be genetically expressed in specific cells of organisms and activated by light. Different tools exist for either activation or inhibition of excitable cells. Some of the most-used tools are channelrhodopsins, which have first been discovered in algae (Nagel *et al.*, 2002, 2003), and ion pumps, which were found in halobacteria (Han *et al.*, 2011). Both can now be genetically expressed in other organisms to depolarize or hyperpolarize cells upon light stimulation. Optogenetics has become widely established in different model organisms, e.g. small nematodes and flies but also mammals such as mice and monkeys (Fenno *et al.*, 2011). *C. elegans* is well suited and established for optogenetic studies (Husson, Gottschalk, & Leifer, 2013; Schmitt, Schultheis, Husson, Liewald, & Gottschalk, 2012). Many physiological processes are conserved across species and can be studied in less complex organisms such as the 1 mm long nematode *C. elegans*. 83% of its genes have human homologs, allowing molecular studies that are of relevance also to human biology (Lai, Chou, Ch'ang, Liu, & Lin, 2000). With 302 neurons, its nervous system is more manageable than that of other animals. Additionally, a single neuron in *C. elegans* can act similarly to brain regions in mammals (Altun and Hall, 2011). Due to the nematode's transparency, optogenetic experiments can be conducted in a non-invasive manner (Husson *et al.*, 2013). *C. elegans* was the first animal in which optogenetics was established (Husson *et al.*, 2013; Nagel *et al.*, 2003).

However, there are still limitations that hinder the complete realization of the potential of optogenetics. In particular, long-term optogenetic experiments have rarely been conducted (Schultheis, Liewald, Bamberg, Nagel, & Gottschalk, 2011). In a standard experiment the neuronal manipulation only lasts for seconds or minutes. While it is true that some reactions and neuronal signals are fast acting, to manipulate physiology in the long term, one typically has to manipulate biological processes for days or even longer. Optogenetic long-term experiments are challenging for several reasons:

1. It is necessary to control the environment of the tested organisms.

2. For high-throughput experiments, many different conditions should be processed in parallel.
3. There is currently no inexpensive device available to account for 1 and 2.

Through optogenetic long-term manipulations, it is possible to investigate how a specific behavior affects organisms systemically (Altun and Hall, 2011; Husson *et al.*, 2013; Lai *et al.*, 2000; Schmitt *et al.*, 2012). Even in *C. elegans* research the above-mentioned challenges in long-term optogenetic studies persist. Due to the development of new rhodopsins, first steps towards long-term optogenetics have been made. These newer genetic tools can continually be activated for minutes (Gengyo-Ando *et al.*, 2017) or even for up to 2 days (Schultheis *et al.*, 2011) after a shorter light pulse. The longest optogenetic lifespan experiment to date lasted 2.5 h (De Rosa *et al.*, 2019). Optogenetic survival assays lasting several days or weeks have not yet been conducted in *C. elegans.*

One additional reason that explains why long-term experiments have rarely been conducted in *C. elegans* is, that blue light, which is often used in optogenetic experiments, is harmful to the worms. Blue light causes a negative phototaxis and prolonged exposure leads to paralysis and death of *C. elegans* (Edwards *et al.*, 2008; Ward, Liu, Feng, & Xu, 2008). Alternative optogenetic actuators have been developed that can be excited with a higher wavelength, thus causing less stress to *C. elegans*. For example, the red-shifted channelrhodopsin (ReaChR) can be used for neuronal activation (Lin, Knutsen, Muller, Kleinfeld, & Tsien, 2013) or ArchT, which hyperpolarizes neurons by pumping out protons, can be used for inhibition (Okazaki, Sudo, & Takagi, 2012). These genetic tools allow the use of yellow to orange light (585-605nm) for excitation.

Increased arousal and decreased sleep affect the survival of *C. elegans* (De Rosa *et al.*, 2019; Wu, Masurat, Preis, & Bringmann, 2018). Many assays that control arousal and sleep deprivation in *C. elegans* build on external stimuli such as tapping mechanisms, the ablation of neurons or mutation (Bringmann, 2019; Driver, Lamb, Wyner, & Raizen, 2013; Hill, Mansfield, Lopez, Raizen, & Van Buskirk, 2014; Schwarz & Bringmann, 2013; Singh, Ju, Walsh, DiIorio, & Hart, 2014; Spies & Bringmann, 2018; Van Buskirk & Sternberg, 2007). Optogenetics activates or inhibits specific neurons and therefore allows the dissection of neuronal mechanisms. ASH is a nociceptor and its activation causes a reverse escape response by activating the second layer RIM interneurons and by inhibiting the sleep neuron RIS (Kaplan & Horvitz, 1993; Maluck *et al.*, 2020). Mechanical tapping or optogenetic RIM activation, which causes a flight response and increase in arousal, shortens the lifespan of adult *C. elegans* (De Rosa *et al.*, 2019). Depolarization of ASH causes a complex response. It activates RIM, therefore triggering release of tyramine and promoting the flight response (De Rosa *et al.*, 2019; Maluck *et al.*, 2020). Additionally, strong RIM activation inhibits the sleep neuron RIS which leads to sleep deprivation (Maluck *et al.*, 2020). RIS is a single neuron that acts as the motor of sleep in *C. elegans.* RIS is active during sleep, its activation induces sleep and its depolarization is homeostatically regulated (Bringmann, 2018; Maluck *et al.*, 2020; Turek, Lewandrowski, & Bringmann, 2013). A more specific experiment for sleep deprivation, in which arousal also gets increased, is hence the inhibition of the sleep neuron RIS through optogenetics (Maluck *et al.*, 2020; Wu *et al.*, 2018).

To solve the problem of long-term optogenetic manipulation, we have developed the OptoGenBox, a simple-to-use stand-alone device, which provides a controllable environment and allows for the execution of complex optogenetic protocols. The total material costs of less than 3500 USD (Table S1) makes it substantially more inexpensive than the use of standard microscope set-ups. The OptoGenBox therefore presents the currently best solution for long-term optogenetic experiments in *C. elegans.*

We successfully tested the OptoGenBox by optogenetically activating the sensory neuron ASH and inhibiting the sleep neuron RIS. Optogenetic activation of ASH or inhibition of RIS in L1 arrested animals both reduced lifespan. Our results show that the OptoGenBox is a valuable tool for long-term optogenetic experiments in *C. elegans*, and potentially also for other small animals.

Results

A device for optogenetic long-term experiments

We developed the OptoGenBox to enable long-term optogenetic experiments in *C. elegans* (Figure 1). Worms were kept in a temperature-controlled environment and illuminated with orange light from the bottom (Figure 2(A)). For this, the OptoGenBox was built as a $70 \times 70 \times 90$cm large device that is programmable via a touch display (Figure 2(B)). The inside consists of a 22×22cm sized experimentation area partitioned into 13 cells (Figure 2(C)). Each cell can hold small plates with nematode growth medium or microfluidic chambers (Bringmann, 2011; Turek, Besseling, & Bringmann, 2015) with a diameter of 3.5 cm, and can thus fit up to 100 worms. Worm plates are placed on 4 mm thick glass (B270), which was polished on the bottom (400 polish) to homogeneously distribute the LED light throughout the worm plate (Figure 3). 6 LEDs are distributed throughout an LED module (Figure 4) 7.4 mm below the glass to illuminate the worms from the bottom. An aluminum casing keeps external light out and creates optically isolated cells. Furthermore, the box is temperature controlled through Peltier devices and protected from external disturbances via foam and an acrylic case (Figure S1). The LED intensities of all 13 cells were measured with a light voltmeter (ThorLabs PM100A) and calibrated through the software while setting up the system to assure equal light intensities between the cells. The temperature for all cells is uniform and can only be determined when no experiments are running. Each cell contains environmental sensors for light intensity and air quality, and temperature recordings are carried out for each cell. Humidity stays constant in the closed plastic dishes that contain the microfluidic devices (Turek *et al.*, 2015) and thus humidity measurements are not necessary when the microfluidic devices are used. Nevertheless, sensors are

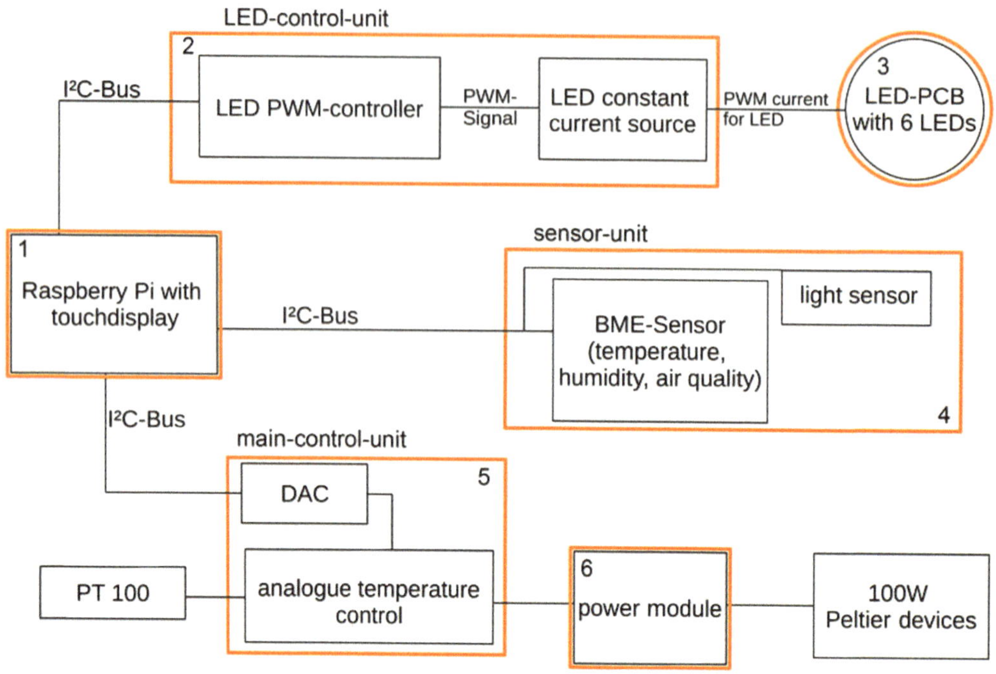

Figure 1. Functional scheme of the OptoGenBox. The box consists of several printed circuit boards (PCBs in coloured outlines), that are connected and controlled by a Raspberry Pi computer.

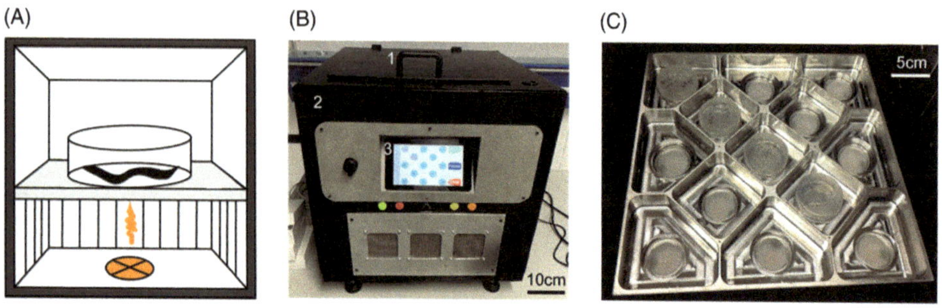

Figure 2. The OptoGenBox is a device for long-term optogenetic experiments in *C. elegans*. (A) Worms are kept in a controlled environment and illuminated with orange light. (B) The outside of the OptoGenBox. (1) opening handle, (2) exterior case and (3) touch screen. (C) The inside of the OptoGenBox is comprised of 13 groupable or separately programmable cells.

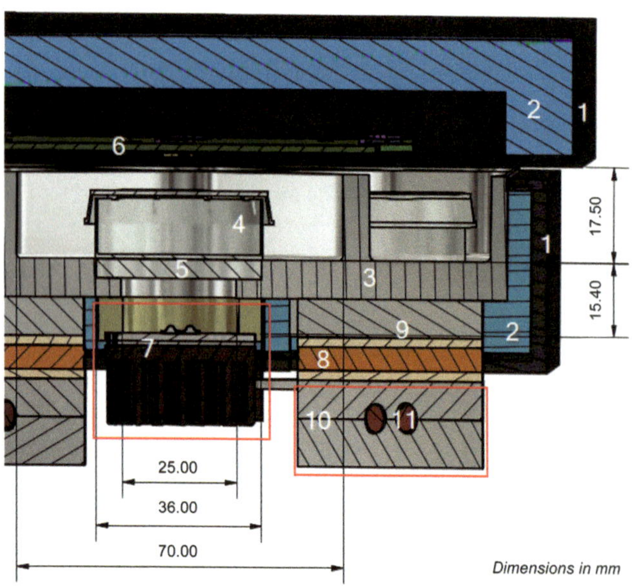

Figure 3. Cross-section of the OptoGenBox incubator. (1) plastic cover, (2) insulating foam, (3) incubator inlet (4) worm plates or microfluidic devices, (5) sanded glass, (6) sensor PCB (in the lid), (7) LED module, (8) mounting bracket Peltier device, (9) Peltier device covered with thermal pads, (10) heat pipe brackets and (11) heat pipe.

included to monitor humidity inside the device in case other types of samples need to be used.

The researcher can easily program the optogenetic protocol through the touch screen. The system is written in Python and implemented on a Raspberry Pi (Figure S2). To start an experiment the exact cells can be selected individually for each experiment and then the optogenetic protocol can be defined (Figure S3). The experimenter can choose how many cycles should run with how much time (hours or minutes) in light and how much time in darkness and can define the light intensity for the experimentation area (between 2-40mW) during the light times. LEDs can be programmed to be either on or off for minutes or hours. The minimum continuous amount of time for a light cycle is hence 1 min and the maximum is 25 h. The same holds for dark phases. The maximum number of cycles is 5000. Theoretically, worms could get illuminated for up to 2083 days. The temperature can only be chosen for the entire OptoGenBox and not individual cells between 15–25 °C (Figure S4).

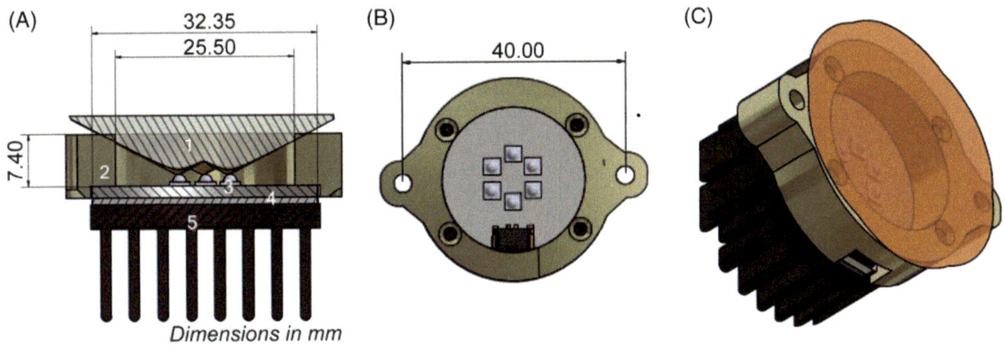

Figure 4. LED module. (A) Cross section of an LED module. (1) calculated light beam (LED current at 50%), (2) distance ring to mount the module, (3) the IMS-PCB with single LEDs on it, (4) the thermal pad and (5) the heat sink. (B) Top view of the LED module. (C) Side view of the LED module with the light distribution.

While one experiment can include up to 13 cells, light intensities and optogenetic protocols of individual or groups of cells can also be programmed separately to allow for parallel experiments (Figure S5). The total material costs of less than 3500 USD (Table S1) make it much less expensive than microscopic set-ups, which one could also use for optogenetic long-term experiments. All code is freely available (https://gitlab.gwdg.de/psapir/inkubator). The OptoGenbox presents an inexpensive and user-friendly tool to conduct optogenetic long-term experiments.

Optogenetic ASH activation in the OptoGenBox triggers an escape response

The sensory neuron ASH is known to promote reverse escape locomotion upon different harmful stimuli (Kaplan & Horvitz, 1993; Zheng, Brockie, Mellem, Madsen, & Maricq, 1999). To test for the functionality of the box, we developed an escape essay in which we optogenetically activated ASH and tested for its effects on behavior. ReaChR was genetically expressed in worms under the *sra-6* promoter to cause ASH activation upon addition of ATR (Wu *et al.*, 2018). For the experiment, a small plate was prepared with a small lawn of bacteria of the *E. coli* strain OP50 as a food source on one half of the plate and an opaque sticky tape, which caused an area of shade in the OptoGenBox, on the other half (Figure 5(A)). Worms without any optogenetic activation were expected to mostly assemble by the food. On the contrary, after ASH activation worms were expected to not gather at the food but to either distribute throughout the plate or gather in the shade, where the activation is interrupted. An optogenetic protocol was run for one hour and the distribution of worms was counted. Indeed, an average of 80% of the control worms gathered by the food. Only around 20% of the ASH-activated *C. elegans* could be counted at the food drop. This significant decrease in worms at the food drop confirms that the worms show an escape response upon ASH activation. Worms did not aggregate in the shade caused by the sticky tape but mostly distributed across the plate. This could potentially be explained by a remaining low light intensity of 0.02mW in the shade (outside the shade there was an intensity of 10mW, so 0.2% of the light intensity could be measured above the sticky tape), which may have still been sufficient for ASH activation and hence an escape response of the worm. The low light intensity in the shade could perhaps be caused by light reflections. Neither the worms in which ASH was activated nor control worms were able to flee from the plate (Figure 5(B)). These results demonstrate the functionality of the OptoGenBox.

Increased arousal and decreased sleep by optogenetic manipulations shortens the lifespan of arrested L1 larvae

Increased arousal and sleep deprivation has been shown to shorten the lifespan in *C. elegans* (De Rosa *et al.*, 2019; Wu *et al.*, 2018). We wanted to test if an increase in arousal or inhibition of sleep can affect the survival of arrested L1 larvae. We therefore conducted experiments in which arousal gets increased or sleep is reduced through different optogenetic manipulations.

The optogenetic manipulations were achieved by treating transgenic worms carrying the optogenetic tool with ATR. Since a toxicity of ATR could not be excluded we first investigated the effects of ATR on the wild type. Two rounds of experiments confirmed that the addition of ATR without optogenetic manipulation did not lead to a significant reduction of survival (Figure 6(A)). Hence, any lifespan phenotypes in our optogenetic experiments can be attributed to the optogenetic manipulations and not the treatment with ATR.

To test for survival phenotypes upon increased arousal, we conducted two experiments, a first experiment in which optogenetic activation of a nociceptive neuron causes an escape response and increases arousal and a second experiment in which optogenetic inhibition of a sleep neuron causes sleep deprivation.

For the optogenetic activation experiment, we used the ASH::ReaChR strain as described before (Wu *et al.*, 2018). All-trans retinal (ATR) was present throughout the L1 arrest lifespan to ensure functionality of the optogenetic tool. Control worms were used that carried the ReaChR transgene but did not receive ATR. In both rounds of the experiment, animals in which ASH was activated died significantly earlier than control worms (Figure 6(B)).

Next, we tested how sleep deprivation caused by the inhibition of the sleep neuron RIS affects survival in L1 arrest. We expressed ArchT under the *flp-11* promoter so

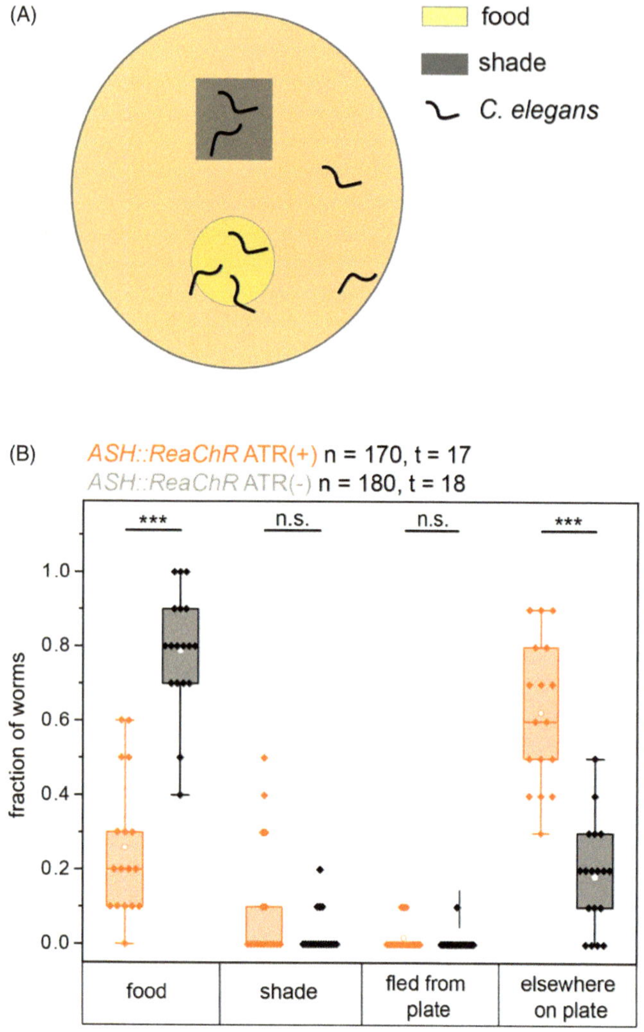

Figure 5. ASH activation in the OptoGenBox caused an escape response. (A) Preparation of the experimental plate. A small NGM plate is prepared with a drop of food (*E. coli* OP50). An opaque sticky tape is used to block the stimulating light. (B) After ASH activation through ReaChR and ATR, worms did not stay on the food drop but distributed throughout the plate. Neither ASH-activated nor control animals fled from the plate. $^{***}p < 0.001$, Kolmogorov Smirnov Test.

that it was specifically expressed in RIS and all-trans retinal was supplemented (Wu *et al.*, 2018). Again, control worms for comparison did not receive ATR treatment. Optogenetic sleep deprivation led to a small but significant reduction of survival in arrested L1 animals by 8.3% (Figure 6(C)).

Discussion

Here we developed the OptoGenBox as a device for optogenetic long-term experiments. The OptoGenBox combines a controlled environment and allows for parallel processing of many experiments for *C. elegans* and perhaps other small animal models. With material costs of less than 3500 USD it is rather inexpensive. While there exist lower cost alternatives such as the DART system for *Drosophila* (Faville, Kottler, Goodhill, Shaw, & Van Swinderen, 2015), the DART system does not allow for parallel processing and

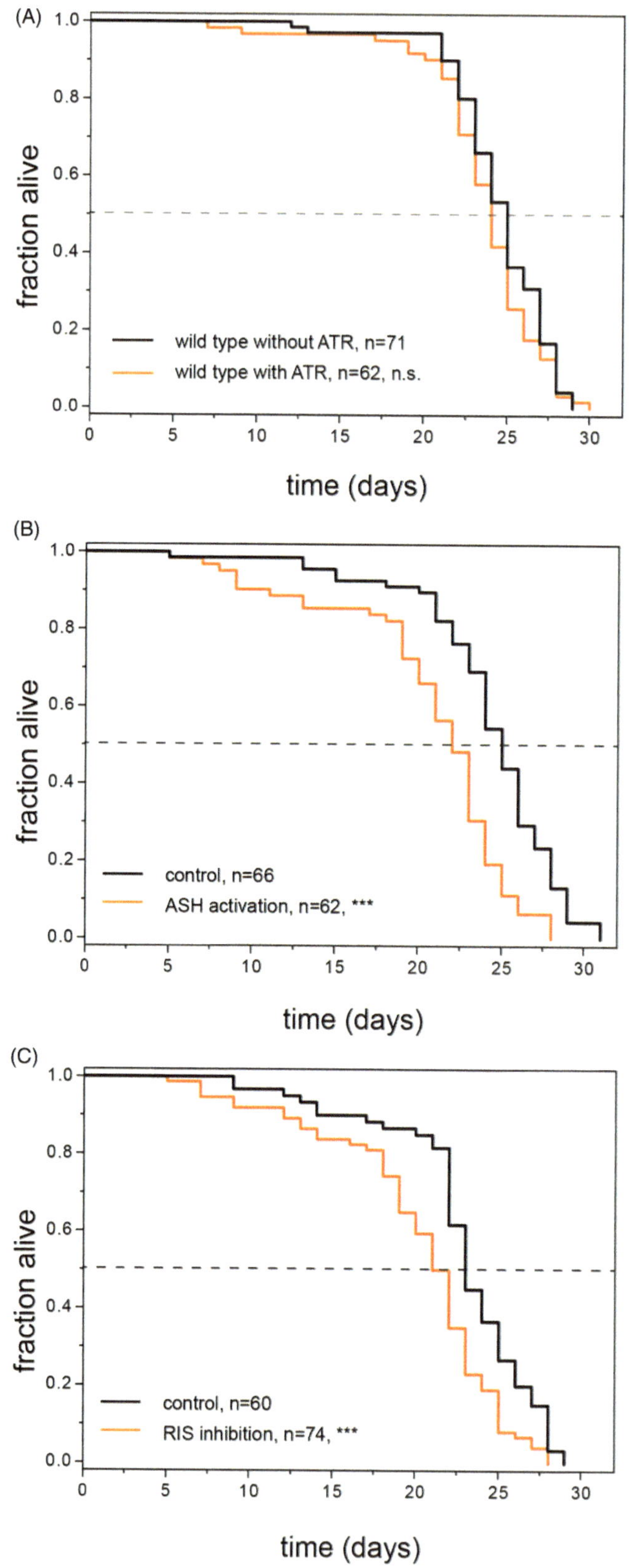

Figure 6. An increase in arousal and sleep deprivation reduces the lifespan of arrested L1 *C. elegans*. (A) All-trans retinal (ATR) did not affect survival of wild-type arrested L1 larvae. The graph includes data from 2 replicates. n.s. $p = 0.2$, Logrank test. (B) ASH activation causes a reduction in lifespan compared to control animals without the addition of ATR. The graph includes data from 2 replicates. $^{***}p = 9.996*10^{-7}$, Logrank test. (C) RIS inhibition causes a reduction in lifespan compared to control animals without the addition of ATR. The graph includes data from 2 replicates. $^{***}p = 3.4*10^{-4}$, Logrank test.

temperature control. Hence, the OptoGenBox currently presents the best solution to allow for parallel optogenetic long-term experiments. Experiments in the OptoGenBox can last up to several weeks. The device allows for optogenetic long-term experiments in a highly controlled environment. The OptoGenBox is not equipped with an imaging system. For performing measurements on the worms, the samples containing the worms thus need to be taken out of the system, which could perturb the measurements. However, an imaging system could be added to the device in the future.

We could demonstrate that different optogenetic manipulations that increase arousal or inhibit sleep have a detrimental effect on *C. elegans*. The activation of the nociceptor ASH led to a reduced survival in L1 arrest. While ASH activation also leads to an inhibition of the sleep neuron RIS (Maluck *et al.*, 2020), the lifespan shortage of ASH activated animals cannot solely be accounted for by sleep inhibition(Maluck *et al.*, 2020). More likely, the reduced survival upon ASH activation is caused by the inhibition of cytoprotective mechanisms through the activation of RIM and release of tyramine as has been previously described (De Rosa *et al.*, 2019).

The optogenetic inhibition of RIS presents a very specific and therefore suitable experiment to investigate the effects of sleep deprivation on *C. elegans*. The shortened survival upon RIS inhibition confirms that sleep plays an essential role in arrested L1 worms as has been previously demonstrated with *aptf-1(gk794)* mutants in which RIS is not functional and with worms in which RIS was genetically ablated (Wu *et al.*, 2018). However, the previously reported phenotypes with *aptf-1(gk794)* mutants were stronger, having a reduction of lifespan of approximately 40% compared to the wild type. In comparison, in the lifespans in which RIS was optogenetically inhibited, the reduction of lifespan was rather small (around 8.3%). There might be several reasons for these differences. The previously reported stronger lifespan effects were obtained in liquid cultures whereas during the optogenetic experiments, worms were kept isolated in microfluidic devices, making a direct comparison impossible. Furthermore, genetic sleep deprivation by a loss of functional APTF-1 can be presumed to lead to more severe effects than temporally-restricted optogenetic sleep deprivation. The advantages of optogenetics are that behavior can be controlled with temporal precision. Instead of completely depriving the worms of sleep it is possible to study the effects of periodic sleep deprivation. For the results presented here, a long light phase (11 h) was followed by only a short dark phase (1 h) in each cycle throughout the lifespan. This is a rather long optogenetic stimulation phase, in which neurons could perhaps get desensitized as desensitization has been shown before in optogenetic experiments in *C. elegans* (Bergs *et al.*, 2018; Berndt, Yizhar, Gunaydin, Hegemann, & Deisseroth, 2009). It is possible that shorter intervals of light/dark phases might be even more effective for optogenetic sleep deprivation in future experiments.

In experiments with worms in which RIS function was impaired, it was shown that sleep counteracts aging phenotypes (Wu *et al.*, 2018). It would be interesting to see how aging phenotypes progress when RIS is inhibited optogenetically. Additionally, how exactly sleep counteracts aging and causes premature death needs further investigation.

Conclusions

With the newly developed OptoGenBox, we have mostly investigated how an increase in arousal and a loss of sleep affects survival in L1 arrest. However, many other questions could be answered with our device. Optogenetics is a method that cannot only be utilized for depolarizing or hyperpolarizing neurons but also any other type of cell such as epidermal or muscle cells. Silencing of body wall muscles for example leads to an inhibition of feeding (Takahashi & Takagi, 2017) and photoablation of epidermal cells causes paralysis in *C. elegans* (Xu & Chisholm, 2016). The OptoGenBox should allow for many optogenetic long-term experiments in *C. elegans* and potentially also other small animals. Long-term optogenetics should thus help understand how behavior affects systemic physiology in the long term.

Methods

Development of the OptoGenBox

The OptoGenBox consists of several parts to allow for orange light illumination and temperature control (Figure 1–3). The user can select the cells and set the exact light level through the touch display of the raspberry pi computer. Signals from and to the raspberry pi are transferred by an inter-integrated circuit bus (I^2C bus). There are four LED controllers that address the LEDs of cells, which the researcher previously chose. The LED controllers convert the set illumination level into a pulse width modulated (PWM) signal. This signal allows a constant current through the LEDs and their current source so that the selected cell gets illuminated. The PWM current finally supplies 6 single high brightness LEDs on one single LED-PCB. A light sensor for each cell gives feedback to the raspberry pi about the activation and wavelengths of the LEDs.

A digital to analogue converter (DAC) connects the digital temperature signal, set by the user, with the analogue temperature control unit. This unit gets the actual value from a PT 100 temperature sensor located at the bottom of the chamber and regulates the power output for six 100 W Peltier devices. With this closed control loop the OptoGenBox can operate at a constant temperature between 15–25 °C

An additional temperature measuring takes part by several evenly placed environmental sensors located in the lid. These sensors measure temperature, air quality (based on gas measurements, 0–50 is excellent air) and humidity. The obtained temperature is displayed on the screen.

The system is built on several printed circuit boards (PCBs), which are separated by function. These different PCBs are: the LED-controlling PCB, the environment measuring PCB, the analogue temperature-control PCB, an analogue power module PCB and an overall supplying PCB.

Assembly of the OptoGenBox

The OptoGenBox consists of a few electronic units (Figure 1), which are: (1) the raspberry pi inclusive the touch display, (2) LED-control-units, (3) LED modules, (4) a sensor-unit, (5) a main-control-unit and (6) a DC/DC-power-supply-unit, These units were specifically produced for the OptoGenBox (except the raspberry pi with its display). Furthermore, all PCBs were assembled manually at the Max Planck Institute for Biophysical Chemistry (MPI-bpC). The bare PCBs were produced by different distributors available in Germany (market compliant).

To assemble a PCB, a soldering iron was sufficient for most PCBs. However, for some PCBs, a reflow-oven was used either because it was required or for a more reliable and time efficient soldering procedure.

Reflow soldering

Reflow soldering requires a special set of tools, which consists of a disposing tool for the soldering paste, a placing machine (not necessary, but facilitates the procedure), and an oven that heats up to at least 270 °C.

Hand soldering

Hand soldering doesn't require as specific tools as reflow soldering but requires more skills from the executing person. To produce reliable PCBs, different types of soldering tips are recommended and a set of tweezers should be available.

After PCB assembly, the PCBs were connected. For different types of signals, different connectors and cables were selected. Every connector has its special crimping tool so that in total four crimping pliers were used. Additionally, a set of screwdrivers and pliers should be available. A digital multi-meter was utilized to adjust the LED voltage and to tune the analogue temperature control circuit.

Mechanical assembly

The components of the OptoGenBox were placed in a modified case originally built for a water-cooled PC system (Figure S7). In the lower tier, all of the AC/DC power supplies and the temperature-control-unit are placed. The incubator sits in the upper tier of the case. The main-control-unit and the raspberry-pi are placed around the incubator (Figure 2).

The incubator itself is assembled in the following manner:

> The outside of the incubator is a plastic cover (Figure 3, number 1) around an insulating foam material (Figure 3, number 2). These two materials provide for a stable temperature environment in the incubator inlet (Figure 3, number 3). The worm plates or microfluidic devices (Figure 3, number 4) can be placed on a one-side sanded glass (Figure 3, number 5) in the inlet. The lid of the incubator contains the sensor-unit (Figure 3, number 6). The LED modules (Figure 3, number 7) and the Peltier devices (Figure 3, number 8) are placed in cut-outs beneath the inlet and each mounted with two screws. This construction makes it possible to change the pre-assembled LED modules. While exchanging the LEDs, one has to pay attention to match the current and voltage to the new LED type for ideal light results. Matching the electrical parameters can be done via already implemented options on the LED-control-unit and DC/DC-power-supply-units.
>
> For an optimized thermal solution, the Peltier devices are clamped with thermal pads between two brackets. One bracket (Figure 3, number 9) is directly attached to the inlet. The other is a two-piece bracket (Figure 3, number 10) clamping the Peltier devices and holding the heat pipes (Figure 3, number 11). The heat pipes transport the emerging heat when the device is cooling the incubator. The elements holding the heat pipes can be assembled separately. The heat pipes were manually bend from a straight pipe to fit in the shape that was needed. All bracket parts were specifically designed for the OptoGenBox. The LED control units were attached to the plastic cover (with standard bolts and screws) and then wired with the 13 LED modules.

With the LED modules and the Peltier devices attached to the insulated, covered aluminium inlet, it was installed on fitting brackets in the upper tier of the modified PC case. The LED-control-unit was wired to the main control unit and to the power supply for the LEDs at the DC/DC power supply unit

Fans were installed on both sides of the case to avoid a cushion of heat beneath the incubator and to create a constant airflow so the LED modules, heatsinks for Peltier devices and the electronics would not get damaged by elevated temperature.

The LED module

One LED module consists of six high-power LEDs (Osram Opto Semiconductors LCY-CLBP Series) with a peak wavelength at approximately 590 nm with 80 lm each (Figure 4). In order to reach a maximum light power of 40mW we placed 6 LEDs in a circle with a diameter of 8.4 mm. The individual LED modules were calibrated after the installation to have the same light intensities. At 10mW, the light intensity difference between the center and the periphery of the experimentation area was measured to be 0.04mW (0.4% difference). The PCB of the LED module is an IMS-Core PCB, (insulated metal substrate) to absorb most of the thermal energy and conduct it through a thermal pad to the attached round heat sink away from the temperature-controlled area.

The lid

The lid is made of insulating foam material covered with plastic. To locate the necessary sensors at the designated position, the lid got a fitting cut-out. In this cut-out, an overall covering PCB with a pair of sensors (light & environment) for each individual chamber was placed. It is directly attached to the plastic that covers the aluminium inlet from above, aligned to small holes so that the light can be detected and measured. Through a separated hole the air-quality is measured. This PCB and the plastic, on which it is mounted, could be modified to add several other functions as for example an IR-camera with an integrated light source.

C. elegans maintenance

Worms were grown at 20 °C on Nematode Growth Medium (NGM) plates. The plates were seeded with *E. coli* OP50, which served as food for the worms (Brenner, 1974). The following strains were used for this study:

HBR974 *goeIs232(psra-6::ReaChr::mKate2-unc-54–3'utr, unc-119(+))*

HBR1463 *goeIs307(pflp-11::ArchT::SL2mKate2-unc-54–3'utr, unc-119(+))*

N2 wild type (Bristol) (Brenner, 1974)

Escape assay

Late L4 stage worms were picked onto NGM plates with 0.2 mM all-*trans* retinal (ATR, Sigma Aldrich). Control late L4 stage worms were picked onto NGM plates without ATR. 9.6 cm^2 large NGM plates were prepared for the experiment by placing a 1 cm^2 opaque tape on the bottom of one side of the plate and a drop of *E. coli* OP50 on the other side (Figure 5(A)). After 4 h, 10 young adult worms were picked into the food drop of the experimental plate for each trial.

The experimental plates were then placed in the OptoGenBox and stimulated with 10mW orange light for 1 h at 20 °C. After one hour the plates were removed from the OptoGenBox and the distribution of worms on the plates was counted.

Lifespan assay

It was shown before that sleep is important for the survival of *C. elegans* by counteracting aging phenotypes. However, non-sleeping *aptf-1*(gk794) mutants only have a reduced lifespan when worms starve upon hatching and arrest in the first larval stage (L1 arrest) and not when they are adults (Wu *et al.*, 2018). For this reason, we conducted our experiments with L1 arrested animals.

Worms were kept in microfluidic devices as previously described (Bringmann, 2011; Turek *et al.*, 2015). A PDMS mold was used as a stamp to cast 110x110x10µm cuboids into a hydrogel. The hydrogel consisted of 3% agarose dissolved in S-Basal (Stiernagle, 2006). Eggs were transferred from a growing plate to a plate without food and then picked into chambers without transferring food. Between 29 and 45 worms were in one microfluidic device housed in individual chambers.

For optogenetic activation or inhibition, chambers were replenished with 10 µl of 10 mM all-trans-retinal (ATR, Sigma Aldrich) every 3–4 days. Control chambers did not receive ATR. To avoid fungal contamination, 20 µl of 10 µg/ml nystatin was pipetted to each chamber 2–4 times throughout the lifespan. Additionally, 20 µl of sterile water was added every 2 days until day 15 of the lifespan and then each day to counteract the agarose drying out over time. In the beginning of the lifespan experiment, worms were counted every second day, in the later stages of the survival assay they were counted every day. A worm was counted as dead if it didn't move for 2 min under stimulation with a blue light LED. This was necessary to distinguish dead from sleeping worms.

The worms were placed in the OptoGenBox and illuminated with 10mW for 11 h to attain a long continuous neuronal manipulation. This was followed by 1 h of darkness to allow the optogenetic tools to recover without giving too much time to sleep homeostasis processes, which initiate a deeper and prolonged sleep upon sleep deprivation. This protocol was repeated until all worms were dead. The temperature of the incubator was set to 20 °C.

Statistics

Sample sizes were determined empirically based on previous studies. The researcher was not blinded since the addition of ATR is easily detectable. Conditions in the escape assay were compared with the Kolmogorov Smirnov Test. The lifespans were compared with the Logrank test.

Acknowledgements

The authors thank the CGC, which is funded by NIH Office of Research Infrastructure Programs (P40 OD010440), for the N2 strain. The mechanics workshop at the MPI BPC provided us with valuable advice for the design and parts of the OptoGenBox. The authors also like to thank Juliane Haase for assisting with laboratory work.

Author contributions

IB and FJ designed the OptoGenBox. IB designed, performed and analyzed the experiments and wrote the manuscript. FJ built the hardware of the OptoGenBox and contributed to the manuscript. PS programmed the software of the OptoGenBox. HB acquired funding, conceived the project, supervised the work, and edited the manuscript.

Disclosure statement

The authors declare that they have no competing interest.

Funding

This work was funded by the Max Planck Society (Max Planck Research Group), a European Research Council Starting Grant (ID: 637860, SLEEPCONTROL), and the University of Marburg.

References

Altun, Z.F., & Hall, D.H. (2011). Nervous system, general description. https://doi.org/doi:10.3908/wormatlas.1.18

Bergs, A., Schultheis, C., Fischer, E., Tsunoda, S.P., Erbguth, K., Husson, S.J., ... Liewald, J.F. (2018). Rhodopsin optogenetic toolbox v2.0 for light-sensitive excitation and inhibition in Caenorhabditis elegans. *PLoS One*, *13*(2), e0191802. doi:10.1371/journal.pone.0191802

Berndt, A., Yizhar, O., Gunaydin, L.A., Hegemann, P., & Deisseroth, K. (2009). Bi-stable neural state switches. *Nature Neuroscience*, *12*(2), 229–234. doi:10.1038/nn.2247

Brenner, S. (1974). The genetics of Caenorhabditis elegans. *Genetics*, *77*(1), 71–94. doi:10.1002/cbic.200300625

Bringmann, H. (2011). Agarose hydrogel microcompartments for imaging sleep- and wake-like behavior and nervous system development in Caenorhabditis elegans larvae. *Journal of Neuroscience Methods, 201*(1), 78–88. doi:10.1016/j.jneumeth.2011.07.013

Bringmann, H. (2018). Sleep-active neurons: Conserved motors of sleep. *Genetics, 208*(4), 1279–1289. doi:10.1534/genetics.117.300521

Bringmann, H. (2019). Genetic sleep deprivation: using sleep mutants to study sleep functions. *EMBO Reports, 20* (3). e46807. doi:10.15252/embr.201846807

De Rosa, M.J., Veuthey, T., Florman, J., Grant, J., Blanco, M.G., Andersen, N., ... Alkema, M.J. (2019). The flight response impairs cytoprotective mechanisms by activating the insulin pathway. *Nature, 573*(7772), 135–138. doi:10.1038/s41586-019-1524-5

Driver, R.J., Lamb, A.L., Wyner, A.J., & Raizen, D.M. (2013). DAF-16/FOXO regulates homeostasis of essential sleep-like behavior during larval transitions in C. elegans. *Current Biology, 23*(6), 501–506. doi:10.1016/j.cub.2013.02.009

Edwards, S.L., Charlie, N.K., Milfort, M.C., Brown, B.S., Gravlin, C.N., Knecht, J.E., & Miller, K.G. (2008). A novel molecular solution for ultraviolet light detection in Caenorhabditis elegans. *PLoS Biology, 6*(8), e198. doi:10.1371/journal.pbio.0060198

Faville, R., Kottler, B., Goodhill, G.J., Shaw, P.J., & Van Swinderen, B. (2015). How deeply does your mutant sleep? Probing arousal to better understand sleep defects in Drosophila. *Scientific Reports, 5*, 8454. doi:10.1038/srep08454

Fenno, L., Yizhar, O., & Deisseroth, K. (2011). The development and application of optogenetics. *Annual Review of Neuroscience, 34*(1), 389–412. doi:10.1146/annurev-neuro-061010-113817

Gengyo-Ando, K., Kagawa-Nagamura, Y., Ohkura, M., Fei, X., Chen, M., Hashimoto, K., & Nakai, J. (2017). A new platform for long-term tracking and recording of neural activity and simultaneous optogenetic control in freely behaving Caenorhabditis elegans. *Journal of Neuroscience Methods, 286*, 56–68. doi:10.1016/j.jneumeth.2017.05.017

Han, X., Chow, B. Y., Zhou, H., Klapoetke, N. C., Chuong, A., Rajimehr, R., Yang, A., Baratta, M. V., Winkle, J., Desimone, R., & Boyden, E. S. (2011). A high-light sensitivity optical neural silencer: development and application to optogenetic control of non-human primate cortex. *Frontiers in systems neuroscience, 5*, 18. https://doi.org/10.3389/fnsys.2011.00018

Hill, A.J., Mansfield, R., Lopez, J.M.N.G., Raizen, D.M., & Van Buskirk, C. (2014). Cellular stress induces a protective sleep-like state in C. elegans. *Current Biology, 24*(20), 2399–2405. doi:10.1016/j.cub.2014.08.040

Husson, S.J., Gottschalk, A., & Leifer, A.M. (2013). Optogenetic manipulation of neural activity in C. elegans: From synapse to circuits and behaviour. *Biology of the Cell, 105*(6), 235–250. doi:10.1111/boc.201200069

Kaplan, J.M., & Horvitz, H.R. (1993). A dual mechanosensory and chemosensory neuron in Caenorhabditis elegans. *Proceedings of the National Academy of Sciences of the United States of America, 90*(6), 2227–2231. doi:10.1073/pnas.90.6.2227

Lai, C.H., Chou, C.Y., Ch'ang, L.Y., Liu, C.S., & Lin, W. (2000). Identification of novel human genes evolutionarily conserved in *Caenorhabditis elegans* by comparative proteomics. *Genome Research, 10*(5), 703–713. doi:10.1101/gr.10.5.703

Lin, J.Y., Knutsen, P.M., Muller, A., Kleinfeld, D., & Tsien, R.Y. (2013). ReaChR: a red-shifted variant of channelrhodopsin enables deep transcranial optogenetic excitation. *Nature Neuroscience, 16*(10), 1499–1508. doi:10.1038/nn.3502

Maluck, E., Busack, I., Besseling, J., Masurat, F., Turek, M., Busch, K.E., & Bringmann, H. (2020). A wake-active locomotion circuit depolarizes a sleep-active neuron to switch on sleep. *PLoS Biology, 18*(2), e3000361. Retrieved from: . doi:10.1371/journal.pbio.3000361

Nagel, G., Ollig, D., Fuhrmann, M., Kateriya, S., Musti, A.M., Bamberg, E., & Hegemann, P. (2002). Channelrhodopsin-1: a light-gated proton channel in green algae. *Science (New York, N.Y.), 296*(5577), 2395–2398. doi:10.1126/science.1072068

Nagel, G., Szellas, T., Huhn, W., Kateriya, S., Adeishvili, N., Berthold, P., ... Bamberg, E. (2003). Channelrhodopsin-2, a directly light-gated cation-selective membrane channel. *Proceedings of the National Academy of Sciences of the United States of America, 100*(24), 13940–13945. doi:10.1073/pnas.1936192100

Okazaki, A., Sudo, Y., & Takagi, S. (2012). Optical silencing of c. elegans cells with arch proton pump. *PLoS One, 7*(5), e35370. doi:10.1371/journal.pone.0035370

Schmitt, C., Schultheis, C., Husson, S.J., Liewald, J.F., & Gottschalk, A. (2012). Specific expression of channelrhodopsin-2 in single neurons of *Caenorhabditis elegans. PLoS One, 7*(8), e43164. doi:10.1371/journal.pone.0043164

Schultheis, C., Liewald, J.F., Bamberg, E., Nagel, G., & Gottschalk, A. (2011). Optogenetic long-term manipulation of behavior and animal development. *PLoS One, 6*(4), e18766. doi:10.1371/journal.pone.0018766

Schwarz, J., & Bringmann, H. (2013). Reduced sleep-like quiescence in both hyperactive and hypoactive mutants of the galphaq gene egl-30 during lethargus in *Caenorhabditis elegans. PLoS One, 8*(9), e75853. doi:10.1371/journal.pone.0075853

Singh, K., Ju, J.Y., Walsh, M.B., DiIorio, M.A., & Hart, A.C. (2014). Deep conservation of genes required for both *Drosophila melanogaster* and *Caenorhabditis elegans* sleep includes a role for dopaminergic signaling. *Sleep, 37*(9), 1439–1451. doi:10.5665/sleep.3990

Spies, J., & Bringmann, H. (2018). Automated detection and manipulation of sleep in C. elegans reveals depolarization of a sleep-active neuron during mechanical stimulation-induced sleep deprivation. *Scientific reports, 8*(1), 9732. https://doi.org/10.1038/s41598-018-28095-5

Stiernagle, T. (2006). Maintenance of C. elegans. *WormBook : The Online Review of.* C. Elegans Biology, doi:10.1895/wormbook.1.101.1

Takahashi, M., & Takagi, S. (2017). Optical silencing of body wall muscles induces pumping inhibition in Caenorhabditis elegans. *PLoS Genetics, 13*(12), e1007134. doi:10.1371/journal.pgen.1007134

Turek, M., Besseling, J., & Bringmann, H. (2015). Agarose microchambers for long-term calcium imaging of caenorhabditis elegans. *J Vis Exp*, (100), e52742. doi:10.3791/52742

Turek, M., Lewandrowski, I., & Bringmann, H. (2013). An AP2 transcription factor is required for a sleep-active neuron to induce sleep-like quiescence in C. elegans. *Curr. Biol, 23*(22), 2215–2223. doi:10.1016/j.cub.2013.09.028

Van Buskirk, C., & Sternberg, P.W. (2007). Epidermal growth factor signaling induces behavioral quiescence in Caenorhabditis elegans. *Nature Neuroscience, 10*(10), 1300–1307. doi:10.1038/nn1981

Ward, A., Liu, J., Feng, Z., & Xu, X.Z.S. (2008). Light-sensitive neurons and channels mediate phototaxis in C. elegans. *Nature Neuroscience, 11*(8), 916–922. doi:10.1038/nn.2155

Wu, Y., Masurat, F., Preis, J., & Bringmann, H. (2018). Sleep counteracts aging phenotypes to survive starvation-induced developmental arrest in C. elegans. *Current Biology, 28*(22), 3610–3624.e8. doi:10.1016/j.cub.2018.10.009

Xu, S., & Chisholm, A.D. (2016). Highly efficient optogenetic cell ablation in C. Elegans using membrane-targeted miniSOG. *Scientific Reports, 6*, 21271. doi:10.1038/srep21271

Zheng, Y., Brockie, P.J., Mellem, J.E., Madsen, D.M., & Maricq, A.V. (1999). Neuronal control of locomotion in C. elegans is modified by a dominant mutation in the GLR-1 ionotropic glutamate receptor. *Neuron, 24*(2), 347–361. doi:10.1016/S0896-6273(00)80849-1

Part VI

Survival, aging and disease

Neuromodulators: an essential part of survival

Joy Alcedo and Veena Prahlad

ABSTRACT
The coordination between the animal's external environment and internal state requires constant modulation by chemicals known as neuromodulators. Neuromodulators, such as biogenic amines, neuropeptides and cytokines, promote organismal homeostasis. Over the past several decades, *Caenorhabditis elegans* has grown into a powerful model organism that allows the elucidation of the mechanisms of action of neuromodulators that are conserved across species. In this perspective, we highlight a collection of articles in this issue that describe how neuromodulators optimize *C. elegans* survival.

An animal receives multiple environmental stimuli, some of which have the potential to disrupt metabolism and overall physiology. To survive environmental stressors, an animal must transition between a range of internal states and behaviors to identify new set points at which its physiological processes function optimally, thereby regaining homeostasis. One mechanism by which all organisms, including *Caenorhabditis elegans*, integrate changes in their external environments with their internal states is through the secretion of chemicals known as neuromodulators, which allow the animal to best exploit its niche and prioritize survival. This perspective introduces a series of articles in this collection that highlight the role of these chemicals in survival programs, aging and disease.

What are neuromodulators?

Neuromodulators were discovered as brain chemicals that transform a neuron's intrinsic excitability or synaptic dynamics (see Bargmann, 2012; Bargmann & Marder, 2013; Marder, 2012; Taghert & Nitabach, 2012, for excellent reviews on neuromodulator function). In contrast to classical neurotransmitters, diverse members of this class of chemicals, such as monoamines, neuropeptides, and cytokines, can be released extrasynaptically from neural sources (Bargmann, 2012; Bargmann & Marder, 2013; Bentley *et al.*, 2016; Marder, 2012; Taghert & Nitabach, 2012). They can also be released from non-neural sources (Marder, 2012; Taghert & Nitabach, 2012). Neuromodulators act locally in a paracrine manner or act hormonally at neural or non-neural targets far from their site of release (Bargmann, 2012; Bargmann & Marder, 2013; Hobert, 2013; Marder, 2012; Schafer, 2006; Taghert & Nitabach, 2012). They can modify the outputs of anatomically defined neural circuits or alter the composition of these circuits to generate entirely different outputs (Bargmann, 2012; Bargmann & Marder, 2013; Marder, 2012; Taghert & Nitabach, 2012). To add to their complexity, a neuromodulator may promote one response by enhancing one cell's activity and/or repressing the activity of another (Bargmann & Marder, 2013; Marder, 2012). Then, due to a change in local cell environments, that same neuromodulator may promote a second or opposite response by affecting the activities of other cells that now express the appropriate receptors (Bargmann & Marder, 2013; Di Giovangiulio *et al.*, 2015; Marder, 2012; Schafer, 2006). This multiplicity of effects by neuromodulators has made their study particularly challenging.

Neuromodulators in *C. elegans* survival

Thanks to the pioneering work of Brenner, Sulston, and others, the worm *C. elegans* has grown into a powerful experimental system to study the effects of neuromodulators on all aspects of animal physiology. *Caenorhabditis elegans* expresses all the major classes of neuromodulators, which include the biogenic amines (serotonin, dopamine, octopamine, and tyramine; Bentley *et al.*, 2016), neuropeptides (short peptides that are processed post-translationally from precursor proteins; reviewed by Hobert, 2013; Li & Kim, 2008), and cytokines [such as TGF-β and the interleukin IL-

17 (Bargmann, 2012; Chen *et al.*, 2017)]. The worm's extraordinary tractability to forward and reverse genetics allows the easy manipulation of neuromodulators and their receptors in specific cells and visualization of the subsequent changes in cellular properties, behavior, and physiology. The secretion of neuromodulators by neural or non-neural tissues into the worm pseudocoelomic cavity also facilitates the study of the systemic effects of these chemicals—how they mediate communication between neural and non-neural cells. The above advantages of *C. elegans* has yielded a wealth of information that allows us to understand the impact of neuromodulators on its biology. Indeed, the worm's food choices, its decision to forage, mate, or reproduce, its metabolism or responses to threats and competition, its developmental programs and longevity are but some processes influenced by neuromodulators and amenable to experimental manipulation (Aprison & Ruvinsky, 2019; Banerjee, Bhattacharya, Gorczyca, Collins, & Francis, 2017; Beets, Temmerman, Janssen, & Schoofs, 2013; Bhattacharya & Francis, 2015; Cermak *et al.*, 2020; Ezcurra, Walker, Beets, Swoboda, & Schafer, 2016; Ghosh *et al.*, 2016; Ishita, Chihara, & Okumura, 2020; Kagawa-Nagamura, Gengyo-Ando, Ohkura, & Nakai, 2018; Ringstad, 2017; Schafer, 2006; Wu *et al.*, 2019; Zang, Ho, & Ringstad, 2017).

A common thread throughout this issue

The profound influence of neuromodulators on behavior, metabolism, and overall physiology is a common thread throughout this issue (Cheon, Hwang, & Kim, 2020; Honer *et al.*, 2020; Kim & Flavell, 2020; Kim, Lee, Kim, & Lee, 2020; Liang, McKinnon, & Rankin, 2020; Liu & Zhang, 2020; Muirhead & Srinivasan, 2020; Prahlad, 2020; Srinivasan, 2020; Takeishi, Takagaki, & Kuhara, 2020; Yang, Lee, Yim, & Lee, 2020). In this perspective, we will focus on the roles of monoamines and neuropeptides in *C. elegans* survival.

Monoamine modulators

Caenorhabditis elegans synthesizes four monoamine neuromodulators—octopamine (OA), tyramine (TA), dopamine (DA), and serotonin (5-HT)—but lack histamine, epinephrine, and norepinephrine, which are found in vertebrates (Bentley *et al.*, 2016; Chase & Koelle, 2007). The major source, and, in some cases, the only source, of these monoamine modulators are neurons. *Caenorhabditis elegans* mutants that lack key biosynthetic enzymes for each of the bioamine neuromodulators are viable, allowing *C. elegans* to serve as a powerful discovery platform to understand neuromodulator function. These bioactive monoamine synthesis mutants exert pleiotropic effects on *C. elegans* internal states, thereby affecting behavior (see, *e.g.*, Cermak *et al.*, 2020; Ghosh *et al.*, 2016; Schafer, 2006).

Octopamine and tyramine. OA and TA are considered the functional equivalent of epinephrine and norepinephrine in invertebrates (Li *et al.*, 2017). OA and TA are best characterized in orchestrating the transition between the foraging state, which is elicited by lack of food, and the dwelling state, which denotes food availability. TA is present in low abundance and is synthesized by the enzyme tyrosine decarboxylase (TDC-1) in the RIM-1 motor neurons, gonadal sheath cells, and the uv1 neuroendocrine cells (Alkema, Hunter-Ensor, Ringstad, & Horvitz, 2005; Chase & Koelle, 2007). OA is synthesized from TA by the enzyme tyramine β-hydroxylase (TBH-1) in RIC interneurons and the gonadal sheath cells (Alkema *et al.*, 2005; Chase & Koelle, 2007; Horvitz, Chalfie, Trent, Sulston, & Evans, 1982). Food deprivation results in the release of OA by the RIC neurons (Churgin, McCloskey, Peters, & Fang-Yen, 2017; Roeder, 2020; Suo, Culotti, & Van Tol, 2009). The released OA acts via the G protein-coupled receptors (GPCRs) SER-3 and SER-6 in SIA neurons to promote roaming behaviors that increase the probability of finding food (Churgin *et al.*, 2017; Suo *et al.*, 2009). When food becomes available, TA promotes reduced locomotion to allow feeding (Churgin *et al.*, 2017).

Caenorhabditis elegans is a bacterivore, and the bacteria encountered by the animal range from highly nutritious to poorly nutritious and outright pathogenic (see Kim & Flavell, 2020; this issue). Interestingly, OA also suppresses aversive behaviors (Guo *et al.*, 2015; Mills *et al.*, 2012) to prioritize feeding. OA allows *C. elegans* to tolerate low-quality or detrimental bacterial food sources by modulating bacteria-elicited innate immune responses (Sellegounder, Yuan, Wibisono, Liu, & Sun, 2018; Suo *et al.*, 2009). Consequently, OA mediates a shift towards attraction to a greater range of foods, like altering the valence of the response to CO_2 levels that typically signify food (Rengarajan, Yankura, Guillermin, Fung, & Hallem, 2019). In this issue, Srinivasan (2020) discusses how RIC neuron-secreted OA coordinates food availability with lipolytic activity by signaling through intestinal SER-3 receptors to activate the intestinal lipases LIPS-6 and ATGL-1.

Remarkably, *C. elegans* is subject to signaling not only from its self-synthesized OA, but also from OA or OA-like compounds secreted by certain bacteria or other *C. elegans*, respectively. In this issue, Kim and Flavell (2020) highlight the recent findings from the Sengupta lab (O'Donnell, Fox, Chao, Schroeder, & Sengupta, 2020) on how OA produced by commensal bacteria alters *C. elegans* behavior and internal state. Cheon *et al.* (2020; this issue) and Muirhead and Srinivasan (2020; this issue) also review how starved larvae produce the OA-like small molecule osas#9, an ascaroside component of the worm-secreted pheromone blend, which is then sensed by nociceptive ASH neurons in adults to initiate their avoidance behavior (Chute *et al.*, 2019).

Dopamine. The DA neurons in *C. elegans* were initially identified by Sulston and coworkers (Sulston, Dew, & Brenner, 1975), using the catecholamine-specific technique of formaldehyde-induced fluorescence (FIF). DA is synthesized in eight neurons (ADEL/R, CEPDL/R, CEPVL/R, and PDEL/R) in hermaphrodites and in six additional neurons (R5AL/R, R7AL/R, R9AL/R) in males by the tyrosine hydroxylase CAT-2, which catalyzes the rate-limiting step in dopamine synthesis (Lints & Emmons, 1999; Sulston, Dew, & Brenner, 1975). As in other animals, *C. elegans* DA plays key roles in coordinating motor programs with the reward system during foraging, feeding, and egg laying (Ardiel *et al.*, 2016; Bettinger & McIntire, 2004; Chase & Koelle, 2007; Cermak *et al.*, 2020; Qin & Wheeler, 2007; Rivard *et al.*, 2010; Sanyal *et al.*, 2004; Sawin, Ranganathan, & Horvitz, 2000; Suo *et al.*, 2019). DA is released upon sensing food (Oranth *et al.*, 2018) to initiate the slowing of movement in the presence of food (Sawin *et al.*, 2000). Thus, DA counteracts OA-induced hyperactivity (Luedtke, O'Connor, Holden-Dye, & Walker, 2010; Rengarajan, Yankura, Guillermin, Fung, & Hallem, 2019). Similarly, DA works antagonistically to OA in switching the responses to CO_2: DA promotes aversion to CO_2 in the fed state and OA promotes attraction in the starved state (Rengarajan *et al.*, 2019). As in mammalian neurodegenerative models, *C. elegans* DA neurons appear more susceptible to degeneration upon expression of disease-associated aggregation-prone proteins, such as α-synuclein (Mor *et al.*, 2017). In this collection, the Rankin lab focuses on how *C. elegans* serves as a powerful model in which to study neurodegeneration (Liang, McKinnon, & Rankin, 2020).

Serotonin. The rate-limiting enzyme tryptophan hydroxylase, TPH-1, synthesizes 5-HT in eight to ten neurons in hermaphrodites (ADFL/R, NSML/R, HSNL/R, ASGL/R upon hypoxia, and rarely in AIM and RIH) and in more neurons in males (CP0 to CP06 and the B-type ray neurons R1BL/R, R3BL/R, and R9BL/R; Hare & Loer, 2004; Loer & Kenyon, 1993; Loer & Rivard, 2007; Pocock & Hobert, 2010; Serrano-Saiz *et al.*, 2017). Release of 5-HT from each of these neurons performs different functions, either because of its co-release with other neurotransmitters (Srinivasan, 2020; this issue) or because the acute versus chronic availability of 5-HT exerts different effects on target tissues (Prahlad, 2020; this issue). In the worm, 5-HT can mimic food and favorable conditions or signal stress, based upon the duration and site of release (Avery & You, 2012; Chase & Koelle, 2007; Cruz-Corchado, Ooi, Das, & Prahlad, 2020; Curran & Chalasani, 2012; Ishita, Chihara, & Okumura, 2020; Rankin, 2006). For instance, 5-HT can promote recovery from the developmental arrest known as dauer that forms in response to early life stress (Cassada & Russell, 1975; Mylenko *et al.*, 2016; see also Yang, Lee *et al.*, 2020), by mimicking food signals that promote growth and differentiation (Srinivasan, 2020; this issue). Alternatively, 5-HT can activate behavioral avoidance responses or stress-responsive transcription programs (Prahlad, 2020; this issue). Notably, these opposing effects resemble what is observed during the administration of 5-HT modulators for the treatment of neuropsychiatric disorders in humans: an acute increase in 5-HT availability causes increased anxiety; chronic treatment leads to antidepressant effects (Sharp & Cowen, 2011).

Neuropeptides

The *C. elegans* genome contains more than 120 genes that encode neuropeptide precursor proteins, and these proteins are processed to more than 250 neuropeptides. Most of their receptors belong to the large GPCR family but can also include ion channels and receptor kinases (for more extensive reviews on neuropeptides and their receptors, see Hobert, 2013; Li & Kim, 2008). *Caenorhabditis elegans* has the FMRFamide-like peptides (FLPs; Li, Kim, & Nelson, 1999), insulin-like peptides (ILPs; Pierce *et al.*, 2001), and the non-FLP, non-ILP neuropeptides called NLPs (Nathoo, Moeller, Westlund, & Hart, 2001). Like the biogenic amines, neuropeptides have also been extensively studied in *C. elegans* and are implicated in behaviors and physiological mechanisms that modulate homeostasis and survival.

FMRFamide-like peptides. A prominent example of a worm FLP-dependent pathway is neuropeptide Y signaling, which is represented by the FLP-21 peptide ligand and its associated GPCR, NPR-1 (Rogers et al., 2003). FLP-21 and NPR-1 are required for avoidance responses to noxious stimuli and loss of pathway activity compromises survival (Glauser *et al.*, 2011; Reddy, Andersen, Kruglyak, & Kim, 2009; Styer et al., 2008). In this issue, Kim and Flavell (2020) review how this pathway can alter *C. elegans* behavior in response to bacterial metabolites in the animal's natural environment. Other FLP genes also modulate longevity and metabolism. In this collection, Kim *et al.* (2020) describe the role of *flp-6* in increasing survival at high temperatures, but *flp-6* also intriguingly exhibits an opposite role in survival at lower temperatures (Chen *et al.*, 2016). Srinivasan (2020; this issue) discusses how FLP-17 coordinates environmental oxygen levels with intestinal fat metabolism. Yang *et al.* (2020; this issue) refer to findings by the Sternberg lab (Lee *et al.*, 2017), where peptides encoded by two *flp* genes, *flp-10* and *flp-17*, facilitate a dispersal behavior adopted by dauers in migrating to environments that support better survival.

Insulin-like peptides. ILP signaling has long been associated with survival (see Kenyon, 2010; and references therein). The worm ILP receptor DAF-2, which is a receptor tyrosine kinase (Kimura, Tissenbaum, Liu, & Ruvkun, 1997), promotes reproductive growth and inhibits dauer arrest (Riddle, Swanson, & Albert, 1981). The downregulation of DAF-2 activity doubles *C. elegans* lifespan (Kenyon, Chang, Gensch, Rudner, & Tabtiang, 1993), a discovery that ushered the birth of a field—the genetics of aging. Like DAF-2 (Gems *et al.*, 1998), at least some of the worm ILPs (Hobert, 2013; Li & Kim, 2008) have pleiotropic functions (Fernandes de Abreu *et al.*, 2014), which might be a consequence of their ILP-to-ILP network organization, where one ILP regulates multiple ILPs (Fernandes de Abreu *et al.*, 2014). Many of the ILP functions typify neuromodulator functions. For example, there are ILPs that sometimes behave like the DAF-2 receptor in one context and opposite from DAF-2 in another context (Fernandes de Abreu *et al.*, 2014). The articles in this collection discuss the roles of ILPs in temperature-sensing (see Takeishi, Takagaki, & Kuhara, 2020), in context-dependent avoidance behaviors (see Cheon, Hwang, & Kim, 2020; Kim & Flavell, 2020), in neuroprotection (see Liang, McKinnon, & Rankin, 2020), the dauer program (see Yang *et al.*, 2020), and longevity (see Kim *et al.*, 2020).

Non-FLP, non-ILP neuropeptides. NLPs comprise a heterogeneous group of neuropeptides, but are again involved in diverse physiological processes (Li & Kim, 2008; Hobert, 2013), from sleep behaviors (see Honer, Buscemi *et al.*, 2020; this issue) to neurodegeneration (Lezi *et al.*, 2018) and longevity (Park, Link, & Johnson, 2010). Similar to FLPs and ILPs, NLPs can amplify or dampen signaling at specific synapses (Chalasani *et al.*, 2010; Hapiak *et al.*, 2013; Macosko *et al.*, 2009), thereby shaping circuit connectivities and behaviors. The three classes of neuropeptides, the FLPs, ILPs, and NLPs, are also known to work together through feedforward or feedback mechanisms to maintain homeostasis at both the circuit level and the organismal level (Chalasani *et al.*, 2010; Chen, Chen *et al.*, 2016).

Coda

The dysregulation of neuromodulator activities can lead to disease. Indeed, numerous studies in mammalian systems implicate neuromodulator dysfunction in neurodegenerative diseases, such as Alzheimer's disease, Huntington's disease and Parkinson's disease, where impaired neuromodulator signaling often preempt disease symptoms (Du, Pang, & Hannan, 2013; Elsworthy & Aldred, 2019; Ohno, Shimizu, Tokudome, Kunisawa, & Sasa, 2015; Politis & Niccolini, 2015). *Caenorhabditis elegans* expresses many orthologs of neurodegenerative disease-associated genes and their study in the worm have contributed to our understanding of the above human diseases (see Liang, McKinnon, & Rankin, 2020; this issue). Understanding the role of neuromodulators in worm neurodegeneration will likely add to our understanding of human neurodegenerative disorders.

To conclude, we would like to highlight an important question. How does a neuromodulator modify a physiological response to a stimulus? This question circles back to experiments performed in the 1960s. Injection of an abdominal ganglion extract from one *Aplysia* into another *Aplysia* elicited the cessation of locomotor and feeding behavior, followed by the stereotyped head-waving behavior that facilitated egg laying in the second animal (Kupfermann, 1967; Strumwasser, Jacklet, & Alvarez, 1969; Toevs & Brackenbury, 1969). These experiments demonstrated that diverse modulatory substances could act centrally and peripherally to change the physiological state of an animal completely. It would be interesting to learn the rules and constraints by which different cocktails of neuromodulators achieve such a dramatic switch in physiological responses to environmental stimuli. Ultimately, the complete identification of the interacting modulators, their receptors and sites of action should allow us to address this question. We posit that *C. elegans* is an ideal system to achieve this goal.

Acknowledgments

The authors apologize to the authors whose work we were unable to cite due to space constraints and the daunting scale of the *C. elegans* literature in neuromodulation. The authors would also like to thank the two reviewers for their valuable critiques of this perspective.

Disclosure statement

No potential conflict of interest was reported by the author(s).

Funding

This work was supported by NIH [R01 GM108962] to J. A. and [R01 AG060616 and R01 AG050653] to V. P.

References

Alkema, M.J., Hunter-Ensor, M., Ringstad, N., & Horvitz, H.R. (2005). Tyramine functions independently of octopamine in the *Caenorhabditis elegans* nervous system. *Neuron*, *46*(2), 247–260. doi: 10.1016/j.neuron.2005.02.024

Aprison, E.Z., & Ruvinsky, I. (2019). Dynamic regulation of adult-specific functions of the nervous system by signaling from the reproductive system. *Current Biology : CB*, *29*(23), 4116–4123.e4113. doi: 10.1016/j.cub.2019.10.011

Ardiel, E.L., Giles, A.C., Yu, A.J., Lindsay, T.H., Lockery, S.R., & Rankin, C.H. (2016). Dopamine receptor DOP-4 modulates habituation to repetitive photoactivation of a *C. elegans* polymodal nociceptor. *Learning & Memory (Cold Spring Harbor, N.Y.)*, *23*(10), 495–503. doi:10.1101/lm.041830.116

Avery, L. and You, Y.J. C. elegans feeding, WormBook, ed. The C. elegans Research Community, WormBook, doi/10.1895/wormbook.1.150.1

Banerjee, N., Bhattacharya, R., Gorczyca, M., Collins, K.M., & Francis, M.M. (2017). Local neuropeptide signaling modulates serotonergic transmission to shape the temporal organization of C. elegans egg-laying behavior. *PLoS Genetics*, *13*(4), e1006697. doi:10.1371/journal.pgen.1006697

Bargmann, C.I. (2012). Beyond the connectome: How neuromodulators shape neural circuits. *BioEssays: News and Reviews in Molecular, Cellular and Developmental Biology*, *34*(6), 458–465. doi:10.1002/bies.201100185

Bargmann, C.I., & Marder, E. (2013). From the connectome to brain function. *Nature Methods*, *10*(6), 483–490. doi:10.1038/nmeth.2451

Beets, I., Temmerman, L., Janssen, T., & Schoofs, L. (2013). Ancient neuromodulation by vasopressin/oxytocin-related peptides. *Worm*, *2*(2), e24246. doi:10.4161/worm.24246

Bentley, B., Branicky, R., Barnes, C.L., Chew, Y.L., Yemini, E., Bullmore, E.T., … Schafer, W.R. (2016). The multilayer connectome of *Caenorhabditis elegans*. *PLoS Computational Biology*, *12*(12), e1005283. doi:10.1371/journal.pcbi.1005283

Bettinger, J.C., & McIntire, S.L. (2004). State-dependency in *C. elegans*. *Genes Brain, and Behavior*, *3*(5), 266–272. doi:10.1111/j.1601-183X.2004.00080.x

Bhattacharya, R., & Francis, M.M. (2015). In the proper context: Neuropeptide regulation of behavioral transitions during food searching. *Worm*, *4*(3), e1062971. doi:10.1080/21624054.2015.1062971

Cassada, R.C., & Russell, R.L. (1975). The dauerlarva, a post-embryonic developmental variant of the nematode *Caenorhabditis elegans*. *Developmental Biology*, 46(2), 326–342. doi:10.1016/0012-1606(75)90109-8

Cermak, N., Yu, S.K., Clark, R., Huang, Y.C., Baskoylu, S.N., & Flavell, S.W. (2020). Whole-organism behavioral profiling reveals a role for dopamine in state-dependent motor program coupling in *C. elegans*. *eLife*, 9, e57093. doi:10.7554/eLife.57093

Chalasani, S.H., Kato, S., Albrecht, D.R., Nakagawa, T., Abbott, L.F., & Bargmann, C.I. (2010). Neuropeptide feedback modifies odor-evoked dynamics in *Caenorhabditis elegans* olfactory neurons. *Nature Neuroscience, 13*(5), 615–621. doi:10.1038/nn.2526

Chase, D.L., & Koelle, M.R. (2007). Biogenic amine neurotransmitters in *C. elegans. WormBook*, 1–15. doi:10.1895/wormbook.1.132.1

Chen, C., Itakura, E., Nelson, G.M., Sheng, M., Laurent, P., Fenk, L.A., … de Bono, M. (2017). IL-17 is a neuromodulator of *Caenorhabditis elegans* sensory responses. *Nature, 542*(7639), 43–48. doi:10.1038/nature20818

Chen, Y.C., Chen, H.J., Tseng, W.C., Hsu, J.M., Huang, T.T., Chen, C.H., & Pan, C.L. (2016). A *C. elegans* thermosensory circuit regulates longevity through crh-1/CREB-dependent flp-6 neuropeptide signaling . *Developmental Cell, 39*(2), 209–223. doi:10.1016/j.devcel.2016.08.021

Cheon, Y.-J., Hwang, H., & Kim, K. (2020). Plasticity of pheromone-mediated avoidance behavior in *C. elegans. Journal of Neurogenetics, 34.*doi: 10.1080/01677063.2020.1802723

Churgin, M.A., McCloskey, R.J., Peters, E., & Fang-Yen, C. (2017). Antagonistic serotonergic and octopaminergic neural circuits mediate food-dependent locomotory behavior in *Caenorhabditis elegans. Journal of Neuroscience: The Official Journal of the Society for Neuroscience, 37*(33), 7811–7823. doi:10.1523/JNEUROSCI.2636-16.2017

Chute, C.D., DiLoreto, E.M., Zhang, Y.K., Reilly, D.K., Rayes, D., Coyle, V.L., … Srinivasan, J. (2019). Co-option of neurotransmitter signaling for inter-organismal communication in *C. elegans. Nature Communications, 10*(1), 3186. doi:10.1038/s41467-019-11240-7

Cruz-Corchado, J., Ooi, F.K., Das, S., & Prahlad, V. (2020). Global transcriptome changes that accompany alterations in serotonin levels in *Caenorhabditis elegans. G3 (Bethesda, Md.), 10*(4), 1225–1246. doi:10.1534/g3.120.401088

Curran, K.P., & Chalasani, S.H. (2012). Serotonin circuits and anxiety: What can invertebrates teach us? *Invertebrate Neuroscience, 12*(2), 81–92. doi:10.1007/s10158-012-0140-y

Di Giovangiulio, M., Verheijden, S., Bosmans, G., Stakenborg, N., Boeckxstaens, G.E., & Matteoli, G. (2015). The neuromodulation of the intestinal immune system and its relevance in inflammatory bowel disease. *Frontiers in Immunology, 6*, 590. doi:10.3389/fimmu.2015.00590

Du, X., Pang, T., & Hannan, A. (2013). A tale of two maladies? Pathogenesis of depression with and without the Huntington's disease gene mutation. *Frontiers in Neurology, 4*, 81. doi:10.3389/fneur.2013.00081

Elsworthy, R.J., & Aldred, S. (2019). Depression in Alzheimer's disease: An alternative role for selective serotonin reuptake inhibitors? *Journal of Alzheimer's Disease, 69*(3), 651–661. doi:10.3233/JAD-180780

Ezcurra, M., Walker, D.S., Beets, I., Swoboda, P., & Schafer, W.R. (2016). Neuropeptidergic signaling and active feeding state inhibit nociception in *Caenorhabditis elegans. Journal of Neuroscience : The Official Journal of the Society for Neuroscience, 36*(11), 3157–3169. doi:10.1523/JNEUROSCI.1128-15.2016

Fernandes de Abreu, D.A., Caballero, A., Fardel, P., Stroustrup, N., Chen, Z., Lee, K., … Ch'ng, Q. (2014). An insulin-to-insulin regulatory network orchestrates phenotypic specificity in development and physiology. *PLoS Genetics, 10*(3), e1004225. doi:10.1371/journal.pgen.1004225

Gems, D., Sutton, A.J., Sundermeyer, M.L., Albert, P.S., King, K.V., Edgley, M.L., … Riddle, D.L. (1998). Two pleiotropic classes of daf-2 mutation affect larval arrest, adult behavior, reproduction and longevity in *Caenorhabditis elegans. Genetics, 150*(1), 129–155.

Ghosh, D.D., Sanders, T., Hong, S., McCurdy, L.Y., Chase, D.L., Cohen, N., … Nitabach, M.N. (2016). Neural architecture of hunger-dependent multisensory decision making in *C. elegans. Neuron, 92*(5), 1049–1062. doi:10.1016/j.neuron.2016.10.030

Glauser, D. A., Chen, W. C., Agin, R., Macinnis, B. L., Hellman, A. B., Garrity, P. A., Goodman, M. B. (2011). Heat avoidance is regulated by transient receptor potential (TRP) channels and a neuropeptide signaling pathway in Caenorhabditis elegans. Genetics, 188(1), 91–103. doi:10.1534/genetics.111.127100

Guo, M., Wu, T.-H., Song, Y.-X., Ge, M.-H., Su, C.-M., Niu, W.-P., … Wu, Z.-X. (2015). Reciprocal inhibition between sensory ASH and ASI neurons modulates nociception and avoidance in *Caenorhabditis elegans. Nature Communications, 6*, 5655. doi:10.1038/ncomms6655

Hapiak, V., Summers, P., Ortega, A., Law, W.J., Stein, A., & Komuniecki, R. (2013). Neuropeptides amplify and focus the monoaminergic inhibition of nociception in *Caenorhabditis elegans. Journal of Neuroscience, 33*(35), 14107–14116. doi:10.1523/JNEUROSCI.1324-13.2013

Hare, E.E., & Loer, C.M. (2004). Function and evolution of the serotonin-synthetic bas-1 gene and other aromatic amino acid decarboxylase genes in *Caenorhabditis. BMC Evolutionary Biology, 4*, 24. doi:10.1186/1471-2148-4-24

Hobert, O. (2013). The neuronal genome of *Caenorhabditis elegans. WormBook*, 1–106. doi:10.1895/wormbook.1.161.1

Honer, M., Buscemi, K., Barrett, N., Riazati, N., Orlando, G., & Nelson, M.D. (2020). Orcokinin neuropeptides regulate sleep in *Caenorhabditis elegans. Journal of Neurogenetics, 34.*doi: 10.1080/01677063.2020.1830084

Horvitz, H.R., Chalfie, M., Trent, C., Sulston, J.E., & Evans, P.D. (1982). Serotonin and octopamine in the nematode *Caenorhabditis elegans. Science (New York, N.Y.), 216*(4549), 1012–1014. doi:10.1126/science.6805073

Ishita, Y., Chihara, T., & Okumura, M. (2020). Serotonergic modulation of feeding behavior in *Caenorhabditis elegans* and other related nematodes. *Neuroscience Research, 154*, 9–19. doi:10.1016/j.neures.2019.04.006

Kagawa-Nagamura, Y., Gengyo-Ando, K., Ohkura, M., & Nakai, J. (2018). Role of tyramine in calcium dynamics of GABAergic neurons and escape behavior in *Caenorhabditis elegans. Zoological Letters, 4*, 19. doi:10.1186/s40851-018-0103-1

Kenyon, C., Chang, J., Gensch, E., Rudner, A., & Tabtiang, R. (1993). A *C. elegans* mutant that lives twice as long as wild type. *Nature, 366*(6454), 461–464. doi:10.1038/366461a0

Kenyon, C.J. (2010). The genetics of ageing. *Nature, 464*(7288), 504–512. doi:10.1038/nature08980

Kim, B., Lee, J., Kim, Y., & Lee, S.-J.V. (2020). Regulatory systems that mediate the effects of temperature on the lifespan of *Caenorhabdities elegans. Journal of Neurogenetics, 34.* doi: 10.1080/01677063.2020.1781849

Kim, D.H., & Flavell, S.W. (2020). Host-microbe interactions and the behavior of *Caenorhabditis elegans. Journal of Neurogenetics, 34.*doi: 10.1080/01677063.2020.1802724

Kimura, K.D., Tissenbaum, H.A., Liu, Y., & Ruvkun, G. (1997). *daf-2*, an insulin receptor-like gene that regulates longevity and diapause in *Caenorhabditis elegans. Science (New York, N.Y.), 277*(5328), 942–946. doi:10.1126/science.277.5328.942

Kupfermann, I. (1967). Stimulation of egg laying: Possible neuroendocrine function of bag cells of abdominal ganglion of *Aplysia californica. Nature, 216*(5117), 814–815. doi:10.1038/216814a0

Lee, J.S., Shih, P.Y., Schaedel, O.N., Quintero-Cadena, P., Rogers, A.K., & Sternberg, P.W. (2017). FMRFamide-like peptides expand the behavioral repertoire of a densely connected nervous system. *Proceedings of the National Academy of Sciences of the United States of America, 114*(50), E10726–E10735. doi:10.1073/pnas.1710374114

Lezi, E., Zhou, T., Koh, S., Chuang, M., Sharma, R., Pujol, N., … Yan, D. (2018). An antimicrobial peptide and its neuronal receptor regulate dendrite degeneration in aging and infection. *Neuron, 97*(1), 125–138.e125. doi:10.1016/j.neuron.2017.12.001

Li, C., & Kim, K. (2008). Neuropeptides. *WormBook*, 1–36. doi:10.1895/wormbook.1.142.1

Li, C., Kim, K., & Nelson, L.S. (1999). FMRFamide-related neuropeptide gene family in *Caenorhabditis elegans. Brain Research, 848*(1–2), 26–34. doi:10.1016/S0006-8993(99)01972-1

Li, Y., Tiedemann, L., von Frieling, J., Nolte, S., El-Kholy, S., Stephano, F., … Roeder, T. (2017). The role of monoaminergic neurotransmission for metabolic control in the fruit fly *Drosophila melanogaster*. *Frontiers in Systems Neuroscience*, *11*, 60. doi:10.3389/fnsys.2017.00060

Liang, J.J.H., McKinnon, I.A., & Rankin, C.H. (2020). The contribution of *C. elegans* neurogenetics to understanding neurodegenerative diseases. *Journal of Neurogenetics*, *34*.doi: 10.1080/01677063.2020.1803302

Lints, R., & Emmons, S.W. (1999). Patterning of dopaminergic neurotransmitter identity among *Caenorhabditis elegans* ray sensory neurons by a TGFb family signaling pathway and a Hox gene. *Development*, *126*(24), 5819–5831.

Liu, H., & Zhang, Y. (2020). What can a worm learn in a bacteria-rich habitat? *Journal of Neurogenetics*, *34*. doi: 10.1080/01677063.2020.1829614

Loer, C.M., & Kenyon, C.J. (1993). Serotonin-deficient mutants and male mating behavior in the nematode *Caenorhabditis elegans*. *The Journal of Neuroscience : The Official Journal of the Society for Neuroscience*, *13*(12), 5407–5417. doi:10.1523/JNEUROSCI.13-12-05407.1993

Loer, C.M., & Rivard, L. (2007). Evolution of neuronal patterning in free-living rhabditid nematodes I: Sex-specific serotonin-containing neurons. *Journal of Comparative Neurology*, *502*(5), 736–767. doi:10.1002/cne.21288

Luedtke, S., O'Connor, V., Holden-Dye, L., & Walker, R.J. (2010). The regulation of feeding and metabolism in response to food deprivation in *Caenorhabditis elegans*. *Invertebrate Neuroscience*, *10*(2), 63–76. doi:10.1007/s10158-010-0112-z

Macosko, E.Z., Pokala, N., Feinberg, E.H., Chalasani, S.H., Butcher, R.A., Clardy, J., & Bargmann, C.I. (2009). A hub-and-spoke circuit drives pheromone attraction and social behaviour in *C. elegans*. *Nature*, *458*(7242), 1171–1175. doi:10.1038/nature07886

Marder, E. (2012). Neuromodulation of neuronal circuits: Back to the future. *Neuron*, *76*(1), 1–11. doi:10.1016/j.neuron.2012.09.010

Mills, H., Wragg, R., Hapiak, V., Castelletto, M., Zahratka, J., Harris, G., … Komuniecki, R. (2012). Monoamines and neuropeptides interact to inhibit aversive behaviour in *Caenorhabditis elegans*. *EMBO Journal*, *31*(3), 667–678. doi:10.1038/emboj.2011.422

Mor, D.E., Tsika, E., Mazzulli, J.R., Gould, N.S., Kim, H., Daniels, M.J., … Ischiropoulos, H. (2017). Dopamine induces soluble α-synuclein oligomers and nigrostriatal degeneration. *Nature Neuroscience*, *20*(11), 1560–1568. doi:10.1038/nn.4641

Muirhead, C.S., & Srinivasan, J. (2020). Small molecule signals mediate social behaviors in *C. elegans*. *Journal of Neurogenetics*, *34*. doi: 10.1080/01677063.2020.1808634

Mylenko, M., Boland, S., Penkov, S., Sampaio, J.L., Lombardot, B., Vorkel, D., … Kurzchalia, T.V. (2016). NAD + Is a food component that promotes exit from dauer diapause in *Caenorhabditis elegans*. *PLoS One*, *11*(12), e0167208. doi:10.1371/journal.pone.0167208

Nathoo, A.N., Moeller, R.A., Westlund, B.A., & Hart, A.C. (2001). Identification of neuropeptide-like protein gene families in *Caenorhabditis elegans* and other species. *Proceedings of the National Academy of Sciences of the United States of America*, *98*(24), 14000–14005. doi:10.1073/pnas.241231298

O'Donnell, M.P., Fox, B.W., Chao, P.H., Schroeder, F.C., & Sengupta, P. (2020). A neurotransmitter produced by gut bacteria modulates host sensory behaviour. *Nature*, *583*(7816), 415–420. doi:10.1038/s41586-020-2395-5

Ohno, Y., Shimizu, S., Tokudome, K., Kunisawa, N., & Sasa, M. (2015). New insight into the therapeutic role of the serotonergic system in Parkinson's disease. *Progress in Neurobiology*, *134*, 104–121. doi:10.1016/j.pneurobio.2015.09.005

Oranth, A., Schultheis, C., Tolstenkov, O., Erbguth, K., Nagpal, J., Hain, D., … Gottschalk, A. (2018). Food sensation modulates locomotion by dopamine and neuropeptide signaling in a distributed neuronal network. *Neuron*, *100*(6), 1414–1428. e1410. doi:10.1016/j.neuron.2018.10.024

Park, S.K., Link, C.D., & Johnson, T.E. (2010). Life-span extension by dietary restriction is mediated by NLP-7 signaling and coelomocyte endocytosis in *C. elegans*. *FASEB Journal: Official Publication of the Federation of American Societies for Experimental Biology*, *24*(2), 383–392. doi:10.1096/fj.09-142984

Pierce, S.B., Costa, M., Wisotzkey, R., Devadhar, S., Homburger, S.A., Buchman, A.R., … Ruvkun, G. (2001). Regulation of DAF-2 receptor signaling by human insulin and *ins-1*, a member of the unusually large and diverse *C. elegans* insulin gene family. *Genes & Development*, *15*(6), 672–686. doi:10.1101/gad.867301

Pocock, R., & Hobert, O. (2010). Hypoxia activates a latent circuit for processing gustatory information in C. elegans. *Nature Neuroscience* *13*(5), 610–614. doi: 10.1038/nn.2537

Politis, M., & Niccolini, F. (2015). Serotonin in Parkinson's disease. *Behavioural Brain Research*, *277*, 136–145. doi:10.1016/j.bbr.2014.07.037

Prahlad, V. (2020). The discovery and consequences of the central role of the nervous system in the control of protein homeostasis. *Journal of Neurogenetics*, *34*.doi: 10.1080/01677063.2020.1771333

Qin, J., & Wheeler, A.R. (2007). Maze exploration and learning in *C. elegans*. *Lab on a Chip*, *7*(2), 186–192. doi:10.1039/b613414a

Rankin, C.H. (2006). Nematode behavior: The taste of success, the smell of danger!. *Current Biology*, *16*(3), R89–R91. doi:10.1016/j.cub.2006.01.025

Reddy, K. C., Andersen, E. C., Kruglyak, L., & Kim, D. H. (2009). A polymorphism in npr-1 is a behavioral determinant of pathogen susceptibility in *C. elegans*. Science, 323(5912), 382-384. doi:10.1126/science.1166527

Rengarajan, S., Yankura, K.A., Guillermin, M.L., Fung, W., & Hallem, E.A. (2019). Feeding state sculpts a circuit for sensory valence in *Caenorhabditis elegans*. *Proceedings of the National Academy of Sciences of the United States of America*, *116*(5), 1776–1781. doi:10.1073/pnas.1807454116

Riddle, D.L., Swanson, M.M., & Albert, P.S. (1981). Interacting genes in nematode dauer larva formation. *Nature*, *290*(5808), 668–671. doi: 10.1038/290668a0

Ringstad, N. (2017). Neuromodulation: The fevered mind of the worm. *Current Biology*, *27*(8), R315–R317. doi:10.1016/j.cub.2017.03.005

Rivard, L., Srinivasan, J., Stone, A., Ochoa, S., Sternberg, P.W., & Loer, C.M. (2010). A comparison of experience-dependent locomotory behaviors and biogenic amine neurons in nematode relatives of *Caenorhabditis elegans*. *BMC Neuroscience*, *11*, 22. doi:10.1186/1471-2202-11-22

Roeder, T. (2020). The control of metabolic traits by octopamine and tyramine in invertebrates. *The Journal of Experimental Biology*, *223*(7), jeb194282. doi:10.1242/jeb.194282

Rogers, C., Reale, V., Kim, K., Chatwin, H., Li, C., Evans, P., & de Bono, M. (2003). Inhibition of Caenorhabditis elegans social feeding by FMRFamide-related peptide activation of NPR-1. Nat Neurosci, 6(11), 1178–1185. doi:10.1038/nn1140

Sanyal, S., Wintle, R.F., Kindt, K.S., Nuttley, W.M., Arvan, R., Fitzmaurice, P., … Van Tol, H.H.M. (2004). Dopamine modulates the plasticity of mechanosensory responses in *Caenorhabditis elegans*. *The EMBO Journal*, *23*(2), 473–482. doi:10.1038/sj.emboj.7600057

Sawin, E.R., Ranganathan, R., & Horvitz, H.R. (2000). *C. elegans* locomotory rate is modulated by the environment through a dopaminergic pathway and by experience through a serotonergic pathway. *Neuron*, *26*(3), 619–631. doi:10.1016/S0896-6273(00)81199-X

Schafer, W.F. (2006). Genetics of egg-laying in worms. *Annual Review of Genetics*, *40*, 487–509. doi:10.1146/annurev.genet.40.110405.090527

Sellegounder, D., Yuan, C.H., Wibisono, P., Liu, Y., & Sun, J. (2018). Octopaminergic signaling mediates neural regulation of innate immunity in *Caenorhabditis elegans*. *mBio*, *9*(5), e01645. doi:10.1128/mBio.01645-18

Serrano-Saiz, E., Pereira, L., Gendrel, M., Aghayeva, U., Bhattacharya, A., Howell, K., … Hobert, O. (2017). A neurotransmitter atlas of the *Caenorhabditis elegans* male nervous system reveals sexually dimorphic neurotransmitter usage. *Genetics*, *206*(3), 1251–1269. doi:10.1534/genetics.117.202127

Sharp, T., & Cowen, P.J. (2011). 5-HT and depression: Is the glass half-full? *Current Opinion in Pharmacology*, *11*(1), 45–51. doi:10.1016/j.coph.2011.02.003

Srinivasan, S. (2020). Neuroendocrine control of lipid metabolism: Lessons from *C. elegans*. *Journal of Neurogenetics*, *34*.doi: 10.1080/01677063.2020.1777116

Strumwasser, F., Jacklet, J.W., & Alvarez, R.B. (1969). A seasonal rhythm in the neural extract induction of behavioral egg-laying in Aplysia. *Comparative Biochemistry and Physiology*, *29*(1), 197–206. doi:10.1016/0010-406X(69)91735-6

Styer, K. L., Singh, V., Macosko, E., Steele, S. E., Bargmann, C. I., & Aballay, A. (2008). Innate immunity in Caenorhabditis elegans is regulated by neurons expressing NPR-1/GPCR. Science, 322(5900), 460–464. doi:10.1126/science.1163673.

Sulston, J., Dew, M., & Brenner, S. (1975). Dopaminergic neurons in the nematode *Caenorhabditis elegans*. *Journal of Comparative Neurology*, *163*(2), 215–226. doi:10.1002/cne.901630207

Suo, S., Culotti, J.G., & Van Tol, H.H. (2009). Dopamine counteracts octopamine signalling in a neural circuit mediating food response in *C. elegans*. *The EMBO Journal*, *28*(16), 2437–2448. doi:10.1038/emboj.2009.194

Suo, S., Harada, K., Matsuda, S., Kyo, K., Wang, M., Maruyama, K., … Tsuboi, T. (2019). Sexually dimorphic regulation of behavioral states by dopamine in *Caenorhabditis elegans*. *Journal of Neuroscience : The Official Journal of the Society for Neuroscience*, *39*(24), 4668–4683. doi:10.1523/JNEUROSCI.2985-18.2019

Taghert, P.H., & Nitabach, M.N. (2012). Peptide neuromodulation in invertebrate model systems. *Neuron*, *76*(1), 82–97. doi:10.1016/j.neuron.2012.08.035

Takeishi, A., Takagaki, N., & Kuhara, A. (2020). Temperature signaling underlying thermotaxis and cold tolerance in *Caenorhabditis elegans*. *Journal of Neurogenetics*, *34*.doi: 10.1080/01677063.2020.1734001

Toevs, L.A., & Brackenbury, R.W. (1969). Bag cell-specific proteins and the humoral control of egg laying in *Aplysia californica*. *Comparative Biochemistry and Physiology*, *29*(1), 207–216. doi:10.1016/0010-406X(69)91736-8

Wu, T., Duan, F., Yang, W., Liu, H., Caballero, A., Fernandes de Abreu, D.A., … Zhang, Y. (2019). Pheromones modulate learning by regulating the balanced signals of two insulin-like peptides. *Neuron*, *104*(6), 1095–1109. e1095. doi:10.1016/j.neuron.2019.09.006

Yang, H., Lee, B.Y., Yim, H., & Lee, J. (2020). Neurogenetics of nictation, a dispersal strategy in nematodes. *Journal of Neurogenetics*, *34*.doi: 10.1080/01677063.2020.1788552

Zang, K.E., Ho, E., & Ringstad, N. (2017). Inhibitory peptidergic modulation of *C. elegans* serotonin neurons is gated by T-type calcium channels. *eLife*, *6*, e22771. doi:10.7554/eLife.22771

Neuroendocrine control of lipid metabolism: lessons from *C. elegans*

Supriya Srinivasan

ABSTRACT
This review article highlights our efforts to decode the role of the nervous system in regulating intestinal lipid metabolism in *Caenorhabditis elegans*. Capitalizing on the prescient and pioneering work of Sydney Brenner and John Sulston in establishing *C. elegans* as an immensely valuable model system, we have uncovered critical roles for oxygen sensing, population density sensing and food sensing in orchestrating the balance between storing lipids and utilizing them for energy in the intestine, the major organ for lipid metabolism in this model system. Our long-term goal is to reveal the integrative mechanisms and regulatory logic that underlies the complex relationship between genes, environment and internal state in the regulation of energy and whole-body physiology.

Introduction

A striking feature of humans around the world is the incredible diversity seen in aspects of physiology such as stature and body weight. When it comes to metabolism, there is a perception that we are, in the 21st century, in an environment that predisposes us towards obesity and the accumulation of body fat because of increased dietary intake and reduced energy expenditure. However, obesity and leanness have always co-existed (Davenport, 1923). This fact is perhaps most easily visualized in ancient and modern art. From the sculpture of the Venus of Willendorf dating back to about 20,000 B.C., to art in the second millennium between the 1300s and 1800s, human obesity has been depicted, in co-existence with leanness (Woodhouse, 2008). In modern parlance, that is, in the post McClintock and Mendelian eras, body weight is a heritable trait. Indeed, genome-wide association studies and twin studies place the heritability of body weight between 0.4–0.7 (Walley, Blakemore, & Froguel, 2006) commensurate with that of height, suggesting that there is a strong genetic basis that underlies the regulation of body fat storage.

General observations of human physiology began to be codified in the scientific literature in the late 1800s, in the works of Claude Bernard, Walter Cannon, Hetherington and others, who were influenced by engineering control theory, and promoted the idea of homeostasis and the body's defense against a so-called set point (Bernard, 1927; Cannon, 1939). This idea was originally used to explain the narrow range within which parameters such as body temperature and blood glucose are maintained, but later also included body fat control and the concept of energy balance, which posits that animals placed on a high calorie diet would resist a shift in body fat stores, and return to a homeostatic set point (Figure 1(A)). However experimental observations in the 20th century did not agree with the set point theory, because animals with lesions in their hypothalami, ovariectomized animals, and those bearing spontaneous mutations in what would later become known as the leptin signaling pathway, rapidly gained weight with no return to a homeostatic set point (Ravussin, Leibel, & Ferrante, 2014; Wade & Gray, 1979; Wirtshafter & Davis, 1977). These experiments and other mathematical modeling approaches in the field soon gave rise to a 'settling point' model (Figure 1(B)), which could sufficiently account for increases in body fat stores when placed on a high-calorie diet (Wirtshafter & Davis, 1977). However this model suggested that upon returning to a regular diet, animals would reduce body weight indefinitely, which was also not been borne out by experimental observations. The 'hybrid model' (Figure 1(C)), incorporates our current understanding of the roles of the major anabolic hormones leptin and insulin, which are sufficient to explain dramatic weight gains on a high calorie diet (Tam, Fukumura, & Jain, 2009). However, a strong body of evidence in both humans and rodents shows that once the weight is gained, the efficacy of weight loss is profoundly variable. Factors that control this variability are postulated to exist, but are poorly understood (Ravussin *et al.*, 2014). Our hypothesis is that both environmental influences and genetic factors will influence the many hormones that combinatorially regulate the physiology underlying body weight control (Figure 1(D)).

Understanding the fundamental biology of fat metabolism is of critical importance to human health given that a host of illnesses including cardiovascular disease, diabetes, metastatic cancers and perhaps even neurodegeneration have metabolic derangements as a root cause. Additionally, the

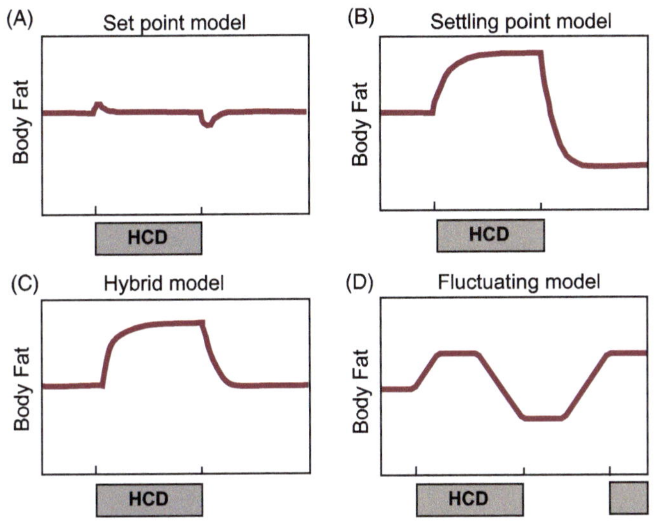

Figure 1. Theories of body weight control. (A) The 'set point' model posits that an internal homeostatic sensor can detect deviations from a predetermined set point, and restore body weight back to homeostasis. HCD: high-calorie diet. (B) The settling point model suggests that body weight will largely be a reflection of the environment, such that a HCD will increase body fat, and withdrawal will decrease body fat to below its original state. (C) The hybrid model suggests that anabolic hormones (such as leptin, insulin) will drive body fat gain under conditions of a HCD, but that restoration to a normal diet will bring body fat back to a preset baseline value. How catabolism is affected in these conditions remains unknown. (D) We propose a fluctuating model, in which body fat gain and loss occur as a result of external influences (both sensory and metabolic) and internal physiological state. There is no predetermined set point *per se*.

prevalence of metabolic diseases has exceeded 30% of the US population and projections suggest a further increase (Ward *et al.*, 2019). Given the complexity of the problem, it is clear that new approaches, orthogonal to the conventional view of energy balance as simple relationship between food intake and energy expenditure, are critical.

Developing *Caenorhabditis elegans* as a model system to study fat metabolism

Capitalizing on the pioneering work of Sydney Brenner and John Sulston in establishing *C. elegans* as a model system (Coulson, Sulston, Brenner, & Karn, 1986; Sulston & Brenner, 1974), sequencing of the genome (C. elegans Sequencing Consortium, 1998; Waterston & Sulston, 1995) and the use of sophisticated molecular genetics including RNAi (Fire *et al.*, 1998; Kamath & Ahringer, 2003), several groups established that the nervous system plays an instrumental role in regulating whole body lipid metabolism (Greer, Perez, Van Gilst, Lee, & Ashrafi, 2008; Mak, Nelson, Basson, Johnson, & Ruvkun, 2006; Srinivasan *et al.*, 2008). Other groups working on genes controlling the lipid composition of *C. elegans* deciphered the polyunsaturated fatty acid synthesis pathway, a novel class of signaling lipids called ascarosides and their many roles in *C. elegans* physiology (Ludewig & Schroeder, 2013; Watts & Browse, 2002; Zhu & Han, 2014), as well as roles for the major nutrient sensors (TOR, AMPK) in regulating lipid metabolism (Narbonne & Roy, 2009; Ristow & Zarse, 2010; Soukas, Kane, Carr, Melo, & Ruvkun, 2009). Several groups including our own have established the use of reporter assays using fluorescent proteins that would allow us, in the living worm, to capture metabolic state information (Walker *et al.*, 2011) (Noble, Stieglitz, & Srinivasan, 2013). Additional and ongoing developments in measuring food intake (Ding, Romenskyy, Sarkisyan, & Brown, 2020; Wu *et al.*, 2019), energy expenditure (Koopman *et al.*, 2016; Srinivasan *et al.*, 2008), fat content and composition (Srinivasan, 2015) have rendered *C. elegans* an exceedingly sophisticated model system in which to uncover the fundamental features of body fat control.

In wild-type *C. elegans* feeding on *Escherichia coli*, the intestine is the primary site for fat synthesis, accumulation and metabolism (Srinivasan, 2015). In reproducing adults, fertilized embryos are a second site. Under certain conditions including food deprivation, fat deposition has also been noted in the germline, presumably to ensure survival of progeny in the face of longer-term food deprivation (Lynn *et al.*, 2015). Electron micrographs of the intestinal cells show that they contain large lipid deposits (Srinivasan, 2015), which from mass spectrometry studies primarily contain storage triglycerides (Ding *et al.*, 2013), as seen in mammals. Additional studies have shown that beta-oxidation, a central process by which stored fats are converted to energy in the mitochondria, regulate the relationship between reduced mitochondrial activity and increased lifespan (Durieux, Wolff, & Dillin, 2011).

Neurobiology of fat metabolism

In 2010 when I started my lab at TSRI, we began with the notion that, in contrast to the early discoveries of fat-accumulating hormones (such as leptin and insulin), genes and hormones regulating the conversion of stored fat to energy had lagged behind. To discover such mechanisms, we began conducting forward-genetic, candidate- and RNAi-based screens. We were also curious to explore more deeply and systematically, the role of the nervous system in intestinal fat metabolism, an endeavor that would have been Herculean without the pioneering work of John Sulston in having deciphered the complete wiring diagram of the *C. elegans* nervous system. Also invaluable is the deep knowledge base that had come from the work of Cori Bargmann and others in deciphering the major roles of the chemosensory neurons (Bargmann, 2006), and the remarkable ability to monitor and manipulate any neuron of choice in living worms (Chalasani *et al.*, 2007).

Oxygen sensing

To date, we have screened an estimated 30% of the neuronal genome (Hobert, 2013) for roles in body fat control. One of our early genetic screens was for the family of null mutants of the Gα family of G proteins (Jansen *et al.*, 1999), that transduce sensory signals from the environment via G protein coupled receptors (GPCRs). Many Gα proteins were expressed predominantly in neurons, and null mutants were available for the viable 19 of 21 genes. Our first 'hit' was a Gα protein called GPA-8, which is orthologous to the mammalian gustducin family of G proteins that regulate intracellular cGMP. GPA-8 is expressed in four oxygen sensing neurons of *C. elegans* called AQR, PQR and the bilateral URX(L/R). These neurons had previously been studied for their role in sensing atmospheric oxygen (de Bono & Bargmann, 1998; Gray *et al.*, 2004). We defined a role for *gpa-8* in the URX neurons in regulating the metabolic response to oxygen-sensing in the following way: worms fasted at 21% oxygen metabolized more than 80% of their intestinal fat stores for energy, whereas those fasted at 10% oxygen did not. This effect was largely due to the presence of the soluble guanylate cyclase *gcy-36*, which was also the oxygen sensor in the URX neurons. Through molecular-genetic and live Ca^{++} imaging approaches, we found that in the URX neurons, *gpa-8* which had emerged in our body fat screen, functions as a negative regulator of *gcy-36*, and modulates the cGMP-gated calcium channel, *tax-4* (Witham *et al.*, 2016). Interestingly, this work also revealed that the internal fat status of the intestine influences URX resting state, suggesting that an internal homeostatic signal regulates the extent to which environmental oxygen regulates intestinal fat metabolism (Ringstad, 2016).

In a set of parallel studies, we uncovered a role for the BAG neurons, which are sensors of low atmospheric oxygen (Zimmer *et al.*, 2009). BAG neurons detect and respond to 5–10% oxygen via the soluble guanylyl cyclase *gcy-33*, which had emerged from our screen of the soluble guanylyl cyclase family in *C. elegans*. It had been previously established that BAG and URX neurons are tonic sensors of oxygen (Zimmer *et al.*, 2009). In contemplating the relationship between BAG and URX neurons with respect to fat metabolism in the intestine, we discovered that BAG neurons function as tonic repressors of URX resting state (Hussey *et al.*, 2018). When exposed to low oxygen, BAG neurons are activated, and a peptide called FLP-17 is secreted from these neurons and acts on its cognate GPCR EGL-6 (Ringstad & Horvitz, 2008), which is necessary and sufficient in the URX neurons (Hussey *et al.*, 2018). BAG activity via *flp-17-egl-6* signaling serves to repress the resting state and activity of the URX neurons, limiting fat oxidation when environmental oxygen levels are low (Hussey *et al.*, 2018). Thus, fluctuations in environmental oxygen, sensed via the chemosensory nervous system, are an important driver of intestinal fat metabolism (Figure 2(A)).

Population density sensing

The most potent hit from our Gα family screen was *gpa-3*, whose absence led to a profound decrease in intestinal fat stores. *gpa-3* is expressed in several pairs of chemosensory neurons (Jansen *et al.*, 1999), however we found that it is necessary and sufficient in the ADL neurons, and functions to negatively regulate the adenylyl cyclase ACY-1 which produces the second messenger cAMP. ADL neurons detect and respond to the ascaroside pheromone ascr#3, via the TRPV channel *osm-9* (Jang *et al.*, 2012). Because the effect of *gpa-3* mutants was fully suppressed in the *gpa-3;osm-9* mutants, we wondered whether the pheromone ascr#3, which indicates population density and structure, might be another salient environmental cue that regulates intestinal fat metabolism. A series of genetic and reconstitution studies showed that pheromone-mediated regulation of cAMP signaling in ADL neurons regulates a metabolic signal to trigger the conversion of stored fats to energy in the intestine (Figure 2(B)). Why would population density and pheromones regulate intestinal lipid metabolism? As an animal encounters a new patch of food, it must adjust its metabolism to reflect its environment. A food patch that contains other worms must be shared, whereas a similar patch without other worms reflects a relatively greater amount of food for a single entering worm (Hussey *et al.*, 2017). We speculate that pheromone-sensing via ADL provides a salient 'denominator' to evaluate relative food availability to best optimize metabolic rate.

Food sensing and serotonin

In an independent line of research, our lab has been interested in the role of the neuromodulator serotonin (5-hydroxytryptamine, 5HT) because of its ancient roles in modulating behavior and physiology across species. In *C. elegans*, 5HT is synthesized by the rate-limiting enzyme tryptophan hydroxylase (tph-1), which is found predominantly in the NSM, ADF, HSN and a few other neurons (Sawin, Ranganathan, & Horvitz, 2000; Sze, Victor, Loer, Shi, & Ruvkun, 2000). Work from a number of groups has shown that 5HT production at each of these sites has distinct roles: NSM neurons gauge food entering via the pharynx and accordingly modulate behavioral responses to food (Rhoades *et al.*, 2019); ADF neurons detect beneficial and pathogenic bacteria, and alter feeding behavior (Cunningham *et al.*, 2012) and pathogen avoidance (Zhang, Lu, & Bargmann, 2005); HSN neurons regulate the rate of egg-laying in proportion to food access (Brewer, Olson, Collins, & Koelle, 2019; Hardaker, Singer, Kerr, Zhou, & Schafer, 2001). Interestingly 5HT from both NSM and ADF neurons also transduces heat stress, detected by the AFD neurons, to the germline (Tatum *et al.*, 2015).

We had found that neuronal 5HT was a potent modulator of fat metabolism in the intestine: loss of endogenous 5HT via genetic ablation of *tph-1* increased intestinal fat stores (Noble *et al.*, 2013), whereas ablation of *mod-5* (the 5HT-specific reuptake transporter) which increases synaptic 5HT, decreased intestinal fat stores (Srinivasan *et al.*, 2008). Through RNAi-based screens, we had found that in the intestine, neuronal 5HT led to increased mitochondrial beta-oxidation and the conversion of stored fats to energy (Srinivasan *et al.*, 2008). In trying to understand the mechanisms by which neuronal 5HT is relayed to the intestine, we uncovered the neural circuit for 5HT-mediated fat loss (Figure 2(C)). We defined a role for the ADF chemosensory neurons from which 5HT signaling is required for intestinal fat loss, and interestingly, a second role for the URX neurons from which MOD-1, a 5HT-gated chloride channel, regulates intestinal fat metabolism. We also found that the RIC neuron in which octopamine (the invertebrate ortholog of adrenaline) is synthesized (Alkema, Hunter-Ensor,

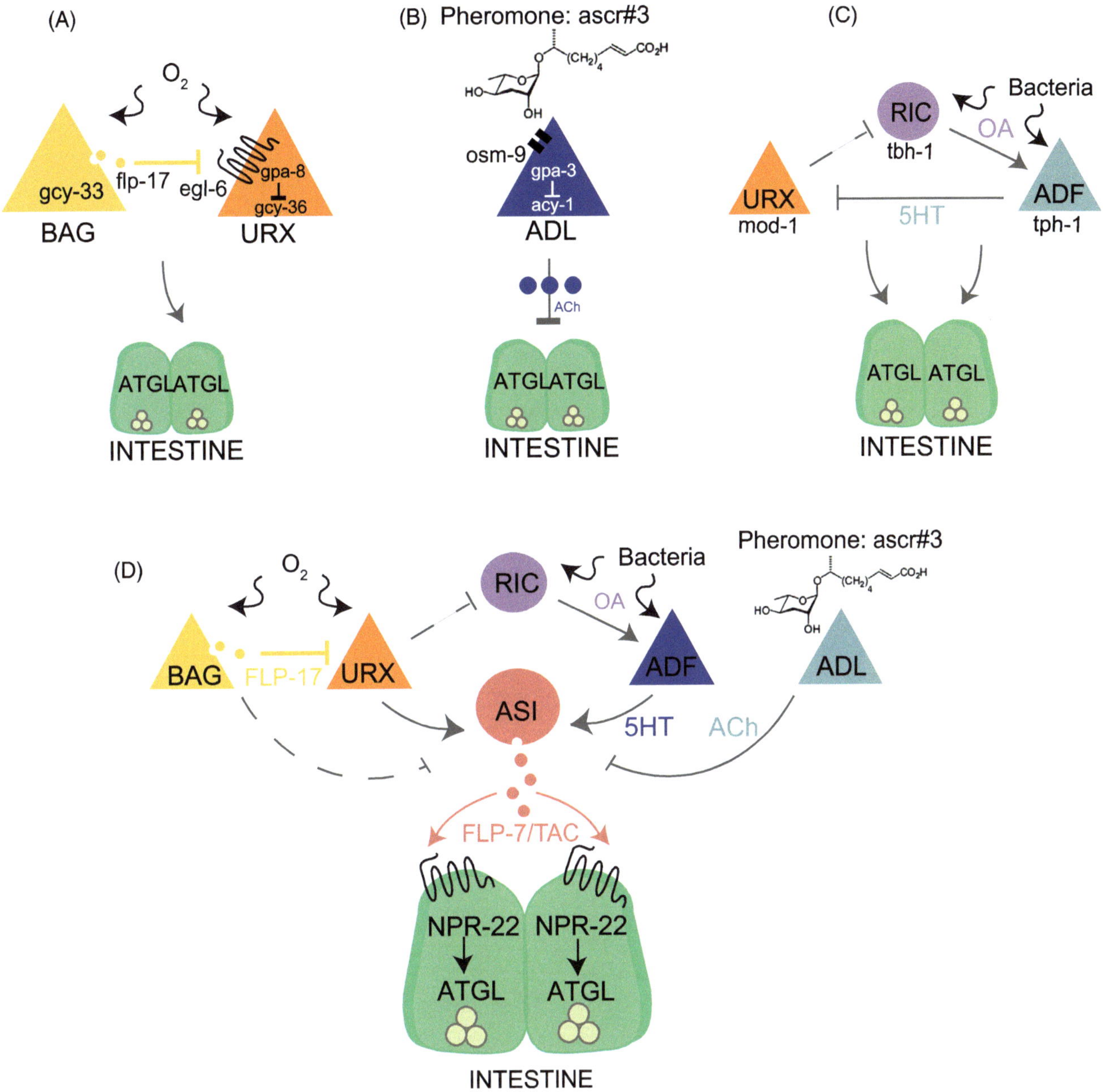

Figure 2. Neuroendocrine control of lipid metabolism. We have found that the *C. elegans* nervous system plays a profound role in regulating body fat stores, stored predominantly in the intestine. (A) Oxygen sensing from the environment plays an important role in regulating intestinal lipid metabolism. The URX neurons, which detect environmental oxygen, drive fat oxidation in the intestine. A Gα protein called GPA-8, controls the resting state of the URX neurons, thus regulating the rate and extent to which they can be activated by environmental oxygen. On the other hand, the BAG neurons negatively regulate the URX neurons by secreting the FLP-17 peptide under conditions of low oxygen, which binds to its cognate receptor EGL-6, necessary and sufficient in the URX neurons. Thus, oxygen sensing in the nervous system regulates intestinal fat metabolism in *C. elegans*. (B) Population-density-sensing is another driver of intestinal fat metabolism. The ascaroside pheromone ascr#3 is secreted by adults and L1 larvae. High ascr#3 concentrations within a given patch of food drive ATGL-1-dependent fat oxidation in the intestine. This process occurs via ADL neuron-mediated detection of ascr#3, and in which the Gα protein GPA-3 plays a critical sensory role. ADL communicates via downstream cholinergic neurons to the intestine to relay a signal for fat oxidation. (C) The serotonergic neural circuit for intestinal fat oxidation detects food availability in the environment, and is amplified by octopamine signaling from the RIC neurons. (D) Current model for the neuroendocrine control of intestinal lipid metabolism in *C. elegans*. The FLP-7 peptide is necessary and sufficient in the ASI neurons, and is secreted in proportion to fluctuations in oxygen, bacteria and pheromones. The FLP-7 receptor NPR-22 is necessary and sufficient in the intestine, and when activated, increases expression of the conserved triglyceride lipase ATGL-1 in the intestine. ATGL-1 then converts stored triglyceride lipids to free fatty acids, which are oxidized in the mitochondria for usable energy in the form of ATP. Thus, fluctuations in the environment, sensed and decoded by the nervous system, are a major driver of fat metabolism in *C. elegans*.

Ringstad, & Horvitz, 2005; Horvitz, Chalfie, Trent, Sulston, & Evans, 1982), and the AWB sensory neurons in which the SER-6 octopamine receptor is sufficient, together modulate ADF *tph-1* expression. Ultimately, food presence is relayed via the serotonergic circuit and amplified via octopamine, to drive the rate and extent of fat loss in the intestine (Noble *et al.*, 2013). The 5HT circuit exerts its effects in the intestine via regulating the essential and highly conserved triglyceride lipase called Adipocyte Triglyceride Lipase (ATGL-1), which is expressed predominantly in the intestine (Noble *et al.*, 2013). One notable feature of the 5HTergic fat regulatory circuit is that neither 5HT synthesis, nor the receptor that governs fat loss, nor the genes that underlie the amplifying effects of octopamine are expressed in the intestine, where the metabolic effects of 5HT occur. This observation echoed our work on oxygen sensing and pheromone sensing, in which genes necessary and sufficient in the nervous system showed large fluctuations in fat modulation in the intestinal cells. Because the *C. elegans* intestine is not innervated, we hypothesized that an unknown endocrine factor may relay sensory information from the nervous system, to the gut (Srinivasan, 2015).

Neuroendocrine communication and the tachykinin brain-to-gut signaling axis

In an effort to uncover such an endocrine signal, we conducted a screen of the known neuropeptide genes in *C. elegans*, for suppressors of 5HT-mediated fat oxidation. Our most potent suppressor was a neuropeptide called FLP-7, which we found is secreted from the ASI neurons in an AMPK-dependent manner, in response to fluctuations in neural 5HT circuit signaling (Palamiuc *et al.*, 2017). We then found that the FLP-7 receptor, a GPCR called NPR-22 (Mertens, Clinckspoor, Janssen, Nachman, & Schoofs, 2006; Mertens *et al.*, 2004), is necessary and sufficient in the intestine to drive ATGL-1-mediated fat oxidation. In *the flp-7;npr-22* double mutant, restoration of neither gene rescued 5HT-mediated fat oxidation without the presence of the other, suggesting that FLP-7 and NPR-22 function as a true ligand-receptor pair in vivo, and thus defining the neuroendocrine axis for 5HT-mediated lipid metabolism (Figure 2(D)). Notably, this ligand-receptor pair does not alter other 5HT-dependent behaviors including locomotion, reproduction and food intake (Palamiuc *et al.*, 2017). For global modulators such as 5HT, the use of distinct peptides for each output may be a predominant strategy to achieve phenotypic selectivity. Our ongoing work suggests that the tachykinin signaling axis may represent the final common pathway that integrates oxygen sensing, population density sensing and food sensing from the nervous system, to regulate the rate and extent of fat metabolism in the intestine (Figure 2(D)).

Open questions

A remarkable feature of the tachykinin neuroendocrine pathway is that all of the neuronal effects on global fat metabolism occur without appreciable changes in food intake (Hussey *et al.*, 2017; Hussey *et al.*, 2018; Noble *et al.*, 2013; Palamiuc *et al.*, 2017; Srinivasan *et al.*, 2008; Witham *et al.*, 2016), suggesting that the regulation of body fat is indeed distinct from food intake alone. The tachykinin neuropeptides, first defined by Substance P were originally identified more than 80 years ago (Guillemin, 2013), but have not previously been associated with lipid metabolism. The tachykinin peptides and receptors have been predominantly associated with inflammation in mammals (Steinhoff, von Mentzer, Geppetti, Pothoulakis, & Bunnett, 2014), which has mechanistic ties to metabolic dysfunction, however the role of tachykinins in mammalian metabolism remain unknown.

By our estimates, approximately 70% of the neuronal genome remains unexplored with respect to metabolic control (Hobert, 2013). Although we have identified the roles of a handful of neurons that transmit environmental cues to regulate intestinal metabolism, the majority remain unknown in this regard. We are optimistic that the use of advanced chemo- and optogenetic tools will hasten this process. In closing, the prescient studies of John Sulston and Sydney Brenner will continue to initiate, inform and strengthen our understanding of the myriad ways in which the nervous system regulates internal physiology. In relation to models that best describe our current understanding of energy regulation, we favor one in which hormones and neuroendocrine factors such as the serotonin-tachykinin system, which fluctuate in proportion to the external environment as well as internal state, regulate body fat stores. How such factors are integrated singly and in combination to orchestrate energy states within the body, is a fascinating unanswered in biology and medicine.

Acknowledgments

I thank all current and previous members of the Srinivasan Laboratory for their dedication and sincere efforts to uncover the biology of body fat metabolism using the *C. elegans* model system. Work discussed in this review was funded by the NIH, and I am thankful for their support in this endeavor. We also thank the CGC, from whom we have requested many *C. elegans* strains.

Disclosure statement

No potential conflict of interest was reported by the author(s).

References

Alkema, M.J., Hunter-Ensor, M., Ringstad, N., & Horvitz, H.R. (2005). Tyramine functions independently of octopamine in the *Caenorhabditis elegans* nervous system. *Neuron*, *46*(2), 247–260. doi:10.1016/j.neuron.2005.02.024

Bargmann, C.I. (2006). *Chemosensation in C. elegans. WormBook*, ed. The *C. elegans* Research Community, *WormBook*, doi/10.1895/wormbook.1.7.1, http://www.wormbook.org.

Bernard, C. (1927). *An introduction to the study of experimental medicine*. New York, NY: MacMillan and Co.

Brewer, J.C., Olson, A.C., Collins, K.M., & Koelle, M.R. (2019). Serotonin and neuropeptides are both released by the HSN command neuron to initiate *Caenorhabditis elegans* egg laying. *PLoS Genetics*, *15*(1), e1007896. doi:10.1371/journal.pgen.1007896

Cannon, W. (1939). *The wisdom of the body*. New York, NY: Norton & Company.

C. elegans Sequencing Consortium. (1998). Genome sequence of the nematode *C. elegans*: A platform for investigating biology. *Science*, *282*, 2012–2018. doi:10.1126/science.282.5396.2012

Chalasani, S.H., Chronis, N., Tsunozaki, M., Gray, J.M., Ramot, D., Goodman, M.B., & Bargmann, C.I. (2007). Dissecting a circuit for olfactory behaviour in Caenorhabditis elegans. *Nature*, *450*(7166), 63–70. doi:10.1038/nature06292

Coulson, A., Sulston, J., Brenner, S., & Karn, J. (1986). Toward a physical map of the genome of the nematode Caenorhabditis elegans. *Proceedings of the National Academy of Sciences of the United States of America*, *83*(20), 7821–7825. doi:10.1073/pnas.83.20.7821

Cunningham, K.A., Hua, Z., Srinivasan, S., Liu, J., Lee, B.H., Edwards, R.H., & Ashrafi, K. (2012). AMP-activated kinase links serotonergic signaling to glutamate release for regulation of feeding behavior in *C. elegans*. *Cell Metabolism*, *16*(1), 113–121. doi:10.1016/j.cmet.2012.05.014

Davenport, C.B. (1923). Body build and its inheritance. *Proceedings of the National Academy of Sciences of the United States of America*, *9*(7), 226–230. doi:10.1073/pnas.9.7.226

de Bono, M., & Bargmann, C.I. (1998). Natural variation in a neuropeptide Y receptor homolog modifies social behavior and food response in *C. elegans*. *Cell*, *94*(5), 679–689. doi:10.1016/S0092-8674(00)81609-8

Ding, S.S., Romenskyy, M., Sarkisyan, K.S., & Brown, A.E.X. (2020). Measuring *Caenorhabditis elegans* spatial foraging and food intake using bioluminescent bacteria. *Genetics*, *214*(3), 577–587. doi:10.1534/genetics.119.302804

Ding, Y., Zhang, S., Yang, L., Na, H., Zhang, P., Zhang, H., … Huo, C. (2013). Isolating lipid droplets from multiple species. *Nature Protocols*, *8*(1), 43–51. doi:10.1038/nprot.2012.142

Durieux, J., Wolff, S., & Dillin, A. (2011). The cell-non-autonomous nature of electron transport chain-mediated longevity. *Cell*, *144*(1), 79–91. doi:10.1016/j.cell.2010.12.016

Fire, A., Xu, S., Montgomery, M.K., Kostas, S.A., Driver, S.E., & Mello, C.C. (1998). Potent and specific genetic interference by double-stranded RNA in Caenorhabditis elegans. *Nature*, *391*(6669), 806–811. doi:10.1038/35888

Gray, J.M., Karow, D.S., Lu, H., Chang, A.J., Chang, J.S., Ellis, R.E., … Bargmann, C.I. (2004). Oxygen sensation and social feeding mediated by a C. elegans guanylate cyclase homologue. *Nature*, *430*(6997), 317–322. doi:10.1038/nature02714

Greer, E.R., Perez, C.L., Van Gilst, M.R., Lee, B.H., & Ashrafi, K. (2008). Neural and molecular dissection of a *C. elegans* sensory circuit that regulates fat and feeding. *Cell Metabolism*, *8*(2), 118–131. doi:10.1016/j.cmet.2008.06.005

Guillemin, R. (2013). A conversation with Roger Guillemin. Interview by Greg Lemke. *Annual Review of Physiology*, *75*, 1–22. doi:10.1146/annurev-physiol-082712-104641

Hardaker, L.A., Singer, E., Kerr, R., Zhou, G., & Schafer, W.R. (2001). Serotonin modulates locomotory behavior and coordinates egg-laying and movement in *Caenorhabditis elegans*. *Journal of Neurobiology*, *49*(4), 303–313. doi:10.1002/neu.10014

Hobert, O. (2013). The neuronal genome of *Caenorhabditis elegans*. *WormBook, ed. The C. elegans Research Community, WormBook*, doi/10.1895/wormbook.1.7.1, http://www.wormbook.org.

Horvitz, H.R., Chalfie, M., Trent, C., Sulston, J.E., & Evans, P.D. (1982). Serotonin and octopamine in the nematode Caenorhabditis elegans. *Science*, *216*(4549), 1012–1014. doi:10.1126/science.6805073

Hussey, R., Littlejohn, N.K., Witham, E., Vanstrum, E., Mesgarzadeh, J., Ratanpal, H., & Srinivasan, S. (2018). Oxygen-sensing neurons reciprocally regulate peripheral lipid metabolism via neuropeptide signaling in *Caenorhabditis elegans*. *PLoS Genetics*, *14*(3), e1007305. doi:10.1371/journal.pgen.1007305

Hussey, R., Stieglitz, J., Mesgarzadeh, J., Locke, T.T., Zhang, Y.K., Schroeder, F.C., & Srinivasan, S. (2017). Pheromone-sensing neurons regulate peripheral lipid metabolism in *Caenorhabditis elegans*. *PLoS Genetics*, *13*(5), e1006806. doi:10.1371/journal.pgen.1006806

Jang, H., Kim, K., Neal, S.J., Macosko, E., Kim, D., Butcher, R.A., … Sengupta, P. (2012). Neuromodulatory state and sex specify alternative behaviors through antagonistic synaptic pathways in *C. elegans*. *Neuron*, *75*(4), 585–592. doi:10.1016/j.neuron.2012.06.034

Jansen, G., Thijssen, K.L., Werner, P., van der Horst, M., Hazendonk, E., & Plasterk, R.H. (1999). The complete family of genes encoding G proteins of Caenorhabditis elegans. *Nature Genetics*, *21*(4), 414–419. doi:10.1038/7753

Kamath, R.S., & Ahringer, J. (2003). Genome-wide RNAi screening in *Caenorhabditis elegans*. *Methods*, *30*(4), 313–321. doi:10.1016/S1046-2023(03)00050-1

Koopman, M., Michels, H., Dancy, B.M., Kamble, R., Mouchiroud, L., Auwerx, J., … Houtkooper, R.H. (2016). A screening-based platform for the assessment of cellular respiration in *Caenorhabditis elegans*. *Nature Protocols*, *11*(10), 1798–1816. doi:10.1038/nprot.2016.106

Ludewig, A.H., & Schroeder, F.C. (2013). Ascaroside signaling in *C. elegans*. *WormBook*, 1–22. doi:10.1895/wormbook.1.155.1

Lynn, D.A., Dalton, H.M., Sowa, J.N., Wang, M.C., Soukas, A.A., & Curran, S.P. (2015). Omega-3 and -6 fatty acids allocate somatic and germline lipids to ensure fitness during nutrient and oxidative stress in *Caenorhabditis elegans*. *Proceedings of the National Academy of Sciences of the United States of America*, *112*(50), 15378–15383. doi:10.1073/pnas.1514012112

Mak, H.Y., Nelson, L.S., Basson, M., Johnson, C.D., & Ruvkun, G. (2006). Polygenic control of *Caenorhabditis elegans* fat storage. *Nature Genetics*, *38*(3), 363–368. doi:10.1038/ng1739

Mertens, I., Clinckspoor, I., Janssen, T., Nachman, R., & Schoofs, L. (2006). FMRFamide related peptide ligands activate the *Caenorhabditis elegans* orphan GPCR Y59H11AL.1. *Peptides*, *27*(6), 1291–1296. doi:10.1016/j.peptides.2005.11.017

Mertens, I., Vandingenen, A., Meeusen, T., Janssen, T., Luyten, W., Nachman, R.J., … Schoofs, L. (2004). Functional characterization of the putative orphan neuropeptide G-protein coupled receptor C26F1.6 in *Caenorhabditis elegans*. *FEBS Letters*, *573*(1–3), 55–60. doi:10.1016/j.febslet.2004.07.058

Narbonne, P., & Roy, R. (2009). *Caenorhabditis elegans* dauers need LKB1/AMPK to ration lipid reserves and ensure long-term survival. *Nature*, *457*(7226), 210–214. doi:10.1038/nature07536

Noble, T., Stieglitz, J., & Srinivasan, S. (2013). An integrated serotonin and octopamine neuronal circuit directs the release of an endocrine signal to control *C. elegans* body fat. *Cell Metabolism*, *18*(5), 672–684. doi:10.1016/j.cmet.2013.09.007

Palamiuc, L., Noble, T., Witham, E., Ratanpal, H., Vaughan, M., & Srinivasan, S. (2017). A tachykinin-like neuroendocrine signalling axis couples central serotonin action and nutrient sensing with peripheral lipid metabolism. *Nature Communications*, *8*, 14237. doi:10.1038/ncomms14237

Ravussin, Y., Leibel, R.L., & Ferrante, A.W., Jr. (2014). A missing link in body weight homeostasis: The catabolic signal of the overfed state. *Cell Metabolism*, *20*(4), 565–572. doi:10.1016/j.cmet.2014.09.002

Rhoades, J.L., Nelson, J.C., Nwabudike, I., Yu, S.K., McLachlan, I.G., Madan, G.K., … Flavell, S.W. (2019). ASICs mediate food responses in an enteric serotonergic neuron that controls foraging behaviors. *Cell*, *176*(1–2), 85.e14–97.e14. doi:10.1016/j.cell.2018.11.023

Ringstad, N. (2016). A controlled burn: Sensing oxygen to tune fat metabolism. *Cell Reports*, *14*(7), 1569–1570. doi:10.1016/j.celrep.2016.02.015

Ringstad, N., & Horvitz, H.R. (2008). FMRFamide neuropeptides and acetylcholine synergistically inhibit egg-laying by *C. elegans*. *Nature Neuroscience*, *11*(10), 1168–1176. doi:10.1038/nn.2186

Ristow, M., & Zarse, K. (2010). How increased oxidative stress promotes longevity and metabolic health: The concept of mitochondrial hormesis (mitohormesis). *Experimental Gerontology*, *45*(6), 410–418. doi:10.1016/j.exger.2010.03.014

Sawin, E.R., Ranganathan, R., & Horvitz, H.R. (2000). *C. elegans* locomotory rate is modulated by the environment through a dopaminergic pathway and by experience through a serotonergic pathway. *Neuron*, *26*(3), 619–631. doi:10.1016/S0896-6273(00)81199-X

Soukas, A.A., Kane, E.A., Carr, C.E., Melo, J.A., & Ruvkun, G. (2009). Rictor/TORC2 regulates fat metabolism, feeding, growth, and life span in *Caenorhabditis elegans*. *Genes & Development*, *23*(4), 496–511. doi:10.1101/gad.1775409

Srinivasan, S. (2015). Neuroendocrine control of body fat in *Caenorhabditis elegans. Annual Reviews in Physiology, 77(1), 161–178.*

Srinivasan, S., Sadegh, L., Elle, I.C., Christensen, A.G., Faergeman, N.J., & Ashrafi, K. (2008). Serotonin regulates *C. elegans* fat and feeding through independent molecular mechanisms. *Cell Metabolism*, *7*(6), 533–544. doi:10.1016/j.cmet.2008.04.012

Steinhoff, M.S., von Mentzer, B., Geppetti, P., Pothoulakis, C., & Bunnett, N.W. (2014). Tachykinins and their receptors: Contributions to physiological control and the mechanisms of disease. *Physiological Reviews*, *94*(1), 265–301. doi:10.1152/physrev.00031.2013

Sulston, J.E., & Brenner, S. (1974). The DNA of *Caenorhabditis elegans*. *Genetics*, *77*(1), 95–104.

Sze, J.Y., Victor, M., Loer, C., Shi, Y., & Ruvkun, G. (2000). Food and metabolic signalling defects in a *Caenorhabditis elegans* serotonin-synthesis mutant. *Nature*, *403*(6769), 560–564. doi:10.1038/35000609

Tam, J., Fukumura, D., & Jain, R.K. (2009). A mathematical model of murine metabolic regulation by leptin: Energy balance and defense of a stable body weight. *Cell Metabolism*, *9*(1), 52–63. doi:10.1016/j.cmet.2008.11.005

Tatum, M.C., Ooi, F.K., Chikka, M.R., Chauve, L., Martinez-Velazquez, L.A., Steinbusch, H.W.M., … Prahlad, V. (2015). Neuronal serotonin release triggers the heat shock response in *C. elegans* in the absence of temperature increase. *Current Biology*, *25*(2), 163–174. doi:10.1016/j.cub.2014.11.040

Wade, G.N., & Gray, J.M. (1979). Gonadal effects on food intake and adiposity: A metabolic hypothesis. *Physiology & Behavior*, *22*(3), 583–593. doi:10.1016/0031-9384(79)90028-3

Walker, A.K., Jacobs, R.L., Watts, J.L., Rottiers, V., Jiang, K., Finnegan, D.M., … Niebergall, L.J. (2011). A conserved SREBP-1/phosphatidylcholine feedback circuit regulates lipogenesis in metazoans. *Cell*, *147*(4), 840–852. doi:10.1016/j.cell.2011.09.045

Walley, A.J., Blakemore, A.I., & Froguel, P. (2006). Genetics of obesity and the prediction of risk for health. *Human Molecular Genetics*, *15*(2), R124–R130. doi:10.1093/hmg/ddl215

Ward, Z.J., Bleich, S.N., Cradock, A.L., Barrett, J.L., Giles, C.M., Flax, C., … Gortmaker, S.L. (2019). Projected U.S. state-level prevalence of adult obesity and severe obesity. *New England Journal of Medicine*, *381*(25), 2440–2450. doi:10.1056/NEJMsa1909301

Waterston, R., & Sulston, J. (1995). The genome of Caenorhabditis elegans. *Proceedings of the National Academy of Sciences of the United States of America*, *92*(24), 10836–10840. doi:10.1073/pnas.92.24.10836

Watts, J.L., & Browse, J. (2002). Genetic dissection of polyunsaturated fatty acid synthesis in *Caenorhabditis elegans*. *Proceedings of the National Academy of Sciences of the United States of America*, *99*(9), 5854–5859. doi:10.1073/pnas.092064799

Wirtshafter, D., & Davis, J.D. (1977). Set points, settling points, and the control of body weight. *Physiology & Behavior*, *19*(1), 75–78. doi:10.1016/0031-9384(77)90162-7

Witham, E., Comunian, C., Ratanpal, H., Skora, S., Zimmer, M., & Srinivasan, S. (2016). *C. elegans* body cavity neurons are homeostatic sensors that integrate fluctuations in oxygen availability and internal nutrient reserves. *Cell Reports*, *14*(7), 1641–1654. doi:10.1016/j.celrep.2016.01.052

Woodhouse, R. (2008). Obesity in art: A brief overview. *Frontiers of Hormone Research*, *36*, 271–286. doi:10.1159/000115370

Wu, Z., Isik, M., Moroz, N., Steinbaugh, M.J., Zhang, P., & Blackwell, T.K. (2019). Dietary restriction extends lifespan through metabolic regulation of innate immunity. *Cell Metabolism*, *29*(5), 1192–1205. doi:10.1016/j.cmet.2019.02.013

Zhang, Y., Lu, H., & Bargmann, C.I. (2005). Pathogenic bacteria induce aversive olfactory learning in *Caenorhabditis elegans*. *Nature*, *438*(7065), 179–184. doi:10.1038/nature04216

Zhu, H., & Han, M. (2014). Exploring developmental and physiological functions of fatty acid and lipid variants through worm and fly genetics. *Annual Review of Genetics*, *48*, 119–148. doi:10.1146/annurev-genet-041814-095928

Zimmer, M., Gray, J.M., Pokala, N., Chang, A.J., Karow, D.S., Marletta, M.A., … Bargmann, C.I. (2009). Neurons detect increases and decreases in oxygen levels using distinct guanylate cyclases. *Neuron*, *61*(6), 865–879. doi:10.1016/j.neuron.2009.02.013

The discovery and consequences of the central role of the nervous system in the control of protein homeostasis

Veena Prahlad

ABSTRACT
Organisms function despite wide fluctuations in their environment through the maintenance of homeostasis. At the cellular level, the maintenance of proteins as functional entities at target expression levels is called protein homeostasis (or proteostasis). Cells implement proteostasis through universal and conserved quality control mechanisms that surveil and monitor protein conformation. Recent studies that exploit the powerful ability to genetically manipulate specific neurons in *C. elegans* have shown that cells within this metazoan lose their autonomy over this fundamental survival mechanism. These studies have uncovered novel roles for the nervous system in controlling how and when cells activate their protein quality control mechanisms. Here we discuss the conceptual underpinnings, experimental evidence and the possible consequences of such a control mechanism.

PRELUDE: Whether the detailed examination of parts of the nervous system and their selective perturbation is sufficient to reconstruct how the brain generates behavior, mental disease, music and religion remains an open question. Yet, Sydney Brenner's development of *C. elegans* as an experimental organism and his faith in the bold reductionist approach that 'the understanding of wild-type behavior comes best after the discovery and analysis of mutations that alter it', has led to discoveries of unexpected roles for neurons in the biology of organisms.

Introduction

Most protein-based biological mechanisms proceed optimally only within a narrow range of environmental conditions (Fields, Dong, Meng, & Somero, 2015; Hofmann & Somero, 1995; Somero, 1995). Despite this, organisms thrive in a variety of ecological niches. A mechanism by which organisms function despite wide fluctuations in their environment is through the maintenance of homeostasis (from Greek: ὅμοιος homoeos, 'similar' and στάσις stasis, 'standing still'), whereby the internal milieu is maintained close to constant in the face of external perturbations. This concept, first described by Claude Bernard in 1865, has provided the foundation for our understanding of macroscopic physiological processes such as the maintenance of core body temperature, pH of blood (Bernard, 1965), regulation of food intake, body mass, and energy metabolism, all of which are under homeostatic control. More recently, the concept of homeostasis has been extended to mechanisms that occur at the sub-cellular level, such as the maintenance of protein function in cells, and has become popularly known as proteostasis (Balch, Morimoto, Dillin, & Kelly, 2008; Gidalevitz, Ben-Zvi, Ho, Brignull, & Morimoto, 2006; Powers, Morimoto, Dillin, Kelly, & Balch, 2009). In general, there appear to be two distinct kinds of strategies to achieve homeostasis. The first are servomechanisms (Leow, 2007), whereby error-sensing negative feedback loops correct the performance of a system to maintain some specific feature, called a *regulated variable*, constant or close to some set-point (Figure 1(A)). Such a mechanism requires an error detector that senses deviations from the set-point, and a mechanism to perform the necessary error corrections (often called the servomotor, as this terminology grew from mechanical devices such as steam engines). While servomechanisms can precisely maintain the regulated variable at the set-point, they are triggered only *after* the set-point, and homeostasis, have been perturbed. An alternate mechanism by which biological systems achieve homeostasis is through the activation of anticipatory mechanisms. These mechanisms, termed Cephalic mechanisms, are predictive, being implemented prior to the actual perturbation of the system. For instance, temperature homeostasis in an organism is maintained through mechanisms that *prevent* the core body temperature from deviating above or below a certain range through launching preemptive/anticipatory adjustments such as sweating, *before* core body temperatures are perturbed. The essence of such mechanisms is that, by definition, the command for the correction of the deviation comes from centers that can predict the oncoming perturbation using associated information, prior to themselves becoming

(A)

Definitions	
Homeostatic control mechanism	Mechanism by which the internal milieu of a system is maintained close to constant in the face of external perturbations. *The levels of 'misfolded' proteins within a cell are held homeostatic.*
Regulated Variable	A feature or variable of a system that is maintained at constant levels (near set-point) by homeostatic control mechanisms. *Protein conformation is a regulated variable being maintained at some set-point within cells.*
Set-point	An optimal value of the regulated variable that is maintained by homeostatic control mechanisms. *Deviations in the levels of 'misfolded' proteins from a set-point value activates homeostatic control mechanisms.*
Controlled Variable	The physiologic processes that maintain the regulated variable at a set-point. *The level of misfolded proteins is maintained at some set-point by the activity of controlled variables: protein biogenesis and protein degradation.*
Controller	The component of the homeostatic control mechanism that monitors the value of regulated variable and detects deviations from the set-point to trigger changes in the controlled variables. *Molecular chaperones are CONTROLLERS that surveil the folding state of proteins and adjust the controlled variables— protein biogenesis and protein degradation.*
Plant	Component of the homeostatic control mechanism that implements the changes triggered by the controller. *Ribosomes are PLANTS for protein biogenesis; the proteasome and autophagosomes are PLANTS for degradation.*

(B)

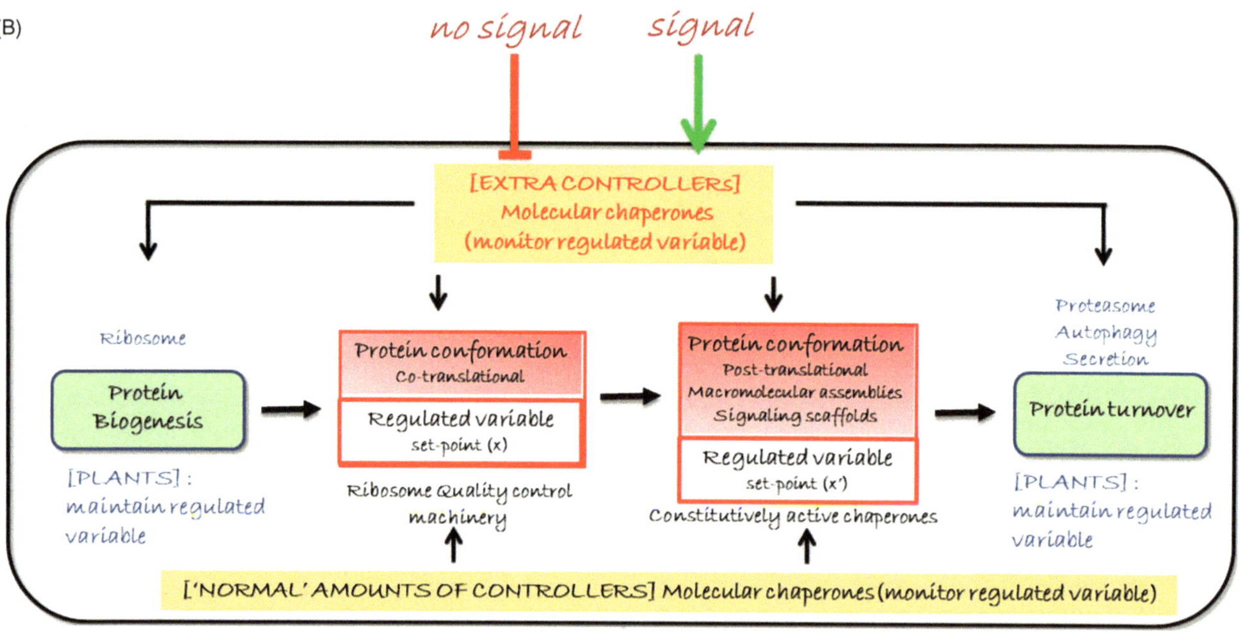

Figure 1. (A) A brief definition of control theory terminology in relation to its use to describe proteostasis. (B) A simplified model of proteostasis. Protein conformation is the *regulated variable* that is maintained at a set-point through the regulated activities of protein biogenesis and protein degradation (controlled variables). CONTROLLERS (here mainly, molecular chaperones) that surveil and monitor protein conformation are present at specific amounts during the normal activities of the cells. They direct the activity of the *controlled variables* through PLANTS (e.g. ribosomes, proteasome, autophagy etc.) to generate homeostatic levels of the *regulated variable* (namely low levels of misfolded proteins). Extra CONTROLLERS can be generated in response to signals that change the levels of the *regulated variable*. However, as explained in the text, increasing the levels of the CONTROLLERS is under neuronal control in *C. elegans*.

perturbed. Thus, cephalic mechanisms can prevent that the system be perturbed in the first place (Lechan & Fekete, 2006; Mattes, 1997); however, they are not fail-safe as they can be 'fooled' and triggered even in the absence of any perturbations to homeostasis.

Until about a decade ago, the mechanisms by which cells and organisms maintained protein homeostasis were best studied in isolated cells in culture or unicellular organisms such as yeast and bacteria (Akerfelt, Morimoto, & Sistonen, 2010; Balch *et al.*, 2008; Goff & Goldberg, 1987; Goldberg, 1971, 1972; Hightower, 1980; Lindquist, 1986; Morano, Liu, & Thiele, 1998; Neef, Turski, & Thiele, 2010; Ritossa, 1996). In these systems, protein homeostasis is implemented mainly through feed-back loops that act as servomechanisms (Morimoto, 1998; Voellmy, 1994, 1996). In the last decade, however, experiments that may not have been so readily feasible were it not for the ability to manipulate neurons in *C. elegans* (Avery, Bargmann, & Horvitz, 1993; Chalfie *et al.*,

1985; Walrond, Kass, Stretton, & Donmoyer, 1985; White, Southgate, Thomson, & Brenner, 1983, 1986), demonstrated that in this metazoan the nervous system exerts neuroendocrine control over the conserved cellular machinery that maintains protein quality control. Such control enables the animal to trigger these deeply conserved survival and defense mechanisms in anticipation of a threat. Here we briefly highlight how the genetic manipulation of *C. elegans* neurons enabled these insights and discuss some of the implications of the existence of cephalic control over cellular protein homeostasis. We focus mainly on the conceptual framework, and would like to direct interested readers to more thorough reviews for details (Miles, Scherz-Shouval, & van Oosten-Hawle, 2019; Wolff, Weissman, & Dillin, 2014).

Proteostasis

A mammalian cell synthesizes on average ~10,000 proteins at any given time, most of which can assume multiple conformations, many of which are unstructured, occur in multiple copies, have widely varying half-lives, and are in the continual process of being modified, transported and degraded (Jayaraj, Hipp, & Hartl, 2020; Mogk, Bukau, & Kampinga, 2018; Tyedmers, Mogk, & Bukau, 2010; Wolff *et al.*, 2014). Despite this complexity, cells largely maintain a target expression level of any particular protein, and protein concentrations do not vary erratically. Moreover, cells can detect the presence of protein aggregates and non-native protein conformations (Geiler-Samerotte *et al.*, 2011; Vendruscolo, Knowles, & Dobson, 2011), and respond in specific and conserved ways to restore conformation or trigger degradation of the abnormal species (Bednarska, Schymkowitz, Rousseau, & Van Eldere, 2013; Hipp, Park, & Hartl, 2014; Labbadia & Morimoto, 2015; Mattoo & Goloubinoff, 2014; Miller *et al.*, 2015; Silverman *et al.*, 2015; Vendruscolo *et al.*, 2011; Walther *et al.*, 2015). Since the conformation of proteins drives their stability and function, this ability of cells to maintain proteins as functional entities at specific concentrations occurs through mechanisms that surveil protein conformation (Ananthan, Goldberg, & Voellmy, 1986; Goff & Goldberg, 1985, 1987; Goldberg, 1972; Kandror, Busconi, Sherman, & Goldberg, 1994; Knowles, Gunn, Hanson, & Ballard, 1975; Knowles & Ballard, 1976, 1978; Prouty, Karnovsky, & Goldberg, 1975). These mechanisms are collectively called proteostasis (Balch *et al.*, 2008; Gidalevitz *et al.*, 2006). Exciting studies have demonstrated the existence of interconnected networks of surveillance and quality control mechanisms that monitor protein quality during normal protein synthesis. For instance, ribosome stalling that can occur due to stochastic perturbations of translation, mutations, or a number of biotic or abiotic fluctuations in the environment will lead to the aborted synthesis of a nascent polypeptide chains. These aborted polypeptides are recognized by the cellular surveillance machinery and modified by the addition of carboxy-terminal alanine and threonine residues (CAT tails) (Brandman *et al.*, 2012; Kostova *et al.*, 2017; Sitron, Park, & Brandman, 2017; Sitron, Park, Giafaglione, & Brandman, 2020; Sitron & Brandman, 2019). This, in turn, leads to their aggregation and targeted degradation through the sequestration of a group of specialized proteins called molecular chaperones that recognize the aberrant conformation of the CAT-tailed polypeptides (Sitron *et al.*, 2017, 2020; Sitron & Brandman, 2019). The sequestration of molecular chaperones is also thought to activate a conserved transcription factor to increase the amounts of molecular chaperone proteins available to surveil and prevent the build-up of aberrant polypeptides. Translating these concepts into homeostatic control theory (Figure 1(A,B)) protein conformation is the *regulated variable*, being maintained at some *set-point* (optimal flux, optimal concentration etc.) by the activity of *controlled variables*: protein biogenesis and protein degradation (green boxes). The *regulated variable* is maintained by the activity of PLANTS: ribosomes for protein biogenesis and proteases, the proteasome, and autophagy pathways for degradation. The *set-point* for misfolded proteins is monitored and maintained through the activity of CONTROLLERS, the best known of which are molecular chaperones that surveil the folding state of proteins. If the CONTROLLER detects deviations from the *set-point*, signals to the PLANTS change the flux of proteins through the system restoring the *regulated variable* to its acceptable set-point (Goldstein & Kopin, 2017; Iberall & Cardon, 1964).

As with most biological processes, the levels of proteins in a cell, and therefore also misfolded proteins, varies with the types of protein synthesized, physiological needs and environmental changes. Accordingly, *controlled variables* i.e. the flux of proteins through the PLANTS, and even the concentration of PLANTS and CONTROLLERS required for maintaining protein homeostasis vary. The best understood regulatory mechanisms by which a sudden increase in the *regulated variable*, i.e. misfolded protein, impacts the PLANTS and CONTROLLERS, are the so-called heat shock response (HSR) in the cytoplasm and the unfolded protein response in the endoplasmic reticulum (Gomez-Pastor, Burchfiel, & Thiele, 2018; Vihervaara, Duarte, & Lis, 2018). First discovered as a heat-inducible transcriptional response when *Drosophila busckii* cells were accidently exposed to high temperatures (De Maio, Santoro, Tanguay, & Hightower, 2012; Ritossa, 1996), the HSR is a highly conserved, universal transcription program executed in response to protein damage (Akerfelt *et al.*, 2010; Ananthan *et al.*, 1986; Morimoto, 1998; Voellmy, 1994). The HSR is mediated by the heat shock transcription factor 1 (HSF1) whose activity results in the rapid expression of more molecular chaperones (CONTROLLERS) to help identify, and refold or degrade the damaged proteins. Molecular chaperones accomplish this by triggering changes in the *controlled variables*, typically promoting a decrease in protein biogenesis and an increase in degradation of the damaged proteins (Goldberg, 1971; Hightower, 1980; Lindquist, 1986; Morimoto, 1998). Similarly, the Unfolded Protein Response of the Endoplasmic Reticulum (UPR^{ER}) is a conserved cellular mechanism to detect and regulate proteostasis within the lumen of the endoplasmic reticulum (ER) if protein processing stalls due to the increase in aberrant

ER proteins (Gething & Sambrook, 1992; Kozutsumi, Segal, Normington, Gething, & Sambrook, 1988; Munro & Pelham, 1986, 1987; Pelham & Munro, 1993). UPR^{ER} acts through the activation of all or a subset of three stress signal transducers found within the ER membrane, Inositol-requiring enzyme 1 (IRE1), Activating transcription factor 6 (ATF6), and protein kinase R-like endoplasmic reticulum kinase (PERK), which also increase the molecular chaperones (CONTROLLERS) that act in the ER and modulate the *controlled variables* (Mori, 2009; Mori, Kawahara, Yoshida, Yanagi, & Yura, 1996; Walter, 2010; Yoshida, Matsui, Yamamoto, Okada, & Mori, 2001). Parallel mechanisms exist for ensuring protein homeostasis in mitochondria (Haynes & Ron, 2010). Until recently, activation of the transcription factors that increase the amount of CONTROLLERS (HSF1 or IRE1, ATF6 and XBP1) was thought to be *solely* triggered by the individual cells in which protein homeostasis was perturbed. Thus, both HSF1 and the transcriptional activators of the UPR^{ER} are typically inhibited unless the flux of protein misfolding exceeds some set-point or threshold because they themselves are under the negative inhibition of the CONTROLLERS. Upon sensing an increase in protein misfolding, the CONTROLLERS are kinetically titrated away, dis-inhibiting (or permitting the activation of) the transcription factors, increasing the amounts of CONTROLLERS, and restoring homeostasis (Bernales, Papa, & Walter, 2006). Indeed, this is what is thought to occur during normal protein synthesis upon ribosome stalling. Under these conditions, the aborted synthesis of the nascent aggregation prone polypeptide titrates away the negative inhibition on HSF1 activity to promote an increase in chaperone gene transcription. This elegant model thus lay the foundation for the proteostasis network in isolated cells.

Proteostasis in *C. elegans*

C. elegans possesses dedicated neurons to sense threats in its environment and can detect environmental changes well before they are severe enough to disrupt protein homeostasis (Bargmann, 1998; Gally & Bessereau, 2003; Goodman & Sengupta, 2019; Hobert, 2005). One such sensory system that senses temperature change is the thermosensory circuitry, which involves at least three pairs of sensory neurons (AFD, AWC and ASI), that are extraordinarily sensitive to temperature changes, and a layer of interneurons (AIZ and AIY) that process and integrate the temperature information (Beverly, Anbil, & Sengupta, 2011; Clark, Biron, Sengupta, & Samuel, 2006; Goodman & Sengupta, 2018, 2019). It had been shown, once all the neurons of *C. elegans* had been catalogued and mapped, that AFD neurons can detect changes as small as 0.05 °C above ambient temperature; hence, they could arguably be excited by temperature increments well below those that cause cellular damage (Beverly *et al.*, 2011; Clark *et al.*, 2006; Colosimo *et al.*, 2004). With such specialized machinery to sense the environment, had metazoans with a nervous system evolved mechanisms to activate their protein homeostasis machinery prior to experiencing temperatures where their proteins would misfold? This was the question that motivated the first experiments that demonstrated that, in addition to the elegant servomechanism of proteostasis that existed in individual cells, in *C. elegans* there also exists cephalic control over protein homeostasis (Prahlad, Cornelius, & Morimoto, 2008). In the first experiments we used mutations that impaired the thermosensory capacity of the AFD neurons, and asked whether chaperones would be induced upon a transient increase in temperatures, if the animal did not sense the temperature upshift. Indeed, mutations affecting the ability of thermosensory AFD neurons to sense heat also delayed HSF1 dependent molecular chaperone induction throughout *C. elegans* upon heat shock (Prahlad *et al.*, 2008) (Figure 2(A)). However, mutations that impair neurons throughout the lifetime of animals could arguably have reset physiology and metabolism such that animals had altered the flux of proteins through the PLANTS or re-set their thresholds for tolerating misfolded proteins. Therefore, the more convincing demonstration that proteostasis was under cephalic control came from experiments using optogenetics to excite the AFD neurons in the absence of heat. Such a perturbation mimicked the sensation of heat and was sufficient to simulate a HSR, activate HSF1 and upregulate HSPs in distal tissues (Figure 2(A)) (Tatum *et al.*, 2015). We found that the signal for AFD-dependent upregulation of HSPs was the ancient bioamine, serotonin that was released upon the excitation of the AFD neurons. Serotonin is best studied for its roles in mood, anxiety and cognition, and in almost all organisms real or perceived threats cause the release of neural serotonin (Cruz-Corchado, Ooi, Das, & Prahlad, 2020; Tatum *et al.*, 2015). In *C. elegans,* the use of powerful genetic tools including neuronal mutants that lack all serotonin, had shown that serotonin played a pivotal role in learning and memory (Chase & Koelle, 2007; Curran & Chalasani, 2012). These studies have thus immediately linked anxiety, experience, learning—aspects that are fundamental to cephalic processes—to cellular changes in transcription and protein quality control. Indeed, in following experiments we have shown that *C. elegans* can 'learn', through prior experience of a threat, to enhance chaperone expression were it to encounter the specific threat in its environment (Ooi & Prahlad, 2017) (Figure 2(B)). Moreover, the release of serotonin that occurs upon the sensing of temperature change has a rapid effect on chromatin accessibility in the germline of animals conferring stress protection on future offspring (Das *et al.*, 2020) (Figure 2(C)).

Analogous neurohormonal signaling mechanisms have since been discovered for the CONTROLLERS of proteostasis in other cellular compartments in *C. elegans* (Berendzen *et al.*, 2016; Durieux, Wolff, & Dillin, 2011; Frakes *et al.*, 2020; Taylor, Berendzen, & Dillin, 2014; Taylor & Dillin, 2013; Zhang *et al.*, 2018) (Figure 2(A)). Thus, the constitutive ectopic expression of XBP1 in the nervous system—in neurons or supporting glial cells—or the activation of the mitochondrial UPR in neurons induces a systemic UPR^{ER} or UPR^{MT} respectively, in other tissues (Durieux *et al.*, 2011; Frakes *et al.*, 2020; Taylor *et al.*, 2014). In addition, it appears that neurons may not be the only

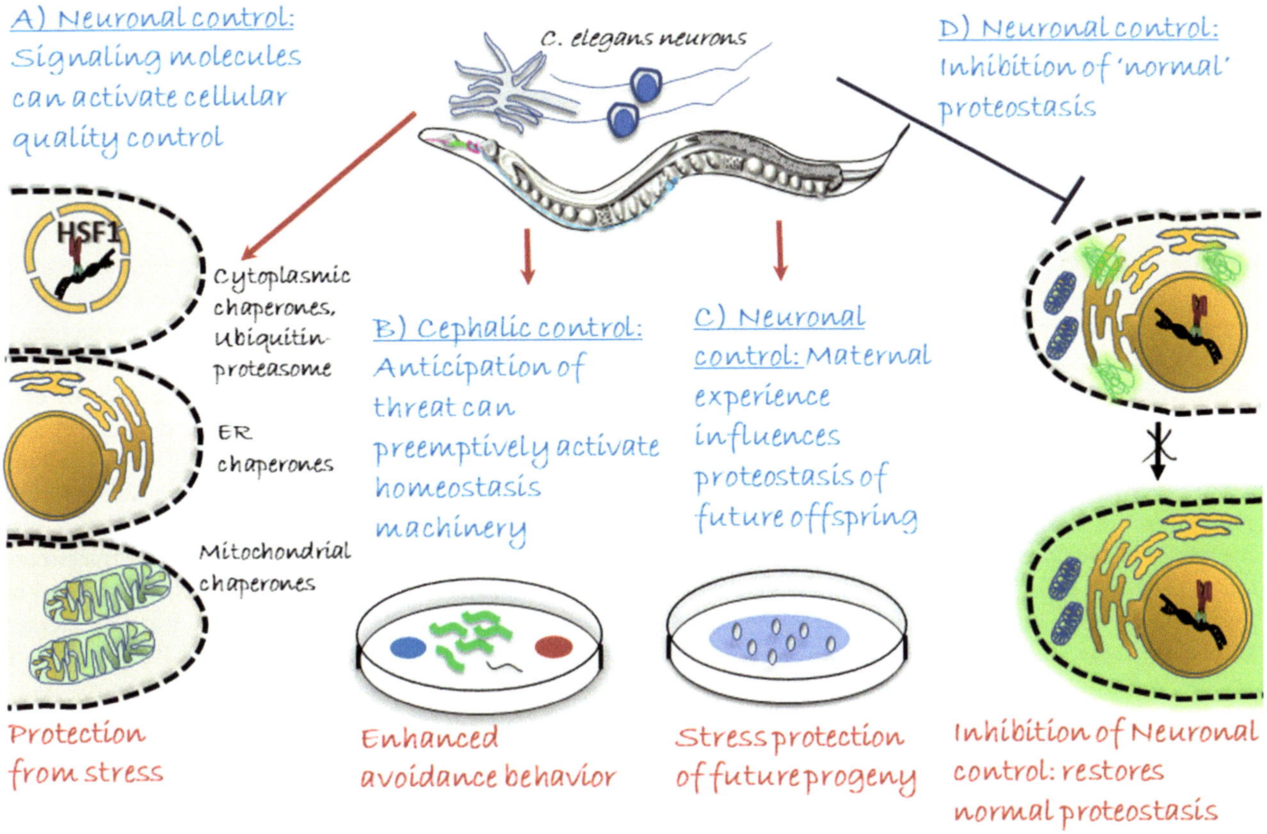

Figure 2. Consequences of neuronal and cephalic control over homeostasis in *C. elegans*. (A) Neuronal circuits through serotonergic, dopaminergic, WNT, FLP-2 and octopaminergic signaling can activate transcription factors that are the master regulators of chaperones, and degradation machinery such as the ubiquitin-proteasome system in distal tissue of C. elegans. (B) Prior experience of an environmental threat can enhance the expression of chaperones upon actual encounter with the threat. (C) Maternal release of serotonin upon stress encounter increases the stress tolerance of progeny. (D) Neuronal control enables rapid response to stimuli at the cost of the normal proteostasis machinery.

sentinels of harm: activation of the heat shock response in one tissue results in the systemic induction of this stress response (Miles *et al.*, 2019; O'Brien *et al.*, 2018; van Oosten-Hawle, Porter, & Morimoto, 2013). Thus, it appears that in a multicellular organism, changes in proteostasis in one tissue can feedback onto protein metabolism in another tissue to systemically increase the quantity of the molecular chaperones or the CONTROLLERS that monitor set points for misfolded proteins.

Many studies have also shown, directly or indirectly, that the autonomy of cells over the *controlled variables*, protein biogenesis and protein degradation, is also outsourced to neurons (Ben-Gedalya & Cohen, 2012). The nervous system controls growth, metabolism and aging in response to nutrient and energy status of the animal (Broughton & Partridge, 2009; Burkewitz *et al.*, 2015; Kawli, Wu, & Tan, 2010; Minnerly, Zhang, Parker, Kaul, & Jia, 2017; Weir *et al.*, 2017; Wolkow, Kimura, Lee, & Ruvkun, 2000; Zhang *et al.*, 2019). As a consequence, dietary restriction activates the conserved cap'n'collar transcription factor, SKN-1 in two neurons (called the ASI neurons) (Bishop & Guarente, 2007; Schmeisser *et al.*, 2013) to increase metabolic activity (Bishop & Guarente, 2007; Schmeisser *et al.*, 2013). SKN-1 is also a key regulator of the abundance of the ubiquitin-proteasome machinery within the animal (Lehrbach & Ruvkun, 2016; Li *et al.*, 2011; Steinbaugh *et al.*, 2015). Dietary restriction is thus known to increase lifespan through (amongst other mechanisms) the increase in proteasome activity and the E3 ubiquitin ligase WWP-1 (Carrano, Liu, Dillin, & Hunter, 2009). Remarkably, the smell of food is sufficient to reverse the benefits of dietary restriction and a recent study demonstrated rapid changes in protein degradation rates in response to food odors sensed by olfactory neurons (Finger *et al.*, 2019). Similarly changes in the levels of the neuromodulator dopamine control the flux of proteins by modulating protein degradation through the ubiquitin-proteasome system (Joshi, Matlack, & Rongo, 2016). Moreover, since growth and differentiation of cells within metazoa such as *C. elegans* occur in a coordinated manner, it is perhaps not surprising that neurons are not the only cell-type that exerts hierarchical control over the proteostasis machinery of another cell: cells also manage the proteostasis control systems of each other. For instance, germline stem cells limit the capacity of muscle cells, and perhaps other cells, to respond to heat stress in multiple tissues by inhibiting the amounts of PLANTS, in this case the regulatory particle non-ATPase 6 (RPN6) proteasome subunit that these distal cells are allowed to express (Sinha & Rae, 2014; Vilchez *et al.*, 2012). Similarly, specialized cells called coelomocytes play a key role in the clearance of aggregation-prone transthyretin expressed in *C. elegans* muscle (Madhivanan *et al.*, 2018), and protein degradation in the muscle, albeit driven by neuronal activity, is the integrated outcome of signals from neurons and other tissues such as hypodermis and intestine. In turn, distal cells modulate *controlled variables* to feedback on neurons and the rest of the animal (Chikka, Anbalagan, Dvorak, Dombeck, & Prahlad, 2016; Dalton & Curran, 2018; Minnerly *et al.*, 2017).

While neuronal networks confer the ability to respond rapidly to environmental threats, we and others have found that this ability comes at the expense of the 'normal' proteostasis machinery (Berendzen *et al.*, 2016; Cao & Aballay, 2016; David, 2013; El-Ami *et al.*, 2014; Imanikia, Ozbey, Krueger, Casanueva, & Taylor, 2019; Joshi *et al.*, 2016; Maman *et al.*, 2013; Moll, Ben-Gedalya, Reuveni, & Cohen, 2016; O'Brien *et al.*, 2018; Prahlad & Morimoto, 2011; Ray, Zhang, Rentas, Caldwell, & Caldwell, 2014; Roitenberg *et al.*, 2018; Taylor & Dillin, 2013) (Figure 2(D)). For example, the AFD-dependent circuity accelerates the activation of HSF1 upon sensing environmental stress but prevents HSF1 from being activated when cells chronically express aggregation prone proteins (Prahlad & Morimoto, 2011). Consequently, the activity of the AFD-thermosensory neurons needs to be inhibited in order to allow cells in the metazoan to activate HSF1 and increase the expression of CONTROLLERS needed to respond adequately to protein aggregation. Similarly, in wild-type animals, the presence of a G protein coupled receptor expressed on chemosensory neurons (*gtr-1*) is required for animals to increase chaperone expression upon exposure to heat, but its presence inhibits the 'normal' proteostasis machinery from rescuing the proteotoxicity of Alzheimer's-disease-linked aggregative peptide $A\beta_{3-42}$ (Maman *et al.*, 2013). A similar phenomenon is seen in studies on aging, a process that results in the systemic loss of protein quality control and the accumulation of misfolded proteins in various tissues: sensory neuronal function while allowing animals to respond to their environment, invariably inhibits the proteostasis machinery from acting on misfolded proteins (Boccitto, Lamitina, & Kalb, 2012; Broughton & Partridge, 2009; Burkewitz *et al.*, 2015; Byrne, Wilhelm, & Richly, 2017; Jiang *et al.*, 2015; Kumsta *et al.*, 2014; Stein & Murphy, 2012; Taylor & Dillin, 2013; Yin, Liu, Yuan, Jiang, & Cai, 2014; Zhang *et al.*, 2019). In these cases, the loss of normal neuronal control is required to allow protein misfolding to activate the normal proteostasis machinery, increase CONTROLLERS, and recalibrate the flow of proteins through the PLANTS to restore health and function (Figures 1 and 2). Why might such a seemingly counter-intuitive mechanism have evolved? We believe this is because neuronal signals have higher physiological priority because they orchestrate the protective response to life threatening environmental changes. Therefore, in a metazoan, the ability of neurons to antagonize the mechanisms that could normally be used by the proteostasis machinery of individual cells allows neurons to disengage this very same machinery for use upon the sensing of a threat. The control over the activation of transcription factors that are responsible for the expression of CONTROLLERS in all compartments of cells allows neurons to play this dominant role. In addition, such a control mechanism also forces the normal proteostasis machinery—the CONTROLLERS and the PLANTS involved in protein biogenesis—to function within a range of stringent set points under normal physiological conditions even through the regulated variable, protein misfolding, itself may not be well defined. Indeed, experiments in *C. elegans* have shown that the proteostasis machinery operates within a narrow range which is not easily perturbed (Gidalevitz *et al.*, 2006).

It is interesting to speculate on what could happen were cell intrinsic and cell extrinsic control mechanisms in disagreement. In human diseases such as Alzheimer's disease and Parkinson's disease, and in animal models of these diseases, cells accumulate misfolded and aggregated proteins but fail to *naturally* activate their cell-autonomous proteostasis response. This suggests that in the absence of external stimuli, the cell-extrinsic control dominates, preventing the increase in CONTROLLERS and PLANTS even at the cost of increasing the cellular burden of misfolded proteins. Indeed, experimentally increasing CONTROLLERS such as molecular chaperones in the affected cells, by transfection or other means, ameliorates misfolding and disease pathology. However, it is also telling that the cell intrinsic increase in the abundance of CONTROLLERS and PLANTS occurring in the apparent absence of cell extrinsic control is a *prerequisite* for cancer maintenance and malignancy in mammals (Dai, Whitesell, Rogers, & Lindquist, 2007; Gaglia *et al.*, 2020; Mendillo *et al.*, 2012; Scherz-Shouval *et al.*, 2014). Since the most obvious function of cephalic control over a community of cells, where each cell has autonomous control over its own proteostasis machinery, is the maintenance of harmony and co-ordination required for multicellular function, one might image that an anarchy of cell intrinsic control over that exerted by neurons may result in defects in development and differentiation, and other such organismal functions.

Is what we have learnt from *C. elegans* more generally true?

The ability to manipulate *C. elegans* neurons has uncovered a novel role for the nervous system in the control of protein quality control mechanisms. Dysfunction of proteostasis is responsible for the devastating age-associated neurodegenerative diseases and other disorders prevalent in our societies today. Therefore, that need to understand the extent to which neurohormonal signaling pathways influence cellular proteostasis in mammals is urgent. Interventions to activate protective protein quality control mechanisms in the treatment of protein conformational diseases have, to date, been based largely on the assumption that protein misfolding, and its subsequent correction, are cell autonomous processes: namely the cells that need to be targeted are the cells within which protein misfolding occurs (Balch *et al.*, 2008; Powers *et al.*, 2009). However, if protein folding homeostasis in all multicellular organisms is under the cell non-autonomous regulation of sensory neuronal modalities, another method for treatment of neurodegenerative diseases could involve modulation of neurosensory systems: a method that can be used alone, or in combination with small molecules to effectively target protein misfolding. Evidence for such control in mammals does exist, and is growing. Prior to the studies in *C. elegans*, restraint stress had been shown to induce expression of HSP70 mRNA in the adrenal cortex of the rat through the activity of the by the hypothalamic-pituitary-adrenal (HPA) axis (Blake, Buckley, & Buckley, 1993; Blake, Udelsman, Feulner, Norton, & Holbrook, 1991; Fawcett, Sylvester, Sarge, Morimoto, & Holbrook, 1994). Notably, the control of the proteostasis machinery of cells by the nervous system shares features with the systemic inhibition of acute inflammation seen in mammals (Tracey, 2002) and invertebrates including *C. elegans* (Anderson, Laurenson-Schafer, Partridge, Hodgkin, & McMullan, 2013; Singh & Aballay, 2012; Styer *et al.*, 2008; Sun, Liu, & Aballay, 2012; Sun, Singh, Kajino-Sakamoto, & Aballay, 2011; Yu, Zhi, Wu, Jing, & Wang, 2018), and treatments to modulate inflammatory processes are leveraging the existence of neuronal control. In mammals cholinergic activation within the efferent vagus nerve suppresses chronic inflammation, restricting the innate immune response to remain an acute response to local insults, and not flare up into chronic inflammation (Tracey, 2002). Moreover, as with the inflammatory response, recent experiments show that increased activity of the CONTROLLERS and PLANTS is protective only under a controlled regimen, and can be detrimental if unchecked (Styer *et al.*, 2008; Sun *et al.*, 2011). In mouse models of Alzheimer's Disease and Prion Disease it was the sustained activation of proteostasis machinery, rather than its dysfunction that accelerated neurodegeneration (Ma *et al.*, 2013; Moreno *et al.*, 2012). Protein misfolding within the cell, if detected and responded to by the CONTROLLERS elicits not repair, but also a change in the activity of the PLANTS such as a transient attenuation of protein translation through the phosphorylation of the α-subunit of eukaryotic initiation factor-2 (eIF2α) (Harding *et al.*, 2003), presumably to lower the burden of nascent polypeptides to an already stressed proteome (Ma *et al.*, 2013; Moreno *et al.*, 2012). This attenuation of translation has been shown to be toxic to neurons.

Thus, in retrospect, the finding that proteostasis is systemically controlled in a metazoan could perhaps have been anticipated. The maintenance of protein homeostasis ensures the presence of adequate amounts of functional proteins in a cell and is therefore directly responsible for cell growth, differentiation and function. It is possible that the nervous system, because of its role in regulating multicellular outcomes such as aging, behavior, metabolism, longevity and reproduction, also determines the set-points of misfolded proteins in the different cells of an organism maintaining them within some 'optimal' range. A better understanding of the mechanisms responsible for determining these set-points could go a long-way in treating protein conformation diseases.

Acknowledgements

We thank the members of V.P. laboratory and Dr. Tali Gidalevitz (Drexel University) for useful comments. We apologize to all our colleagues whose work we may have omitted to cite.

Disclosure statement

No potential conflict of interest was reported by the author(s).

Funding

This work was supported by NIH R01 AG 050653 (V.P.).

References

Akerfelt, M., Morimoto, R.I., & Sistonen, L. (2010). Heat shock factors: Integrators of cell stress, development and lifespan. *Nature Reviews. Molecular Cell Biology*, *11*(8), 545–555. doi:10.1038/nrm2938

Ananthan, J., Goldberg, A.L., & Voellmy, R. (1986). Abnormal proteins serve as eukaryotic stress signals and trigger the activation of heat shock genes. *Science (New York, N.Y.)*, *232*(4749), 522–524. doi:10.1126/science.3083508

Anderson, A., Laurenson-Schafer, H., Partridge, F.A., Hodgkin, J., & McMullan, R. (2013). Serotonergic chemosensory neurons modify the *C. elegans* immune response by regulating G-protein signaling in epithelial cells. *PLoS Pathogens*, *9*(12), e1003787. doi:10.1371/journal.ppat.1003787

Avery, L., Bargmann, C.I., & Horvitz, H.R. (1993). The *Caenorhabditis elegans* unc-31 gene affects multiple nervous system-controlled functions. *Genetics*, *134*(2), 455–464. Retrieved from http://www.ncbi.nlm.nih.gov/pubmed/8325482

Balch, W.E., Morimoto, R.I., Dillin, A., & Kelly, J.W. (2008). Adapting proteostasis for disease intervention. *Science (New York, N.Y.)*, *319*(5865), 916–919. doi:10.1126/science.1141448

Bargmann, C.I. (1998). Neurobiology of the *Caenorhabditis elegans* genome. *Science (New York, N.Y.)*, *282*(5396), 2028–2033. doi:10.1126/science.282.5396.2028

Bednarska, N.G., Schymkowitz, J., Rousseau, F., & Van Eldere, J. (2013). Protein aggregation in bacteria: the thin boundary between functionality and toxicity. *Microbiology (Reading, England)*, *159*(Pt 9), 1795–1806. doi:10.1099/mic.0.069575-0

Ben-Gedalya, T., & Cohen, E. (2012). Quality control compartments coming of age. *Traffic (Copenhagen, Denmark)*, *13*(5), 635–642. doi:10.1111/j.1600-0854.2012.01330.x

Berendzen, K.M., Durieux, J., Shao, L.-W., Tian, Y., Kim, H.-E., Wolff, S., … Dillin, A. (2016). Neuroendocrine coordination of mitochondrial stress signaling and proteostasis. *Cell*, *166*(6), 1553–1563 e1510. doi:10.1016/j.cell.2016.08.042

Bernales, S., Papa, F.R., & Walter, P. (2006). Intracellular signaling by the unfolded protein response. Annual Review of Cell and Developmental Biology, *22*, 487–508. doi:10.1146/annurev.cellbio.21.122303.120200

Bernard, C. (1965). An introduction to the study of experimental medicine. *Medical Journal of Australia*, *1*(4), 119–120. Retrieved from http://www.ncbi.nlm.nih.gov/pubmed/14248653

Beverly, M., Anbil, S., & Sengupta, P. (2011). Degeneracy and neuromodulation among thermosensory neurons contribute to robust thermosensory behaviors in *Caenorhabditis elegans*. *The Journal of Neuroscience : The Official Journal of the Society for Neuroscience*, *31*(32), 11718–11727. doi:10.1523/JNEUROSCI.1098-11.2011

Bishop, N.A., & Guarente, L. (2007). Two neurons mediate diet-restriction-induced longevity in *C. elegans*. *Nature*, *447*(7144), 545–549. doi:10.1038/nature05904

Blake, M.J., Buckley, D.J., & Buckley, A.R. (1993). Dopaminergic regulation of heat shock protein-70 expression in adrenal gland and aorta. *Endocrinology*, *132*(3), 1063–1070. doi:10.1210/endo.132.3.8095012

Blake, M.J., Udelsman, R., Feulner, G.J., Norton, D.D., & Holbrook, N.J. (1991). Stress-induced heat shock protein 70 expression in adrenal cortex: An adrenocorticotropic hormone-sensitive, age-dependent response. *Proceedings of the National Academy of Sciences of the United States of America*, *88*(21), 9873–9877. doi:10.1073/pnas.88.21.9873

Boccitto, M., Lamitina, T., & Kalb, R.G. (2012). Daf-2 signaling modifies mutant SOD1 toxicity in *C. elegans*. *PLoS One*, *7*(3), e33494. doi:10.1371/journal.pone.0033494

Brandman, O., Stewart-Ornstein, J., Wong, D., Larson, A., Williams, C.C., Li, G.-W., ... Weissman, J.S. (2012). A ribosome-bound quality control complex triggers degradation of nascent peptides and signals translation stress. *Cell*, *151*(5), 1042–1054. doi:10.1016/j.cell.2012.10.044

Broughton, S., & Partridge, L. (2009). Insulin/IGF-like signalling, the central nervous system and aging. *The Biochemical Journal*, *418*(1), 1–12. doi:10.1042/BJ20082102

Burkewitz, K., Morantte, I., Weir, H.J.M., Yeo, R., Zhang, Y., Huynh, F.K., ... Mair, W.B. (2015). Neuronal CRTC-1 governs systemic mitochondrial metabolism and lifespan via a catecholamine signal. *Cell*, *160*(5), 842–855. doi:10.1016/j.cell.2015.02.004

Byrne, J., Wilhelm, T., & Richly, H. (2017). Inhibition of neuronal autophagy mediates longevity. *Aging*, *9*(9), 1953–1954. doi:10.18632/aging.101297

Cao, X., & Aballay, A. (2016). Neural inhibition of dopaminergic signaling enhances immunity in a cell-non-autonomous manner. Current Biology, *26*(17), 2329–2334. doi:10.1016/j.cub.2016.06.036

Carrano, A.C., Liu, Z., Dillin, A., & Hunter, T. (2009). A conserved ubiquitination pathway determines longevity in response to diet restriction. *Nature*, *460*(7253), 396–399. doi:10.1038/nature08130

Chalfie, M., Sulston, J.E., White, J.G., Southgate, E., Thomson, J.N., & Brenner, S. (1985). The neural circuit for touch sensitivity in *Caenorhabditis elegans*. *The Journal of Neuroscience : The Official Journal of the Society for Neuroscience*, *5*(4), 956–964. doi:10.1523/JNEUROSCI.05-04-00956.1985

Chase, D.L., & Koelle, M.R. (2007). Biogenic amine neurotransmitters in C. elegans. *WormBook*, pp. 1–15. doi:10.1895/wormbook.1.132.1

Chikka, M.R., Anbalagan, C., Dvorak, K., Dombeck, K., & Prahlad, V. (2016). The mitochondria-regulated immune pathway activated in the *C. elegans* intestine is neuroprotective. *Cell Reports*, *16*(9), 2399–2414. doi:10.1016/j.celrep.2016.07.077

Clark, D.A., Biron, D., Sengupta, P., & Samuel, A.D. (2006). The AFD sensory neurons encode multiple functions underlying thermotactic behavior in *Caenorhabditis elegans*. *The Journal of Neuroscience : The Official Journal of the Society for Neuroscience*, *26*(28), 7444–7451. doi:10.1523/JNEUROSCI.1137-06.2006

Colosimo, M.E., Brown, A., Mukhopadhyay, S., Gabel, C., Lanjuin, A.E., Samuel, A.D., & Sengupta, P. (2004). Identification of thermosensory and olfactory neuron-specific genes via expression profiling of single neuron types. *Current Biology : CB*, *14*(24), 2245–2251. doi:10.1016/j.cub.2004.12.030

Cruz-Corchado, J., Ooi, F.K., Das, S., & Prahlad, V. (2020). Global transcriptome changes that accompany alterations in serotonin levels in *Caenorhabditis elegans*. *G3 (Bethesda)*, *10*, 1225–1246. doi:10.1534/g3.120.401088

Curran, K.P., & Chalasani, S.H. (2012). Serotonin circuits and anxiety: What can invertebrates teach us? *Invertebrate Neuroscience : IN*, *12*(2), 81–92. doi:10.1007/s10158-012-0140-y

Dai, C., Whitesell, L., Rogers, A.B., & Lindquist, S. (2007). Heat shock factor 1 is a powerful multifaceted modifier of carcinogenesis. *Cell*, *130*(6), 1005–1018. doi:10.1016/j.cell.2007.07.020

Dalton, H.M., & Curran, S.P. (2018). Hypodermal responses to protein synthesis inhibition induce systemic developmental arrest and AMPK-dependent survival in *Caenorhabditis elegans*. *PLoS Genetics*, *14*(7), e1007520. doi:10.1371/journal.pgen.1007520

Das, S., Ooi, F.K., Cruz Corchado, J., Fuller, L.C., Weiner, J.A., & Prahlad, V. (2020). Serotonin signaling by maternal neurons upon stress ensures progeny survival. eLife 2020;9:e55246. doi:10.7554/eLife.55246

David, R. (2013). Protein metabolism: Proteostasis goes global. Nature Reviews Molecular Cell Biology, *14*(8), 461. doi:10.1038/nrm3626

De Maio, A., Santoro, M.G., Tanguay, R.M., & Hightower, L.E. (2012). Ferruccio Ritossa's scientific legacy 50 years after his discovery of the heat shock response: A new view of biology, a new society, and a new journal. *Cell Stress Chaperones*, *17*(2), 139–143. doi:10.1007/s12192-012-0320-z

Durieux, J., Wolff, S., & Dillin, A. (2011). The cell-non-autonomous nature of electron transport chain-mediated longevity. *Cell*, *144*(1), 79–91. doi:10.1016/j.cell.2010.12.016

El-Ami, T., Moll, L., Carvalhal Marques, F., Volovik, Y., Reuveni, H., & Cohen, E. (2014). A novel inhibitor of the insulin/IGF signaling pathway protects from age-onset, neurodegeneration-linked proteotoxicity. *Aging Cell*, *13*(1), 165–174. doi:10.1111/acel.12171

Fawcett, T.W., Sylvester, S.L., Sarge, K.D., Morimoto, R.I., & Holbrook, N.J. (1994). Effects of neurohormonal stress and aging on the activation of mammalian heat shock factor 1. *The Journal of Biological Chemistry*, *269*(51), 32272–32278. Retrieved from http://www.ncbi.nlm.nih.gov/entrez/query.fcgi?cmd=Retrieve&db=PubMed&dopt=Citation&list_uids=7798227

Fields, P.A., Dong, Y., Meng, X., & Somero, G.N. (2015). Adaptations of protein structure and function to temperature: There is more than one way to 'skin a cat'. *The Journal of Experimental Biology*, *218*(Pt 12), 1801–1811. doi:10.1242/jeb.114298

Finger, F., Ottens, F., Springhorn, A., Drexel, T., Proksch, L., Metz, S., ... Hoppe, T. (2019). Olfaction regulates organismal proteostasis and longevity via microRNA-dependent signaling. *Nature Metabolism*, *1*(3), 350–359. doi:10.1038/s42255-019-0033-z

Frakes, A.E., Metcalf, M.G., Tronnes, S.U., Bar-Ziv, R., Durieux, J., Gildea, H.K., ... Dillin, A. (2020). Four glial cells regulate ER stress resistance and longevity via neuropeptide signaling in C. elegans. *Science (New York, N.Y.)*, *367*(6476), 436–440. doi:10.1126/science.aaz6896

Gaglia, G., Rashid, R., Yapp, C., Joshi, G.N., Li, C.G., Lindquist, S.L., ... Santagata, S. (2020). HSF1 phase transition mediates stress adaptation and cell fate decisions. *Nature Cell Biology*, *22*(2), 151–158. doi:10.1038/s41556-019-0458-3

Gally, C., & Bessereau, J.L. (2003). [*C. elegans*: of neurons and genes]. *Medecine Sciences : M/S*, *19*(6–7), 725–734. doi:10.1051/medsci/20031967725

Geiler-Samerotte, K.A., Dion, M.F., Budnik, B.A., Wang, S.M., Hartl, D.L., & Drummond, D.A. (2011). Misfolded proteins impose a dosage-dependent fitness cost and trigger a cytosolic unfolded protein response in yeast. *Proceedings of the National Academy of Sciences of the United States of America*, *108*(2), 680–685. doi:10.1073/pnas.1017570108

Gething, M.J., & Sambrook, J. (1992). Protein folding in the cell. *Nature*, *355*(6355), 33–45. doi:10.1038/355033a0

Gidalevitz, T., Ben-Zvi, A., Ho, K.H., Brignull, H.R., & Morimoto, R.I. (2006). Progressive disruption of cellular protein folding in models of polyglutamine diseases. *Science (New York, N.Y.)*, *311*(5766), 1471–1474. doi:10.1126/science.1124514

Goff, S.A., & Goldberg, A.L. (1985). Production of abnormal proteins in *E. coli* stimulates transcription of lon and other heat shock genes. *Cell*, *41*(2), 587–595. doi:10.1016/S0092-8674(85)80031-3

Goff, S.A., & Goldberg, A.L. (1987). An increased content of protease La, the lon gene product, increases protein degradation and blocks growth in *Escherichia coli*. *The Journal of Biological Chemistry*, *262*(10), 4508–4515. Retrieved from http://www.ncbi.nlm.nih.gov/pubmed/3549709

Goldberg, A.L. (1971). A role of aminoacyl-tRNA in the regulation of protein breakdown in *Escherichia coli*. *Proceedings of the National Academy of Sciences of the United States of America*, *68*(2), 362–366. doi:10.1073/pnas.68.2.362

Goldberg, A.L. (1972). Correlation between rates of degradation of bacterial proteins in vivo and their sensitivity to proteases. *Proceedings of the National Academy of Sciences of the United States of America*, *69*(9), 2640–2644. doi:10.1073/pnas.69.9.2640

Goldstein, D.S., & Kopin, I.J. (2017). Homeostatic systems, biocybernetics, and autonomic neuroscience. *Autonomic Neuroscience : Basic & Clinical*, *208*, 15–28. doi:10.1016/j.autneu.2017.09.001

Gomez-Pastor, R., Burchfiel, E.T., & Thiele, D.J. (2018). Regulation of heat shock transcription factors and their roles in physiology and disease. *Nature Reviews. Molecular Cell Biology*, *19*(1), 4–19. doi:10.1038/nrm.2017.73

Goodman, M.B., & Sengupta, P. (2018). The extraordinary AFD thermosensor of *C. elegans*. *Pflugers Archiv : European Journal of Physiology*, *470*(5), 839–849. doi:10.1007/s00424-017-2089-5

Goodman, M.B., & Sengupta, P. (2019). How *Caenorhabditis elegans* senses mechanical stress, temperature, and other physical stimuli. *Genetics*, *212*(1), 25–51. doi:10.1534/genetics.118.300241

Harding, H.P., Zhang, Y., Zeng, H., Novoa, I., Lu, P.D., Calfon, M., … Ron, D. (2003). An integrated stress response regulates amino acid metabolism and resistance to oxidative stress. *Molecular Cell*, *11*(3), 619–633. doi:10.1016/S1097-2765(03)00105-9

Haynes, C.M., & Ron, D. (2010). The mitochondrial UPR - protecting organelle protein homeostasis. *Journal of Cell Science*, *123*(Pt 22), 3849–3855. doi:10.1242/jcs.075119

Hightower, L.E. (1980). Cultured animal cells exposed to amino acid analogues or puromycin rapidly synthesize several polypeptides. *Journal of Cellular Physiology*, *102*(3), 407–427. doi:10.1002/jcp.1041020315

Hipp, M.S., Park, S.H., & Hartl, F.U. (2014). Proteostasis impairment in protein-misfolding and -aggregation diseases. *Trends in Cell Biology*, *24*(9), 506–514. doi:10.1016/j.tcb.2014.05.003

Hobert, O. (2005). Specification of the nervous system. In *WormBook* (pp. 1–19). WormBook, ed. The C. elegans Research Community, doi:10.1895/wormbook.

Hofmann, G., & Somero, G. (1995). Evidence for protein damage at environmental temperatures: seasonal changes in levels of ubiquitin conjugates and hsp70 in the intertidal mussel *Mytilus trossulus*. *The Journal of Experimental Biology*, *198*(Pt 7), 1509–1518. Retrieved from http://www.ncbi.nlm.nih.gov/entrez/query.fcgi?cmd=Retrieve&db=PubMed&dopt=Citation&list_uids=9319406

Iberall, A.S., & Cardon, S.Z. (1964). Control in Biological systems-a physical review. *Annals of the New York Academy of Sciences*, *117*, 445–518. doi:10.1111/j.1749-6632.1964.tb48202.x

Imanikia, S., Ozbey, N.P., Krueger, C., Casanueva, M.O., & Taylor, R.C. (2019). Neuronal XBP-1 activates intestinal lysosomes to improve proteostasis in *C. elegans*. *Current Biology : CB*, *29*(14), 2322–2338 e2327. doi:10.1016/j.cub.2019.06.031

Jayaraj, G.G., Hipp, M.S., & Hartl, F.U. (2020). Functional modules of the proteostasis network. *Cold Spring Harbor Perspectives in Biology*, *12*(1), a033951. doi:10.1101/cshperspect

Jiang, H.C., Hsu, J.M., Yen, C.P., Chao, C.C., Chen, R.H., & Pan, C.L. (2015). Neural activity and CaMKII protect mitochondria from fragmentation in aging *Caenorhabditis elegans* neurons. *Proceedings of the National Academy of Sciences of the United States of America*, *112*(28), 8768–8773. doi:10.1073/pnas.1501831112

Joshi, K.K., Matlack, T.L., & Rongo, C. (2016). Dopamine signaling promotes the xenobiotic stress response and protein homeostasis. *The EMBO Journal*, *35*(17), 1885–1901. doi:10.15252/embj.201592524

Kandror, O., Busconi, L., Sherman, M., & Goldberg, A.L. (1994). Rapid degradation of an abnormal protein in Escherichia coli involves the chaperones GroEL and GroES. *The Journal of Biological Chemistry*, *269*(38), 23575–23582. Retrieved from http://www.ncbi.nlm.nih.gov/pubmed/7916344

Kawli, T., Wu, C., & Tan, M.W. (2010). Systemic and cell intrinsic roles of Gqalpha signaling in the regulation of innate immunity, oxidative stress, and longevity in *Caenorhabditis elegans*. *Proceedings of the National Academy of Sciences of the United States of America*, *107*(31), 13788–13793. doi:10.1073/pnas.0914715107

Knowles, S.E., & Ballard, F.J. (1976). Selective control of the degradation of normal and aberrant proteins in Reuber H35 hepatoma cells. *The Biochemical Journal*, *156*(3), 609–617. doi:10.1042/bj1560609

Knowles, S.E., & Ballard, F.J. (1978). Effects of amino acid analogues on protein synthesis and degradation in isolated cells. *The British Journal of Nutrition*, *40*(2), 275–287. doi:10.1079/bjn19780123

Knowles, S.E., Gunn, J.M., Hanson, R.W., & Ballard, F.J. (1975). Increased degradation rates of protein synthesized in hepatoma cells in the presence of amino acid analogues. *The Biochemical Journal*, *146*(3), 595–600. doi:10.1042/bj1460595

Kostova, K.K., Hickey, K.L., Osuna, B.A., Hussmann, J.A., Frost, A., Weinberg, D.E., & Weissman, J.S. (2017). CAT-tailing as a fail-safe mechanism for efficient degradation of stalled nascent polypeptides. *Science (New York, N.Y.)*, *357*(6349), 414–417. doi:10.1126/science.aam7787

Kozutsumi, Y., Segal, M., Normington, K., Gething, M.J., & Sambrook, J. (1988). The presence of malfolded proteins in the endoplasmic reticulum signals the induction of glucose-regulated proteins. *Nature*, *332*(6163), 462–464. doi:10.1038/332462a0

Kumsta, C., Ching, T.-T., Nishimura, M., Davis, A.E., Gelino, S., Catan, H.H., … Hansen, M. (2014). Integrin-linked kinase modulates longevity and thermotolerance in *C. elegans* through neuronal control of HSF-1. *Aging Cell*, *13*(3), 419–430. doi:10.1111/acel.12189

Labbadia, J., & Morimoto, R.I. (2015). The biology of proteostasis in aging and disease. *Annual Review of Biochemistry*, *84*, 435–464. doi:10.1146/annurev-biochem-060614-033955

Lechan, R.M., & Fekete, C. (2006). The TRH neuron: A hypothalamic integrator of energy metabolism. *Progress in Brain Research*, *153*, 209–235. doi:10.1016/S0079-6123(06)53012-2

Lehrbach, N.J., & Ruvkun, G. (2016). Proteasome dysfunction triggers activation of SKN-1A/Nrfl by the aspartic protease DDI-1. *eLife*, *5*, 7554. doi:10.7554/eLife.17721

Leow, M.K. (2007). A mathematical model of pituitary-thyroid interaction to provide an insight into the nature of the thyrotropin-thyroid hormone relationship. *Journal of Theoretical Biology*, *248*(2), 275–287. doi:10.1016/j.jtbi.2007.05.016

Li, X., Matilainen, O., Jin, C., Glover-Cutter, K.M., Holmberg, C.I., & Blackwell, T.K. (2011). Specific SKN-1/Nrf stress responses to perturbations in translation elongation and proteasome activity. *PLoS Genetics*, *7*(6), e1002119pdoi:10.1371/journal.pgen.1002119

Lindquist, S. (1986). The heat-shock response. *Annual Review of Biochemistry*, *55*, 1151–1191. doi:10.1146/annurev.bi.55.070186.005443

Ma, T., Trinh, M.A., Wexler, A.J., Bourbon, C., Gatti, E., Pierre, P., … Klann, E. (2013). Suppression of eIF2α kinases alleviates Alzheimer's disease-related plasticity and memory deficits . *Nature Neuroscience*, *16*(9), 1299–1305. doi:10.1038/nn.3486

Madhivanan, K., Greiner, E.R., Alves-Ferreira, M., Soriano-Castell, D., Rouzbeh, N., Aguirre, C.A., … Encalada, S.E. (2018). Cellular clearance of circulating transthyretin decreases cell-nonautonomous proteotoxicity in *Caenorhabditis elegans*. *Proceedings of the National Academy of Sciences of the United States of America*, *115*(33), E7710–E7719. doi:10.1073/pnas.1801117115

Maman, M., Carvalhal Marques, F., Volovik, Y., Dubnikov, T., Bejerano-Sagie, M., & Cohen, E. (2013). A neuronal GPCR is critical for the induction of the heat shock response in the nematode C. elegans. *The Journal of Neuroscience : The Official Journal of the Society for Neuroscience*, *33*(14), 6102–6111. doi:10.1523/JNEUROSCI.4023-12.2013

Mattes, R.D. (1997). Physiologic responses to sensory stimulation by food: Nutritional implications. *Journal of the American Dietetic Association*, *97*(4), 406–413. doi:10.1016/S0002-8223(97)00101-6

Mattoo, R.U., & Goloubinoff, P. (2014). Molecular chaperones are nanomachines that catalytically unfold misfolded and alternatively folded proteins. *Cellular and Molecular Life Sciences : Cmls*, *71*(17), 3311–3325. doi:10.1007/s00018-014-1627-y

Mendillo, M.L., Santagata, S., Koeva, M., Bell, G.W., Hu, R., Tamimi, R.M., … Lindquist, S. (2012). HSF1 drives a transcriptional program distinct from heat shock to support highly malignant human cancers. *Cell*, *150*(3), 549–562. doi:10.1016/j.cell.2012.06.031

Miles, J., Scherz-Shouval, R., & van Oosten-Hawle, P. (2019). Expanding the organismal proteostasis network: Linking systemic stress signaling with the innate immune response. *Trends in Biochemical Sciences*, *44*(11), 927–942. doi:10.1016/j.tibs.2019.06.009

Miller, S.B.M., Ho, C.-T., Winkler, J., Khokhrina, M., Neuner, A., Mohamed, M.Y.H., … Bukau, B. (2015). Compartment-specific aggregases direct distinct nuclear and cytoplasmic aggregate deposition. *The EMBO Journal*, *34*(6), 778–797. doi:10.15252/embj.201489524

Minnerly, J., Zhang, J., Parker, T., Kaul, T., & Jia, K. (2017). The cell non-autonomous function of ATG-18 is essential for neuroendocrine regulation of *Caenorhabditis elegans* lifespan. *PLoS Genetics, 13*(5), e1006764. doi:10.1371/journal.pgen.1006764

Mogk, A., Bukau, B., & Kampinga, H.H. (2018). Cellular handling of protein aggregates by disaggregation machines. *Molecular Cell, 69*(2), 214–226. doi:10.1016/j.molcel.2018.01.004

Moll, L., Ben-Gedalya, T., Reuveni, H., & Cohen, E. (2016). The inhibition of IGF-1 signaling promotes proteostasis by enhancing protein aggregation and deposition. *FASEB Journal : Official Publication of the Federation of American Societies for Experimental Biology, 30*(4), 1656–1669. doi:10.1096/fj.15-281675

Morano, K.A., Liu, P.C., & Thiele, D.J. (1998). Protein chaperones and the heat shock response in Saccharomyces cerevisiae. *Current Opinion in Microbiology, 1*(2), 197–203. doi:10.1016/S1369-5274(98)80011-8

Moreno, J.A., Radford, H., Peretti, D., Steinert, J.R., Verity, N., Martin, M.G., ... Mallucci, G.R. (2012). Sustained translational repression by eIF2α-P mediates prion neurodegeneration . *Nature, 485*(7399), 507–511. doi:10.1038/nature11058

Mori, K. (2009). Signalling pathways in the unfolded protein response: Development from yeast to mammals. *Journal of Biochemistry, 146*(6), 743–750. doi:10.1093/jb/mvp166

Mori, K., Kawahara, T., Yoshida, H., Yanagi, H., & Yura, T. (1996). Signalling from endoplasmic reticulum to nucleus: Transcription factor with a basic-leucine zipper motif is required for the unfolded protein-response pathway. *Genes to Cells : Devoted to Molecular & Cellular Mechanisms, 1*(9), 803–817. doi:10.1046/j.1365-2443.1996.d01-274.x

Morimoto, R.I. (1998). Regulation of the heat shock transcriptional response: Cross talk between a family of heat shock factors, molecular chaperones, and negative regulators. *Genes & Development, 12*(24), 3788–3796. doi:10.1101/gad.12.24.3788

Munro, S., & Pelham, H.R. (1986). An Hsp70-like protein in the ER: Identity with the 78 kd glucose-regulated protein and immunoglobulin heavy chain binding protein. *Cell, 46*(2), 291–300. doi:10.1016/0092-8674(86)90746-4

Munro, S., & Pelham, H.R. (1987). A C-terminal signal prevents secretion of luminal ER proteins. *Cell, 48*(5), 899–907. doi:10.1016/0092-8674(87)90086-9

Neef, D.W., Turski, M.L., & Thiele, D.J. (2010). Modulation of heat shock transcription factor 1 as a therapeutic target for small molecule intervention in neurodegenerative disease. *PLoS Biology, 8*(1), e1000291. doi:10.1371/journal.pbio.1000291

O'Brien, D., Jones, L.M., Good, S., Miles, J., Vijayabaskar, M.S., Aston, R., ... van Oosten-Hawle, P. (2018). A PQM-1-mediated response triggers transcellular chaperone signaling and regulates organismal proteostasis. *Cell Reports, 23*(13), 3905–3919. doi:10.1016/j.celrep.2018.05.093

Ooi, F.K., & Prahlad, V. (2017). Olfactory experience primes the heat shock transcription factor HSF-1 to enhance the expression of molecular chaperones in *C. elegans. Science Signaling, 10*(501). eaan4893. doi:10.1126/scisignal.aan4893

Pelham, H.R., & Munro, S. (1993). Sorting of membrane proteins in the secretory pathway. *Cell, 75*(4), 603–605. doi:10.1016/0092-8674(93)90479-A

Powers, E.T., Morimoto, R.I., Dillin, A., Kelly, J.W., & Balch, W.E. (2009). Biological and chemical approaches to diseases of proteostasis deficiency. Annual Review of Biochemistry, *78*, 959–991. doi:10.1146/annurev.biochem.052308.114844

Prahlad, V., Cornelius, T., & Morimoto, R.I. (2008). Regulation of the cellular heat shock response in *Caenorhabditis elegans* by thermosensory neurons. *Science (New York, N.Y.), 320*(5877), 811–814. doi:10.1126/science.1156093

Prahlad, V., & Morimoto, R.I. (2011). Neuronal circuitry regulates the response of *Caenorhabditis elegans* to misfolded proteins. *Proceedings of the National Academy of Sciences of the United States of America, 108*(34), 14204–14209. doi:10.1073/pnas.1106557108

Prouty, W.F., Karnovsky, M.J., & Goldberg, A.L. (1975). Degradation of abnormal proteins in *Escherichia coli.* Formation of protein inclusions in cells exposed to amino acid analogs. *The Journal of Biological Chemistry, 250*(3), 1112–1122. Retrieved from http://www.ncbi.nlm.nih.gov/pubmed/1089651

Ray, A., Zhang, S., Rentas, C., Caldwell, K.A., & Caldwell, G.A. (2014). RTCB-1 mediates neuroprotection via XBP-1 mRNA splicing in the unfolded protein response pathway. *The Journal of Neuroscience : The Official Journal of the Society for Neuroscience, 34*(48), 16076–16085. doi:10.1523/JNEUROSCI.1945-14.2014

Ritossa, F. (1996). Discovery of the heat shock response. *Cell Stress & Chaperones, 1*(2), 97–98. doi:10.1379/1466-1268(1996)001<0097:DOTHSR>2.3.CO;2

Roitenberg, N., Bejerano-Sagie, M., Boocholez, H., Moll, L., Marques, F.C., Golodetzki, L., ... Cohen, E. (2018). Modulation of caveolae by insulin/IGF-1 signaling regulates aging of *Caenorhabditis elegans. EMBO Reports, 19*(8). e45673. doi:10.15252/embr.201745673

Scherz-Shouval, R., Santagata, S., Mendillo, M.L., Sholl, L.M., Ben-Aharon, I., Beck, A.H., ... Lindquist, S. (2014). The reprogramming of tumor stroma by HSF1 is a potent enabler of malignancy. *Cell, 158*(3), 564–578. doi:10.1016/j.cell.2014.05.045

Schmeisser, S., Priebe, S., Groth, M., Monajembashi, S., Hemmerich, P., Guthke, R., ... Ristow, M. (2013). Neuronal ROS signaling rather than AMPK/sirtuin-mediated energy sensing links dietary restriction to lifespan extension. *Molecular Metabolism, 2*(2), 92–102. doi:10.1016/j.molmet.2013.02.002

Silverman, R.M., Cummings, E.E., O'Reilly, L.P., Miedel, M.T., Silverman, G.A., Luke, C.J., ... Pak, S.C. (2015). The aggregation-prone intracellular serpin SRP-2 fails to transit the ER in *Caenorhabditis elegans. Genetics, 200*(1), 207–219. doi:10.1534/genetics.115.176180

Singh, V., & Aballay, A. (2012). Endoplasmic reticulum stress pathway required for immune homeostasis is neurally controlled by arrestin-1. *The Journal of Biological Chemistry, 287*(40), 33191–33197. doi:10.1074/jbc.M112.398362

Sinha, A., & Rae, R. (2014). A functional genomic screen for evolutionarily conserved genes required for lifespan and immunity in germline-deficient C. elegans. *PLoS One, 9*(8), e101970. doi:10.1371/journal.pone.0101970

Sitron, C.S., & Brandman, O. (2019). CAT tails drive degradation of stalled polypeptides on and off the ribosome. Nature Structural & Molecular Biology, *26*(6), 450–459. doi:10.1038/s41594-019-0230-1

Sitron, C.S., Park, J.H., & Brandman, O. (2017). Asc1, Hel2, and Slh1 couple translation arrest to nascent chain degradation. *RNA (New York, N.Y.), 23*(5), 798–810. doi:10.1261/rna.060897.117

Sitron, C.S., Park, J.H., Giafaglione, J.M., & Brandman, O. (2020). Aggregation of CAT tails blocks their degradation and causes proteotoxicity in *S. cerevisiae. PLoS One, 15*(1), e0227841. doi:10.1371/journal.pone.0227841

Somero, G.N. (1995). Proteins and temperature. Annual Review of Physiology, *57*, 43–68. doi:10.1146/annurev.ph.57.030195.000355

Stein, G.M., & Murphy, C.T. (2012). The intersection of aging, longevity pathways, and learning and memory in *C. elegans. Frontiers in Genetics, 3*, 259. doi:10.3389/fgene.2012.00259

Steinbaugh, M.J., Narasimhan, S.D., Robida-Stubbs, S., Moronetti Mazzeo, L.E., Dreyfuss, J.M., Hourihan, J.M., ... Blackwell, T.K. (2015). Lipid-mediated regulation of SKN-1/Nrf in response to germ cell absence. eLife 2015;4:e07836. doi:10.7554/eLife.07836

Styer, K.L., Singh, V., Macosko, E., Steele, S.E., Bargmann, C.I., & Aballay, A. (2008). Innate immunity in *Caenorhabditis elegans* is regulated by neurons expressing NPR-1/GPCR. *Science (New York, N.Y.), 322*(5900), 460–464. doi:10.1126/science.1163673

Sun, J., Liu, Y., & Aballay, A. (2012). Organismal regulation of XBP-1-mediated unfolded protein response during development and immune activation. *EMBO Reports, 13*(9), 855–860. doi:10.1038/embor.2012.100

Sun, J., Singh, V., Kajino-Sakamoto, R., & Aballay, A., (2011). Neuronal GPCR controls innate immunity by regulating noncanonical unfolded protein response genes. *Science (New York, N.Y.), 332*(6030), 729–732. doi:10.1126/science.1203411

Tatum, M.C., Ooi, F.K., Chikka, M.R., Chauve, L., Martinez-Velazquez, L.A., Steinbusch, H.W.M., ... Prahlad, V. (2015). Neuronal serotonin release triggers the heat shock response in *C. elegans* in the absence of temperature increase. Current Biology : CB, *25*(2), 163–174. doi:10.1016/j.cub.2014.11.040

Taylor, R.C., Berendzen, K.M., & Dillin, A. (2014). Systemic stress signalling: Understanding the cell non-autonomous control of proteostasis. *Nature Reviews. Molecular Cell Biology*, *15*(3), 211–217. doi: 10.1038/nrm3752

Taylor, R.C., & Dillin, A. (2013). XBP-1 is a cell-nonautonomous regulator of stress resistance and longevity. *Cell*, *153*(7), 1435–1447. doi: 10.1016/j.cell.2013.05.042

Tracey, K.J. (2002). The inflammatory reflex. *Nature*, *420*(6917), 853–859. doi:10.1038/nature01321

Tyedmers, J., Mogk, A., & Bukau, B. (2010). Cellular strategies for controlling protein aggregation. *Nature Reviews. Molecular Cell Biology*, *11*(11), 777–788. doi:10.1038/nrm2993

van Oosten-Hawle, P., Porter, R.S., & Morimoto, R.I. (2013). Regulation of organismal proteostasis by transcellular chaperone signaling. *Cell*, *153*(6), 1366–1378. doi:10.1016/j.cell.2013.05.015

Vendruscolo, M., Knowles, T.P., & Dobson, C.M. (2011). Protein solubility and protein homeostasis: a generic view of protein misfolding disorders. *Cold Spring Harbor Perspectives in Biology*, *3*(12), a010454. doi:10.1101/cshperspect.a010454

Vihervaara, A., Duarte, F.M., & Lis, J.T. (2018). Molecular mechanisms driving transcriptional stress responses. *Nature Reviews. Genetics*, *19*(6), 385–397. doi:10.1038/s41576-018-0001-6

Vilchez, D., Morantte, I., Liu, Z., Douglas, P.M., Merkwirth, C., Rodrigues, A.P.C., ... Dillin, A. (2012). RPN-6 determines *C. elegans* longevity under proteotoxic stress conditions. *Nature*, *489*(7415), 263–268. doi:10.1038/nature11315

Voellmy, R. (1994). Transduction of the stress signal and mechanisms of transcriptional regulation of heat shock/stress protein gene expression in higher eukaryotes. *Critical Reviews in Eukaryotic Gene Expression*, *4*(4), 357–401. Retrieved from http://www.ncbi.nlm.nih.gov/pubmed/7734836

Voellmy, R. (1996). Sensing stress and responding to stress. *EXS*, *77*, 121–137. doi:10.1007/978-3-0348-9088-5_9

Walrond, J.P., Kass, I.S., Stretton, A.O., & Donmoyer, J.E. (1985). Identification of excitatory and inhibitory motoneurons in the nematode Ascaris by electrophysiological techniques. *The Journal of Neuroscience : The Official Journal of the Society for Neuroscience*, *5*(1), 1–8. doi:10.1523/JNEUROSCI.05-01-00001.1985

Walter, P. (2010). Walking along the serendipitous path of discovery. *Molecular Biology of the Cell*, *21*(1), 15–17. doi:10.1091/mbc.E09-08-0662

Walther, D.M., Kasturi, P., Zheng, M., Pinkert, S., Vecchi, G., Ciryam, P., ... Hartl, F.U. (2015). Widespread Proteome Remodeling and Aggregation in Aging *C. elegans*. *Cell*, *161*(4), 919–932. doi:10.1016/j.cell.2015.03.032

Weir, H.J., Yao, P., Huynh, F.K., Escoubas, C.C., Goncalves, R.L., Burkewitz, K., ... Mair, W.B. (2017). Dietary restriction and AMPK increase lifespan via mitochondrial network and peroxisome remodeling. *Cell Metabolism*, *26*(6), 884–896 e885. doi:10.1016/j.cmet.2017.09.024

White, J.G., Southgate, E., Thomson, J.N., & Brenner, S. (1983). Factors that determine connectivity in the nervous system of *Caenorhabditis elegans*. *Cold Spring Harbor Symposia on Quantitative Biology*, *48 Pt 2*, 633–640. doi:10.1101/sqb.1983.048.01.067

White, J.G., Southgate, E., Thomson, J.N., & Brenner, S. (1986). The structure of the nervous system of the nematode *Caenorhabditis elegans*. *Philosophical Transactions of the Royal Society of London. Series B, Biological Sciences*, *314*(1165), 1–340. doi:10.1098/rstb.1986.0056

Wolff, S., Weissman, J.S., & Dillin, A. (2014). Differential scales of protein quality control. *Cell*, *157*(1), 52–64. doi:10.1016/j.cell.2014.03.007

Wolkow, C.A., Kimura, K.D., Lee, M.S., & Ruvkun, G. (2000). Regulation of *C. elegans* life-span by insulinlike signaling in the nervous system. *Science (New York, N.Y.)*, *290*(5489), 147–150. doi:10.1126/science.290.5489.147

Yin, J.A., Liu, X.J., Yuan, J., Jiang, J., & Cai, S.Q. (2014). Longevity manipulations differentially affect serotonin/dopamine level and behavioral deterioration in aging *Caenorhabditis elegans*. *The Journal of Neuroscience : The Official Journal of the Society for Neuroscience*, *34*(11), 3947–3958. doi:10.1523/JNEUROSCI.4013-13.2014

Yoshida, H., Matsui, T., Yamamoto, A., Okada, T., & Mori, K. (2001). XBP1 mRNA is induced by ATF6 and spliced by IRE1 in response to ER stress to produce a highly active transcription factor. *Cell*, *107*(7), 881–891. doi:10.1016/S0092-8674(01)00611-0

Yu, Y., Zhi, L., Wu, Q., Jing, L., & Wang, D. (2018). NPR-9 regulates the innate immune response in *Caenorhabditis elegans* by antagonizing the activity of AIB interneurons. *Cellular & Molecular Immunology*, *15*(1), 27–37. doi:10.1038/cmi.2016.8

Zhang, Y., Lanjuin, A., Chowdhury, S.R., Mistry, M., Silva-García, C.G., Weir, H.J., ... Mair, W.B. (2019). Neuronal TORC1 modulates longevity via AMPK and cell nonautonomous regulation of mitochondrial dynamics in *C. elegans*. eLife 2019;8:e49158 doi:10.7554/eLife.49158

Zhang, Q., Wu, X., Chen, P., Liu, L., Xin, N., Tian, Y., & Dillin, A. (2018). The Mitochondrial unfolded protein response is mediated cell-non-autonomously by retromer-dependent Wnt signaling. *Cell*, *174*(4), 870–883 e817. doi:10.1016/j.cell.2018.06.029

OPEN ACCESS

Host-microbe interactions and the behavior of *Caenorhabditis elegans*

Dennis H. Kim and Steven W. Flavell

ABSTRACT

Microbes are ubiquitous in the natural environment of *Caenorhabditis elegans*. Bacteria serve as a food source for *C. elegans* but may also cause infection in the nematode host. The sensory nervous system of *C. elegans* detects diverse microbial molecules, ranging from metabolites produced by broad classes of bacteria to molecules synthesized by specific strains of bacteria. Innate recognition through chemosensation of bacterial metabolites or mechanosensation of bacteria can induce immediate behavioral responses. The ingestion of nutritive or pathogenic bacteria can modulate internal states that underlie long-lasting behavioral changes. Ingestion of nutritive bacteria leads to learned attraction and exploitation of the bacterial food source. Infection, which is accompanied by activation of innate immunity, stress responses, and host damage, leads to the development of aversive behavior. The integration of a multitude of microbial sensory cues in the environment is shaped by experience and context. Genetic, chemical, and neuronal studies of *C. elegans* behavior in the presence of bacteria have defined neural circuits and neuromodulatory systems that shape innate and learned behavioral responses to microbial cues. These studies have revealed the profound influence that host-microbe interactions have in governing the behavior of this simple animal host.

Microbiology

Caenorhabditis elegans lives in a microbe-rich environment that defines the ecology and has shaped the evolution of the organism (Schulenburg & Félix, 2017). The basic dichotomy for *C. elegans* in its interactions with this microbial community is that bacteria are an essential source of nutrition but may also be pathogenic and cause infection and death. Ecological survey and sequence analysis of bacteria isolated from the natural environment from which *C. elegans* are isolated has revealed a community of *Proteobacteria*, *Bacteroidetes*, *Firmicutes*, and *Actinobacteria* (Samuel, Rowedder, Braendle, Félix, & Ruvkun, 2016), with notable enrichment of alpha-*Proteobacteria* genera such as *Ochrobactrum* and *Pseudomonas* in close association with *C. elegans*, likely colonizing the intestine (Dirksen *et al.*, 2016). These associated bacteria may benefit the *C. elegans* host as a food source and in other ways. At the same time, the characterization of *C. elegans* strains in the wild has uncovered a broad range of pathogenic microorganisms, including bacteria, fungi, and viruses, which can cause sickness and death (Schulenburg & Félix, 2017).

The experimental study of *C. elegans* has typically involved laboratory cultivation on monoaxenic lawns of *Escherichia coli* OP50 seeded on agar plates supplemented with cholesterol (Brenner, 1974). Genetic and metabolomic characterization of alternative bacterial food sources for *C. elegans*, such as *Comamonas*, *Bacillus subtilis* and mutants of *E. coli*, has defined conserved requirements for micronutrients (Qi, Kniazeva, & Han, 2017; Watson *et al.*, 2014), novel mechanisms of host co-option of bacterial siderophores for iron acquisition (Qi & Han, 2018), and metabolic determinants of bacteria that can influence complex phenotypes such as lifespan (Han *et al.*, 2017; Qi & Han, 2018; Saiki *et al.*, 2008; Virk *et al.*, 2012). Semi-defined axenic media has been developed for growth and cultivation of *C. elegans*, but live bacteria support optimal growth and development (Lenaerts, Walker, Van Hoorebeke, Gems, & Vanfleteren, 2008).

Diverse bacteria are pathogenic to *C. elegans*. The human opportunistic pathogen *Pseudomonas aeruginosa*, which resides in soil and water, was shown to infect an evolutionarily diverse range of hosts, including *C. elegans* (Rahme *et al.*, 1995; Tan, Mahajan-Miklos, & Ausubel, 1999). *Bacillus thurigiensis* produces a crystal pore-forming toxin that is highly toxic to *C. elegans* upon ingestion (Marroquin, Elyassnia, Griffitts, Feitelson, & Aroian, 2000) and has attracted interest as a potential biocontrol method for nematodes that are pathogenic to animals and plants. The coryneform bacterium *Microbacterium nematophilum* causes a distinct mode of infection and host response through adherence to the rectal and post-anal cuticle, likely reflecting a natural infection of *C. elegans* (Hodgkin, Kuwabara, & Corneliussen, 2000). A broad range of bacterial species, including environmental isolates and human pathogens, exhibit increased virulence towards *C. elegans*, compared with the survival of *C. elegans* on *E. coli* OP50 (Couillault & Ewbank, 2002). Experimental conditions and host status play an important role when considering the pathogenicity of a bacterial strain. As an illustration of this point, even *E. coli* OP50 can be considered as pathogenic in the presence of richer growth media or when colonizing aging or feeding-defective animals (Garigan *et al.*, 2002; Garsin *et al.*, 2001; Herndon *et al.*, 2002; Kumar *et al.*, 2019). Pathogenicity can also be altered by co-ingestion of non-pathogenic bacteria (Montalvo-Katz, Huang, Appel, Berg, & Shapira, 2013; Samuel *et al.*, 2016). For example, factors secreted by *Enterococcus faecium* are protective against *Salmonella* pathogenesis (Rangan *et al.*, 2016). The range of bacteria that can be pathogenic to *C. elegans* in the laboratory setting has been expanded by molecular engineering, for example by having *E. coli* strains carry plasmids expressing specific toxins of *B. thurigiensis* (Wei *et al.*, 2003) or *P. aeruginosa* (McEwan, Kirienko, & Ausubel, 2012), or even RNAi clones targeting essential *C. elegans* genes (Kamath *et al.*, 2003).

Innate recognition

The innate recognition of bacteria by *C. elegans* is evident from observations of behavioral phenotypes of *C. elegans* in the presence of bacteria and its metabolites (Figure 1). Behaviors such as feeding (Avery & Horvitz, 1990), defecation (Thomas, 1990), egg-laying (Trent, Tsuing, & Horvitz, 1983), and locomotion (Sawin, Ranganathan, & Horvitz, 2000) are affected by the presence of bacteria. Insights into the molecular cues sensed by *C. elegans* have come from the characterization of chemotaxis behaviors of *C. elegans*, which demonstrate that *C. elegans* propagated in the laboratory are attracted to a broad range of volatile organic molecules that are produced by bacterial metabolism and may serve as food cues (Bargmann, Hartwieg, & Horvitz, 1993; Bargmann & Horvitz, 1991; Ward, 1973). The genome of *C. elegans* encodes an expanded family of over 1000 chemoreceptor genes (Bargmann, 1998). Genetic analysis of chemotactic responses to diacetyl, which is produced by many bacteria, identified ODR-10 as a chemoreceptor for diacetyl (Sengupta, Chou, & Bargmann, 1996). The chemical characterization of natural bacterial isolates that may serve as food for *C. elegans* in the wild has further revealed a number of volatile organic compounds that attract *C. elegans* (Worthy *et al.*, 2018a). Bacterial food cues modulate dauer entry and exit (Golden & Riddle, 1984), and fatty acids derived from bacteria cause dauer larvae, which do not ingest bacteria, to exit dauer diapause (Kaul *et al.*, 2014). In addition to chemosensation, *C. elegans* also detect the presence of bacteria through mechanosensation. Four classes of ciliated dopaminergic neurons are required for an innate slowing response that *C. elegans* display upon encountering a bacterial food source (Sawin *et al.*, 2000). These neurons express mechanically-sensitive ion channels, display neuronal responses to mechanical forces, and also drive slowing in response to microbeads that are similar in size to bacteria (Kang, Gao, Schafer, Xie, & Xu, 2010; Sawin *et al.*, 2000).

Gradients of molecular oxygen and carbon dioxide are generated by bacterial metabolism, and *C. elegans* exhibits robust detection and behavioral responses to these gases (Bretscher, Busch, & de Bono, 2008, Bretscher *et al.*, 2011; Chang, Chronis, Karow, Marletta, & Bargmann, 2006;

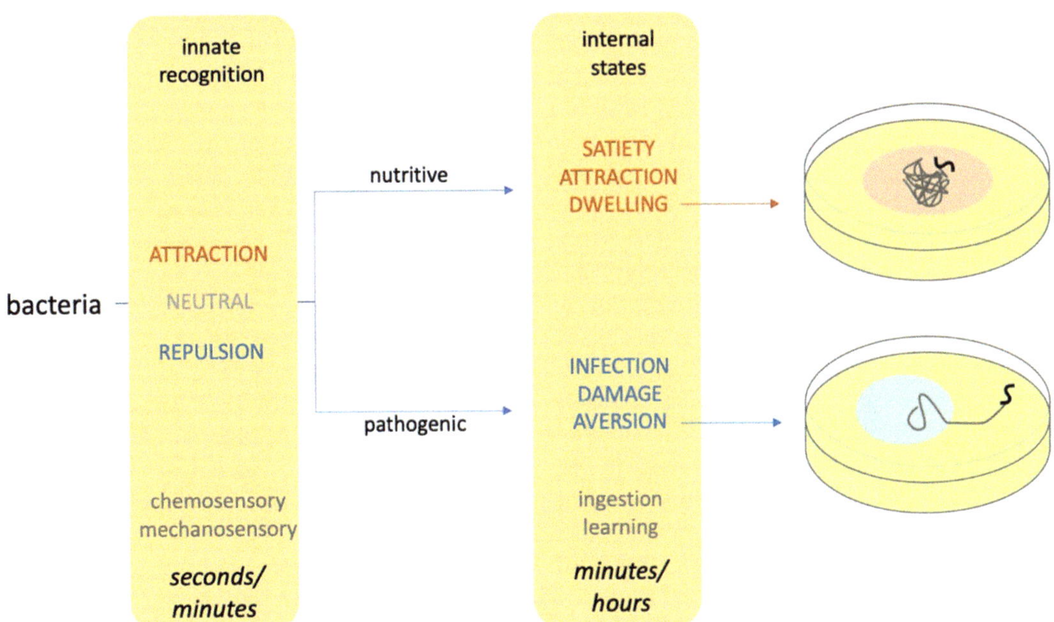

Figure 1. *C. elegans* behavioral responses to bacteria unfold over multiple time scales. *C. elegans* consume diverse bacteria that differ in their nutritive qualities and their pathogenicity. Innate recognition of bacteria allows animals to generate rapid behavioral responses to bacterial odors and textures (left). After bacteria are ingested, animals undergo internal state changes that underlie long-lasting behavioral changes (middle). These long-lasting changes include alterations in their foraging strategies (right) and learned changes in bacterial preference.

Cheung, Cohen, Rogers, Albayram, & de Bono, 2005; Gray *et al.*, 2004; Hallem & Sternberg, 2008), which can drive the behavior of *C. elegans* in a microbial environment. *C. elegans* detects molecular oxygen with cytosolic guanylyl cyclase receptors expressed in the AQR, PQR, URX, and BAG neurons (Cheung *et al.*, 2005; Gray *et al.*, 2004; Zimmer *et al.*, 2009) and exhibits a peaked preference for oxygen levels around ~8% in the absence of bacterial food (Gray *et al.*, 2004). Multiple sensory neurons including BAG and AFD are responsive to carbon dioxide levels (Bretscher *et al.*, 2008, 2011; Hallem & Sternberg, 2008), which may also serve as a cue that modulates behavioral responses to bacterial density. Ambient laboratory oxygen levels (~21%) would be expected to drive *C. elegans* into a bacterial lawn, where oxygen levels can be substantially lower (Gray *et al.*, 2004; Reddy, Hunter, Bhatla, Newman, & Kim, 2011). However, the wild-type laboratory strain, N2, exhibits aerotaxis behavior that is altered in the presence of bacteria, such that the peaked preference for ~8% O_2 levels is lost, with an increased attraction to higher levels of molecular oxygen. The altered aerotaxis behavior of N2 *C. elegans* in the presence of bacteria is caused by allele differences in the *npr-1* gene, encoding a neuropeptide receptor (Chang *et al.*, 2006; Cheung *et al.*, 2005; de Bono & Bargmann, 1998; Gray *et al.*, 2004), and the *glb-5* gene, which encodes a neural globin (McGrath *et al.*, 2009; Persson *et al.*, 2009). The N2 strain carries a laboratory-acquired neomorphic 215 V allele of *npr-1*, whereas natural isolates of *C. elegans* carry an ancestral 215 F allele with reduced NPR-1 activity (de Bono & Bargmann, 1998; McGrath *et al.*, 2009). Bacteria induce NPR-1-dependent differences in not only aerotaxis behavior, but a number of different behavioral phenotypes including differences in CO_2 avoidance (Bretscher *et al.*, 2008; Hallem & Sternberg, 2008), aggregation in feeding behavior, and roaming versus dwelling locomotion on bacterial lawns (Cheung *et al.*, 2005; Gray *et al.*, 2004). Altered behavioral responses to gradients of molecular oxygen also underlie the NPR-1-dependent avoidance of *P. aeruginosa* lawns (Reddy, Andersen, Kruglyak, & Kim, 2009). Of note, behaviors dependent on NPR-1 in the N2 strain can be attenuated in the presence of mucoid Gram-negative bacterial strains, which have an altered surface due to the overproduction of an exopolysaccharide coat (Reddy *et al.*, 2011). The pervasive influence that bacteria can have on the behavior of *C. elegans* is underscored by the modulation of behavior by NPR-1-dependent signaling circuitry (Macosko *et al.*, 2009).

In addition to the innate responsiveness to molecules and mechanical cues that are produced by many bacteria, *C. elegans* also detects molecules that enable discrimination among specific bacterial species. Food choice assays and microdroplet-based assays have enabled the sensitive monitoring of head-turning behaviors to bacterial odors, which demonstrated that *C. elegans* can innately distinguish between odors emanating from *E. coli* OP50 and *P. aeruginosa* PA14 (Ha *et al.*, 2010). *C. elegans* also exhibit chemotactic responses to autoinducer molecules produced by *P. aeruginosa* and *Vibrio cholerae* that mediate quorum sensing (Beale, Li, Tan, & Rumbaugh, 2006; Werner, Perez, Ghosh, Semmelhack, & Bassler, 2014). Innate recognition of pathogenic *Serratia marcescens* has been shown to include the detection of volatile cues (Glater, Rockman, & Bargmann, 2014; Worthy, Rojas, Taylor, & Glater, 2018b), as well as Serrawettin W2, a surfactant-like lipodepsipentapeptide, which acts as a chemical repellent of *C. elegans* (Pradel *et al.*, 2007).

In vivo calcium imaging methods that directly measure activation of individual sensory neurons in response to bacterial cues has corroborated observations from behavioral assays and enabled the dissection of neuronal circuitry mediating innate recognition and preference. The activation of AWB and AWC chemosensory neurons, which mediate repulsive and attractive responses to volatile chemicals, was demonstrated and defined components of a circuit that mediates innate preference for *P. aeruginosa* PA14-conditioned media over *E. coli* OP50-conditioned media (Ha *et al.*, 2010). Activation of the ASH neurons, in response to surfactant-like dodecanoic acid that is secreted by *Streptomyces*, was observed to be dependent on the SRB-6 olfactory receptor, defining a molecular mechanism for innate recognition of a repellant that induces immediate avoidance behavior (Tran *et al.*, 2017). Comprehensive analyses of sensory responses in *C. elegans* have revealed at least 10 classes of sensory neurons that respond to nutritive *E. coli* supernatants (Zaslaver *et al.*, 2015). The ASJ neurons were shown to be activated by both *E. coli* supernatants (Zaslaver *et al.*, 2015), as well as nitric oxide (Hao *et al.*, 2018) and phenazine-1-carboxamide (Meisel, Panda, Mahanti, Schroeder, & Kim, 2014), which are produced by *P. aeruginosa* PA14.

Neuronal activation is accompanied by the rapid induction of gene transcription (Yap & Greenberg, 2018), and ASJ activation is accompanied by the induction of *daf-7*/TGF-beta transcription in the ASJ neurons within six minutes of exposure to *P. aeruginosa* and specifically its secondary metabolites, phenazine-1-carboxamide and the siderophore pyochelin (Meisel *et al.*, 2014). DAF-7 is necessary for *C. elegans* avoidance of *P. aeruginosa*, and increased DAF-7 activity in the ASJ neurons induced by innate recognition of *P. aeruginosa* metabolites alters aerotaxis behavior to promote avoidance behavior (Meisel *et al.*, 2014). Genetic analysis has defined the involvement of distinct cyclic-GMP-dependent signaling pathways in the ASJ neurons that couple the recognition of *P. aeruginosa* metabolites to the selective transcription of *daf-7* (Park, Meisel, & Kim, 2020). The modulation of avoidance behavior induced by infection (discussed further below) by DAF-7 in response to innate recognition in the ASJ neurons of *P. aeruginosa* metabolites contrasts with the immediate attractive or repulsive responses that are induced by activation of the ASH neurons (Tran *et al.*, 2017).

The characterization of *C. elegans* sensory responses to microbe-derived molecules in its environment suggest these host animals recognize molecules produced by broad classes of bacteria as well as molecules that are highly specific to bacterial strains. Many molecules are attractive to *C. elegans*, as might be anticipated for cues of bacterial food, but the diverse repertoire of molecules that can elicit sensory responses also suggests the ability to recognize and respond to specific strains of bacteria, depending on the physiological context.

Bacterial food and internal state

The ingestion of nutritive bacteria can influence the subsequent behavior of *C. elegans* (Figure 1). Associative learning paradigms that involve pairing either the presence or absence of nutritive, non-pathogenic *E. coli* bacteria with environmental conditions such as temperature result in a learned preference for the conditions associated with the fed, not starved, state (Hedgecock & Russell, 1975; Mori & Ohshima, 1995; Torayama, Ishihara, & Katsura, 2007). In addition, *C. elegans* that are fed non-pathogenic bacteria with a wide range of nutritive qualities – 'good food' versus 'bad food' that differ in their ease of ingestion and metabolic factors – can also exhibit a learned change in preference that is calibrated for better versus poorer food (Shtonda & Avery, 2006). These learned changes in behavior can last from minutes to days, depending on the conditioning protocol, thus reflecting a stable change in preference.

The foraging behaviors of *C. elegans* are also strongly influenced by past and present bacterial feeding conditions. While feeding on *E. coli* OP50, *C. elegans* alternate between active 'roaming' states and inactive 'dwelling' states in which they either explore or exploit their food source, respectively (Fujiwara, Sengupta, & McIntire, 2002). The proportion of time that *C. elegans* spends in roaming versus dwelling states is controlled by satiety levels, chemosensory inputs, and internal sensing of bacterial food ingestion (Ben Arous, Laffont, & Chatenay, 2009; Fujiwara *et al.*, 2002; Shtonda & Avery, 2006). Animals that have been deprived of bacterial food exhibit an 'enhanced slowing response' upon encountering a bacterial lawn, suggesting an important role for satiety state in these modes of food exploration (Sawin *et al.*, 2000). The acute ingestion of bacterial food also plays a central role in food exploration: animals that are exposed to a lawn of bacteria rendered largely inedible as a result of pharmacological treatment, spend almost all of their time in the roaming state (Ben Arous *et al.*, 2009).

Serotonin signaling has been shown to have a key role in mediating the effects of the ingestion of nutritive bacteria on behavior. Serotonin biosynthesis and the serotonergic NSM neurons were shown to be required for the enhanced slowing response (Sawin *et al.*, 2000) and for maintenance of the dwelling state (Flavell *et al.*, 2013). NSM neurons extend a sensory dendrite to the surface of the pharyngeal lumen and are acutely activated upon bacterial food ingestion (Rhoades *et al.*, 2019). Feeding-dependent NSM activation requires the acid-sensing ion channels DEL-3 and DEL-7 that localize to the NSM sensory dendrite and appear to mediate detection of a heat-stable bacterial component, connecting microbial recognition in the *C. elegans* alimentary canal to the modification of feeding behaviors.

While serotonin appears to promote states of slow locomotion, the neuropeptide pigment dispersing factor (PDF) is required for sustained roaming states (Flavell *et al.*, 2013) and for mate search behaviors (Barrios, Ghosh, Fang, Emmons, & Barr, 2012), in which male *C. elegans* leave a bacterial food source to search for a mating partner (Lipton, Kleemann, Ghosh, Lints, & Emmons, 2004). PDF signaling appears to be antagonized by serotonin release (Flavell *et al.*, 2013) and, in males, it promotes expression of *daf-7* in ASJ neurons (Hilbert & Kim, 2018), which promotes mate searching behavior (Hilbert & Kim, 2017). *daf-7* expression in ASJ is also positively regulated by satiety, thus allowing it to serve as a signal that integrates multiple internal cues to promote exploration. Increased motion through bacterial lawns during roaming states also activates dopamine signaling via the dopaminergic PDE neurons that appear to integrate the presence of food with the animal's own motion (Cermak *et al.*, 2020). Activation of dopaminergic neurons during roaming elevates egg-laying rates, allowing animals disperse their eggs across bacterial food sources.

Metabolic changes arising from bacterial food ingestion also influence foraging. Rictor/TORC2 signaling in intestinal cells promotes *daf-7* expression in ASI neurons and elevates roaming behavior in a PDF-dependent manner (O'Donnell *et al.*, 2018). The ETS-5 transcription factor functions in ASG and BAG sensory neurons to limit intestinal fat storage and promote PDF-dependent roaming (Juozaityte *et al.*, 2017). Interestingly, the effects of *ets-5* on roaming can be reversed by altering intestinal fat storage. In addition to ASG and BAG, the URX and ASI sensory neurons also impact fat storage (Palamiuc *et al.*, 2017; Witham *et al.*, 2016). URX has also been shown to detect the mobilization of peripheral fat stores, suggesting bi-directional communication (Witham *et al.*, 2016). Genetic analysis of the sterol response element binding protein pathway for fat metabolism has also suggested a critical role for fat metabolism in the regulation of food-induced quiescence behaviors in *C. elegans* (Hyun *et al.*, 2016). Reduced food intake can also alter the production of other metabolites that act on the nervous system. For example, 2 h of fasting reduces levels of kynurenic acid, which alters NMDA signaling and downstream serotonin signaling to impact feeding (Lemieux *et al.*, 2015). Our current understanding of how specific species of bacteria might alter metabolic state to impact behavior remains more limited. However, a recent study showed that *Providencia* bacteria in the *C. elegans* gut produce the neurotransmitter tyramine, which is converted to octopamine by the *C. elegans* host to alter aversive sensory responses (O'Donnell *et al.*, 2020), which were previously shown to be regulated by feeding state (Chao, Komatsu, Fukuto, Dionne, & Hart, 2004). A large number of neuroactive metabolites are produced by non-pathogenic bacteria, suggesting that other bacterial species-specific signals produced in the gut may similarly influence behavior.

The chemosensory and mechanosensory detection of bacteria can also influence *C. elegans* foraging states. When animals are removed from an *E. coli* food source, they exhibit a 'local search' state where they display a high frequency of high-angle turns for ~15 min, before switching to a 'global search' state where they dramatically reduce turning (Gray,

Hill, & Bargmann, 2005; Hills, Brockie, & Maricq, 2004; Wakabayashi, Kitagawa, & Shingai, 2004). The frequency of turning during the local search state depends on the density of the *E. coli* food lawn from which animals were removed (López-Cruz *et al.*, 2019). In this case, chemosensory and mechanosensory neurons are required for bacterial food detection, as depletion of glutamate from both populations of sensory neurons abolishes local search (López-Cruz *et al.*, 2019). Together, these studies of internal states reveal that the *C. elegans* nervous system surveys the past and present levels of non-pathogenic bacteria through multiple sensory modalities in order to change behavior over long time scales.

Infection and internal state

Infection following the ingestion of *P. aeruginosa* PA14, which is not only a nutritive food source for *C. elegans,* but also highly pathogenic (Tan *et al.*, 1999), causes an aversive learned response that is distinct from the effect of feeding on nutritive non-pathogenic bacteria. The initial innate preference of *C. elegans* for *P. aeruginosa* PA14 over *E. coli* OP50 was found to be reversed after feeding on *P. aeruginosa* PA14, with a subsequent preference for *E. coli* OP50 and aversion to *P. aeruginosa* PA14 (Ha *et al.*, 2010; Zhang, Lu, & Bargmann, 2005).

P. aeruginosa can kill *C. elegans* through multiple modes of toxicity that depend on bacterial strain and experimental conditions, including rapid toxicity from the secretion of diffusible toxins over the course of minutes (Darby, Cosma, Thomas, & Manoil, 1999; Kirienko *et al.*, 2013; Mahajan-Miklos, Tan, Rahme, & Ausubel, 1999), or the development of an intestinal infection associated with intralumenal, extracellular bacterial proliferation and distention of the intestinal lumen with effacement of epithelial cells (Irazoqui *et al.*, 2010; Tan *et al.*, 1999) over the course of several hours (Tan *et al.*, 1999). Infection of *C. elegans* by *P. aeruginosa* induces the activation of host innate immunity (Kim & Ewbank, 2018). The host response integrates innate immunity with cellular stress response pathways, such as the endoplasmic reticulum Unfolded Protein Response (Richardson, Kooistra, & Kim, 2010) and mitochondrial stress pathways (Pellegrino *et al.*, 2014), as well as responses to exogenous, toxin-mediated effects on mRNA translation (Dunbar, Yan, Balla, Smelkinson, & Troemel, 2012; McEwan *et al.*, 2012). The widespread induction of stress-activated signaling pathways, immune and stress-responsive genes, and morphological changes to the host reflect the disruption of normal physiology and homeostasis caused by infection with pathogenic bacteria. The broad activation of cellular stress responses with infection has itself been proposed to be a mechanism by which immune defense is activated (Liu, Samuel, Breen, & Ruvkun, 2014; Pukkila-Worley, 2016; Reddy, Dunbar, Nargund, Haynes, & Troemel, 2016). Whereas specific microbial cues for the activation of innate immunity in *C. elegans* remain elusive, evidence points to a key role for host damage resulting from infection in activating innate immunity. The PMK-1 p38 mitogen-activated protein kinase pathway is activated by both bacterial infection (Fletcher, Tillman, Butty, Levine, & Kim, 2019; Kim *et al.*, 2002; Troemel *et al.*, 2006) and pore-forming toxin activity (Huffman *et al.*, 2004). In the epidermal response to fungal infection, host damage may be signaled by the endogenously produced metabolite 4-hydroxyphenyllactic acid, which acts through GPCR signaling to activate PMK-1 and antifungal immunity (Zugasti *et al.*, 2014).

The evolutionarily ancient role for host damage in the activation of innate immunity parallels its apparent role in the development of aversive behavior. Consistent with the kinetics of infection and modified choice behaviors, *C. elegans* exhibits avoidance of a lawn of pathogenic bacteria following bacterial infection (Melo & Ruvkun, 2012; Pradel *et al.*, 2007; Pujol *et al.*, 2001; Reddy *et al.*, 2009; Schulenburg & Müller, 2004). Whereas morphological changes such as intestinal distention accompany the development of pathogen infection (Tan *et al.*, 1999), such changes are only correlative, and host damage in the absence of such changes has been shown to be sufficient for *C. elegans* lawn avoidance behavior. *C. elegans* avoids bacteria such as *Bacillus thurigiensis* (Schulenburg & Müller, 2004), which produces a pore-forming toxin that rapidly kills *C. elegans* over a time scale of minutes without associated intestinal proliferation of bacteria (Marroquin *et al.*, 2000). The development of a lawn avoidance response is also observed in the presence of *Microbacterium nematophilum*, which causes a distinct mode of infection resulting in the induction of a rectal swelling response and sickness (McMullan, Anderson, & Nurrish, 2012; Yook & Hodgkin, 2007). Moreover, *E. coli* strain HT115, which is used for feeding RNAi-based experiments and is a nutritious food source of *C. elegans* can be engineered to be toxic to *C. elegans* through the expression of dsRNA targeting genes that are essential for viability of *C. elegans* (Kamath *et al.*, 2003). Aversion to the *E. coli* HT115 lawn develops over the time course that RNAi exerts toxic effects on the host, and even the addition of abiotic toxins to the lawn can also induce *C. elegans* to leave a bacterial lawn (Melo & Ruvkun, 2012). Notably, the subsequent lawn aversive behavior is not specific for *E. coli* HT115 only but also observed in response to other *E. coli* and even other bacterial species.

In addition to this generalized aversive response, exposure to pathogenic *P. aeruginosa* PA14 also causes a learned change in bacterial preference, where the preference for *P. aeruginosa* PA14 over *E. coli* OP50 is reversed. This change can be elicited by 4 h of *P. aeruginosa* PA14 exposure in adulthood or 12 h of exposure during the L1 larval stage (Jin, Pokala, & Bargmann, 2016; Zhang *et al.*, 2005). In addition, 24 h of PA14 exposure beginning at the L4 larval stage can also reduce PA14 preference in progeny for up to four generations later (Moore, Kaletsky, & Murphy, 2019). The molecular and circuit mechanisms by which exposure to *P. aeruginosa* elicits a change in *C. elegans* preference have been carefully examined. Infection with *P. aeruginosa* induces the increased transcription of *tph-1* from the ADF neurons, and *tph-1* mutants are defective for learned aversive choice behavior following *P. aeruginosa* PA14 infection

(Zhang *et al.*, 2005). *tph-1* mutants exhibit increased susceptibility to killing by *P. aeruginosa* compared to wild-type, a difference that is abrogated when animals are constrained such that they cannot avoid the *P. aeruginosa* lawn (Shivers, Kooistra, Chu, Pagano, & Kim, 2009).

Circuit-level studies have localized the site of learning within sensorimotor circuits that underlies the learned change in bacterial preference. AWB and AWC olfactory neurons detect *P. aeruginosa* PA14 and *E. coli* OP50 odors, but their sensory responses to these cues are not altered after *P. aeruginosa* PA14 exposure (Ha *et al.*, 2010). In addition, a subset of neurons in the downstream circuitry, such as AIY, AIZ, and AIB neurons, are required for navigation towards food sources in naïve animals, but not required for learned changes. In contrast, RIA and SMD neurons are not required for naïve food choice, but are required for the learned change in food preference after *P. aeruginosa* PA14 exposure (Jin *et al.*, 2016; Zhang *et al.*, 2005). These results suggest that RIA and SMD neurons are likely modulated during *P. aeruginosa* PA14 exposure. This modulation appears to require serotonergic signaling from ADF (Ha *et al.*, 2010) and *ins-6* and *ins-7* insulin-like peptides whose expression also changes after learning (Chen *et al.*, 2013). Thus, *P. aeruginosa* PA14 elicits changes in neuroendocrine signaling that impact specific nodes in the sensorimotor circuit to alter food preference. Interestingly, neuroendocrine signals from the gut are also critical for food aversion, as the dynamic expression of an intestinal insulin, *ins-11*, modulates aversive responses to *P. aeruginosa* (Lee & Mylonakis, 2017). In addition, changes in neuroendocrine signaling that modulate learning may be accompanied by changes in neuroendocrine signaling that alter innate behaviors, as is observed in the modulation of *P. aeruginosa* PA14 avoidance behavior by dynamic *daf-7* expression (Meisel *et al.*, 2014).

The observations that *C. elegans* exhibits a preference for one bacterial species over another, and that this preference can be changed based on experience, suggest that *C. elegans* can discriminate among bacterial species, which may be enabled in part by the diversity of microbial ligands that it can recognize. At the same time, the observations that *C. elegans* will leave a lawn of bacteria following infection and damage, and that the subsequent aversive response may not be specific for the bacteria causing the damage, suggest that a behavioral state characterized by a more general aversion to bacteria may also develop. This distinction underscores differences in assays for aversive behavior following infection. Lawn-leaving behavior may be influenced by changes in internal states that confer both specific and non-specific responses to bacteria. It is also possible that non-specific lawn aversion determinants, such as differential oxygen or carbon dioxide levels, may act differentially on the attraction or repulsion of *C. elegans* to particular bacteria to also influence the choice of *C. elegans* between two different bacterial strains.

Summary

The behavioral responses of *C. elegans* animals to microbial cues in their environment rely on the innate recognition of a vast and diverse set of bacterial sensory cues. Behavioral responses to these molecular and mechanosensory cues are influenced by experience and context, endowing *C. elegans* with a great deal of flexibility in how it responds and adapts to its microbial environment. Changes in satiety, tissue damage resulting from pathogenic infection, and recognition of specific bacterial metabolites can alter how the neural circuits in this animal process subsequent microbial cues. Nutritive bacteria can elicit a range of adaptive behavioral changes including learned attraction and stable switches to exploitative foraging behaviors. Pathogenic bacteria can elicit generalized responses, like bacterial aversion, as well as highly specific changes, like learned avoidance of harmful food sources. Thus, evolutionary ancient mechanisms to sense bacteria, as well as host damage arising from pathogenic bacteria, triggers not only cell-autonomous innate immune responses, but also organism-wide behavioral responses that are controlled by a nervous system that can flexibly respond to microbial sensory cues. These studies of host-microbe interactions in *C. elegans* may ultimately inform our understanding of how microbes impact nervous system function in more complex animals (Li & Liberles, 2015; Yang & Chiu, 2017).

Acknowledgements

We are grateful to the editors for the opportunity to write in honor of the memories of Sydney Brenner and John Sulston, who had central roles in the development of *Caenorhabditis elegans* as a simple organism for the molecular genetic analysis of behavior. We thank members of our research groups over the years who have joined us in studying host-microbe interactions of *C. elegans*.

Disclosure statement

No potential conflict of interest was reported by the author(s).

Funding

D.H.K. acknowledges funding from NIH GM084477. S.W.F acknowledges funding from NIH NS104892.

References

Avery, L., & Horvitz, H.R. (1990). Effects of starvation and neuroactive drugs on feeding in *Caenorhabditis elegans*. *The Journal of Experimental Zoology*, *253*(3), 263–270. doi:10.1002/jez.1402530305

Bargmann, C.I. (1998). Neurobiology of the *Caenorhabditis elegans* genome. *Science (New York, N.Y.)*, *282*(5396), 2028–2033. doi:10.1126/science.282.5396.2028

Bargmann, C.I., Hartwieg, E., & Horvitz, H.R. (1993). Odorant-selective genes and neurons mediate olfaction in *C. elegans*. *Cell*, *74*(3), 515–527. doi:10.1016/0092-8674(93)80053-H

Bargmann, C.I., & Horvitz, H.R. (1991). Chemosensory neurons with overlapping functions direct chemotaxis to multiple chemicals in *C. elegans*. *Neuron*, *7*(5), 729–742. doi:10.1016/0896-6273(91)90276-6

Barrios, A., Ghosh, R., Fang, C., Emmons, S.W., & Barr, M.M. (2012). PDF-1 neuropeptide signaling modulates a neural circuit for mate-searching behavior in *C. elegans*. *Nature Neuroscience*, *15*(12), 1675–1682. doi:10.1038/nn.3253

Beale, E., Li, G., Tan, M.-W., & Rumbaugh, K.P. (2006). *Caenorhabditis elegans* senses bacterial autoinducers. *Applied and Environmental Microbiology, 72*(7), 5135–5137. doi:10.1128/AEM.00611-06

Ben Arous, J., Laffont, S., & Chatenay, D. (2009). Molecular and sensory basis of a food related two-state behavior in *C. elegans. PLoS One., 4*(10), e7584. doi:10.1371/journal.pone.0007584

Brenner, S. (1974). The genetics of *Caenorhabditis elegans. Genetics, 77*(1), 71–94.

Bretscher, A.J., Busch, K.E., & de Bono, M. (2008). A carbon dioxide avoidance behavior is integrated with responses to ambient oxygen and food in *Caenorhabditis elegans. Proceedings of the National Academy of Sciences, 105* (23), 8044–8049. doi:10.1073/pnas.0707607105

Bretscher, A.J., Kodama-Namba, E., Busch, K.E., Murphy, R.J., Soltesz, Z., Laurent, P., & de Bono, M. (2011). Temperature, oxygen, and salt-sensing neurons in *C. elegans* are carbon dioxide sensors that control avoidance behavior. *Neuron, 69*(6), 1099–1113. doi:10.1016/j.neuron.2011.02.023

Cermak, N., Yu, S.K., Clark, R., Huang, Y.-C., Baskoylu, S.N., & Flavell, S.W. (2020). Whole-organism behavioral profiling reveals a role for dopamine in state-dependent motor program coupling in *C. elegans. eLife, 9.* doi:10.7554/eLife.57093

Chang, A.J., Chronis, N., Karow, D.S., Marletta, M.A., & Bargmann, C.I. (2006). A distributed chemosensory circuit for oxygen preference in *C. elegans. PLoS Biology, 4*(9), e274. doi:10.1371/journal.pbio.0040274

Chao, M.Y., Komatsu, H., Fukuto, H.S., Dionne, H.M., & Hart, A.C. (2004). Feeding status and serotonin rapidly and reversibly modulate a *Caenorhabditis elegans* chemosensory circuit. *Proceedings of the National Academy of Sciences of the United States of America, 101*(43), 15512–15517. doi:10.1073/pnas.0403369101

Chen, Z., Hendricks, M., Cornils, A., Maier, W., Alcedo, J., & Zhang, Y. (2013). Two insulin-like peptides antagonistically regulate aversive olfactory learning in *C. elegans. Neuron, 77*(3), 572–585. doi:10.1016/j.neuron.2012.11.025

Cheung, B.H.H., Cohen, M., Rogers, C., Albayram, O., & de Bono, M. (2005). Experience-dependent modulation of *C. elegans* behavior by ambient oxygen. Current Biology: CB, *15*(10), 905–917. doi:10.1016/j.cub.2005.04.017

Couillault, C., & Ewbank, J.J. (2002). Diverse bacteria are pathogens of *Caenorhabditis elegans. Infection and Immunity, 70*(8), 4705–4707. doi:10.1128/iai.70.8.4705-4707.2002

Darby, C., Cosma, C.L., Thomas, J.H., & Manoil, C. (1999). Lethal paralysis of *Caenorhabditis elegans* by *Pseudomonas aeruginosa. Proceedings of the National Academy of Sciences of the United States of America, 96*(26), 15202–15207. doi:10.1073/pnas.96.26.15202

de Bono, M., & Bargmann, C.I. (1998). Natural variation in a neuropeptide Y receptor homolog modifies social behavior and food response in *C. elegans. Cell, 94*(5), 679–689. doi:10.1016/S0092-8674(00)81609-8

Dirksen, P., Marsh, S.A., Braker, I., Heitland, N., Wagner, S., Nakad, R., … Rosenstiel, P. (2016). The native microbiome of the nematode *Caenorhabditis elegans*: Gateway to a new host-microbiome model. *BMC Biology, 14*, 38. doi:10.1186/s12915-016-0258-1

Dunbar, T.L., Yan, Z., Balla, K.M., Smelkinson, M.G., & Troemel, E.R. (2012). *C. elegans* detects pathogen-induced translational inhibition to activate immune signaling. *Cell Host Microbe, 11*(4), 375–386. doi:10.1016/j.chom.2012.02.008

Flavell, S.W., Pokala, N., Macosko, E.Z., Albrecht, D.R., Larsch, J., & Bargmann, C.I. (2013). Serotonin and the neuropeptide PDF initiate and extend opposing behavioral states in *C. elegans. Cell, 154*(5), 1023–1035. doi:10.1016/j.cell.2013.08.001

Fletcher, M., Tillman, E.J., Butty, V.L., Levine, S.S., & Kim, D.H. (2019). Global transcriptional regulation of innate immunity by ATF-7 in *C. elegans. PLoS Genetics, 15*(2), e1007830. doi:10.1371/journal.pgen.1007830

Fujiwara, M., Sengupta, P., & McIntire, S.L. (2002). Regulation of body size and behavioral state of *C. elegans* by sensory perception and the EGL-4 cGMP-dependent protein kinase. *Neuron, 36*(6), 1091–1102. doi:10.1016/S0896-6273(02)01093-0

Garigan, D., Hsu, A.-L., Fraser, A.G., Kamath, R.S., Ahringer, J., & Kenyon, C. (2002). Genetic analysis of tissue aging in *Caenorhabditis elegans*: A role for heat-shock factor and bacterial proliferation. *Genetics, 161*(3), 1101–1112.

Garsin, D.A., Sifri, C.D., Mylonakis, E., Qin, X., Singh, K.V., Murray, B.E., … Ausubel, F.M. (2001). A simple model host for identifying Gram-positive virulence factors. *Proceedings of the National Academy of Sciences of the United States of America, 98*(19), 10892–10897. doi:10.1073/pnas.191378698

Glater, E.E., Rockman, M.V., & Bargmann, C.I. (2014). Multigenic natural variation underlies *Caenorhabditis elegans* olfactory preference for the bacterial pathogen *Serratia marcescens. G3 (Bethesda, Md.), 4*(2), 265–276. doi:10.1534/g3.113.008649

Golden, J.W., & Riddle, D.L. (1984). The *Caenorhabditis elegans* dauer larva: Developmental effects of pheromone, food, and temperature. *Developmental Biology, 102*(2), 368–378. doi:10.1016/0012-1606(84)90201-X

Gray, J.M., Hill, J.J., & Bargmann, C.I. (2005). A circuit for navigation in *Caenorhabditis elegans. Proceedings of the National Academy of Sciences of the United States of America, 102*(9), 3184–3191. doi:10.1073/pnas.0409009101

Gray, J.M., Karow, D.S., Lu, H., Chang, A.J., Chang, J.S., Ellis, R.E., … Bargmann, C.I. (2004). Oxygen sensation and social feeding mediated by a *C. elegans* guanylate cyclase homologue. *Nature, 430*(6997), 317–322. doi:10.1038/nature02714

Ha, H., Hendricks, M., Shen, Y., Gabel, C.V., Fang-Yen, C., Qin, Y., … Zhang, Y. (2010). Functional organization of a neural network for aversive olfactory learning in *Caenorhabditis elegans. Neuron, 68*(6), 1173–1186. doi:10.1016/j.neuron.2010.11.025

Hallem, E.A., & Sternberg, P.W. (2008). Acute carbon dioxide avoidance in *Caenorhabditis elegans. Proceedings of the National Academy of Sciences of the United States of America, 105*(23), 8038–8043. doi:10.1073/pnas.0707469105

Han, S., Schroeder, E.A., Silva-García, C.G., Hebestreit, K., Mair, W.B., & Brunet, A. (2017). Mono-unsaturated fatty acids link H3K4me3 modifiers to *C. elegans* lifespan. *Nature, 544*(7649), 185–190. doi:10.1038/nature21686

Hao, Y., Yang, W., Ren, J., Hall, Q., Zhang, Y., & Kaplan, J.M. (2018). Thioredoxin shapes the *C. elegans* sensory response to Pseudomonas produced nitric oxide. *eLife, 7.* doi:10.7554/eLife.36833

Hedgecock, E.M., & Russell, R.L. (1975). Normal and mutant thermotaxis in the nematode *Caenorhabditis elegans. Proceedings of the National Academy of Sciences of the United States of America, 72*(10), 4061–4065. doi:10.1073/pnas.72.10.4061

Herndon, L.A., Schmeissner, P.J., Dudaronek, J.M., Brown, P.A., Listner, K.M., Sakano, Y., … Driscoll, M. (2002). Stochastic and genetic factors influence tissue-specific decline in ageing *C. elegans. Nature, 419*(6909), 808–814. doi:10.1038/nature01135

Hilbert, Z.A., & Kim, D.H. (2017). Sexually dimorphic control of gene expression in sensory neurons regulates decision-making behavior in *C. elegans. eLife, 6.* doi:10.7554/eLife.21166

Hilbert, Z.A., & Kim, D.H. (2018). PDF-1 neuropeptide signaling regulates sexually dimorphic gene expression in shared sensory neurons of *C. elegans. eLife, 7.* doi:10.7554/eLife.36547

Hills, T., Brockie, P.J., & Maricq, A.V. (2004). Dopamine and glutamate control area-restricted search behavior in *Caenorhabditis elegans.* The Journal of Neuroscience : The Official journal of the Society for Neuroscience, *24*(5), 1217–1225. doi:10.1523/JNEUROSCI.1569-03.2004

Hodgkin, J., Kuwabara, P.E., & Corneliussen, B. (2000). A novel bacterial pathogen, *Microbacterium nematophilum*, induces morphological change in the nematode *C. elegans. Current Biology: CB, 10*(24), 1615–1618. doi:10.1016/S0960-9822(00)00867-8

Huffman, D.L., Abrami, L., Sasik, R., Corbeil, J., van der Goot, F.G., & Aroian, R.V. (2004). Mitogen-activated protein kinase pathways defend against bacterial pore-forming toxins. *Proceedings of the National Academy of Sciences of the United States of America, 101*(30), 10995–11000. doi:10.1073/pnas.0404073101

Hyun, M., Davis, K., Lee, I., Kim, J., Dumur, C., & You, Y.-J. (2016). Fat metabolism regulates satiety behavior in *C. elegans*. *Scientific Reports, 6*, 24841. doi:10.1038/srep24841

Irazoqui, J.E., Troemel, E.R., Feinbaum, R.L., Luhachack, L.G., Cezairliyan, B.O., & Ausubel, F.M. (2010). Distinct pathogenesis and host responses during infection of *C. elegans* by *P. aeruginosa* and *S. aureus*. *PLoS Pathogens, 6*, e1000982. doi:10.1371/journal.ppat.1000982

Jin, X., Pokala, N., & Bargmann, C.I. (2016). Distinct circuits for the formation and retrieval of an imprinted olfactory memory. *Cell, 164*(4), 632–643. doi:10.1016/j.cell.2016.01.007

Juozaityte, V., Pladevall-Morera, D., Podolska, A., Nørgaard, S., Neumann, B., & Pocock, R. (2017). The ETS-5 transcription factor regulates activity states in *Caenorhabditis elegans* by controlling satiety. *Proceedings of the National Academy of Sciences of the United States of America, 114*(9), E1651–E1658. doi:10.1073/pnas.1610673114

Kamath, R.S., Fraser, A.G., Dong, Y., Poulin, G., Durbin, R., Gotta, M., ... Sohrmann, M. (2003). Systematic functional analysis of the *Caenorhabditis elegans* genome using RNAi. *Nature, 421*(6920), 231–237. doi:10.1038/nature01278

Kang, L., Gao, J., Schafer, W.R., Xie, Z., & Xu, X.Z.S. (2010). *C. elegans* TRP family protein TRP-4 is a pore-forming subunit of a native mechanotransduction channel. *Neuron, 67*(3), 381–391. doi:10.1016/j.neuron.2010.06.032

Kaul, T.K., Rodrigues, P.R., Ogungbe, I.V., Kapahi, P., and Gill, M.S. (2014). Bacterial Fatty Acids Enhance Recovery from the Dauer Larva in Caenorhabditis elegans. Plos One 9(1): e86979.

Kim, D.H., & Ewbank, J.J. (2018). Signaling in the innate immune response. *WormBook : The Online Review of* C. elegans *Biology, 2018*, 1–35. doi:10.1895/wormbook.1.83.2

Kim, D.H., Feinbaum, R., Alloing, G., Emerson, F.E., Garsin, D.A., Inoue, H., ... Tan, M.-W. (2002). A conserved p38 MAP kinase pathway in *Caenorhabditis elegans* innate immunity. *Science (New York, N.Y.), 297*(5581), 623–626. doi:10.1126/science.1073759

Kirienko, N.V., Kirienko, D.R., Larkins-Ford, J., Wählby, C., Ruvkun, G., & Ausubel, F.M. (2013). Pseudomonas aeruginosa disrupts *Caenorhabditis elegans* iron homeostasis, causing a hypoxic response and death. *Cell Host & Microbe, 13*(4), 406–416. doi:10.1016/j.chom.2013.03.003

Kumar, S., Egan, B.M., Kocsisova, Z., Schneider, D.L., Murphy, J.T., Diwan, A., & Kornfeld, K. (2019). Lifespan extension in *C. elegans* caused by bacterial colonization of the intestine and subsequent activation of an innate immune response. *Developmental Cell, 49*(1), 100–117.e6. doi:10.1016/j.devcel.2019.03.010

Lee, K., & Mylonakis, E. (2017). An intestine-derived neuropeptide controls avoidance behavior in *Caenorhabditis elegans*. *Cell Reports, 20*(10), 2501–2512. doi:10.1016/j.celrep.2017.08.053

Lemieux, G.A., Cunningham, K.A., Lin, L., Mayer, F., Werb, Z., & Ashrafi, K. (2015). Kynurenic acid is a nutritional cue that enables behavioral plasticity. *Cell, 160*(1–2), 119–131. doi:10.1016/j.cell.2014.12.028

Lenaerts, I., Walker, G.A., Van Hoorebeke, L., Gems, D., & Vanfleteren, J.R. (2008). Dietary restriction of *Caenorhabditis elegans* by axenic culture reflects nutritional requirement for constituents provided by metabolically active microbes. *The Journals of Gerontology. Series A, Biological Sciences and Medical Sciences, 63*(3), 242–252. doi:10.1093/gerona/63.3.242

Li, Q., & Liberles, S.D. (2015). Aversion and attraction through olfaction. *Current Biology : CB, 25*(3), R120–R129. doi:10.1016/j.cub.2014.11.044

Lipton, J., Kleemann, G., Ghosh, R., Lints, R., & Emmons, S.W. (2004). Mate searching in *Caenorhabditis elegans*: A genetic model for sex drive in a simple invertebrate. *The Journal of Neuroscience: The Official Journal of the Society for Neuroscience, 24*(34), 7427–7434. doi:10.1523/JNEUROSCI.1746-04.2004

Liu, Y., Samuel, B.S., Breen, P.C., & Ruvkun, G. (2014). *Caenorhabditis elegans* pathways that surveil and defend mitochondria. *Nature, 508*(7496), 406–410. doi:10.1038/nature13204

López-Cruz, A., Sordillo, A., Pokala, N., Liu, Q., McGrath, P.T., & Bargmann, C.I. (2019). Parallel multimodal circuits control an innate foraging behavior. *Neuron, 102*(2), 407–419.e8. doi:10.1016/j.neuron.2019.01.053

Macosko, E.Z., Pokala, N., Feinberg, E.H., Chalasani, S.H., Butcher, R.A., Clardy, J., & Bargmann, C.I. (2009). A hub-and-spoke circuit drives pheromone attraction and social behaviour in *C. elegans*. *Nature, 458*(7242), 1171–1175. doi:10.1038/nature07886

Mahajan-Miklos, S., Tan, M.W., Rahme, L.G., & Ausubel, F.M. (1999). Molecular mechanisms of bacterial virulence elucidated using a *Pseudomonas aeruginosa-Caenorhabditis elegans* pathogenesis model. *Cell, 96*(1), 47–56. doi:10.1016/S0092-8674(00)80958-7

Marroquin, L.D., Elyassnia, D., Griffitts, J.S., Feitelson, J.S., & Aroian, R.V. (2000). *Bacillus thuringiensis* (Bt) toxin susceptibility and isolation of resistance mutants in the nematode *Caenorhabditis elegans*. *Genetics, 155*(4), 1693–1699.

McEwan, D.L., Kirienko, N.V., & Ausubel, F.M. (2012). Host translational inhibition by *Pseudomonas aeruginosa* Exotoxin A Triggers an immune response in *Caenorhabditis elegans*. *Cell Host & Microbe, 11*(4), 364–374. doi:10.1016/j.chom.2012.02.007

McGrath, P.T., Rockman, M.V., Zimmer, M., Jang, H., Macosko, E.Z., Kruglyak, L., & Bargmann, C.I. (2009). Quantitative mapping of a digenic behavioral trait implicates globin variation in *C. elegans* sensory behaviors. *Neuron, 61*(5), 692–699. doi:10.1016/j.neuron.2009.02.012

McMullan, R., Anderson, A., & Nurrish, S. (2012). Behavioral and immune responses to infection require Gαq- RhoA signaling in *C. elegans*. *PLoS Pathogens, 8*(2), e1002530. doi:10.1371/journal.ppat.1002530

Meisel, J.D., Panda, O., Mahanti, P., Schroeder, F.C., & Kim, D.H. (2014). Chemosensation of bacterial secondary metabolites modulates neuroendocrine signaling and behavior of *C. elegans*. *Cell, 159*(2), 267–280. doi:10.1016/j.cell.2014.09.011

Melo, J.A., & Ruvkun, G. (2012). Inactivation of conserved *C. elegans* genes engages pathogen- and xenobiotic-associated defenses. *Cell, 149*(2), 452–466. doi:10.1016/j.cell.2012.02.050

Montalvo-Katz, S., Huang, H., Appel, M.D., Berg, M., & Shapira, M. (2013). Association with soil bacteria enhances p38-dependent infection resistance in *Caenorhabditis elegans*. *Infection and Immunity, 81*(2), 514–520. doi:10.1128/IAI.00653-12

Moore, R.S., Kaletsky, R., & Murphy, C.T. (2019). Piwi/PRG-1 argonaute and TGF-β mediate transgenerational learned pathogenic avoidance. *Cell, 177*(7), 1827–1841.e12. doi:10.1016/j.cell.2019.05.024

Mori, I., & Ohshima, Y. (1995). Neural regulation of thermotaxis in *Caenorhabditis elegans*. *Nature, 376*(6538), 344–348. doi:10.1038/376344a0

O'Donnell, M.P., Chao, P.-H., Kammenga, J.E., & Sengupta, P. (2018). Rictor/TORC2 mediates gut-to-brain signaling in the regulation of phenotypic plasticity in *C. elegans*. *PLoS Genetics, 14*(2), e1007213. doi:10.1371/journal.pgen.1007213

O'Donnell, M.P., Fox, B.W., Chao, P.-H., Schroeder, F.C., & Sengupta, P. (2020). A neurotransmitter produced by gut bacteria modulates host sensory behaviour. *Nature, 583*(7816), 415–420. doi:10.1038/s414586-020-2395-5

Palamiuc, L., Noble, T., Witham, E., Ratanpal, H., Vaughan, M., & Srinivasan, S. (2017). A tachykinin-like neuroendocrine signalling axis couples central serotonin action and nutrient sensing with peripheral lipid metabolism. *Nature Communications, 8*, 14237. doi:10.1038/ncomms14237

Park, J., Meisel, J.D., & Kim, D.H. (2020). Immediate activation of chemosensory neuron gene expression by bacterial metabolites is selectively induced by distinct cyclic-GMP-dependent pathways in *C. elegans*. *PLoS Genetics 16*(8): e1008505. doi:10.1371/journal.pgen.1008505.

Pellegrino, M.W., Nargund, A.M., Kirienko, N.V., Gillis, R., Fiorese, C.J., & Haynes, C.M. (2014). Mitochondrial UPR-regulated innate immunity provides resistance to pathogen infection. *Nature, 516*(7531), 414–417. doi:10.1038/nature13818

Persson, A., Gross, E., Laurent, P., Busch, K.E., Bretes, H., & de Bono, M. (2009). Natural variation in a neural globin tunes oxygen sensing in wild *Caenorhabditis elegans*. *Nature*, *458*(7241), 1030–1033. doi: 10.1038/nature07820

Pradel, E., Zhang, Y., Pujol, N., Matsuyama, T., Bargmann, C.I., & Ewbank, J.J. (2007). Detection and avoidance of a natural product from the pathogenic bacterium Serratia marcescens by *Caenorhabditis elegans*. *Proceedings of the National Academy of Sciences of the United States of America*, *104*(7), 2295–2300. doi:10.1073/pnas.0610281104

Pujol, N., Link, E.M., Liu, L.X., Kurz, C.L., Alloing, G., Tan, M.W., … Ewbank, J.J. (2001). A reverse genetic analysis of components of the Toll signaling pathway in *Caenorhabditis elegans*. *Current Biology: CB*, *11*(11), 809–821. doi:10.1016/S0960-9822(01)00241-X

Pukkila-Worley, R. (2016). Surveillance immunity: An emerging paradigm of innate defense activation in *Caenorhabditis elegans*. *PLoS Pathogens*, *12*(9), e1005795. doi:10.1371/journal.ppat.1005795

Qi, B., & Han, M. (2018). Microbial siderophore enterobactin promotes mitochondrial iron uptake and development of the host via interaction with ATP synthase. *Cell*, *175*(2), 571–582.e11. doi:10.1016/j.cell.2018.07.032

Qi, B., Kniazeva, M., & Han, M. (2017). A vitamin-B2-sensing mechanism that regulates gut protease activity to impact animal's food behavior and growth. *eLife*, *6*. doi:10.7554/eLife.26243

Rahme, L.G., Stevens, E.J., Wolfort, S.F., Shao, J., Tompkins, R.G., & Ausubel, F.M. (1995). Common virulence factors for bacterial pathogenicity in plants and animals. *Science (New York, N.Y.)*, *268*(5219), 1899–1902. doi:10.1126/science.7604262

Rangan, K.J., Pedicord, V.A., Wang, Y.-C., Kim, B., Lu, Y., Shaham, S., … Hang, H.C. (2016). A secreted bacterial peptidoglycan hydrolase enhances tolerance to enteric pathogens. *Science (New York, N.Y.)*, *353*(6306), 1434–1437. doi:10.1126/science.aaf3552

Reddy, K.C., Andersen, E.C., Kruglyak, L., & Kim, D.H. (2009). A polymorphism in npr-1 is a behavioral determinant of pathogen susceptibility in *C. elegans*. *Science (New York, N.Y.)*, *323*(5912), 382–384. doi:10.1126/science.1166527

Reddy, K.C., Dunbar, T.L., Nargund, A.M., Haynes, C.M., & Troemel, E.R. (2016). The *C. elegans* CCAAT-enhancer-binding protein gamma is required for surveillance immunity. *Cell Reports*, *14*(7), 1581–1589. doi:10.1016/j.celrep.2016.01.055

Reddy, K.C., Hunter, R.C., Bhatla, N., Newman, D.K., & Kim, D.H. (2011). *Caenorhabditis elegans* NPR-1-mediated behaviors are suppressed in the presence of mucoid bacteria. *Proceedings of the National Academy of Sciences of the United States of America*, *108*(31), 12887–12892. doi:10.1073/pnas.1108265108

Rhoades, J.L., Nelson, J.C., Nwabudike, I., Yu, S.K., McLachlan, I.G., Madan, G.K., … Flavell, S.W. (2019). ASICs mediate food responses in an enteric serotonergic neuron that controls foraging behaviors. *Cell*, *176*(1–2), 85–97.e14. doi:10.1016/j.cell.2018.11.023

Richardson, C.E., Kooistra, T., & Kim, D.H. (2010). An essential role for XBP-1 in host protection against immune activation in *C. elegans*. *Nature*, *463*(7284), 1092–1095. doi:10.1038/nature08762

Saiki, R., Lunceford, A.L., Bixler, T., Dang, P., Lee, W., Furukawa, S., … Clarke, C.F. (2008). Altered bacterial metabolism, not coenzyme Q content, is responsible for the lifespan extension in *Caenorhabditis elegans* fed an Escherichia coli diet lacking coenzyme Q. *Aging Cell*, *7*(3), 291–304. doi:10.1111/j.1474-9726.2008.00378.x

Samuel, B.S., Rowedder, H., Braendle, C., Félix, M.-A., & Ruvkun, G. (2016). *Caenorhabditis elegans* responses to bacteria from its natural habitats. *Proceedings of the National Academy of Sciences of the United States of America*, *113*(27), E3941–3949. doi:10.1073/pnas.1607183113

Sawin, E.R., Ranganathan, R., & Horvitz, H.R. (2000). *C. elegans* locomotory rate is modulated by the environment through a dopaminergic pathway and by experience through a serotonergic pathway. *Neuron*, *26*(3), 619–631. doi:10.1016/S0896-6273(00)81199-X

Schulenburg, H., & Félix, M.-A. (2017). The natural biotic environment of *Caenorhabditis elegans*. *Genetics*, *206*(1), 55–86. doi:10.1534/genetics.116.195511

Schulenburg, H., & Müller, S. (2004). Natural variation in the response of *Caenorhabditis elegans* towards Bacillus thuringiensis. *Parasitology*, *128*(Pt 4), 433–443. doi:10.1017/s003118200300461x

Sengupta, P., Chou, J.H., & Bargmann, C.I. (1996). odr-10 encodes a seven transmembrane domain olfactory receptor required for responses to the odorant diacetyl. *Cell*, *84*(6), 899–909. doi:10.1016/S0092-8674(00)81068-5

Shivers, R.P., Kooistra, T., Chu, S.W., Pagano, D.J., & Kim, D.H. (2009). Tissue-specific activities of an immune signaling module regulate physiological responses to pathogenic and nutritional bacteria in *C. elegans*. *Cell Host & Microbe*, *6*(4), 321–330. doi:10.1016/j.chom.2009.09.001

Shtonda, B.B., & Avery, L. (2006). Dietary choice behavior in *Caenorhabditis elegans*. *The Journal of Experimental Biology*, *209*(Pt 1), 89–102. doi:10.1242/jeb.01955

Tan, M.W., Mahajan-Miklos, S., & Ausubel, F.M. (1999). Killing of *Caenorhabditis elegans* by Pseudomonas aeruginosa used to model mammalian bacterial pathogenesis. *Proceedings of the National Academy of Sciences of the United States of America*, *96*(2), 715–720. doi:10.1073/pnas.96.2.715

Thomas, J.H. (1990). Genetic analysis of defecation in *Caenorhabditis elegans*. *Genetics*, *124*, 855–872.

Torayama, I., Ishihara, T., & Katsura, I. (2007). *Caenorhabditis elegans* integrates the signals of butanone and food to enhance chemotaxis to butanone. *The Journal of Neuroscience : The Official Journal of the Society for Neuroscience*, *27*(4), 741–750. doi:10.1523/JNEUROSCI.4312-06.2007

Tran, A., Tang, A., O'Loughlin, C.T., Balistreri, A., Chang, E., Coto Villa, D., … VanHoven, M.K. (2017). *C. elegans* avoids toxin-producing Streptomyces using a seven transmembrane domain chemosensory receptor. *eLife*, *6*. doi:10.7554/eLife.23770

Trent, C., Tsuing, N., & Horvitz, H.R. (1983). Egg-laying defective mutants of the nematode *Caenorhabditis elegans*. *Genetics*, *104*(4), 619–647.

Troemel, E.R., Chu, S.W., Reinke, V., Lee, S.S., Ausubel, F.M., & Kim, D.H. (2006). p38 MAPK regulates expression of immune response genes and contributes to longevity in *C. elegans*. *PLoS Genetics*, *2*(11), e183. doi:10.1371/journal.pgen.0020183

Virk, B., Correia, G., Dixon, D.P., Feyst, I., Jia, J., Oberleitner, N., … Ward, J. (2012). Excessive folate synthesis limits lifespan in the *C. elegans*: *E. coli* aging model. *BMC Biology*, *10*(1), 67. doi:10.1186/1741-7007-10-67

Wakabayashi, T., Kitagawa, I., & Shingai, R. (2004). Neurons regulating the duration of forward locomotion in *Caenorhabditis elegans*. *Neuroscience Research*, *50*(1), 103–111. doi:10.1016/j.neures.2004.06.005

Ward, S. (1973). Chemotaxis by the nematode *Caenorhabditis elegans*: Identification of attractants and analysis of the response by use of mutants. *Proceedings of the National Academy of Sciences of the United States of America*, *70*(3), 817–821. doi:10.1073/pnas.70.3.817

Watson, E., MacNeil, L.T., Ritter, A.D., Yilmaz, L.S., Rosebrock, A.P., Caudy, A.A., & Walhout, A.J.M. (2014). Interspecies systems biology uncovers metabolites affecting *C. elegans* gene expression and life history traits. *Cell*, *156*(4), 759–770. doi:10.1016/j.cell.2014.01.047

Wei, J.-Z., Hale, K., Carta, L., Platzer, E., Wong, C., Fang, S.-C., & Aroian, R.V. (2003). Bacillus thuringiensis crystal proteins that target nematodes. *Proceedings of the National Academy of Sciences of the United States of America*, *100*(5), 2760–2765. doi:10.1073/pnas.0538072100

Werner, K.M., Perez, L.J., Ghosh, R., Semmelhack, M.F., & Bassler, B.L. (2014). *Caenorhabditis elegans* recognizes a bacterial quorum-sensing signal molecule through the AWCON neuron. The Journal of Biological Chemistry, *289*(38), 26566–26573. doi:10.1074/jbc.M114.573832

Witham, E., Comunian, C., Ratanpal, H., Skora, S., Zimmer, M., & Srinivasan, S. (2016). *C. elegans* body cavity neurons are homeostatic sensors that integrate fluctuations in oxygen availability and internal nutrient reserves. *Cell Reports*, *14*(7), 1641–1654. doi:10.1016/j.celrep.2016.01.052

Worthy, S.E., Haynes, L., Chambers, M., Bethune, D., Kan, E., Chung, K., … Glater, E.E. (2018a). Identification of attractive odorants released by preferred bacterial food found in the natural habitats of *C. elegans*. *PloS One*, *13*(7), e0201158. doi:10.1371/journal.pone.0201158

Worthy, S.E., Rojas, G.L., Taylor, C.J., & Glater, E.E. (2018b). Identification of odor blend used by *Caenorhabditis elegans* for pathogen recognition. *Chemical Senses*, *43*(3), 169–180. doi:10.1093/chemse/bjy001

Yang, N.J., & Chiu, I.M. (2017). Bacterial signaling to the nervous system through toxins and metabolites. Journal of Molecular Biology, *429*(5), 587–605. doi:10.1016/j.jmb.2016.12.023

Yap, E.-L., & Greenberg, M.E. (2018). Activity-regulated transcription: Bridging the gap between neural activity and behavior. *Neuron*, *100*(2), 330–348. doi:10.1016/j.neuron.2018.10.013

Yook, K., & Hodgkin, J. (2007). Mos1 mutagenesis reveals a diversity of mechanisms affecting response of *Caenorhabditis elegans* to the bacterial pathogen Microbacterium nematophilum. *Genetics*, *175*(2), 681–697. doi:10.1534/genetics.106.060087

Zaslaver, A., Liani, I., Shtangel, O., Ginzburg, S., Yee, L., & Sternberg, P.W. (2015). Hierarchical sparse coding in the sensory system of *Caenorhabditis elegans*. *Proceedings of the National Academy of Sciences of the United States of America*, *112*(4), 1185–1189. doi:10.1073/pnas.1423656112

Zhang, Y., Lu, H., & Bargmann, C.I. (2005). Pathogenic bacteria induce aversive olfactory learning in *Caenorhabditis elegans*. *Nature*, *438*(7065), 179–184. doi:10.1038/nature04216

Zimmer, M., Gray, J.M., Pokala, N., Chang, A.J., Karow, D.S., Marletta, M.A., … Bargmann, C.I. (2009). Neurons detect increases and decreases in oxygen levels using distinct guanylate cyclases. *Neuron*, *61*(6), 865–879. doi:10.1016/j.neuron.2009.02.013

Zugasti, O., Bose, N., Squiban, B., Belougne, J., Kurz, C.L., Schroeder, F.C., … Ewbank, J.J. (2014). Activation of a G protein-coupled receptor by its endogenous ligand triggers the innate immune response of *Caenorhabditis elegans*. *Nature Immunology*, *15*(9), 833–838. doi:10.1038/ni.2957

Neurogenetics of nictation, a dispersal strategy in nematodes

Heeseung Yang, Bo Yun Lee, Hyunsoo Yim and Junho Lee

ABSTRACT
Nictation is a behaviour in which a nematode stands on its tail and waves its head in three dimensions. This activity promotes dispersal of dauer larvae by allowing them to attach to other organisms and travel on them to a new niche. In this review, we describe our understanding of nictation, including its diversity in nematode species, how it is induced by environmental factors, and neurogenetic factors that regulate nictation. We also highlight the known cellular and signalling factors that affect nictation, for example, IL2 neurons, insulin/IGF-1 signalling, TGF-β signalling, FLP neuropeptides and piRNAs. Elucidation of the mechanism of nictation will contribute to increased understanding of the conserved dispersal strategies in animals.

Dispersal and phoresy in animals

Dispersal is a crucial survival strategy for many animal species and phoresy is a type of dispersal strategy by which an animal attaches to another organism with higher mobility and uses that organism to facilitate dispersal. Allowing small organisms the opportunity to overcome their limited mobility, phoresy also enables these organisms to increase the distance and efficiency with which they disperse. Various phoretic interactions in nature include the attachment of the ostracod to the frog (Lopez, Rodrigues, & Rios, 1999), the blister beetle larvae to the bee (Saul-Gershenz & Millar, 2006), and mites to the nostrils of birds (Proctor & Owens, 2000).

Nictation is a specific behaviour that promotes phoretic interactions between nematodes and carrier animals (Lee *et al.*, 2011). Nictation involves nematodes standing on their tails and waving their bodies in three dimensions, increasing the probability that they will encounter and climb onto the bodies of other organisms, such as isopods, slugs and snails (Frézal & Felix, 2015). In *Caenorhabditis elegans*, nictation is unique to a specialised larval stage called the dauer. *C. elegans* larvae enter the dauer stage as an alternative developmental stage when they are exposed to harsh conditions, such as high temperatures, high population densities and food scarcity (Cassada & Russell, 1975; Golden & Riddle, 1984a). Having increased stress resistance, dauers can survive several months without feeding, thus, this stage promotes long-term survival under harsh conditions (Cassada & Russell, 1975; Klass & Hirsh, 1976).

Nictation facilitates the dispersal of dauers to a new niche. In nature, with the population increasing rapidly when food is plentiful and rapidly decreasing when food is depleted, a *C. elegans* population has a boom-and-bust life strategy (Felix & Braendle, 2010). Nictation allows the increasing number of dauers in this harsh environment to spread to a new niche via exploiting a more mobile host. Once they enter a more suitable environment for development, dauers begin feeding and developing again so that the population can increase. In the wild, *C. elegans* are usually collected in the dauers stage, suggesting that a large proportion of *C. elegans* exists as dauers in nature (Barriere & Felix, 2005). Dispersal by nictation in the dauer stage appears to have a great impact on the survival and evolution of nematode species.

Diversity of nictation behaviour in nematodes

Nictation in multiple nematode species

Nictation occurs among many nematode species, including other *Caenorhabditis* species, such as *Caenorhabditis remanei, Caenorhabditis briggssae, Caenorhabditis japonica* and *Caenorhabditis plicata*, and in other genera such as *Pristionchus pacificus*, but *C. elegans* is the species in which nictation has been most intensely studied (Brown, D'Anna, & Sommer, 2011; Kiontke & Sudhaus, 2006). The first observed nictation was in parasitic nematodes (Reed & Wallace, 1965), where nictation is specific to the infective juvenile stage. Nictation by infective larvae is a form of parasitic behaviour as it enables larvae to attach to hosts during infection. Parasitism in the nematode has evolved independently from the free-living nematodes multiple times (Holterman *et al.*, 2006). Based on the similarities between the infective stage of parasitic nematodes and the dauer stage of free-living nematodes, the dauer hypothesis suggests that dauers could be a pre-adaptation of parasitism (Crook, 2014; Hotez, Hawdon, & Schad, 1993). Various parasitic species in *Steinernema, Strongyloides, Heterorhabditis, Nippostrongylus, Ancylostoma, Necator, Heligmosomoides Mermis* show nictation behaviour (Campbell & Gaugler, 1993; Castelletto *et al.*, 2014; Gans & Burr, 1994; Granzer & Haas, 1991; Hernandez & Sukhdeo, 1995; Ishibashi & Kondo, 1990).

Diverse modes of nictation

Different forms of nictation behaviour have been reported (Figure 1). As a simple example, nictating dauers may wave or stay stationary. Nictation can also occur as a group; in *C. elegans* and *C. briggssae*, it was observed that thousands of dauers aggregate and stand as a group on a point-structured surface (Felix & Duveau, 2012). Group nictation is thought to increase the possibility of attaching to other structures or hosts, with another example of group nictation being observed in *P. pacificus* (Penkov *et al.*, 2014), which secrete long-chain polyunsaturated wax esters that form an adhesive layer that covers the surface of their bodies. Many dauers nictate in aggregates, which are called 'dauer towers' (Penkov *et al.*, 2014). Group nictation of *C. elegans* and *P. pacificus* exhibit different characteristics. Group nictation of *C. elegans* is relatively less stable than that of *P. pacificus*, and *C. elegans* dauers do not aggregate in water while *P. pacificus* dauers do (Penkov *et al.*, 2014). Group nictation may also occur in other nematode species. Preliminary observations in *Ditylenchus myceliophagous* and rhabditids on mushrooms showed an aggregated standing structure (Hesling, 1966; Staniland, 1957), although *Ditylenchus* swarming contains all larval stages rather than dauers only. Future studies are needed to elucidate the different properties of swarming and group nictation of dauers.

Tube-waving is another example of nictation. Tube-waving species do not completely shed the second juvenile cuticle at the dauer stage. These species can carry the cuticle, retract into the cuticle after nictation, and extend out of it when they begin nictation. Tube wavers can nictate on flat surfaces, whereas other nematodes require rough structures. Tube waving has been observed in *Rhabditella* spp., *Cephaloboides* sp. and *Auanema* spp. (Kanzaki *et al.*, 2017; Kiontke, 1999; Osche, 1954; Sudhaus, 2011). Studies suggest that tube wavers form a clade, evolving only once (Kanzaki *et al.*, 2017).

Some nematodes also show specific jumping behaviours, as well as the standard nictation (Reed & Wallace, 1965). Upon standing, the parasitic nematode *Steinernema* spp., such as *Steinernema carpocapsae* and *Steinernema scapterisci* (Campbell & Kaya, 1999, 2000), bend their body into a loop before jumping forward with their entire bodies stretched (Reed & Wallace, 1965). On average, jumps can be nine-body lengths long and seven-body lengths high (Campbell & Kaya, 1999). This behaviour further enhances the probability that the nematode will attach to hosts.

Mechanisms underlying these diverse processes are largely unknown and they may or may not have identical neurogenetic bases. Future work is needed to elucidate the regulatory mechanisms and evolution of the diverse modes of nictation.

Intraspecific quantitative diversity of nictation

Nictation behaviour can vary within species. Various wild isolates of *C. elegans* have different proportions of individuals that undergo nictation (nictation ratios) (Lee, Yang, *et al.*, 2017). Efforts to find genetic factors that contribute to nictation variability between the N2 reference strain (from Bristol) and the CB4856 isolate (from Hawaii) using quantitative trait locus (QTL) mapping led to the discovery that natural variations in a piRNA-rich region underlie the nictation diversity (Lee, Yang, *et al.*, 2017). *C. elegans* isolates from Orsay in France also show differences in nictation ratio (Richaud, Zhang, Lee, Lee, & Felix, 2018).

Environmental factors, host sensing and nictation

Nictation is affected by several environmental factors. For example, nictation in *C. elegans* was observed only on rough

Figure 1. Diverse modes of nictation in nematode species. *C. elegans* shows individual or group nictation. Different nictation phenotypes such as tube-waving and jumping are also observed in other nematode species.

surfaces of agar plates that bear three-dimensional structures, but not on flat surfaces. One possibility is that tactile cues sensed by mechanosensory neurons in dauers promote nictation. High temperature and air movement also seem to induce nictation. Near the normal breeding temperature (15–25 °C), the nictation ratio in *C. elegans* dauers increases as temperature increases (H. Yang and J. Lee, unpublished data). Nictation of *Ancylostoma caninum* is also induced by high temperatures (Granzer & Haas, 1991).

Since the nictation ratio appears higher in certain nematodes in covered Petri dishes than in open dishes, it is possible that humidity also affects nictation (Ishibashi & Kondo, 1990). Indeed, other studies have shown that nictation of *A. caninum* is stimulated by humidity (Granzer & Haas, 1991). However, nictation is not observed when the surface is excessively moist, possibly because the behaviour is inhibited by surface tension (Ishibashi & Kondo, 1990). Because dauers do not nictate well on moist chips, agar micro-dirt chips with micro-structures are dried for 90 min before use (Lee, Lee, Choi, Park, & Lee, 2015).

Nematodes must sense nearby hosts to nictate, which facilitates their attachment to the hosts. Thus, the enhanced nictation induced by increased temperature and air movement might be related to the presence of hosts. For example, air movement that contains host volatile cues (Campbell & Kaya, 1999), like CO_2 (Hallem *et al.*, 2011), induces nictation and jumping behaviour in infective juveniles of *S. carpocapsae*. BAG neurons are CO_2-sensing neurons *in C.* elegans and this function is conserved in several parasitic nematodes (Hallem *et al.*, 2011). Ablation of BAG neurons reduces CO_2 attraction in infective juveniles of several nematodes and in *C. elegans* dauers, as well as the CO_2-evoked jumping in *S. carpocapsae* (Hallem *et al.*, 2011). In *C. elegans*, other developmental stages show CO_2 avoidance, whereas dauers show CO_2 attraction (Hallem *et al.*, 2011). CO_2 attraction in dauers may facilitate interactions with hosts (Hallem *et al.*, 2011), and further studies are needed to assess the effects of CO_2 on nictation in *C. elegans*.

Host sensing is intensively studied in parasitic nematodes. The nictation responses for different host-specific chemicals differ between nematode species, as they interact with diverse hosts. CO_2 may be an example of a general cue signalling the presence of a host.

Nictation assay in the laboratory

Whereas most behaviours of nematodes are studied in two dimensions, nictation is a behaviour that occurs in three dimensions. Therefore, the study of nictation in the lab requires a specific assay. In the laboratory, *C. elegans* are typically grown on agar plates. However, the flat agar does not provide structures that favour nictation, so nictation is observed on fungal hyphae in contaminated plates or after sand or glass beads are sprayed onto the agar to form a rough surface (Baird, 1999; Gaugler & Bilgrami, 2004). However, detailed studies of the behaviour require methods that are more sophisticated to yield consistent results.

The micro-dirt assay was developed to overcome the inconvenience and inconsistency of previous methods (Lee *et al.*, 2011). Micro-dirt chips are agar chips with a dense array of finely shaped pillars. Agar chips are made using a polydimethylsiloxane mould and their microstructure mimics the dirt structure of the natural environment of free-living nematodes (Park *et al.*, 2008). Micro-dirt chips allow nictation, where its duration and ratio can then be quantified. In an individual assay, each individual dauer is observed for one minute and the fraction of time it nictates is measured. In a population assay, a population of dauers is observed and the ratio of nictating dauers at a specific time point can also be measured. Typically, the average nictation ratio of N2 dauers measured by the individual assay is about 0.2–0.4 (Lee *et al.*, 2015), but this ratio can be affected by experimental conditions. Nictation ratios that include quiescent dauers may be different as only moving dauers are typically counted in this assay.

Neurons, genes and signalling pathways that influence *C. elegans* nictation

Several studies that focus on the regulatory mechanisms of nictation have been enabled by the development of a micro-dirt chip assay and recent advances in genetics. We describe here the known regulatory factors of nictation, although the details are not yet fully understood.

IL2 neurons

IL2 neurons are a set of six ciliated inner labial neurons. Acetylcholine transmission and intact cilia are required in IL2 neurons for nictation to occur in *C. elegans* (Lee *et al.*, 2011). Although micro-dirt chip and flat plate surfaces are identical in their chemical compositions, *C. elegans* dauers are able to nictate on a micro-dirt chip, but not on a flat plate (Lee *et al.*, 2011). This suggests that the physical properties of the micro-dirt chip, which might be sensed by IL2 neurons, may directly elicit *C. elegans* nictation (Lee *et al.*, 2011).

A TRPV channel protein, OSM-9, is expressed in several sensory neurons, including IL2 neurons. OSM-9 is involved in mechanosensory, osmosensory and olfactory responses (Colbert, Smith, & Bargmann, 1997). Only *osm-9* mutants showed reduced nictation ratio among mutations in several mechanoreceptor genes examined (Lee *et al.*, 2011). These studies suggest that OSM-9 may act as a mechanoreceptor for the initiation of nictation, although future studies are necessary to further investigate this process.

Various tissues and cells, such as epithelia, muscles and neurons, undergo morphological changes during dauer development (Albert & Riddle, 1983; Cassada & Russell, 1975; Dixon, Alexander, Chan, & Roy, 2008). IL2 neurons also change their morphology during dauer development, which may regulate nictation. In non-dauers, the inner labial channels are located at the nose tip and IL2 neurons extend to the end of the subcuticle (Albert & Riddle, 1983). In dauers, however, the channels are located at the outer edge of the lip, and the ends of IL2 neurons are retracted (Albert & Riddle, 1983). In addition, quadrant (dorsal and ventral) IL2s undergo dauer-specific dendrite arborisation and axon remodelling, where additional dendrites are observed and the axons branch and thicken (Schroeder *et al.*, 2013) (Figure 2). UNC-86 is a transcription factor that promotes arborisation of IL2 neurons during dauer development and is involved in determining IL2 cell fate (Schroeder *et al.*, 2013; Shaham & Bargmann, 2002). KPC-1 is a serine-type endopeptidase that has cell-autonomous action to promote the arborisation of IL2 neurons, where *kpc-1* is specifically expressed during the dauer stage (Schroeder *et al.*, 2013). The fact that *kpc-1* or *unc-86* mutants have defective nictation, which is restored by the expression of *kpc-1* or *unc-86* in IL2 neurons, suggests that dauer-specific IL2 arborisation regulates nictation (Schroeder *et al.*, 2013).

Remodelling of gap junctions also occurs during dauer development (Bhattacharya, Aghayeva, Berghoff, & Hobert, 2019). In lateral IL2 neurons, INX-1a, INX-2, UNC-7 and UNC-9 are expressed in non-dauer larvae, but not in dauers, which instead express CHE-7 (Bhattacharya *et al.*, 2019). In quadrant IL2s, INX-2 is expressed in non-dauers, but not in dauers (Bhattacharya *et al.*, 2019). Although differences in the types of innexin proteins expressed in IL2 neurons may induce changes in the neural circuitry in the dauer stage, and thereby affect nictation behaviour, no neuronal cells that form gap junctions with IL2 neurons in the dauer stage have been identified to date.

Signalling pathways regulating nictation

Insulin/IGF-1 signalling (IIS) pathway

Insulin signalling pathways and their components are well conserved in *C. elegans*. In *C. elegans*, the insulin/IGF-1 signalling (IIS) pathway regulates longevity (Kenyon, Chang, Gensch, Rudner, & Tabtiang, 1993), metabolism (Kimura, Tissenbaum, Liu, & Ruvkun, 1997), development (Kimura *et al.*, 1997), and behaviour (Dillon, Holden-Dye, O'Connor, & Hopper, 2016; Kodama *et al.*, 2006) by distinguishing the nutritional status of individuals. *daf-2* is so far the only identified insulin/IGF-1 receptor (Kimura *et al.*, 1997) for the 40 *C. elegans* insulin-like peptides (Duret, Guex, Peitsch, & Bairoch, 1998; Gregoire, Chomiki, Kachinskas, & Warden, 1998; Husson, Mertens, Janssen, Lindemans, & Schoofs, 2007; Li, Kennedy, & Ruvkun, 2003; Pierce *et al.*, 2001). Insulin signalling inhibits expression of DAF-16/FOXO downstream genes by phosphorylating DAF-16 (Lin, Hsin, Libina, & Kenyon, 2001; Paradis & Ruvkun, 1998).

IIS inhibits nictation in *C. elegans,* since dauers with mutations in *daf-2* or other factors of the IIS pathway generally have a higher nictation ratio than N2 dauers (Lee, Lee, Kim, Lim & Lee, 2017). IIS regulates nictation in the early larval stages, before dauer entry (Lee, Lee, *et al.*, 2017). Cell-specific rescue experiments showed that IIS can regulate nictation in ASI and ASJ sensory neurons that are also known to regulate dauer development and recovery (Bargmann & Horvitz, 1991; Lee, Lee, *et al.*, 2017). The data described above show that IIS acts for both dauer development and nictation. More detailed analysis, however, reveals distinct regulation modes of IIS for dauer development and nictation. First, *daf-16*, the key downstream regulator of IIS, plays somewhat distinct roles in dauer formation and nictation. While the *daf-16(null); daf-2* double mutants that expressed *daf-16a* or *daf-16d/f* isoforms fully or nearly fully recovered dauer formation capability, those isoforms alone did not rescue nictation capability (Kwon, Narasimhan, Yen, & Tissenbaum, 2010; Lee, Lee, *et al.*, 2017), suggesting that several *daf-16* isoforms, not a single *daf-16* isoform, are required for nictation. Secondly, mutations in the insulin-like peptide genes *ins-28* or *ins-34* lowered nictation ratio, with little or no effect on dauer formation (Fernandes de Abreu *et al.*, 2014; Lee, Lee, *et al.*, 2017), suggesting that ligands for the DAF-2 receptor may be different for dauer development and nictation.

In mammals, p21-activated kinase (PAK), *pak1*, is activated by FOXO and regulates brain neuronal polarity (de la Torre-Ubieta *et al.*, 2010). In *C. elegans*, PAK genes *pak-1* and *max-2* are involved in the axonal guidance of ventral cord commissural motor neurons during development (Lucanic, Kiley, Ashcroft, L'Etoile, & Cheng, 2006). Consistent with the idea that the actin remodeller MAX-2 acts downstream of IIS to regulate nictation, the *max-2* mutant nictation phenotype is epistatic to that of the *daf-2* mutant phenotype (Lee, Lee, *et al.*, 2017). The *max-2* mutation suppresses the high nictation ratio caused by the *daf-2* mutation, where the *max-2; daf-2* double mutant resembles the *max-2* single mutant (Lee, Lee, *et al.*, 2017).

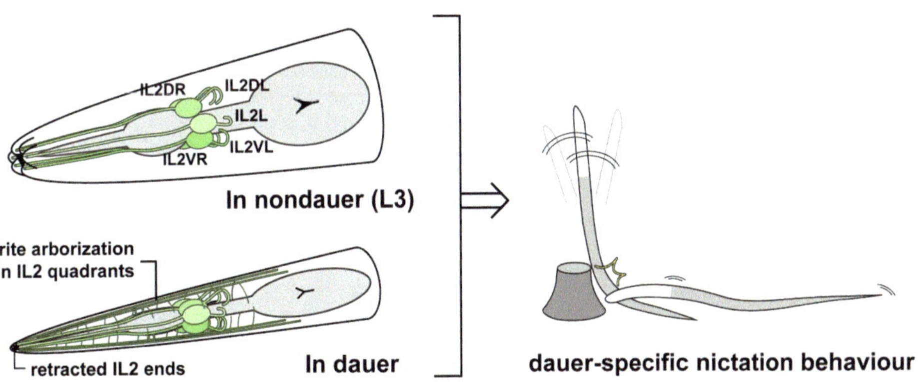

Figure 2. IL2 arborization in *C. elegans* dauer. Quadrant IL2s undergo dauer-specific dendrite arborisation regulating nictation behaviour. More arborisation and axonal remodelling are present in a full dauer. This is not indicated due to simplification for the figure.

TGF-β signalling

TGF-β binds to heteromers of type-I and type-II receptors to activate Smad to regulate the expression of downstream genes (Shi & Massague, 2003). *C. elegans* have five TGF-β-related genes, of which DAF-7 contributes to regulation of dauer development (Golden & Riddle, 1984b; Ren *et al.*, 1996). DAF-1 is a type-I receptor, and DAF-4 is a type-II receptor (Estevez *et al.*, 1993; Georgi, Albert, & Riddle, 1990). *C. elegans* have several R-Smads, among which DAF-8 and DAF-14, together with DAF-7, inhibit dauer development through inhibition of the DAF-3/co-Smad and DAF-5/Sno-Ski transcription factors (da Graca *et al.*, 2004; Inoue & Thomas, 2000; Park, Estevez, & Riddle, 2010; Patterson, Koweek, Wong, Liu, & Ruvkun, 1997).

TGF-β signalling through DAF-7 is involved in nictation behaviour (Lee, Lee, *et al.*, 2017). In dauers that have a mutation in *daf-7* or its downstream factors *daf-1*, *daf-4*, *daf-8* and *daf-14*, the nictation ratio is dramatically reduced, and this phenotype is suppressed by mutations in *daf-3* (Lee, Lee, *et al.*, 2017). As in the case of dauer formation, DAF-1 regulates nictation in RIM/RIC interneurons (Greer, Perez, Van Gilst, Lee, & Ashrafi, 2008; Lee, Lee, *et al.*, 2017).

TGF-β signalling and IIS inhibit dauer formation such that if either pathway has a low level of activity, dauer formation is induced (Gottlieb & Ruvkun, 1994; Patterson & Padgett, 2000; Thomas, Birnby, & Vowels, 1993). However, during nictation, IIS acts opposite from TGF-β signalling (Lee, Lee, *et al.*, 2017). Double mutants of *age-1*/PI3K, a component of IIS, and *daf-7* also exhibited low nictation ratio, similar to that of the *daf-7* single mutant (Lee, Lee, *et al.*, 2017). Unlike dauer development (Vowels & Thomas, 1992), these findings suggest that TGF-β signalling is epistatic to IIS in nictation regulation, although the mechanism is not yet fully understood. It is conceivable that TGF-β signalling and IIS may respond to the environmental conditions and reduced activity of either pathway may indicate that the environment is not sufficient for reproduction, thus leading to the induction of the development of dauers. On the other hand, dauers induced by reduced IIS have a high nictation ratio and dauers induced by reduced TGF-β signalling have a low nictation ratio, suggesting that these two signalling pathways are reduced due to differing environmental conditions (Figure 3). Future studies are required to determine whether the levels of TGF-β signalling correlate with the nictation ratio, which may allow the determination of which environmental conditions these pathways affect.

Neuropeptides

In the dauer stage, expression of neuropeptides and G protein-coupled receptors is increased (Lee, Shie, *et al.*, 2017). SBT-1 is a chaperone of the proprotein convertase EGL-3/PC2, an enzyme that cleaves pre-neuropeptides (Lindberg, Tu, Muller, & Dickerson, 1998). Dauer formation is reduced and nictation duration is increased in *sbt-1* mutants (Lee, Shie, *et al.*, 2017). Also, while wild-type dauers are attracted to CO_2, *sbt-1* mutant dauers display CO_2 repulsion and avoidance (Lee, Shie, *et al.*, 2017). Of the neuropeptides that increase in dauer larvae, *flp-10* and *flp-17* are expressed in BAG neurons (Kim & Li, 2004; Smith, Martinez-Velazquez, & Ringstad, 2013). Dauers of these mutants display CO_2 avoidance and reduced nictation duration (Lee, Shie, *et al.*, 2017). Neuropeptides also regulate nictation behaviour in parasites (Morris *et al.*, 2017). Knockdown of *flp-21* in the infective juveniles of *S. carpocapsae* decreases hyperactive nictation, chemotaxis on host volatiles, dispersals and jumping behaviours (Morris *et al.*, 2017).

piRNA

QTL mapping revealed that a piRNA-rich region in chromosome IV contributes to the nictation differences between the N2 and CB4856 strains (Lee, Yang, *et al.*, 2017). N2 has a higher nictation ratio than CB4856 and the N2-specific piRNA profiles are likely to promote nictation. *prg-1*/Piwi mutations in the N2 background reduced the nictation ratio, whereas *prg-1* mutations in the CB4856 did not affect nictation (Lee, Yang, *et al.*, 2017). Although PRG-1 is expressed mainly in the gonads, a small amount of PRG-1 is also expressed in somatic cells (Kim *et al.*, 2018). PRG-1

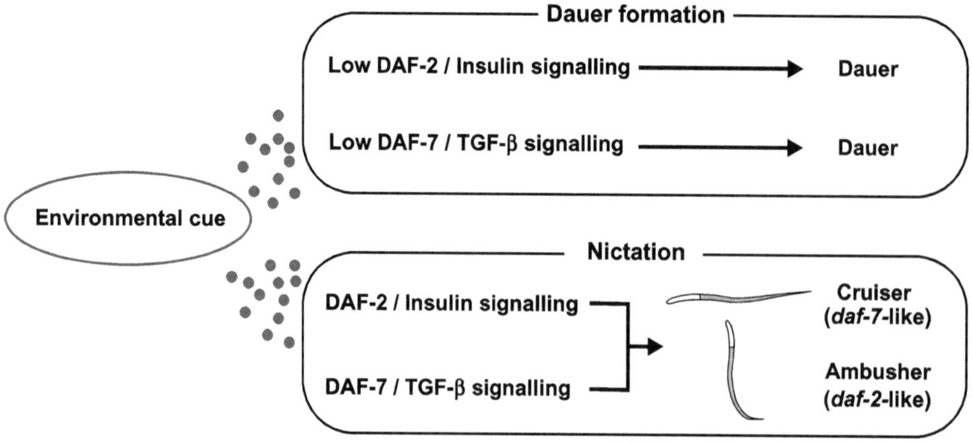

Figure 3. Physiological significance of nictation regulation by the TGF-β and insulin/IGF-1 signalling pathways. Reduced activity of either pathway, TGF-β or insulin/IGF-1 signalling, induces dauers. In contrast, reduced activity of TGF-β signalling lowers the nictation ratio and reduced activity of insulin/IGF-1 signalling increases the nictation ratio, suggesting that the two signalling pathways may represent different environmental qualities. The 'cruiser' and 'ambusher' dauers indicate the dauers that prefer cruising and nictation, respectively, for dispersal to another niche.

regulates axon regeneration of touch receptor neurons in a cell-autonomous manner (Kim *et al.*, 2018). PRG-1 might act in somatic cells to alter piRNA expression and thereby regulate nictation.

Concluding remarks and perspectives

Recent studies have led to increased understanding of nictation from a variety of viewpoints, including the cellular basis, signalling pathways and natural variation involved in nictation. Also, studies on IL2 neurons revealed neuroplasticity and behavioural differentiation. However, many questions about nictation remain. Nictation occurs in dauers, with the exception of adult females in *Mermis nigrescens* (Gans & Burr, 1994). Future studies with focus on developmental plasticity are needed to investigate why this behaviour is restricted. Dauer-specific muscle arm extensions or gap junction plasticity may be possible reasons (Bhattacharya *et al.*, 2019; Dixon *et al.*, 2008). Future studies may lead to a circuit-level understanding of the nictation behaviour, including neurons downstream from IL2, as IL2 neurons are responsible for nictation. Studies on the environmental and host cues affecting nictation can increase understanding of nematode ecology in nature. The study of the conservation and diversification of nictation regulatory mechanisms in nematode species may be of interest, including tube-waving nematodes and jumping nematodes. Elucidation of nictation regulatory mechanisms will contribute to a deeper understanding of dispersal strategies in animals.

Acknowledgements

The authors thank members of Lee laboratory for discussion. The authors thank Jun Kim for help with Figure 1.

Disclosure statement

No potential conflict of interest was reported by the author(s).

Funding

This work was supported by the Samsung Science and Technology Foundation [Project Number SSTF-BA1501-04]. H. Yang, B. Lee and H. Yim were supported by BK21 program.

References

Albert, P.S., & Riddle, D.L. (1983). Developmental alterations in sensory neuroanatomy of the *Caenorhabditis elegans* dauer larva. *The Journal of Comparative Neurology*, *219*(4), 461–481. doi:10.1002/cne.902190407

Baird, S.E. (1999). Natural and experimental associations of *Caenorhabditis remanei* with Trachelipus rathkii and other terrestrial isopods. *Nematology*, *1*(5), 471–475. doi:10.1163/156854199508478

Bargmann, C.I., & Horvitz, H.R. (1991). Control of larval development by chemosensory neurons in *Caenorhabditis elegans*. *Science (New York, N.Y.)*, *251*(4998), 1243–1246. doi:10.1126/science.2006412

Barriere, A., & Felix, M.A. (2005). High local genetic diversity and low outcrossing rate in *Caenorhabditis elegans* natural populations. *Current Biology*, *15*(13), 1176–1184. doi:10.1016/j.cub.2005.06.022

Bhattacharya, A., Aghayeva, U., Berghoff, E.G., & Hobert, O. (2019). Plasticity of the electrical connectome of *C. elegans*. *Cell*, *176*(5), 1174–1189.e16. doi:10.1016/j.cell.2018.12.024

Brown, F.D., D'Anna, I., & Sommer, R.J. (2011). Host-finding behaviour in the nematode *Pristionchus pacificus*. *Proceedings. Biological Sciences*, *278*(1722), 3260–3269. doi:10.1098/rspb.2011.0129

Campbell, J.F., & Gaugler, R. (1993). Nictation behavior and its ecological implications in the host search strategies of entomopathogenic nematodes (heterorhabditidae and steinernematidae). *Behaviour*, *126* (3-4), 155–169. doi:10.1163/156853993X00092

Campbell, J.F., & Kaya, H.K. (1999). How and why a parasitic nematode jumps. *Nature*, *397*(6719), 485–486. doi:10.1038/17254

Campbell, J.F., & Kaya, H.K. (2000). Influence of insect associated cues on the jumping behavior of entomopathogenic nematodes (*Steinernema* spp.). *Behaviour*, *137*(5), 591–609. doi:10.1163/156853900502231

Cassada, R.C., & Russell, R.L. (1975). The dauerlarva, a post-embryonic developmental variant of the nematode *Caenorhabditis elegans*. *Developmental Biology*, *46*(2), 326–342. doi:10.1016/0012-1606(75)90109-8

Castelletto, M.L., Gang, S.S., Okubo, R.P., Tselikova, A.A., Nolan, T.J., Platzer, E.G., … Hallem, E.A. (2014). Diverse host-seeking behaviors of skin-penetrating nematodes. *PLoS Pathogens*, *10*(8), e1004305. doi:10.1371/journal.ppat.1004305

Colbert, H.A., Smith, T.L., & Bargmann, C.I. (1997). OSM-9, a novel protein with structural similarity to channels, is required for olfaction, mechanosensation, and olfactory adaptation in *Caenorhabditis elegans*. *The Journal of Neuroscience*, *17*(21), 8259–8269. doi:10.1523/JNEUROSCI.17-21-08259.1997

Crook, M. (2014). The dauer hypothesis and the evolution of parasitism: 20 years on and still going strong. International Journal for Parasitology, *44*(1), 1–8. doi:10.1016/j.ijpara.2013.08.004

da Graca, L.S., Zimmerman, K.K., Mitchell, M.C., Kozhan-Gorodetska, M., Sekiewicz, K., Morales, Y., & Patterson, G.I. (2004). DAF-5 is a Ski oncoprotein homolog that functions in a neuronal TGFβ pathway to regulate *C. elegans* dauer development. *Development*, *131*(2), 435–446. doi:10.1242/dev.00922

de la Torre-Ubieta, L., Gaudilliere, B., Yang, Y., Ikeuchi, Y., Yamada, T., DiBacco, S., … Bonni, A. (2010). A FOXO-Pak1 transcriptional pathway controls neuronal polarity. *Genes & Development*, *24*(8), 799–813. doi:10.1101/gad.1880510

Dillon, J., Holden-Dye, L., O'Connor, V., & Hopper, N.A. (2016). Context-dependent regulation of feeding behaviour by the insulin receptor, DAF-2, in *Caenorhabditis elegans*. *Invertebrate Neuroscience*, *16*(2), 4. doi:10.1007/s10158-016-0187-2

Dixon, S.J., Alexander, M., Chan, K.K., & Roy, P.J. (2008). Insulin-like signaling negatively regulates muscle arm extension through DAF-12 in *Caenorhabditis elegans*. *Developmental Biology*, *318*(1), 153–161. doi:10.1016/j.ydbio.2008.03.019

Duret, L., Guex, N., Peitsch, M.C., & Bairoch, A. (1998). New insulin-like proteins with atypical disulfide bond pattern characterized in *Caenorhabditis elegans* by comparative sequence analysis and homology modeling. *Genome Research*, *8*(4), 348–353. doi:10.1101/gr.8.4.348

Estevez, M., Attisano, L., Wrana, J.L., Albert, P.S., Massague, J., & Riddle, D.L. (1993). The daf-4 gene encodes a bone morphogenetic protein receptor controlling *C. elegans* dauer larva development. *Nature*, *365*(6447), 644–649. doi:10.1038/365644a0

Felix, M.A., & Braendle, C. (2010). The natural history of *Caenorhabditis elegans*. Current Biology, *20*(22), R965–R969. doi:10.1016/j.cub.2010.09.050

Felix, M.A., & Duveau, F. (2012). Population dynamics and habitat sharing of natural populations of *Caenorhabditis elegans* and *C. briggsae*. *BMC Biology*, *10*(1), 59. doi:10.1186/1741-7007-10-59

Fernandes de Abreu, D.A., Caballero, A., Fardel, P., Stroustrup, N., Chen, Z., Lee, K., … Ch'ng, Q. (2014). An insulin-to-insulin regulatory network orchestrates phenotypic specificity in development and physiology. *PLoS Genetics*, *10*(3), e1004225. doi:10.1371/journal.pgen.1004225

Frézal, L., & Felix, M.-A. (2015). The natural history of model organisms: C. *elegans* outside the Petri dish. *eLife, 4*, e05849. doi:10.7554/eLife.05849

Gans, C., & Burr, A.H.J. (1994). Unique locomotory mechanism of *Mermis nigrescens*, a large nematode that crawls over soil and climbs through vegetation. *Journal of Morphology, 222*(2), 133–148. doi:10.1002/jmor.1052220203

Gaugler, R., & Bilgrami, A.L. (2004). *Nematode behaviour*. Wallingford: CABI.

Georgi, L.L., Albert, P.S., & Riddle, D.L. (1990). daf-1, a *C. elegans* gene controlling dauer larva development, encodes a novel receptor protein kinase. *Cell, 61*(4), 635–645. doi:10.1016/0092-8674(90)90475-T

Golden, J.W., & Riddle, D.L. (1984a). The *Caenorhabditis elegans* dauer larva: Developmental effects of pheromone, food, and temperature. *Developmental Biology, 102*(2), 368–378. doi:10.1016/0012-1606(84)90201-X

Golden, J.W., & Riddle, D.L. (1984b). A pheromone-induced developmental switch in *Caenorhabditis elegans*: Temperature-sensitive mutants reveal a wild-type temperature-dependent process. *Proceedings of the National Academy of Sciences of the United States of America, 81*(3), 819–823. doi:10.1073/pnas.81.3.819

Gottlieb, S., & Ruvkun, G. (1994). daf-2, daf-16 and daf-23: Genetically interacting genes controlling Dauer formation in *Caenorhabditis elegans. Genetics, 137*(1), 107–120.

Granzer, M., & Haas, W. (1991). Host-finding and host recognition of infective *Ancylostoma caninum* larvae. *International Journal for Parasitology, 21*(4), 429–440. doi:10.1016/0020-7519(91)90100-L

Greer, E.R., Perez, C.L., Van Gilst, M.R., Lee, B.H., & Ashrafi, K. (2008). Neural and molecular dissection of a *C. elegans* sensory circuit that regulates fat and feeding. *Cell Metabolism, 8*(2), 118–131. doi:10.1016/j.cmet.2008.06.005

Gregoire, F.M., Chomiki, N., Kachinskas, D., & Warden, C.H. (1998). Cloning and developmental regulation of a novel member of the insulin-like gene family in *Caenorhabditis elegans. Biochemical and Biophysical Research Communications, 249*(2), 385–390. doi:10.1006/bbrc.1998.9164

Hallem, E.A., Dillman, A.R., Hong, A.V., Zhang, Y., Yano, J.M., DeMarco, S.F., & Sternberg, P.W. (2011). A sensory code for host seeking in parasitic nematodes. Current Biology, *21*(5), 377–383. doi:10.1016/j.cub.2011.01.048

Hernandez, A.D., & Sukhdeo, M.V. (1995). Host grooming and the transmission strategy of *Heligmosomoides polygyrus. The Journal of Parasitology, 81*(6), 865–869. doi:10.2307/3284031

Hesling, J. (1966). Preliminary experiments on the control of mycophagous eelworms in mushroom beds, with a note on their swarming. *Plant Pathology, 15*(4), 163–166. doi:10.1111/j.1365-3059.1966.tb00342.x

Holterman, M., van der Wurff, A., van den Elsen, S., van Megen, H., Bongers, T., Holovachov, O., … Helder, J. (2006). Phylum-wide analysis of SSU rDNA reveals deep phylogenetic relationships among nematodes and accelerated evolution toward crown Clades. *Molecular Biology and Evolution, 23*(9), 1792–1800. doi:10.1093/molbev/msl044

Hotez, P., Hawdon, J., & Schad, G.A. (1993). Hookworm larval infectivity, arrest and amphiparatenesis: the *Caenorhabditis elegans* Daf-c paradigm. *Parasitology Today (Personal ed.), 9*(1), 23–26. doi:10.1016/0169-4758(93)90159-D

Husson, S.J., Mertens, I., Janssen, T., Lindemans, M., & Schoofs, L. (2007). Neuropeptidergic signaling in the nematode *Caenorhabditis elegans. Progress in Neurobiology, 82*(1), 33–55. doi:10.1016/j.pneurobio.2007.01.006

Inoue, T., & Thomas, J.H. (2000). Targets of TGF-beta signaling in *Caenorhabditis elegans* dauer formation. *Developmental Biology, 217*(1), 192–204. doi:10.1006/dbio.1999.9545

Ishibashi, N., & Kondo, E. (1990). *Behavior of infective juveniles (entomopathogenic nematodes in biological control* (pp. 139–150). Boca Raton, FL: CRC Press.

Kanzaki, N., Kiontke, K., Tanaka, R., Hirooka, Y., Schwarz, A., Muller-Reichert, T., … Pires-daSilva, A. (2017). Description of two three-gendered nematode species in the new genus Auanema (Rhabditina) that are models for reproductive mode evolution. *Scientific Reports, 7*(1), 11135. doi:10.1038/s41598-017-09871-1

Kenyon, C., Chang, J., Gensch, E., Rudner, A., & Tabtiang, R. (1993). A *C. elegans* mutant that lives twice as long as wild type. *Nature, 366*(6454), 461–464. doi:10.1038/366461a0

Kim, K., & Li, C. (2004). Expression and regulation of an FMR famide-related neuropeptide gene family in *Caenorhabditis elegans*. Journal of Comparative Neurology, *475*(4), 540–550. doi:10.1002/cne.20189

Kim, K.W., Tang, N.H., Andrusiak, M.G., Wu, Z., Chisholm, A.D., & Jin, Y. (2018). A neuronal piRNA pathway inhibits axon regeneration in *C. elegans. Neuron, 97*(3), 511–519 e516. doi:10.1016/j.neuron.2018.01.014

Kimura, K.D., Tissenbaum, H.A., Liu, Y., & Ruvkun, G. (1997). daf-2, an insulin receptor-like gene that regulates longevity and diapause in *Caenorhabditis elegans. Science (New York, N.Y.), 277*(5328), 942–946. doi:10.1126/science.277.5328.942

Kiontke, K. (1999). The Rhabditis (Rhabditella) octopleura species complex and descriptions of three new species. *Russian Journal of Nematology, 7*(2), 71–94.

Kiontke, K., & Sudhaus, W. (2006). Ecology of *Caenorhabditis* species. *WormBook*, 9, 1–14. doi:10.1895/wormbook.1.37.1

Klass, M., & Hirsh, D. (1976). Non-ageing developmental variant of *Caenorhabditis elegans. Nature, 260*(5551), 523–525. doi:10.1038/260523a0

Kodama, E., Kuhara, A., Mohri-Shiomi, A., Kimura, K.D., Okumura, M., Tomioka, M., … Mori, I. (2006). Insulin-like signaling and the neural circuit for integrative behavior in *C. elegans. Genes & Development, 20*(21), 2955–2960. doi:10.1101/gad.1479906

Kwon, E.S., Narasimhan, S.D., Yen, K., & Tissenbaum, H.A. (2010). A new DAF-16 isoform regulates longevity. *Nature, 466*(7305), 498–502. doi:10.1038/nature09184

Lee, D., Lee, H., Choi, M-k., Park, S., & Lee, J. (2015). Nictation assays for *Caenorhabditis* and other nematodes. *Bioprotocol, 5*, e1433. doi:10.21769/BioProtoc.1433

Lee, D., Lee, H., Kim, N., Lim, D.S., & Lee, J. (2017). Regulation of a hitchhiking behavior by neuronal insulin and TGF-β signaling in the nematode *Caenorhabditis elegans. Biochemical and Biophysical Research Communications, 484*(2), 323–330. doi:10.1016/j.bbrc.2017.01.113

Lee, D., Yang, H., Kim, J., Brady, S., Zdraljevic, S., Zamanian, M., … Lee, J. (2017). The genetic basis of natural variation in a phoretic behavior. *Nature Communications, 8*(1), 273. doi:10.1038/s41467-017-00386-x

Lee, H., Choi, M.K., Lee, D., Kim, H.S., Hwang, H., Kim, H., … Lee, J. (2011). Nictation, a dispersal behavior of the nematode *Caenorhabditis elegans*, is regulated by IL2 neurons. *Nature Neuroscience, 15*(1), 107–112. doi:10.1038/nn.2975

Lee, J.S., Shih, P.Y., Schaedel, O.N., Quintero-Cadena, P., Rogers, A.K., & Sternberg, P.W. (2017). FMRFamide-like peptides expand the behavioral repertoire of a densely connected nervous system. *Proceedings of the National Academy of Sciences of the United States of America, 114*(50), E10726–E10735. doi:10.1073/pnas.1710374114

Li, W., Kennedy, S.G., & Ruvkun, G. (2003). daf-28 encodes a *C. elegans* insulin superfamily member that is regulated by environmental cues and acts in the DAF-2 signaling pathway. *Genes & Development, 17*(7), 844–858. doi:10.1101/gad.1066503

Lin, K., Hsin, H., Libina, N., & Kenyon, C. (2001). Regulation of the *Caenorhabditis elegans* longevity protein DAF-16 by insulin/IGF-1 and germline signaling. *Nature Genetics, 28*(2), 139–145. doi:10.1038/88850

Lindberg, I., Tu, B., Muller, L., & Dickerson, I.M. (1998). Cloning and functional analysis of *C. elegans* 7B2. *DNA and Cell Biology, 17*(8), 727–734. doi:10.1089/dna.1998.17.727

Lopez, L.C.S., Rodrigues, P.J.F.P., & Rios, R.I. (1999). Frogs and snakes as phoretic dispersal agents of bromeliad ostracods (Limnocytheridae: Elpidium) and annelids (Naididae: Dero). *Biotropica, 31*(4), 705–708. doi:10.1111/j.1744-7429.1999.tb00421.x

Lucanic, M., Kiley, M., Ashcroft, N., L'Etoile, N., & Cheng, H.J. (2006). The *Caenorhabditis elegans* P21-activated kinases are differentially required for UNC-6/netrin-mediated commissural motor axon guidance. *Development (Cambridge, England), 133*(22), 4549–4559. doi:10.1242/dev.02648

Morris, R., Wilson, L., Sturrock, M., Warnock, N.D., Carrizo, D., Cox, D., … Dalzell, J.J. (2017). A neuropeptide modulates sensory perception in the entomopathogenic nematode Steinernema carpocapsae. *PLoS Pathogens, 13*(3), e1006185. doi:10.1371/journal.ppat.1006185

Osche, G. (1954). Über verhalten und morphologie der dauerlarven freilebender nematoden. *Zoologischer Anzeiger, 152*, 65–73.

Paradis, S., & Ruvkun, G. (1998). *Caenorhabditis elegans* Akt/PKB transduces insulin receptor-like signals from AGE-1 PI3 kinase to the DAF-16 transcription factor. *Genes & Development, 12*(16), 2488–2498. doi:10.1101/gad.12.16.2488

Park, D., Estevez, A., & Riddle, D.L. (2010). Antagonistic Smad transcription factors control the dauer/non-dauer switch in *C. elegans*. *Development (Cambridge, England), 137*(3), 477–485. doi:10.1242/dev.043752

Park, S., Hwang, H., Nam, S.W., Martinez, F., Austin, R.H., & Ryu, W.S. (2008). Enhanced *Caenorhabditis elegans* locomotion in a structured microfluidic environment. *PloS One, 3*(6), e2550. doi:10.1371/journal.pone.0002550

Patterson, G.I., & Padgett, R.W. (2000). TGFβ-related pathways: Roles in *Caenorhabditis elegans* development. *Trends in Genetics, 16*(1), 27–33. doi:10.1016/S0168-9525(99)01916-2

Patterson, G.I., Koweek, A., Wong, A., Liu, Y., & Ruvkun, G. (1997). The DAF-3 Smad protein antagonizes TGF-β-related receptor signaling in the *Caenorhabditis elegans* dauer pathway. *Genes & Development, 11*(20), 2679–2690. doi:10.1101/gad.11.20.2679

Penkov, S., Ogawa, A., Schmidt, U., Tate, D., Zagoriy, V., Boland, S., … Kurzchalia, T.V. (2014). A wax ester promotes collective host finding in the nematode *Pristionchus pacificus*. Nature Chemical Biology, *10*(4), 281–285. doi:10.1038/nchembio.1460

Pierce, S.B., Costa, M., Wisotzkey, R., Devadhar, S., Homburger, S.A., Buchman, A.R., … Ruvkun, G. (2001). Regulation of DAF-2 receptor signaling by human insulin and ins-1, a member of the unusually large and diverse *C. elegans* insulin gene family. *Genes & Development, 15*(6), 672–686. doi:10.1101/gad.867301

Proctor, H., & Owens, I.I. (2000). Mites and birds: Diversity, parasitism and coevolution. *Trends in Ecology & Evolution, 15*(9), 358–364. doi:10.1016/S0169-5347(00)01924-8

Reed, E.M., & Wallace, H.R. (1965). Leaping locomotion by an insect-parasitic nematode. *Nature, 206*(4980), 210–210. doi:10.1038/206210a0

Ren, P., Lim, C.S., Johnsen, R., Albert, P.S., Pilgrim, D., & Riddle, D.L. (1996). Control of *C. elegans* larval development by neuronal expression of a TGF-beta homolog. *Science (New York, N.Y.), 274*(5291), 1389–1391. doi:10.1126/science.274.5291.1389

Richaud, A., Zhang, G., Lee, D., Lee, J., & Felix, M.A. (2018). The local coexistence pattern of selfing genotypes in *Caenorhabditis elegans* natural metapopulations. *Genetics, 208*(2), 807–821. doi:10.1534/genetics.117.300564

Saul-Gershenz, L.S., & Millar, J.G. (2006). Phoretic nest parasites use sexual deception to obtain transport to their host's nest. *Proceedings of the National Academy of Sciences of the United States of America, 103*(38), 14039–14044. doi:10.1073/pnas.0603901103

Schroeder, N.E., Androwski, R.J., Rashid, A., Lee, H., Lee, J., & Barr, M.M. (2013). Dauer-specific dendrite arborization in *C. elegans* is regulated by KPC-1/Furin. *Current Biology : CB, 23*(16), 1527–1535. doi:10.1016/j.cub.2013.06.058

Shaham, S., & Bargmann, C.I. (2002). Control of neuronal subtype identity by the *C. elegans* ARID protein CFI-1. *Genes & Development, 16*(8), 972–983. doi:10.1101/gad.976002

Shi, Y., & Massague, J. (2003). Mechanisms of TGF-beta signaling from cell membrane to the nucleus. *Cell, 113*(6), 685–700. doi:10.1016/S0092-8674(03)00432-X

Smith, E.S., Martinez-Velazquez, L., & Ringstad, N. (2013). A chemoreceptor that detects molecular carbon dioxide. *The Journal of Biological Chemistry, 288*(52), 37071–37081. doi:10.1074/jbc.M113.517367

Staniland, L. (1957). The swarming of Rhabditid eelworms in mushroom houses. *Plant Pathology, 6*(2), 61–62. doi:10.1111/j.1365-3059.1957.tb00775.x

Sudhaus, W. (2011). Phylogenetic systematisation and catalogue of paraphyletic "Rhabditidae" (Secernentea, Nematoda). *Journal of Nematode Morphology and Systematics, 14*(2), 113–178.

Thomas, J.H., Birnby, D.A., & Vowels, J.J. (1993). Evidence for parallel processing of sensory information controlling dauer formation in *Caenorhabditis elegans*. *Genetics, 134*(4), 1105–1117.

Vowels, J.J., & Thomas, J.H. (1992). Genetic analysis of chemosensory control of dauer formation in *Caenorhabditis elegans*. *Genetics, 130*(1), 105–123.

Regulatory systems that mediate the effects of temperature on the lifespan of *Caenorhabditis elegans*

Byounghun Kim, Jongsun Lee, Younghun Kim and Seung-Jae V. Lee

ABSTRACT

Temperature affects animal physiology, including aging and lifespan. How temperature and biological systems interact to influence aging and lifespan has been investigated using model organisms, including the nematode *Caenorhabditis elegans*. In this review, we discuss mechanisms by which diverse cellular factors modulate the effects of ambient temperatures on aging and lifespan in *C. elegans*. *C. elegans* thermosensory neurons alleviate lifespan-shortening effects of high temperatures via sterol endocrine signaling and probably through systemic regulation of cytosolic proteostasis. At low temperatures, *C. elegans* displays a long lifespan by upregulating the cold-sensing TRPA channel, lipid homeostasis, germline-mediated prostaglandin signaling, and autophagy. In addition, co-chaperone p23 amplifies lifespan changes affected by high and low temperatures. Our review summarizes how external temperatures modulate *C. elegans* lifespan and provides information regarding responses of biological processes to temperature changes, which may affect health and aging at an organism level.

Introduction

Aging is a gradual functional and structural decline in biological systems, leading to decreases in reproductive capacity and increased mortality. Aging is influenced by various genetic and environmental factors. Several model organisms, such as budding yeast, *Caenorhabditis elegans*, *Drosophila melanogaster*, and mice, have been used to determine how biological factors modulate aging and lifespan. The roundworm *C. elegans*, a small transparent nematode with a lifespan of only 2–3 weeks, is an invaluable model organism for aging research (Son, Altintas, Kim, Kwon, & Lee, 2019). By performing EMS mutagenesis and defining cell lineages for *C. elegans* (Brenner, 1974; Sulston, Schierenberg, White, & Thomson, 1983), Sydney Brenner and John Sulston established *C. elegans* as a model organism for developmental genetics. Their contributions provide the foundation for subsequent breakthrough discoveries in the biology of aging, using *C. elegans* as a primary model. Various genetic factors, including those acting in insulin/IGF-1 signaling (IIS) and target of rapamycin (TOR) signaling pathways, have been extensively studied using *C. elegans* (DiLoreto & Murphy, 2015; Kenyon, 2010a; Lee *et al.*, 2015). Many reports also show how *C. elegans* lifespan is affected by environmental factors, including temperature, diet, and various external stresses.

Temperature is an important environmental factor, which affects lifespan and aging. The 'rate-of-living' theory asserts that the faster the metabolism, the shorter the lifespan (Pearl, 1928). Therefore, it was widely accepted that chemical reactions are facilitated as temperature rises, leading to increased metabolic rates and consequently short lifespan. This scenario is particularly plausible for ectotherms, whose body temperatures are subject to changes in environmental temperatures. Interestingly, however, many studies using ectotherms, including *C. elegans*, indicate that genetic factors modulate lifespan changes in response to external temperatures (Jeong, Artan, Seo, & Lee, 2012; Xiao, Liu, & Xu, 2015).

In this review, we summarize regulatory systems that modulate lifespan in response to environmental temperatures, focusing on studies using *C. elegans*. We describe established mechanisms by which thermosensory neurons modulate endocrine signaling, in turn influencing lifespan and proteostasis. We also discuss temperature-sensing cation channels, co-chaperones, and lipid signaling/homeostasis systems, which affect lifespan at different temperatures. Many aspects of the regulation of *C. elegans* lifespan and aging are conserved in mammals, and therefore these studies may provide useful information regarding the effects of environmental temperatures on human aging and health.

Thermosensory neurons counter short lifespan at high temperatures via modulating neuroendocrine signaling

Adaptive behaviors for maintaining homeostasis under changing environmental conditions are essential for survival. *C. elegans* is sensitive to changes in temperature and can discriminate temperature differences as small as 0.05 °C

(Hedgecock & Russell, 1975; Luo, Clark, Biron, Mahadevan, & Samuel, 2006; Mori, 1999; Ramot, MacInnis, & Goodman, 2008). Amphid sensory neurons, AFD, AWC, and ASI, which were first identified by Brenner *et al.* (Ward, Thomson, White, & Brenner, 1975; White, Southgate, Thomson, & Brenner, 1986), act as thermosensory neurons required for normal thermotaxis (Beverly, Anbil, & Sengupta, 2011; Biron, Wasserman, Thomas, Samuel, & Sengupta, 2008; Kuhara *et al.*, 2008; Mori & Ohshima, 1995). Among thermosensory neurons, AFD neurons that form a neural circuit with AIY, AIZ, and RIA interneurons are the primary thermosensory neurons (Goodman & Sengupta, 2019). The responses to temperatures sensed by the AFD neurons are divided into a fast response in minutes and a slow response in hours (Goodman & Sengupta, 2019; Yu *et al.*, 2014). For the fast response, temperature rise activates AFD neuron-specific guanylyl cyclases, GCY-8, GCY-18, and GCY-23 (Inada *et al.*, 2006; Wasserman, Beverly, Bell, & Sengupta, 2011), leading to upregulation of intracellular cGMP levels and subsequent opening of TAX-2/TAX-4 cation channels (Satterlee, Ryu, & Sengupta, 2004). The slow response causes nuclear localization of CMK-1, which upregulates the expression of the guanylyl cyclase genes, *gcy-8*, *gcy-18*, and *gcy-23* (Yu *et al.*, 2014).

Interestingly, the thermosensory neural circuit modulates temperature-dependent changes in longevity and stress responses at high temperatures (Figure 1; Chen *et al.*, 2016; Lee & Kenyon, 2009). The AFD thermosensory neurons actively counteract the life-shortening effects of high temperatures through a sterol endocrine signaling pathway (Lee & Kenyon, 2009). Specifically, AFD neurons upregulate DAF-9/cytochrome P450, which is required for producing a sterol that regulates the activity of DAF-12/nuclear receptor (Lee & Kenyon, 2009). In particular, *daf*-9 expression in XXX neurosecretory cells and hypodermis is important for maintaining lifespan at a high temperature, 25 °C (Lee & Kenyon, 2009). This neuroendocrine process acts in a systemic, tissue-nonautonomous fashion (Lee & Kenyon, 2009). Lifespan regulation by thermosensory neurons acts independently of chemosensory neurons, which regulate longevity at lower temperatures (Alcedo & Kenyon, 2004; Apfeld & Kenyon, 1999; Artan *et al.*, 2016; Lee & Kenyon, 2009). Overall this finding challenges the conventional view that the rate of living of ectotherms is elevated at high temperatures solely by increased chemical reaction rates. Instead, this report suggests that thermosensation actively regulates lifespan at higher temperatures, acting through sterol hormonal signaling.

A follow-up paper provides additional detail for the role of AFD neurons in translating temperature information into lifespan maintenance (Chen *et al.*, 2016) (Figure 1). *crh-1*, which encodes a homolog of the cyclic AMP-responsive element-binding protein (CREB) transcription factor, acts in AFD neurons and is essential for normal response to temperature (Nishida, Sugi, Nonomura, & Mori, 2011). CRH-1/CREB is phosphorylated at its serine 48 residue by CMK-1, Ca^{2+}/calmodulin-dependent protein kinase I (CaMKI) (Satterlee *et al.*, 2004), in AFD neurons in response to warm temperatures (Chen *et al.*, 2016). This event subsequently increases expression and secretion of FLP-6, an FMRF amide neuropeptide that acts on AIY interneurons (Chen *et al.*, 2016); the AIY neurons form a circuit with AFD neurons, required for maintaining lifespan at high temperatures (Lee & Kenyon, 2009). CREB-dependent upregulation of FLP-6 in turn increases DAF-9-mediated sterol hormone signaling in XXX neurosecretory cells (Chen *et al.*, 2016). In addition, FLP-6 signaling downregulates *ins-7* (Chen *et al.*, 2016), which encodes an agonistic insulin-like peptide that decreases lifespan (Murphy, Lee, & Kenyon, 2007; Murphy *et al.*, 2003). Thus, the thermosensory AFD/AIY neural circuit employs multiple neuroendocrine systems, such as FLP-6, INS-7, and sterol, for protecting worms from substantially reduced lifespan at high temperatures.

AFD thermosensory neurons regulate heat shock responses in a cell-nonautonomous manner

The heat shock response (HSR) helps protect organisms from detrimental effects of misfolded and aggregated proteins in the cytosol caused by proteotoxic stress, including heat (Morimoto, 2011). Heat shock factor 1 (HSF-1), a major transcription factor that induces genes encoding heat shock proteins (HSPs), including cytosolic molecular chaperones, is upregulated by HSR and ameliorates cellular damage (Barna, Csermely, & Vellai, 2018). In *C. elegans*, HSF-1 promotes longevity and resistance against various stressors (Altintas, Park, & Lee, 2016) and is also required for maintenance of normal lifespan at high temperatures (Lee & Kenyon, 2009).

Interestingly, HSR in somatic cells is nonautonomously modulated by the thermosensory neural circuit in *C. elegans* (Figure 1; Prahlad, Cornelius, & Morimoto, 2008; Prahlad & Morimoto, 2011; Tatum *et al.*, 2015), in addition to conventional cell-autonomous regulation. Functional thermosensory AFD neurons and AIY interneurons are required for HSF-1-dependent heat shock-induced cytosolic chaperones in non-neuronal tissues, suggesting intercellular regulation of HSR by thermosensory neurons (Prahlad *et al.*, 2008). The thermosensory neural circuit is also responsible for normal response to chronic protein misfolding in the absence of heat shock, and this activity is dependent on dense core vesicles (Prahlad & Morimoto, 2011). The activation of AFD neurons increases the release of serotonin from serotonergic neurons, which acts as an endocrine signal (Tatum *et al.*, 2015). This signal, in turn, activates the metabotropic serotonin receptor/SER-1 in the intestine, leading to the induction of HSF-1-regulated molecular chaperones (Tatum *et al.*, 2015). Thus, the AFD/AIY thermosensory neural circuit regulates protein homeostasis of somatic cells through neuroendocrine serotonin signaling at high temperatures.

The thermosensitive TRPA channel extends lifespan in response to cold temperatures

The first transient receptor potential (TRP) was identified in Drosophila (Minke, Wu, & Pak, 1975; Montell, 2011; Montell & Rubin, 1989). Subsequent discoveries of TRP family members

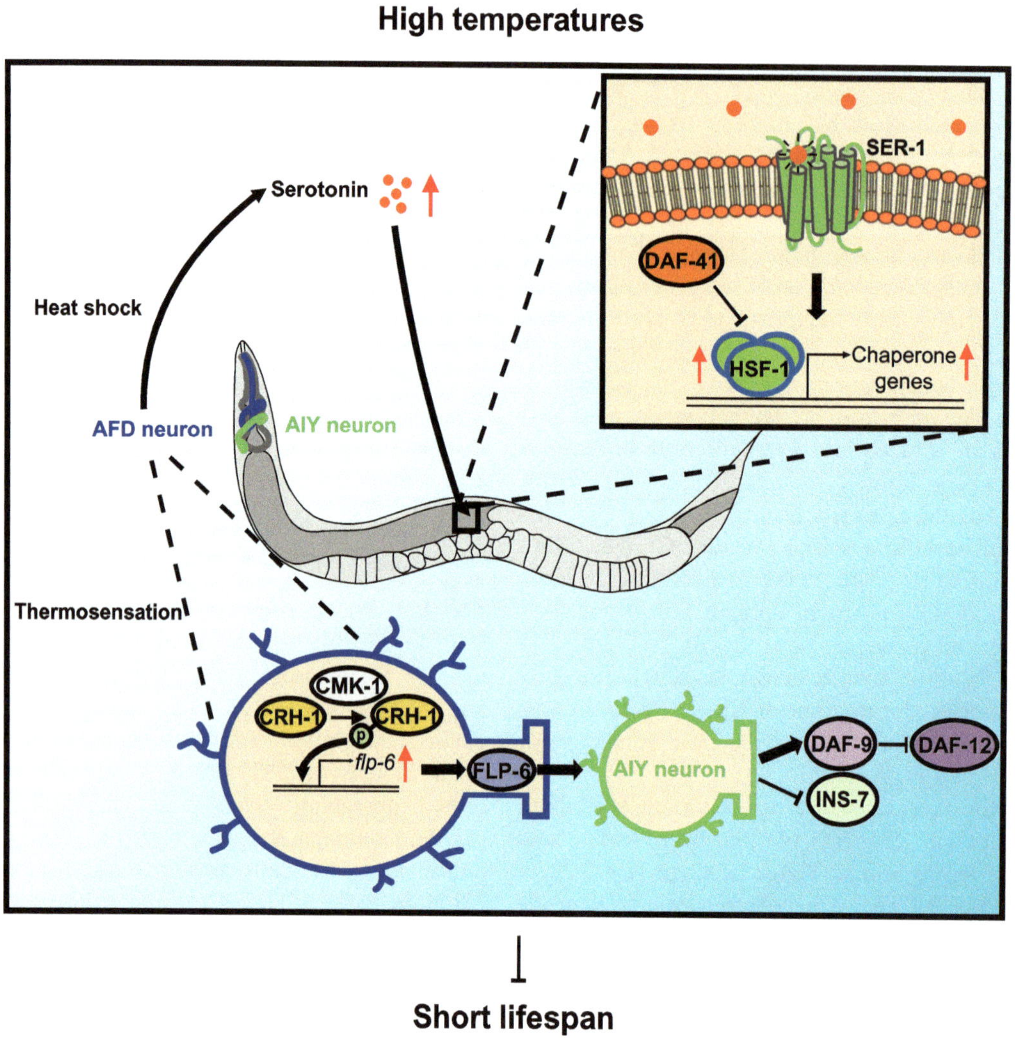

Figure 1. Thermosensory AFD neurons regulate lifespan at high temperatures and heat shock responses via endocrine signaling. At high temperatures, AFD neurons, the major thermosensory neurons in *C. elegans*, increase the activity of Ca^{2+}/calmodulin-dependent protein kinase 1 (CMK-1), which leads to the phosphorylation and activation of cyclic AMP-responsive element-binding protein (CREB)/CRH-1. This activation results in upregulation of FMRFamide neuropeptide/FLP-6, in turn elevating the level of insulin-like peptide 7 (INS-7) and altering sterol signaling composed of DAF-9/cytochrome P450 and DAF-12/nuclear receptor through AIY interneurons. This process is required for preventing shortening of lifespan at high ambient temperatures. Under heat shock conditions, the activation of AFD thermosensory neurons leads to release of serotonin from serotonergic neurons. Serotonin receptor SER-1 mediates serotonin signaling and cell-nonautonomous activation of heat shock factor 1 (HSF-1) in the intestine. These activities are required for inducing cytosolic chaperone genes to proper heat shock response at an organismal level. HSF-1 also acts downstream of DAF-41/co-chaperone p23 for modulating lifespan at high temperatures.

in multiple species indicate that functions of TRP channels are conserved across phyla, including *C. elegans* (Venkatachalam & Montell, 2007; Xiao & Xu, 2011). TRP channels play roles as molecular sensors in many cells, including neurons (Venkatachalam & Montell, 2007). For example, TRPA1 (TRP-ankyrin 1) is a cold sensor in mammals (Karashima *et al.*, 2009; Story *et al.*, 2003). *C. elegans* TRPA-1 is expressed in various tissues, including sensory neurons (Dupuy *et al.*, 2007; Kindt *et al.*, 2007; Xiao & Xu, 2011), and plays a role in cold sensation (Chatzigeorgiou *et al.*, 2010).

Thermosensitive TRPA-1 enhances longevity at cold temperatures (Figure 2; Xiao *et al.*, 2013). TRPA-1 mediates Ca^{2+} influx in response to cold temperatures, subsequently activating PKC-2/Ca^{2+}-sensitive protein kinase C (Xiao *et al.*, 2013). This activation leads to long lifespan by activation of DAF-16/FOXO, a major pro-longevity transcription factor that acts downstream of IIS (DiLoreto & Murphy, 2015; Kenyon, 2010a; Lee *et al.*, 2015), through phosphorylation by SGK-1/serum and glucocorticoid-regulated kinase 1 (Xiao *et al.*, 2013) and glutamate and serotonin neuroendocrine signals (Zhang *et al.*, 2018). Interestingly, cold exposure at an adult stage extends adult lifespan by upregulation of TRPA-1, but cold exposure during larval developmental stages shortens lifespan (Zhang *et al.*, 2015). This developmental stage-specific longevity response to low temperatures is mediated by DAF-16/FOXO, which differentially regulates gene expression in a stage-dependent manner (Zhang *et al.*, 2015). Thus, TRPA-1 appears to act as a cold sensor, which operates differentially at different developmental stages and transmits low-temperature signals to altering animal physiology for promoting longevity.

Co-chaperone p23 augments effects of high and low temperatures on *C. elegans* lifespan

In addition to molecular chaperones, co-chaperones promote proper protein folding in response to proteotoxic stresses

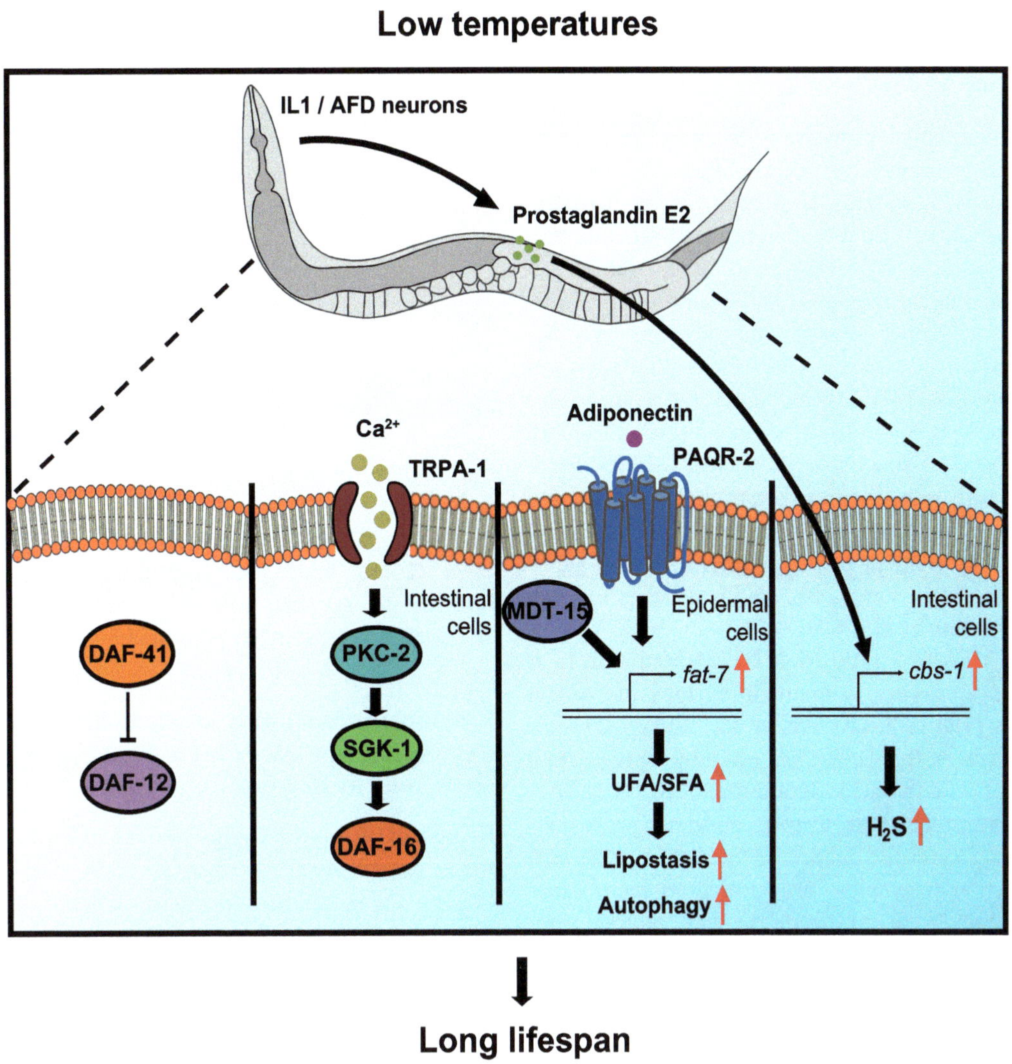

Figure 2. Factors that mediate longevity in response to low temperatures. A schematic model showing how various factors that act in temperature-sensing and lipid signaling pathways promote long lifespan in *C. elegans* at low temperatures. DAF-41/co-chaperone p23 downregulates DAF-12/nuclear receptor sterol signaling. Cold-sensitive TRPA-1/transient receptor potential-ankyrin 1 increases Ca^{2+} influx in cells, leading to the activation of DAF-16/FOXO via PKC-2/a Ca^{2+}-sensitive protein kinase C and SGK-1/serum and the glucocorticoid-regulated kinase 1 cascade to mediate cold-induced longevity. MDT-15/Mediator 15 and adiponectin receptor/PAQR-2 increase the expression of *fat-7*, which encodes a fatty acid desaturase required for increasing unsaturated fatty acid (UFA) to saturated fatty acid (SFA) ratios. This alteration leads to long lifespan by enhancing lipostasis and autophagy at low temperatures. Cold-sensing IL1 and AFD neurons stimulate the release of prostaglandin E2 (PGE2) from germline and increase the expression of cystathionine β-synthase-1 (*cbs-1*) in the intestinal cells, for the release of hydrogen sulfide (H_2S) that contributes to low temperature-induced longevity.

(Chen & Klionsky, 2011; Genest, Wickner, & Doyle, 2019). For example, co-chaperone p23 forms a complex with heat shock protein 90 (HSP90) and stabilizes the interaction between HSP90 and its client proteins, including steroid receptor and HSF-1 (Ali *et al.*, 2006; Freeman, Felts, Toft, & Yamamoto, 2000; Freeman & Yamamoto, 2002; Johnson & Toft, 1994; Zou, Guo, Guettouche, Smith, & Voellmy, 1998). p23 also acts as a chaperone independently of HSP90, by interacting with various additional proteins (Echtenkamp *et al.*, 2011). Knockout of p23 in mice leads to perinatal lethality (Grad *et al.*, 2006; Lovgren, Kovarova, & Koller, 2007; Nakatani *et al.*, 2007), and therefore its role in the physiology of adult mammals, such as aging and longevity, remains unknown.

In *C. elegans*, *daf-41*, which encodes the sole ortholog of p23, confers differential effects on lifespan at high and low temperatures (Figures 1 and 2; Horikawa, Sural, Hsu, & Antebi, 2015). Specifically, loss-of-function mutations in *daf-41/p23* decrease longevity at low temperatures, while extending short lifespan at high temperatures (Horikawa *et al.*, 2015). Thus, DAF-41/p23 appears to amplify differences in lifespan between low and high temperatures. At low temperatures, DAF-41/p23 increases lifespan by inhibiting DAF-12/nuclear receptor sterol signaling (Horikawa *et al.*, 2015). At high temperatures, DAF-41/p23 decreases lifespan via downregulating DAF-16/FOXO and HSF-1 (Horikawa *et al.*, 2015). Intriguingly, as described above, DAF-12/nuclear receptor and HSF-1 act downstream of thermosensory neurons for longevity and stress response at high temperatures (Chen *et al.*, 2016; Lee & Kenyon, 2009). In contrast, DAF-16/FOXO acts downstream of TRPA-1 for promoting low temperature-induced longevity (Xiao *et al.*, 2013). Therefore, elucidating mechanisms by which DAF-41/p23 amplifies lifespan changes at different temperatures is an important topic for future research.

MDT-15/Mediator 15 maintains lipid and protein homeostasis by enhancing UFA/SFA ratio and increases lifespan at low temperatures

Regulation of the relative proportion of unsaturated fatty acids (UFAs) and saturated fatty acids (SFAs) is important for lipid homeostasis (Holthuis & Menon, 2014). In ectotherms, lipid membrane fluidity is regulated at different temperatures by altering UFA/SFA ratios. For example, at low temperatures, increasing the UFA/SFA ratio in *C. elegans* is crucial for proper growth and development (Ma *et al.*, 2015; Svensk *et al.*, 2013). A key factor that regulates the UFA/SFA ratio is MDT-15/Mediator 15, a transcriptional co-regulator that promotes lipid homeostasis (Taubert, Van Gilst, Hansen, & Yamamoto, 2006). MDT-15/Mediator 15 acts together with transcription factors that regulate lipid metabolism, such as nuclear hormone receptor-49 (NHR-49) and sterol regulatory element-binding protein (SREBP)/SBP-1 (Taubert *et al.*, 2006; Van Gilst, Hadjivassiliou, Jolly, & Yamamoto, 2005; Yang *et al.*, 2006).

Regulation of UFA/SFA ratio by MDT-15/Mediator 15 is a key process for *C. elegans* in maintaining long lifespan at low temperatures (Figure 2; Lee, An, *et al.*, 2019). MDT-15/Mediator 15 increases the UFA/SFA ratio by upregulating the expression of a fatty acid desaturase, *fat-7* (Lee, An, *et al.*, 2019). This upregulation, in turn, alleviates proteotoxicity via enhancing cytosolic chaperone-mediated proteostasis systems, which may employ biological lipid membranes for proper function and contribute to low temperature-induced longevity (Lee, An, *et al.*, 2019). Importantly, addition of oleic acid, a mono-UFA enriched in olive oil, suppresses defects in proteostasis and short lifespan caused by mutations in *mdt-15* at low temperatures (Lee, An, *et al.*, 2019). Thus, maintenance of lipid fluidity by increasing UFA/SFA ratio is crucial for long lifespan at low temperatures. These results highlight the emerging importance of lipid homeostasis in animal health and longevity.

PAQR-2 acts as a thermal sensor and extends longevity by enhancing autophagy at low temperatures

Autophagy, a process that degrades damaged organelles and dysfunctional components in cytosol, regulates multiple biological and pathological processes, including development, metabolism and stress resistance, and cancer and neurodegenerative diseases (Parzych & Klionsky, 2014; Yang & Klionsky, 2020). Autophagy also plays an important role in longevity of diverse organisms (Hansen, Rubinsztein, & Walker, 2018; Wong, Kumar, Mills, & Lapierre, 2020). Mutations in genes that are crucial for autophagy decrease lifespan in model organisms ranging from nematodes to mammals (Hansen *et al.*, 2018). Conversely, overexpression of several autophagy regulators increases lifespan in multiple species (Aparicio, Hansen, Walker, & Kumsta, 2020; Hansen *et al.*, 2018). In *C. elegans*, autophagy is a common effector process acting downstream of various longevity regimens. Such regimens include dietary restriction, reduced IIS, decreased TOR signaling, and impaired mitochondrial respiration (Hansen *et al.*, 2018).

A recent study indicates that autophagy is also activated by low temperatures, and this activation contributes to longevity in *C. elegans* (Figure 2; Chen *et al.*, 2019). Low temperatures are likely sensed by signaling from PAQR-2, the *C. elegans* adiponectin receptor homolog that increases UFA/SFA ratio at low temperatures (Svensk *et al.*, 2013), although a direct temperature sensor for membrane fluidity remains unknown. PAQR-2 leads to upregulation of poly-UFAs, γ-linolenic acid and arachidonic acid (Chen *et al.*, 2019). This upregulation increases autophagy, particularly in the epidermis, and contributes to low temperature-induced longevity (Chen *et al.*, 2019). Thus, sensing low temperatures by PAQR-2 promotes cold-induced longevity by increasing poly-UFAs and autophagy-mediated proteostasis. Overall, this study (Chen *et al.*, 2019) and the report by Lee, An, *et al.* (2019) independently demonstrate the crucial role of homeostatic regulation of lipid composition in proteostasis and longevity at low temperatures.

Prostaglandin signals increase lifespan by decreasing age-associated exhaustion of germline stem cells at low temperatures

Reproductive capacity negatively correlates with longevity in many organisms (Partridge, Gems, & Withers, 2005). In *C. elegans*, the germline exacerbates somatic aging, and germline removal extends lifespan via multiple longevity signaling systems, including DAF-16/FOXO, and sterol signaling components, DAF-9/cytochrome P450 and DAF-12/nuclear receptor, at normal and high temperatures (Antebi, 2013; Kenyon, 2010b). This phenomenon appears to be conserved in flies and perhaps also in humans (Flatt *et al.*, 2008; Min, Lee, & Park, 2012). Therefore, findings regarding regulation of longevity by germline signaling may lead to insights into human longevity.

In contrast to responses at normal and high temperatures, the germline in *C. elegans* is required for longevity at low temperatures (Figure 2; Lee, Noormohammadi, *et al.*, 2019). Low temperatures sensed by cold-sensing IL1 and AFD neurons (Mori & Ohshima, 1995; Zhang *et al.*, 2018) increase the release of prostaglandin E2 (PGE2), a lipophilic hormone, from germline stem cells. This release leads to upregulation of cystathionine β-synthase-1 (*cbs-1*) in the intestine (Lee, Noormohammadi, *et al.*, 2019). CBS-1 then increases the production of hydrogen sulfide (H_2S), which maintains cellular homocysteine levels (Vozdek, Hnizda, Krijt, Kostrouchova, & Kozich, 2012) and induces longevity at low temperatures (Lee, Noormohammadi, *et al.*, 2019). Overall, this study demonstrates the novel role of germline lipophilic hormonal signals in cold-induced longevity in *C. elegans*.

Conclusions and perspectives

Environmental temperatures affect lifespan in various organisms, in particular ectotherms. In this review, we discussed studies that investigate relationships between ambient temperatures and biological systems, which affect lifespan in *C. elegans*. Thermosensory neurons in *C. elegans* modulate changes in lifespan and stress responses at different temperatures by influencing endocrine signaling and proteostasis. At high temperatures, thermosensory AFD neurons prevent lifespan shortening via sterol endocrine signaling. Further, AFD neurons increase serotonin release and enhance cytosolic chaperone responses under heat shock conditions. The TRPA-1 channel senses low temperatures and contributes to cold-induced longevity by upregulating DAF-16/FOXO. In addition, co-chaperone DAF-41/p23 amplifies the effects of high and low temperatures on lifespan changes. Various factors in lipid metabolism pathways modulate lifespan in response to low temperatures; these include MDT-15/Mediator 15, which maintains lipid homeostasis, PAQR-2, which promotes longevity by enhancing autophagy, and prostaglandin signals from germline to intestine, which regulates homocysteine levels.

Despite the identification of diverse cellular factors that modulate temperature-dependent lifespan changes in *C. elegans*, many questions remain unresolved. In particular, how all the identified factors that modulate temperature-dependent lifespan changes communicate in altering animal physiology remain unclear. The following are important specific questions: how do endocrine signaling molecules, including sterol, prostaglandin, serotonin, glutamate, insulin-like peptide, and FMRFamide neuropeptide, discriminate and coordinate upstream and downstream signaling that affects lifespan at different temperatures? What is the specific contribution of thermosensory neurons and molecular temperature sensors to changes in lifespan and aging rates at various temperatures? For example, AIY interneurons are crucial for thermotaxis and modulate lifespan at high temperatures (Chen *et al.*, 2016; Lee & Kenyon, 2009). However, the contribution of AIZ and RIA interneurons to thermotaxis remains controversial (Luo *et al.*, 2014; Mori & Ohshima, 1995), and the roles of these two interneurons in aging and longevity are unknown. Although cell-specific promoters are currently not available for AIZ neurons, it will be interesting to test whether disruption of these neurons modulates lifespan changes at different temperatures. In addition, many features of molecular mechanisms described in this review remain unclear. Specifically, identification of the receptor for FLP-6, which regulates DAF-9 and DAF-12 sterol hormone signaling in response to increases in temperatures, will be critical. Further, defining how lipid composition regulated by MDT-15 and PAQR-2 contributes to cytosolic chaperone- and autophagy-mediated proteostasis will be important. In addition, identifying transcription factors, which upregulate intestinal *cbs-1* in response to germline PGE-2, will help clarify the mechanisms of cold-induced longevity. Perhaps most importantly, investigation of conserved molecular temperature sensors and longevity-promoting factors, including TRPA-1, FOXO, and autophagy regulators, may be crucial to defining physiological responses to temperature changes in other organisms, including mammals.

Reducing core body temperature in mice results in an extended lifespan (Conti *et al.*, 2006). Several reports also suggest a correlation between increases in lifespan and decreases in body temperature in humans (Protsiv, Ley, Lankester, Hastie, & Parsonnet, 2020; Roth *et al.*, 2002; Simonsick, Meier, Shaffer, Studenski, & Ferrucci, 2016). These studies raise the possibility that low body temperature may underlie longevity in mammals. Further, the reproductive capacity and growth rate of wild *C. elegans* strains correlate with thermal preference (Anderson, Albergotti, Ellebracht, Huey, & Phillips, 2011), suggesting that temperature-dependent lifespan regulation is not a passive process, but an evolutionarily adaptive mechanism for fitness and reproduction. Although many aspects regarding the relationship between *C. elegans* lifespan regulation and ambient temperatures will be different from regulation in homeotherms, several factors that regulate longevity response to temperature changes, such as autophagy, FOXO, and chaperone systems, are evolutionarily conserved. Therefore, we believe that studies using *C. elegans* described in this review provide valuable information regarding the effects of temperatures on aging and longevity in mammals, including humans.

Acknowledgements

The authors thank all Lee laboratory members for help and discussion.

Disclosure statement

The authors report no conflict of interest.

Funding

This study is supported by the Korean Government (MSICT) through the National Research Foundation of Korea (NRF) (NRF-2019R1A3B2067745) to S-J.V.L.

References

Alcedo, J., & Kenyon, C. (2004). Regulation of *C. elegans* longevity by specific gustatory and olfactory neurons. *Neuron*, *41*(1), 45–55. doi:10.1016/S0896-6273(03)00816-X

Ali, M.M.U., Roe, S.M., Vaughan, C.K., Meyer, P., Panaretou, B., Piper, P.W., … Pearl, L.H. (2006). Crystal structure of an Hsp90–nucleotide–p23/Sba1 closed chaperone complex. *Nature*, *440*(7087), 1013–1017. doi:10.1038/nature04716

Altintas, O., Park, S., & Lee, S.J. (2016). The role of insulin/IGF-1 signaling in the longevity of model invertebrates, *C. elegans* and *D. melanogaster*. *BMB Reports*, *49*(2), 81–92. doi:10.5483/bmbrep.2016.49.2.261

Anderson, J.L., Albergotti, L., Ellebracht, B., Huey, R.B., & Phillips, P.C. (2011). Does thermoregulatory behavior maximize reproductive fitness of natural isolates of *Caenorhabditis elegans*? *BMC Evolutionary Biology*, *11*, 157. doi:10.1186/1471-2148-11-157

Antebi, A. (2013). Regulation of longevity by the reproductive system. *Experimental Gerontology*, *48*(7), 596–602. doi:10.1016/j.exger.2012.09.009

Aparicio, R., Hansen, M., Walker, D.W., & Kumsta, C. (2020). The selective autophagy receptor SQSTM1/p62 improves lifespan and proteostasis in an evolutionarily conserved manner. *Autophagy, 16*(4), 772–774. doi:10.1080/15548627.2020.1725404

Apfeld, J., & Kenyon, C. (1999). Regulation of lifespan by sensory perception in *Caenorhabditis elegans*. *Nature, 402*(6763), 804–809. doi: 10.1038/45544

Artan, M., Jeong, D.-E., Lee, D., Kim, Y.-I., Son, H.G., Husain, Z., … Lee, S.-J.V. (2016). Food-derived sensory cues modulate longevity via distinct neuroendocrine insulin-like peptides. *Genes & Development, 30*(9), 1047–1057. doi:10.1101/gad.279448.116

Barna, J., Csermely, P., & Vellai, T. (2018). Roles of heat shock factor 1 beyond the heat shock response. *Cellular and Molecular Life Sciences, 75*(16), 2897–2916. doi:10.1007/s00018-018-2836-6

Beverly, M., Anbil, S., & Sengupta, P. (2011). Degeneracy and neuromodulation among thermosensory neurons contribute to robust thermosensory behaviors in *Caenorhabditis elegans*. *The Journal of Neuroscience, 31*(32), 11718–11727. doi:10.1523/jneurosci.1098-11.2011

Biron, D., Wasserman, S., Thomas, J.H., Samuel, A.D., & Sengupta, P. (2008). An olfactory neuron responds stochastically to temperature and modulates *Caenorhabditis elegans* thermotactic behavior. *Proceedings of the National Academy of Sciences of the United States of America, 105*(31), 11002–11007. doi:10.1073/pnas.0805004105

Brenner, S. (1974). The genetics of *Caenorhabditis elegans*. *Genetics, 77*(1), 71–94.

Chatzigeorgiou, M., Yoo, S., Watson, J.D., Lee, W.-H., Spencer, W.C., Kindt, K.S., … Schafer, W.R. (2010). Specific roles for DEG/ENaC and TRP channels in touch and thermosensation in *C. elegans* nociceptors. *Nature Neuroscience, 13*(7), 861–868. doi:10.1038/nn.2581

Chen, Y., & Klionsky, D.J. (2011). The regulation of autophagy – Unanswered questions. *Journal of Cell Science, 124*(Pt 2), 161–170. doi:10.1242/jcs.064576

Chen, Y.C., Chen, H.J., Tseng, W.C., Hsu, J.M., Huang, T.T., Chen, C.H., & Pan, C.L. (2016). A *C. elegans* thermosensory circuit regulates longevity through crh-1/CREB-dependent flp-6 neuropeptide signaling. *Developmental Cell, 39*(2), 209–223. doi:10.1016/j.devcel.2016.08.021

Chen, Y.L., Tao, J., Zhao, P.J., Tang, W., Xu, J.P., Zhang, K.Q., & Zou, C.G. (2019). Adiponectin receptor PAQR-2 signaling senses low temperature to promote *C. elegans* longevity by regulating autophagy. *Nature Communications, 10*(1), 2602. doi:10.1038/s41467-019-10475-8

Conti, B., Sanchez-Alavez, M., Winsky-Sommerer, R., Morale, M.C., Lucero, J., Brownell, S., … Bartfai, T. (2006). Transgenic mice with a reduced core body temperature have an increased life span. *Science (New York, N.Y.), 314*(5800), 825–828. doi:10.1126/science.1132191

DiLoreto, R., & Murphy, C.T. (2015). The cell biology of aging. *Molecular Biology of the Cell, 26*(25), 4524–4531. doi:10.1091/mbc.E14-06-1084

Dupuy, D., Bertin, N., Hidalgo, C.A., Venkatesan, K., Tu, D., Lee, D., … Vidal, M. (2007). Genome-scale analysis of in vivo spatiotemporal promoter activity in *Caenorhabditis elegans*. *Nature Biotechnology, 25*(6), 663–668. doi:10.1038/nbt1305

Echtenkamp, F.J., Zelin, E., Oxelmark, E., Woo, J.I., Andrews, B.J., Garabedian, M., & Freeman, B.C. (2011). Global functional map of the p23 molecular chaperone reveals an extensive cellular network. *Molecular Cell, 43*(2), 229–241. doi:10.1016/j.molcel.2011.05.029

Flatt, T., Min, K.-J., D'Alterio, C., Villa-Cuesta, E., Cumbers, J., Lehmann, R., … Tatar, M. (2008). Drosophila germ-line modulation of insulin signaling and lifespan. *Proceedings of the National Academy of Sciences of the United States of America, 105*(17), 6368–6373. doi:10.1073/pnas.0709128105

Freeman, B.C., Felts, S.J., Toft, D.O., & Yamamoto, K.R. (2000). The p23 molecular chaperones act at a late step in intracellular receptor action to differentially affect ligand efficacies. *Genes & Development, 14*(4), 422–434.

Freeman, B.C., & Yamamoto, K.R. (2002). Disassembly of transcriptional regulatory complexes by molecular chaperones. *Science (New York, N.Y.), 296*(5576), 2232–2235. doi:10.1126/science.1073051

Genest, O., Wickner, S., & Doyle, S.M. (2019). Hsp90 and Hsp70 chaperones: Collaborators in protein remodeling. *The Journal of Biological Chemistry, 294*(6), 2109–2120. doi:10.1074/jbc.REV118.002806

Goodman, M.B., & Sengupta, P. (2019). How *Caenorhabditis elegans* senses mechanical stress, temperature, and other physical stimuli. *Genetics, 212*(1), 25–51. doi:10.1534/genetics.118.300241

Grad, I., McKee, T.A., Ludwig, S.M., Hoyle, G.W., Ruiz, P., Wurst, W., … Picard, D. (2006). The Hsp90 cochaperone p23 is essential for perinatal survival. *Molecular and Cellular Biology, 26*(23), 8976–8983. doi:10.1128/mcb.00734-06

Hansen, M., Rubinsztein, D.C., & Walker, D.W. (2018). Autophagy as a promoter of longevity: Insights from model organisms. *Nature Reviews. Molecular Cell Biology, 19*(9), 579–593. doi:10.1038/s41580-018-0033-y

Hedgecock, E.M., & Russell, R.L. (1975). Normal and mutant thermotaxis in the nematode *Caenorhabditis elegans*. *Proceedings of the National Academy of Sciences of the United States of America, 72*(10), 4061–4065. doi:10.1073/pnas.72.10.4061

Holthuis, J.C., & Menon, A.K. (2014). Lipid landscapes and pipelines in membrane homeostasis. *Nature, 510*(7503), 48–57. doi:10.1038/nature13474

Horikawa, M., Sural, S., Hsu, A.L., & Antebi, A. (2015). Co-chaperone p23 regulates *C. elegans* lifespan in response to temperature. *PLoS Genetics, 11*(4), e1005023. doi:10.1371/journal.pgen.1005023

Inada, H., Ito, H., Satterlee, J., Sengupta, P., Matsumoto, K., & Mori, I. (2006). Identification of guanylyl cyclases that function in thermosensory neurons of *Caenorhabditis elegans*. *Genetics, 172*(4), 2239–2252. doi:10.1534/genetics.105.050013

Jeong, D.E., Artan, M., Seo, K., & Lee, S.J. (2012). Regulation of lifespan by chemosensory and thermosensory systems: Findings in invertebrates and their implications in mammalian aging. *Frontiers in Genetics, 3*, 218. doi:10.3389/fgene.2012.00218

Johnson, J.L., & Toft, D.O. (1994). A novel chaperone complex for steroid receptors involving heat shock proteins, immunophilins, and p23. *The Journal of Biological Chemistry, 269*(40), 24989–24993.

Karashima, Y., Talavera, K., Everaerts, W., Janssens, A., Kwan, K.Y., Vennekens, R., … Voets, T. (2009). TRPA1 acts as a cold sensor in vitro and in vivo. *Proceedings of the National Academy of Sciences of the United States of America, 106*(4), 1273–1278. doi:10.1073/pnas.0808487106

Kenyon, C.J. (2010a). The genetics of ageing. *Nature, 464*(7288), 504–512. doi:10.1038/nature08980

Kenyon, C.J. (2010b). A pathway that links reproductive status to lifespan in *Caenorhabditis elegans*. *Annals of the New York Academy of Sciences, 1204*, 156–162. doi:10.1111/j.1749-6632.2010.05640.x

Kindt, K.S., Viswanath, V., Macpherson, L., Quast, K., Hu, H., Patapoutian, A., & Schafer, W.R. (2007). *Caenorhabditis elegans* TRPA-1 functions in mechanosensation. *Nature Neuroscience, 10*(5), 568–577. doi:10.1038/nn1886

Kuhara, A., Okumura, M., Kimata, T., Tanizawa, Y., Takano, R., Kimura, K.D., … Mori, I. (2008). Temperature sensing by an olfactory neuron in a circuit controlling behavior of *C. elegans*. *Science (New York, N.Y.), 320*(5877), 803–807. doi:10.1126/science.1148922

Lee, D., An, S.W.A., Jung, Y., Yamaoka, Y., Ryu, Y., Goh, G.Y.S., … Lee, S.-J.V. (2019). MDT-15/MED15 permits longevity at low temperature via enhancing lipidostasis and proteostasis. *PLoS Biology, 17*(8), e3000415. doi:10.1371/journal.pbio.3000415

Lee, H.J., Noormohammadi, A., Koyuncu, S., Calculli, G., Simic, M.S., Herholz, M., … Vilchez, D. (2019). Prostaglandin signals from adult germ stem cells delay somatic aging of *Caenorhabditis elegans*. *Nature Metabolism, 1*(8), 790–810. doi:10.1038/s42255-019-0097-9

Lee, S.J., & Kenyon, C. (2009). Regulation of the longevity response to temperature by thermosensory neurons in *Caenorhabditis elegans*. *Current Biology, 19*(9), 715–722. doi:10.1016/j.cub.2009.03.041

Lee, Y., An, S.W.A., Artan, M., Seo, M., Hwang, A.B., Jeong, D.-E., … Lee, S.-J.V. (2015). Genes and pathways that influence longevity in *Caenorhabditis elegans*. In N. Mori & I. Mook-Jung (Eds.), *Aging mechanisms: Longevity, metabolism, and brain aging* (pp. 123–169). Tokyo: Springer.

Lovgren, A.K., Kovarova, M., & Koller, B.H. (2007). cPGES/p23 is required for glucocorticoid receptor function and embryonic growth but not prostaglandin E2 synthesis. *Molecular and Cellular Biology, 27*(12), 4416–4430. doi:10.1128/mcb.02314-06

Luo, L., Clark, D.A., Biron, D., Mahadevan, L., & Samuel, A.D. (2006). Sensorimotor control during isothermal tracking in *Caenorhabditis elegans. The Journal of Experimental Biology, 209*(Pt 23), 4652–4662. doi:10.1242/jeb.02590

Luo, L., Cook, N., Venkatachalam, V., Martinez-Velazquez, L.A., Zhang, X., Calvo, A.C., … Samuel, A.D.T. (2014). Bidirectional thermotaxis in *Caenorhabditis elegans* is mediated by distinct sensorimotor strategies driven by the AFD thermosensory neurons. *Proceedings of the National Academy of Sciences of the United States of America, 111*(7), 2776–2781. doi:10.1073/pnas.1315205111

Ma, D.K., Li, Z., Lu, A.Y., Sun, F., Chen, S., Rothe, M., … Horvitz, H.R. (2015). Acyl-CoA dehydrogenase drives heat adaptation by sequestering fatty acids. *Cell, 161*(5), 1152–1163. doi:10.1016/j.cell.2015.04.026

Min, K.J., Lee, C.K., & Park, H.N. (2012). The lifespan of Korean eunuchs. *Current Biology: CB, 22*(18), R792–R793. doi:10.1016/j.cub.2012.06.036

Minke, B., Wu, C., & Pak, W.L. (1975). Induction of photoreceptor voltage noise in the dark in *Drosophila* mutant. *Nature, 258*(5530), 84–87. doi:10.1038/258084a0

Montell, C. (2011). The history of TRP channels, a commentary and reflection. *Pflugers Archiv: European Journal of Physiology, 461*(5), 499–506. doi:10.1007/s00424-010-0920-3

Montell, C., & Rubin, G.M. (1989). Molecular characterization of the *Drosophila* trp locus: A putative integral membrane protein required for phototransduction. *Neuron, 2*(4), 1313–1323. doi:10.1016/0896-6273(89)90069-X

Mori, I. (1999). Genetics of chemotaxis and thermotaxis in the nematode *Caenorhabditis elegans. Annual Review of Genetics, 33*, 399–422. doi:10.1146/annurev.genet.33.1.399

Mori, I., & Ohshima, Y. (1995). Neural regulation of thermotaxis in *Caenorhabditis elegans. Nature, 376*(6538), 344–348. doi:10.1038/376344a0

Morimoto, R.I. (2011). The heat shock response: Systems biology of proteotoxic stress in aging and disease. *Cold Spring Harbor Symposia on Quantitative Biology, 76*, 91–99. doi:10.1101/sqb.2012.76.010637

Murphy, C.T., Lee, S.J., & Kenyon, C. (2007). Tissue entrainment by feedback regulation of insulin gene expression in the endoderm of *Caenorhabditis elegans. Proceedings of the National Academy of Sciences of the United States of America, 104*(48), 19046–19050. doi:10.1073/pnas.0709613104

Murphy, C.T., McCarroll, S.A., Bargmann, C.I., Fraser, A., Kamath, R.S., Ahringer, J., … Kenyon, C. (2003). Genes that act downstream of DAF-16 to influence the lifespan of *Caenorhabditis elegans. Nature, 424*(6946), 277–283. doi:10.1038/nature01789

Nakatani, Y., Hokonohara, Y., Kakuta, S., Sudo, K., Iwakura, Y., & Kudo, I. (2007). Knockout mice lacking cPGES/p23, a constitutively expressed PGE2 synthetic enzyme, are peri-natally lethal. *Biochemical and Biophysical Research Communications, 362*(2), 387–392. doi:10.1016/j.bbrc.2007.07.180

Nishida, Y., Sugi, T., Nonomura, M., & Mori, I. (2011). Identification of the AFD neuron as the site of action of the CREB protein in *Caenorhabditis elegans* thermotaxis. *EMBO Reports, 12*(8), 855–862. doi:10.1038/embor.2011.120

Partridge, L., Gems, D., & Withers, D.J. (2005). Sex and death: What is the connection? *Cell, 120*(4), 461–472. doi:10.1016/j.cell.2005.01.026

Parzych, K.R., & Klionsky, D.J. (2014). An overview of autophagy: Morphology, mechanism, and regulation. *Antioxidants & Redox Signaling, 20*(3), 460–473. doi:10.1089/ars.2013.5371

Pearl, R. (1928). Experiments on Longevity. *The Quarterly Review of Biology, 3*(3), 391–407. doi:10.1086/394311

Prahlad, V., & Morimoto, R.I. (2011). Neuronal circuitry regulates the response of *Caenorhabditis elegans* to misfolded proteins. *Proceedings of the National Academy of Sciences of the United States of America, 108*(34), 14204–14209. doi:10.1073/pnas.1106557108

Prahlad, V., Cornelius, T., & Morimoto, R.I. (2008). Regulation of the cellular heat shock response in *Caenorhabditis elegans* by thermosensory neurons. *Science (New York, N.Y.), 320*(5877), 811–814. doi:10.1126/science.1156093

Protsiv, M., Ley, C., Lankester, J., Hastie, T., & Parsonnet, J. (2020). Decreasing human body temperature in the United States since the industrial revolution. *eLife, 9*, e49555. doi:10.7554/eLife.49555

Ramot, D., MacInnis, B.L., & Goodman, M.B. (2008). Bidirectional temperature-sensing by a single thermosensory neuron in *C. elegans. Nature Neuroscience, 11*(8), 908–915. doi:10.1038/nn.2157

Roth, G.S., Lane, M.A., Ingram, D.K., Mattison, J.A., Elahi, D., Tobin, J.D., … Metter, E.J. (2002). Biomarkers of caloric restriction may predict longevity in humans. *Science (New York, N.Y.), 297*(5582), 811. doi:10.1126/science.1071851

Satterlee, J.S., Ryu, W.S., & Sengupta, P. (2004). The CMK-1 CaMKI and the TAX-4 cyclic nucleotide-gated channel regulate thermosensory neuron gene expression and function in *C. elegans.* Current Biology, *14*(1), 62–68. doi:10.1016/j.cub.2003.12.030

Simonsick, E.M., Meier, H.C.S., Shaffer, N.C., Studenski, S.A., & Ferrucci, L. (2016). Basal body temperature as a biomarker of healthy aging. *Age (Dordrecht, Netherlands), 38*(5–6), 445–454. doi:10.1007/s11357-016-9952-8

Son, H.G., Altintas, O., Kim, E.J.E., Kwon, S., & Lee, S.V. (2019). Age-dependent changes and biomarkers of aging in *Caenorhabditis elegans. Aging Cell, 18*(2), e12853. doi:10.1111/acel.12853

Story, G.M., Peier, A.M., Reeve, A.J., Eid, S.R., Mosbacher, J., Hricik, T.R., … Patapoutian, A. (2003). ANKTM1, a TRP-like channel expressed in nociceptive neurons, is activated by cold temperatures. *Cell, 112*(6), 819–829. doi:10.1016/S0092-8674(03)00158-2

Sulston, J.E., Schierenberg, E., White, J.G., & Thomson, J.N. (1983). The embryonic cell lineage of the nematode *Caenorhabditis elegans. Developmental Biology, 100*(1), 64–119. doi:10.1016/0012-1606(83)90201-4

Svensk, E., Stahlman, M., Andersson, C.H., Johansson, M., Boren, J., & Pilon, M. (2013). PAQR-2 regulates fatty acid desaturation during cold adaptation in *C. elegans. PLoS Genetics, 9*(9), e1003801. doi:10.1371/journal.pgen.1003801

Tatum, M.C., Ooi, F.K., Chikka, M.R., Chauve, L., Martinez-Velazquez, L.A., Steinbusch, H.W.M., … Prahlad, V. (2015). Neuronal serotonin release triggers the heat shock response in *C. elegans* in the absence of temperature increase. Current Biology, *25*(2), 163–174. doi:10.1016/j.cub.2014.11.040

Taubert, S., Van Gilst, M.R., Hansen, M., & Yamamoto, K.R. (2006). A mediator subunit, MDT-15, integrates regulation of fatty acid metabolism by NHR-49-dependent and -independent pathways in *C. elegans. Genes & Development, 20*(9), 1137–1149. doi:10.1101/gad.1395406

Van Gilst, M.R., Hadjivassiliou, H., Jolly, A., & Yamamoto, K.R. (2005). Nuclear hormone receptor NHR-49 controls fat consumption and fatty acid composition in *C. elegans. PLoS Biology, 3*(2), e53. doi:10.1371/journal.pbio.0030053

Venkatachalam, K., & Montell, C. (2007). TRP channels. Annual Review of Biochemistry, *76*, 387–417. doi:10.1146/annurev.biochem.75.103004.142819

Vozdek, R., Hnizda, A., Krijt, J., Kostrouchova, M., & Kozich, V. (2012). Novel structural arrangement of nematode cystathionine β-synthases: Characterization of *Caenorhabditis elegans* CBS-1. *The Biochemical Journal, 443*(2), 535–547. doi:10.1042/bj20111478

Ward, S., Thomson, N., White, J.G., & Brenner, S. (1975). Electron microscopical reconstruction of the anterior sensory anatomy of the nematode *Caenorhabditis elegans.?2UU. The Journal of Comparative Neurology, 160*(3), 313–337. doi:10.1002/cne.901600305

Wasserman, S.M., Beverly, M., Bell, H.W., & Sengupta, P. (2011). Regulation of response properties and operating range of the AFD thermosensory neurons by cGMP signaling. *Current Biology: CB, 21*(5), 353–362. doi:10.1016/j.cub.2011.01.053

White, J.G., Southgate, E., Thomson, J.N., & Brenner, S. (1986). The structure of the nervous system of the nematode *Caenorhabditis elegans. Philosophical Transactions of the Royal Society of London.*

Series B, Biological Sciences, *314*(1165), 1–340. doi:10.1098/rstb.1986.0056

Wong, S.Q., Kumar, A.V., Mills, J., & Lapierre, L.R. (2020). Autophagy in aging and longevity. *Human Genetics*, *139*(3), 277–290. doi:10.1007/s00439-019-02031-7

Xiao, R., Liu, J., & Xu, X.Z. (2015). Thermosensation and longevity. *Journal of Comparative Physiology. A, Neuroethology, Sensory, Neural, and Behavioral Physiology*, *201*(9), 857–867. doi:10.1007/s00359-015-1021-8

Xiao, R., & Xu, X.Z. (2011). *C. elegans* TRP channels. *Advances in Experimental Medicine and Biology*, *704*, 323–339. doi:10.1007/978-94-007-0265-3_18

Xiao, R., Zhang, B., Dong, Y., Gong, J., Xu, T., Liu, J., & Xu, X.Z. (2013). A genetic program promotes *C. elegans* longevity at cold temperatures via a thermosensitive TRP channel. *Cell*, *152*(4), 806–817. doi:10.1016/j.cell.2013.01.020

Yang, F., Vought, B.W., Satterlee, J.S., Walker, A.K., Jim Sun, Z.-Y., Watts, J.L., ... Näär, A.M. (2006). An ARC/mediator subunit required for SREBP control of cholesterol and lipid homeostasis. *Nature*, *442*(7103), 700–704. doi:10.1038/nature04942

Yang, Y., & Klionsky, D.J. (2020). Autophagy and disease: Unanswered questions. *Cell Death and Differentiation*, *27*(3), 858–871. doi:10.1038/s41418-019-0480-9

Yu, Y.V., Bell, H.W., Glauser, D., Van Hooser, S.D., Goodman, M.B., & Sengupta, P. (2014). CaMKI-dependent regulation of sensory gene expression mediates experience-dependent plasticity in the operating range of a thermosensory neuron. *Neuron*, *84*(5), 919–926. doi:10.1016/j.neuron.2014.10.046

Zhang, B., Gong, J., Zhang, W., Xiao, R., Liu, J., & Xu, X.Z.S. (2018). Brain-gut communications via distinct neuroendocrine signals bidirectionally regulate longevity in *C. elegans*. *Genes & Development*, *32*(3–4), 258–270. doi:10.1101/gad.309625.117

Zhang, B., Xiao, R., Ronan, E.A., He, Y., Hsu, A.L., Liu, J., & Xu, X.Z. (2015). Environmental temperature differentially modulates *C. elegans* longevity through a thermosensitive TRP channel. *Cell Reports*, *11*(9), 1414–1424. doi:10.1016/j.celrep.2015.04.066

Zou, J., Guo, Y., Guettouche, T., Smith, D.F., & Voellmy, R. (1998). Repression of heat shock transcription factor HSF1 activation by HSP90 (HSP90 complex) that forms a stress-sensitive complex with HSF1. *Cell*, *94*(4), 471–480. doi:10.1016/S0092-8674(00)81588-3

The contribution of *C. elegans* neurogenetics to understanding neurodegenerative diseases

Joseph J. H. Liang, Issa A. McKinnon and Catharine H. Rankin

ABSTRACT
Since *Caenorhabditis elegans* was first introduced as a genetic model organism by Sydney Brenner, researchers studying it have made significant contributions in numerous fields including investigations of the pathophysiology of neurodegenerative diseases. The simple anatomy, optical transparency, and short life-span of this small nematode together with the development and curation of many openly shared resources (including the entire genome, cell lineage and the neural map of the animal) allow researchers using *C. elegans* to move their research forward rapidly in an immensely collaborative community. These resources have allowed researchers to use *C. elegans* to study the cellular processes that may underlie human diseases. Indeed, many disease-associated genes have orthologs in *C. elegans,* allowing the effects of mutations in these genes to be studied in relevant and reproducible neuronal cell-types at single-cell resolution *in vivo*. Here we review studies that have attempted to establish genetic models of specific human neurodegenerative diseases (ALS, Alzheimer's Disease, Parkinson's Disease, Huntington's Disease) in *C. elegans* and what they have contributed to understanding the molecular and genetic underpinnings of each disease. With continuous advances in genome engineering, research conducted using this small organism first established by Brenner, Sulston and their contemporaries will continue to contribute to the understanding of human nervous diseases.

Introduction

When Sydney Brenner first chose the roundworm *Caenorhabditis elegans* as an organism in which to thoroughly investigate the genetic regulation of development and of the nervous system, it would have been unimaginable to envision the landscape of research that this simple roundworm would contribute to in the present day. Exploiting the small and tractable nature of the transparent nematode, Brenner and his colleagues began efforts that would eventually culminate in determining the invariant cell lineage (Sulston, Schierenberg, White, & Thomson, 1983), a fully mapped nervous system connectome (White, Southgate, Thomson, & Brenner, 1986) and a completely sequenced genome (The *C. elegans* Sequencing Consortium, 1998). In particular, Sulston's foresight in seeing the significance of mapping the *C. elegans* genome led to the establishment of the nematode as the advanced genetic model organism it is today.

Consistent with Sulston's involvement in the battle against patenting the human genome, all of this work was made freely available to facilitate open science. Therefore, contributions that came out of Sydney Brenner's laboratory pioneered an immensely collaborative research culture that persists in the current *C. elegans* research community. Indeed, it is the works of Sydney Brenner, John Sulston, and their colleagues that seeded the establishment of *C. elegans* as an ideal model organism for studying the genetics of many key biological processes.

Advantages of C. elegans *in nervous system disease research*

With the discovery of the molecular pathways of programmed cell death (Hengartner & Horvitz, 1994; Sulston & Horvitz, 1977), and the striking degree of conservation of the apoptosis pathway across phyla, the idea that one could learn about mammalian cellular processes from a microscopic invertebrate organism became much more widely accepted. As the molecular pathways of cell death were mapped (Cook *et al.*, 2019), it did not take long for researchers to start using the same *C. elegans* techniques to delineate the molecular mechanisms underlying human neurodegenerative diseases. In many of these diseases, *C. elegans* research has contributed significantly to understanding gene function, identifing genetic pathways and determining how the pathological proteins behave *in vivo*.

Researchers using *C. elegans* to understand the mechanisms of human neurodegenerative diseases exploit the many advantages of this invertebrate model. Due to its relatively short generation cycle and life-span compounded with cheap maintenance costs, research conducted with this animal can be relatively rapid and inexpensive. *Caenorhabditis elegans* have a small nervous system composed of only 302 neurons that make up an invariant neuronal network, allowing for reductionist studies where cellular processes and mechanisms can be tracked and studied at the level of individual neurons (White *et al.*, 1986). Additionally, its transparent body that allows for visualization of all cell types at all stages of its condition *in vivo* is exceptionally useful in the context of research into neurodegenerative disorders – neuronal cell death and protein aggregation can be easily detected and quantified in transgenic strains expressing fluorescent proteins under tissue-specific or neuron-specific promoters (Chalfie, Tu, Euskirchen, Ward, & Prasher, 1994).

Advantages for neuroscientists include the knowledge that major neurotransmitter systems are conserved in *C. elegans*, and around 52.6% of human disease-related genes have a *C. elegans* ortholog (Kim, Underwood, Greenwald, & Shaye, 2018), suggesting that most important biochemical pathways (including many of those implicated in disease) are conserved across evolution. Perhaps the most vital characteristic of the nematode in the scope of research is its suitability for genetic experimental approaches that are not feasible in more complex models. Unbiased genetic screens can reveal genes essential to biological processes of interest and, in respect to neurodegenerative diseases, thus far have led to the discovery of conserved genes and pathways involved in synaptic vesicle trafficking, apoptosis and translational regulation (Figure 1). Because the functions of many disease genes discovered through genome sequencing of patients are often not fully understood, the low cost and availability of many genetic tools to investigate these candidate genes in *C. elegans* allow for rapid insights into gene function. Indeed, having a rich background in the development of tools for genetic manipulation means that researchers using *C. elegans* can employ a wide array of genome engineering techniques that are less practical in other model organisms.

Despite all its benefits, there are also challenges for those conducting neurogenetic research with *C. elegans*. Although *C. elegans* researchers typically enjoy the luxury of having an exhaustive and advanced toolset for genetic and biological manipulations, some techniques come with nuanced complications. For example, a popular method to disrupt gene function in *C. elegans* has been to silence the target gene using RNA interference (Fire *et al.*, 1998). Regardless of advances in the field to address this issue (Calixto, Chelur, Topalidou, Chen, & Chalfie, 2010), the nervous system remains relatively refractory to RNAi (Asikainen, Vartiainen, Lakso, Nass, & Wong, 2005; Kamath *et al.*, 2003; Timmons, Tabara, Mello, & Fire, 2003). Recently, developments in protein-degradation approaches to induce protein knock-down in specific cells have yielded exciting and promising results. One new method for degrading proteins *in vivo* repurposes the endogenous *C. elegans* protein degradation mechanism to remove proteins tagged with a small zinc-finger domain, ZF1 (Armenti, Lohmer, Sherwood, & Nance, 2014). Several adaptations have allowed this approach to target GFP-tagged proteins (Wang *et al.*, 2017) and are inducible optogenetically (Hermann, Liewald, & Gottschalk, 2015) or via administration of a biologically inert chemical (Martinez *et al.*, 2020; Zhang, Ward, Cheng, & Dernburg, 2015). The degron approach is likely to become the gold standard for inducible knockdown of protein function in the coming years as it addresses the problems researchers face with RNAi in *C. elegans* neurons. As libraries of driver lines for degrons and GAL4-UAS are currently being made for *C. elegans*, it is likely that high-throughput cell-specific protein degradation experiments like those routinely done in *Drosophila* will soon be possible. It is also important to remember that the researcher must be cognisant of the insights that can and cannot be gleaned from modelling mechanisms of disease in *C. elegans*. As many features of the human organ systems (ex. a circulatory system) and some key features of vertebrate neurons (ex. myelinated neurons) are lacking in *C. elegans*, some aspects of human physiology simply cannot be modeled with the worm.

Ongoing research has indicated that many neurodegenerative diseases show strong genetic influences. While this is particularly true for familial types of these diseases where genetics plays a larger role, research in cellular pathways that these genes are involved in may also yield an understanding of sporadic occurrences of these diseases. Here, an overview of the landscape of neurodegenerative disease research done using the nematode will be presented. Contributions made by the *C. elegans* research community to understand the effects of genes involved in Amyotrophic Lateral Sclerosis (ALS), Alzheimer's Disease (AD), Parkinson's Disease (PD) and Huntington's Disease (HD) specifically on the nervous system will be reviewed (Table 1) as they present a majority of the *C. elegans* literature on neurodegenerative disorders to date. The majority of the mutations in these disease-associated genes are autosomal dominant mutations, as a single copy of these variations can cause disease.

Amyotrophic lateral sclerosis

ALS is a neurodegenerative disease affecting approximately 1–2 out of every 100,000 individuals. Patients with ALS typically present with a progressive and selective loss of glutamatergic and cholinergic motor neurons in the brain and spinal cord (Turner *et al.*, 2013). There can be both familial (10% of all cases) and more common sporadic forms of ALS: the former applies to cases where there is more than one occurrence of the disease in the family, and the latter to *de novo* cases where there is no family history of the disease (Turner *et al.*, 2013). Both types, familial ALS (fALS) and sporadic ALS (sALS) have been investigated in *C. elegans* (Therrien & Parker, 2014). Since the advent of gene sequencing technologies, a number of genetic risk variants in human patients have been implicated in ALS and investigated in the worm. Here we will discuss research on the genes and the associated pathways currently believed to play major roles in fALS and sALS: SOD1, TDP-4, FUS,

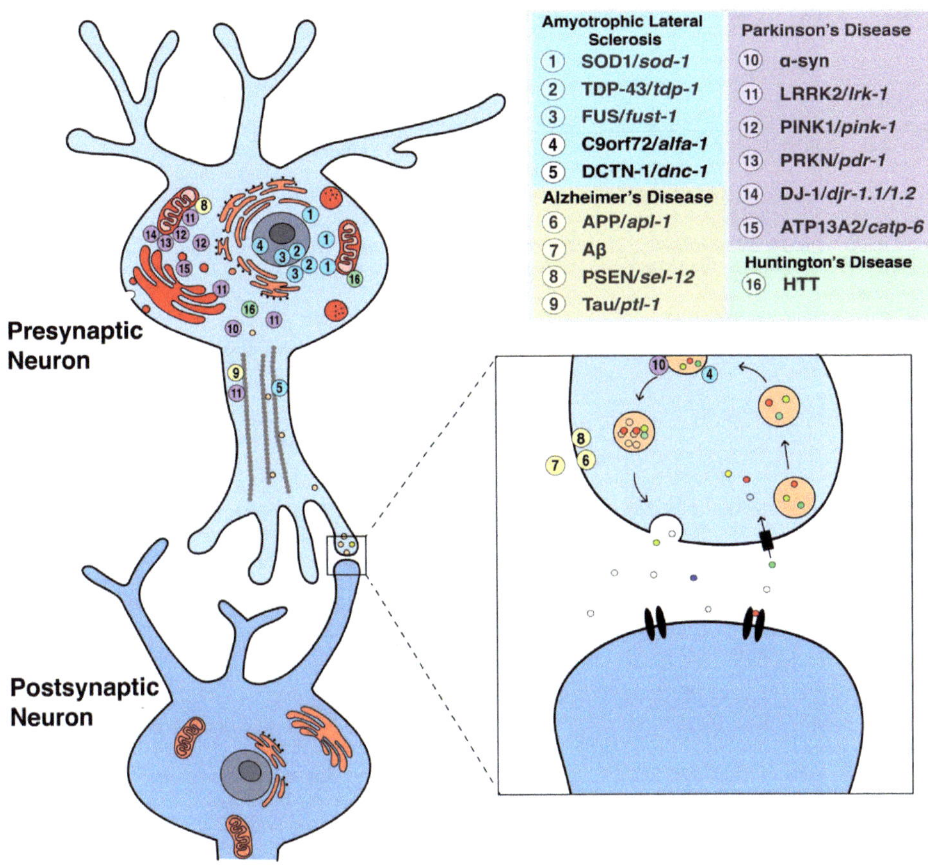

Figure 1. Genes associated with neurodegenerative diseases and their worm orthologs. Many genes associated with nervous diseases have worm homologs and have been well studied in *C. elegans*. SOD1 (1) has been shown to regulate super-oxide species in the cytoplasm and in the mitochondria and can mediate endoplasmic ER stress signaling. TDP-43 (2) and FUS (3) have nucleocytoplasmic shuttling properties and have been shown to mediate many aspects of gene regulation. A GGGGCC expansion in the non-coding region of C9orf72 (4) is the most frequent genetic cause of ALS in Europe and North America. C9ORF72 has been shown to localize in the presynaptic regions and the dipeptide repeat (DPR) protein product of the hexanucleotide expansion were shown to localize to the nucleus, though the exact functions and mechanisms of toxicity of C9ORF72 and DPRs are still under investigation. Dynactin-1 (DCTN1; 5) is a vital component of dynactin, a multimember complex important for retrograde transport. In motor neurons of sALS patients, DCTN1 is downregulated in motor neurons from an early disease stage. Amyloid precursor protein, or APP (6) is a transmembrane protein that can be cleaved and processed through multiple pathways. When APP is processed via the amyloidogenic pathway, it is cleaved by γ-secretase to secrete pathogenic Aβ (7) to the extracellular environment. Presenilin (PSEN, 8) is an important component of γ-secretase but is also shown to regulate mitochondrial processes that may induce neurodegeneration characteristic of AD. Tau (9) is important in mediating microtubule assembly in the axons, an important process for vesicle transport in neurons. α-synuclein (10) is the predominant component in Lewy Bodies seen in affected tissues of PD patients and has been postulated to interact with synaptic vesicles. LRRK2 (11) is a large protein with multiple domains of which the kinase domain is the most associated to PD pathogenesis. LRRK2 has been hypothesized to mediate many cellular processes, from mediating vesicle trafficking to mitochondrial stress response. PINK1 (12) and PRKN (13) mediate the mitophagy and mitochondria unfolded protein response in a process that LRRK2 (11) may also be involved in. DJ-1 (14) is another protein that has been shown to interact in a pathway parallel to the mitophagy pathway of PINK1 and PRKN, and is also involved in UPR. ATP13A2 (15) localizes to intracellular vesicles and when dysfunctional can enhance the accumulation of α-synuclein. Huntingtin (Htt; 16) is an ubiquitously expressed protein of and interacts with many partners – it has been implicated in many cellular processes from vesicle trafficking to regulating transcription. In many *C. elegans* models of neurodegenerative diseases caused by these genes, neurodegeneration has been observed.

C9ORF72 and DCTN1. Mutations in these genes predominantly result in autosomal dominant ALS.

SOD1 models to investigate familial ALS (fALS)

The discovery of the gene encoding superoxide dismutase (SOD1) in human patients in 1993 made it the first ALS gene to be identified (Rosen *et al.*, 1993). Mutations in SOD1 contribute to approximately 20% of all fALS patients (Cudkowicz *et al.*, 1997). SOD1 is an enzyme that catalyzes the conversion of harmful free superoxide radicals that are formed as by-products of mitochondrial respiration (e.g. O_{2-}) to more inert molecules in cells. SOD1 is synthesized in the cytoplasm and can then be transported to the mitochondria for maturation, where it remains to catalyze free radical conversion (Figure 1; Tafuri, Ronchi, Magri, Comi, & Corti, 2015). Pathogenic mutations in SOD1 can alter its protein structure and change interactions with its co-factors, effectively decreasing or hyper-activating its function (Kaur, McKeown, & Rashid, 2016). The *C. elegans* ortholog SOD-1 is structurally (~50% amino acid identity) and functionally similar to human SOD1. Loss-of-function (LOF) *sod-1* mutants showed increased superoxide radical levels, shorter life-span and were more sensitive to environmental stressors. In contrast, *sod-1* over-expression increased the levels of H_2O_2, a catalytic by-product of normal SOD1 enzymatic reactions (Yanase, Onodera, Tedesco, Johnson, & Ishii, 2009). This was also seen in a SOD1 (G93A) variant mouse model that showed a heightened production of H_2O_2, suggesting that this increase in H_2O_2 may be a pathway to ALS-associated neurodegeneration for GOF mutations (Yim

Table 1. A summary of *C. elegans* research on human neurodegenerative diseases.

Disease model, gene	Expression	Phenotype	Reference
Amyotrophic lateral sclerosis (ALS)			
SOD1/sod-1	Muscle-driven over-expression of human WT, A4V, G37R, G93A SOD1	More sensitive to paraquat-induced oxidative stress, slowed degradation of mutant proteins, protein aggregation.	Oeda *et al.* (2001)
SOD1/sod-1	Pan-neuronal over-expression of human WT, G85R SOD1-YFP fusion	Age-dependent locomotion defect.	Wang *et al.* (2009)
SOD1/sod-1	Muscle-driven (*unc-54*) over-expression of human WT, G85R, G93A, 127X SOD1-YFP fusion	Aggregation only in mutant SOD1, toxicity on body wall muscles.	Gidalevitz *et al.* (2009)
SOD1/sod-1	Single copy knock-in expression of human WT, A4V, H71Y, L84V, G85R, G93A	Accelerated aggregation of SOD1 and neurodegeneration under paraquat-driven oxidative stress, neurodegeneration also seen in loss-of-function variants.	Baskoylu *et al.* (2018)
TARDBP/tdp-1	*tdp-1(ok803 and ok781)* loss of function mutant	Locomotion defect, increased sensitivity to oxidative and osmotic stresses. Toxicity rescued by *TARDBP* over-expression. Decreased lifespan in *daf-2(e1370)* background.	Liachko *et al.* (2010), Zhang *et al.* (2012), Vaccaro, Tauffenberger, Ash, *et al.* (2012)
TARDBP/tdp-1	Pan-neuronal over-expression of mutant TDP-43	Uncoordinated movement and GABAergic motor neuron degeneration, elevated phosphorylation levels.	Ash *et al.* (2010), Liachko *et al.* (2010)
TARDBP/tdp-1	Pan-neuronal over-expression of TDP-43 C′ fragments	Abnormal synaptic transmission, however no neuronal loss.	Zhang *et al.* (2011)
TARDBP/tdp-1	Over-expression of wildtype and mutant (A315T) human TDP-43 in GABAergic motor neurons using the *unc-47* promoter	Age-dependent paralysis, impaired synaptic transmission, neurodegeneration with A315T TDP-43 over-expression. Drugs supressing ER stress response rescue phenotypes	Vaccaro, Tauffenberger, Aggad, *et al.* (2012), Vaccaro *et al.* (2013)
FUS/fust-1	Over-expression of mutant and truncated FUS variants throughout the nervous system	Motor deficits seen only in aggregation-causing variants, and cannot be rescued by expression of WT FUS.	Murakami *et al.* (2012)
FUS/fust-1	Single-copy knock-in of R524S and P525L equivalent mutations into *fust-1*	Impaired neuronal and muscular autophagy, intact neuronal ubiquitin proteasome pathway.	Baskoylu *et al.* (2019)
FUS/fust-1	Over-expression of wildtype and mutant (S57Δ) human FUS in GABAergic motor neurons using the *unc-47* promoter	Age-dependent paralysis, impaired synaptic transmission, neurodegeneration with S57Δ FUS over-expression.	Vaccaro, Tauffenberger, Aggad, *et al.* (2012)
C9orf72/alfa-1	*alfa-1(ok3062)* loss of function mutant	Age-dependent locomotion defect and motor neuron degeneration phenotype. Sensitive to osmotic stress. Enhances TDP-43(A315T) induced paralysis. Abnormal release of yolk blobs in embryo associated with impaired endolysosomal pathway.	Therrien *et al.* (2013), Corrionero and Horvitz (2018)
C9orf72/alfa-1	Over-expression of dipeptide repeat proteins (DPRs) in motor neurons using *unc-47* promoter.	Arginine containing DPRs induce age-dependent tneurodegeneration.	Rudich *et al.* (2017)
Dynactin-1/dnc-1	*dnc-1(or404ts)* loss-of-function mutants and motor-neuron specific RNAi using the *acr-2* promoter	Increase in size and number of protein aggregates in neuronal processes, reduced locomotion and reduce lifespan in *dnc-1(or404ts)* mutants. Motor neuron axon degeneration, increased number of autophagosomes and defective axonal transport in RNAi transgenic animals.	Koushika *et al.* (2004), Ikenaka *et al.* (2013)
Alzheimer's disease (AD)			
Aβ	Aβ over-expression in body wall muscle (*unc-54*)	Intracellular Aβ accumulation, formation of Aβ deposits in muscle cells, age-dependent paralysis phenotype as a result of neurodegeneration. Rescuing deposit-forming phenotype did not reduce toxicity.	Link (1995), Fay *et al.* (1998)
Aβ	Temperature-dependent pan-neuronal and muscle-specific over-expression of Aβ	Paralysis, neurodegeneration, impaired synaptic transmission.	Link *et al.* (2003)
Aβ	Pan-neuronal expression of full length $A\beta_{1\text{-}42}$	Age dependent deficits in neuromuscular behaviours and sensorimotor function. Mitochondrial dysfunction.	Fong *et al.* (2016)
APP/*apl-1*	*apl-1(tm385)* loss-of-function mutation, RNAi	Lethality, defective molting rescued by *apl-1* over-expression but not human *APP*, *APLP1*, or *APLP2*.	Wiese *et al.* (2010), Hornsten *et al.* (2007)
APP/*apl-1*	*apl-1(yn5)* gain-of-function mutation, *apl-1* over-expression	Incompletely penetrant L1 lethality, defects in brood size, movement and viability. Lethality partially rescued by *sel-12* knockdown.	Hornsten *et al.* (2007)
PSEN/sel-12	*sel-12(ar131)* loss-of-function mutant	Egg-laying phenotype, rescued by human PSEN1/PSEN2 over-expression.	Levitan and Greenwald (1995), Levitan *et al.* (1996)
PSEN/sel-12	Over-expression of $PSEN1_{s169del}$, clinical variant with intact Notch signalling	Defective γ-secretase signalling, neurodegeneration observed.	Zhang *et al.* (2020)
PSEN/sel-12	*sel-12(ar131, ty11 and ok2078)* loss-of-function mutant animals	Impaired mitochondria function and morphology, rescued by inhibiting calcium release from the ER, neurodegeneration in mechanosensory neurons – rescued by a *sel-12* variant with constitutively active γ-secretase activity.	Sarasija and Norman (2015), Sarasija *et al.* (2018)

(continued)

Table 1. Continued.

Disease model, gene	Expression	Phenotype	Reference
Tau/*ptl-1*	*ptl-1(ok621 and tm543)* loss-of-function	Incompletely penetrant lethality during embryogenesis, shortened lifespan for surviving animals. Impairment in sensing mechanical stimuli.	Chew *et al.* (2013), Gordon *et al.* (2008)
Tau/*ptl-1*	Pan-neuronal (*aex*-3) over-expression of human WT and mutant tau	Accumulation of insoluble phosphorylated tau, age-dependent neurodegeneration, greater toxicity in mutant tau.	Kraemer *et al.* (2003)
Parkinson's Disease (PD)			
α-synuclein	Pan-neuronal (*aex-3*), dopaminergic (*dat-1*), motor-neuron-specific (*unc-30, acr-2*) over-expression of WT and A53T α-synuclein	Age-dependent DA-neuron neurodegeneration and locomotion deficits.	Lakso *et al.* (2003), Cao *et al.* (2005)
α-synuclein	over-expression of WT, A53T and A30P in DA neurons	α-synuclein accumulation in cell bodies and neurites, but no neurodegeneration in CEP.	Kuwahara *et al.* (2006)
α-synuclein	Intestinal over-expression of A53T, A30P, and Δ1-32 ($P_{ges\text{-}1}$)	Increased mitochondrial fragmentation and hyperactivated UPR^{MT} and mitochondrial stress.	Martinez, Peterson, *et al.* (2017)
LRRK2/lrk-1	Loss-of-function *lrk-1(km17 and tm1898)* mutant animals	Defects in polarized sorting of SV's, deficits in dopamine-specific behaviours, defective synaptic vesicle trafficking (GFP::RAB-3). Decreases *pink-1* deficits in mitochondrial-mediated stress response.	Sakaguchi-Nakashima *et al.* (2007), Choudhary *et al.* (2017), Sämann *et al.* (2009), Martinez, Peterson, *et al.* (2017)
LRRK2/lrk-1	Pan-neuronal and DA neuron-specific over-expression of human WT or mutant LRRK2	Decreased dopamine levels, impaired DA-dependent behaviours, progressive degeneration in DA neurons. Heightened Kinase activity in disease variant (G2019S).	Saha *et al.* (2009), Liu *et al.* (2011), Yao *et al.* (2013)
PRKN/pdr-1	Loss-of-function *pdr-1(gk448, lg103, tm598, lg101, tm395)* mutant animals	Reduction in DA-dependent behaviours. Increased sensitivity to drug and metal-induced stresses. Accumulation of dysfunctional mitochondria and deficiencies in oxidative phosphorylation. Protein aggregation and increased sensitivity to proteotoxic stress.	Cooper *et al.* (2017), Springer *et al.* (2005)
PINK1/pink-1	Loss-of-function *pink-1(ok3538, tm1779)* mutant animals	Increased sensitivity to multiple stresses. Reduction in DA-dependent behaviours. Altered mitochondria morphology and deficiencies in oxidative phosphorylation.	Sämann *et al.* (2009), Cooper *et al.* (2017)
DJ-1/djr-1.1, djr-1.2	*djr-1.1(tm918), djr-1.2(tm1346)* loss-of-function deletion mutant animals	Increased sensitivity to oxidative stress, increased mitochondrial fragmentation and decreased ability to generate energy. No neurodegeneration or impairments in DA-dependent behaviours.	Cooper *et al.* (2017), Lee *et al.* (2012)
ATP13A2/catp-6	Loss-of-function *catp-6(unknown)* mutant animals	Accelerated DA neuron loss, deficits in DA-dependent behaviour, decreased rate of movement and increased sensitivity to multiple stresses. Can influence accumulation of α-synuclein.	Cooper and Van Raamsdonk (2018)
ATP13A2/catp-6	Over-expression of *catp-6* using the *dat-1* promoter, RNAi in body wall muscle cells over-expressing α-syn	No change in dopaminergic neurons. Co-expression with α-syn partially rescued α-syn induced neurodegeneration. RNAi enhanced misfolding of over-expressed α-syn.	Gitler *et al.* (2009)
Huntington's disease			
Htt	Over-expression of exon-1 fragments under $P_{osm\text{-}10}$ with varying length PolyQ repeats (Htn-Q2, Htn-Q23, Htn-Q95, Htn-Q150)	Longer PolyQ repeats (Htn-Q95, Htn-Q150) induced ASH neuron dysfunction and neurodegeneration, formation of Htt-positive cytoplasmic aggregates.	Faber *et al.* (2002)
Polyglutamate fragments	Over-expression of PolyQ-GFP fusion proteins in body wall muscles and neurons	Longer repeats induced protein aggregation and toxicity, and induced activation of heat shock proteins, disrupted protein homeostasis.	Brignull *et al.* (2006), Satyal *et al.* (2000), Morley *et al.* (2002), Gidalevitz *et al.* (2006)

et al., 1996). Interestingly, the life-span phenotypes observed may be independent of SOD-1 catalase activity and may instead be due to the protein's role in endoplasmic reticulum ER stress signalling, as the longevity phenotype observed in *sod-1* over-expression animals could be partially suppressed by inactivating mediators of the ER stress response (Cabreiro *et al.*, 2011).

Oeda *et al.* (2001) were the first to express human wild-type and fALS-associated (A4V, G37R, G93A) variants of SOD1 in *C. elegans*. They investigated whether animal survival or behaviour was affected by these human WT and disease-variant proteins when they were induced by heat shock ($p_{hsp16\text{-}2}$) or muscle-driven ($p_{myo\text{-}2}$) expression. While there were no observable changes in life-span or behaviour, animals expressing fALS-associated SOD1 were more sensitive to paraquat-induced oxidative stress. Most notably, oxidative stress slowed the degradation of these mutant proteins, leading to the accumulation of aggregates in muscle cells. Gidalevitz, Krupinski, Garcia, and Morimoto (2009) also investigated SOD1 aggregation by studying wild-type and mutant SOD1-YFP fusion proteins (G85R, G93A, 127X) in the body wall muscles using the *unc-54* promoter. They only saw aggregation of mutant SOD1 proteins that exerted mild toxic effects and led to muscle dysfunction. A separate study by Wang and colleagues (2009) showed that pan-neuronal expression of mutant SOD1 (G85R/G85R-YFP), but not wild-type SOD1, led to a clear progressive locomotion defect the severity of which correlated with increasing SOD1 aggregation. Both of these findings support the standing hypothesis that mutant SOD1 proteins are aggregation-prone – and that they potentially induce cellular damage by overwhelming the limited protein-folding machinery within the cell. A pan-neuronal G85R-YFP SOD1 transgenic model developed by Wang *et al.* (2009) was used in an RNAi screen using a sensitized background to identify modifier genes. Here, they discovered genes encoding chaperone components of SOD1 (TGF-β, SUMO and Topoisomerase I) that when knocked down greatly aggravated formation G85R SOD1 aggregates.

While these over-expression experiments present promising findings, mounting evidence from mammalian studies suggests that SOD1 over-expression models may not be ideal for studying mechanisms of these diseases. First, over-expression models are unable to determine the effects of loss-of-function SOD1 on ALS pathogenesis – recent knock-in mouse models have shown that both loss- and gain-of-function of SOD1 may contribute to disease symptoms (Joyce *et al.*, 2015; Şahin *et al.*, 2017). Interestingly, it has been reported that over-expression of even wild-type SOD1 protein has yielded disease-like phenotypes in rodent models (Graffmo *et al.*, 2013), presenting a newfound but necessary challenge to distinguish the impact of fALS-associated mutations from the effects of simple protein over-expression in these models. To address these issues, Baskoylu *et al.* (2018) complemented previous over-expression models by generating single-copy disease-associated mutations in conserved residues in SOD1 knock-in models in *C. elegans* (A4V, H71Y, L84V, G85R, G93A). In addition to showing that mutant SOD1 accelerated aggregation of SOD1 under paraquat-driven oxidative stress (which was consistent with previous work), it was also found that loss of *sod-1* function could also induce neurodegeneration following oxidative stress. More importantly, they discovered that SOD1 LOF mutations caused selective degeneration of glutamatergic neurons, while GOF mutations preferentially impacted cholinergic neurons (Baskoylu *et al.*, 2018). These findings are the first to demonstrate that neuron populations producing different neurotransmitters could show differential sensitivity to *SOD-1* alleles, and more work has yet to be done to show that this phenomenon occurs in the human ALS patient population. The fact that a very common disease allele with a large patient population, A4V SOD1, is a GOF mutation that leads to cholinergic spinal motor neuron degeneration and spares glutamatergic neurons in human patients suggests this hypothesis may be one worth pursuing. While more work is required to investigate the mechanisms by which different neuron types can have differential susceptibility to LOF or GOF mutations of SOD1, the answer may lie in the unique transcriptome and metabolic profiles of each neuron type. Current advancements in cell-type-specific and single-cell transcriptomics leave researchers hopeful and excited for new insights.

TDP-43 models

Pathways of sALS have also been investigated with *C. elegans*, particularly the TDP-43 and FUS pathways. TDP-43 was identified in 2006 as a major constituent of ALS aggregates; ALS patients express aggregates containing a hyperphosphorylated form of TDP-43 and the C-terminus cleaved fragment was found in the most affected regions of the brain and spinal cord (Neumann *et al.*, 2006). So far, more than 40 mutations in *TARDBP* (which encodes the TDP-43 protein) have been linked to ALS, most of them located in the protein-protein interaction domain at the C-terminus region (Al-Chalabi *et al.*, 2012). Although the exact functions of TDP-43 are still unclear, it has been shown that the DNA and RNA-binding protein participates in regulatory aspects of transcription and translation as it shuttles from the nucleus to the cytoplasm in a highly regulated manner (Figure 1, Ling, Polymenidou, & Cleveland, 2013). Work done with *C. elegans* has contributed to the current understanding that both the loss and gain of function of TDP-43 can cause ALS and that the functions of this protein are tightly controlled in the healthy cell (Mejzini *et al.*, 2019). *TARDBP* has a *C. elegans* ortholog called *tdp-1* that is primarily a nuclear protein expressed in most tissues including body wall muscles, pharynx and neurons. *tdp-1* LOF mutants exhibit several phenotypes including locomotion defects and higher sensitivity to oxidative and osmotic stresses (Liachko, Guthrie, & Kraemer, 2010; Zhang, Hwang, Hao, Talbot, & Wang, 2012). TDP-1 seems to be functionally conserved because over-expression of human TDP-43 rescued phenotypes observed in *C. elegans tdp-1* loss-of-function mutants (Zhang *et al.*, 2012). However, the worm ortholog does not share similarities with the human gene in all of its functional domains (*tdp-1* lacks the glycine-rich domain found in human TDP-43).

Presently, *C. elegans* ALS models expressing human TDP-43 are primarily based on over-expression of TDP-43 and are likely to represent a gain-of-function. Ash *et al.* developed the first TDP-43 ALS model in *C. elegans*, showing that over-expression of human TDP-43 and its worm ortholog *tdp-1* caused uncoordinated movement and degeneration of GABAergic motor neurons (Ash *et al.*, 2010). These results and observations were confirmed by another study expressing mutant TDP-43 pan-neuronally (Liachko *et al.*, 2010). These phenotypes were also highly correlated with elevated protein phosphorylation levels (Liachko *et al.*, 2010). Expressing only the C-terminus of human mutant TDP-43, aggregates of which were found in human patients, was sufficient to cause abnormal synaptic transmission, albeit with no neuronal loss (Zhang, Mullane, Periz, & Wang, 2011).

Interestingly, multiple studies found that *C. elegans* GABAergic motor neurons were particularly sensitive to TDP-43 perturbations and underwent neurodegeneration but the mechanism for this selective degradation remains unknown (Liachko *et al.*, 2010; Vaccaro, Tauffenberger, Aggad, *et al.*, 2012). However, two conserved cellular responses have been shown to play a role TDP-43 mediated toxicity. One the one hand, the ER-mediated unfolded protein response (UPR^{ER}) was activated in the GABA-ergic motor neuron over-expression model and drugs that activate this response yielded neuroprotective effects (Vaccaro *et al.*, 2013). On the other hand, the innate antimicrobial immune pathway mediated by *tir-1* was activated in the same model, and TDP-43-induced paralysis and neurodegeneration were reduced in mutants that inhibit the *TIR-1* pathway (Vérièpe, Fossouo, & Parker, 2015). It is possible that in addition to activating the unfolded protein response, misfolded TDP-43 protein is also recognized as a pathogen and elicits an immune response as a host defense mechanism. While more work has to be done to show that these conserved pathways are indeed implicated in ALS neuropathology, these findings suggest that the above response pathways can be interesting targets for pharmacological interventions to reduce neurodegeneration.

Several studies have shown that wild-type *TDP-1*/TDP-43 can contribute to neurodegeneration caused by the mutant protein, suggesting that TDP-43 induced toxicity may be explained by a dominant-negative mechanism (e.g. protein aggregation induced toxicity). The paralysis and motor neuron degeneration caused by mutant TDP-43 over-expression in only GABA-ergic motor neurons was reduced by the deletion of endogenous *tdp-1* (Vaccaro, Tauffenberger, Ash, *et al.*, 2012), although results from experiments done with pan-neuronal over-expression of human TDP-43 transgenes were less consistent (Ash *et al.*, 2010; Liachko *et al.*, 2010; Zhang *et al.*, 2012). In two separate experiments, pan-neuronal over-expression of the wildtype human TDP-43 in wildtype worms caused uncoordinated locomotion phenotypes (Ash *et al.*, 2010; Zhang *et al.*, 2011) as profound as the other full-length, ALS-linked alleles of TDP-43 (Q331K and M337V; Zhang *et al.*, 2011). Taken together, these findings were consistent with studies in other model organisms (Estes *et al.*, 2011; Wils *et al.*, 2010; Xu *et al.*, 2010) that demonstrated that wildtype TDP-43 may have neurotoxic effects and control of TDP-43 levels is important for cell survival. However, more work has yet to be done to elucidate the exact mechanisms through which TDP-43 can induce cytotoxicity in a dominant-negative manner. Findings from other TDP-43 aggregation models, both *in vitro* and *in vivo*, showed a low amount of aggregation could be observed when there was a high level of neurotoxicity, weakening the emphasis on aggregation itself as the vehicle for toxicity (Hergesheimer *et al.*, 2019). That being said, accumulating data show that increasing TDP-43 clearance can decrease aggregation and improve neuron survival and motor symptoms in ALS models (Hergesheimer *et al.*, 2019). These studies reinforce this perspective and support this approach as a therapeutic strategy. As of now, the consensus remains that TDP-43 aggregation leads to neurotoxicity – though the understanding on how exactly toxicity is induced is incomplete.

FUS models

Identification of TDP-43 as a causative factor for ALS led to the examination of related RNA-binding proteins in mammals for possible involvement in ALS pathogenesis. As a result, FUS was identified as another major causative factor for ALS (Kwiatkowski *et al.*, 2009). Like TDP-43, FUS contains an RNA-binding domain and a glycine-rich domain and can bind to both DNA and RNA. FUS is also involved in many of the same RNA processing activities as TDP-43 as it undergoes nucleocytoplasmic shuttling (Figure 1). Additionally, FUS is well conserved and has a *C. elegans* ortholog, *fust-1*. Over-expressing several mutants and truncated variants of human FUS throughout the wild-type worm nervous system showed that motor deficits only occurred in FUS variants that caused aggregation and could not be rescued by wild-type FUS, suggesting a gain-of-function mechanism (Murakami *et al.*, 2012). These phenotypes were replicated by Vaccaro, Tauffenberger, Aggad, *et al.* (2012) another study in which wildtype or disease-associated S57Δ FUS was expressed only in GABA-ergic motor neurons. Baskoylu *et al.* (2019) inserted equivalent patient mutations for R524S and P525L into *fust-1* and observed that these mutations impaired neuronal and muscular autophagy but left the neuronal ubiquitin-proteasome pathway intact. These findings replicated FUS over-expression experiments conducted in other model organisms to support the hypothesis that autophagy dysfunction may play a role in ALS FUS pathogenesis. Similar to TDP-43, FUS-associated neurodegeneration has been associated with the *TIR-1* mediated immune pathway in a GABAergic neuron-specific over-expression model, suggesting that the misfolded protein is potentially recognized as a pathogen (Vérièpe *et al.*, 2015).

C9ORF72 models

A recent landmark discovery in ALS research was the identification of a hexanucleotide GGGGCC (G_4C_2) repeat expansion in the *C9orf72* gene as the most frequent genetic cause of ALS in Europe and North America (DeJesus-Hernandez *et al.*, 2011; Renton *et al.*, 2011). In a healthy neuron, C9ORF72 is postulated to act as a guanine nucleotide exchange factor for Rab proteins and regulate endosomal trafficking and autophagy (Balendra & Isaacs, 2018). Therrien, Rouleau, Dion, and Parker (2013) were the first to explore functions of *C9orf72* in *C. elegans* using its ortholog, *alfa-1*. *alfa-1(ok3062)* deletion mutants exhibit age-dependent locomotion and motor neuron degeneration phenotype and are sensitive to osmotic stress (Therrien *et al.*, 2013). Additionally, the *alfa-1(ok3062)* deletion mutation enhanced TDP-43^{A315T}-induced age-dependent paralysis (Therrien *et al.*, 2013). A further study of *alfa-1(ok3062)* in embryos further elucidated its role in lysosome reformation and the maintenance of lysosomal homeostasis (Corrionero & Horvitz, 2018). Here, *alfa-1(ok3062)* mutant embryos were shown to have an abnormal release of yolk into the extra-embryonic fluid due to an impaired endolysosomal pathway; Over-expression of a codon-optimized long isoform of the human *C9orf72* (*ceC9orf72*) partially rescued this phenotype, suggesting functional conservation of the protein in this pathway and validating the potential for this model to be adapted to a future drug screen (Corrionero & Horvitz, 2018).

Another intriguing aspect of C9orf72 in ALS is the G_4C_2 repeats. Despite not being in the coding region of *C9orf72*, the G_4C_2 repeats can be translated into dipeptide repeat proteins (DPRs). To study DPR-induced toxicity, Rudich *et al.* (2017) expressed 50 repeats of four different GFP-tagged DPRs (GA, PA, GR and PR) under the motor neuron-specific *unc-47* promoter. Arginine-containing DPRs (GR, PR) are particularly toxic and are localized to the nucleus to induce toxicity (Rudich *et al.*, 2017). Further, mutations in *daf-16* and *hsf-1* that exhibit accelerated rates of aging enhanced toxicity, and the age-dependent deficits in locomotion were reduced by mutations in *daf-2* that exhibit delayed aging (Rudich *et al.*, 2017). Many of these findings were in line with DPR studies done in other model organisms (Balendra & Isaacs, 2018), suggesting a level of conservation in DPR-mediated toxicity and promising exciting discoveries in future nematode studies.

Dynactin models

Altered vesicular transport is also implicated in ALS pathologies. Slowed cargo transport is suggested to be an early pre-symptomatic phenotype observed in a mouse model of ALS (Williamson & Cleveland, 1999), and mutations in the human and mouse dynein-dynactin complex replicate phenotypes seen in ALS (Puls *et al.*, 2003). In addition, a *Drosophila* study reported the downregulation of *dynactin-1* and impairment in the fusion of autophagosomes in a TDP-43-depleted model (Xia *et al.*, 2016). Indeed, in characterizing the motor neuron-specific gene expression profiles of sALS patients, *dynactin-1(DCTN1)* was downregulated in motor neurons from an early disease stage (Jiang *et al.*, 2005; Jiang *et al.*, 2007). *DCTN1* is a vital component of dynactin, a multi-member complex associated with dynein in its role as a molecular motor for retrograde transport. An early study on the worm ortholog of DCTN1, *dnc-1*, and other components in the complex had interesting parallels with other ALS disease models (Koushika *et al.*, 2004). Of note, an increase in size and number of protein aggregates in neuronal processes, reduced locomotion and a reduced lifespan were observed in *C. elegans* dynein complex mutants (Koushika *et al.*, 2004). Ikenaka *et al.* (2013) further investigated the role of *dnc-1* with motor neuron-specific RNAi and observed an increased number of autophagosomes, a remarkable pathological feature of ALS patients. The group had since utilized this model as a behavior-based drug screening system to identify drugs that can ameliorate *dnc-1* mutant phenotypes (Ikenaka *et al.*, 2019).

At this point, the increasing use of next-generation sequencing of family pedigrees as well as large cohorts of ALS patients has led to the discovery of many other genetic risk factors associated with ALS (Mejzini *et al.*, 2019). Investigations into these genes have found many cellular mechanisms to be implicated in ALS pathogenesis, which include RNA processing, protein homeostasis, mitochondrial dysfunction and oxidative stress (Mejzini *et al.*, 2019). Despite all the research so far, a definitive understanding of causative pathogenic mechanisms in ALS remains incomplete. The temporal etiology of ALS is a challenge to current researchers, with downstream effects and potential causes feeding back into each other. A more advanced understanding of ALS genetics will hopefully yield a future with more personalized and effective treatment strategies, and it will be fascinating to see how the ongoing *C. elegans* research contributes to this effort.

Alzheimer's disease

AD is a neurodegenerative disease characterized by a progressive loss of memory due to the loss of neurons in the cerebral cortex, primarily but not exclusively at the frontotemporal association cortex. Most cases of AD occur sporadically in people aged 60 years or older without a clear inheritance pattern (late-onset AD). In 5% of the cases, AD symptoms appear earlier and are linked with autosomal dominant mutations in specific genes (familial or early-onset, fAD). Pathologically, patients with either form of AD feature the presence of extra-neuronal amyloid plaques (also called senile plaques) and intraneuronal neurofibrillary tangles (Kidd, 1964). Where senile plaques are shown to consist primarily of the amyloid-β (Aβ) peptide that originates from its cleavage precursor APP (amyloid precursor protein), the neurofibrillary tangles are composed of the microtubule binding protein tau. Both of these molecules have been extensively studied in *C. elegans* (Alexander, Marfil, & Li, 2014).

Amyloid-β models

A dominant, but recently challenged hypothesis for the cause of AD is the Aβ hypothesis (Hardy & Higgins, 1992; Karran & De Strooper, 2016; Selkoe & Hardy, 2016). In mammals, APP is part of a family of proteins that also includes APLP1 and APLP2: all three proteins have a large extracellular region containing conserved E1 and E2 domains, a single transmembrane domain and a small cytosolic domain (Kang *et al.*, 1987). This family of proteins is required for viability and brain development – they have essential and redundant Aβ-independent functions during development. Notably, out of the three proteins, only APP contains the Aβ domain. While the normal function of Aβ is not well understood, it has been shown that the cleaved protein is involved in a variety of cellular activities including activating kinase enzymes (Tabaton, Zhu, Perry, Smith, & Giliberto, 2010), protecting against oxidative stress (Zou, Gong, Yanagisawa, & Michikawa, 2002), functioning as a transcription factor (Maloney & Lahiri, 2011) and having a pro-inflammatory activity (Figure 1; Kagan *et al.*, 2012).

In *C. elegans*, there is only one APP ortholog, *apl-1*. APL-1 has many similarities to human APP – it shares homologous large extracellular domains (E1 and E2) and a cytosolic domain. However, unlike human APP, the nematode ortholog does not contain the Aβ sequence. Therefore, investigations of Aβ in *C. elegans* have been primarily carried out by over-expressing human Aβ in worms. The first attempt to use *C. elegans* to understand Aβ toxicity over-expressed human $A\beta_{1-42}$ using the body wall muscle promoter *unc-54* and saw Aβ accumulating intracellularly and forming amyloid deposits in muscle cells, causing an age-dependent paralysis phenotype (Link, 1995). Engineered substitutions of human Aβ in the same model that ameliorated the deposit-forming phenotype could not reduce toxicity, suggesting that amyloid itself is not the toxic agent (Fay, Fluet, Johnson, & Link, 1998). Immunoprecipitation studies in this model identified possible interactions with chaperone proteins, suggesting that the toxicity effect is primarily caused by the perturbation of protein homeostasis (Fonte *et al.*, 2002). In another study, *C. elegans* was engineered to have pan-neuronal or muscle-specific expression of $A\beta_{1-42}$ only when exposed to a temperature upshift (from 20 to 23 °C vs 16 °C; Link *et al.*, 2003). Using this system, Link and his colleagues were able to confirm that Aβ expression induced a deleterious phenotype: worms that had wild-type movement became paralyzed upon the temperature shift. Further studies using this model found a role for autophagy in countering Aβ toxicity. Accumulation of autophagosomes was observed in a transgenic line overexpressing human $A\beta_{1-42}$ in the body-wall muscles with the *myo-3* promoter, and in mutants, with impaired lysosomal action, both the accumulation of autophagosomes and the toxic effects of the human $A\beta_{1-42}$ transgene was exacerbated (Florez-McClure, Hohsfield, Fonte, Bealor, & Link, 2007). A micro-array analysis with the same model identified *aip-1*, a conserved regulator of proteasome function (Hassan, Merin, Fonte, & Link, 2009). Over-expressing *aip-1* reduced both Aβ accumulation and the severity of the paralysis phenotype induced by $A\beta_{1-42}$ over-expression (Hassan *et al.*, 2009). That said, whether *aip*-1 is involved in autophagy or another protein degradation pathway remains unclear and requires further investigation. Interestingly, over-expressing the human ortholog of AIP-1 (AIRAPL) also reduced Aβ toxicity in *C. elegans* (Hassan *et al.*, 2009), suggesting that designing drugs to activate AIRAPL may be a promising therapeutic strategy. Aside from producing and secreting Aβ, *apl-1* is also involved in many molecular signalling pathways. The extensive work done on uncovering the signalling pathways of *apl-1* in *C. elegans* has been discussed in detail in another review (Alexander *et al.*, 2014). The insulin signalling pathway has also been implicated in the similar muscle-driven $A\beta_{1-42}$ model but using the *unc-54* promoter. The clearance of $A\beta_{1-42}$ aggregation involves disaggregation and proteolysis, which were shown to be separate cellular activities (Bieschke, Cohen, Murray, Dillin, & Kelly, 2009). The disaggregase activity which appears to be the primary mechanism for $A\beta_{1-42}$ aggregate clearance in the *C. elegans* model was shown to be upregulated by the HSF-1 transcription factor that was in turn down-regulated by the insulin growth factor-1 receptor signalling pathway that also influences aging (Bieschke *et al.*, 2009). This finding was consistent with a previous study that showed that decreased insulin-receptor signalling caused by the *daf-2(e1370)* mutation suppresses $A\beta_{1-42}$ toxicity through an autophagy mechanism (Florez-McClure *et al.*, 2007). An important caveat must accompany these studies: The *C. elegans* models of Aβ described here were designed to over-express the $A\beta_{1-42}$ signal peptide. However, McColl *et al.* (2009) found that animals engineered to express this signal peptide construct actually harboured a truncated $A\beta_{3-42}$ protein due to a post-translational cleavage site at position 3 of $A\beta_{1-42}$, and this may well be the case for the other transgenic *C. elegans* expressing this peptide. Thus, studies must confirm which form of the protein is being produced, and if it is $A\beta_{3-42}$ researchers can refocus on understanding the toxic effects and processing of this peptide.

In a separate model, Fong *et al.* (2016) developed a strain with pan-neuronal full-length $A\beta_{1-42}$ expression and observed age-dependent deficits in neuromuscular behaviours and sensorimotor function. More interestingly, it was shown in this model that mitochondrial dysfunction can be an early event in AD pathogenesis as reduced energy metabolism preceded the global metabolic failure that coincided with the behavioural deficits observed in 12-day old animals (Fong *et al.*, 2016). Further studies with the same model identified that $A\beta_{1-42}$ reduced the activity of alpha-ketoglutarate, a rate-limiting step in the Krebs cycle and that this effect could be ameliorated by applying Metformin, an anti-diabetic drug and metabolic modular that improves substrate availability in the Krebs cycle (Teo *et al.*, 2019). Taken together, these studies point to metabolic dysfunction as an early and causative phenomenon in Aβ-induced AD pathology and may be a promising target for treatment and intervention.

Presenilin models

In humans, APP is cleaved by several secretases to yield its downstream products, and its products come from two primary cleavage pathways. In the non-amyloidogenic pathway, APP is cleaved first by α-secretase to release extracellular sAPPα. The remaining fragment is then cleaved by the γ-secretase to yield more of its components. In the amyloidogenic pathway, *β*-secretase first cleaves APP to release the sAPPβ fragment and the remaining fragment is cleaved by γ-secretase to yield Aβ in the extracellular space as one of its components (Figure 1). In *C. elegans*, no *β*-secretase activity that cleaves APP has been detected, suggesting that APL-1 is processed by the α/γ-secretase processing pathway. As α-secretase cleavage releases the extracellular fragment and subsequent cleavage with γ-secretase liberates the intracellular domain, more interest lies in the characterization of γ-secretase. Of the four core proteins that make up γ-secretase, presenilins (human PSEN1 and PSEN2) are the most studied due to its proteolytic function. Indeed, pathogenic mutations in PSEN1 account for a majority of cases of fAD (Janssen *et al.*, 2003).

Studies of presenilin are well-founded in *C. elegans* research: the first cellular functions of presenilin were determined by their homology to the *C. elegans* protein, SEL-12. Levitan and Greenwald (1995) showed that *sel-12* was involved in the Notch pathway to mediate cell fate decisions during development. Over-expression of both human PSEN1 and PSEN2 rescued a Notch-dependent egg-laying phenotype shown by *sel-12(ar171)* – suggesting a conserved function between human presenilin and *C. elegans sel-12* (Levitan *et al.*, 1996). Because of presenilin's role in Notch signalling, it was hypothesized that the notch pathway may also play a role in AD pathogenesis. Indeed, fAD-associated mutant PSEN1 proteins have also been shown to decrease Notch1 cleavage and Notch signalling in addition to its impacts on APP processing and Aβ generation, and over-expression of fAD-associated human PSEN1 variants (M146L, H163R, L286V and C410Y) in the *sel-12(ar131)* mutant background did not rescue the Notch-dependent egg-laying phenotype (Levitan *et al.*, 1996). However, this hypothesis has been questioned by Zhang *et al.* (2020) who demonstrated that a novel PSEN1 mutation identified in an fAD patient, $PSEN1_{s169del}$, altered APP processing and Aβ generation without affecting Notch1 cleavage and Notch signalling (overexpression of this variant in a *sel-12(ok2078)* mutant background rescued the egg-laying phenotype), suggesting that Notch signalling may not always be critical for AD pathogenesis. However, follow-up studies on Notch signalling's role in AD pathology associated neurodegeneration should be done before definite conclusions can be made. Hornsten *et al.* (2007) made an interesting observation bringing together *sel-12* and *apl-1*. When *sel-12* activity was reduced in transgenic worms with APL-1 over-expression, the lethality phenotype observed in *apl-1* GOF mutants was partially rescued – suggesting that *sel-12* may regulate APL-1 cleavage and/or trafficking.

Some work has been done to suggest that *sel-12* may contribute to neurodegeneration through its role in mediating mitochondrial function. Sarasija and Norman (2015) observed that *sel-12* loss-of-function mutations affect mitochondria morphology and function in a pathway independent of both Notch signalling and γ-secretase (Sarasija & Norman, 2015). The mitochondrial defects were attributed to the role of *sel-12* in mediating calcium release from the endoplasmic reticulum – inhibiting ER calcium release rescued these defects (Sarasija & Norman, 2015). Furthermore, these mitochondrial defects caused aberrant neuron morphology and neurodegeneration in mechanosensory neurons that could be rescued by a variant of *sel-12* that had constitutively inactive γ-secretase activity (D226A; Sarasija *et al.*, 2018). These neurodegeneration phenotypes were rescued with a supplement of the reactive oxygen species (ROS) scavenger MitoTEMPO, suggesting that the neurodegeneration was induced by mitochondrially generated ROS species (Sarasija *et al.*, 2018). This work has suggested a potential second pathway to neurodegeneration in PSEN pathogenesis in AD.

Tau models

In addition to human Aβ and the processing of its precursor protein APP, *C. elegans* researchers have also studied tau pathology. Tau is a major component of neurofibrillary tangles that appear prominently in AD and other neurodegenerative diseases. Tau belongs to a family of microtubule-associated protein (MAP) that share characteristic orthologous domains with each other (Goedert, Spillantini, Jakes, Rutherford, & Crowther, 1989). Tau is a predominant MAP expressed in axons and is responsible for promoting microtubule assembly and stability (Figure 1). Phosphorylation of tau regulates its ability to bind microtubules and has been shown to induce a conformational change that favours tubulin assembly. However, hyperphosphorylation of tau can impair its ability to bind with microtubules and lead to their disassembly instead (Bramblett *et al.*, 1993). In addition, hyperphosphorylated tau can self-aggregate into paired helical filaments, which can form the intracellular neurofibrillary tangles seen in AD patients (Alonso, Zaidi, Novak, Grundke-Iqbal, & Iqbal, 2001).

The *C. elegans* protein orthologous to tau is PTL-1: PTL-1 exists as two isoforms, differing in the number of tandem repeats (PTL-1a has five and PTL-1b has four). *ptl-1* has a high level of sequence homology with mammalian tau and shares a homologous microtubule-binding region in the C-terminal (McDermott, Aamodt, & Aamodt, 1996). Loss of *ptl-1* resulted in an incompletely penetrant lethality during embryogenesis and animals that escaped lethality showed normal development but a shortened life-span (Chew, Fan, Götz, & Nicholas, 2013). Interestingly, although the overall integrity of the microtubule structure appeared unaffected at the light microscope level, there was a significant reduction in the mutant's sensitivity to mechanical stimuli (Gordon *et al.*, 2008). These touch defects were enhanced in tubulin mutants (α- and *β*-tubulin), suggesting that the absence of PTL-1 disrupts mechanosensation independent of tubulin (Gordon *et al.*, 2008). Animals expressing a truncated form of PTL-1 that is missing the C-terminal microtubule-binding repeats had touch sensitivity like that of wild-type animals, showing that the N-terminal domain of PTL-1 is sufficient for its involvement in regulating this response to gentle touch (Chew *et al.*, 2013).

Kraemer *et al.* (2003) were the first to establish a *C. elegans* tauopathy model by expressing human wild-type and mutant tau pan-neuronally using the *aex-3* promoter. Observations from this model replicated key clinical pathologies: there was accumulation of insoluble phosphorylated tau, age-dependent neurodegeneration and greater toxicity in worms expressing mutant tau than wild-type tau. In addition, this model presented a clear organism-specific phenotype in an uncoordinated movement that has since been used for both classical forward and reverse genetic screens, yielding interesting findings. Loss-of-function mutations in *sut-1* and *sut-2* suppressed the tau-induced uncoordinated phenotype, leading to the hypothesis that these two genes may be involved in the pathological pathway activated by tau (Kraemer & Schellenberg, 2007). While there is currently no human protein orthologous to *sut-1*, *sut-2* is orthologous to human ZC3H14 – ZC3H14 has been shown to regulate Poly(A) tail length and expression of synaptic proteins critical for brain function in the mouse brain (Rha *et al.*, 2017). More work is required to identify the target mRNAs to which ZC3H14 binds, but the findings from *sut-1* and *sut-2* foreshadow a potential role in the tau pathway.

Increased phosphorylation of tau is considered a contributing factor to neurodegeneration AD pathology. Phosphorylation of tau was also explored in *C. elegans*. Brandt, Gergou, Wacker, Fath, and Hutter (2009) generated transgenic strains overexpressing either wild-type human tau or a pseudo-hyperphosphorylated (PHP) tau pan-neuronally. In the *C. elegans* body, wild-type human tau was hyperphosphorylated to a level similar to that observed in AD patients, and it also induced motor neuron defects that were reflected in an accelerated uncoordinated phenotype (Brandt *et al.*, 2009). Only PHP tau expressing animals showed a progressive tau protein aggregation (indicative of the insoluble characteristic of hyperphosphorylated tau seen in mammalian systems). Interestingly, neurodegeneration was not seen in transgenic animals expressing either WT or PHP tau (Brandt *et al.*, 2009).

There has been evidence in rodents that suggest a link between Aβ toxicity and tau hyperphosphorylation. Aβ upregulated the expression of a regulator of Calcineurin, a phosphatase that dephosphorylates tau (Lloret *et al.*, 2011), and transgenic rats exposed to or over-expressing Aβ peptides showed an increase in tau phosphorylation (Oliveira, Henriques, Martins, Rebelo, & da Cruz e Silva, 2015; Takashima *et al.*, 1996). Recently, two transgenic *C. elegans* studies investigated this interaction. In pan-neuronal transgenic models over-expressing $A\beta_{1-42}$ (P_{snb-1}::$A\beta_{1-42}$) and either pro-aggregating or anti-aggregating tau (P_{rab-3}::*tau*), Wang, Saar, Leung, Chen, and Wong (2018) observed exacerbated phenotypes in a host of behavioral and lifespan assays compared to transgenic animals expressing just the $A\beta_{1-42}$ or the tau transgene, as well as marked neurodegeneration of dopaminergic neurons. Here, pro-aggregating tau elicited a much larger effect than the anti-aggregating tau, suggesting this synergistic toxicity may be aggregation-related (Wang *et al.*, 2018). Transcriptomic analysis of this model revealed that these strains resulted in a large number of different up-regulated and down-regulated genes (Wang *et al.*, 2018). Benbow, Strovas, Darvas, Saxton, and Kraemer (2020) generated a different transgenic *C. elegans* model over-expressing both $A\beta_{1-42}$ (P_{snb-1}::$A\beta_{1-42}$) and human tau (P_{aex-3}::tau) pan-neuronally and found that these animals exhibited exacerbated swimming dysfunction and a reduced number of dye-filled sensory neurons (indicative of neurodegeneration) that was more severe than in animals harbouring either the $A\beta_{1-42}$ or tau transgene alone. Interestingly, there were no changes to the levels of tau phosphorylation or aggregation in this model, suggesting that at least in the *C. elegans* model, $A\beta_{1-42}$ and tau can induce toxicity in neurons additively but not by mediating tau phosphorylation (Benbow *et al.*, 2020).

Today, in addition to the continued efforts to understand more about the cellular mechanisms behind these major risk factors, an emphasis has been placed on the discovery of additional genetic risk factors of AD. The development of large international consortia for Azheimer's disease research, new genotyping technologies and falling costs of next-generation sequencing has drastically increased both the scale and quality of discovering new genetic risk factors (Bellenguez, Grenier-Boley, & Lambert, 2020). In the last 10 years, more than 40 AD-associated genes have been identified by GWAS and sequencing projects (Dourlen, Kilinc, Malmanche, Chapuis, & Lambert, 2019). However, the discovery of new risk factors begets the follow-up efforts to exhaustively characterize functional and pathogenic variants to further understanding the etiology of the disease. Given the advantages of the model organism, it is anticipated that work done with *C. elegans* will make important contributions in this direction in the future.

Parkinson's disease

Parkinson's disease (PD) is a neurodegenerative disorder of the central nervous system that impacts more than 10 million patients globally. Patients with PD suffer from progressive impairment of motor and cognitive functions over time. Pathologically, this disease is characterized by the loss of dopaminergic (DA) neurons in the substantia nigra pars compacta (SNc) and the accumulation of Lewy bodies (LB) in PD patient brains (Fahn & Sulzer, 2004; Holdorff, Rodrigues e Silva, & Dodel, 2013). The most abundant component of LBs is α-synuclein (α-syn), a protein normally found in presynaptic terminals and nuclei of neurons. Genetic analyses of human patients have identified a number of other genes that may be involved in PD pathogenesis in addition to the gene encoding α-syn, *SNCA* (Nalls *et al.*, 2014). While there is no *C. elegans SNCA* ortholog, the nematode genome encodes orthologs of a number of these other genetic risk factors (Martinez, Caldwell, & Caldwell, 2017). Numerous groups have investigated the molecular pathways of these genes in *C. elegans* (Martinez, Caldwell, *et al.*, 2017). Here, *C. elegans* genetic models of α-syn, LRRK2 and genes involved in the mitophagy pathway, will be described. Mutations in *SNCA* and *LRRK2* are largely responsible for autosomal dominant forms of PD, while mutations in *PINK1*, *Parkin*, *DJ-1* and *ATP13A2* account for cases of PD that display an autosomal recessive mode of inheritance. Note that *C. elegans* researchers have also modeled PD in a pharmacological or drug/metal-induced neurodegeneration context – this body of work has been covered in another recent review (Cooper & Van Raamsdonk, 2018).

α-synuclein models

Despite the absence of a *C. elegans* α-syn ortholog, a plethora of *C. elegans* PD models have been generated by over-expressing the human gene in the worm to examine α-syn toxicity as LBs are the neuropathological hallmark of the disease. α-synuclein is a presynaptic neuronal protein with functions that remain largely uncharacterized (Figure 1). However, evidence suggests that α-syn can control the supply of synaptic vesicles in the neuron terminals and regulate dopamine release. In its normal state, α-syn is present abundantly in the brain and especially in the cytosol of DA neurons, where it has been shown to interact with membranes of synaptic vesicles containing dopamine (Bellani *et al.*, 2010). PD-causing mutations in α-syn have been suspected to induce misfolding and accelerate aggregation compared to the WT protein. Lakso *et al.* (2003) studied α-syn toxicity in *C. elegans* by expressing human wild-type and mutant (A53T) α-syn-GFP fusion proteins in all neurons (*aex-3* promoter), dopamine neurons (*dat-1*), or motor neurons (*unc-30* for GABA-ergic and *acr-2* for cholinergic) and examining dopaminergic neuron loss. Pan-neuronal over-expression of wild-type and A53T α-syn both induced degradation of DA neurons and locomotion deficits at a comparable level. However, although DA neuron-specific over-expression did induce DA neuron loss, doing so did not induce locomotor deficits, and cholinergic neuron-specific over-expression of α-syn appeared to have no effect on DA neuron loss or locomotor function, regardless of the α-syn genotype (Lakso *et al.*, 2003). Interestingly, over-expressing α-syn in GABAergic neurons induced DA neuron toxicity. These results suggest that regulation of α-syn expression may play a role in neuronal toxicity, and toxic damage inflicted by α-syn on dopamine neurons may originate from a different neuron population (Lakso *et al.*, 2003). DA neuron degeneration induced by α-syn toxicity was confirmed in another study over-expressing human α-synuclein in only DA neurons – here, Cao, Gelwix, Caldwell, and Caldwell (2005) determined that the neurodegeneration was progressive and age-dependent. In this latter model, neuroprotective roles of ER-associated proteins with chaperone-like activity was found, suggesting a link between α-syn toxicity and ER-Golgi vesicular trafficking (Cao *et al.*, 2005; Cooper *et al.*, 2006). Interestingly, findings from another study appear to partially contradict these effects of α-syn on neurodegeneration: Kuwahara *et al.* (2006) saw the accumulation of α-syn in cell bodies and neurites of DA neurons after over-expressing wild-type or mutant (A53T and A30P) α-syn, but observed that there was no significant loss of the dopaminergic neuron CEP cell bodies.

The molecular pathways of synaptic endocytosis are also implicated in α-syn toxicity. Using the same pan-neuronal α-syn model described above, Kuwahara *et al.* (2008) carried out a large scale RNAi feeding screen involving 1673 genes in a neuronal RNAi sensitive background (*eri-1*) that implicated genes involved in synaptic endocytosis in α-syn toxicity. Of the eleven genes that enhanced transgenic α-syn induced phenotypes, four of them (*apa-2*, *aps-2*, *eps-8*, and *rab-7*) are involved in endocytosis, and two of them (*apa-2*, *aps-2*) are subunits of the AP-2 complex (Kuwahara *et al.*, 2008), which recruits clathrin and cargo receptors to initiate endocytosis. RNAi knockdown of another synaptic endocytosis gene, *unc-11*, also enhanced transgenic α-syn induced touch insensitivity phenotypes in P_{mec-7}::α-syn transgenic models, further emphasizing a connection between endocytosis and α-syn toxicity (Kuwahara *et al.*, 2008). Findings from studies in other model organisms also suggest the existence of a pathogenic link between endocytosis and α-syn. Namely, α-syn inhibits activity of an PLD2, an upstream activator of AP-2 (Ahn *et al.*, 2002), and that normal α-syn is required for the formation and/or maintenance of a reserve pool or presynaptic vesicles, a process that relies on endocytosis heavily (Cabin *et al.*, 2002). As endocytosis and exocytosis are two closely linked processes especially in the context of proper functioning at neuronal synapses, it should be no surprise that α-syn would be implicated in both rather than one or the other. The exact roles of α-syn in both of these processes are yet to be fully understood and there remains much to be explored in future research (Huang *et al.*, 2019).

LRRK2 models

Mutations in the leucine-rich repeat kinase 2 (LRRK2/PARK8) gene have also been implicated in the neuropathology of autosomal-dominant Parkinson's disease (Paisán-Ruíz *et al.*, 2004). In humans, LRRK2 is expressed in many organs including the brain, gastrointestinal tract, muscle tissues and is highly expressed in the lung and kidney. LRRK2 is a large protein with multiple conserved domains (including a GTPase domain and a kinase domain). While the protein has been shown to be involved in mitochondrial function, autophagy, cytoskeletal remodeling, protein synthesis and vesicle transport, additional functions of LRRK2 are still under investigation (Figure 1; Wallings, Manzoni, & Bandopadhyay, 2015). PD-causing mutations in LRRK2 typically center around the GTPase and kinase domains and induce hyperactivity of the protein. One of the most common PD-causing mutations, G2019S, is located in the kinase domain and augments its activity towards its intended substrates.

The *C. elegans* ortholog of *LRRK2*, *lrk-1*, is expressed broadly in many cell types including neurons, muscles, hypodermis and intestine and the protein has been observed to associate with the Golgi apparatus (Fukuzono *et al.*, 2016). Mutant animals with LOF mutations in *lrk-1* exhibit defects in polarized sorting of synaptic vesicles and dopamine-specific behaviours (Sakaguchi-Nakashima, Meir, Jin, Matsumoto, & Hisamoto, 2007). Transgenic models over-expressing wild-type and mutant human LRRK2 have been established in *C. elegans*. In models over-expressing these transgenes pan-neuronally (Saha *et al.*, 2009) and specifically in the DA neurons (Liu *et al.*, 2011), researchers saw that both wild-type and mutant LRRK2 decreased dopamine levels, impaired dopamine-dependent behaviours and caused progressive degeneration in DA neurons. Another over-expression model of LRRK2 in only DA neurons found that the heightened kinase activity of LRRK2(G2019S) may be key to this development of age-dependent neurodegeneration as treatment with kinase inhibitors arrested neurodegeneration in these transgenic worms (Yao *et al.*, 2013).

Another role of LRRK2 investigated in *C. elegans* is its involvement in regulating synaptic vesicle protein trafficking. Synaptic vesicle proteins are essential for neurotransmission, and *lrk-1* has been shown to localize in the Golgi and regulate trafficking and distribution of synaptic vesicles in presynaptic boutons (Wallings *et al.*, 2015). By observing the vesicles in PLM neurons using the synaptic vesicle marker GFP::RAB-3, Choudhary *et al.* (2017) observed that *lrk-1* mutant animals were defective in one form of synaptic vesicle trafficking and that this phenotype could be restored by the over-expression of *lrk-1*. In addition, Choudhary *et al.* (2017) found that the localization of an integral regulator of synaptic vesicle size (*unc-101*) is mediated by *lrk-1*. These findings provide support for the hypothesis that neurodegeneration seen in PD patients arises from a chronic impairment in neurotransmitter transport mechanisms that are mediated by LRRK2. One possible way to explain this phenomenon is that diseased neurons over-exert themselves as a compensatory mechanism to make up for the reduced efficiency in neurotransmitter transport and become more susceptible to metabolic damage as a result.

Role of PD genes in mitochondria

Accumulating evidence suggests that mitochondria-related processes are also important in PD pathology. Indeed, several PD-associated genes have functions involving mitochondria. Genes encoding the phosphatase and tensin homolog (PTEN)-induced putative kinase 1 (PINK1) and E3 ubiquitin ligase (Parkin) are both crucial regulators of mitochondrial autophagy (mitophagy) and have been heavily implicated in PD (Figure 1; Chen, Turnbull, & Reeve, 2019). Both of these genes have *C. elegans* orthologs (*pink-1* and *pdr-1*, respectively), and these genes have been shown to mediate mitochondrial morphology, mitophagic potential and mitochondrial accumulation. LOF mutations in these genes render animals more susceptible to neurodegeneration after oxidative damage (UVC and 6-OHDA; Hartman *et al.*, 2019). Although LOF mutations in *pink-1* and *pdr-1* do not cause neurodegeneration without any drug-induced perturbations, they do display deficits in dopamine-dependent behaviour, suggesting some level of neuronal dysfunction (Cooper *et al.*, 2017). These genes may also mediate neurodegeneration through the mitochondrial unfolded protein response (UPR^{MT}; Cooper *et al.*, 2017). Protein folding environments of the mitochondrial matrix can be prone to dysregulation as a result of inherent genetic inefficiencies or environmental factors (i.e. mitochondrial stress), upon which the UPR^{MT} is activated to restore homeostasis to the proper protein-folding environment. Impairment in this compensatory mechanism can induce a necrotic form of neurodegeneration in neurons that undergo excessive stress (Martinez, Peterson, *et al.*, 2017). Experiments conducted by Cooper *et al.* (2017) showed that *pink-1* and *pdr-1* mutants accumulate dysfunctional mitochondria with age and present hyperactive UPR^{MT} and that UPR^{MT} can have neuroprotective properties when there are dysfunctional mitochondria: in an *atfs-1* LOF mutant background that has impaired UPR^{MT}, *pink-1* and *pdr-1* LOF mutants showed increased DA neuron degeneration. This finding is consistent with an earlier study done on *pdr-1* where Springer, Hoppe, Schmidt, and Baumeister (2005) noted that an in-frame deletion variant of *pdr-1(lg103)* is hypersensitive to the toxicity of α-syn over-expression – this was attributed to its role in the UPR pathway.

In addition to showing that *pink-1* and *pdr-1* play a role in this pathway, Martinez *et al.* also showed that α-synuclein plays a big role in mediating the UPR^{MT}, and that disease-associated variant of α-synuclein increased mitochondrial fragmentation (a sign of poor mitochondrial health) and hyperactivated the UPR^{MT}, an indicator of mitochondrial stress (Martinez, Peterson, *et al.*, 2017). LRRK2 has also been shown to be involved in mitochondria function in the same pathways *as pink-1*. In a series of experiments with *lrk-1* and *pink-1* mutant animals, Sämann *et al.* (2009) showed that *lrk-1(tm1898)* supresses *pink-1* phenotypes observed in *pink-1(tm1779)* mutants including sensitivity to paraquat-induced oxidative stress and mitochondria cristae deformations. This antagonistic relationship was also observed in UPR^{MT} experiments done by Martinez, Peterson, *et al.* (2017): *lrk-1(tm1898)* mutants attenuated the toxic effect of disrupted ATFS-1-mediated UPR^{MT}, a phenotype which was abolished in *pink-1(tm1779)* mutants. Taken together, findings from these studies provide strong evidence that PD-associated genes play a vital role in regulating important cellular pathways involving the mitochondria. However, whether the mitochondrial dysfunction seen in these studies is a cause or consequence of PD pathogenesis is yet unknown.

Studies on DJ-1 and ATP13A2

Caenorhabditis elegans researchers have also investigated the role of *DJ-1* and *ATP13A2* in PD pathogenesis. *DJ-1* is another PD-associated gene that has implications in mitochondrial function, encoding an oxidative stress sensor that protects neurons from oxidative damage (Figure 1; Canet-Avilés *et al.*, 2004). There are two orthologs of *DJ-1* in *C. elegans*, one expressed primarily in the intestine (*djr-1.1*) and another primarily in neurons (*djr1.2*; Lee *et al.*, 2012). Although LOF mutations in these genes either by themselves or together do not cause DA neuron degeneration or alter dopamine-dependent behaviours, *djr-1.1* mutants show increased sensitivity to oxidative stress, mitochondrial fragmentation and a decreased metabolic output (Cooper *et al.*, 2017; Lee *et al.*, 2012). In mammals, *ATP13A2* encodes a lysosomal P-type ATPase transporter localized to acidic vesicles and mutations in this gene are linked to a unique form of early-onset parkinsonism (Figure 1; Ramirez *et al.*, 2006; Ramonet *et al.*, 2012). In a multi-organism study conducted with *C. elegans*, yeast and mice, Gitler *et al.* (2009) showed that mutations in *ATP13A2* (*catp-6* is the nematode ortholog) can increase the accumulation of α-syn in cells.

At present, the genetic landscape of PD remains a complicated one. Since the development of genome-wide association studies (GWAS), the number of genes implicated in PD has bloomed to at least 90 risk loci (Bandres-Ciga, Diez-Fairen, Kim, & Singleton, 2020). Yet, there remains a high number of familial and early-onset cases with no known genetic causes, and the validation of novel genes associated with PD remains challenging. Evidence suggests that in addition to PD-associated genes converging on common pathways, various disease mechanisms can exist on the same locus and influence disease pathology through different biological effects of one gene (Bandres-Ciga *et al.*, 2020). In addition to pathway analysis with epistatic experiments, parsing out the different functions of each gene remain a focus to those studying PD. Another area of particular interest in recent years is the selective vulnerability of neuron populations in PD. Although the loss of DA neurons at the SNc is the canonical pathological feature in PD patients, a large body of evidence has shown that non-DA neuron populations in other regions of the brain also undergo neuron loss (Giguère, Burke Nanni, & Trudeau, 2018). *Caenorhabditis elegans* models are well-equipped to undertake further investigations in this direction.

Huntington's disease and other polyglutamine diseases

Huntington's disease (HD) is an autosomal dominant neurodegenerative disorder characterized by affective, cognitive, behavioural and motor dysfunctions due to neuron loss in the striatal part of the basal ganglia (Reiner, Dragatsis, & Dietrich, 2011). HD is the most frequent autosomal-dominant disorder with a prevalence of 5–10 per 100,000 in most countries of European descent (Reiner *et al.*, 2011). This disease is caused by the pathological expansion of CAG trinucleotide repeats (which codes for the amino acid glutamine (Q) in the N terminus of the huntingtin (Htt) protein, a ubiquitously expressed protein at both the tissue and subcellular levels (Figure 1; Reiner *et al.*, 2011). It interacts with many partners and has been shown to traffic vesicles, mediate endocytosis, vesicle recycling and endosomal trafficking, among many more other functions (Saudou & Humbert, 2016). The age of onset and severity of the disease is directly correlated with the number of polyglutamine (polyQ) repeats; having 35 or more glutamines invariably results in HD neurodegeneration (Bates, 2003). Neurodegeneration in HD patients is postulated to be caused by aggregated N-terminal fragments of mutant Htt in the cytoplasm and inclusion bodies in the affected tissues (Bauer & Nukina, 2009; DiFiglia *et al.*, 1997). However, the molecular and cellular mechanisms that lead to neuronal dysfunction in HD and other polyQ disorders are largely unknown. While not studied extensively in *C. elegans*, the Htt gene is also hypothesized to increase mitochondria-associated oxidative stress HD patients (Zheng, Winderickx, Franssens, & Liu, 2018). Note that while both neuron-specific and muscle-specific *C. elegans* models have been established, only neuron-based models will be discussed here.

Although *C. elegans* does not have an Htt ortholog, the expression of parts of human Htt in *C. elegans* has been widely used to model several aspects of polyQ cytotoxicity. Faber, Voisine, King, Bates, and Hart (2002) drove expression of huntingtin exon 1 GFP-tagged fragment with various polyQ lengths in specific sensory neurons of *C. elegans* using the *osm-10* promoter and saw that transgenic animals expressing fragments with longer polyQ repeats (Htn-Q150 and Htn-Q95) showed sensory neuron dysfunction and eventual cell death. Consistent with clinical observations, the formation of Htt-positive cytoplasmic aggregates was seen and increased with age in the sensory and axonal processes of these neurons in transgenic worms expressing Htn-Q150. This transgenic model was subsequently used in an enhancer screen to identify the exonuclease polyQ-enhancer gene (*pqe-1*) where loss-of-function mutations in *pqe-1* dramatically enhanced ASH neurodegeneration in Htn-Q150 expressing transgenic animals – suggesting that *pqe-1* may protect from Htn-Q150 toxicity by competing for proteins interacting with Ht-Q150 (Faber *et al.*, 2002). Although currently limited work has been done to investigate the role of the human ortholog of *pqe-1*, REXO1, this finding suggests that REXO1 and other exonucleases may play a role in HD pathogenesis. Another HD neuronal model was developed by expressing the first 57 amino acids of human Htt with normal and expanded polyQ tracts fused to GFP in the six touch receptor neurons using the *mec-7* promoter (Parker *et al.*, 2001). In addition to observing that mechanosensory defects correlated with increasing polyQ expansion, various morphological abnormalities were seen in Q128 expressing animals. However, no apoptotic features were observed in this animal model. This model was used to discover a protective role of the *C. elegans* Htt-interacting protein 1 ortholog (*hipr-1*) and other synaptic endocytosis proteins from polyQ toxicity (Parker *et al.*, 2007).

The role of histone deacetylases in PolyQ toxicity especially in the context of HD has been extensively studied in *C. elegans* models. Over-expression of a class III HDAC, *sir-2.1*, rescued the mechanosensory defects of animals overexpressing Htn-Q128 in the touch neurons (Parker *et al.*, 2005). Bates, Victor, Jones, Shi, and Hart (2006) built on this finding by showing that multiple HDACs could ameliorate ASH-specific Htn-Q150 induced cytotoxicity. Here, loss-of-function mutations or RNAi of any of a majority, but not all, of *C. elegans* HDAC genes (*hda-1, hda-2, hda-4, hda-5, hda-6, hda-10, hda-11, sir-2.1, sir2.2)* worsened Htn-Q150 induced degeneration (Bates *et al.*, 2006). However, pharmacological experiments investigating the effects of increasing or decreasing acetylation on PolyQ toxicity yielded rather conflicting results. Different classes of HDACs appeared to have different effects on Htn-Q150 toxicity. While decreasing acetylation by activating class III HDACs rescued Htn-Q128 induced toxicity, increasing acetylation by inhibiting class I and II HDACs reduced Htn-Q150 toxicity (Bates *et al.*, 2006; Parker *et al.*, 2005). Increasing acetylation of mutant Htt at K444 by HDAC1 knockdown or over-expression of an upstream regulator also led to increased Htt clearance and neuroprotective effects (Jeong *et al.*, 2009). Taken together, these results suggest that the different classes of HDACs likely work via distinct pathways and have specific targets that impact PolyQ toxicity: some targets may be neuroprotective when acetylated; others may be neurotoxic.

Modeling polyglutamine toxicity

HD is but just one member of a family of diseases sharing toxic polyglutamine repeats. To date, including HD a total of nine polyQ neurodegenerative disorders have been described (Fan *et al.*, 2014). The Morimoto group established a model aimed towards a more general understanding of polyQ repeat toxicity by expressing short (Q-19) and long (Q-82) polyQ-GFP fusion proteins in *C. elegans* body wall muscle cells and neurons (Brignull, Moore, Tang, & Morimoto, 2006; Satyal *et al.*, 2000). Further work with this model showed that there is a narrow threshold of polyQ repeat size for induction of aggregation and toxicity: expression of anything more than 35–40 polyQ repeats resulted in the aggregate formation and signs of cellular stress (expression of heat shock proteins; Morley, Brignull, Weyers, & Morimoto, 2002). Additionally, by using this model to observe physiological changes the group showed that the formation of polyQ aggregates generally disrupted protein homeostasis (Gidalevitz, Ben-Zvi, Ho, Brignull, & Morimoto, 2006). Results from these studies support a general 'chaperone depletion' hypothesis proposing that the aggregating protein toxicity comes from competition for limited components of the protein homeostasis machinery. Indeed, an RNAi screen was able to identify a large set of protein folding or protein degradation genes that increased polyQ aggregation when their expression was knocked down (Nollen *et al.*, 2004).

Unlike other diseases discussed here, HD and other polyglutamine diseases are unique in the sense that they are monogenic. Without the need for advanced genomic discovery, more effort can be concentrated on discovering novel therapeutic strategies for polyglutamine diseases. Currently, many drugs targeting the pathogenic effects of mutant HTT (mHTT) or poly-Q are in clinical trials (Kumar *et al.*, 2020). Exploring new mechanisms for therapeutic targets and understanding mechanisms of polyglutamine action continues to be a promising venture for *C. elegans* researchers studying this family of diseases.

Discussion

Researchers developing *C. elegans* models to study the genetic underpinnings of neurodegenerative diseases continue to pave the way to important medical insights. In each of the diseases covered, *C. elegans* researchers have contributed significantly to the understanding of the genetic pathways that cause disease and have aided in finding potential drug targets to treat these diseases. These neurodegenerative diseases seem to share common characteristics – *C. elegans* models highlighted in this review show *in vivo* that dysfunctional and misfolded proteins lead to detrimental aggregation and disrupt important cellular processes. As a result of a catastrophic failure of such important processes, neurons undergo unrecoverable damage and begin to die. However, as researchers continue to make new discoveries, we are reminded of what is yet a mystery. For some of these diseases, even the mechanism of neurodegeneration is up for debate. In the literature, neurodegeneration can be classified into three separate types characterized by separate genetic mechanisms: programmed cell death, autophagic cell death and necrotic cell death (Figure 2; Green & Llambi, 2015). Unique features of these different neurodegeneration pathways can be used in addition to genetic markers to determine how neurons die in *C. elegans*. For example, apoptosis features the condensation and fragmentation of nuclear DNA and blebbing of the neuron processes, while necrotic death has been associated with increased Ca^{2+} release from the ER, inflammation and a rupture of the damaged cell. Autophagic cell death, on the other hand, is identified by the presence of autophagic vacuoles and lysosomes. However, features from multiple pathways can be detected simultaneously in both animal models and human patients with these neurodegenerative diseases – indicating that the pathways to neuron death in this disease may involve

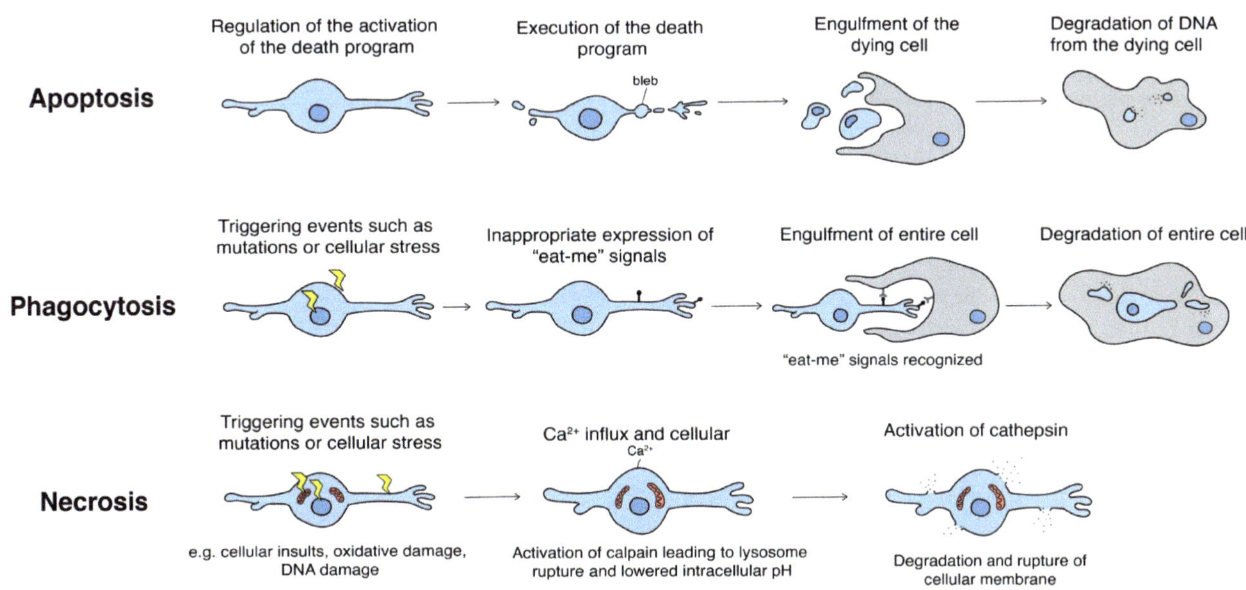

Figure 2. Neurodegeneration is categorized into three major types. Apoptosis, or programmed cell death, results from activation of cellular pathways that triggers a gradual and tidy degradation of the affected neuron. Phagocytosis is an induced engulfment of an affected neuron prior to the neuron undergoing its own degrading process. Necrosis occurs where the neuron undergoes volatile degeneration due to inflammation or oxidative stress and causes the most negative impact to its neighboring environment. Each of these pathways have key molecular components that can help characterize one from the other. However, in neurodegeneration seen in nervous diseases, markers and features of multiple pathways test are present – more work has yet to be done to distinguish these pathways in diseases in much more resolution.

multiple simultaneous mechanisms and be more complicated than originally thought (Sarasija *et al.*, 2018; Sawa, Tomoda, & Bae, 2003).

Many of these diseases are considered multifactorial in nature, the etiology of each of these diseases is distinct from the others and specific populations of neurons and specific molecular pathways are targeting by unknown mechanisms. What causes these diseases to target specific neuronal populations, and why do these pathways not influence other neurons? Much of what triggers the onset of these neurodegenerative diseases remain largely unknown.

Yet, current research also converges on key common features in these diseases. For example, mitochondrial processes are implicated in many of these disorders – indeed, genes associated with the diseases covered here appear to interact with the mitochondria through vital pathways (Figure 1). Another factor that these diseases all share is their association with the aging process. Age is the largest risk factor for the development of many neurodegenerative diseases including PD, AD and ALS and is an exacerbating factor to their progression (Mattson & Magnus, 2006). With age comes the gradual accumulation of damage and stress to cells that can induce cellular alterations that can lead to a decrement of a neuron's condition over time (Mattson & Magnus, 2006). Notably, neurons are particularly susceptible to the accumulated damage from aging, due to their non-replenishing nature and high metabolic load. To this end, numerous studies have been conducted on wild-type *C. elegans* to elucidate the effects of aging on neurodegeneration and the molecular mechanisms behind the aging process, yet more is to be done (Chen, Chen, Jiang, Chen, & Pan, 2013).

With the advent of more advanced genome engineering technology, it is becoming clear that findings from over-expression models only reveal a part of a larger picture. As covered in this review, transgenic models traditionally over-express human WT and disease-associated mutant genes in a tissue-specific manner. However, it is difficult to adjudicate the validity of over-expression models in loss-of-function variants of the disease, and there is increasing evidence where WT genes can be pathogenic when over-expressed. Furthermore, especially in autosomal dominant forms of these illnesses, patients with these nervous system disorders suffer from deleterious effects of the mutation in only one copy of the gene(s). With CRISPR-Cas-9, it is now possible to study these diseases more realistically in the context of knock-in or gene-to-gene replacement models (Dickinson, Ward, Reiner, & Goldstein, 2013; Friedland *et al.*, 2013). In the future, single-copy knock-in models such as the one designed by Baskoylu *et al.* (2018, 2019) for FUS and SOD1 will be the benchmark for investigating the intricacies of the molecular pathways in question. In fact, targeted human gene replacement *C. elegans* models have been established for ASD-associated genes in a similar manner (McDiarmid *et al.*, 2018). Many of these disease-associated genes with multiple variants of varying significance present as great candidates for CRISPR-mediated genome engineering experiments, and it is inevitable that going in this direction will yield new insights into these disorders.

Conclusion

Caenorhabditis elegans researchers continue to exploit the many advantages of using the tiny transparent worm to examine many facets to neurodegeneration: from the genetic pathways leading to neuronal death to how different disease-associated molecular pathways can induce neuronal atrophy. Indeed, this model organism first established by Brenner, Sulston and their contemporaries has blossomed into a well-understood and tractable system that continues to benefit many fields of research. Yet, the most meaningful legacy this cohort leaves behind remains undoubtedly the passionately collaborative spirit that led to the continued curation of openly shared resources that are the foundation of the thriving *C. elegans* research community today.

Acknowledgements

The authors would like to thank Troy A. McDiarmid and Alex J. Yu for helpful comments to the manuscript. They would also like to thank Lexi D. Kepler for help making the figures.

Disclosure statement

No potential conflict of interest was reported by the author(s).

Funding

This work was supported by a Natural Sciences and Engineering Research Council of Canada under the Canada Graduate Scholarships – Canadian Graduate Scholarship Masters to J.L. as well as by the Natural Sciences and Engineering Research Council of Canada under project grant [NSERC RGPIN-2019–05558] and Canadian Institutes of Health Research under project grant [CIHR PJT 165947] to C.H.R.

References

Ahn, B.-H., Rhim, H., Kim, S.Y., Sung, Y.-M., Lee, M.-Y., Choi, J.-Y., ... Min, D.S. (2002). alpha-synuclein interacts with phospholipase D isozymes and inhibits pervanadate-induced phospholipase D activation in human embryonic kidney-293 cells. *The Journal of Biological Chemistry*, *277*(14), 12334–12342. doi:10.1074/jbc.M110414200

Al-Chalabi, A., Jones, A., Troakes, C., King, A., Al-Sarraj, S., & Van Den Berg, L.H. (2012). The genetics and neuropathology of amyotrophic lateral sclerosis. *Acta Neuropathologica*, *124*(3), 339–352. doi:10.1007/s00401-012-1022-4

Alexander, A.G., Marfil, V., & Li, C. (2014). Use of *Caenorhabditis elegans* as a model to study Alzheimer's disease and other neurodegenerative diseases. *Frontiers in Genetics*, *5*, 279. doi:10.3389/fgene.2014.00279

Alonso, A.D.C., Zaidi, T., Novak, M., Grundke-Iqbal, I., & Iqbal, K. (2001). Hyperphosphorylation induces self-assembly of tau into tangles of paired helical filaments/straight filaments. *Proceedings of the National Academy of Sciences of the United States of America*, *98*(12), 6923–6928. doi:10.1073/pnas.121119298

Armenti, S.T., Lohmer, L.L., Sherwood, D.R., & Nance, J. (2014). Repurposing an endogenous degradation system for rapid and targeted depletion of *C. elegans* proteins. *Development*, *141*(23), 4640–4647. doi:10.1242/dev.115048

Ash, P.E.A., Zhang, Y.-J., Roberts, C.M., Saldi, T., Hutter, H., Buratti, E., ... Link, C.D. (2010). Neurotoxic effects of TDP-43 overexpression in *C. elegans*. *Human Molecular Genetics*, *19*(16), 3206–3218. doi:10.1093/hmg/ddq230

Asikainen, S., Vartiainen, S., Lakso, M., Nass, R., & Wong, G. (2005). Selective sensitivity of *Caenorhabditis elegans* neurons to RNA interference. *Neuroreport*, *16*(18), 1995–1999. doi:10.1097/00001756-200512190-00005

Balendra, R., & Isaacs, A.M. (2018). C9orf72-mediated ALS and FTD: Multiple pathways to disease. *Nature Reviews Neurology*, *14*(9), 544–558. doi:10.1038/s41582-018-0047-2

Bandres-Ciga, S., Diez-Fairen, M., Kim, J.J., & Singleton, A.B. (2020). Genetics of Parkinson's disease: An introspection of its journey towards precision medicine. *Neurobiology of Disease*, *137*, 104782. doi:10.1016/j.nbd.2020.104782

Baskoylu, S.N., Chapkis, N., Unsal, B., Lins, J., Schuch, K., Simon, J., & Hart, A.C. (2019). Disrupted Autophagy and Neuronal Dysfunction in C. elegans Knock-in Models of FUS Amyotrophic Lateral Sclerosis. *BioRxiv*, *7*, 799932. doi:10.1101/799932

Baskoylu, S.N., Yersak, J., O'Hern, P., Grosser, S., Simon, J., Kim, S., … Hart, A.C. (2018). Single copy/knock-in models of ALS SOD1 in *C. elegans* suggest loss and gain of function have different contributions to cholinergic and glutamatergic neurodegeneration. *PLoS Genetics*, *14*(10), e1007682. doi:10.1371/journal.pgen.1007682

Bates, E.A., Victor, M., Jones, A.K., Shi, Y., & Hart, A.C. (2006). Differential contributions of *Caenorhabditis elegans* histone deacetylases to huntingtin polyglutamine toxicity. *Journal of Neuroscience*, *26*(10), 2830–2838. doi:10.1523/JNEUROSCI.3344-05.2006

Bates, G. (2003). Huntingtin aggregation and toxicity in Huntington's disease. *The Lancet*, *361*(9369), 1642–1644. doi:10.1016/S0140-6736(03)13304-1

Bauer, P.O., & Nukina, N. (2009). The pathogenic mechanisms of polyglutamine diseases and current therapeutic strategies. *Journal of Neurochemistry*, *110*(6), 1737–1765. doi:10.1111/j.1471-4159.2009.06302.x

Bellani, S., Sousa, V.L., Ronzitti, G., Valtorta, F., Meldolesi, J., & Chieregatti, E. (2010). The regulation of synaptic function by alpha-synuclein. *Communicative & Integrative Biology*, *3*(2), 106–109. doi:10.4161/cib.3.2.10964

Bellenguez, C., Grenier-Boley, B., & Lambert, J.C. (2020). Genetics of Alzheimer's disease: Where we are, and where we are going. *Current Opinion in Neurobiology*, *61*, 40–48. doi:10.1016/j.conb.2019.11.024

Benbow, S.J., Strovas, T.J., Darvas, M., Saxton, A., & Kraemer, B.C. (2020). Synergistic toxicity between tau and amyloid drives neuronal dysfunction and neurodegeneration in transgenic *C. elegans*. *Human Molecular Genetics*, *29*(3), 495–505. doi:10.1093/hmg/ddz319

Bieschke, J., Cohen, E., Murray, A., Dillin, A., & Kelly, J.W. (2009). A kinetic assessment of the *C. elegans* amyloid disaggregation activity enables uncoupling of disassembly and proteolysis. *Protein Science: A Publication of the Protein Society*, *18*(11), 2231–2241. doi:10.1002/pro.234

Bramblett, G.T., Goedert, M., Jakes, R., Merrick, S.E., Trojanowski, J.Q., & Lee, V.M.Y. (1993). Abnormal tau phosphorylation at Ser396 in Alzheimer's disease recapitulates development and contributes to reduced microtubule binding. *Neuron*, *10*(6), 1089–1099. doi:10.1016/0896-6273(93)90057-X

Brandt, R., Gergou, A., Wacker, I., Fath, T., & Hutter, H. (2009). A *Caenorhabditis elegans* model of tau hyperphosphorylation: Induction of developmental defects by transgenic overexpression of Alzheimer's disease-like modified tau. *Neurobiology of Aging*, *30*(1), 22–33. doi:10.1016/j.neurobiolaging.2007.05.011

Brignull, H.R., Moore, F.E., Tang, S.J., & Morimoto, R.I. (2006). Polyglutamine proteins at the pathogenic threshold display neuron-specific aggregation in a pan-neuronal *Caenorhabditis elegans* model. *The Journal of Neuroscience: The Official Journal of the Society for Neuroscience*, *26*(29), 7597–7606. doi:10.1523/JNEUROSCI.0990-06.2006

Cabin, D.E., Shimazu, K., Murphy, D., Cole, N.B., Gottschalk, W., McIlwain, K.L., … Nussbaum, R.L. (2002). Synaptic vesicle depletion correlates with attenuated synaptic responses to prolonged repetitive stimulation in mice lacking α-synuclein. *The Journal of Neuroscience: The Official Journal of the Society for Neuroscience*, *22*(20), 8797–8807. doi:10.1523/JNEUROSCI.22-20-08797.2002

Cabreiro, F., Ackerman, D., Doonan, R., Araiz, C., Back, P., Papp, D., … Gems, D. (2011). Increased life span from overexpression of superoxide dismutase in *Caenorhabditis elegans* is not caused by decreased oxidative damage. *Free Radical Biology & Medicine*, *51*(8), 1575–1582. doi:10.1016/j.freeradbiomed.2011.07.020

Calixto, A., Chelur, D., Topalidou, I., Chen, X., & Chalfie, M. (2010). Enhanced neuronal RNAi in C. elegans using SID-1. *Nature Methods*, *7*(7), 554–559. doi:10.1038/nmeth.1463

Canet-Avilés, R.M., Wilson, M.A., Miller, D.W., Ahmad, R., McLendon, C., Bandyopadhyay, S., … Cookson, M.R. (2004). The Parkinson's disease protein DJ-1 is neuroprotective due to cysteine-sulfinic acid-driven mitochondrial localization. *Proceedings of the National Academy of Sciences of the United States of America*, *101*(24), 9103–9108. doi:10.1073/pnas.0402959101

Cao, S., Gelwix, C.C., Caldwell, K.A., & Caldwell, G.A. (2005). Torsin-mediated protection from cellular stress in the dopaminergic neurons of Caenorhabditis elegans. *The Journal of Neuroscience: The Official Journal of the Society for Neuroscience*, *25*(15), 3801–3812. doi:10.1523/JNEUROSCI.5157-04.2005

Chalfie, M., Tu, Y., Euskirchen, G., Ward, W.W., & Prasher, D.C. (1994). Green fluorescent protein as a marker for gene expression. *Science*, *263*(5148), 802–805. doi:10.1126/science.8303295

Chen, C., Turnbull, D.M., & Reeve, A.K. (2019). Mitochondrial dysfunction in Parkinson's disease—cause or consequence? *Biology*, *8*(2), 38. doi:10.3390/biology8020038

Chen, C.H., Chen, Y.C., Jiang, H.C., Chen, C.K., & Pan, C.L. (2013). Neuronal aging: Learning from *C. elegans*. *Journal of Molecular Signaling*, *8* (1), 10–14. doi:10.1186/1750-2187-8-14

Chew, Y.L., Fan, X., Götz, J., & Nicholas, H.R. (2013). PTL-1 regulates neuronal integrity and lifespan in *C. elegans*. *Journal of Cell Science*, *126*(9), 2079–2091. doi:10.1242/jcs.jcs124404

Choudhary, B., Kamak, M., Ratnakaran, N., Kumar, J., Awasthi, A., Li, C., … Koushika, S.P. (2017). UNC-16/JIP3 regulates early events in synaptic vesicle protein trafficking via LRK18 1/LRRK2 and AP complexes. *PLoS Genetics*, *13*(11), 1–25. doi:10.1371/journal.pgen.1007100

Cook, S.J., Jarrell, T.A., Brittin, C.A., Wang, Y., Bloniarz, A.E., Yakovlev, M.A., … Emmons, S.W. (2019). Whole-animal connectomes of both *Caenorhabditis elegans* sexes. *Nature*, *571*(7763), 63–71. doi:10.1038/s41586-019-1352-7

Cooper, A.A., Gitler, A.D., Cashikar, A., Haynes, C.M., Hill, K.J., Bhullar, B., … Lindquist, S. (2006). Alpha-synuclein blocks ER-golgi traffic and Rab1 rescues neuron loss in Parkinson's models. *Science*, *313*(5785), 324–329. doi:10.1126/science.1129462

Cooper, J.F., Machiela, E., Dues, D.J., Spielbauer, K.K., Senchuk, M.M., & Van Raamsdonk, J.M. (2017). Activation of the mitochondrial unfolded protein response promotes longevity and dopamine neuron survival in Parkinson's disease models. *Scientific Reports*, *7*(1), 16441. doi:10.1038/s41598-017-16637-2

Cooper, J.F., & Van Raamsdonk, J.M. (2018). Modeling Parkinson's disease in *C. elegans*. *Journal of Parkinson's Disease*, *8*(1), 17–32. doi:10.3233/JPD-171258

Corrionero, A., & Horvitz, H.R. (2018). A C9orf72 ALS/FTD ortholog acts in endolysosomal degradation and lysosomal homeostasis. *Current Biology*, *28*(10), 1522.e5–1535.e5. doi:10.1016/j.cub.2018.03.063

Cudkowicz, M.E., McKenna-Yasek, D., Sapp, P.E., Chin, W., Geller, B., Hayden, D.L., … Brown, R.H. (1997). Epidemiology of mutations in superoxide dismutase in amyotrophic lateral sclerosis. *Annals of Neurology*, *41*(2), 210–221. doi:10.1002/ana.410410212

DeJesus-Hernandez, M., Mackenzie, I.R., Boeve, B.F., Boxer, A.L., Baker, M., Rutherford, N.J., … Rademakers, R. (2011). Expanded GGGGCC hexanucleotide repeat in noncoding region of C9ORF72 causes chromosome 9p-linked FTD and ALS. *Neuron*, *72*(2), 245–256. doi:10.1016/j.neuron.2011.09.011

Dickinson, D.J., Ward, J.D., Reiner, D.J., & Goldstein, B. (2013). Engineering the *Caenorhabditis elegans* genome using Cas9-triggered homologous recombination. *Nature Methods*, *10*(10), 1028–1034. doi:10.1038/nmeth.2641

DiFiglia, M., Sapp, E., Chase, K.O., Davies, S.W., Bates, G.P., Vonsattel, J.P., & Aronin, N. (1997). Aggregation of huntingtin in neuronal intranuclear inclusions and dystrophic neurites in brain. *Science, 277*(5334), 1990–1993. doi:10.1126/science.277.5334.1990

Dourlen, P., Kilinc, D., Malmanche, N., Chapuis, J., & Lambert, J.C. (2019). The new genetic landscape of Alzheimer's disease: From amyloid cascade to genetically driven synaptic failure hypothesis? *Acta Neuropathologica, 138*(2), 221–236. doi:10.1007/s00401-019-02004-0

Estes, P.S., Boehringer, A., Zwick, R., Tang, J.E., Grigsby, B., & Zarnescu, D.C. (2011). Wild-type and A315T mutant TDP-43 exert differential neurotoxicity in a *Drosophila* model of ALS. *Human Molecular Genetics, 20*(12), 2308–2321. doi:10.1093/hmg/ddr124

Faber, P.W., Voisine, C., King, D.C., Bates, E.A., & Hart, A.C. (2002). Glutamine/proline-rich PQE-1 proteins protect *Caenorhabditis elegans* neurons from huntingtin polyglutamine neurotoxicity. *Proceedings of the National Academy of Sciences of the United States of America, 99*(26), 17131–17136. doi:10.1073/pnas.262544899

Fahn, S., & Sulzer, D. (2004). Neurodegeneration and neuroprotection in Parkinson disease. *NeuroRx: The Journal of the American Society for Experimental Neurotherapeutics, 1*(1), 139–154. doi:10.1602/neurorx.1.1.139

Fan, H.-C., Ho, L.-I., Chi, C.-S., Chen, S.-J., Peng, G.-S., Chan, T.-M., … Harn, H.-J. (2014). Polyglutamine (PolyQ) diseases: Genetics to treatments. *Cell Transplantation, 23*(4–5), 441–458. doi:10.3727/096368914X678454

Fay, D.S., Fluet, A., Johnson, C.J., & Link, C.D. (1998). In vivo aggregation of beta-amyloid peptide variants. *Journal of Neurochemistry, 71*(4), 1616–1625. doi:10.1046/j.1471-4159.1998.71041616.x

Fire, A., Xu, S., Montgomery, M.K., Kostas, S.A., Driver, S.E., & Mello, C.C. (1998). Potent and specific genetic interference by double-stranded RNA in *Caenorhabditis elegans. Nature, 391*(6669), 806–811. doi:10.1038/35888

Florez-McClure, M.L., Hohsfield, L.A., Fonte, G., Bealor, M.T., & Link, C.D. (2007). Decreased insulin-receptor signaling promotes the autophagic degradation of beta-amyloid peptide in *C. elegans. Autophagy, 3*(6), 569–580. doi:10.4161/auto.4776

Fong, S., Teo, E., Ng, L.F., Chen, C.-B., Lakshmanan, L.N., Tsoi, S.Y., … Gruber, J. (2016). Energy crisis precedes global metabolic failure in a novel *Caenorhabditis elegans* Alzheimer disease model. *Scientific Reports, 6*(1), 33781. doi:10.1038/srep33781

Fonte, V., Kapulkin, W.J., Kapulkin, V., Taft, A., Fluet, A., Friedman, D., & Link, C.D. (2002). Interaction of intracellular beta amyloid peptide with chaperone proteins. *Proceedings of the National Academy of Sciences of the United States of America, 99*(14), 9439–9444. doi:10.1073/pnas.152313999

Friedland, A.E., Tzur, Y.B., Esvelt, K.M., Colaiácovo, M.P., Church, G.M., & Calarco, J.A. (2013). Heritable genome editing in *C. elegans* via a CRISPR-Cas9 system. *Nature Methods, 10*(8), 741–743. doi:10.1038/nmeth.2532

Fukuzono, T., Pastuhov, S.I., Fukushima, O., Li, C., Hattori, A., Iemura, S-i., … Hisamoto, N. (2016). Chaperone complex BAG2-HSC70 regulates localization of *Caenorhabditis elegans* leucine-rich repeat kinase LRK-1 to the Golgi. *Genes to Cells: Devoted to Molecular & Cellular Mechanisms, 21*(4), 311–324. doi:10.1111/gtc.12338

Gidalevitz, T., Ben-Zvi, A., Ho, K.H., Brignull, H.R., & Morimoto, R.I. (2006). Progressive disruption of cellular protein folding in models of polyglutamine diseases. *Science, 311*(5766), 1471–1474. doi:10.1126/science.1124514

Gidalevitz, T., Krupinski, T., Garcia, S., & Morimoto, R.I. (2009). Destabilizing protein polymorphisms in the genetic background direct phenotypic expression of mutant SOD1 toxicity. *PLoS Genetics, 5*(3), e1000399. doi:10.1371/journal.pgen.1000399

Giguère, N., Burke Nanni, S., & Trudeau, L.-E. (2018). On cell loss and selective vulnerability of neuronal populations in Parkinson's disease. *Frontiers in Neurology, 9*, 455. doi:10.3389/fneur.2018.00455

Gitler, A.D., Chesi, A., Geddie, M.L., Strathearn, K.E., Hamamichi, S., Hill, K.J., … Lindquist, S. (2009). Alpha-synuclein is part of a diverse and highly conserved interaction network that includes PARK9 and manganese toxicity. *Nature Genetics, 41*(3), 308–315. doi:10.1038/ng.300

Goedert, M., Spillantini, M.G., Jakes, R., Rutherford, D., & Crowther, R.A. (1989). Multiple isoforms of human microtubule-associated protein tau: Sequences and localization in neurofibrillary tangles of Alzheimer's disease. *Neuron, 3*(4), 519–526. doi:10.1016/0896-6273(89)90210-9

Gordon, P., Hingula, L., Krasny, M.L., Swienckowski, J.L., Pokrywka, N.J., & Raley-Susman, K.M. (2008). The invertebrate microtubule-associated protein PTL-1 functions in mechanosensation and development in *Caenorhabditis elegans. Development Genes and Evolution, 218*(10), 541–551. doi:10.1007/s00427-008-0250-z

Graffmo, K.S., Forsberg, K., Bergh, J., Birve, A., Zetterström, P., Andersen, P.M., … Brännström, T. (2013). Expression of wild-type human superoxide dismutase-1 in mice causes amyotrophic lateral sclerosis. *Human Molecular Genetics, 22*(1), 51–60. doi:10.1093/hmg/dds399

Green, D.R., & Llambi, F. (2015). Cell death signaling. *Cold Spring Harbor Perspectives in Biology, 7*(12), a006080. doi:10.1101/cshperspect.a006080

Hardy, J.A., & Higgins, G.A. (1992). Alzheimer's disease: The amyloid cascade hypothesis. *Science, 256*(5054), 184–185. doi:10.1126/science.1566067

Hartman, J.H., Gonzalez-Hunt, C., Hall, S.M., Ryde, I.T., Caldwell, K.A., Caldwell, G.A., & Meyer, J.N. (2019). Genetic defects in mitochondrial dynamics in *Caenorhabditis elegans* impact ultraviolet C radiation- and 6-hydroxydopamine-induced neurodegeneration. *International Journal of Molecular Sciences, 20*(13), 3202. doi:10.3390/ijms20133202

Hassan, W.M., Merin, D.A., Fonte, V., & Link, C.D. (2009). AIP-1 ameliorates beta-amyloid peptide toxicity in a *Caenorhabditis elegans* Alzheimer's disease model. *Human Molecular Genetics, 18*(15), 2739–2747. doi:10.1093/hmg/ddp209

Hengartner, M.O., & Horvitz, R.H. (1994). Programmed cell death in *Caenorhabditis elegans. Current Opinion in Genetics & Development, 4*(4), 581–586. doi:10.1016/0959-437X(94)90076-F

Hergesheimer, R.C., Chami, A.A., de Assis, D.R., Vourc'h, P., Andres, C.R., Corcia, P., … Blasco, H. (2019). The debated toxic role of aggregated TDP-43 in amyotrophic lateral sclerosis: A resolution in sight? *Brain: A Journal of Neurology, 142*(5), 1176–1194. doi:10.1093/brain/awz078

Hermann, A., Liewald, J.F., & Gottschalk, A. (2015). A photosensitive degron enables acute light-induced protein degradation in the nervous system. *Current Biology* , *25*(17), R749–R750. doi:10.1016/j.cub.2015.07.040

Holdorff, B., Rodrigues e Silva, A.M., & Dodel, R. (2013). Centenary of Lewy bodies (1912-2012). *Journal of Neural Transmission, 120*(4), 509–516. doi:10.1007/s00702-013-0984-2

Hornsten, A., Lieberthal, J., Fadia, S., Malins, R., Ha, L., Xu, X., … Li, C. (2007). APL-1, a *Caenorhabditis elegans* protein related to the human beta-amyloid precursor protein, is essential for viability. *Proceedings of the National Academy of Sciences of the United States of America, 104*(6), 1971–1976. doi:10.1073/pnas.0603997104

Huang, M., Wang, B., Li, X., Fu, C., Wang, C., & Kang, X. (2019). A-synuclein: A multifunctional player in exocytosis, endocytosis, and vesicle recycling. *Frontiers in Neuroscience, 13*, 28–28. doi:10.3389/fnins.2019.00028

Ikenaka, K., Kawai, K., Katsuno, M., Huang, Z., Jiang, Y.-M., Iguchi, Y., … Sobue, G. (2013). dnc-1/dynactin 1 knockdown disrupts transport of autophagosomes and induces motor neuron degeneration. *PLoS One, 8*(2), e54511. doi:10.1371/journal.pone.0054511

Ikenaka, K., Tsukada, Y., Giles, A.C., Arai, T., Nakadera, Y., Nakano, S., … Mori, I. (2019). A behavior-based drug screening system using a *Caenorhabditis elegans* model of motor neuron disease. *Scientific Reports, 9*(1), 1–10. doi:10.1038/s41598-019-46642-6

Janssen, J.C., Beck, J.A., Campbell, T.A., Dickinson, A., Fox, N.C., Harvey, R.J., … Collinge, J. (2003). Early onset familial Alzheimer's disease: Mutation frequency in 31 families. *Neurology, 60*(2), 235–239. doi:10.1212/01.WNL.0000042088.22694.E3

Jeong, H., Then, F., Melia, T.J., Mazzulli, J.R., Cui, L., Savas, J.N., … Krainc, D. (2009). Acetylation targets mutant huntingtin to autophagosomes for degradation. *Cell, 137*(1), 60–72. doi:10.1016/j.cell.2009.03.018

Jiang, Y.-M., Yamamoto, M., Kobayashi, Y., Yoshihara, T., Liang, Y., Terao, S., … Sobue, G. (2005). Gene expression profile of spinal motor neurons in sporadic amyotrophic lateral sclerosis. *Annals of Neurology, 57*(2), 236–251. doi:10.1002/ana.20379

Jiang, Y.-M., Yamamoto, M., Tanaka, F., Ishigaki, S., Katsuno, M., Adachi, H., … Sobue, G. (2007). Gene expressions specifically detected in motor neurons (dynactin 1, early growth response 3, acetyl-CoA transporter, death receptor 5, and cyclin C) differentially correlate to pathologic markers in sporadic amyotrophic lateral sclerosis. *Journal of Neuropathology and Experimental Neurology, 66*(7), 617–627. doi:10.1097/nen.0b013e318093ece3

Joyce, P.I., Mcgoldrick, P., Saccon, R.A., Weber, W., Fratta, P., West, S.J., … Acevedo-Arozena, A. (2015). A novel SOD1-ALS mutation separates central and peripheral effects of mutant SOD1 toxicity. *Human Molecular Genetics, 24*(7), 1883–1897. doi:10.1093/hmg/ddu605

Kagan, B.L., Jang, H., Capone, R., Teran Arce, F., Ramachandran, S., Lal, R., & Nussinov, R. (2012). Antimicrobial properties of amyloid peptides. *Molecular Pharmaceutics, 9*(4), 708–717. doi:10.1021/mp200419b

Kamath, R.S., Fraser, A.G., Dong, Y., Poulin, G., Durbin, R., Gotta, M., … Ahringer, J. (2003). Systematic functional analysis of the *Caenorhabditis elegans* genome using RNAi. *Nature, 421*(6920), 231–237. doi:10.1038/nature01278

Kang, J., Lemaire, H.G., Unterbeck, A., Salbaum, J.M., Masters, C.L., Grzeschik, K.H., … Müller-Hill, B. (1987). The precursor of Alzheimer's disease amyloid A4 protein resembles a cell-surface receptor. *Nature, 325*(6106), 733–736. doi:10.1038/325733a0

Karran, E., & De Strooper, B. (2016). The amyloid cascade hypothesis: Are we poised for success or failure? *Journal of Neurochemistry, 139*, 237–252. doi:10.1111/jnc.13632

Kaur, S.J., McKeown, S.R., & Rashid, S. (2016). Mutant SOD1 mediated pathogenesis of amyotrophic lateral sclerosis. *Gene, 577*(2), 109–118. doi:10.1016/j.gene.2015.11.049

Kidd, M. (1964). Alzheimer's disease-an electron microscopical study. *Brain: a Journal of Neurology, 87*, 307–320. doi:10.1093/brain/87.2.307

Kim, W., Underwood, R.S., Greenwald, I., & Shaye, D.D. (2018). Ortholist 2: A new comparative genomic analysis of human and *Caenorhabditis elegans* genes. *Genetics, 210*(2), 445–461. doi:10.1534/genetics.118.301307

Koushika, S.P., Schaefer, A.M., Vincent, R., Willis, J.H., Bowerman, B., & Nonet, M.L. (2004). Mutations in *Caenorhabditis elegans* cytoplasmic dynein components reveal specificity of neuronal retrograde cargo. *The Journal of Neuroscience: The Official Journal of the Society for Neuroscience, 24*(16), 3907–3916. doi:10.1523/JNEUROSCI.5039-03.2004

Kraemer, B.C., & Schellenberg, G.D. (2007). SUT-1 enables tau-induced neurotoxicity in *C. elegans*. *Human Molecular Genetics, 16*(16), 1959–1971. doi:10.1093/hmg/ddm143

Kraemer, B.C., Zhang, B., Leverenz, J.B., Thomas, J.H., Trojanowski, J.Q., & Schellenberg, G.D. (2003). Neurodegeneration and defective neurotransmission in a *Caenorhabditis elegans* model of tauopathy. *Proceedings of the National Academy of Sciences of the United States of America, 100*(17), 9980–9985. doi:10.1073/pnas.1533448100

Kumar, A., Kumar, V., Singh, K., Kumar, S., Kim, Y.-S., Lee, Y.-M., & Kim, J.-J. (2020). Therapeutic advances for Huntington's disease. *Brain Sciences, 10*(1), 43. doi:10.3390/brainsci10010043

Kuwahara, T., Koyama, A., Gengyo-Ando, K., Masuda, M., Kowa, H., Tsunoda, M., … Iwatsubo, T. (2006). Familial Parkinson mutant alpha-synuclein causes dopamine neuron dysfunction in transgenic *Caenorhabditis elegans*. *The Journal of Biological Chemistry, 281*(1), 334–340. doi:10.1074/jbc.M504860200

Kuwahara, T., Koyama, A., Koyama, S., Yoshina, S., Ren, C.-H., Kato, T., … Iwatsubo, T. (2008). A systematic RNAi screen reveals involvement of endocytic pathway in neuronal dysfunction in alpha-synuclein transgenic *C. elegans*. *Human Molecular Genetics, 17*(19), 2997–3009. doi:10.1093/hmg/ddn198

Kwiatkowski, T.J., Bosco, D.A., Leclerc, A.L., Tamrazian, E., Vanderburg, C.R., Russ, C., … Brown, R.H. (2009). Mutations in the FUS/TLS gene on chromosome 16 cause familial amyotrophic lateral sclerosis. *Science, 323*(5918), 1205–1208. doi:10.1126/science.1166066

Lakso, M., Vartiainen, S., Moilanen, A.-M., Sirviö, J., Thomas, J.H., Nass, R., … Wong, G. (2003). Dopaminergic neuronal loss and motor deficits in *Caenorhabditis elegans* overexpressing human alpha-synuclein. *Journal of Neurochemistry, 86*(1), 165–172. doi:10.1046/j.1471-4159.2003.01809.x

Lee, J-y., Song, J., Kwon, K., Jang, S., Kim, C., Baek, K., … Park, C. (2012). Human DJ-1 and its homologs are novel glyoxalases. *Human Molecular Genetics, 21*(14), 3215–3225. doi:10.1093/hmg/dds155

Levitan, D., Doyle, T.G., Brousseau, D., Lee, M.K., Thinakaran, G., Slunt, H.H., … Greenwald, I. (1996). Assessment of normal and mutant human presenilin function in *Caenorhabditis elegans*. *Proceedings of the National Academy of Sciences of the United States of America, 93*(25), 14940–14944. doi:10.1073/pnas.93.25.14940

Levitan, D., & Greenwald, I. (1995). Facilitation of lin-12-mediated signalling by sel-12, a *Caenorhabditis elegans* S182 Alzheimer's disease gene. *Nature, 377*(6547), 351–354. doi:10.1038/377351a0

Liachko, N.F., Guthrie, C.R., & Kraemer, B.C. (2010). Phosphorylation promotes neurotoxicity in a *Caenorhabditis elegans* model of TDP-43 proteinopathy. *The Journal of Neuroscience: The Official Journal of the Society for Neuroscience, 30*(48), 16208–16219. doi:10.1523/JNEUROSCI.2911-10.2010

Ling, S.C., Polymenidou, M., & Cleveland, D.W. (2013). Converging mechanisms in ALS and FTD: Disrupted RNA and protein homeostasis. *Neuron, 79*(3), 416–438. doi:10.1016/j.neuron.2013.07.033

Link, C.D. (1995). Expression of human beta-amyloid peptide in transgenic *Caenorhabditis elegans*. *Proceedings of the National Academy of Sciences of the United States of America, 92*(20), 9368–9372. doi:10.1073/pnas.92.20.9368

Link, C.D., Taft, A., Kapulkin, V., Duke, K., Kim, S., Fei, Q., … Sahagan, B.G. (2003). Gene expression analysis in a transgenic *Caenorhabditis elegans* Alzheimer's disease model. *Neurobiology of Aging, 24*(3), 397–413. doi:10.1016/S0197-4580(02)00224-5

Liu, Z., Hamamichi, S., Lee, B.D., Yang, D., Ray, A., Caldwell, G.A., … Dawson, V.L. (2011). Inhibitors of LRRK2 kinase attenuate neurodegeneration and Parkinson-like phenotypes in *Caenorhabditis elegans* and Drosophila Parkinson's disease models. *Human Molecular Genetics, 20*(20), 3933–3942. doi:10.1093/hmg/ddr312

Lloret, A., Badia, M.-C., Giraldo, E., Ermak, G., Alonso, M.-D., Pallardó, F.V., … Viña, J. (2011). Amyloid-β toxicity and tau hyperphosphorylation are linked via RCAN1 in Alzheimer's disease. *Journal of Alzheimer's Disease, 27*(4), 701–709. doi:10.3233/JAD-2011-110890

Maloney, B., & Lahiri, D.K. (2011). The Alzheimer's amyloid β-peptide (Aβ) binds a specific DNA Aβ-interacting domain (AβID) in the APP, BACE1, and APOE promoters in a sequence-specific manner: Characterizing a new regulatory motif. *Gene, 488*(1–2), 1–12. doi:10.1016/j.gene.2011.06.004

Martinez, B.A., Caldwell, K.A., & Caldwell, G.A. (2017). *C. elegans* as a model system to accelerate discovery for Parkinson disease. *Current Opinion in Genetics & Development, 44*, 102–109. doi:10.1016/j.gde.2017.02.011

Martinez, B.A., Petersen, D.A., Gaeta, A.L., Stanley, S.P., Caldwell, G.A., & Caldwell, K.A. (2017). Dysregulation of the mitochondrial unfolded protein response induces non-apoptotic dopaminergic neurodegeneration in *C. elegans* models of Parkinson's disease. *The Journal of Neuroscience: The Official Journal of the Society for Neuroscience, 37*(46), 11085–11100. doi:10.1523/JNEUROSCI.1294-17.2017

Martinez, M.A.Q., Kinney, B.A., Medwig-Kinney, T.N., Ashley, G., Ragle, J.M., Johnson, L., … Matus, D.Q. (2020). Rapid degradation of *Caenorhabditis elegans* proteins at single-cell resolution with a synthetic auxin. *G3, 10*(1), 267–280. doi:10.1534/g3.119.400781

Mattson, M.P., & Magnus, T. (2006). Ageing and neuronal vulnerability. *Nature Reviews. Neuroscience*, *7*(4), 278–294. doi:10.1038/nrn1886

McColl, G., Roberts, B.R., Gunn, A.P., Perez, K.A., Tew, D.J., Masters, C.L., … Bush, A.I. (2009). The *Caenorhabditis elegans* A beta 1-42 model of Alzheimer disease predominantly expresses A beta 3-42. *The Journal of Biological Chemistry*, *284*(34), 22697–22702. doi:10.1074/jbc.C109.028514

McDermott, J.B., Aamodt, S., & Aamodt, E. (1996). ptl-1, a *Caenorhabditis elegans* gene whose products are homologous to the tau microtubule-associated proteins. *Biochemistry*, *35*(29), 9415–9423.

McDiarmid, T.A., Au, V., Loewen, A., Liang, J.J.H., Mizumoto, K., Moerman, D.G., & Rankin, C.H. (2018). CRISPR-Cas9 human gene replacement and phenomic characterization in *Caenorhabditis elegans* to understand the functional conservation of human genes and decipher variants of uncertain significance. *Disease Models and Mechanisms*, *11*, 1–15. 10.1101/369249

Mejzini, R., Flynn, L.L., Pitout, I.L., Fletcher, S., Wilton, S.D., & Akkari, P.A. (2019). ALS genetics, mechanisms, and therapeutics: Where are we now? *Frontiers in Neuroscience*, *13*, 1310–1327. doi:10.3389/fnins.2019.01310

Morley, J.F., Brignull, H.R., Weyers, J.J., & Morimoto, R.I. (2002). The threshold for polyglutamine-expansion protein aggregation and cellular toxicity is dynamic and influenced by aging in *Caenorhabditis elegans*. *Proceedings of the National Academy of Sciences of the United States of America*, *99*(16), 10417–10422. doi:10.1073/pnas.152161099

Murakami, T., Yang, S.-P., Xie, L., Kawano, T., Fu, D., Mukai, A., … St George-Hyslop, P. (2012). ALS mutations in FUS cause neuronal dysfunction and death in *Caenorhabditis elegans* by a dominant gain-of-function mechanism. *Human Molecular Genetics*, *21*(1), 1–9. doi:10.1093/hmg/ddr417

Nalls, M.A., Pankratz, N., Lill, C.M., Do, C.B., Hernandez, D.G., Saad, M., … Singleton, A.B. (2014). Large-scale meta-analysis of genome-wide association data identifies six new risk loci for Parkinson's disease. *Nature Genetics*, *46*(9), 989–993. doi:10.1038/ng.3043

Neumann, M., Sampathu, D.M., Kwong, L.K., Truax, A.C., Micsenyi, M.C., Chou, T.T., … Lee, V.M.-Y. (2006). Phosphorylated TDP-43 in frontotemporal lobar degeneration and amyotrophic lateral sclerosis. *Science*, *314*(5796), 130–133. doi:10.1126/science.1134108

Nollen, E.A.A., Garcia, S.M., van Haaften, G., Kim, S., Chavez, A., Morimoto, R.I., & Plasterk, R.H.A. (2004). Genome-wide RNA interference screen identifies previously undescribed regulators of polyglutamine aggregation. *Proceedings of the National Academy of Sciences of the United States of America*, *101*(17), 6403–6408. doi:10.1073/pnas.0307697101

Oeda, T., Shimohama, S., Kitagawa, N., Kohno, R., Imura, T., Shibasaki, H., & Ishii, N. (2001). Oxidative stress causes abnormal accumulation of familial amyotrophic lateral sclerosis-related mutant SOD1 in transgenic *Caenorhabditis elegans*. *Human Molecular Genetics*, *10*(19), 2013–2023. doi:10.1093/hmg/10.19.2013

Oliveira, J.M., Henriques, A.G., Martins, F., Rebelo, S., & da Cruz e Silva, O.A.B. (2015). Amyloid-β modulates both AβPP and tau phosphorylation. *Journal of Alzheimer's Disease*, *45*(2), 495–507. doi:10.3233/JAD-142664

Paisán-Ruíz, C., Jain, S., Evans, E.W., Gilks, W.P., Simón, J., van der Brug, M., … Singleton, A.B. (2004). Cloning of the gene containing mutations that cause PARK8-linked Parkinson's disease. *Neuron*, *44*(4), 595–600. doi:10.1016/j.neuron.2004.10.023

Parker, J.A., Arango, M., Abderrahmane, S., Lambert, E., Tourette, C., Catoire, H., & Néri, C. (2005). Resveratrol rescues mutant polyglutamine cytotoxicity in nematode and mammalian neurons. *Nature Genetics*, *37*(4), 349–350. doi:10.1038/ng1534

Parker, J.A., Connolly, J.B., Wellington, C., Hayden, M., Dausset, J., & Neri, C. (2001). Expanded polyglutamines in *Caenorhabditis elegans* cause axonal abnormalities and severe dysfunction of PLM mechanosensory neurons without cell death. *Proceedings of the National Academy of Sciences of the United States of America*, *98*(23), 13318–13323. doi:10.1073/pnas.231476398

Parker, J.A., Metzler, M., Georgiou, J., Mage, M., Roder, J.C., Rose, A.M., … Néri, C. (2007). Huntingtin-interacting protein 1 influences worm and mouse presynaptic function and protects *Caenorhabditis elegans* neurons against mutant polyglutamine toxicity. *The Journal of Neuroscience: The Official Journal of the Society for Neuroscience*, *27*(41), 11056–11064. doi:10.1523/JNEUROSCI.1941-07.2007

Puls, I., Jonnakuty, C., LaMonte, B.H., Holzbaur, E.L.F., Tokito, M., Mann, E., … Fischbeck, K.H. (2003). Mutant dynactin in motor neuron disease. *Nature Genetics*, *33*(4), 455–456. doi:10.1038/ng1123

Ramirez, A., Heimbach, A., Gründemann, J., Stiller, B., Hampshire, D., Cid, L.P., … Kubisch, C. (2006). Hereditary Parkinsonism with dementia is caused by mutations in ATP13A2, encoding a lysosomal type 5 P-type ATPase. *Nature Genetics*, *38*(10), 1184–1191. doi:10.1038/ng1884

Ramonet, D., Podhajska, A., Stafa, K., Sonnay, S., Trancikova, A., Tsika, E., … Moore, D.J. (2012). PARK9-associated ATP13A2 localizes to intracellular acidic vesicles and regulates cation homeostasis and neuronal integrity. *Human Molecular Genetics*, *21*(8), 1725–1743. doi:10.1093/hmg/ddr606

Reiner, A., Dragatsis, I., & Dietrich, P. (2011). Genetics and neuropathology of Huntington's disease. *International Review of Neurobiology*, *98*, 325–372. doi:10.1016/B978-0-12-381328-2.00014-6

Renton, A.E., Majounie, E., Waite, A., Simón-Sánchez, J., Rollinson, S., Gibbs, J.R., … Traynor, B.J. (2011). A hexanucleotide repeat expansion in C9ORF72 is the cause of chromosome 9p21-linked ALS-FTD. *Neuron*, *72*(2), 257–268. doi:10.1016/j.neuron.2011.09.010

Rha, J., Jones, S.K., Fidler, J., Banerjee, A., Leung, S.W., Morris, K.J., … Corbett, A.H. (2017). The RNA-binding protein, ZC3H14, is required for proper poly(A) tail length control, expression of synaptic proteins, and brain function in mice. *Human Molecular Genetics*, *26*(19), 3663–3681. doi:10.1093/hmg/ddx248

Rosen, D.R., Siddique, T., Patterson, D., Figlewicz, D.A., Sapp, P., Hentati, A., … Deng, H.X. (1993). Mutations in Cu/Zn superoxide dismutase gene are associated with familial amyotrophic lateral sclerosis. *Nature*, *362*(6415), 59–62. doi:10.1038/362059a0

Rudich, P., Snoznik, C., Watkins, S.C., Monaghan, J., Pandey, U.B., & Lamitina, S.T. (2017). Nuclear localized C9orf72-associated arginine-containing dipeptides exhibit age-dependent toxicity in *C. elegans*. *Human Molecular Genetics*, *26*(24), 4916–4928. doi:10.1093/hmg/ddx372

Saha, S., Guillily, M.D., Ferree, A., Lanceta, J., Chan, D., Ghosh, J., … Wolozin, B. (2009). LRRK2 modulates vulnerability to mitochondrial dysfunction in *Caenorhabditis elegans*. *The Journal of Neuroscience: The Official Journal of the Society for Neuroscience*, *29*(29), 9210–9218. doi:10.1523/JNEUROSCI.2281-09.2009

Şahin, A., Held, A., Bredvik, K., Major, P., Achilli, T.-M., Kerson, A.G., … Reenan, R. (2017). Human SOD1 ALS mutations in a Drosophila knock-in model cause severe phenotypes and reveal dosage-sensitive gain- and loss-of-function components. *Genetics*, *205*(2), 707–723. doi:10.1534/genetics.116.190850

Sakaguchi-Nakashima, A., Meir, J.Y., Jin, Y., Matsumoto, K., & Hisamoto, N. (2007). LRK-1, a *C. elegans* PARK8-related kinase, regulates axonal-dendritic polarity of SV proteins. *Current Biology*, *17*(7), 592–598. doi:10.1016/j.cub.2007.01.074

Sämann J., Hegermann J., von Gromoff E., Eimer S., Baumeister R. & Schmidt E. (2009) Caenorhabditis elegans LRK-1 and PINK-1 Act Antagonistically in Stress Response and Neurite Outgrowth. *Journal of Biological Chemistry*, *284*(24), 16482–16491.

Sarasija, S., Laboy, J.T., Ashkavand, Z., Bonner, J., Tang, Y., & Norman, K.R. (2018). Presenilin mutations deregulate mitochondrial Ca^{2+} homeostasis and metabolic activity causing neurodegeneration in *Caenorhabditis elegans*. *eLife*, *7*, 1–30. doi:10.7554/eLife.33052

Sarasija, S., & Norman, K.R. (2015). A γ-Secretase independent role for presenilin in calcium homeostasis impacts mitochondrial function and morphology in *Caenorhabditis elegans. Genetics, 201*(4), 1453–1466. doi:10.1534/genetics.115.182808

Satyal, S.H., Schmidt, E., Kitagawa, K., Sondheimer, N., Lindquist, S., Kramer, J.M., & Morimoto, R.I. (2000). Polyglutamine aggregates alter protein folding homeostasis in *Caenorhabditis elegans. Proceedings of the National Academy of Sciences of the United States of America, 97*(11), 5750–5755. doi:10.1073/pnas.100107297

Saudou, F., & Humbert, S. (2016). The biology of huntingtin. *Neuron, 89*(5), 910–926. doi:10.1016/j.neuron.2016.02.003

Sawa, A., Tomoda, T., & Bae, B.-I. (2003). Mechanisms of neuronal cell death in Huntington's disease. *Cytogenetic and Genome Research, 100*(1–4), 287–295. doi:10.1159/000072864

Selkoe, D.J., & Hardy, J. (2016). The amyloid hypothesis of Alzheimer's disease at 25 years. *EMBO Molecular Medicine, 8*(6), 595–608. doi:10.15252/emmm.201606210

Springer, W., Hoppe, T., Schmidt, E., & Baumeister, R. (2005). A *Caenorhabditis elegans* Parkin mutant with altered solubility couples alpha-synuclein aggregation to proteotoxic stress. *Human Molecular Genetics, 14*(22), 3407–3423. doi:10.1093/hmg/ddi371

Sulston, J.E., & Horvitz, H.R. (1977). Post-embryonic cell lineages of the nematode, *Caenorhabditis elegans. Developmental Biology, 56*(1), 110–156. doi:10.1016/0012-1606(77)90158-0

Sulston, J.E., Schierenberg, E., White, J.G., & Thomson, J.N. (1983). The embryonic cell lineage of the nematode *Caenorhabditis elegans. Developmental Biology, 100*(1), 64–119. doi:10.1016/0012-1606(83)90201-4

Tabaton, M., Zhu, X., Perry, G., Smith, M.A., & Giliberto, L. (2010). Signaling effect of amyloid-beta(42) on the processing of AbetaPP. *Experimental Neurology, 221*(1), 18–25. doi:10.1016/j.expneurol.2009.09.002

Tafuri, F., Ronchi, D., Magri, F., Comi, G.P., & Corti, S. (2015). SOD1 misplacing and mitochondrial dysfunction in amyotrophic lateral sclerosis pathogenesis. *Frontiers in Cellular Neuroscience, 9*, 336–312. doi:10.3389/fncel.2015.00336

Takashima, A., Noguchi, K., Michel, G., Mercken, M., Hoshi, M., Ishiguro, K., & Imahori, K. (1996). Exposure of rat hippocampal neurons to amyloid β peptide (25–35) induces the inactivation of phosphatidyl inositol-3 kinase and the activation of tau protein kinase I/glycogen synthase kinase-3β. *Neuroscience Letters, 203*(1), 33–36. doi:10.1016/0304-3940(95)12257-5

Teo, E., Ravi, S., Barardo, D., Kim, H.-S., Fong, S., Cazenave-Gassiot, A., … Gruber, J. (2019). Metabolic stress is a primary pathogenic event in transgenic *Caenorhabditis elegans* expressing pan-neuronal human amyloid beta. *eLife, 8*, 1–25. doi:10.7554/eLife.50069

The *C. elegans* Sequencing Consortium. (1998). Genome sequence of the nematode *C. elegans*: A platform for investigating biology. *Science, 282*(5396), 2012–2018. 10.1126/science.282.5396.2012

Therrien, M., & Parker, J.A. (2014). Worming forward: Amyotrophic lateral sclerosis toxicity mechanisms and genetic interactions in *Caenorhabditis elegans. Frontiers in Genetics, 5*, 85. doi:10.3389/fgene.2014.00085

Therrien, M., Rouleau, G.A., Dion, P.A., & Parker, J.A. (2013). Deletion of C9ORF72 results in motor neuron degeneration and stress sensitivity in *C. elegans. PLoS One, 8*(12), e83450. doi:10.1371/journal.pone.0083450

Timmons, L., Tabara, H., Mello, C.C., & Fire, A.Z. (2003). Inducible systemic RNA silencing in *Caenorhabditis elegans. Molecular Biology of the Cell, 14*(7), 2972–2983. doi:10.1091/mbc.e03-01-0858

Turner, M.R., Hardiman, O., Benatar, M., Brooks, B.R., Chio, A., de Carvalho, M., … Kiernan, M.C. (2013). Controversies and priorities in amyotrophic lateral sclerosis. *The Lancet Neurology, 12*(3), 310–322. doi:10.1016/S1474-4422(13)70036-X

Vaccaro, A., Patten, S.A., Aggad, D., Julien, C., Maios, C., Kabashi, E., … Parker, J.A. (2013). Pharmacological reduction of ER stress protects against TDP-43 neuronal toxicity in vivo. *Neurobiology of Disease, 55*, 64–75. doi:10.1016/j.nbd.2013.03.015

Vaccaro, A., Tauffenberger, A., Aggad, D., Rouleau, G., Drapeau, P., & Parker, J.A. (2012). Mutant TDP-43 and FUS cause age-dependent paralysis and neurodegeneration in *C. elegans. PLoS One, 7*(2), e31321. doi:10.1371/journal.pone.0031321

Vaccaro, A., Tauffenberger, A., Ash, P.E.A., Carlomagno, Y., Petrucelli, L., & Parker, J.A. (2012). TDP-1/TDP-43 regulates stress signaling and age-dependent proteotoxicity in *Caenorhabditis elegans. PLoS Genetics, 8*(7), e1002806. doi:10.1371/journal.pgen.1002806

Vérièpe, J., Fossouo, L., & Parker, J.A. (2015). Neurodegeneration in *C. elegans* models of ALS requires TIR-1/Sarm1 immune pathway activation in neurons. *Nature Communications, 6.* doi:10.1038/ncomms8319

Wallings, R., Manzoni, C., & Bandopadhyay, R. (2015). Cellular processes associated with LRRK2 function and dysfunction. *The FEBS Journal, 282*(15), 2806–2826. doi:10.1111/febs.13305

Wang, C., Saar, V., Leung, K.L., Chen, L., & Wong, G. (2018). Human amyloid β peptide and tau co-expression impairs behavior and causes specific gene expression changes in *Caenorhabditis elegans. Neurobiology of Disease, 109*, 88–101. doi:10.1016/j.nbd.2017.10.003

Wang, J., Farr, G.W., Hall, D.H., Li, F., Furtak, K., Dreier, L., & Horwich, A.L. (2009). An ALS-linked mutant SOD1 produces a locomotor defect associated with aggregation and synaptic dysfunction when expressed in neurons of *Caenorhabditis elegans. PLoS Genetics, 5*(1), e1000350. doi:10.1371/journal.pgen.1000350

Wang, S., Tang, N.H., Lara-Gonzalez, P., Zhao, Z., Cheerambathur, D.K., Prevo, B., … Oegema, K. (2017). A toolkit for GFP-mediated tissue-specific protein degradation in *C. elegans. Development, 144*(14), 2694–2701. doi:10.1242/dev.150094

White, J.G., Southgate, E., Thomson, J.N., & Brenner, S. (1986). The structure of the nervous system of the nematode *Caenorhabditis elegans. Philosophical Transactions of the Royal Society of London. Series B, Biological Sciences, 314*(1165), 1–340. doi:10.1098/rstb.1986.0056

Wiese M., Antebi A., Zheng H. (2010). Intracellular Trafficking and Synaptic Function of APL-1 in Caenorhabditis elegans. *PLoS One, 5*(9), e12790.

Williamson, T.L., & Cleveland, D.W. (1999). Slowing of axonal transport is a very early event in the toxicity of ALS-linked SOD1 mutants to motor neurons. *Nature Neuroscience, 2*(1), 50–56. doi:10.1038/4553

Wils, H., Kleinberger, G., Janssens, J., Pereson, S., Joris, G., Cuijt, I., … Kumar-Singh, S. (2010). TDP-43 transgenic mice develop spastic paralysis and neuronal inclusions characteristic of ALS and frontotemporal lobar degeneration. *Proceedings of the National Academy of Sciences of the United States of America, 107*(8), 3858–3863. doi:10.1073/pnas.0912417107

Xia, Q., Wang, H., Hao, Z., Fu, C., Hu, Q., Gao, F., … Wang, G. (2016). TDP-43 loss of function increases TFEB activity and blocks autophagosome-lysosome fusion. *The EMBO Journal, 35*(2), 121–142. doi:10.15252/embj.201591998

Xu, Y.-F., Gendron, T.F., Zhang, Y.-J., Lin, W.-L., D'Alton, S., Sheng, H., … Petrucelli, L. (2010). Wild-type human TDP-43 expression causes TDP-43 phosphorylation, mitochondrial aggregation, motor deficits, and early mortality in transgenic mice. *The Journal of Neuroscience: The Official Journal of the Society for Neuroscience, 30*(32), 10851–10859. doi:10.1523/JNEUROSCI.1630-10.2010

Yanase, S., Onodera, A., Tedesco, P., Johnson, T.E., & Ishii, N. (2009). SOD-1 deletions in *Caenorhabditis elegans* alter the localization of intracellular reactive oxygen species and show molecular compensation. *The Journals of Gerontology. Series A, Biological Sciences and Medical Sciences, 64*(5), 530–539. doi:10.1093/gerona/glp020

Yao, C., Johnson, W.M., Gao, Y., Wang, W., Zhang, J., Deak, M., … Chen, S.G. (2013). Kinase inhibitors arrest neurodegeneration in cell and *C. elegans* models of LRRK2 toxicity. *Human Molecular Genetics, 22*(2), 328–344. doi:10.1093/hmg/dds431

Yim, M.B., Kang, J.H., Yim, H.S., Kwak, H.S., Chock, P.B., & Stadtman, E.R. (1996). A gain-of-function of an amyotrophic lateral sclerosis-associated Cu,Zn-superoxide dismutase mutant: An enhancement of free radical formation due to a decrease in Km for hydrogen peroxide. *Proceedings of the National Academy of Sciences of the United States of America, 93*(12), 5709–5714. doi:10.1073/pnas.93.12.5709

Zhang, L., Ward, J.D., Cheng, Z., & Dernburg, A.F. (2015). The auxin-inducible degradation (AID) system enables versatile conditional protein depletion in *C. elegans*. *Development, 142*(24), 4374–4384. doi:10.1242/dev.129635

Zhang, S., Cai, F., Wu, Y., Bozorgmehr, T., Wang, Z., Zhang, S., ... Song, W. (2020). A presenilin-1 mutation causes Alzheimer disease without affecting Notch signaling. *Molecular Psychiatry, 25* (3), 603–611. doi:10.1038/s41380-018-0101-x

Zhang, T., Hwang, H.-Y., Hao, H., Talbot, C., & Wang, J. (2012). *Caenorhabditis elegans* RNA-processing protein TDP-1 regulates protein homeostasis and life span. *The Journal of Biological Chemistry, 287*(11), 8371–8382. doi:10.1074/jbc.M111.311977

Zhang, T., Mullane, P.C., Periz, G., & Wang, J. (2011). TDP-43 neurotoxicity and protein aggregation modulated by heat shock factor and insulin/IGF-1 signaling. *Human Molecular Genetics, 20*(10), 1952–1965. doi:10.1093/hmg/ddr076

Zheng, J., Winderickx, J., Franssens, V., & Liu, B. (2018). A mitochondria-associated oxidative stress perspective on Huntington's disease. *Frontiers in Molecular Neuroscience, 11*, 329–310. doi:10.3389/fnmol.2018.00329

Zou, K., Gong, J.S., Yanagisawa, K., & Michikawa, M. (2002). A novel function of monomeric amyloid β-protein serving as an antioxidant molecule against metal-induced oxidative damage. *The Journal of Neuroscience: The Official Journal of the Society for Neuroscience, 22*(12), 4833–4841. doi:10.1523/JNEUROSCI.22-12-04833.2002

Part VII
Worm photo and art gallery

A journey to 'tame a small metazoan organism', seen through the artistic eyes of *C. elegans* researchers

Eleni Gourgou, Alexandra R. Willis, Sebastian Giunti, Maria J. De Rosa, Amanda G. Charlesworth, Mirella Hernandez Lima, Elizabeth Glater, Sonja Soo, Bianca Pereira, Kübra Akbaş, Anushka Deb, Madhushree Kamak, Mark W. Moyle, Annika Traa, Aakanksha Singhvi, Surojit Sural and Eugene Jennifer Jin

ABSTRACT

In the following pages, we share a collection of photos, drawings, and mixed-media creations, most of them especially made for this JoN issue, manifesting *C. elegans* researchers' affection for their model organism and the founders of the field. This is a celebration of our community's growth, flourish, spread, and bright future. Descriptions provided by the contributors, edited for space.[1]

[1]See page 11 for a complete list of authors and credits.

(A) (B) (C) (D) (E) (F)

Figure 1. S. Brenner, J. Sulston, and colleagues. J. Sulston's FRS celebration party, MRC coffee room, 1986. Individuals are named from left to right. (A) J. Sulston and S. Brenner; (B) J. Sulston, Alan Coulson, S. Brenner; (C) Peter Lawrence, J. Sulston, Alan Coulson, S. Brenner; (D) Jim Priess, Peter Lawrence, Jonathan Hodgkin, J. Sulston; (E) Richard Durbin, Maria Leptin, Nichol Thomson, Jim Priess, Peter Lawrence; (F) Cynthia Kenyon, Michael Shen.

Figure 2. S. Brenner and J. Sulston, as artistic inspiration. (A) J. Sulston and S. Brenner; drawing; (B) Vitruvian Worm; drawing, mixed media.

Figure 3. 'When you blow bubbles in the wind you never know where they will go. When Sydney Brenner started us down the incredible path of researching *C. elegans* he could never have imagined how much they would teach us all. Represented here are only a fraction of some of the amazing things we have learned from Worms', drawing.

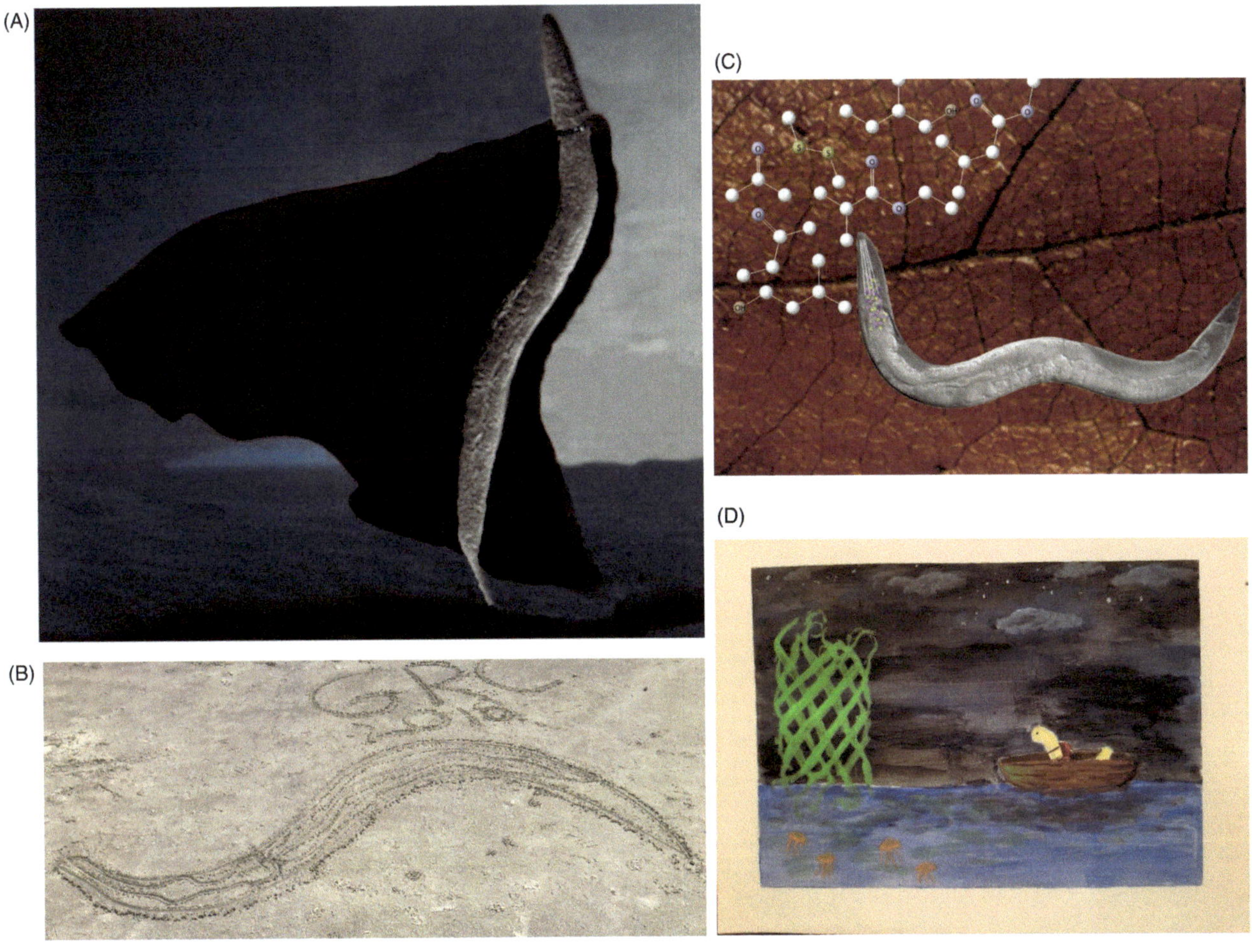

Figure 4. Just a few of *C. elegans*' many adventures. (A) '*Caenorhabditis elegans*, a transparent mystical nematode that has provided insight into the mysteries of science, exposing truths that transcend human understanding'. (B) A worm drawn in the sand, March 2018 Sleep Gordon Research Conference, Galveston, Texas, USA. (C) '*C. elegans* uses chemosensation to distinguish among different species of bacteria, their major food source. Work from the Glater lab and others have identified volatile organic compounds that *C. elegans* uses to recognize and detect different bacterial species', mixed media; (D) 'Journey of the elegant glow', acrylic on paper: 'As the first model organism to express Green Fluorescent Protein (GFP), *C. elegans* provide a powerful tool for scientific discoveries'.

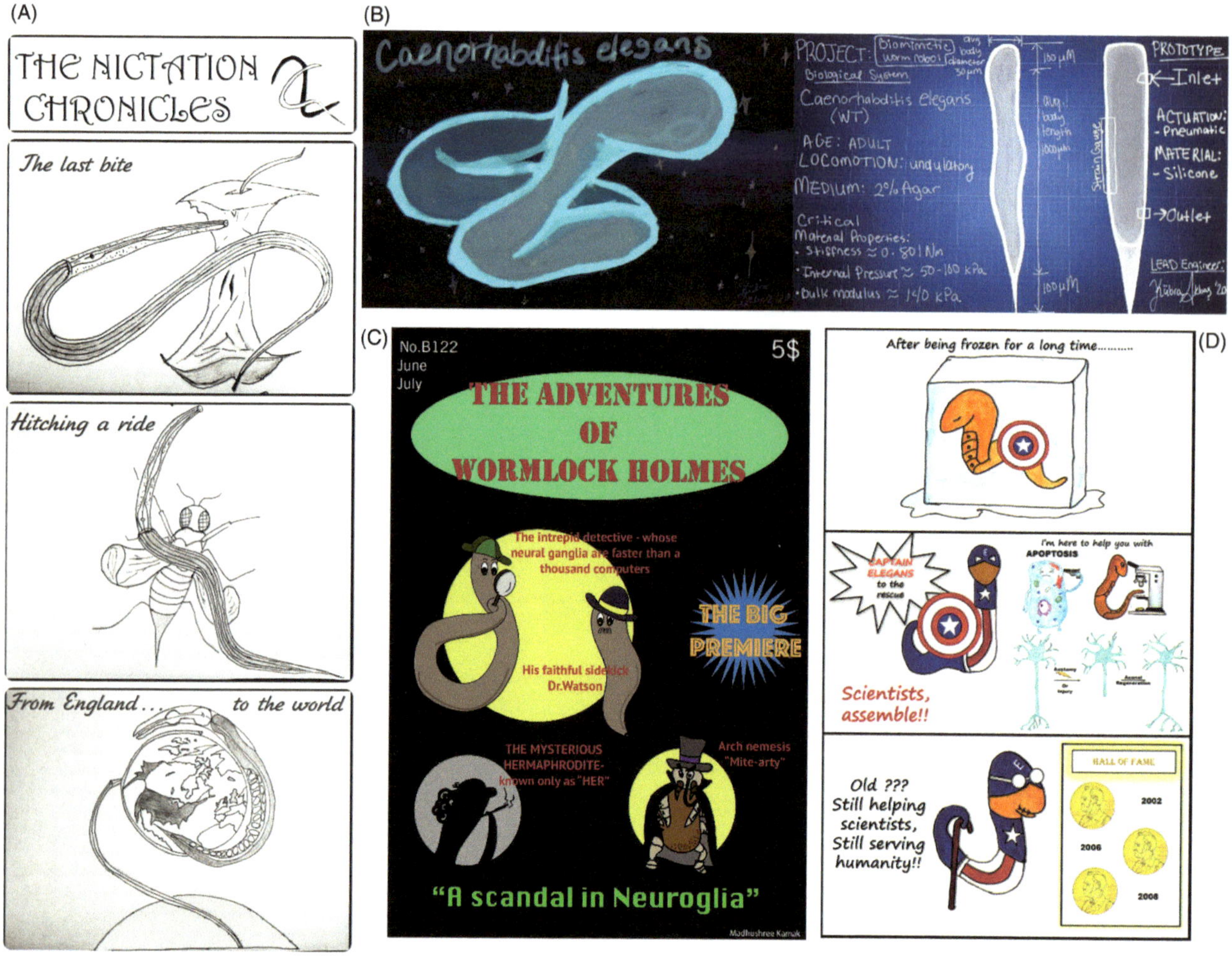

Figure 5. *C. elegans* in graphic novels. (A) The Nictation Chronicles, comic strip. (B) How an engineer-roboticist envisions *C. elegans*. Left: Star-gazing glowing worm: a metaphorical representation of how *C. elegans* can be considered a building block in the universe due to its use as a model organism in many biological studies, especially those pertaining to biomedical applications. Right: Notes on a worm-inspired soft robot: a mechanical and robotic approach to *C. elegans*, where the inspiration for a potential pneumatically-actuated soft robot is outlined. (C) The Adventures of the Wormlock Holmes-A scandal in Neuroglia. (D) Captain Elegans.

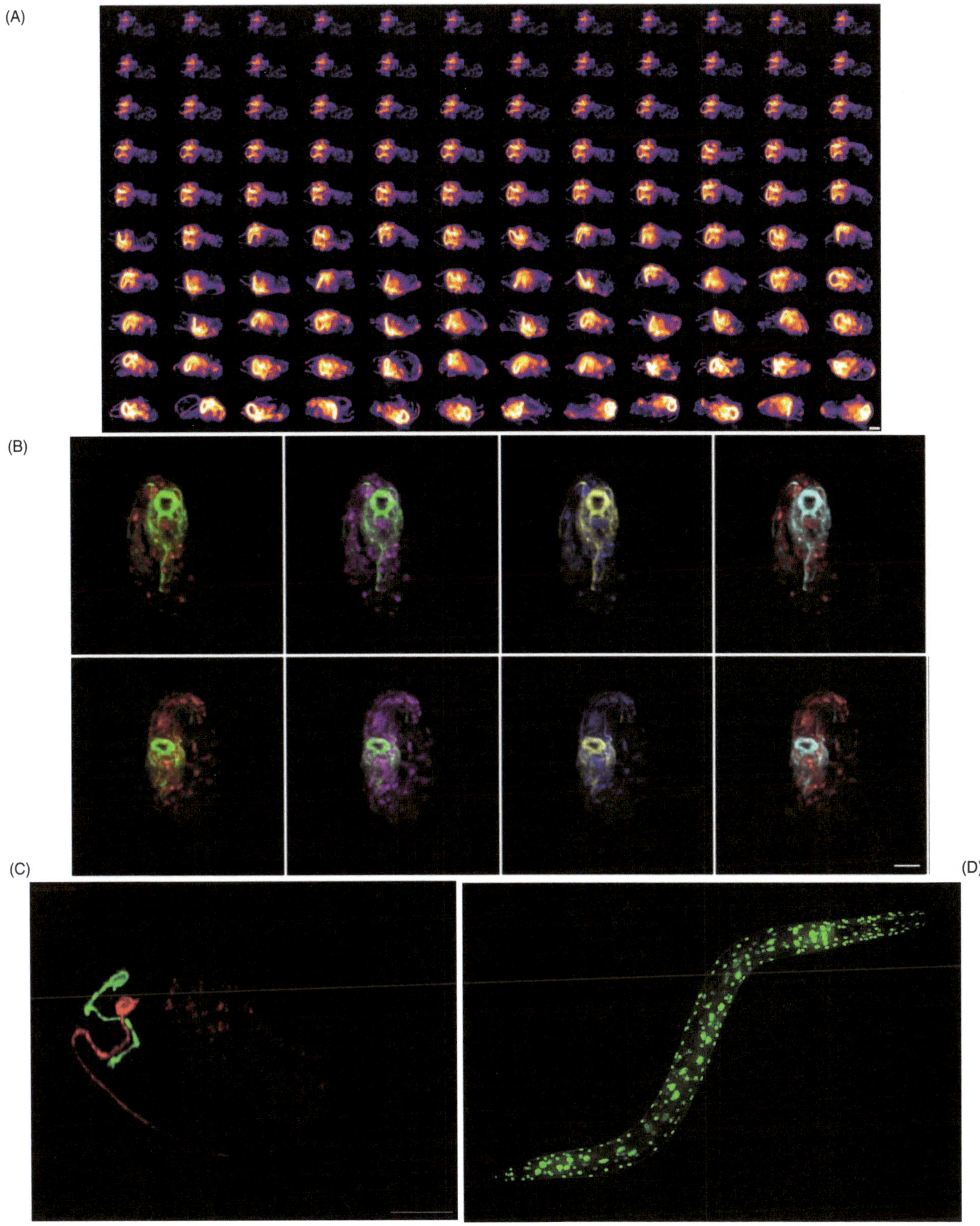

Figure 6. *C. elegans* microscopy, Part I. (A) Two-hour time-lapse sequence of the developing *C. elegans* nervous system, *rab-3p::membrane tethered GFP* transgene. The bright, omega-like structure (Ω) is the nerve ring neuropil (the *C. elegans* brain). (B) *C. elegans* neurodevelopment. *rgef-1p::membrane tethered GFP* transgene. Nuclei were simultaneously imaged by using *mCherry::histone* transgene. Circular ring structure corresponds to the nerve ring neuropil of the animal. (C) Developing neurons in *C. elegans* embryo, *hlh-16p::membrane tethered GFP* transgene. Both neurites are in close proximity at the nerve ring neuropil. Neurons have been pseudocolored red and green with Photoshop to highlight individual outgrowth dynamics. (D) Nuclear localization of DAF-16::GFP. This image depicts a DAF-16::GFP reporter strain, which is used to monitor the nuclear localization of DAF-16, after having been exposed to a 37 °C heat stress for four hours. (Color images available in the electronic version of this photo gallery.)

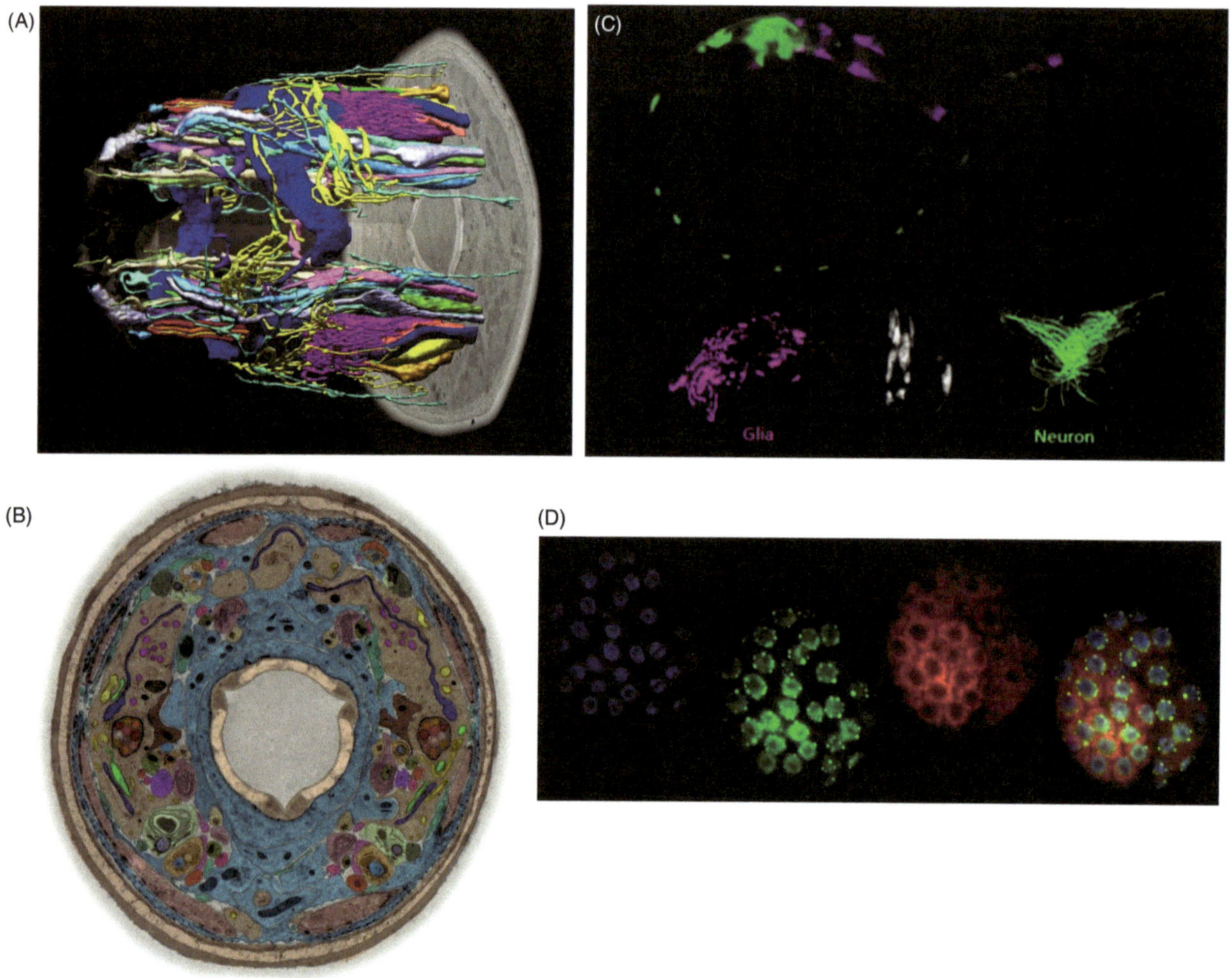

Figure 7. *C. elegans* microscopy, Part II. Panels (A) and (B): Graphical model of high-resolution system 3D reconstruction of anterior sensory endings and TEM cross-section. 3D reconstruction of the cilia and dendritic endings of anterior sensory neurons modeled from 166 thin serial sections with superimposed example TEM cross-section of a high-pressure frozen/freeze-substituted (HPF-FS) *C. elegans* hermaphrodite animal. (C) *C. elegans* as a model to study single glia-neuron interactions, mixed media. (D) Fluorescent staining of subcellular structures in the germ line of an adult hermaphrodite. DAPI stains DNA (blue), Alexa Fluor Plus 488-conjugated antibody labels histone H4 (green) and MitoTracker Orange stains mitochondria (red). (Color images available in the electronic version of this photo gallery.)

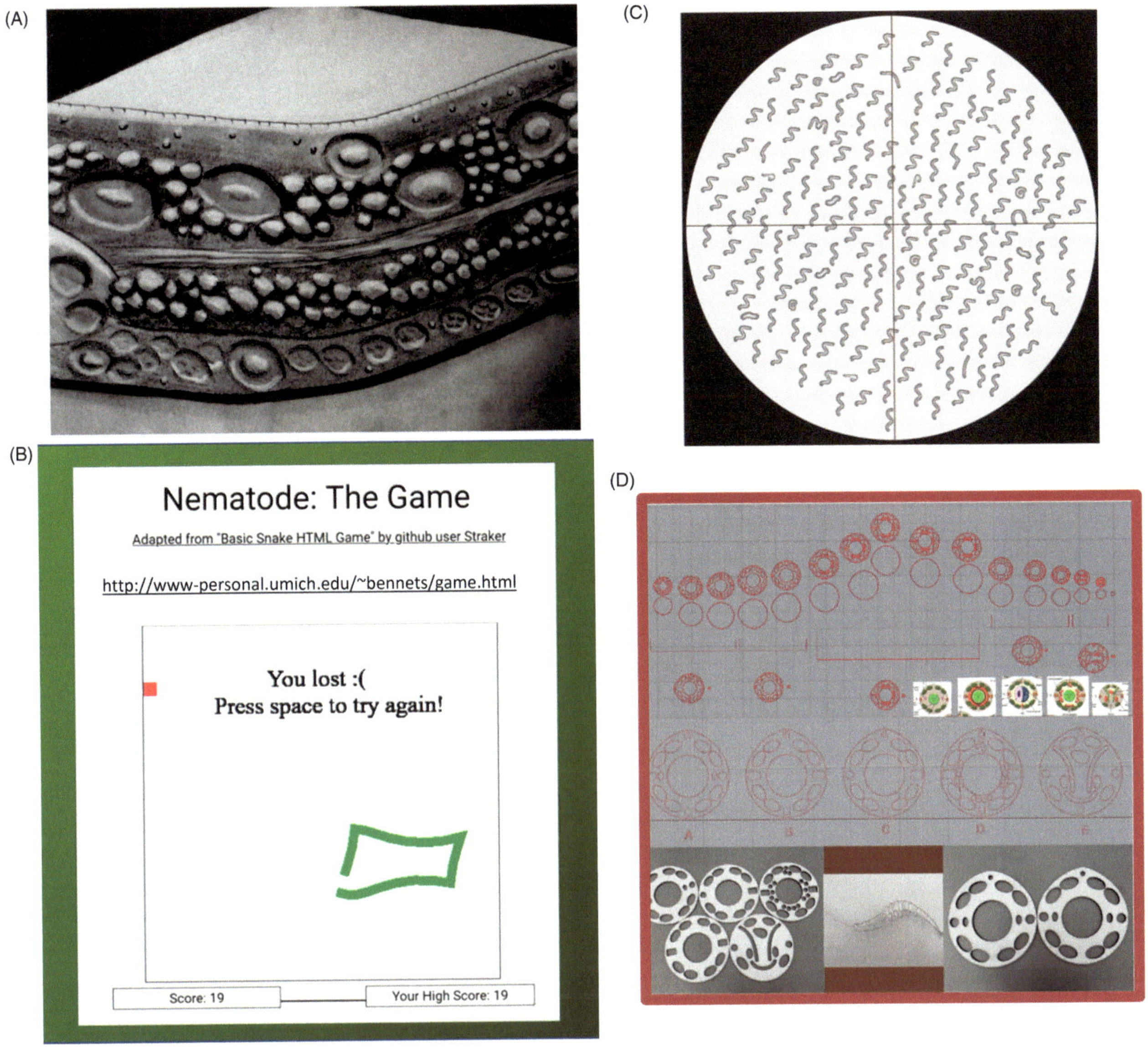

Figure 8. *C. elegans* mixed media art. (A) 'Transparency hit by light': P cell divisions under Nomarski microscope to determine developmental time, in honor of J. Sulston's work on cell lineages, showing cells posterior of gonad in a late L1 worm, after P9's 4th cell division. Anterior is left and ventral is down. P cells are positions along the ventral cord, and I cells are surrounded by gut granules. Graphite pencil drawing based on a Nomarski microscopy image. (B) 'Opening to neurobiology': view through a microscope eyepiece illustrating S. Brenner's forward genetic screen for uncoordinated worms. In the field of view, are coiled, kinky, dumpy and omega-shaped mutant worms. Paralyzed worms are depicted as straight or less curvy worms. Created using Adobe Illustrator. (C) *C. elegans* is the star of a video game: *Nematode, The Game*! It is like 'Snake', the nematode moves in a sinusoidal way and pirouettes off in a random direction every once in a while, sometimes into itself. A perfect blend of skill and luck! (Play here: http://www-personal.umich.edu/~bennets/game.html..) (D) *C. elegans* body cross sections translated in CAD (Computer Aided Design-Rhino). Used to make cardboard or acrylic discs, combine them with metallic wire and build a worm prototype.

Figure 9. *C. elegans* embroidery, textile, fabric and gastronomy. (A) 'Wormville', hand embroidery. Depicts *C. elegans* living in what would be their imaginary own little village. Participated in Worm Art show in worm meeting 2019 in LA, where it won an award. (B) Front part of a T-shirt, made for Javier Apfeld by his labmates, for his graduation from Cynthia Kenyon's lab, mosaic representation of *C. elegans* lifespan curve. (C) Front part of a T-shirt, made for Jen Whangbo by her labmates at the Kenyon Lab, referring to her research on Q neuroblast migration. (D) Worm nervous system cake.

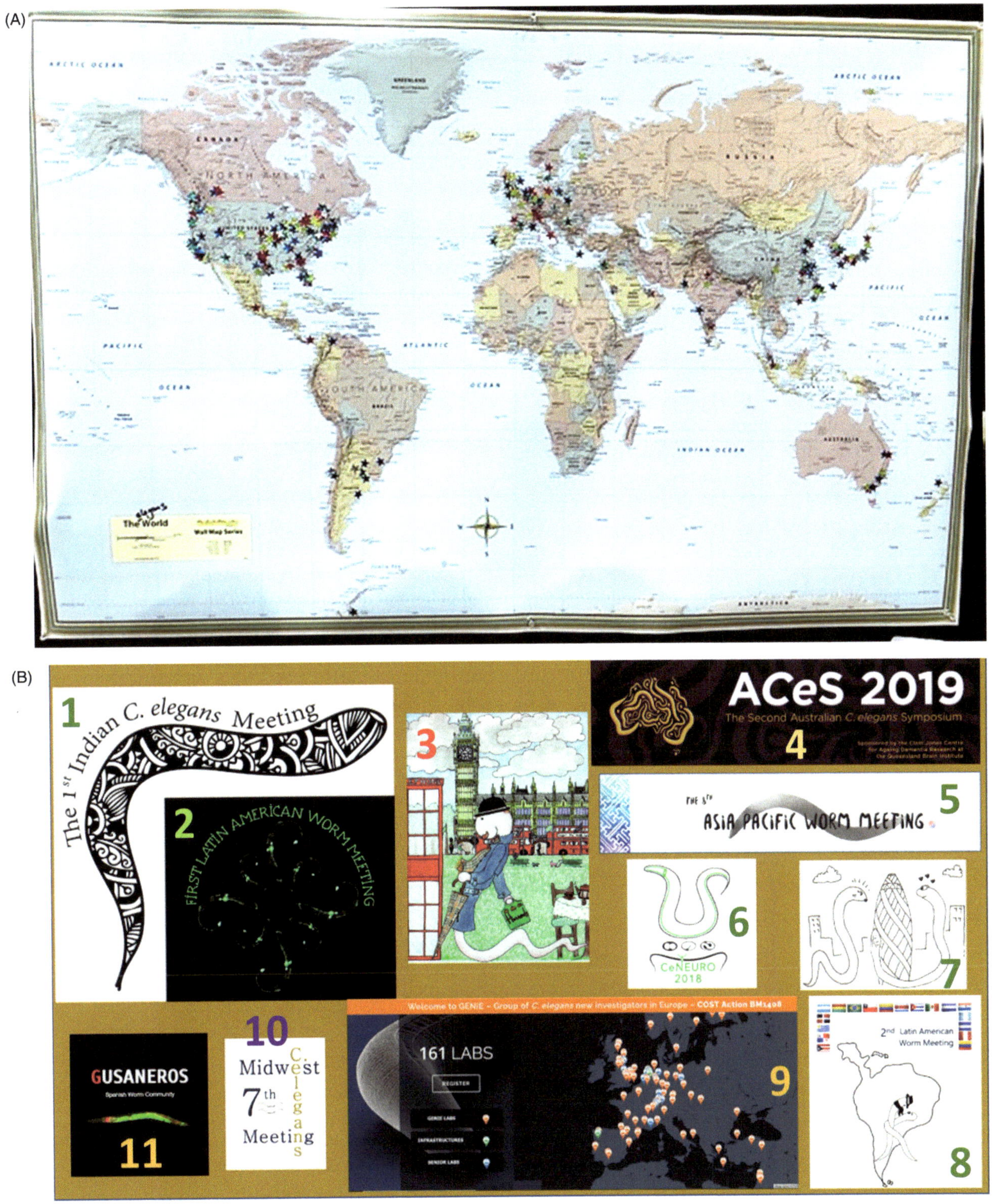

Figure 10. *C. elegans* thrives around the world. (A) A world map, featuring locations of *C. elegans* labs, collective work by attendants of the 2019 International *C. elegans* Meeting, California. This initiative was part of the beloved Worm Art Show, organized for years by Ahna Skop. (B) A collection of worm meetings logos from around the globe, featuring: 1. 1st Indian *C. elegans* meeting logo, 2. 1st Latin American *C. elegans* meeting logo, to signify the visit of Martin Chalfie, hence the GFP. 3. 1st UK *C. elegans* meeting logo, 4. 2nd Australian *C. elegans* meeting banner, 5. 8th Asia Pacific worm meeting logo, 6. CeNeuro 2018 logo (topic meeting, USA), 7. 2nd UK worm meeting, featuring the Gherkin tower, 8. 2nd Latin American Worm Meeting, 9. Website screenshot of GENiE, group of *C. elegans* investigators in Europe and neighboring areas, locations of labs highlighted. 10. 7th Midwest *C. elegans* meeting logo (regional, USA), 11. Gusaneros, a Spanish-speaking worm community.

Credits

We are grateful to all these talented people, who enthusiastically contributed to this gallery.

Figure 1. Photos taken by Tabitha Doniach, generously contributed by Scott Emmons (Einstein College of Medicine, USA), with the kind assistance of Jonathan Hodgkin (University of Oxford, UK).

Figure 2. (A) Alexandra R. Willis (Reinke Lab, University of Toronto, Canada); (B) Sebastián Giunti and María José De Rosa (Rayes and De Rosa Labs, Instituto de Investigaciones Bioquímicas Bahía Blanca - INIBIBB, Argentina).

Figure 3. Amanda G. Charlesworth (Claycomb Lab, University of Toronto, Canada).

Figure 4. (A) Mirella Hernandez Lima (Truttmann Lab, University of Michigan, USA); (B) Henrik Bringmann (Max Planck Institute, Germany), photo contributed by David Raizen (University of Pennsylvania, USA); (C) Elizabeth Glater (Pomona College, USA), Luisa Scott (University of Texas at Austin, USA), and Madeleine Huong Le (Flansburgh Architects, USA); (D) Sonja Soo (Van Raamsdonk Lab, McGill University, Canada).

Figure 5. (A) Bianca Pereira, (Alcedo Lab, Wayne State University, USA); (B) Kübra Akbaş (Coppélia Research Lab-Mummolo Group, New Jersey Institute of Technology, USA); (C) Madhushree Kamak; (D) Anushka Deb; (C) and (D): Koushika Lab, DBS-TIFR, India.

Figure 6. (A) and (B): Mark W. Moyle; (C) Javier Marquina-Solis, Mark W. Moyle; (A)-(C): Colón-Ramos Lab, Yale University, USA. Collaborators: Hari Shroff, Zhirong Bao, William Mohler; (D) Annika Traa (Van Raamsdonk Lab, McGill University, Canada).

Figure 7. (A) and (B): David Doroquez, Cristina Berciu, James R Anderson, Piali Sengupta, Daniela Nicastro (Sengupta and Nicastro Labs, Brandeis University, USA, published in *eLife*, PMID: 24668170); (C) Aakanksha Singhvi, Stephan Raiders and Maria Purice (Singhvi Lab, Fred Hutchinson Cancer Research Center, USA); (D) Surojit Sural (Hsu Lab, University of Michigan, USA).

Figure 8. (A) and (B): Eugene Jennifer Jin, (Y. Jin Lab, University of California, San Diego, USA); (C) Bennet Sakelaris (Gourgou and Booth Groups, University of Michigan, USA); (D) Manali Desai, Jiwen Chen, Richard Wall, Fee Christoph, Melinda Li (Gourgou Group, University of Michigan, USA).

Figure 9. (A) Hala Tamim El Jarkass (Reinke Lab, University of Toronto, Canada); (B) Contributed by J. Apfeld (Northeastern University, USA); (C) Contributed by J. Whangbo (Harvard Medical School, USA); (D) Lindsey Lopes and Jessica Schwartz (Raizen Lab, University of Pennsylvania, USA).

Figure 10. (A) Photo contributed by Ahna Skop (University of Wisconsin, USA). (B) 1. Madhushree Kamak (Koushika lab, DBS-TIFR, India); 2. Andrea Calixto (Universidad Mayor, Chile, and Centro Interdisciplinario de Neurociencia de Valparaiso, Chile), and Ines Carrera (Institut Pasteur de Montevideo, Uruguay); 3. Giovanna Lalli (UK Dementia Research Institute), we thank Giovanni Lesa for help; 4. Logo owned by The University of Queensland, Australia; 5. Kyoung-Hye Yoon (Yonsei University, Republic of Korea), we thank Seung-Jae V. Lee for help; 6. Tari Tan (Harvard Medical School, USA), we thank Denise Ferkey for help; 7. David Gems (University College London, UK); 8. Permission provided by organizers; 9. From GENiE website; 10. Permission provided by organizers; 11. We thank Julian Ceron Madrigal for help.

ORCID

Eleni Gourgou http://orcid.org/0000-0003-1561-5545

Index